Comprehensive
Chemistry

Comprehensive Chemistry

JOHN HICKS, M.A. (Cantab.)

Headmaster, The Howard School, Gillingham, Kent
Formerly Senior Chemistry Master,
Merchant Taylors' School, Middlesex

REVISED SECOND EDITION

SI Units

M

First edition 1963
Reprinted (with minor amendments) 1964
Reprinted 1965, 1967, 1968

Second edition 1970
Revised second edition 1971
Reprinted (with corrections) 1972, 1973, (twice), 1975

Published by
THE MACMILLAN PRESS LTD
London and Basingstoke
Associated companies in New York Dublin
Melbourne Johannesburg and Madras

SBN 333 13189 4 (cased edition)
333 14299 3 (low-priced edition)
333 09045 4 (paper edition)

Printed in Hong Kong by
BRIGHT SUN PRINTING PRESS CO., LTD.

Drawings by William A. Keel

Preface to the Revised Second Edition

In this edition the whole book, text and examination questions, has been rewritten in SI units. It has also been very extensively altered to conform to the recent recommendations of the Royal Society and the Association for Science Education in their respective reports *Symbols, Signs, and Abbreviations for British Scientific Publications* and *SI Units, Signs, Symbols, and Abbreviations for Use in School Science*. These two publications offer the most authoritative and up-to-date guidance available on approved international practice and have been followed meticulously.

Outdated textbooks are often blamed for hindering educational advance; it is hoped that the provision of a book which implements these important recommendations rigorously will give powerful impetus to them and facilitate their rapid adoption in schools and colleges throughout the world.

J. H.

January 1971

Preface to the Second Edition

The distinctive features of the first edition, the broad unified approach and the emphasis upon the principles of the subject, have been preserved, but the opportunity has been taken of making many important additions to the text and bringing the contents thoroughly up-to-date. Amongst the topics which have been added in this edition or given a much fuller and deeper treatment are kinetics, energetics and entropy, chromatography, ionization energy, atomic orbitals, oxidation numbers, hydrides, nuclear chemistry, natural gas, alicyclic compounds, amino-acids and proteins, and carbon fibres. Two additional chapters have been provided, one dealing with polymers and polymerizations and the other with the mechanisms and types of organic reactions. Appendices have been added on chemical nomenclature and on SI units and the text has been extensively altered in these respects. A list of Nobel prizewinners in Chemistry has been included and certain diagrams have been redrawn to incorporate glass-jointed apparatus.

These substantial alterations and additions, which amount altogether to some 120 extra pages, take account of all the changes recently made or proposed in the Advanced Level syllabuses of the various examining boards including the Cambridge (Alternative T) syllabus, the new syllabus B of the Joint Matriculation Board, the new London University B syllabus, and the syllabus of the West African Examinations Council. This edition also covers most of the ground of the Nuffield A Level Chemistry Course, for which it should prove very useful as an accompaniment.

I am grateful to the many students and teachers who have written from all over the world expressing their appreciation of the book and am always glad to receive further suggestions on ways of improving it.

J. H.

January 1970

Preface to the First Edition

This book covers the Chemistry required by the various examining boards up to the Advanced and Special (Scholarship) levels of the G.C.E. Written also with the needs of the Colleges of Technology in mind, it should prove valuable to students taking National Certificate courses and courses in Applied Chemistry. In addition, it is hoped that the book will be useful to those studying other subjects such as Engineering, Pharmacy, or Medicine, who require a broad survey of Chemistry at this level, and to those taking Chemistry as a subsidiary subject at a University.

The book departs from the practice of its contemporaries in several important ways. As its title is intended to suggest, it includes physical, general, inorganic and organic Chemistry in a single volume. Not only does this result in an appreciable economy, but it has the further advantage of giving a unified approach to the subject which is emphasized here by the use of numerous cross references. The book aims to provide complete coverage of the new examination syllabuses at A and S levels. These new syllabuses, and the recent suggestions of the Association for Science Education (formerly the Science Masters' Association) which inspired them, represent substantial changes in content and approach, changes which will be welcomed generally by students and teachers alike because of their greater relevance to modern needs. The dependence upon petroleum, for example, as a source of organic chemicals has been duly emphasized, and elements such as lithium, beryllium, boron, fluorine, silicon, and titanium have been given a place commensurate with their present importance. To meet the complaints which have been voiced that Chemistry textbooks are completely out of touch with modern industrial practice, every effort has been made to obtain up-to-date information on the manufactures and uses of the chemicals considered and to reflect the enormous advances made in these respects in recent years. The sign conventions used in electrochemistry and thermochemistry are those recommended by the International Union of Pure and Applied Chemistry.

The contents of this book are, however, not limited to coverage of examination syllabuses. For instance, the concept of electronegativity has been discussed in a general way and so have the simplest ideas on, atomic and ionic radii, overpotential, resonance, hydrogen bonding, energetics, and molecular shape. Experience has shown that topics such as these are well within the grasp of the student at this stage and that even an elementary study of them stimulates interest and gives a clearer understanding of the subject.

Although knowledge sufficient to pass O level Chemistry or Physics-with-Chemistry is assumed, special attention has been paid to the needs of the weaker student who aims only at securing a pass in the subject at A level or in the O.N.C. examination. Numerous worked examples have been included in the text, and wherever possible the treatment within a chapter has been so ordered that the most difficult parts appear at the end. Care has been taken to reduce the mathematical content to a minimum, so that it can be followed by anyone with a knowledge of O level Mathematics. Tables are provided displaying the reactions of the principal organic compounds, and the properties of the elements in each periodic group are summarized in a comparative way; these features should prove particularly useful during revision.

My thanks are due to the Examining Authorities listed on page 853 for permission to reproduce more than 300 questions taken from their previous papers. These questions have been carefully selected from a much greater number because they test an understanding of the principles and concepts of the subject and of their application, and do not merely demand memorization of factual material. Since this sort of examination question seems likely to supersede the older type in the next few years, the provision of these questions should considerably enhance the value of the book.

I am extremely grateful to Mr. J. V. Westwood, M.Sc., F.R.I.C., of Sir John Cass College, London, for reading the whole of the manuscript. Not only has he made countless constructive criticisms of the text, but he has also given valuable guidance on its suitability for use in Colleges of Technology. I am indebted to Dr. John Bradley, M.A., M.Sc., of the Department of Education of Hull University, who will be well known to readers of the *School Science Review*, for useful suggestions relating to some of the earlier chapters. Lastly, I am happy to acknowledge the immense debt I owe to my wife Margaret for her invaluable contribution. She has painstakingly prepared the entire manuscript in longhand and has assisted with the diagrams, the checking of the proofs, and the indexing. Without her constant help and encouragement this book could never have been written.

J. H.

January 1963

Contents

1

The Atomic and Molecular Theories

1.1. Introduction. Chemistry involves a systematic study of the substances of the universe, of their properties and their reactions with each other to form new substances with different properties, and of the conditions governing those reactions. Were this all, then chemistry would be little more than a massive catalogue of facts and recipes and a collection of techniques for discovering them. Chemistry is also concerned with classifying the data it has acquired and where possible condensing them into a series of precise generalizations called laws. *These laws are statements which summarize in an ideal form the results of a large number of separate experiments or observations.* A law can be verified by performing experiments and showing that the results conform to the law within the known limits of experimental error. Because a law holds true for all investigated cases, it is assumed to be a reliable guide to cases not yet investigated experimentally, at least until evidence to the contrary is obtained. These scientific laws differ radically from the laws of a country or of a game; the latter are entirely arbitrary (i.e. they are drawn up by general agreement within a community or group to express what is desirable or expedient), whereas the laws of science are summaries of what has always been found to happen in Nature whether they suit man's interests or not.

Nor is chemistry concerned only with the discovery of laws. Like other experimental sciences its aim is to find *the simplest and most plausible explanation of these laws*—to invent a *theory*, as it is called—*which makes clear the relationship between one law and another*. Such a theory is an imaginative idea or set of ideas which links together a host of apparently unrelated facts into one harmonious whole. The theory must fit all the established facts and if a new fact becomes known which is in conflict with it, then the theory must be abandoned or modified to take account of it. A number of these theories will be encountered in this book—the atomic theory, the kinetic theory, the ionic theory, and the electronic theory of valency are four outstanding examples.

It is an important feature of scientific theories that they should not only provide a rational explanation of facts already known, but that they should also lead to a further advance in our knowledge by suggesting new experiments which are likely to prove fruitful and new facts which have hitherto remained undiscovered. Successful predictions of this kind tend to increase our confidence in the theory because they increase the range of phenomena to which it is known to apply. It follows that in science the value of a theory lies not so much in its truth (no theory is regarded as the absolute or ultimate truth, but rather as the best approximation to that truth which can be reached on the evidence so far available), as in its usefulness in unifying our knowledge and extending it. Only if this is realized can the importance of, for example, Dalton's atomic theory be fully appreciated.

From time to time a scientist puts forward a supposition called a *hypothesis* to serve as a basis for reasoning and further experiment. Such a proposal is a sort of inspired guess; at the time it often has little evidence to support it and is, in any case, by its very nature incapable of *direct* verification. Such a hypothesis is then examined critically to see what deductions can be drawn from it and how it can lead to new advances in our knowledge and understanding. Only if these deductions and the results which follow from them are in complete agreement with one another and with experimental fact is the hypothesis regarded as justified and acceptable.

In this chapter no attempt is made to provide a rigorous and detailed historical account of the development of the atomic and molecular theories as this is available in books on the history of chemistry, but the subject is treated here in such a way as to illustrate the meaning of the terms law, theory, and hypothesis and emphasize the workings of the scientific method. We take up the story at the close of the eighteenth century, when the extensive use of the balance in chemistry was providing the quantitative data which gave rise to the laws of chemical combination.

1.2. The Law of Conservation of Mass. This was first formally stated in 1789 by Lavoisier, who found that *in any chemical reaction the total mass of the products is always equal to the total mass of the reactants.* The word 'total' needs stressing here, as many reactions appear at first sight to proceed with a loss or gain in mass because one or more of the substances is a gas whose mass is omitted (e.g. the combustion of coal to ash involves an apparent loss in mass and the conversion of magnesium into magnesium oxide by heating in air a gain in mass).

The law was verified to one part in ten million by Landolt between 1890 and 1907. He chose fifteen reactions between liquids which involved only small heat changes and which could be performed in closed vessels shaped as in Fig. 1.1. Two such vessels of identical volume were counterbalanced against each other (to avoid buoyancy errors arising from changes in the density of the air during the experiment), then one vessel was inverted to mix the reactants and any change in mass was noted. Landolt's experiments showed that there was a very small loss in mass during the reaction, but if the apparatus was left to stand for a long time (in some cases as long as three weeks), then the final mass was always the same as the initial one within the limits of his experimental errors. He attributed the temporary change in mass to two side effects of the heat evolved by the reaction—to the evaporation of some of the moisture condensed upon the outside surface of the glass vessel and to slight expansion of the glass leading to an increase in the mass of air displaced and hence in the uplift on the vessel. An assessment of the experimental errors involved is always an important part of any quantitative scientific experiment. Landolt estimated his errors by doing what are called 'control experiments', i.e. he carried out numerous blank experiments using the same vessels but with substances which did not react together in order to find the range of random mass changes which resulted.

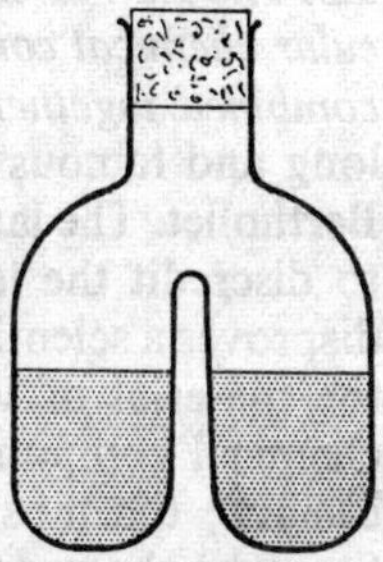

FIG. 1.1 Landolt's Flask

Working with even greater accuracy in 1912, Manley verified the law to one part in one hundred million for the reaction between solutions of barium chloride and sodium sulphate.

This law has sometimes been stated in the form that matter can neither be created nor destroyed in a chemical reaction, but on the basis of a much wider range of scientific evidence it is now believed that mass and energy are interchangeable and that a finite but extremely small loss in mass occurs in every chemical reaction in which energy is evolved. According to this theory, put forward by Einstein, energy E (in joule) is related to mass m (in kilogrammes) by the expression $E = mc^2$, where c, the velocity of light, has the value 3×10^8 m s^{-1}. Thus the change in mass expected during a chemical reaction involving the quantities normally used in the laboratory is of the order of 10^{-14} kg, which is far too small to be detected by even the most sensitive balance. It is explained in chapter 31 that when nuclear transformations occur (e.g. atomic fission or fusion) the energy changes involved are very much greater and the change in mass then becomes appreciable. Thus the law of conservation of mass is seen to be but a part of a much more comprehensive generalization that the total quantity of matter and energy in any isolated system remains constant. As far as chemical reactions are concerned, the law can be made literally true by rewording it thus: *No change in the total mass of all the substances taking part in a chemical reaction has ever been observed.* It is, of course, the implied basis of all gravimetric analysis and of every chemical equation.

1.3. The Law of Constant Composition (or Definite Proportions). *A particular chemical compound, when pure, always contains the same elements combined together in the same proportions by mass.* This law led to a long and famous controversy between Proust, who first stated it, and Berthollet. The latter cited numerous apparent exceptions in his attempt to discredit the law (an exception may 'prove a rule' but it certainly disproves a scientific law), but in every case Proust was able to show that the material of variable composition was a mixture and not a pure chemical compound. The law has been verified on a great many occasions; the most accurate experiments were carried out in 1860 by Stas who showed that silver chloride prepared in four different ways had a composition which varied by less than one part in 100 000.

It should be noted that the converse of this law is certainly not always true. It is often possible for the same elements to combine together in the same proportions by mass to give a number of different chemical compounds, as many organic examples demonstrate.

This law needs modification in two respects in the light of modern knowledge. With the discovery of isotopes (§ 1.14) it was realized that in exceptional circumstances it was possible to obtain samples of a given compound containing different isotopic mixtures, so that the proportions by mass of its constituent elements would differ. This is unlikely to happen in the course of ordinary chemical reactions because the proportions of each isotope in a sample of an element are normally constant, but it does limit the applicability of the law. Secondly, whilst the law applies exactly to gases and liquids, many crystalline solids are now known in which the composition is not constant but tends to vary continuously over a range. Such substances are called *non-stoichiometric compounds*; they are discussed in § 14.10, where explanations are given of their variable composition.

Care must be taken not to define chemical compounds in terms of fixed composition *alone*, for to do so involves arguing in a circle and the whole force of the law is lost. A compound should rather be seen as a substance composed of two or more simpler substances (elements) combined together chemically, so that it displays characteristic physical and chemical properties which are not the same as those of the elements of which it is made.

1.4. Law of Reciprocal Proportions (or Equivalent Proportions). *If two elements X and Y combine together and each also combines with a third* *element Z, then the proportions by mass in which X and Y combine together is that proportion in which they combine with a fixed mass of Z, or some simple multiple or fraction of that proportion.* For example, let *X* and *Y* be hydrogen and oxygen respectively, and let the third element *Z* be carbon, then the following experimental results illustrate the law:

Methane, CH_4	Carbon dioxide, CO_2	Water, H_2O
C : H = 12 : 4	C : O = 12 : 32	H : O = 2 : 16

Thus the proportions of H and O combining with 12 parts of C is 4 : 32, i.e. 1 : 8, which is the proportion found in water. Similarly, analysis of methane and carbon monoxide, CO, would give an H : O ratio of 1 : 4, and a ratio of 1 : 16 is obtained if ethylene, C_2H_4, and carbon dioxide are considered. These proportions are simple multiples or fractions of the proportion of 1 : 8 found in water.

The law grew out of the experimental work of many chemists, of which Richter and Berzelius were the most prominent. Its most accurate deliberate verification was made by Stas half a century later, but since the law is the basis of equivalent weights, every accurate determination of these quantities has indirectly verified it to a high degree of accuracy.

1.5. Dalton's Atomic Theory. To explain these three laws of chemical combination and the results of his work on mixtures of gases (see § 3.5), Dalton put forward the following theory in the first decade of the nineteenth century:

1. All matter is composed of extremely small, indivisible and indestructible particles called *atoms.*

2. The atoms of any one element are all exactly alike in every respect including mass, but are different from the atoms of every other element.

3. When elements form compounds, their atoms combine in simple numerical proportions such as 1 : 1, 2 : 1, 2 : 3, etc.

For many centuries people had suggested that matter was composed of discrete indivisible particles, and in composing his theory Dalton was undoubtedly influenced by their ideas, particularly those of Newton, but the originality of Dalton's theory lay in its quantitative aspect, i.e. in the emphasis it gave to the masses of the atoms and the simple proportions in which they combined. It was this part of the theory which aroused so much interest and which stimulated further speculation and experiment.

It will be seen that Dalton's theory provides a simple explanation of the law of conservation of mass, since according to the theory atoms are indestructible and chemical changes are merely rearrangements of these atoms to form new compounds, the total number of atoms (and therefore their total mass) remaining constant throughout. Moreover, according to Dalton's theory, when elements combine to form a compound they do so in proportions which, for any particular compound, are simple and fixed. Since the atoms of any one element are always the same mass, it follows that the proportions by mass of the various elements present in the compound are also fixed, which explains the law of constant composition. In the same kind of way the theory can be used to explain the law of reciprocal proportions. But the theory did more than just explain the laws that had already been discovered, for Dalton used it to predict a new law, known as the law of multiple proportions (§ 1.6) and showed that this law could be verified experimentally.

In the light of modern knowledge, Dalton's theory must be modified in two respects. Atoms can now be divided by physical means, e.g. by bombardment with very fast-moving particles or by passing an electrical discharge through a rarefied gas. Moreover, a number of radioactive elements are known in which the atoms cannot be prevented from disintegrating of their own accord. The first postulate should be reworded, therefore, that all matter is composed of extremely small particles called atoms, which cannot be divided or destroyed by chemical means. Secondly, the discovery of isotopes has invalidated the second of Dalton's postulates, since many elements are now known to exist as atoms which differ in mass although they have identical chemical properties. Despite these changes the essential ideas of the theory remain and it still provides the fundamental basis of theoretical chemistry.

1.6. The Law of Multiple Proportions. *When two elements combine together to form more than one compound, there is a simple numerical ratio between the different masses of one element which will combine with a fixed mass of the other.*

The law is best illustrated by the simple example provided by the oxides of carbon. Experiment shows that in carbon monoxide 16 parts by mass of oxygen combine with 12 parts by mass of carbon, whereas in carbon dioxide 32 parts by mass of oxygen combine with 12 parts by mass of carbon. Thus the masses of oxygen combining with the same mass of carbon in the two oxides are in the simple ratio of 16 : 32, i.e. 1 : 2.

This law was deduced by Dalton from his atomic theory and was published by him in 1803 with some inaccurate experimental evidence in support of it. In the ensuing years Berzelius verified the law for a large number of cases, thereby greatly increasing confidence in the atomic theory. Since then many highly accurate analyses have confirmed the exactness of the law, although it will be seen that deviations would occur if the isotopic composition of the elements concerned varied between the weighings.

1.7. Gay-Lussac's Law of Combining Volumes. In 1808 Gay-Lussac summarized the results of his experiments on the combination of gases in the following law:

The volumes of gases entering into and formed by a chemical reaction bear a simple numerical relation to one another when measured under the same conditions of temperature and pressure.

For example, experiment shows that 2 volumes of hydrogen combine with 1 volume of oxygen to form 2 volumes of steam, that 1 volume of nitrogen combines with 3 volumes of hydrogen to form 2 volumes of ammonia, that 1 volume of hydrogen combines with 1 volume of chlorine to form 2 volumes of hydrogen chloride, and so on. Very accurate analyses carried out during the period 1890–1920 have shown that the ratios of the combining volumes are not in fact exactly whole numbers. These discrepancies arise from the deviations of the gases concerned from the ideal gas laws of Boyle and Charles (see § 3.12) and they in no way detract from the importance of Gay-Lussac's law.

1.8. The Dalton-Berzelius Hypothesis. In his atomic theory Dalton had proposed that when elements combined their atoms united in small whole number proportions, and in his law Gay-Lussac had discovered that when gases combined the volumes that did so were in small whole number proportions. It was only natural, therefore, that Dalton and Berzelius should seek to correlate these two statements, which, by their very similarity, strongly suggested a connection between numbers of atoms and volumes. They did, in fact, make the simplest possible assumption by proposing the hypothesis that equal volumes of all gases at the same temperature and pressure contain equal numbers of atoms. When this supposition was applied to the experimental results for particular reactions, however, a serious contradiction arose, as the following example demonstrates:

1 vol. of hydrogen + 1 vol. of chlorine
gives 2 vols. of hydrogen chloride

$\therefore$ n atoms of hydrogen + n atoms of chlorine
gives $2n$ compound-atoms of hydrogen chloride
(where n is the number of atoms in 1 volume of the gases).

Dividing throughout by n we get:

1 atom of hydrogen + 1 atom of chlorine gives 2 compound-atoms of hydrogen chloride.

But it is clear that each compound-atom of hydrogen chloride must contain some hydrogen and some chlorine—at least one atom of each, in fact. Hence each of the atoms of hydrogen and of chlorine must have sub-divided during the reaction into at least two atoms, which is contrary to the fundamental postulate of the atomic theory that atoms are chemically indivisible. Provided Gay-Lussac's law holds true (and Dalton at first cast doubts on this, but they were soon dispelled by further experimental verification), then clearly either the Dalton-Berzelius hypothesis or the atomic theory must be abandoned, as they are incompatible. Since the evidence supporting the atomic theory was by that time very extensive, it was rightly decided that obvious though it appeared, the hypothesis was false and must be rejected.

The purpose of including this section has been to emphasize the vulnerability of even the most attractive hypothesis. In science each suggestion of this kind must be subjected to a searching and critical examination to ensure that it (and all the deductions made from it) are completely consistent with our existing knowledge. Where this is not so, the false hypothesis must be immediately discarded, for to retain it is to clutter up the channels of imaginative thought and hinder further progress.

1.9. Avogadro's Hypothesis. In 1811 Avogadro showed that it was possible to correlate Gay-Lussac's law with Dalton's atomic theory by proposing the existence of particles called *molecules*. He defined the molecule as the smallest particle of an element or compound that can exist in the free state, distinguishing it clearly from the atom which was regarded as the smallest particle of an element that can take part in a chemical reaction. His really original contribution was to suggest that atoms of the same element might cling together to form a single particle which was capable of undergoing sub-division when it participated in a chemical change. He further suggested that different elements probably contained different numbers of atoms in their molecules. The only differences between the molecules of an element and those of a compound, therefore, were that all the atoms in the former were alike and that the molecule of an element might, in some cases, consist of only one atom. Avogadro embodied these ideas in the hypothesis:

Equal volumes of all gases under the same conditions of temperature and pressure contain the same number of molecules.

One importance of this hypothesis is that it provides a means of deducing molecular relationships from experimental measurements on the volumes of gases taking part in a reaction. As we shall see in the next section, with its aid we can switch from data on volumes of gases to data on reacting molecules and so gain an understanding of events which, because they are on a molecular scale, are incapable of direct measurement or observation.

1.10. Applications of Avogadro's Hypothesis. When this hypothesis is applied to the experimental results obtained from a study of gaseous reactions, no contradiction arises, as the following example shows:

1 vol. of hydrogen + 1 vol. of chlorine
gives 2 vols. of hydrogen chloride (see § 25.6)

$\therefore$ n mols. of hydrogen + n mols. of chlorine
gives $2n$ mols. of hydrogen chloride

$\therefore$ 1 mol. of hydrogen + 1 mol. of chlorine
gives 2 mols. of hydrogen chloride.

Since each molecule of hydrogen chloride is identical, this means that each molecule of hydrogen and each molecule of chlorine must contain an even number of atoms. Now starting from Avogadro's hypothesis it can be shown by reasoning which is given later that these gases are definitely diatomic, i.e. their molecules each contain two atoms. Once this fact is established we can use Avogadro's hypothesis to determine the composition of hydrogen chloride by continuing the argument as follows:

1 mol. of hydrogen + 1 mol. of chlorine
gives 2 mols. of hydrogen chloride

$\therefore$ 2 atoms of hydrogen + 2 atoms of chlorine
gives 2 mols. of hydrogen chloride

$\therefore$ 1 atom of hydrogen + 1 atom of chlorine
gives 1 mol. of hydrogen chloride

showing that hydrogen chloride has the formula HCl.

In a similar way with the help of Avogadro's hypothesis we can determine the composition of other gaseous compounds once the *atomicity* (i.e. the number of atoms in the molecule) of the common gaseous elements has been settled. This is done for ammonia in § 22.4, for nitric oxide in § 22.11, and so on.

As Avogadro pointed out, one important consequence of his hypothesis is that we can determine the relative masses of the molecules of gaseous substances from a knowledge of their densities. A simple example will make this clear. Fig. 1.2 shows two identical vessels A and B containing two different gases at the same temperature and pressure. Now we know from Avogadro's hypothesis that these two vessels contain an equal number of molecules—let us imagine the number is five as shown. Then if the gas in A weighs twice as much, say, as the gas in B we can conclude that each of the molecules in A must weigh twice as much as each of the molecules in B, since there are equal numbers of each in the two vessels. Thus if the density of hydrogen,

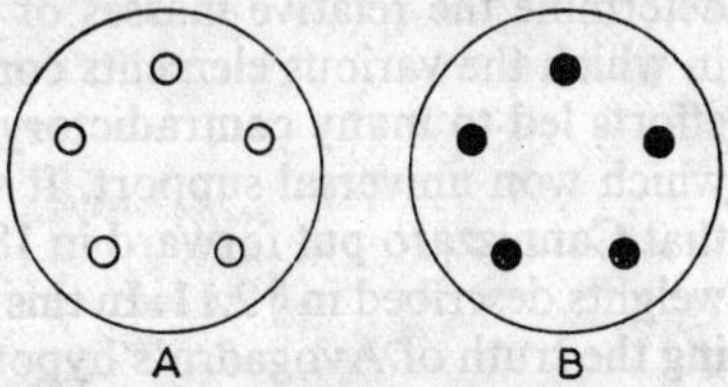

FIG. 1.2 Two Identical Vessels

the lightest gas known, is taken as our standard and the densities of all the other gases are measured and compared with that of hydrogen, then we can determine the masses of their molecules relative to the mass of the molecule of hydrogen. Since this molecule can be shown by Avogadro's hypothesis to contain two atoms of hydrogen, we can determine the mass of each molecule relative to the mass of an atom of hydrogen, i.e. we can find the molecular weight. This argument is summarized in the following steps:

Vapour density

$$= \frac{\text{mass of a volume of a gas}}{\text{mass of an equal volume of hydrogen under the same conditions}} \quad \text{(by definition)}$$

$$= \frac{\text{mass of } n \text{ molecules of the gas}}{\text{mass of } n \text{ molecules of hydrogen}} \quad \text{(by Avogadro's hypothesis)}$$

$$= \frac{\text{mass of 1 molecule of the gas}}{\text{mass of 1 molecule of hydrogen}} \quad \text{(dividing by } n\text{).}$$

But since the hydrogen molecule contains two atoms,

$$\text{Vapour density} = \frac{\text{mass of 1 molecule of the gas}}{\text{mass of 2 atoms of hydrogen}}$$

Now molecular weight is defined as $\dfrac{\text{mass of 1 molecule of a gas}}{\text{mass of 1 atom of hydrogen}}$

From which it follows that molecular weight = 2 × vapour density.

1.11. Cannizzaro's Contribution. It is remarkable but true that Avogadro's hypothesis was ignored and misunderstood for nearly fifty years after it had first been put forward, and this despite the fact that Ampère made a similar proposition in 1813. During this period great confusion prevailed in chemical theory with the emphasis placed upon efforts to determine the relative masses of atoms from the proportions by mass in which the various elements combined in forming compounds. These efforts led to many contradictory systems of 'atomic weights', none of which won universal support. It was against this background of chaos that Cannizzaro put forward in 1858 the method of determining atomic weights described in § 2.11. In this method Cannizzaro started by assuming the truth of Avogadro's hypothesis and used it for determining first the molecular weights of gaseous compounds and then the atomic weights of the elements in them. In this way, by considering a large number of their volatile compounds, he established the diatomicity of hydrogen, oxygen, chlorine, nitrogen, etc., by showing that their atomic weights were one half of their molecular weights. Finally he arrived at a set of reliable atomic weights of these and other elements including carbon, sulphur, phosphorus, and mercury.

Cannizzaro's strong advocacy of Avogadro's hypothesis soon led to its universal acceptance, since the atomic weights obtained were clearly not only consistent amongst themselves but gave rise to a consistent system of 'saturation capacities' or *valencies* for the elements concerned. This was a new and great achievement which made possible a rapid advance in theoretical organic chemistry during the period following 1860 and which solved many outstanding problems concerning the structure of organic compounds. The discovery by Mendeléeff and Lothar Meyer of a periodicity in the properties of the elements when they were arranged in the order of their newly determined atomic weights provided further striking confirmation of the truth of Avogadro's hypothesis and gave rise to the periodic classification of the elements described in chapter 13. Thus, although Avogadro's hypothesis by its very nature defies direct experimental verification, the indirect evidence in its favour is overwhelming and it has consequently become the agreed basis of all modern molecular theories.

1.12. The Avogadro Constant and the Mole. Various masses of substances bearing the same proportion to their molecular weights will contain equal numbers of molecules. Thus if we take that quantity of a substance equal to its molecular weight expressed in gramme, variously called the *gramme-molecular weight*, or the *gramme-molecule*, it will necessarily contain the same number of molecules as a gramme-molecule of any other substance. For example, 32 gramme of oxygen will contain the same number of molecules as 28 gramme of nitrogen, 44 gramme of carbon dioxide, or approximately 2 gramme of hydrogen. This number of molecules in a gramme-molecule is known as the *Avogadro constant*, N_A. Its value can be determined by various physical methods, e.g. from the diffraction of X-rays by crystals (see § 4.5), from the charge on the electron and measurements in electrolysis (see § 8.2), from measurements of radioactivity, from the viscosity of gases, from the surface tension of mono-molecular films, from observations of the Brownian motion (§ 12.3), and from the distribution of particles suspended in a medium. Diverse and independent though these methods are, they give results which are in good agreement with each other, the best value being 6.023×10^{23} mol^{-1}. Despite the vastness of this number a knowledge of it gives to the concept of the molecule a new reality which was not available to the chemists of the nineteenth century.

We shall see in the next section, and also in later chapters, that a great simplification can be made when certain quantities and concentrations are expressed in terms of gramme-molecule (or in the case of crystal molecules in terms of gramme-formula) instead of gramme. For example, as explained in § 2.12, the heat capacity of most solid elements measured at constant pressure at 25°C has the same value of about 26 JK^{-1} when referred to a gramme-atom, and the molar latent heat of vaporization of most liquids has a value of about 88 TJ, where T is the b.p. on the absolute scale (see § 3.3). Similarly, we shall see in § 5.20 that dilute solutions of non-electrolytes of the same gramme-molecular or gramme-formula concentration exhibit the same magnitude of colligative effect. This simplicity arises because in each case the *same number of particles* are involved when one gramme-molecule or one gramme-formula is considered, no matter what the substance is, provided that no association or dissociation occurs.

In SI one of the seven independent basic physical quantities is amount of substance. The basic unit chosen for amount of substance is the mole (symbol mol). One can think and speak of a mole of molecules, atoms, ions, or radicals as the amount of that substance which contains the same number of the specified particles as there are atoms of carbon present in exactly 0.012 kilogramme of the carbon-12 isotope. Thus one mole of any substance always contains the same number of particles (molecules, atoms, or ions, as the case may be), and this number is the Avogadro constant, N_A. It follows that the terms gramme-molecule, gramme-atom, and gramme-ion can now be defined as the masses (expressed in gramme) of one mole of that particular species, and this has led to the widespread use of the mole to replace these other concepts.

The significance of the mole lies in the way it directly relates the actual masses of substances taken, which are measurable, to the numbers of molecules (or atoms or ions) involved, which are not. Thus by use of the mole concept we are able to visualize and calculate what is happening on a molecular scale.

1.13. The Gramme-Molecular Volume. The converse of Avogadro's hypothesis is also true, viz. the same number of molecules of all gases occupy equal volumes under the same conditions of temperature and pressure. It follows that one mole of any gas will occupy the same volume under the same conditions. This volume, known as the *gramme-molecular* (or *molar*) *volume*, is approximately 2.24×10^{-2} m^3 mol^{-1} at the standard temperature and pressure (0°C, 101 325 Nm^{-2} pressure). Like Gay-Lussac's law, Avogadro's hypothesis cannot be exactly true under all conditions owing to the deviations which real gases show from the laws of Boyle and Charles (see § 3.12). The same applies to its converse, of course, so that the gramme-molecular volume does vary very slightly from gas to gas, but any substantial deviation is significant because it suggests that the gas is undergoing association or dissociation.

1.14. Modern Theories of Atomic Structure. Since Dalton's time a great deal of additional experimental evidence has become available to help us build up a picture of the structure of matter. Much of this evidence has come from a study of radioactivity and atomic transmutations and the scattering of α-particles (all three topics are discussed in chapter 31), and from experiments involving the discharge of electricity in rarefied gases. As a result it has been realized that an atom is not to be regarded as a solid particle of matter like a miniature ball-bearing, but that it appears to consist largely of empty space with most of its mass concentrated into a relatively very small central particle called the *nucleus*, which bears a positive charge. This nucleus is believed to be composed of two types of particle, the *proton*, which has unit mass and unit positive charge, and the *neutron*, which has almost the same mass as the proton but, as the name implies, bears no charge.

There is no simple relation between the number of protons and neutrons in an atom; in general (hydrogen is an exception here) there are at least as many neutrons as protons and in the heavier atoms the neutrons are the more numerous. Thus the total charge on the nucleus of an atom is determined by the number of protons it contains. This important quantity is called the *atomic number* of the element. It is the most fundamental property of an atom since it uniquely determines which element it is—all the atoms of any one element have the same atomic number, which differs from the atomic number of any other element. Elements with atomic numbers from 1 to 103 are now known. The atomic number is more than just a theoretical concept and a means of defining elements, however, for it can be measured indirectly by Moseley's method as described in § 13.5. Its importance as the basis of the periodic classification of the elements is made clear in chapter 13.

It is further believed that the nucleus of an atom is surrounded by a number of particles called *electrons*. Experiments have shown that each of these electrons bears unit negative charge but has a mass only about $\frac{1}{1836}$ as great as that of a proton. Since under normal conditions the atom of an element is electrically neutral overall, the number of electrons in it must equal the atomic number. Experiments on the spectra of elements suggest that these electrons are arranged around the nucleus in a definite way rather like the planets in a miniature solar system. This arrangement, which we now believe to be the predominant influence in deciding the chemical properties of an element, is discussed in detail in §§ 13.7, 13.8 and 13.11.

Although the atoms of any one element all contain the same number of protons in their nuclei, it is believed that the number of neutrons present may vary from one atom to another, so that the atoms of an element are not necessarily all the same mass. This fact is best demonstrated experimentally by using the mass spectrograph as described in § 2.17. These various forms of the same element are called *isotopes*, since they are allocated to the same place in the periodic table. Isotopes of the same element differ from each other only in the number of neutrons contained in their nuclei; their atomic numbers and the number of electrons in each of their atoms are exactly the same. Chemically they are identical and it is impossible to separate them by chemical means, but isotopes of the same element do differ slightly in certain physical properties (e.g. density) which depend upon the masses of their atoms. These differences are most marked in the isotopes of the lightest element hydrogen (see § 16.3), because the percentage difference in mass is greatest in their case. The widespread occurrence of isotopes and the relation between atomic weights and isotopic weights are discussed in the next chapter. Some of the uses of radioactive isotopes are given in § 31.10.

In recent years elaborate experiments in physics have led to more advanced theories concerning the composition of the nucleus and the nature of the forces which hold it together. These cohesive forces are vastly stronger than the forces between electrostatic charges, which is why positively charged protons in the nucleus do not fly apart by mutual repulsion. The density of an atomic nucleus is also immense; a typical nucleus, as experiments on the scattering of α-particles show (§ 31.5), has a radius of about 10^{-15} m and accounts for about 99.97% of the mass of the atom, so that its density must exceed the value of 10^{17} kilogramme per cubic metre. The existence of a large number of sub-atomic particles has been proposed to account for these facts, but for our purposes the simple ideas on the nucleus outlined above, which were largely due to Rutherford and his contemporaries, will suffice. In later chapters our interest in atomic structure will focus instead upon the spatial configurations of the electrons and upon the strength of the forces retaining them in the atom.

1.15. Modern Molecular Theories. Just as in the last 160 years the simple atomic theory of Dalton has developed into the elaborate theories of atomic structure which are so useful today, so also has Avogadro's simple concept of the molecule become the basis of more complicated molecular theories. The chief of these in scope and importance is the kinetic theory, which is discussed in detail in chapter 3. Developed in the middle of the nineteenth century to explain the physical properties of gases and the laws summarizing their behaviour, this theory has proved to be a fruitful source of new knowledge.

Once the idea of the molecule had been firmly established, chemists set about determining molecular weights by the various methods described in chapter 5. Later they became increasingly concerned with the structure and shape of molecules. At first their attention was concentrated upon the molecules of organic compounds in an attempt to solve such problems as the structure of benzene (§ 43.3) and the explanation of stereoisomerism (§ 48.2), but with the discovery of the electron and the development of the modern theories of atomic structure they turned their attention to the field of inorganic chemistry as well. Here, aided by the results of X-ray analysis, electron diffraction, spectroscopy, and the measurement of dipole moments, they have used the electronic theory of valency (chapter 14) and the theory of molecular orbitals to provide explanations of a host of puzzling facts and have greatly increased our understanding of the structure of molecules. It is interesting to note, however, that it was not until 1927 that Heitler and London, using the quantum theory and the idea of electron spin, were able to provide a satisfactory solution of a problem that had been one of the greatest stumbling blocks in the early days of Avogadro's hypothesis, viz. why two apparently identical atoms of hydrogen should unite so firmly to form a diatomic molecule.

The success of these molecular theories must not blind us to their limitations. Whilst the concept of the molecule is so valuable when dealing with gaseous substances, care must be exercised not to apply it indiscriminately to liquids and solids. As explained in later chapters, we now believe that many compounds (e.g. hydrogen fluoride, water, ammonia, alcohols, and organic acids) are associated in the liquid state into multiple molecules as a result of hydrogen bonding (§ 14.7), so that to represent them by their simplest formulae can be misleading. Again, many substances are reversibly dissociated into smaller molecules at high temperatures or undergo electrolytic dissociation in solution. As described in chapter 4, a great many solids, particularly the salts of the strongly electropositive metals, are known to consist solely of ions arranged in a regular three-dimensional framework known as the crystal lattice. For such substances the term molecule can have only a limited application since there is no means of clearly defining its extent or shape. When we apply a chemical formula to such substances we do so to convey only the proportions in which the various elements are present without inferring the existence of any actual discrete particles. A similar difficulty arises in considering certain non-volatile, insoluble, crystalline substances (e.g. carbon and silica) which exist as giant-molecules, i.e. the whole crystal is composed of atoms linked into one relatively vast molecule. It follows that the term molecular weight can have no meaning in such cases as these.

Indeed the view of the molecule as a particle of matter capable of leading a separate and independent existence is only really applicable to volatile covalent substances (§ 14.4), but since this category includes nearly all the non-metallic elements and their halides and hydrides as well as the majority of organic compounds, the concept is of paramount importance.

2

Equivalent Weights and Atomic Weights

2.1. Introduction. It follows from the law of reciprocal proportions (§ 1.4) that the masses of different elements which combine with a fixed mass of any one element are also the masses, or simple multiples of those masses, that combine together. Thus if we select a given mass of one particular element as our standard, we can draw up a list of corresponding combining masses for all the other elements. The standard universally chosen in defining *equivalent weights* is 8.0000 parts by mass of oxygen. With the advent of the concept of the mole (§ 1.12), equivalent weights have assumed less importance than hitherto and some chemists now prefer to think entirely in terms of molar rather than normal solutions. However, most examining boards still expect a knowledge of equivalents, and in view of the great historical importance of equivalent weights as a source of atomic weights and their usefulness and convenience in volumetric analysis, their retention in this chapter is felt to be fully justified.

2.2. The Equivalent Weight of an Element. This is defined as *the number of parts by mass of that element which will combine with or displace 8.0000 parts by mass of oxygen, or the equivalent weight of any other element*. For example, in magnesium oxide 24 parts by mass of the metal* are combined with 16 parts by mass of oxygen, so the equivalent weight of magnesium is 12. Similarly, when magnesium dissolves in hydrochloric acid, 24 parts by mass of it displace 2 parts by mass of hydrogen, again giving an equivalent weight of 12.

Two points need emphasizing here. First, an equivalent weight is a pure number—it is the number of parts by mass or the number of gramme. Equivalent weight itself has no units, therefore, but it is often convenient to refer to a *gramme-equivalent* of an element which is *its equivalent weight expressed in gramme*. Thus, in the above example, magnesium has an equivalent weight of 12, and the gramme-equivalent

* In the examples quoted in this section, only approximate figures are given for the sake of simplicity.

of magnesium is 12 gramme. Secondly, certain elements combine together to form more than one compound, so that they necessarily have two or more different equivalent weights, but from the law of multiple proportions (§ 1.6) it follows that these equivalent weights will bear a simple numerical ratio to each other. For example, iron forms an oxide FeO in which 56 parts by mass of iron are combined with 16 parts by mass of oxygen, and an oxide Fe_2O_3 containing 112 parts by mass of iron and 48 parts by mass of oxygen. Thus iron has an equivalent weight of 28 in ferrous oxide and $18\frac{2}{3}$ in ferric oxide, and these are in the ratio of 3 : 2.

2.3. Determination of Equivalent Weights. As explained later in this chapter, this process is of great importance because it provides the basis for the chemical methods of determining atomic weights. Various methods are available, although only a few of them are capable of high accuracy.

1. By Conversion into the Oxide: A known mass of the element is completely converted into its oxide and the mass of the product is found. The oxide may be formed directly by burning the element in excess of air or oxygen (e.g. sulphur or carbon) or it may be prepared indirectly by converting the metal (e.g. copper, lead, or tin) into its nitrate, carbonate, or hydroxide first and then decomposing the product to constant weight by strong heating. If the oxide formed is acidic and gaseous (e.g. carbon dioxide or sulphur dioxide) its mass is found by absorbing it in a solution of caustic alkali which is weighed before and after the reaction. The oxide method is simple and widely applicable, but it is difficult to avoid an error arising from the occlusion of gases by metallic oxides.

Example: 3.000 g of copper were converted into 3.755 g of copper oxide.

∴ 3.000 g of copper combine with 3.755 − 3.000 = 0.755 g of oxygen
∴ 8.000 g of oxygen combine with (3 × 8)/0.755 = 31.79 g of copper
∴ Equivalent weight of copper = 31.79.

2. By Reduction of the Oxide: This is the reverse of the first method. A known mass of pure oxide is reduced by heating it in hydrogen until no further loss in weight occurs, and the mass of the residual metal is found. The method is limited in its application to such metals as lead, copper, and nickel which are only weakly electropositive (§ 10.7).

3. By Displacement of Hydrogen: A known mass of metal is added to an excess of a suitable acid or alkali and the mass of hydrogen

liberated is calculated from the volume corrected to s.t.p.* The method can be used for magnesium, aluminium, zinc, iron, or tin and concentrated hydrochloric acid, and for aluminium or zinc and concentrated potassium hydroxide solution, but it is not very accurate.

Example: When added to an excess of hydrochloric acid, 1.22 g of zinc liberated 458 cm³ of hyrogen at 17°C and 98 660 Nm^{-2} pressure.

$$\therefore \text{ Volume of hydrogen liberated at s.t.p.} = 458 \times \frac{273 \times 98\,660}{290 \times 101\,325} \text{ (see § 3.4)}$$

$$= 420 \text{ cm}^3$$

Now 1000 cm³ of hydrogen at s.t.p. weight 0.09 g

∴ 420 cm³ „ „ „ „ „ (0.09 × 420)/1000 = 0.0378 g

∴ 1.22 g of zinc displace 0.0378 g of hydrogen

$$\therefore \frac{1.22 \times 1.008}{0.0378} = 32.6 \text{ g of zinc displace 1.008 g of hydrogen}$$

∴ Equivalent weight of zinc = 32.6.

4. BY DIRECT COMBINATION WITH HYDROGEN: Although of limited application, this method has been used to obtain very accurate values for chlorine (Edgar, 1908) and for the H : O ratio (Morley, 1895, and Burt and Edgar, 1916). Details of these classic experiments and of the extraordinary precautions taken to minimize errors will be found in more advanced books.

5. BY ANALYSIS OF CHLORATES: This method is capable of the highest accuracy and has been used first by Stas and then by Richards to obtain our best present-day values for the equivalent weights of silver and chlorine. These values are then used as secondary standards for determining the equivalent weights of other elements. The method is best explained by considering the typical example of potassium chlorate. After extensive purification a known mass of this salt is heated to constant weight and the amount of residual potassium chloride is found. From this the mass of potassium chloride which combines with 48 parts by mass of oxygen can be calculated. This quantity of potassium chloride is then added to an excess of silver nitrate solution and the mass of silver chloride precipitated is found. The mass of silver needed to make this mass of silver chloride is then determined; this is the equivalent weight of silver. By subtracting it from the mass of the silver chloride, the equivalent weight of chlorine can be calculated.

6. BY DOUBLE DECOMPOSITION WITH SILVER NITRATE: This method depends upon a knowledge of the accurate equivalent weights of silver and chlorine obtained by the previous method. The element whose

* Standard temperature and pressure, i.e. 0°C and 101 325 Nm^{-2} pressure. (See § 3.3.)

equivalent weight is required is first converted into a pure sample of its chloride. A known mass of this is dissolved in water and added to an excess of silver nitrate solution, and the mass of precipitated silver chloride is found. This method gives good results with a large range of metals, including those too electropositive for the first three methods. It can also be used for finding the equivalent weight of nitrogen by using a solution of ammonium chloride.

Example: When 0.974 gramme of pure sodium chloride were added to an excess of silver nitrate solution, 2.390 gramme of silver chloride were precipitated. Given that the equivalent weights of silver and chlorine are 107.88 and 35.46 respectively, calculate the equivalent weight of sodium.

Let the equivalent weight of sodium be x.

Then $$\frac{x + 35.46}{107.88 + 35.46} = \frac{\text{mass of sodium chloride}}{\text{mass of silver chloride}} = \frac{0.974}{2.390}$$

$$\therefore \quad 2.39(x + 35.46) = 143.34 \times 0.974$$

$$\therefore \; x = \frac{139.6 - 84.7}{2.39} = 22.97$$

∴ Equivalent weight of sodium is 22.97.

7. By Conversion Methods: A known mass of one compound of an element is converted completely into another of its compounds, whose mass is found. Provided the equivalent weights of two of the radicals concerned are known, that of the third can be calculated since the equivalent weight of any compound is the sum of the equivalent weights of its constituent radicals. It will be seen that method 6 is really a special case of this general method in which the chloride radical is common to both compounds.

8. By Electrolysis: A current is passed through two solutions connected in series and the masses of metals deposited at the cathodes are compared. If the equivalent weight of one metal is known, then that of the other can be calculated because by Faraday's second law (§ 8.2) the masses liberated by the same quantity of electricity are proportional to their chemical equivalents. Although this is an accurate method, it is only of limited application.

Example: When a current was passed through solutions of silver nitrate and copper sulphate connected in series, 1.20 gramme of silver and 0.353 gramme of copper were deposited on the respective cathodes. Given that the equivalent weight of silver is 108, calculate the equivalent weight of copper.

Let the equivalent weight of copper be x.

Then $$\frac{1.20}{0.353} = \frac{108}{x}, \text{ giving } x = 31.8.$$

∴ Equivalent weight of copper is 31.8.

9. BY DISPLACEMENT: As explained in § 10.4, metals will displace other metals lower in the electrochemical series from solutions of their salts, e.g. zinc will displace copper from copper sulphate solution. Since the masses of the metals concerned are in the ratio of their equivalent weights, if the equivalent weight of one metal is known, that of the other can be calculated. The method is of no practical importance, however.

2.4. The Equivalent Weight of Acids, Bases, and Salts. The idea of equivalence is not restricted to elements but can be applied to compounds as well. Indeed Cavendish discovered as early as 1766 that definite amounts of certain acids and alkalis were chemically equivalent to each other. The following definitions and examples illustrate how the equivalent weights of various types of compound are calculated.

The equivalent weight of an acid is the number of parts by mass of it which contains 1 part by mass of hydrogen replaceable by a metal.* For example, the molecular weight of sulphuric acid, H_2SO_4, is 98 and each molecule contains two atoms of hydrogen replaceable by a metal. Hence in the reaction represented by the equation $H_2SO_4 + 2NaOH = Na_2SO_4 + 2H_2O$, 2 gramme of hydrogen are replaced in every 98 gramme of acid and the equivalent weight is $\frac{98}{2}$, i.e. 49. However, in acetic acid, $CH_3.COOH$, which has a molecular weight of 60, only one of the hydrogen atoms in each molecule is replaceable by a metal. The equivalent weight is equal to the molecular weight, therefore, despite the fact that each molecule contains four hydrogen atoms. The number of hydrogen atoms in each molecule of an acid replaceable, directly or indirectly, by a metal is called the *basicity* of that acid. From these examples it will be seen that the equivalent weight of an acid is equal to its molecular weight divided by its basicity.

The equivalent weight of a base or carbonate is the number of parts by mass of it which will react with an acid to replace 1 part by mass of hydrogen by a metal. For example, the reaction between sodium hydroxide and hydrochloric acid proceeds according to the equation:

$$\underset{(23+16+1)}{NaOH} + HCl = NaCl + H_2O.$$

Thus 40 parts by mass of sodium hydroxide bring about the replacement of 1 part by mass of hydrogen by sodium and the equivalent weight of sodium hydroxide is therefore 40.

The equivalent weight of a salt (as a salt) is the number of parts by mass of it produced by the replacement of 1 part by mass of hydrogen

* Strictly 1.008 parts by mass, but the approximate value is usually adequate in these definitions.

from its parent acid. For example, potassium chlorate, $KClO_3$, is formed theoretically from chloric acid, $HClO_3$, by replacing one hydrogen atom in each molecule. Its equivalent weight as a salt is therefore equal to its molecular weight, i.e. 122.5.

The equivalent weights of acids, bases, and salts are chiefly determined by volumetric analysis, but special methods do exist for organic acids and amines (§ 32.10).

2.5. The Equivalent Weight of an Oxidizing Agent. This is usually* defined as *the number of parts by mass of it which yields 8 parts by mass of available oxygen* (i.e. oxygen given up to a reducing agent). As an example let us consider potassium permanganate in acid solution, when it behaves in accordance with the following fundamental equation:

$$\underset{2(39+55+4\times16)}{2KMnO_4} \equiv K_2O + 2MnO + 5\bar{O}$$

2×158 g	$\equiv$	5×16 g of available oxygen
316 g	$\equiv$	80 " " " "
$\therefore$ 31.6 g	$\equiv$	8 " " " "

$\therefore$ Equivalent weight of potassium permanganate in acid solution is 31.6.

It should be noted that in neutral or alkaline solution potassium permanganate behaves differently, oxidizing in accordance with the equation:

$$2KMnO_4 \equiv K_2O + 2\,MnO_2 + 3\bar{O}.$$

Under these circumstances 316 gramme of the permanganate yield 48 gramme of available oxygen, so its equivalent weight is $\frac{316}{6}$, i.e. 53.2. Note also that its equivalent weight as a salt is different again, being equal to its molecular weight, i.e. 158.

2.6. The Equivalent Weight of a Reducing Agent. This is usually defined as *the number of parts by mass of it which will combine with 8 parts by mass of available oxygen.* For example, when oxalic acid is acting as a reducing agent it behaves in accordance with the equation:

$$\underset{(2\times1+2\times12+4\times16)}{H_2C_2O_4} + \bar{O} = 2CO_2 + H_2O.$$

Thus 90 gramme of it combine with 16 gramme of available oxygen and the equivalent weight of oxalic acid as a reducing agent is $\frac{90}{2}$, i.e. 45. Similarly when iron(II) sulphate is oxidized to iron(III) sulphate the

* Alternative definitions of the equivalent weights of oxidizing and reducing agents in terms of electron transfer are given in § 15.7.

fundamental change is given by:

$$2FeSO_4 + H_2SO_4 + \bar{O} = Fe_2(SO_4)_3 + H_2O$$
$$2(56+32+4\times16)$$
$$2 \times 152 \text{ g} \equiv 16 \text{ g available oxygen.}$$

Hence 152 gramme of iron(II) sulphate are equivalent to 8 gramme of available oxygen and its equivalent weight is 152.

2.7. Compounds with Several Equivalent Weights. When a compound is hydrated the equivalent weight must be calculated from the formula weight of the hydrate and not from the molecular weight of the anhydrous compound. For example, considering the two examples quoted in the previous section, crystals of oxalic acid having the formula $H_2C_2O_4.2H_2O$ (formula weight 126) have an equivalent weight of $\frac{126}{2}$, i.e. 63 as a reducing agent, and hydrated ferrous sulphate crystals, $FeSO_4.7H_2O$ (formula weight 278), have an equivalent weight of 278 not 152.

It is important to realize that a compound may have more than one equivalent weight depending upon the reactions in which it takes part. In such cases the equation should always be considered in calculating the equivalent weight, and to avoid ambiguity the value should always be stated with respect to acidity, or reducing action, etc. This is well illustrated by potassium hydrogen oxalate, as follows:

As an acid:

$$\begin{matrix} COOK \\ | \\ COOH \end{matrix} + KOH = \begin{matrix} COOK \\ | \\ COOK \end{matrix} + H_2O$$
$$(2\times12+4\times16+39+1)$$

128 gramme contain 1 gramme of replaceable hydrogen

∴ Equivalent weight (as an acid) is 128.

As a reducing agent:

$$\begin{matrix} COOK \\ | \\ COOH \end{matrix} + \bar{O} = 2CO_2 + KOH$$

$$128 \text{ gramme} \equiv 16 \text{ gramme of available oxygen}$$

∴ Equivalent weight (as a reducing agent) is $\frac{128}{2}$, i.e. 64.

2.8. Atomic Weight. As we have seen in § 1.5, Dalton's atomic theory focused attention upon the mass of atoms. Since atoms were far too light to be weighed by even the most sensitive balance, there was no means of finding their actual absolute masses, and chemists concentrated instead upon determining the relative masses of the atoms of the various elements. Some element had to be taken as a standard for this purpose and hydrogen was the obvious choice since its atom was the lightest known. The masses of all the other atoms were expressed, therefore, on a scale on which the weight of the hydrogen atom was one unit. This scale was later abandoned in favour of one using oxygen as the standard instead, so that for many years the atomic weight of an element was defined as the mass of one of its atoms on a scale on which an atom of oxygen weighed 16.0000 units. This change was made for two reasons. Firstly, many atomic weights were derived from equivalent weights which were in turn based upon the combining proportions of the elements with oxygen, so that every time the combining ratio O : H was determined with greater accuracy, all these atomic weights had to be altered. Secondly, with O = 16.000 as standard, many atomic weights approximated more closely to whole numbers. As explained in § 2.18, the discovery in this century that ordinary oxygen consists of a mixture of three types of oxygen atoms differing slightly in mass and known as isotopes led to the abandonment of oxygen as the standard and the adoption in 1961 of the carbon-12 isotope instead. *Atomic weight is now defined, therefore, as the mass of an atom on the scale on which the carbon-12 isotope has a mass of exactly twelve.*

It has already been stressed that atomic weights are not real weights at all, but only relative weights*. They are, therefore, pure numbers with no units. However, since the Avogadro constant, N_A (see § 1.12), is the number of molecules in one gramme-molecule, if the molecular weight and the atomicity of an element are known then the actual mass of its atoms can be calculated. For example, since 6.023×10^{23} molecules of hydrogen, each containing two atoms, has a total mass of 2.016 g, it follows that the mass of each atom of hydrogen must be

$$\frac{2.016}{2 \times 6.023 \times 10^{23}} = 1.673 \times 10^{-24} \text{ g} = 1.673 \times 10^{-27} \text{ kg}$$

i.e. 0.000 000 000 000 000 000 000 001 673 kg. Little wonder that chemists are not concerned with the absolute masses of atoms, but prefer to deal with the much more convenient system of relative weights known as atomic weights in which the mass of any atom is expressed as the number of times it is greater than one twelfth of the mass of an atom of the carbon-12 isotope (see § 2.18).

2.9. Valency. It follows directly from the atomic theory of Dalton that the atomic weight of an element must be an exact multiple of its relative combining weight provided both are measured on the same scale, i.e.

* Hence the alternative term *relative atomic mass.*

atomic weight = $n \times$ combining weight, where n is necessarily an integer because atoms are chemically indivisible. Since the equivalent weight is a multiple of the combining weight, then atomic weight = n' $\times$ equivalent weight, where the integer n' is known as the *valency* of the element. Valency is a measure of the combining capacity of an element. It is best defined numerically as the *number of hydrogen atoms which combine with or are displaced by one atom of the element.* From the definitions of atomic weight and equivalent weight already given, it should be clear that hydrogen is always univalent (i.e. it always has a valency of one) and that oxygen is divalent (i.e. it has a valency of two). Elements with valencies of 3, 4, 5, 6, 7, or 8 are described as trivalent, tetravalent, pentavalent, hexavalent, heptavalent, and octavalent, respectively.

It was seen in § 2.2 that certain elements have more than one equivalent weight. It follows, since the atomic weight of an element is fixed, that such elements also have more than one valency. In the example already considered iron has been shown to have equivalent weights of 28 and $18\frac{2}{3}$. Since its atomic weight is known to be 56, we conclude that its valencies are 2 and 3 respectively. Variable valency is very common in the transition metals (see § 27.2), although it is by no means restricted to those elements.

2.10. Determination of Atomic Weights. There are seven important ways of determining the atomic weight of an element. Four of these are chemical methods which depend upon a knowledge of the equivalent weight of the element. They are, in effect, means of determining which multiple of the equivalent weight is the atomic weight, i.e. they are essentially methods for deciding the valency of the element. Mitscherlich's method does this by direct comparison, but the methods of Dulong and Petit, Cannizzaro, and Mendeléeff do it by establishing the approximate value of the atomic weight. When this is known the valency of the element can be deduced from the expression valency = rough atomic weight $\div$ accurate equivalent weight, since it must be an integer and its value will be obvious even though the atomic weight value is only an approximate one. Once the valency is known it can be multiplied by the accurate equivalent weight to give the atomic weight accurately. The importance of determining the equivalent weights of elements very accurately should now be apparent, since all the chemical methods of determining atomic weight depend directly upon them.

2.11. Cannizzaro's Method. The historical importance of this method has already been described in § 1.11. The method involves the following steps:

1. The molecular weights of a large number of volatile compounds of

the element are found by measuring their vapour densities (by methods described in §§ 5.3–5.5) and multiplying these values by 2. The theoretical justification for this procedure has already been given in § 1.10.

2. The percentage composition by mass of these compounds is determined by chemical analysis.

3. From the molecular weight of a compound and its percentage composition, the relative weight of that particular element in the molecular weight of each of its compounds is calculated.

4. *The assumption is then made that in at least one of the many compounds selected there is only one atom of the chosen element present in each molecule.* This is the more probable the greater the number of compounds considered. Thus the atomic weight of the element is taken as the least weight of it that is found in the molecular weight of any of its compounds, or, more strictly, as the highest common factor of the values obtained.

5. Since vapour density measurements give only approximate values for molecular weights, the atomic weight found by this method is only a rough one, but as explained in the previous section, if the equivalent weight is known accurately then the accurate atomic weight is easily calculated.

6. Although originally restricted to elements which formed a large number of volatile compounds (e.g. H, C, O, N, Cl and S) because molecular weights could only be found from vapour densities, the Cannizzaro method can also be applied to molecular weights determined by the colligative methods described in chapter 5.

By comparing the density of a great many volatile hydrogen compounds with that of hydrogen gas itself, Cannizzaro was able to show conclusively that the least weight of hydrogen found in any of its compounds (which must, by definition, be the atomic weight of hydrogen) was in fact one half of the weight of a hydrogen molecule. This proof that hydrogen is definitely diatomic was one of the first and most important fruits of Cannizzaro's method; its theoretical significance has already been explained in §§ 1.10 and 1.11.

Cannizzaro's method is illustrated for carbon in Table 2.1. From this table it is seen that the rough atomic weight is probably 12, a conclusion which could be confirmed by considering a much larger range of carbon compounds. Since the equivalent weight of carbon is 3.0025 the valency must be four and the accurate atomic weight must be 12.010.

2.12. Dulong and Petit's Rule. In 1819 Dulong and Petit published their rule that *for solid elements the product of the heat capacity and the atomic weight, known as the molar heat capacity, is approximately constant.* Using modern atomic weights, the molar heat capacity of

TABLE 2.1. CANNIZZARO'S METHOD APPLIED TO CARBON

Substance	*Vapour density*	*Approx. mol. wt.*	*% carbon*	*Wt. of carbon in mol. wt.*
Carbon monoxide	14	28	43	12
„ dioxide	22	44	27.3	12
Methane	8	16	75	12
Ethylene	14	28	85.7	24
Acetylene	13	26	92.3	24
Benzene	39	78	92.3	72
Ethyl alcohol	23	46	52.2	24
Acetone	29	58	62.1	36
Diethyl ether	37	74	64.9	48
Carbon disulphide	38	76	15.8	12
Toluene	46	92	91.3	84
Carbon tetrachloride	71	142	84.5	12

most solid elements lies between 25 and 27 joule per mole kelvin, although certain elements (e.g. carbon, silicon, beryllium, and boron),

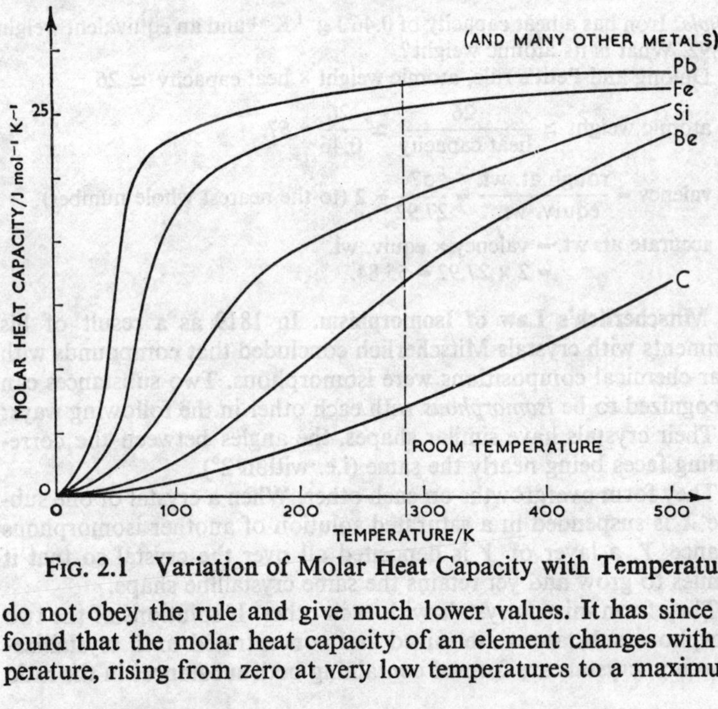

FIG. 2.1 Variation of Molar Heat Capacity with Temperature

do not obey the rule and give much lower values. It has since been found that the molar heat capacity of an element changes with temperature, rising from zero at very low temperatures to a maximum of about 27 J mol^{-1} K^{-1}. As shown in Fig. 2.1, the temperature at which this maximum value is reached varies widely from one element to another, however, and the common exceptions to Dulong and Petit's rule at room temperature do in fact conform if their heat capacity is measured at a high enough temperature.

Although it is now of little practical importance, the method was much used during the nineteenth century for deciding the atomic weights of metals. Before Dulong and Petit's rule could be used in this way its general validity (which up to then had been based only upon the tentative atomic weight values of Berzelius) had to be firmly established. This was done in 1858 by Cannizzaro, who ingeniously applied the rule to solid bromine, iodine, and mercury, and showed that the atomic weights determined with its aid agreed with those he had already found for these three elements by applying his method to their volatile compounds. Having calibrated the rule by demonstrating that it worked correctly in these test cases, Cannizzaro was then able to use it with confidence to determine the atomic weights of a large number of metallic elements for which his own method was not suitable because they did not form many volatile compounds.

Example: Iron has a heat capacity of 0.46 J g^{-1} K^{-1} and an equivalent weight of 27.92. What is its atomic weight?

By Dulong and Petit's rule, atomic weight × heat capacity $\simeq$ 26

$$\therefore \text{ atomic weight} \simeq \frac{26}{\text{heat capacity}} \simeq \frac{26}{0.46} \simeq 57.$$

$$\therefore \text{ valency} = \frac{\text{rough at. wt.}}{\text{equiv. wt.}} = \frac{57}{27.92} = 2 \text{ (to the nearest whole number)}$$

$$\therefore \text{ accurate at. wt.} = \text{valency} \times \text{equiv. wt.} = 2 \times 27.92 = 55.84.$$

2.13. Mitscherlich's Law of Isomorphism. In 1819 as a result of his experiments with crystals Mitscherlich concluded that compounds with similar chemical compositions were isomorphous. Two substances can be recognized to be *isomorphous* with each other in the following ways:

1. Their crystals have similar shapes, the angles between the corresponding faces being nearly the same (i.e. within 2°).

2. They form overgrowths on each other. When a crystal of one substance X is suspended in a saturated solution of another isomorphous substance Y, a layer of Y is deposited all over the crystal so that it continues to grow and yet retains the same crystalline shape.

3. They form mixed crystals with each other. If solutions of the two isomorphous substances are mixed and concentrated until crystallization occurs, crystals are formed containing both substances in the ratio

of their relative proportions in the solution. These crystals are homogeneous (i.e. every sample of them, however small, has the same composition), and are really solid solutions of one substance in the other.

The law of isomorphism, as it is called, is best worded thus: *If two substances are isomorphous, then they have similar chemical formulae.* It follows that corresponding pairs of elements in such substances have the same valency. Amongst the best known examples of series of isomorphous substances are the alums (see § 19.13) and also the sulphates of magnesium, zinc, ferrous iron, nickel, cobalt, and manganese.

As a means of determining atomic weights the method is now only of historic interest and in any case has limited applicability, but it was used to decide the valencies and hence the atomic weights of selenium (potassium sulphate and potassium selenate are isomorphous) and gallium (which was judged to be trivalent because it formed an alum like aluminium, ferric iron, and chromium). However in later years a few exceptions were discovered (calcium carbonate $CaCO_3$ is isomorphous with sodium nitrate $NaNO_3$, and lead sulphide PbS with silver sulphide Ag_2S, whilst sodium and potassium nitrates, $NaNO_3$ and KNO_3, are not isomorphous), and these cast serious doubts on the reliability and validity of the law. It is now known that these exceptions arise because isomorphism only occurs when the ratio of the sizes of the cation and anion are closely similar in the two substances.

2.14. Mendeléeff's Method. In chapter 13 the periodic classification of the elements is described and it is explained there how Mendeléeff used his classification to check the atomic weights of elements. In several cases he altered the values then accepted because they led to serious misplacements of those elements in his table. A description of one of his best known corrections, that of beryllium, is given in § 13.2. Another example was the element indium, which was wrongly regarded as divalent and was given the atomic weight of 75.6. By reference to the periodic table Mendeléeff showed that indium was really trivalent and that its atomic weight was therefore three times its equivalent weight, i.e. 113.4, a value which was subsequently confirmed by other means. The atomic weight of uranium was corrected from 119 to 238 in much the same way.

2.15. From Molecular Weight and Atomicity. If the molecular weight of a gaseous element is determined with great accuracy by the limiting density method described in § 5.2, and its atomicity (i.e. the number of atoms in each molecule) is known, then the accurate atomic weight can be calculated, since atomic weight $= \dfrac{\text{molecular weight}}{\text{atomicity}}$.

Now it is shown in § 3.11 that the atomicity of most gases can readily be deduced from a measurement of γ, the ratio of its specific heats at constant pressure and constant volume. The inert gases form no compounds and their atomic weights cannot be found, therefore, by chemical methods; for them this purely physical method of determining atomic weights is of great importance.

2.16. X-ray Method. It is explained in § 4.5 that X-ray analysis of crystals can be used to measure the size of the unit cell, that small grouping of atoms or ions which, repeated regularly in three dimensions, makes up the structure of the crystal. In most cases X-rays can also be used to determine the number of atoms in this unit cell. If the density of the crystal is measured very accurately then the absolute mass of the unit cell and of each atom in it can be deduced. Thus from a knowledge of the absolute mass of the oxygen atom obtained from our knowledge of the Avogadro number, the atomic weight of an element can be calculated to a high degree of accuracy. The method can be used for very reactive elements (e.g. fluorine) which are difficult to obtain in a pure condition, since it can be applied to their crystalline compounds provided the atomic weights of the other elements involved are accurately known.

2.17. The Mass Spectrograph. This device, the original form of which was designed by Aston in 1919, can be used to determine atomic weights extremely accurately. A simplified diagram of the apparatus is given in Fig. 2.2.

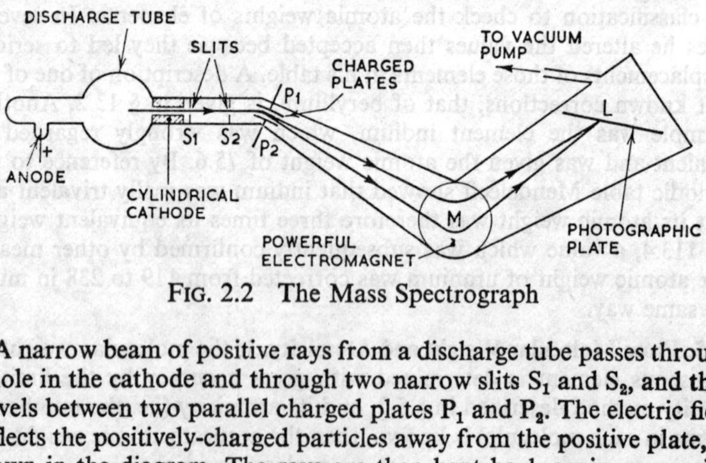

FIG. 2.2 The Mass Spectrograph

A narrow beam of positive rays from a discharge tube passes through a hole in the cathode and through two narrow slits S_1 and S_2, and then travels between two parallel charged plates P_1 and P_2. The electric field deflects the positively-charged particles away from the positive plate, as shown in the diagram. The rays are then bent back again on passing through the field of the powerful electromagnet M. In both cases the amount of deflection varies with the velocity of the particles, the slower ones being deflected more than the faster ones. The two fields are so adjusted that all particles with the same ratio of charge to mass are brought to a focus to give a sharp line on the photographic plate at the point L. Thus if the discharge tube contains atoms of various weights, a series of different lines, known as a mass spectrum, will be produced. Provided some atoms of known mass are present to calibrate the spectrum, the relative masses of all the atoms can be calculated.

2.18. Isotopes and Atomic Weight. If the mass spectrograph method is applied to an element which exists as isotopes, several distinct lines will be produced, each corresponding to one isotopic weight. Now the atomic weight of such an element is the weighted mean of the isotopic weights, i.e. it is the average weight of the isotopes having regard to their relative abundance. For example, chlorine consists of two isotopes weighing approximately 35 and 37 on the atomic weight scale (as explained in § 2.19 the isotopic weights are usually not exact integers owing to the packing fraction or mass defect correction), the lighter isotope being very nearly three times as abundant as the other. The atomic weight of chlorine is 35.457 because this is the average weight of chlorine atoms, about three-quarters of which weigh approximately 35 and the rest approximately 37. It will be noted that the atomic weight is not just the arithmetical average of the various isotopic weights—it would be 36 if it were.

From the preceding paragraph it will be realized that to use the mass spectrograph to determine the atomic weight of an element which exists naturally as isotopes (and most elements are in this category) two distinct sets of measurements have to be made, the weights of the isotopes and their relative abundance. It is possible to calculate atomic weights to an accuracy of 1 in 10 000 by estimating the abundance of the isotopes from the density of the trace produced on the photographic plate of the mass spectrograph, but even more accurate results are obtainable from a *mass spectrometer*. In this apparatus the photographic plate is replaced by a slit and the electric field is varied so as to focus the rays from each isotope onto the slit in turn. An electrometer measures the current caused by the charged particles of each isotope and gives a very accurate indication of the relative abundance of the isotopes. The values for atomic weight obtained in this way are amongst the best available and in conjunction with chemically determined values they appear in the list of internationally accepted atomic weights given at the end of this book.

In 1929 a study of band spectra revealed the hitherto unsuspected fact that ordinary gaseous oxygen is itself a mixture of three different isotopes, 99.76% of it consisting of $^{16}_{8}O$,* 0.04% of $^{17}_{8}O$, and 0.20% of $^{18}_{8}O$. In determining atomic weights by means of the mass spectrograph, $^{16}_{8}O = 16.0000$ is taken as the standard (the physical scale), whereas all chemically determined atomic weights are based upon a standard value of 16.0000 for the mixture of isotopes which makes up ordinary gaseous oxygen. Although the difference between the two scales is too slight to be important in most work, when great precision is required for chemical purposes the atomic weights on the physical scale have to be multiplied by the factor 1.00028, which is the ratio of the atomic weight of ordinary oxygen on the physical scale, 16.0044, to its standard weight of 16.0000 on the chemical scale.

The existence of two different scales of atomic weight has been a source of considerable inconvenience to both chemists and physicists, so it is hardly surprising that attempts have been made to find a new scale acceptable to both. In 1957 the carbon-12 isotope was suggested as the new standard, being taken as exactly twelve units; this was adopted by the International Union of Physics in 1960 and by the International Union of Pure and Applied Chemistry in 1961. *Atomic weight is now defined, therefore, as the weight of an atom on a scale on which the carbon-12 isotope weighs 12.0000 units.* The change from the old chemical scale is only 43 parts in a million, so that no alteration is necessary in most of our existing gravimetric and molar data. Ordinary oxygen becomes 15.9999 on the new scale and oxygen-16 becomes 15.9949. The new values are listed at the end of the book.

The discovery of isotopes sets a limit to the accuracy with which we can expect to determine atomic weights. In general the proportions of the various isotopes of an element in a natural sample are remarkably constant both from place to place on the earth's surface and from time to time. Nevertheless certain well established exceptions (e.g. sulphur, boron, and carbon) make it clear that the relative abundance of the isotopes of an element may vary slightly from one source to another, giving a very small variation in its atomic weight. Moreover, any variation in the isotopic composition of oxygen will cause small discrepancies for all the other elements whose atomic weights are determined by chemical means.

2.19. Prout's Hypothesis. Observing that most atomic weights were approximately whole numbers, in 1815 Prout put forward the startling hypothesis that the atoms of all the elements were really conglomerations of one fundamental unit, the hydrogen atom. The first effect of

* $^{16}_{8}O$ represents an oxygen atom having an atomic number of 8 (subscript) and a mass number or weight of 16 (superscript). This symbolism, or the less desirable alternative of $_{8}O^{16}$, is widely used for distinguishing between isotopes.

this hypothesis was to concentrate attention upon certain fractional atomic weights (e.g. those of chlorine and copper) which were suspected of being in error, but when more accurate determinations showed that this was not the case, the hypothesis was eventually abandoned. Prout's theme of the composite nature of atoms was revived a century later, however, when Aston showed by use of the mass spectrograph that these atomic weights were fractional because they were the average weights of a mixture of isotopes, all of which had approximately integral values. With the discovery of the proton, neutron, and electron, and the accurate determination of their masses, an attempt was made to calculate atomic weights by summing the masses of the particles they contained. In every case it was found that the mass of the atom was less than expected; the discrepancy, which is known as the *mass defect*, arises because some mass has been converted into potential energy in forming the nucleus. The mass defect is a measure, therefore, of the binding energy or stability of a nucleus. Calculation shows that for most elements it amounts to about 1.36×10^{-12} J for each particle, although it is less for very light atoms such as hydrogen and for very heavy atoms (where the atomic number exceeds 83), indicating that the nuclei of these atoms are less stable than the others. This is of great theoretical importance in connection with nuclear fusion and fission (see chapter 31), since it suggests which elements may be used as sources of nuclear energy.

2.20. Conclusion. Although very accurate values of atomic weight are required for certain purposes and their determination is a matter of great importance, on most occasions such great precision is not needed and less exact values suffice. For this reason it is customary in chemical calculations to use the set of approximate atomic weights shown in the table at the end of this book, and this practice will be followed in all the numerical examples considered.

Lest familiarity breed contempt, it is worth pausing occasionally in our everyday use of atomic weights to consider the marvellous achievement whereby man has devised means of determining correctly to five significant figures the relative masses of atoms none of which he has ever seen or weighed directly.

3

The Kinetic Theory

3.1. Introduction. A given substance may exist as a solid, liquid, or gas or in any two of these states or even in all three simultaneously depending on the physical conditions. As a solid it has a definite volume and shape and is said to be rigid, but as a liquid it can flow so that although its volume is fixed once the conditions are specified, it always assumes the shape of the vessel in which it is placed. A gas or vapour has neither definite volume nor definite shape; placed in any vessel it *diffuses*, i.e. it spreads uniformly throughout the whole volume of the container. When solids or liquids are heated they usually expand by an amount which is characteristic of the substance concerned and varies widely from one substance to another. Similarly, solids and liquids undergo contraction by characteristic amounts when compressed. Gases, on the other hand, show remarkable similarity in their behaviour when heated or compressed. The changes in the volume of a gas are very much greater than the corresponding changes produced in solids and liquids and, what is more important, are approximately the same whatever the nature of the gas. It was this common behaviour of gases which stimulated the interest of the early experimenters and led to the discovery of certain general laws applicable to all of them. It also suggested that the physical properties of all gases might be explained on some simple mechanical basis. As we shall see in this chapter, the theory that was eventually devised, the kinetic theory, not only explained the various gas laws, but also provided a new method of determining atomicity and gave valuable insight into the nature of liquids and the processes of melting and evaporation.

3.2. Boyle's Law. This, the first of the gas laws, was discovered in 1662: *The volume of a fixed mass of gas at constant temperature is inversely proportional to the applied pressure.*

Stated in symbols,

$$v \propto \frac{1}{p} \ (T \text{ constant})$$

or pv = constant where $\begin{cases} p = \text{the pressure of the gas} \\ v = \text{the volume of the gas} \\ T = \text{the temperature of the gas} \end{cases}$

Thus if v_1 is the volume at pressure p_1, and v_2 the volume of the same mass of gas at a different pressure p_2, then provided the temperature remains unchanged,

$$p_1v_1 = p_2v_2.$$

Boyle's law is presented graphically in two ways in Fig. 3.1; in (*a*) the volume is plotted against the reciprocal of the pressure giving a straight line through the origin, and in (*b*) the product pv is plotted against the

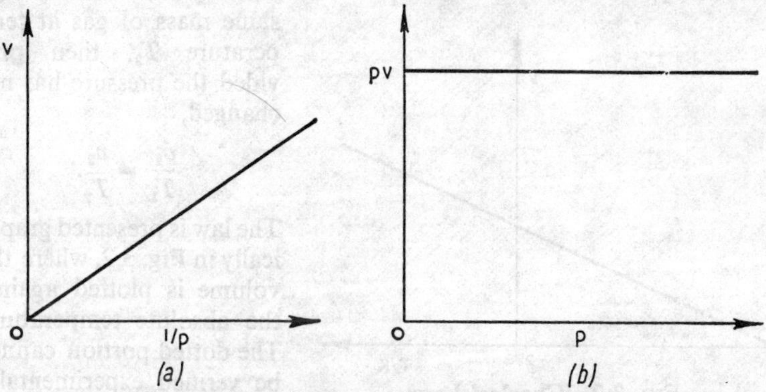

FIG. 3.1 Graphical Representations of Boyle's Law

pressure, when a straight line parallel to the pressure axis is obtained. In both graphs a fixed mass of gas at constant temperature is considered.

We shall see later (§ 3.12) that Boyle's law only holds approximately true in practice and that most gases show slight but definite deviations from it, particularly at high pressures and at temperatures near their liquefaction points.

3.3. Charles' Law. This law was first discovered by Charles and stated later by Gay-Lussac. From their experiments they concluded that all gases expanded by approximately $\frac{1}{273}$ of their volume at 0°C for each C° rise in temperature. As a matter of convenience, therefore, it was decided to introduce a new scale of temperature, the *absolute* or *Kelvin scale*, in which the temperature interval is 1C°, known as the kelvin, symbol K, and the zero is at −273°C.* When temperature is expressed

* Actually −273·15°C, although the approximate value is usually preferred in elementary work. The scale is named after Lord Kelvin because he was the first to define it thermodynamically in a way which was completely independent of any particular substance. On this scale the triple point of water is 273.16 K.

on this scale, in K, then Charles' law can be stated in the familiar form:

The volume of a fixed mass of gas is directly proportional to the absolute temperature, provided the pressure remains constant.

Expressed in symbols,

$$v \propto T \ (p \text{ constant})$$

or $\frac{v}{T}$ = constant where $\begin{cases} v = \text{the volume of the gas} \\ p = \text{the pressure of the gas} \\ T = \text{the } \textit{absolute} \text{ temperature,} \end{cases}$

Thus if v_1 is the volume at temperature T and v_2 the volume of the same mass of gas at temperature T_2, then provided the pressure has not changed,

$$\frac{v_1}{T_1} = \frac{v_2}{T_2}.$$

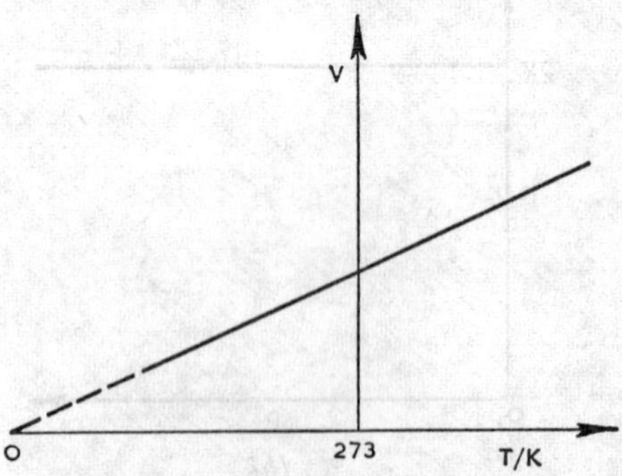

FIG. 3.2 Charles' Law

The law is presented graphically in Fig. 3.2, where the volume is plotted against the absolute temperature. The dotted portion cannot be verified experimentally because gases tend to liquefy at such low temperatures. Like Boyle's law, Charles' law is not obeyed exactly by real gases. The deviations, which are only slight unless the gas is near liquefaction, are explained in § 3.12.

3.4. The Ideal Gas Equation. When Boyle's law and Charles' law are combined, the following relationship is obtained:

$$v \propto \frac{1}{p} \text{ and } v \propto T$$

$\therefore v \propto \frac{T}{p}$ or $pv = kT$ where k is a constant for a fixed mass of one particular gas.

Thus if the volume of a gas is v_1 at pressure p_1 and absolute temperature T_1, and the volume of the same mass of gas is v_2 at pressure p_2 and absolute temperature T_2, then:

$$\frac{p_1 v_1}{T_1} = \frac{p_2 v_2}{T_2}.$$

The derivation of this relationship from the two gas laws sometimes troubles students, so the following treatment, in which the change of a fixed mass of gas from conditions p_1, v_1, T_1, to p_2, v_2, T_2, is achieved in two distinct stages, may be helpful:

First stage (at constant temperature): $p_1\ v_1\ T_1 \rightarrow p_2\ v'\ T_1$

Second stage (at constant pressure): $p_2\ v'\ T_1 \rightarrow p_2\ v_2\ T_2$

$$\text{In stage I, } p_1v_1 = p_2v' \text{ by Boyle's law}$$

$$\text{In stage II, } \frac{v'}{T_1} = \frac{v_2}{T_2} \text{ by Charles' law}$$

$$\therefore p_1\ v_1 = \frac{p_2\ v_2\ T_1}{T_2} \left(\text{since } v' = \frac{v_2\ T_1}{T_2}\right)$$

$$\therefore \frac{p_1\ v_1}{T_1} = \frac{p_2\ v_2}{T_2}.$$

The constant k has a different value for each gas, but if we consider 1 mole of any gas we are always concerned with the same number of molecules and the ratio $\frac{pv}{T}$ then has the same value for every gas,

i.e. $$\frac{PV}{T} = R \quad \text{or} \quad PV = RT$$

where R is called the universal or molar gas constant.
V is the volume occupied by one mole of the gas.
P is the pressure exerted by the gas.
T is the absolute temperature.

The equation $PV = RT$ is known as the *ideal gas equation*, since it summarizes in an ideal form how gases behave. In practice real gases show deviations from this ideal, although these deviations are very small at low pressures and high temperatures and can often be ignored. An *ideal* or *perfect gas* is defined as one which behaves strictly in accord with this equation; no such gas actually exists, but it is often helpful to imagine a hypothetical gas of this kind as it provides a simple approximate guide to the behaviour of real gases.

The constant R can be calculated by substituting typical values of pressure, volume, and temperature, although its value obviously depends upon what units are used.

In SI units, the pressure is expressed in newton per square metre*, the volume in cubic metre, and the temperature in kelvin. Now at standard temperature and pressure, one mole of any gas occupies

* Or in pascal (Pa); 1 Pa $\equiv$ 1 N m^{-2}.

2.24×10^{-2} cubic metre (see § 1.13), so under these conditions we have:

$$P = 101\,325 \text{ N m}^{-2}$$
$$V = 2.24 \times 10^{-2} \text{ m}^3 \text{ mol}^{-1}$$
$$T = 273 \text{ K}$$

$$\therefore R = \frac{PV}{T} = \frac{101\,325 \times 2.24 \times 10^{-2}}{273} \frac{\text{N m}^{-2} \times \text{m}^3 \text{ mol}^{-1}}{\text{K}}$$

$$= 8.31 \frac{\text{N m mol}^{-1}}{\text{K}}$$

$$= 8.31 \text{ J K}^{-1} \text{ mol}^{-1}$$

since the work done when a force of one newton acts over a distance of one metre is one *joule*, denoted by the symbol J.

If desired the ideal gas equation can be stated in a more general way as $PV = nRT$ where n is the number of mole of gas considered.

It is important to notice how the ideal gas equation is derived by the combination of two distinct experimental laws. It is a common practice in science when investigating the way in which a quantity depends upon two variables to find how it varies with each one of them in turn whilst the other is kept constant, and then to deduce the more general relationship. Unfortunately when applying the scientific method to some problems, particularly those in the biological or social fields, it is often extremely difficult to arrange that only one factor varies at a time and that the other variables are all kept constant during the experiment.

3.5. Dalton's Law of Partial Pressures. As a result of his experiments on mixtures of gases Dalton concluded that each gas in a mixture exerts upon the walls of the containing vessel a pressure, known as its *partial pressure*, which is equal to the pressure that the gas would exert if it alone occupied the vessel at that temperature. He summarized his findings in 1801 in the statement that *the total pressure exerted by a mixture of gases is equal to the sum of the partial pressures.*

i.e. $$P = p_A + p_B + p_C + \ldots$$

Although real gases do not obey the law exactly, particularly at high pressures, the deviations at ordinary pressures are so slight that they can usually be ignored.

The partial pressure of a gas is easily calculated if its molecular proportion in the mixture (known as its mole fraction) is known, since it is given by the following expression:

$$p_A = \frac{\text{number of molecules of } A}{\text{total number of molecules in the mixture}} \times P$$

where p_A = the partial pressure of the gas A.
P = the total pressure of the mixture.

For example, if air consists of 21% by volume of oxygen, then the partial pressure of the oxygen when the atmospheric pressure is 101 325 $N\,m^{-2}$ is $\frac{21}{100} \times 101\,325$, i.e. $2.12 \times 10^4\ N\,m^{-2}$, since all the gases in the mixture are under the same conditions and so by Avogadro's hypothesis their molecular proportions will be the same as their volume proportions.

An important application of Dalton's law arises when a gas is collected over water. Under these conditions it may be assumed that the gas is saturated with water vapour and the total pressure acting on the container is the sum of the partial pressures of the gas and the water vapour. The true pressure of the gas is not the atmospheric pressure, therefore, but the atmospheric pressure less the saturated vapour pressure* of water at that temperature (the value of which can be found from tables).

Two other instances of the importance of the idea of partial pressure may be mentioned. We shall see in chapter 6 that the mass of gas which dissolves in a liquid is proportional to the applied pressure. It follows that the partial pressures of the various gases in a mixture will be a governing factor in deciding how much of each gas will dissolve when the mixture is exposed to a liquid. This point is illustrated in § 6.7.

Secondly, when considering the law of mass action in chapter 7 we shall find that the partial pressure of a gas may be taken as a measure of its 'active mass'; in applying the law to gaseous equilibria it is often more convenient to use partial pressures in this way than to consider gramme-molecular concentrations.

3.6. Graham's Law of Gaseous Diffusion. The tendency of gases to diffuse has already been mentioned in this chapter. When placed in an evacuated container a gas rapidly spreads until it is distributed uniformly throughout the whole volume. Diffusion is slower when the container into which the gas is introduced is already occupied by another gas, but eventually the two gases will mix so thoroughly that samples taken from any part will contain exactly the same proportions of each gas. The easiest way to demonstrate that two gases diffuse at different rates is to separate them by some sort of porous diaphragm or partition through which they can both pass, when there will be a temporary difference in pressure between the two sides of the partition. The apparatus shown in Fig. 3.3 illustrates the effect well. Originally the gas A in the porous pot was at atmospheric pressure, so that the liquid

* This term is explained in § 3.16.

in the manometer was at the same height in both limbs. When the pot is lowered into the beaker the gas *A* diffuses out of the pot at a faster rate than gas *B* diffuses in, because it is less dense. This leads to a fall in pressure inside the pot and causes the liquid levels in the manometer to differ as in the Figure. In a short while, however, gas *A* will have distributed itself uniformly throughout both pot and beaker, whereas gas *B* by continuing to diffuse into the pot causes the pressure in it to rise slowly to that of the atmosphere again, as indicated by a levelling of the liquid in the manometer.

The rates of diffusion of two gases can be compared by measuring the time taken for a given volume of each of them to diffuse through a

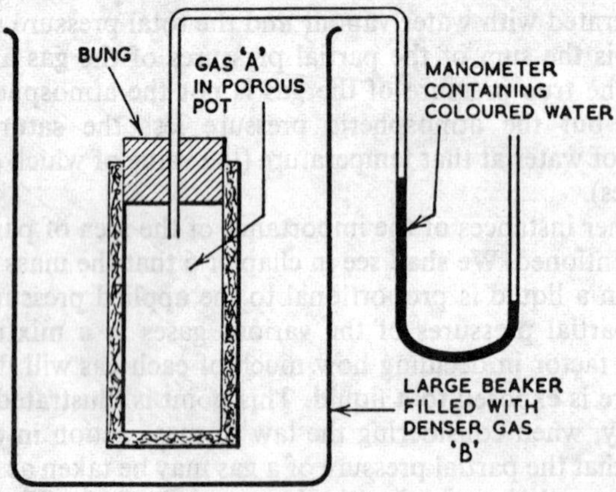

FIG. 3.3 Demonstration of Gaseous Diffusion

porous surface under identical conditions. After experimenting in this way with various gases, Graham summarized his results in 1833 in the following law:

The relative rates of diffusion of different gases under the same conditions are inversely proportional to the square roots of their densities.

Stated in symbols, $$\frac{r_1}{r_2} = \sqrt{\frac{\rho_2}{\rho_1}}$$

where r_1 and r_2 are the rates of diffusion of gases with densities ρ_1 and ρ_2 respectively.

Since both gases are at the same temperature and pressure, the ratio of their densities is also the ratio of their molecular weights, by Avogadro's hypothesis (see § 1.10). The law may be used, therefore, for determining the molecular weight of one gas if the molecular weight of the other is known.

Example: If 100 cm³ of oxygen (mol. wt. = 32) diffuse through a porous plug in 128 seconds, what is the molecular weight of carbon dioxide, 100 cm³ of which take 150 seconds to diffuse through the same plug under the same conditions?

Now $\frac{r_1}{r_2} = \sqrt{\frac{M_2}{M_1}}$ by Graham's law.

But $\frac{t_2}{t_1} = \frac{r_1}{r_2}$ (since the rate of diffusion is inversely proportional to the time taken)

$$\therefore \frac{t_2}{t_1} = \sqrt{\frac{M_2}{M_1}} \quad \text{or} \quad M_2 = \left(\frac{t_2}{t_1}\right)^2 \times M_1.$$

Here $M_1 = 32$, $t_1 = 128$ s, $t_2 = 150$ s,

$$\therefore M_2 = \frac{150 \times 150}{128 \times 128} \times 32 = 43.9.$$

$\therefore$ Mol. wt. of carbon dioxide is 44 approx.

The fact that gases of different density diffuse at different rates can be used to bring about their separation by a process known as *atmolysis*. When this method is used industrially, the mixture of gases is passed under pressure through the inner of a pair of concentric pipes, the dividing wall of which is porous, as in Fig. 3.4. Both gases diffuse through to the outer pipe, but at different rates depending upon their densities, so that after a short while the mixture in the outer pipe is richer in the less dense gas and the mixture in the inner pipe richer in the denser gas than the original mixture. If these enriched mixtures are then passed repeatedly through similar pieces of piping, a progressive separation of the two gases can gradually be achieved, although when the two gases differ only slightly in density a great number of stages is required. Expensive and tedious though it is, this is one of the chief ways in which the two isotopes of uranium, $^{235}_{92}U$ and $^{238}_{92}U$, are separated (as their volatile hexafluorides UF_6) in order to make atomic bombs (see § 31.10). In this particular case the ratio of the densities is 349 : 352, so that the rates of diffusion are in the very low ratio 1.0042 : 1.

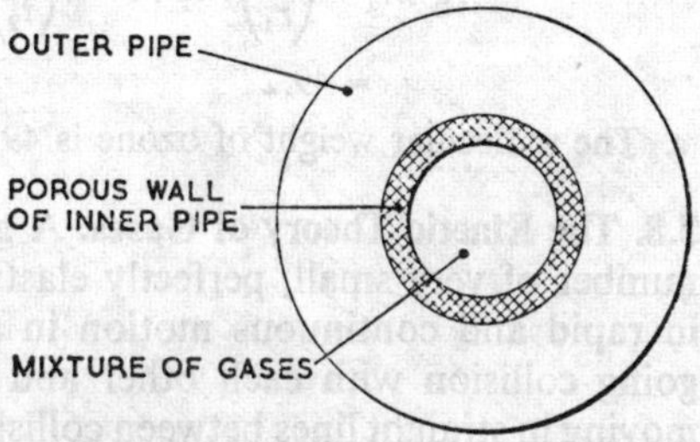

FIG. 3.4 Diffusion Pipes

3.7. Effusion. This phenomenon resembles diffusion but it is concerned with the tendency of a gas to pass through a single very small hole in the wall of its container instead of through a porous plug. Graham studied effusion and discovered that *the relative rates of effusion of different gases under the same conditions are inversely proportional to the square roots of their densities.* This law can be used to determine the molecular weight of ozone, even though it cannot be obtained pure, as the following calculation illustrates:

Example: A certain volume of ozonized oxygen (see § 24.3) effuses through a small hole in 55 seconds, whereas the same volume of a mixture of oxygen and chlorine containing the same proportion of oxygen takes 66 seconds under the same conditions. Calculate the molecular weight of ozone.

Now under these conditions the oxygen in both samples effuses at the same rate irrespective of whether it is mixed with ozone or with chlorine, and since the percentage of oxygen is identical, any difference in the rates of effusion of the two samples is solely attributable to the ozone and the chlorine. In fact, the ratio of the rates is inversely proportional to the square root of the ratio of the molecular weights of these two gases, so that if the ozonized oxygen effuses at rate r_1 in time t_1 and the chlorine–oxygen mixture at rate r_2 in time t_2, and if the molecular weight of ozone is M_1,

then $\frac{r_1}{r_2} = \sqrt{\frac{71}{M_1}}$ and $\frac{r_1}{r_2} = \frac{t_2}{t_1}$

$$\therefore M_1 = \left(\frac{r_2}{r_1}\right)^2 \times 71 = \left(\frac{t_1}{t_2}\right)^2 \times 71 = \left(\frac{55}{66}\right)^2 \times 71$$

$$= 49.2$$

∴ The molecular weight of ozone is 49 and its molecular formula is O_3.

3.8. The Kinetic Theory of Gases. A gas is believed to consist of a vast number of very small, perfectly elastic particles (molecules) which are in rapid and continuous motion in all directions, continually undergoing collision with each other and with the walls of the vessel but moving in straight lines between collisions. It is this bombardment of the containing walls and the change of momentum accompanying it which accounts for the pressure exerted by the gas. In this picture of the nature of a gas it is assumed that the molecules are so small compared with the average distance travelled between consecutive collisions (which is known as the *mean free path*) that the space actually occupied by the molecules and the attractive forces between them are both negligible. The great compressibility of gases and the relatively large volume of gas which is obtained on evaporating a liquid suggest that these assumptions are justified.

The speeds of the various molecules are believed to differ widely and to be constantly changing as a result of their frequent random collisions. Thus at any instant a small proportion will be almost stationary and a similar proportion will be moving at very high speeds indeed, but the majority of the molecules will have speeds between these two extremes. The way in which the molecular velocities are distributed was predicted mathematically by Maxwell in 1860; provided the number of molecules considered is very large (as it always is, even in the smallest volume of gas) then the distribution is as shown in Fig. 3.5. The exact shape of this curve depends upon the temperature (see § 7.12). In our calculations it is convenient to represent the speeds of the molecules by the *root mean square velocity* (the *r.m.s.* velocity) conventionally denoted by $c_{r.m.s.}$ This is not the average velocity,* not is it the commonest velocity, but as the name implies it is the square root of the

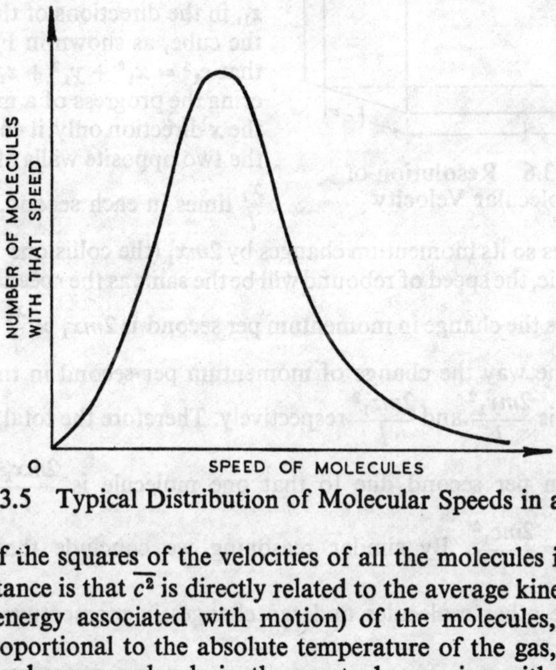

FIG. 3.5 Typical Distribution of Molecular Speeds in a Gas

average of the squares of the velocities of all the molecules in the gas. Its importance is that $\overline{c^2}$ is directly related to the average kinetic energy (i.e. the energy associated with motion) of the molecules, which in turn is proportional to the absolute temperature of the gas, so that if we imagined every molecule in the gas to be moving with the r.m.s. velocity, we should get a true idea of the total kinetic energy and temperature of the gas. The justification for this important assumption will be found in the next section where it is shown that with its aid we are able to deduce correctly the various gas laws already discovered by experiment.

* This point is made clear by considering three molecules with velocities 1, 2, and 3, respectively. Then $\overline{c^2} = (1^2 + 2^2 + 3^2) \div 3 = 4.67$ and its square root is 2.16, whereas the average velocity, $\bar{c}$, is obviously 2.0.

By expressing these ideas mathematically it is possible to derive an expression for the pressure of a gas in terms of the speeds of its molecules. Let us imagine a cubicle vessel with sides l m long which contains n molecules of the gas, each of mass m. Let the velocities of the molecules be c_1, c_2, c_3, etc., and let the pressure of the gas be P. Let us consider for a moment only the molecule with velocity c_1, moving in a random direction across the cube. Its velocity may be resolved into three components x_1, y_1, and z_1, in the directions of the edges of the cube, as shown in Fig. 3.6, so that $c_1^2 = x_1^2 + y_1^2 + z_1^2$. Considering the progress of a molecule in the x direction only, it collides with the two opposite walls of the vessel $\frac{x_1}{l}$ times in each second, and each time it does so its momentum changes by $2mx_1$ (the collisions being perfectly elastic, the speed of rebound will be the same as the speed before impact). Thus the change in momentum per second is $2mx_1 \times \frac{x_1}{l} = \frac{2mx_1^2}{l}$.

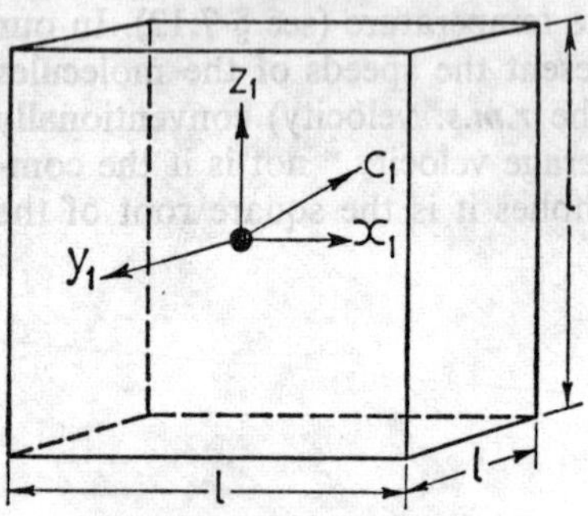

FIG. 3.6 Resolution of Molecular Velocity

In the same way the change of momentum per second in the y and z directions is $\frac{2my_1^2}{l}$ and $\frac{2mz_1^2}{l}$ respectively. Therefore the total change in momentum per second due to that one molecule is $\frac{2mx_1^2}{l} + \frac{2my_1^2}{l} + \frac{2mz_1^2}{l} = \frac{2mc_1^2}{l}$. By similar reasoning we conclude that in each second the other molecules undergo changes in momentum of $\frac{2mc_2^2}{l}$, $\frac{2mc_3^2}{l}$, etc., so the total change in momentum per second of all the molecules is $\frac{2m}{l}(c_1^2 + c_2^2 + c_3^2 + \ldots) = \frac{2mn\overline{c^2}}{l}$ since $\overline{c^2}$, the mean square velocity, is equal to $\frac{(c_1^2 + c_2^2 + c_3^2 + \ldots)}{n}$.

Now this is equal to the total force exerted by the gas upon all six walls of the cube, and the pressure, which is defined as the force acting upon unit area of surface, is obtained therefore by dividing by the total area $6l^2$:

$$\therefore \text{Pressure } P = \frac{2mn\overline{c^2}}{l \times 6l^2} = \frac{1}{3}\frac{mn\overline{c^2}}{l^3} = \frac{1}{3}\frac{mn\overline{c^2}}{V}$$

where V is the volume of the vessel.

$$\therefore PV = \frac{1}{3} mn\overline{c^2}$$

Although this expression has been derived for a cubicle vessel, it also applies to one of any other shape since any vessel may be imagined to be made up of a number of small cubes.

3.9. Deductions from the Kinetic Theory. By using the relationship $PV = \frac{1}{3} mn\overline{c^2}$ it is possible to deduce the various gas laws as follows:

BOYLE'S LAW: For a given mass of gas, m and n are constant. When the temperature of a gas is constant, the average kinetic energy of its molecules $(\frac{1}{2}m\overline{c^2})$ must be constant, so $\overline{c^2}$ must also be constant.

$\therefore$ Under these conditions $PV =$ constant, which is Boyle's law.

CHARLES' LAW: If $PV = \frac{1}{3} mn\overline{c^2}$, then $V = \frac{1}{3}\frac{mn\overline{c^2}}{P}$. Now, as has already been explained, $\overline{c^2}$ is proportional to the absolute temperature, T, so for a fixed mass of gas at constant pressure, m, n, and P are all constant and $V \propto T$, which is Charles' law.

DALTON'S LAW: If we consider, for the sake of simplicity, a mixture of only two gases, one consisting of n_1 molecules of mass m_1 and r.m.s. velocity $\sqrt{\overline{c_1^2}}$, and the other of n_2 molecules of mass m_2 and r.m.s velocity $\sqrt{\overline{c_2^2}}$, then

$$\frac{P_1}{n_1} = \frac{m_1\overline{c_1^2}}{3V_1} \quad \text{and} \quad \frac{P_2}{n_2} = \frac{m_2\overline{c_2^2}}{3V_2} \text{ (since } PV = \frac{1}{3}mn\overline{c^2}\text{).}$$

But the gases in the mixture occupy the same vessel (i.e. $V_1 = V_2$) and are at the same temperature (i.e. $\frac{1}{2}m_1\overline{c_1^2} = \frac{1}{2}m_2\overline{c_2^2}$)

$$\therefore \quad \frac{P_1}{n_1} = \frac{P_2}{n_2} \quad \text{or} \quad P_1 = \frac{n_1}{n_2} \times P_2$$

$$\therefore \quad \text{The total pressure exerted} = P_1 + P_2 = \frac{n_1}{n_2}P_2 + P_2$$

$$= P_2\frac{(n_1 + n_2)}{n_2}$$

$$\therefore\ P_2 = \frac{n_2}{(n_1 + n_2)} \times (P_1 + P_2)$$

i.e. each gas exerts a partial pressure equal to the total pressure multiplied by its mole fraction, which is one way of stating Dalton's law.

GRAHAM'S LAW: Since $PV = \frac{1}{3}mn\overline{c^2}$, $\overline{c^2} = \frac{3PV}{mn}$. Now $\frac{mn}{V}$ is the density of the gas, denoted by ρ.

$$\therefore\ \overline{c^2} = \frac{3P}{\rho} \quad \text{and} \quad c_{\text{r.m.s.}} = \sqrt{\frac{3P}{\rho}}.$$

Now the rates of diffusion and effusion of a gas depend upon the r.m.s. velocity of its molecules—one gas diffuses or effuses faster than another because its molecules move faster and undergo more collisions per second, i.e. $r = k\,.\,c_{\text{r.m.s.}}$. For two different gases with r.m.s. velocities $\sqrt{\overline{c_1^2}}$ and $\sqrt{\overline{c_2^2}}$ and densities ρ_1 and ρ_2 at the same conditions of temperature and pressure,

$$\frac{r_1}{r_2} = \sqrt{\frac{\overline{c_1^2}}{\overline{c_2^2}}} = \sqrt{\frac{3P}{\rho_1}} \div \sqrt{\frac{3P}{\rho_2}} = \sqrt{\frac{\rho_2}{\rho_1}}$$

which is a symbolic statement of Graham's laws of diffusion and effusion.

Expressed in another way, when two gases are at the same temperature, the average kinetic energy of their molecules is the same, so $\frac{1}{2}m_1\overline{c_1^2} = \frac{1}{2}m_2\overline{c_2^2}$ and $\frac{\overline{c_1^2}}{\overline{c_2^2}} = \frac{m_2}{m_1}$.

Hence $$\frac{r_1}{r_2} = \sqrt{\frac{\overline{c_1^2}}{\overline{c_2^2}}} = \sqrt{\frac{m_2}{m_1}} = \sqrt{\frac{M_2}{M_1}}$$

where M_1 and M_2 are the molecular weights of the two gases.

AVOGADRO'S HYPOTHESIS: Since $PV = \frac{1}{3}mn\overline{c^2}$, $n = \frac{3PV}{m\overline{c^2}}$. Consider two gases, one consisting of n_1 molecules of mass m_1 and r.m.s. velocity $\sqrt{\overline{c_1^2}}$, and the other of n_2 molecules of mass m_2 and r.m.s. velocity $\sqrt{\overline{c_2^2}}$.

Then $$n_1 = \frac{3P_1V_1}{m_1\overline{c_1^2}} \quad \text{and} \quad n_2 = \frac{3P_2V_2}{m_2\overline{c_2^2}}.$$

If equal volumes are considered (i.e. $V_1 = V_2$) and if the two gases are at the same temperature (i.e. $\frac{1}{2}m_1\overline{c_1^2} = \frac{1}{2}m_2\overline{c_2^2}$) and pressure (i.e. $P_1 = P_2$), then under these circumstances

$$\frac{3P_1V_1}{m_1\overline{c_1^2}} = \frac{3P_2V_2}{m_2\overline{c_2^2}} \quad \text{and} \quad n_1 = n_2,$$

which is Avogadro's hypothesis.

3.10. Molecular Velocities. From the expression $c_{r.m.s.} = \sqrt{\frac{3P}{\rho}}$ it is possible, by substituting known values of P and ρ, to calculate the r.m.s. velocities of various gases. For example, if we consider hydrogen at s.t.p. and express P in N m^{-2} and ρ in kg m^{-3}, then

$$c_{r.m.s.} = \sqrt{\frac{3 \times 1.013 \times 10^5}{0.09}} = 1838 \text{ m s}^{-1}$$

A hydrogen molecule moving at this speed would travel about 1.15 miles in each second or over 4000 miles in an hour.* Similarly, the r.m.s. velocity of oxygen at s.t.p. can be shown to be 460 m s^{-1} or 0.3 miles s^{-1}. We can also calculate the mean free path of the molecules, i.e. the average distance travelled between consecutive collisions. For oxygen at s.t.p. this is of the order of 10^{-7} m or about 1000 diameters. Of course, many molecules will be moving, at any instant, with even higher speeds, but these figures are quoted to give some idea of how rapid the motion of the molecules really is and what a short distance, on the average, they travel before colliding. It does indeed give a vivid picture of the nature of a gas to realize that at room temperature the molecules in the air are moving about at an average speed of over 1000 miles per hour and that about five thousand million (5 000 000 000) molecular collisions occur in each cubic centimetre of air during every second!

3.11. Specific Heat Capacity and Atomicity. The specific heat capacity of a substance is the quantity of heat, measured in joule, required to raise the temperature of 1 kilogramme of it by 1 K. Because gases have such large coefficients of expansion, its value for a gas depends upon the conditions under which it is measured and it is customary to consider two particular cases—*the specific heat capacity at constant pressure* c_p, and *the specific heat capacity at constant volume*, c_v. The first of these will always be larger than the second because when a gas is heated at constant pressure it expands and does work against the atmosphere, which requires additional energy. In chemistry it is more usual to consider the *molar heat* capacities, i.e. the heat capacities for one mole of the gas, which are denoted by C_{pm} and C_{vm}.

Considering one mole of a gas, we have $PV = \frac{1}{3}mN_A\overline{c^2}$ (since n is the Avogadro constant, N_A, in this case)

$$\therefore \quad PV = \frac{2}{3} \times \frac{1}{2}mN_A\overline{c^2} = \frac{2}{3}E,$$

* Everyday units are used here to give a more vivid impression.

where E is the total kinetic energy of one mole of the gas. But $PV = RT$

$\therefore\ E = \frac{3}{2}RT$, and the energy required to raise T by 1 K is $\frac{3}{2}R = 12.5$ J.

Now this is C_{vm}, the molar heat at constant volume.

Now let one gram-molecule of a gas expand from volume V_1 to volume V_2 at a constant pressure P when the temperature is raised from T_1 to T_2. Then the work done by the gas is $P(V_2 - V_1)$ which is equal to $R(T_2 - T_1)$. For a 1 K rise in temperature, $T_2 - T_1 = 1$ and the work done $= R = 8.3$ joule. So for one mole of a gas,

$$C_{pm} = C_{vm} + 8.3 \text{ J} = 20.8 \text{ J, and the ratio } \frac{C_{pm}}{C_{vm}} = \frac{20.8}{12.5} = 1.67.$$

In this treatment we have assumed that when a gas is heated at constant volume all the energy given to its molecules takes the form of kinetic energy and goes into increasing the speed of the molecules. It was soon realized that whilst this was true for monatomic gases (gases in which the molecule consisted of a single atom), when the atomicity was greater than one some of the acquired energy was distributed in other ways—as rotational energy and, at very high temperatures, as vibrational energy, for example. By making the assumption, known as *the principle of equipartition of energy*, that the energy possessed by a polyatomic gas was divided equally between the various degrees of freedom (i.e. the various means that a gas uses for storing energy), Maxwell and Boltzmann were able to calculate the theoretical value of the ratio of its specific heats, $\frac{C_{pm}}{C_{vm}}$. According to their calculations this ratio, conventionally denoted by γ, should have a value of $\frac{5}{3}$ (i.e. 1.67) for monatomic gases, $\frac{7}{5}$ (1.4) for diatomic gases, and an even lower value for a triatomic gas. When they put these predictions to the test they found that broadly speaking they were confirmed by experiment, as Table 3.1 shows. It is possible to measure γ indirectly by various physical methods, details of which will be found in most textbooks of physics.

The importance of this deduction from the kinetic theory is that it provides a means of determining the atomicity of a gas. If the molecular weight of a gaseous element is known, then its atomic weight can be found as described in § 2.15, since atomic weight = molecular weight ÷ atomicity. The method has been used for finding the atomic weights of the inert gases.

3.12. Deviations from Ideal Gas Behaviour. As described in § 3.4, real gases do not obey the ideal gas laws exactly. Under normal conditions the deviations are small, rarely exceeding 3%, which explains why they were not known to Boyle and Charles, but if a gas is subjected to very

TABLE 3.1. C_{pm}/C_{vm} FOR VARIOUS GASES

Gas	*Molecular Formula*	C_{pm}/C_{vm}
Sodium	Na	1.68
Mercury	Hg	1.67
Argon	Ar	1.67
Helium	He	1.66
Hydrogen	H_2	1.41
Hydrogen chloride	HCl	1.41
Oxygen	O_2	1.40
Nitrogen	N_2	1.40
Carbon monoxide	CO	1.40
Chlorine	Cl_2	1.36
Hydrogen sulphide	H_2S	1.32
Carbon dioxide	CO_2	1.30

high pressure, particularly at low temperature, then much bigger deviations may occur. These were thoroughly investigated by Amagat, who found that the value of the product PV for a fixed mass of gas at constant temperature varies with pressure as shown in Fig. 3.7. These curves, known as *isothermals*, are of three types depending upon the gas considered and the temperature. When a gas is near its liquefaction temperature the value of PV varies as in curve I, decreasing sharply at first as the pressure is increased, and then rising steadily at higher pressures. At a higher temperature the value of PV follows the course of curve II with rising pressure, whilst at a temperature at least three hundred degrees above the liquefaction temperature an isothermal shows a steady rise with pressure as in curve III. For example, at 40°C the gases carbon dioxide, oxygen, and hydrogen have isothermals of type I, II, and III, respectively. The isothermal of a perfect or ideal gas would, of course, be a straight line parallel to the pressure axis, as shown, since PV would have a constant value at any one temperature whatever the pressure applied.

It will be noticed that the deviations from ideal gas behaviour are greatest at high pressures and low temperatures, which are the very conditions when the molecules of a gas are closest together. This suggests two explanations of these deviations in terms of the kinetic theory. Firstly the volumes of the molecules may not always be negligible compared with the volume occupied by the gas, and secondly the molecules may exert appreciable attraction for each other under these conditions.

In using the kinetic theory to explain the ideal gas laws, it was assumed that both the volume occupied by the molecules and their forces of

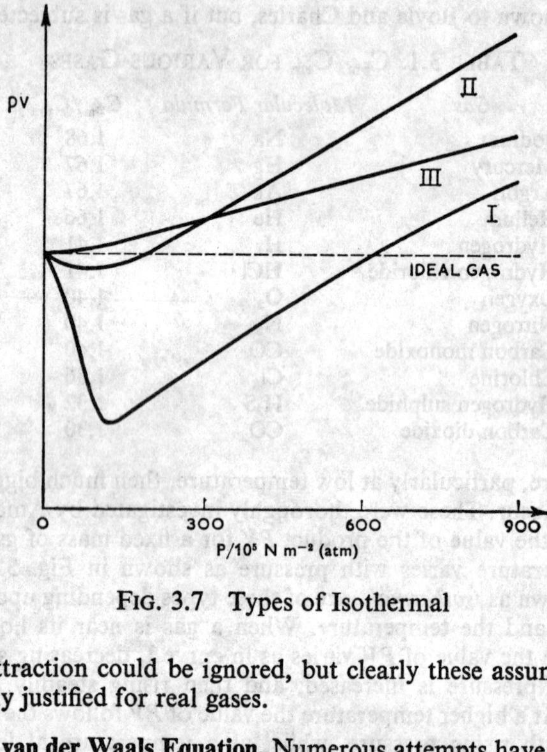

FIG. 3.7 Types of Isothermal

mutual attraction could be ignored, but clearly these assumptions are not strictly justified for real gases.

3.13. The van der Waals Equation. Numerous attempts have been made to devise an equation relating *PV* and *T* which would agree more closely with the behaviour of real gases than the ideal gas equation. The best known of these is the one put forward by van der Waals in 1873:

$$\left(P + \frac{a}{V^2}\right)(V - b) = RT$$

where a and b are different constants for each gas. The term $\frac{a}{V^2}$ takes account of the attractive forces acting between the molecules, and the term b is related to the volume of the molecules. These two constants can be determined from the critical constants of the gas as described in the next section. Other equations have been suggested which claim to represent the behaviour of real gases even more exactly, but their usefulness is limited by their increased complexity.

3.14. Liquefaction. If a gas is cooled strongly or is highly compressed at a sufficiently low temperature, the molecules are brought close enough together for their mutual attraction to cause liquefaction. The gas changes abruptly into a liquid of clearly defined volume, bounded by a surface which confers certain important characteristic properties. Early attempts were made to liquefy all gases by compression, but some (e.g. hydrogen, nitrogen, and oxygen) defied these attempts no matter how highly they were compressed and were known as 'permanent gases' for that reason.

As a result of his experiments with carbon dioxide, Andrews showed that if it was desired to liquefy a gas by applying pressure, then it was necessary first to cool it below a certain temperature. This temperature, known as the *critical temperature* of a gas, *Tc*, is defined as the highest temperature at which it is possible to liquefy that gas by the application of pressure alone. The pressure required to produce liquefaction at the critical temperature is known as the *critical pressure*, *Pc*, and the corresponding volume of one gramme-molecule of the gas (expressed as a fraction of the volume at s.t.p.) as the *critical volume*, *Vc*. It is shown in textbooks of physics that these critical constants are related to the terms *a* and *b* in the van der Waals equation by the expressions:

$$Tc = \frac{8a}{27Rb}, \quad Pc = \frac{a}{27b^2}, \quad Vc = 3b.$$

The so-called permanent gases differ from the others not in kind, therefore, but only in degree in that their critical temperatures are exceptionally low (e.g. *Tc* for nitrogen is −147°C and for hydrogen −240°C), and provided they are cooled sufficiently first, they can be liquefied by compression.

Modern methods of liquefying gases make use of the *Joule-Thomson* effect. When a gas under high pressure flows through a jet or porous plug and expands into a region of lower pressure it is cooled as a result, provided that its temperature beforehand is below a certain value which varies from gas to gas and is known as the *inversion temperature*, *Ti*. The cooling occurs as a result of the rapid expansion of the gas which causes it to do work against the internal attractive forces between the

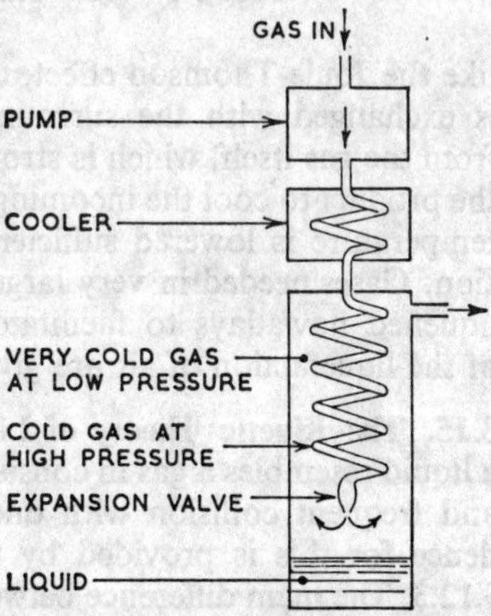

FIG. 3.8 The Hampson–Linde Liquefaction Process

molecules. The inversion temperature for hydrogen is about −80C°. but for most other gases it is considerably higher.

Two main processes for liquefying gases are in use. In the Hampson–Linde process the gas is compressed to 2×10^7 N m^{-2} (200 atm), cooled, and then passed through a valve or plug into a chamber at lower pressure, as in Fig. 3.8. The cold expanded gas is then used to cool the incoming compressed gas so that eventually the temperature falls to a level where liquefaction takes place on expansion. In the other method, the Claude process, some of the compressed gas is used to drive an expansion engine and do external work (see Fig. 3.9). Since,

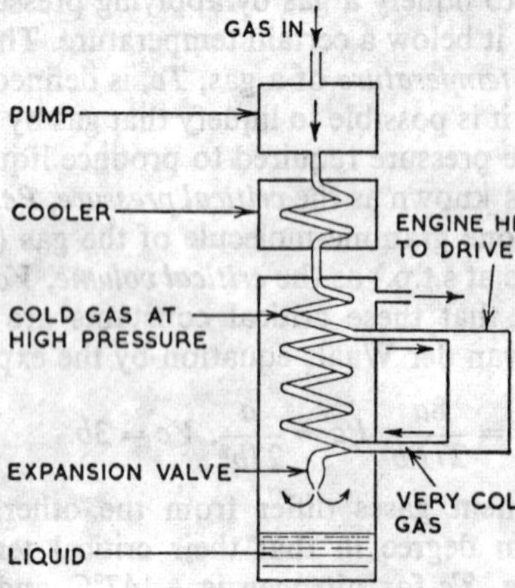

FIG. 3.9 The Claude Liquefaction Process

like the Joule-Thomson effect, this is an adiabatic change (i.e. no heat is exchanged with the surroundings), the energy required is drawn from the gas itself, which is strongly cooled as a consequence. By using the product to cool the incoming gas before it passes through the jet the temperature is lowered sufficiently to cause liquefaction upon expansion. Gases needed in very large quantities for industrial purposes are liquefied nowadays to facilitate their storage and transport. Details of the liquefaction of air are given in § 24.2.

3.15. The Kinetic Theory of Liquids. According to the kinetic theory a liquid resembles a gas in consisting of particles* in continuous motion and frequent collision with one another. The best experimental evidence for this is provided by the Brownian movement described in § 12.3. The main difference between a liquid and a gas is that in a liquid

* These particles are molecules in most liquids, but ions or ion-pairs in molten electrolytes. For simplicity only molecules are considered here.

the molecules are much closer together (e.g. over 1700 cm^3 of steam at 100°C condense into 1 cm^3 of water at the same temperature). As a result the cohesive forces between them are much greater in a liquid than in a gas and they are less free to move about independently of each other. For a molecule in the interior of the liquid this strong inter-molecular attraction has little net effect since it acts upon the molecule equally from all directions, but near the surface of the liquid the situation is very different because the attractive forces there, as shown in Fig. 3.10, are not symmetrical. A molecule at the surface of the liquid will consequently experience a net inward pull which will, in most cases, prevent its escape. This accounts for the *surface tension* of a liquid and for the tendency of its surface to contract to the smallest possible area.

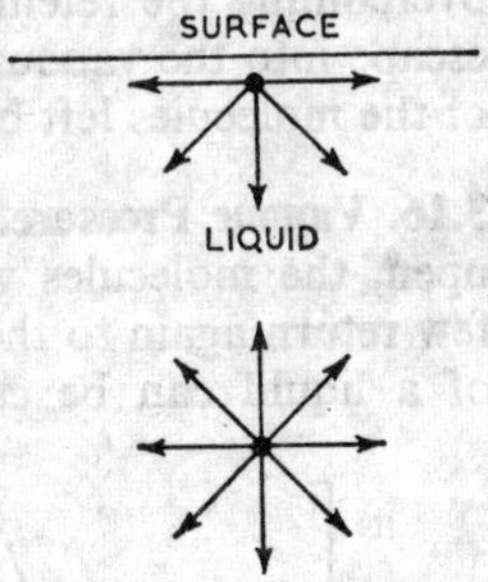

FIG. 3.10 Inter-molecular Attractions in a Liquid

Now whilst the average speed of the molecules in a liquid is determined by the temperature, the individual speeds will differ widely. Although at any instant most of the molecules will be moving at or near the average speed, a small proportion will have acquired very great speeds as a result of their random collisions. Let us imagine one of these fast-moving molecules approaching the surface of the liquid. It will, of course, be subject to a strong inward pull, but if its speed is great enough its momentum will enable it to overcome this retentive force and it will escape from the liquid into the space above.

Since the molecules which escape from the liquid during this process of *evaporation*, as it is called, are the ones with the highest speeds, unless heat is added to the liquid at the same time the average kinetic energy of its remaining molecules (and hence its temperature) will fall. This is why very volatile liquids feel cold to the touch and are sometimes used as local anaesthetics (e.g. ethyl chloride—see § 40.5). It is also the principle underlying most methods of refrigeration. The amount of heat which has to be supplied to prevent the temperature of the liquid from falling during the evaporation of one gramme-molecule of a substance is known as its *molar latent heat of vaporization* because it is 'hidden' energy which does not show itself by causing a rise in temperature. It becomes, in fact, potential energy stored in the vapour by virtue of the separation of the molecules against their forces of mutual attraction, and it is released again when the vapour condenses back into liquid.

It is found experimentally that the molar latent heat of vaporization of a liquid, expressed in joule, is approximately equal to 88*T*, where *T* is the normal boiling point of the liquid on the absolute scale of temperature, i.e. K. This general relationship, which is known as *Trouton's rule*, holds true for most substances except those with very low boiling points, but those liquids in which hydrogen bonding (§ 14.7) occurs such as water, alcohols, and ammonia, have substantially higher values of latent heat than would be expected from the rule.

The fact that it is the fastest-moving molecules which escape from a liquid during evaporation might give the impression that the vapour formed was at a higher temperature than the liquid. This is not so because the molecules are slowed down considerably in the process of overcoming the retentive forces at the surface and when they finally escape into the vapour their average kinetic energy is the same as that of the molecules left behind in the liquid.

3.16. Vapour Pressure. If the process of evaporation takes place in the open, the molecules which escape diffuse away in all directions and few return again to the liquid. Indeed the effective rate of evaporation of a liquid can be considerably increased by removing the vapour

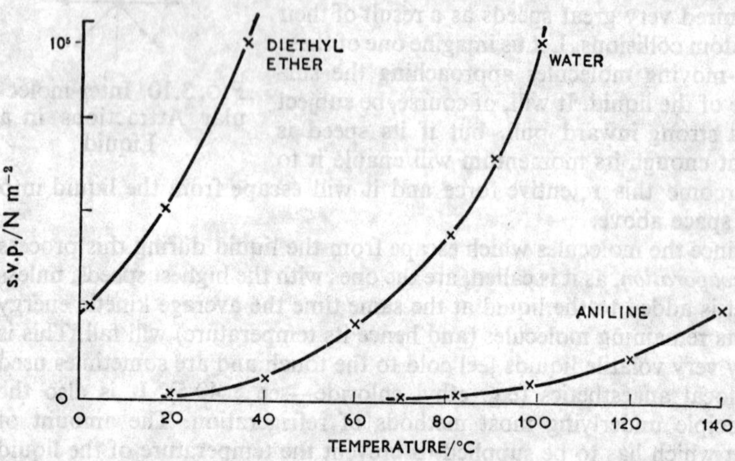

FIG. 3.11 Vapour Pressure Curves

molecules as they escape, as, for example, happens in a desiccator or in a strong draught. When a liquid evaporates in a closed container, however, the molecules which escape from the liquid gradually accumulate in the space above the liquid, colliding frequently with one another and with the walls of the vessel and consequently exerting a pressure, which is known as the *vapour pressure*. After suffering numerous random collisions some of the molecules in the vapour will approach the liquid again. As they do so they experience a strong downward pull towards the surface, owing to the attraction by the molecules in the liquid, causing them to accelerate and re-enter the liquid.

The rate at which molecules condense in this way increases as the vapour pressure rises, whereas the rate of evaporation is constant provided the temperature remains the same. At any one temperature, therefore, a state of dynamic equilibrium will eventually be achieved when the number of molecules condensing into the liquid in a given time just equals the number escaping by evaporation. When this position is reached the vapour pressure will assume a constant value known as the *saturated vapour pressure* (s.v.p.), although the word 'saturated' is frequently omitted here and the pressure is often referred to merely as the vapour pressure of that liquid at that temperature. The s.v.p. depends very much upon the nature of the liquid and the temperature, as shown in Fig. 3.11. It does not depend upon the quantity of liquid present, the area of its surface, or the volume of the vapour.

3.17. Boiling Point. The boiling point of a liquid is that temperature at which its vapour pressure is equal to the external pressure. Thus any

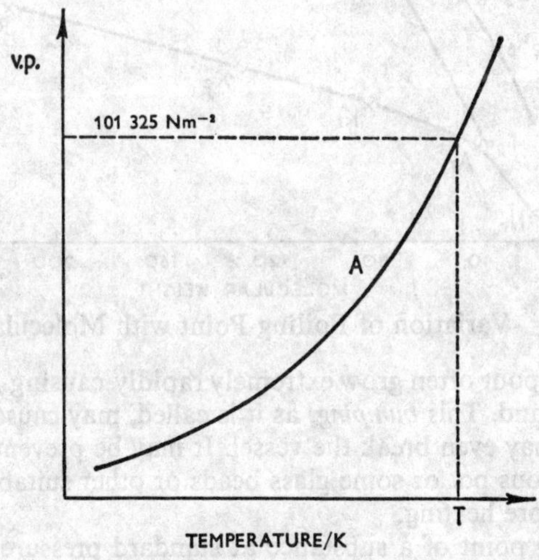

FIG. 3.12 Boiling Point and Vapour Pressure

given liquid has an infinite number of boiling points depending upon the pressure applied. For the sake of convenience one particular external pressure (101 325 N m^{-2}) is arbitrarily chosen as standard so that the volatility of different liquids can readily be compared. Thus the *normal boiling point of a liquid* is defined as *that temperature at which its vapour pressure is equal to 101 325* N m^{-2}. For example, in Fig. 3.12 the liquid A has a boiling point of T.

When heated to its boiling point a liquid not only evaporates rapidly from its surface, but it also vaporizes throughout its whole volume. Small pockets or bubbles of vapour form in the interior of the liquid and grow in size because of their high internal pressure, eventually rising to the surface and escaping. Unless the liquid contains something to act as nuclei for the bubbles of vapour, *superheating* will probably take place, i.e. the temperature of the liquid will rise temporarily above the boiling point. When the superheated liquid eventually boils the

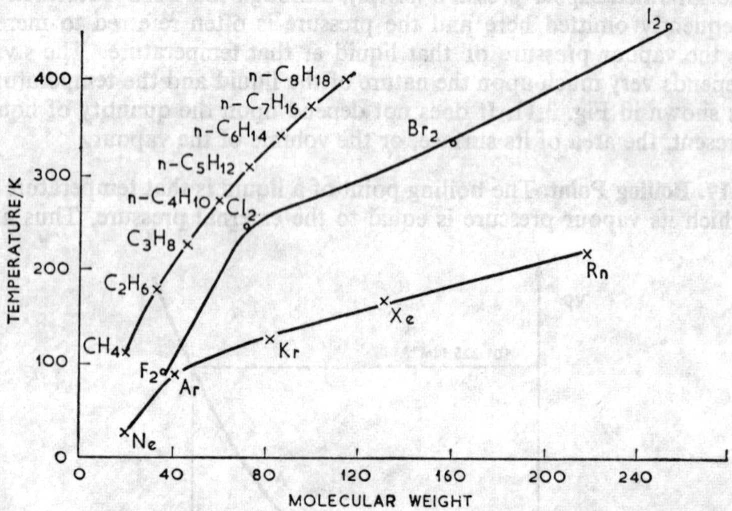

FIG. 3.13 Variation of Boiling Point with Molecular Weight

bubbles of vapour often grow extremely rapidly causing violent agitation of the liquid. This *bumping*, as it is called, may cause the liquid to overflow or may even break the vessel. It may be prevented by adding pieces of porous pot or some glass beads or other suitable material to the liquid before heating.

The boiling point of a substance at standard pressure is dependent on two main factors, its molecular weight and its molecular structure. Obviously the heavier the molecules the lower their average speed at a given temperature and the fewer molecules there are with sufficient energy to escape from the liquid. This explains why volatility generally decreases with rising molecular weight in a homologous series of organic compounds, for example, and in certain groups of the periodic classification (see Fig. 3.13). The influence of molecular structure upon volatility is illustrated not only in the striking differences between

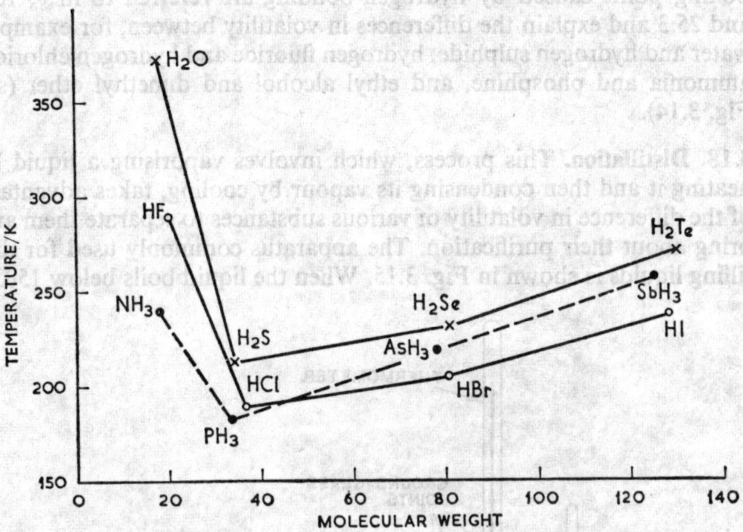

FIG. 3.14 The Effect of Hydrogen Bonding upon Boiling Point

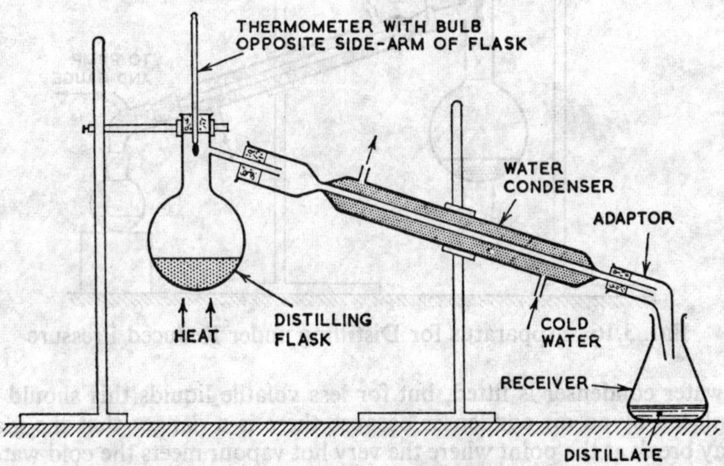

FIG. 3.15 Apparatus for Ordinary Distillation

carbon and nitrogen, and between silicon and phosphorus, where macromolecules are concerned, but also by comparing compounds of the same molecular weight such as *n*-butane (b.p. −1°C) and iso-butane (b.p. −12°C) and acetone (b.p. 56°C). The abnormalities in boiling point caused by hydrogen bonding are referred to in §§ 16.4 and 25.3 and explain the differences in volatility between, for example, water and hydrogen sulphide, hydrogen fluoride and hydrogen chloride, ammonia and phosphine, and ethyl alcohol and dimethyl ether (see Fig. 3.14).

3.18. Distillation. This process, which involves vaporising a liquid by heating it and then condensing its vapour by cooling, takes advantage of the difference in volatility of various substances to separate them and bring about their purification. The apparatus commonly used for distilling liquids is shown in Fig. 3.15. When the liquid boils below 150°C

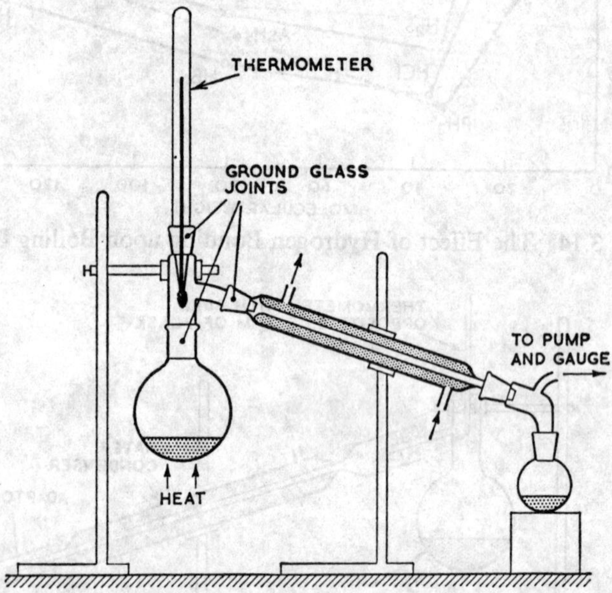

FIG. 3.16 Apparatus for Distilling under Reduced Pressure

a water condenser is fitted, but for less volatile liquids this should be replaced by an air condenser because there is a danger that the glass may break at the point where the very hot vapour meets the cold water. If the liquid is very volatile (e.g. diethyl ether) it may be necessary to use a double surface water condenser or stand the receiver in a cooling bath to ensure as complete condensation of the vapour as possible.

Where a substance decomposes when heated to its normal boiling point it can be *distilled under reduced pressure* in the apparatus shown in Fig. 3.16. By lowering the external pressure in this way it is possible to lower the boiling point of the liquid to a level at which it vaporizes without decomposition. For example, hydrogen peroxide decomposes when heated above 80°C, but it boils at 70°C under a pressure of only 4×10^3 N m^{-2}.

The principles of fractional distillation and of steam distillation and details of the apparatus used for each are given in § 6.10 and § 6.13 respectively.

4

Solids and Solutions

4.1. Introduction. The property of a solid which most clearly distinguishes it from a liquid or gas is its rigidity—every solid has not only a definite volume but also a definite shape. In the solid state almost all pure substances are crystalline, i.e. they have a regular and orderly structure. So-called amorphous powders consist of crystalline material so finely ground that the individual crystals are no longer visible to the naked eye; if they are examined under a microscope or subjected to X-ray analysis their crystalline nature becomes obvious. A few truly amorphous substances do exist which are completely devoid of a regular, ordered structure. These substances, of which glass, pitch, and resins are examples, are best regarded as congealed liquids. When heated they soften gradually over a wide range of temperature (e.g. ordinary soda glass has no sharp or definite melting point and becomes progressively more fluid over the range 500°C–1500°C). They also differ from crystals in not occurring naturally in regular shapes with plane faces, and in not splitting easily along well-defined cleavage planes.

4.2. Crystalline Solids. The most obvious characteristic of a crystal is its regular shape. The crystals of any one substance often show different relative development of their plane faces (i.e. different *habit*), but the angles between their corresponding faces are always the same. Crystals are classified into 32 classes and 7 systems according to the various elements of symmetry which they display in their outward form. The student should consult a textbook on crystallography for details. When a crystal is broken it often fractures along particular planes called cleavage planes. For example, a crystal of calcite breaks up into little rhombohedra when hammered, each a replica of the original crystal. Another characteristic of crystals is that many of their physical properties such as refractive index or coefficient of thermal expansion are *anisotropic*, i.e. they differ in magnitude in different directions in the crystal.

These properties of crystals strongly suggest that their structure is an orderly one, the internal order being reflected in the symmetry of their outward form. This view has been abundantly confirmed by X-ray analysis of crystals since the results obtained can only be explained by assuming that in a crystal the constituent particles are arranged in a rigid, three-dimensional *space lattice*. These particles, which may be atoms, molecules, or ions (atoms or groups of atoms which bear a positive or negative charge as a result of the loss or gain of electrons), are positioned in such a regular way in space that it is possible to represent the structure of the whole crystal in terms of the spatial distribution of a small group of particles known as the *unit cell*. Just as a patterned wallpaper can be built up by repeating a small unit of design many times in the right orientation, so can the unit cell, repeated to an almost infinite extent in all three dimensions, give rise to the space lattice of the crystal. This is illustrated in Fig. 4.1, where a unit cell of

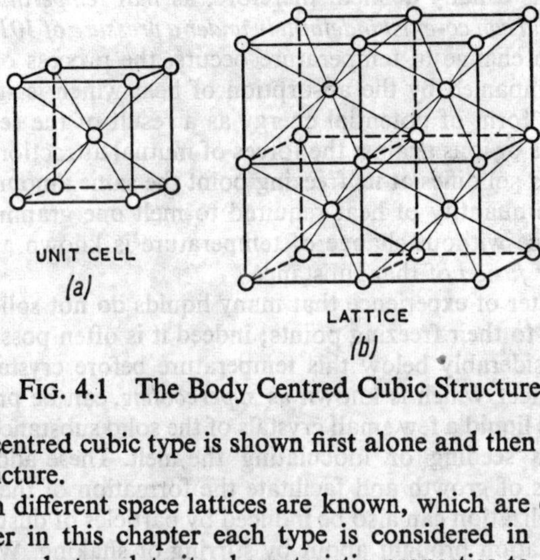

FIG. 4.1 The Body Centred Cubic Structure

the body-centred cubic type is shown first alone and then as part of a crystal structure.

Fourteen different space lattices are known, which are of four main types. Later in this chapter each type is considered in turn and an attempt is made to relate its characteristic properties to its structure.

4.3. Freezing and Melting. In the previous chapter we have seen how the gaseous and liquid states of matter can be described in terms of the random motion of molecules. The kinetic theory can also provide an explanation of the process of freezing or solidification which takes place when a liquid is cooled. As the temperature falls the molecules or ions move, on the average, more and more slowly until eventually a stage is reached when their thermal or kinetic energy is insufficient to overcome the cohesive forces between them. The molecules or ions then settle into fixed positions in a rigid space lattice and as a result of the strong forces of mutual attraction lose virtually all their freedom of translational movement. They are still able to oscillate or vibrate about their mean positions in the lattice and do so to an extent which is proportional to the absolute temperature.

The reverse process of melting takes place when a solid is heated to such an extent that the thermal energy of the particles of which it is composed causes them to break loose from their rigid lattice arrangement and move independently. Since in a particular crystalline solid the particles will, unlike those in a gas, all have about the same amount of energy at any one temperature, the disruptive process will occur in all parts of the crystal at the same temperature and the *melting point* is usually sharp and definite. As explained in chapter 6, at the melting point the solid and liquid can exist in equilibrium with each other. Now experiment shows that the melting point of a substance varies with pressure. It is usually defined, therefore, as *that temperature at which solid and liquid can co-exist indefinitely under a pressure of 101 325* N m^{-2}. Although no change in temperature occurs, the process of melting is always accompanied by the absorption of heat which is stored in the liquid in the form of potential energy as a result of the separation of the molecules or ions against the forces of mutual attraction. Similarly, when a liquid solidifies at its freezing point the same amount of heat is evolved. The quantity of heat required to melt one gramme-molecule of a substance without change of temperature is known as the *molar latent heat of fusion* of that substance.

It is a matter of experience that many liquids do not solidify readily when cooled to their freezing points; indeed it is often possible to cool a liquid considerably below this temperature before crystals begin to form. This effect, which is known as *supercooling*, can be prevented by adding to the liquid a few small crystals of the solid substance, a process referred to as 'seeding' or 'inoculating' the melt. These added crystals act as centres of growth and facilitate the formation of the crystalline solid. Crystallization can also be induced by particles of dust or even by vigorous agitation brought about by stirring or shaking. When supercooling has taken place in a liquid and solidification begins, the latent heat of fusion released causes the temperature to rise rapidly to the melting point and stay there, despite normal heat losses, until all the liquid has solidified. A typical cooling curve for such a substance is shown in Fig. 4.2. Supercooling is commonly encountered when determining molecular weights by the freezing point depression method (see § 5.14) and when investigating eutectic mixtures (§ 6.15).

The factor which exercises the biggest influence upon the melting point of a substance is the strength of the attractive forces acting between the particles in the crystal. As described in later sections of this chapter, these cohesive forces are of several different kinds depending upon the nature of the constituent particles and the type of lattice formed, which accounts for the great differences in melting point displayed by different substances.

When a solid melts it often happens that some of its constituent particles remain grouped in the liquid state in bunches or clusters within which they retain the same orientation towards each other as existed in the crystal. This persistence of localized order in the midst of general disorder is characteristic of many liquids, particularly at temperatures not far above their melting points. At higher temperatures the forces of thermal agitation and the bombardment of the clusters by fast-moving

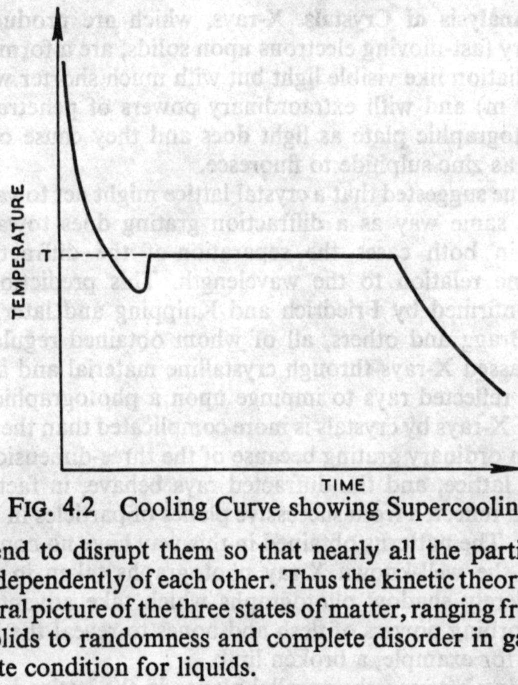

FIG. 4.2 Cooling Curve showing Supercooling

particles tend to disrupt them so that nearly all the particles are free to move independently of each other. Thus the kinetic theory provides us with a general picture of the three states of matter, ranging from complete order in solids to randomness and complete disorder in gases, with an intermediate condition for liquids.

4.4. The Vapour Pressure of Solids. Like liquids, solids also exert a vapour pressure, presumably because molecules at the surface of the lattice tend to be held less firmly than the rest and can occasionally escape into space as a result of their thermal energy. The fact that many solids have smells is evidence of this process. The vapour pressure differs from one solid to another and increases as the temperature of the solid is raised, but it is independent of the amount of solid present. Where the vapour pressure of a solid is so low that it can be ignored, which is frequently the case, the solid is conventionally referred to as being *non-volatile.* Certain solids do exert appreciable vapour pressures, e.g. ice has a v.p. of 613 N m^{-2} at 0°C and iodine vapour exerts a pressure of about 1.2×10^4 N m^{-2} just below its melting point (114°C). Naphthalene, iodine, benzoic acid, and similar substances with relatively high vapour pressures are readily converted directly from the solid state into the vapour when heated gently below their melting points. On exposing their vapours to a cold surface they condense directly to the solid again without passing through the liquid state. This phenomenon, which is known as *sublimation,* can be used to purify these solids. Another aspect of sublimation is discussed in § 6.4.

4.5. X-ray Analysis of Crystals. X-rays, which are produced by the impact of very fast-moving electrons upon solids, are a form of electromagnetic radiation like visible light but with much shorter wavelengths (about 10^{-10} m) and with extraordinary powers of penetration. They affect a photographic plate as light does and they cause certain substances such as zinc sulphide to fluoresce.

In 1912 Laue suggested that a crystal lattice might act towards X-rays in much the same way as a diffraction grating does towards visible light, since in both cases the separation of the diffracting planes bore the same relation to the wavelength. This prediction was immediately confirmed by Friedrich and Knipping and later by W. H. and W. L. Bragg and others, all of whom obtained regular patterns when they passed X-rays through crystalline material and allowed the diffracted or reflected rays to impinge upon a photographic film. The diffraction of X-rays by crystals is more complicated than the diffraction of light by an ordinary grating because of the three-dimensional nature of the space lattice, and the diffracted rays behave, in fact, just as if they had been reflected from successive planes of particles in the crystal, as in Fig. 4.3. The patterns obtained in this way have no connection, of course, with the well-known X-ray photographs taken in a hospital. These are merely shadow photographs which take advantage of the different absorbing powers of flesh and bones to reveal the position of the latter in, for example, a broken limb.

If the spacing between the parallel planes in the lattice is *d*, and the wavelength of the X-rays λ, and if θ is the angle between the incident rays and the lattice planes, then the conditions for the reflected rays from the various parallel planes to be so phased as to reinforce one another and form a ray of maximum intensity is given by the equation:

$$n\lambda = 2d.\sin\theta \quad \text{(Bragg's law)}$$

where n is an integer. Thus if n, λ, and θ are known, the lattice spacing d can be determined. Since in most crystals this spacing is of the order of 10^{-10} metre, it used to be expressed for convenience in the non-SI *ångström unit* (1 Å being 10^{-10} m), but now the *nanometre* (1 nm = 10^{-9} m) is used.

Textbooks of X-ray crystallography should be consulted for details of the two main experimental methods. In one a single large crystal is oscillated or rotated repeatedly in a beam of X-rays so that a series of reflections are obtained from the principal planes in its lattice. More important nowadays is the powder method devised by Debye, Scherrer, and Hull in 1917 in which a mass of finely ground crystalline material with its lattice planes orientated in all directions at random is suspended in a beam of X-rays. In both methods monochromatic X-rays are used

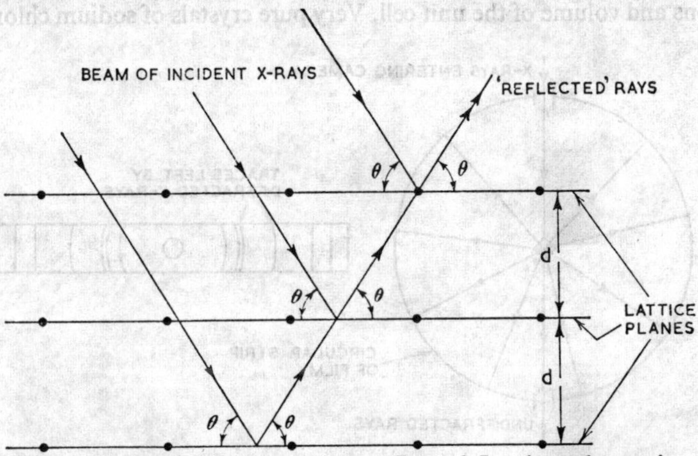

FIG. 4.3 Diffraction of X-rays by a Crystal Lattice, shown in two dimensions only

(i.e. rays of only one wavelength) and the directions of the more important reflected rays are recorded in the form of dots or lines upon a photographic film placed in a strip around the crystal, as shown in Fig. 4.4.

The interpretation of these photographs is not easy, especially when the structure examined is a complicated one, but in skilled hands X-ray analysis can yield a great deal of precise information about the internal structure of a crystal which could not be discovered by any other means. Whereas our previous knowledge of the solid state was mostly deduced indirectly from the external symmetry and physical properties of the crystals and from the way in which they behaved when melted or dissolved, we now have the means of investigating their orderly structure directly. With the help of X-rays we can 'see' the long zig-zag chains of carbon atoms in aliphatic compounds (§ 33.5) and the rings of equidistant carbon atoms in aromatic compounds (§ 43.3). The various silicates can be understood in terms of individual particles, chains, sheets, or networks of SiO_4 groups instead of unlikely and hypothetical silicic acids (§ 20.23), and the differences between diamond and graphite can be explained in terms of their respective lattice structures (see § 20.3). Nor is this valuable technique limited to crystals in the ordinary sense, since most metals and many fibres also possess structures which give significant X-ray patterns.

X-ray analysis also provides an accurate method of determining the Avogadro constant (§ 1.12). If the lattice spacing in a particular crystal is determined by means of X-rays, it is possible to calculate the dimensions and volume of the unit cell. Very pure crystals of sodium chloride,

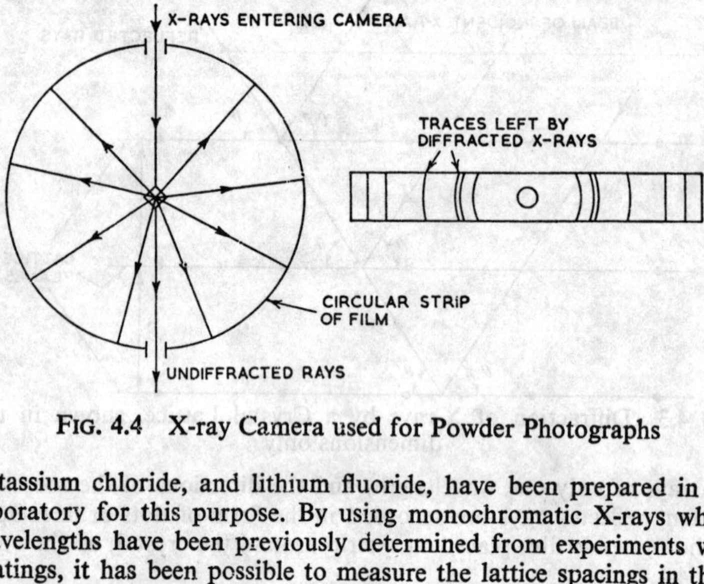

FIG. 4.4 X-ray Camera used for Powder Photographs

potassium chloride, and lithium fluoride, have been prepared in the laboratory for this purpose. By using monochromatic X-rays whose wavelengths have been previously determined from experiments with gratings, it has been possible to measure the lattice spacings in these crystals with great accuracy (for example, the spacing in sodium chloride at 18 °C is found to be 2.81400×10^{-10} m). If the density of the crystal is then determined very accurately, the mass of the unit cell can be found. From a knowledge of the number of atoms or ions in the unit cell it is possible, therefore, to calculate the actual mass of each one and to divide this into the mass of a gramme-molecule to obtain the value of N_A. There is close agreement between such values and those obtained by the electrical method described in § 8.2.

If a beam of fast-moving electrons is passed through a gas at low pressure the electrons are diffracted by the gas in much the same way as X-rays are by a crystalline solid. This process of *electron diffraction*, as it is called, allows us to determine bond lengths and inter-bond angles in gaseous molecules and can give us valuable information about molecular shape, some of which is discussed in § 14.8. Electron diffraction can also be applied to solids (provided the incident beam just skims along the surface), but since the electrons possess little powers of penetration, only the surface of the solid can be investigated.

The intensity of the diffracted electrons and X-rays is greater the higher the atomic number of the element, since it is the electrons in the atom which cause the diffraction in each case. This explains why neither method can be used to locate hydrogen atoms with much precision and why *neutron diffraction*, which principally depends upon scattering of neutrons by the nucleus of the atom, is preferred for this purpose.

4.6. Ionic Lattices. In this type the constituent particles are, as the name implies, *ions*, i.e. charged atoms or groups of atoms which are formed by the transfer of electrons from one atom to another as described in § 14.3. These ions are held rigidly in place in the lattice by the strong electrostatic attraction of their opposite charges. Each ion of one type is surrounded by a number of ions of opposite charge equidistant from it. For example, in a crystal of sodium chloride, around each Na^+ ion

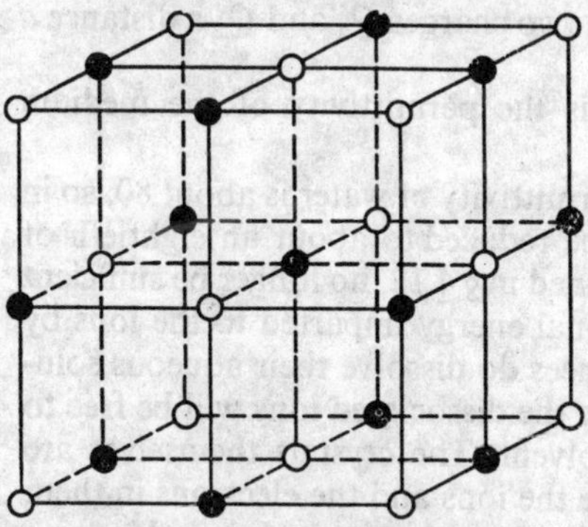

FIG. 4.5 The Structure of Sodium Chloride

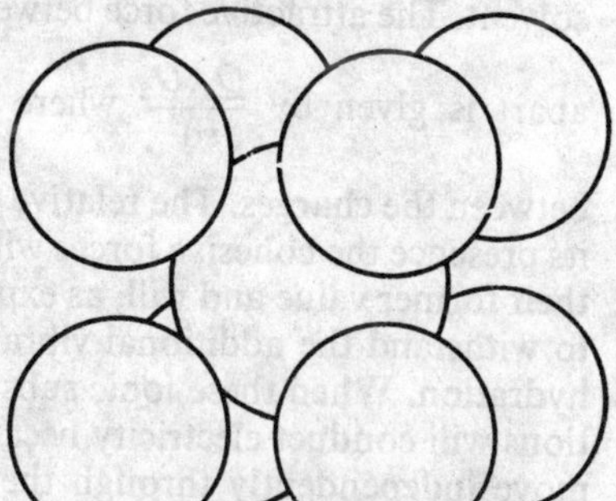

FIG. 4.6 The Structure of Caesium Chloride

are six Cl^- ions at equal distances from it, and each Cl^- ion has 6 Na^+ ions symmetrically disposed around it as in Fig. 4.5. Such a structure is said to have a *co-ordination number* of six. The co-ordination number is determined by geometrical considerations, i.e. by how many of the larger Cl^- ions can be packed around one of the smaller Na^+ ions. Once this is settled, then the number of Na^+ ions around each Cl^- ion is also fixed, since the necessity for electrical neutrality demands equal numbers of each type of ion in the crystal when both ions bear equal charges.

The vital factor, therefore, in deciding which kind of ionic lattice is formed is not the absolute size of the ions but their *relative* sizes, since it is this which ultimately determines the co-ordination number in any particular crystal. Thus all the alkali halides except the chloride, bromide, and iodide of caesium have the same structure as sodium chloride (Fig. 4.5). In these caesium halides, however, the cation is large enough to allow eight of the anions to be arranged around it, giving a cubic structure with a co-ordination number of eight, as in Fig. 4.6. This should be contrasted with the lattices of the sodium and potassium halides, where there is the octahedral or six co-ordinated structure shown in Fig. 4.5.

In these ionic lattices there are no molecules as such, and it is impossible, for example, to detect any pairs of ions corresponding to the formula NaCl since all six Cl^- ions are the same distance away from any one Na^+ ion. If this is clearly understood then it will be appreciated that the properties of substances with an ionic lattice are necessarily the characteristic properties of the ionic or electrovalent bond (§ 14.3). Such substances will be brittle and hard. They will tend to dissolve readily in water and other highly polar solvents because the attraction between the ions will be greatly reduced by the high permittivity of the solvent. The attractive force between two charges Q_1 and Q_2 a distance d apart is given by $\frac{Q_1 . Q_2}{\varepsilon d^2}$ where ε is the permittivity of the medium between the charges. The relative permittivity of water is about 80, so in its presence the cohesive forces will be reduced to about an eightieth of their former value and will, as explained in § 4.12, no longer be sufficient to withstand the additional vibrational energy imparted to the ions by hydration. When these ionic substances do dissolve their aqueous solutions will conduct electricity because the dissociated ions will be free to move independently through the solvent. The crystals themselves are not conductors of electricity because the ions and the electrons in them are held rigidly in position in the lattice, but when melted the substances become good electrical conductors. In general, substances with ionic lattices are non-volatile, i.e. they have high melting points and boiling points and are solid at room temperature, (e.g. sodium chloride melts at 800°C and sodium fluoride at 980°C). Melting such crystals necessitates breaking a large number of strong electrostatic bonds between the ions, so when the ions bear a double charge the thermal energy needed is often very great and melting only takes place at very high temperatures (e.g. calcium oxide melts at over 2500°C and magnesium oxide at about 2800°C). It is for this reason that such substances are used as refractories for lining furnaces.

4.7. Metallic Lattices. This type of lattice consists of atoms so packed together in a rigid framework that some of the outermost electrons are only loosely held. When a potential difference or temperature gradient is applied these electrons tend to migrate through the lattice from one atom to another, temporarily converting the metal atoms into positive ions. This migration of mobile electrons explains the high electrical and thermal conductivity of metals and also their opacity and high lustre. The bonds between the atoms in metals are not spatially directed, and since no limitation is set by requirements of neutrality, geometrical factors alone decide the co-ordination. This leads to highly co-ordinated structures (a co-ordination of twelve, the highest possible, is common), which are dense and close-packed.

Metallic structures are of three main types. Copper, for example, has the *face-centred cubic* structure shown in Fig. 4.7 in which every atom

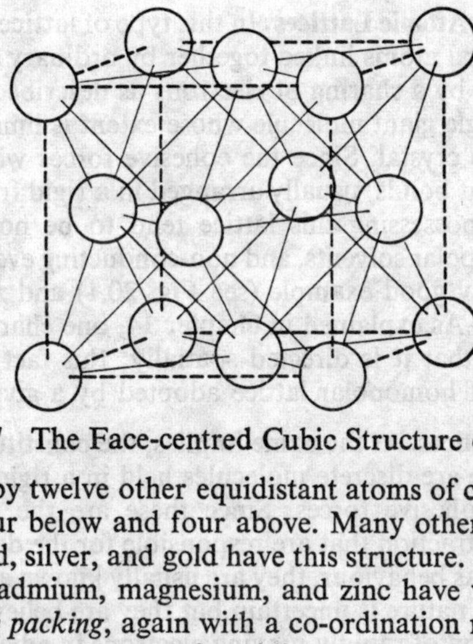

FIG. 4.7 The Face-centred Cubic Structure of Copper

is surrounded by twelve other equidistant atoms of copper, four in the same plane, four below and four above. Many other metals including aluminium, lead, silver, and gold have this structure. Other metals such as beryllium, cadmium, magnesium, and zinc have what is known as *hexagonal close packing*, again with a co-ordination number of twelve, in which six atoms are arranged equidistant from an atom in one plane, with three more above and three below. In such a structure alternate planes of spheres are identical in orientation. The third structure adopted by metals is known as *body-centred cubic* packing; here each metal atom is surrounded by eight equidistant others arranged as though at the corners of a cube. This eight co-ordination type is found in the alkali metals, tungsten, and vanadium. Some metals can exist in more than one type of structure, depending upon the temperature; iron, for example, has a body-centred cubic lattice below 906°C but exists as face-centred cubic structure between 906°C and 1400°C (see § 30.2).

Among the most striking characteristics of metals are their malleability, i.e. the ease with which they can be rolled into thin sheets and foil, and their ductility, i.e. the ease with which they can be drawn into wires. These properties arise because the closely-packed planes of atoms readily glide over one another when pure metals are subjected to a stress. The presence of even small amounts of impurity often completely destroys the malleability of a metal, presumably because the impurity causes imperfections in the lattice which hinder this gliding action of adjacent layers. The great hardness and tensile strength of some metals arises from their tendency to form a large number of small crystals or crystallites with their space lattices orientated at random. A metal can be made poly-crystalline in this way by suitable heat treatment or by the addition of selected impurities, as in alloys, and its physical properties can be altered to give the desired blend of flexibility and hardness.

4.8. Homopolar or Atomic Lattices. In this type of lattice the constituent particles are neutral atoms linked together by ordinary covalent bonds, which are formed by a sharing of electrons as described in § 14.4. This gives rise to a single giant molecule whose extent is limited only by the boundaries of the crystal. Since the cohesive forces within the crystal are strong covalent bonds, usually arranged in a rigid framework structure, substances possessing this lattice tend to be non-volatile, very hard, insoluble in polar solvents, and non-conducting even when molten. Diamond is a very good example (see Fig. 20.1) and zinc blende and silicon are others. As explained in chapter 14, one characteristic of the covalent bond is that it is directed spatially; this fact determines the particular form of homopolar lattice adopted by a given substance.

4.9. Molecular Lattices. As the name implies, the constituent particles in this type of lattice are discrete molecules held in a rigid framework by relatively weak cohesive forces. Since these are the same forces of inter-molecular attraction that are responsible for the deviations of real gases from ideal gas behaviour, they are usually known as *van der Waals forces*. Their exact nature is uncertain but they are believed to originate from the interaction of rapidly moving electrons in adjacent molecules. These feeble van der Waals forces are so easily broken that crystals

TABLE 4.1. MELTING POINTS OF SOME FLUORIDES

Compound	NaF	MgF_2	AlF_3	SiF_4	PF_5	SF_6
Melting Point (°C)	980	1410	1035	−77	−83	−56

containing them are soft and volatile. All the inert gases, for example, have extremely low melting points, and so do hydrogen, fluorine, oxygen, nitrogen, and carbon dioxide. Even molecules of fairly high molecular weight, such as iodine (I_2), sulphur (S_8), and naphthalene ($C_{10}H_8$), melt or sublime on gentle heating. This influence of lattice structure upon volatility is well illustrated in Table 4.1, by contrasting the fluorides of sodium, magnesium, and aluminium, which crystallize in ionic lattices, with those of silicon, phosphorus, and sulphur, which form molecular lattices.

In many molecular lattices the van der Waals forces are not the only ones acting between the particles. Where the molecule is covalent in its bonding and involves directed valencies, its particular shape often results in an asymmetry in the distribution of charge within the molecule which gives it a highly polar character. The extent of this polarity is indicated by its *dipole moment*, which is proportional to the effective charges set up and their distance apart and can be measured by various physical methods. As a result of this polarity the van der Waals forces are often supplemented by other stronger attractive forces such as hydrogen bonding (see § 14.7), which raise the melting point appreciably. The best examples are provided by ice, hydrogen fluoride, and ammonia.

The dipole moment of a molecule may give useful information about its shape. For example, carbon dioxide has zero dipole moment, indicating that the molecule is a linear one in which the polarities of the two carbon-oxygen bonds cancel each other exactly. On the other hand the high dipole moments of water and sulphur dioxide establish that these molecules cannot be linear, whilst the appreciable value for ammonia is inconsistent with a symmetrical planar molecule but could be given by one with a pyramidal shape (see § 14.8).

One type of molecular lattice of special interest is that of graphite, which is described in detail in § 20.3. Here the lattice has a pronounced layer structure which arises from the contrast between the strong covalent bonds between adjacent carbon atoms and the weaker van der Waals type bonds between the layers (see Fig. 20.3).

4.10. Polymorphism. Many substances can exist in more than one form in the solid state; this phenomenon is known as *polymorphism* (or *dimorphism* when only two forms exist). For example, calcium carbonate occurs naturally as calcite and as aragonite, two dissimilar crystalline forms, and anhydrous sodium sulphate can be obtained in crystals with rhombic or monoclinic symmetry depending upon whether the temperature is above or below 234°C. At room temperature mercuric iodide is a scarlet colour, but when it is heated above its transition temperature of 126°C it changes into a brilliant yellow form consisting of orthorhombic crystals. When this form is cooled it tends to persist in the metastable condition until it is disturbed, by scratching with a rod for example, when it rapidly reverts to the red tetragonal variety. Ammonium nitrate can exist in five different forms depending upon the temperature:

$$\text{tetragonal} \overset{-17°C}{\rightleftharpoons} \text{rhombic I} \overset{32°C}{\rightleftharpoons} \text{rhombic II} \overset{84°C}{\rightleftharpoons} \text{rhombohedral} \overset{125°C}{\rightleftharpoons} \text{cubic}$$

Polymorphism is not restricted to compounds and many elements can also exist in more than one form in the same physical state, although in these cases it is usual to refer to the phenomenon as *allotropy* and to the individual forms as *allotropes.* For example, carbon exists as a solid in two very different forms diamond and graphite, yet both consist of pure carbon only. Descriptive accounts of the commonest examples of allotropy are given in the sections dealing with the elements concerned, e.g. carbon in § 20.3, sulphur in § 24.5, phosphorus in § 23.2, oxygen in § 24.3, tin in § 21.2, arsenic in § 23.13, and so on. The principal types of allotropy and the phase relationships involved in each are discussed in § 6.5.

Polymorphism arises from the different ways in which the particles in the crystal can be packed together, one arrangement being more stable than another over a certain range of temperature. When one allotropic form is converted into another there is usually a change in volume (and therefore density) as well as in outward symmetry, and heat is taken in or evolved. As might be expected, one form has a greater intrinsic energy than another owing to the difference in crystal structure. This energy difference can be determined experimentally by the indirect method described in § 11.4.

4.11. Atomic and Ionic Size. Since the Avogadro constant N_A is the number of atoms in a gramme-atom of an element, it is possible to calculate the approximate size of the atom of an element from a knowledge of its density in the solid state. For example, since there are 6.02×10^{23} atoms of aluminium in 0.027 kg of aluminium and its density is 2.70×10^3 kg m^{-3}, then there are 6.02×10^{28} atoms of aluminium in each cubic metre of the metal and each atom of aluminium occupies 1.66×10^{-29} cubic metre. This calculation takes no account, of course, of any space between the atoms in the crystal lattice, which depends upon the packing and will vary in allotropic elements, but even so it does serve to give a rough idea of atomic volume.

A more accurate estimate of atomic size can be made from X-ray and electron diffraction measurements upon solids. The assumption is made that each atom can be regarded as a sphere just in contact with its neighbours, so that the distances between adjacent nuclei are the sum of the atomic radii. Now in the crystal of an element all the atoms are of the same size so the atomic radius is equal to one half of the interatomic distance and can readily be determined by X-ray analysis. Thus the size of atoms of carbon, silicon, phosphorus, sulphur, etc., can be found in this way, and if the further assumption is then made that a given atom has the same size in all of its covalent compounds, it is possible to compile a list of atomic radii based upon measurements of interatomic distances in crystalline compounds.

In much the same way the size of ions can be assessed from X-ray measurements of interionic distances in crystals. The assumption is made that the ions are spherical in shape and that adjacent ions in the crystal lattice just touch one another. The results of these determinations of atomic and ionic size are shown diagrammatically in Table 13.7. Their significance is discussed in the light of our views of atomic structure in § 13.9.

4.12. Solutions of Solids in Liquids. When a solid is left for some time in contact with a suitable liquid, it dissolves, forming a *solution*, i.e. a mixture which is homogeneous in the sense that even the smallest samples of it contain the same proportion of liquid (*solvent*) and solid (*solute*). A solution differs from a compound in that its composition can vary within wide limits and it can readily be separated into its constituents by physical means. This process of dissolving can be explained in terms of lattices in the following way. When a crystal is added to a liquid in which it is soluble, the molecules of the liquid begin to attach themselves to the lattice. Some interaction usually takes place between liquid and solid releasing energy of solvation which causes the affected crystal particles to oscillate vigorously and break loose from the lattice. Provided the energy of solvation exceeds the energy needed to disrupt the lattice and separate its constituent particles, this process will continue all over the surface of the crystal. As explained in § 4.6, dissolution is aided considerably in ionic lattices if the liquid is composed of highly polar molecules (e.g. water) since the solvent will then have a high permittivity and will weaken the cohesive forces between the ions.

If the liquid is not agitated by stirring or shaking, the process of dissolving soon becomes very slow because particles of solute tend to return to the lattice from the concentrated solution around the solid at almost the same rate as they go into solution. On standing for a long while, however, the solute spreads slowly throughout the solution. This process of *diffusion*, which is much slower in liquids than in gases, can be demonstrated by putting some crystals of copper sulphate at the bottom of a gas jar, filling it with water, and leaving it to stand in a place where it will not be disturbed (see Fig. 4.8). After about ten days the solution will have the same depth of blue colour throughout. Thus eventually the whole solution becomes homogeneous and, assuming that an excess of solid is present, a truly *saturated solution* is obtained, i.e. *a solution which contains as much solute as can dissolve at that temperature in the presence of undissolved solid.* Thus when some solid is added to a saturated solution no apparent change occurs because solution and excess solid are in dynamic equilibrium with each other. That this condition is one of continual exchange of particles of solute between solid and solution and not, as it appears, a state of stagnation, can be demonstrated by mixing some solid containing a radioactive isotope with a saturated solution, when the latter soon becomes strongly radioactive by exchange. The amount of solid which dissolves in a specified

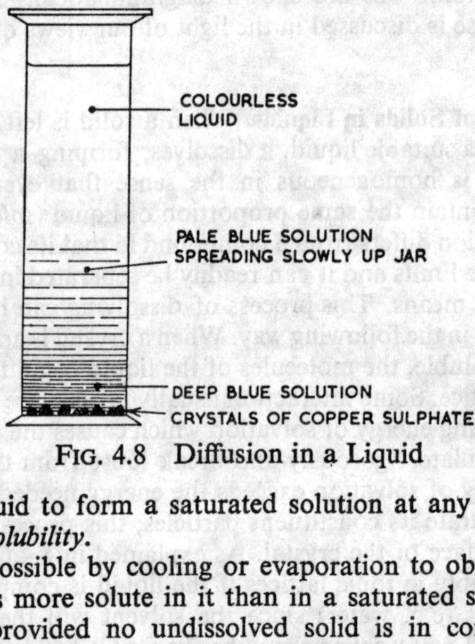

FIG. 4.8 Diffusion in a Liquid

amount of liquid to form a saturated solution at any temperature is known as its *solubility.*

It is often possible by cooling or evaporation to obtain a solution which contains more solute in it than in a saturated solution at that temperature, provided no undissolved solid is in contact with the solution. Such a solution is said to be *supersaturated.* Supersaturation occurs readily, for example, when a nearly saturated aqueous solution of sodium thiosulphate or sodium sulphate is cooled. The addition of a few small crystals or particles of dust, or even violent agitation of the liquid, can cause a supersaturated solution to deposit its excess of dissolved solid as crystals in much the same way as these measures can induce the solidification of a supercooled liquid.

4.13. Concentration of Solutions. The concentration of a solution can be expressed in so many different ways that confusion often arises, and it is important, therefore, to be clear about the various definitions. These are of two main kinds, depending upon whether the quantity of solution is specified by *mass* or by *volume*. For analytical work involving pipettes and burettes it is naturally more convenient to relate the strength of a solution to its volume, but where changes in temperature occur it is preferable to express concentration in terms of a stated mass of solution since the units are then independent of temperature. Mass for mass units have the further advantage that no difficulties arise from the change in volume which occurs on dissolving.

MASS PER VOLUME UNITS:

(1) gramme of solute per cubic decimetre of solution, (g dm^{-3}).
(2) kilogramme of solute per cubic metre of solution, (kg m^{-3}).
(3) mole of solute per cubic decimetre of solution, (mol dm^{-3}). This unit is known as the *molarity* of the solution, denoted by the symbol M.
(4) mole of solute per cubic metre of solution, (mol m^{-3}).
(5) gramme-equivalents of solute per cubic decimetre. This unit is known as the *normality* of the solution, denoted by N.

Conversion from one unit to another is easy, since 1 g $dm^{-3} \equiv 1$ kg m^{-3} and 1 mol $dm^{-3} \equiv 1000$ mol m^{-3}.

MASS PER MASS UNITS:

(1) gramme of solute per 100 gramme of solution. This is often expressed simply as a percentage.
(2) *mole fraction.* This unit, which is of the utmost importance in colligative work (see § 5.20), is the ratio of the number of mole of solute to the total number of mole of solute and solvent in the solution. It is also, of course, equal to the number of molecules of solute divided by the number of molecules of all kinds in the solution.
(3) mole per kilogramme of solvent, (mol kg^{-1}). This unit is known as the *molality* of the solution, *m*.

The best way to show the relationship between these various units is to express the concentration of a given solution in terms of each of them in turn. For this purpose we shall consider a solution of potassium permanganate containing 31.6 gramme of solute per dm^3 of solution. Then clearly it will contain 31.6 kilogramme per cubic metre of solution. Since its molecular weight is 158, our solution contains 31.6/158, i.e. exactly one-fifth of a mole per cubic decimetre, so it is said to be a one-fifth molar or 0.200 M solution. Thus it contains 200 mol m^{-3}. Now the equivalent weight of potassium permanganate when it acts as an oxidizing agent in acid solution is 31.6 (see § 2.5), so that the given solution contains exactly one gramme-equivalent per dm^3, its normality is 1.00, and it is said to be 1.00 N. If the density of our solution is, say, 1.01 g cm^{-3}, then 100 gramme of solution contains

$\frac{100}{101} \times 3.16 = 3.13$ gramme of solute, and it is a 3.13% solution. Now the solution contains 0.20 mole of potassium permanganate per dm^3 and 968.4/18, i.e. 53.8 mole of water per dm^3, so the mole fraction of solute is 0.20/(0.2 + 53.8), i.e. 0.0037. This solution contains 0.20 mole of solute in 968.4 gramme of solvent, so it contains $\frac{0.20 \times 1000}{968.4}$ mole, i.e. 0.206 mole in a kilogramme of solvent, and its molality is therefore 0.206.

4.14. The Determination of the Solubility of Solids. Several methods are in use. All start by preparing a saturated solution of the solid by adding excess of it to the solvent, shaking thoroughly, and allowing the mixture to stand for a while in a stoppered bottle. Since solubility depends upon temperature the mixture should be kept in a thermostat. Periodical shaking is necessary to accelerate dissolution. When the equilibrium condition has been achieved, a volume of the saturated solution is removed by pipette using a filter tip of glass wool or cotton wool (as in Fig. 4.9) to avoid taking up particles of undissolved solute.

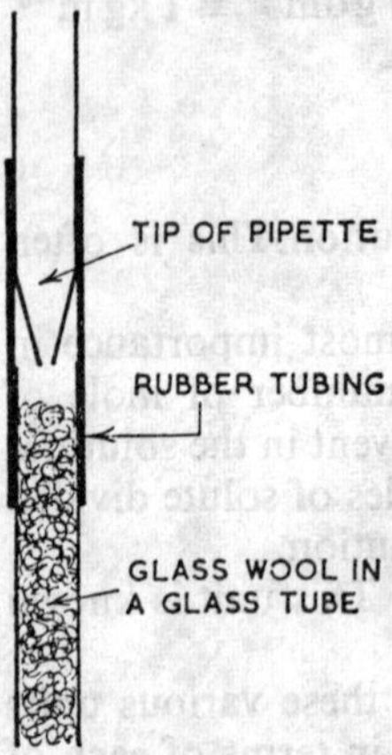

FIG. 4.9 Filter Tip for Pipette

VOLUMETRIC ANALYSIS METHOD: Known volumes of the saturated solution are first weighed and then estimated volumetrically in the usual way. For example, the solubility of a reducing agent (e.g. ammonium oxalate) can be found by titrating it against a standard solution of potassium permanganate, that of an acid (e.g. oxalic acid) by titration with standard alkali, and that of a halide (e.g. potassium chloride) with standard silver nitrate solution.

EVAPORATION METHOD: This method, which is not an accurate one, is virtually restricted to very soluble substances. The saturated solution is transferred to an evaporating basin of known weight and weighed. It is then heated gently to dryness, care being taken to avoid spurting of the liquid and spitting or thermal decomposition of the solid. For aqueous solutions it is best to use a vacuum desiccator to remove the last traces of solvent. Finally the basin and residue are weighed. By subtracting the weight of the residue from the weight of solution taken, the weight of solvent can be found and the solubility can be calculated.

ELECTRICAL METHOD: The best way of finding the solubility of a sparingly soluble salt such as silver chloride, which can be assumed to be completely dissociated into its ions in solution, is to determine the ionic concentration in its saturated solution from its conductivity (§ 8.4). A worked example of the method, which is capable of high accuracy, is given in § 8.8.

4.15. Solubility and Temperature. The solubility of a given solid in a particular solvent depends upon the temperature. With few exceptions (e.g. calcium hydroxide, normal zinc phosphate, and anhydrous sodium

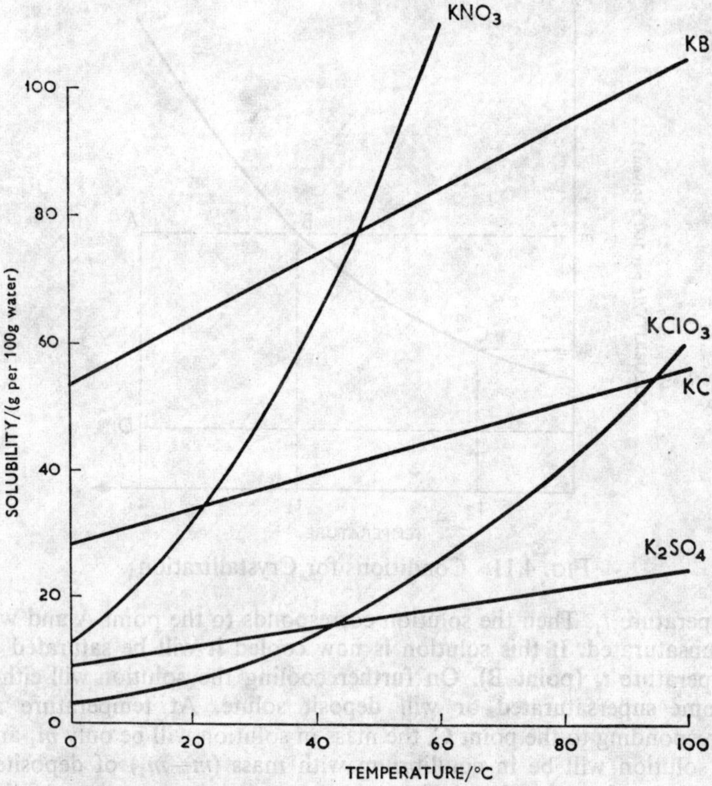

FIG. 4.10 Solubility Curves of Some Potassium Salts

sulphate) the higher the temperature the greater the solubility of the solid, although the effect of temperature is much more marked in some cases than in others. Some typical curves, which are known as *solubility curves*, are shown in Fig. 4.10. Such a curve is constructed by measuring the solubility of the solid at various temperatures using the methods described in the previous section. A solubility curve is, however, much more than just a smooth line joining points determined by experiment, for it separates two areas, the lower one representing the conditions under which unsaturated solutions are formed and the other those conditions under which either supersaturated solutions or saturated solutions in equilibrium with excess of solid are obtained. This is made clear in Fig. 4.11. Let us imagine that a solution of the solid whose solubility curve is shown contains mass m of solute in 100 gramme at

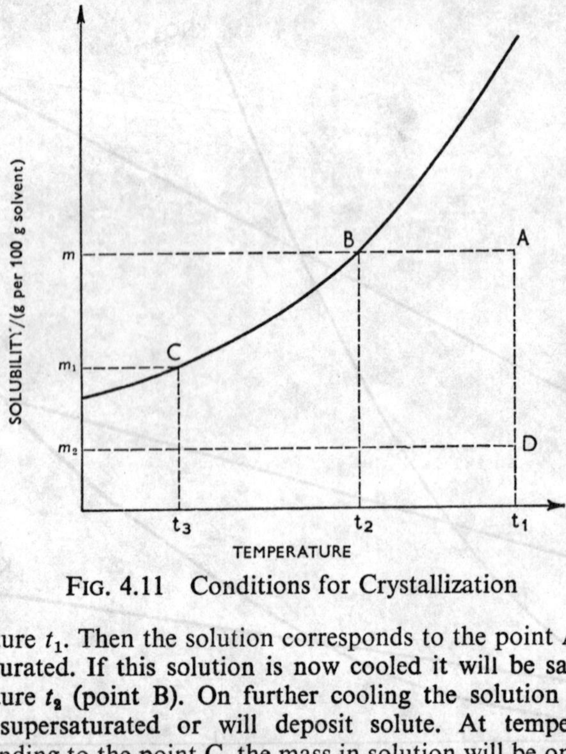

FIG. 4.11 Conditions for Crystallization

temperature t_1. Then the solution corresponds to the point A and will be unsaturated. If this solution is now cooled it will be saturated at temperature t_2 (point B). On further cooling the solution will either become supersaturated or will deposit solute. At temperature t_3, corresponding to the point C, the mass in solution will be only m_1 and the solution will be in equilibrium with mass (m—m_1) of deposited solute. On the other hand if a solution containing mass m_2 of this solid at temperature t_1 (D) is cooled to temperature t_3, no crystallization occurs because the solubility is not exceeded even at the lower temperature. The importance of these considerations will become clear in the next section.

Two other features of solubility curves deserve special mention. A discontinuity (i.e. an abrupt change in direction) in a solubility curve is important because it indicates that two different substances are involved, which usually differ in their degree of hydration. Sodium sulphate is a good example of this (see Fig. 17.7). Secondly, there is a relation between the heat of solution of a substance and the gradient of its solubility curve; this matter is discussed in § 11.7 with the help of Le Chatelier's principle.

4.16. Recrystallization. This is the most important method for purifying a solid, being used extensively in industry and in the laboratory for organic and inorganic substances. The first step consists of choosing a suitable solvent, i.e. one in which the solid is very soluble at high temperatures but only sparingly soluble at room temperature. The solvent is usually selected by trial and error using small samples. The commonest solvents are water, alcohol or methylated spirits, glacial acetic acid, acetone, benzene, petrol, and ether, although the great

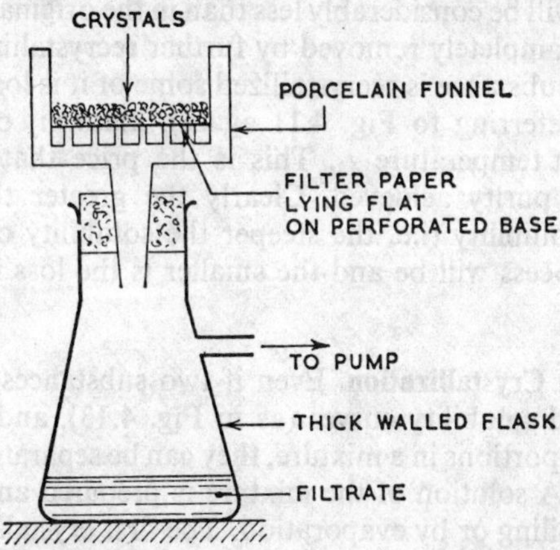

FIG. 4.12 Buchner Funnel and Flask

inflammability of the last two makes them least desirable. Sometimes a mixture of two solvents is chosen.

Having selected a suitable solvent, the crude material is dissolved in the least possible quantity of it at its boiling point. Any insoluble impurities are removed by filtering the hot solution through a fluted filter paper. In order to prevent undue cooling and premature crystallization during filtration, the funnel is usually preheated by passing through it just beforehand a little of the hot pure solvent. The filtered solution is then thoroughly cooled. Since the hot solution was almost saturated, the effect of this cooling is to bring about the crystallization of most of the solute, as described in the previous section. When this process is complete the crystals are filtered off, using a Buchner funnel and flask (see Fig. 4.12) and washed carefully with a little cold solvent to remove traces of impurity adhering to them. Finally the crystals are dried or are subjected to further recrystallization until the desired state of purity is reached. It is not uncommon for a chemical to need repeated recrystallization before it is pure enough for use in medicine or analysis.

It is important to understand exactly how impurities are removed during the process of recrystallization. If they are more soluble in the cold solvent than is the substance being purified, then they will almost certainly remain in solution in the *mother liquor*, as the residual cold solution from which the substance crystallizes is called. Impurities which are less soluble will also probably not be precipitated on cooling because they are unlikely to be present in sufficiently high concentration to form a saturated solution even at the lower temperature. Moreover, even if some impurity is deposited with the pure substance on cooling, its proportion will be considerably less than in the original crude material and it will be completely removed by further recrystallization.

Each time a substance is recrystallized some of it is lost in the mother liquor, since (referring to Fig. 4.11 again), mass m_1 of it remains in solution even at temperature t_3. This is the price that has to be paid for the higher purity achieved. Clearly the greater the temperature coefficient of solubility (i.e. the steeper the solubility curve) the more efficient the process will be and the smaller is the loss incurred in this way.

4.17. Fractional Crystallization. Even if two substances A and B have similarly shaped solubility curves (as in Fig. 4.13), and are present in comparable proportions in a mixture, they can be separated by fractional crystallization. A solution of the mixture is prepared and concentrated gradually by boiling or by evaporation. The first crystals to appear will be predominantly A mixed with some of B, since A is the less soluble of the two. These crystals are removed and represent the first crop or *fraction.* Further concentration of the solution will produce a fraction containing crystals of A and B in about equal proportion, and these are also removed. Finally, as the solution is concentrated even further, a fraction consisting mostly of crystals of B mixed with some of A is obtained. These three fractions are then dissolved in separate portions of fresh solvent and concentrated until each in turn yields three more fractions of crystals. In this way, by repeated crystallization and by amalgamation of similar mixtures of crystals it is possible to obtain samples of A and B of any desired purity, but the process is expensive and tedious unless A and B differ markedly in solubility.

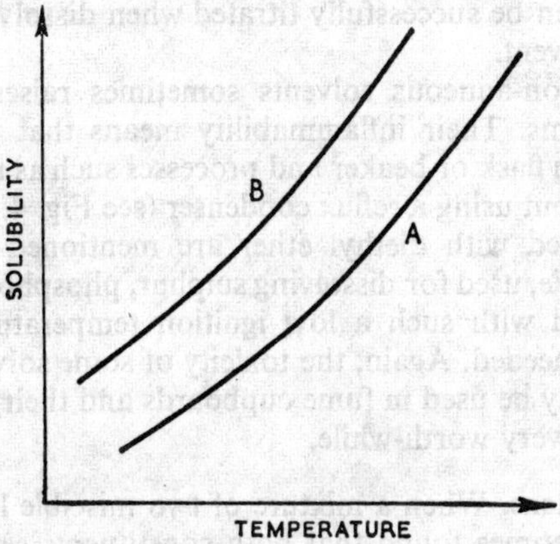

FIG. 4.13 Solubility Curves of A and B

4.18. Non-aqueous Solvents. In elementary work with solutions water is usually chosen as the solvent not only on account of its excellent solvent action, particularly on substances with ionic lattices, but also because of its non-inflammability, non-toxicity, and cheapness. In more advanced work, however, other solvents are often preferred. For example, many organic compounds are insoluble in water but dissolve readily in non-polar solvents such as benzene or ether. In general a substance tends to dissolve readily in solvents with similar structure and polarity, and whereas water is a good solvent for organic substances whose molecules contain hydroxyl groups (e.g. sugars, alcohols, and acids), solvents of a different type (e.g. petrol) are necessary for dissolving long-chain aliphatic compounds such as paraffin wax or fats. Even for ionic compounds water is not always the best solvent and other substances (e.g. liquid ammonia, hydrogen fluoride, and liquid sulphur dioxide) are proving increasingly important. Their use has extended the range of substances to which volumetric analysis can be applied since compounds which are too weakly basic or acidic to give satisfactory end points in aqueous

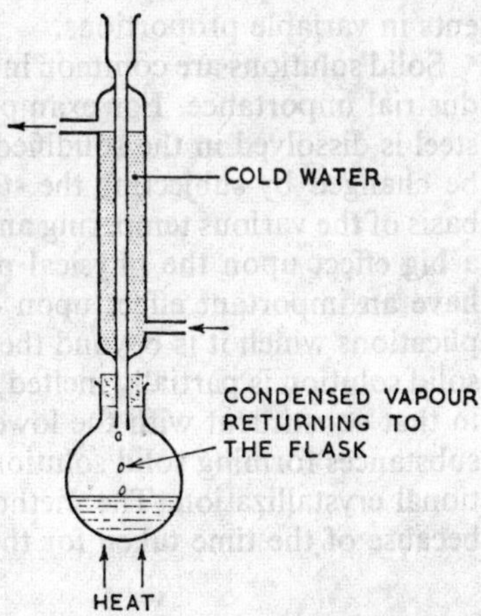

FIG. 4.14 A Reflux Condenser

solution can often be successfully titrated when dissolved in a suitable non-aqueous solvent.

The use of non-aqueous solvents sometimes raises a number of practical problems. Their inflammability means that they cannot be heated in an open flask or beaker and processes such as recrystallization must be carried out using a reflux condenser (see Fig. 4.14). The special dangers associated with diethyl ether are mentioned in chapter 39. Carbon disulphide, used for dissolving sulphur, phosphorus, and iodine, is another liquid with such a low ignition temperature that special precautions are needed. Again, the toxicity of some solvents is so great that they can only be used in fume cupboards and their cost frequently makes their recovery worth-while.

4.19. Solid Solutions. When a mixture of two miscible liquids (§ 6.9) is cooled it is sometimes found that both constituents separate together in a crystalline mass. The proportion of each substance in the solid mixture is usually the same as in the melt. Similarly, when a solid is dissolved in a liquid and the solution is cooled until crystals are deposited, they are sometimes found to be an intimate mixture of solvent and solute. For example, when a solution of iodine in benzene crystallizes, the crystals of benzene are always contaminated with dissolved iodine which is present in about the same proportion as in the liquid solution. In cases such as these *solid solutions* are said to be formed. They are really homogeneous mixtures of two or more solid constituents in variable proportions.

Solid solutions are common in alloys and are often of considerable industrial importance. For example, some of the iron carbide present in steel is dissolved in the solidified iron. The fact that its proportion can be changed by subjecting the steel to particular heat treatments is the basis of the various tempering and quenching processes which have such a big effect upon the physical properties of steel. Solid solutions also have an important effect upon certain phase equilibria, causing complications which it is beyond the scope of this book to discuss. When a solid solution is partially melted, the melt is always richer than the solid in that constituent with the lower melting point, so that in theory two substances forming solid solutions could be separated by repeated fractional crystallization. The method is of little practical value, however, because of the time taken for the system to reach equilibrium.

5

Molecular Weights

5.1 Introduction. As explained in the first two chapters, it is conventional to express the masses of molecules on the same scale as the masses of atoms. It follows that molecular weights, like atomic weights, are only relative weights* and need no units. The main importance of molecular weight lies in the information it gives about the molecular formula of a substance. Even an approximate value will usually suffice to enable us to determine what multiple of the *empirical formula* (the simplest formula which is based upon chemical analysis of the compound and which expresses the proportions in which the various elements are present) should be chosen as the formula to represent a molecule of the substance. This procedure is discussed in greater detail in § 32.3. Once this point is appreciated it will be realized that it is not a serious disadvantage that many of the methods available for determining molecular weights are incapable of high accuracy. It also explains why a molecular weight is usually expressed only to the nearest integer and why in most cases it is quite adequate to define it as the mass of the molecule relative to the mass of one atom of hydrogen or as the mass of 2.24×10^{-2} m^3 of its vapour of s.t.p.

The experimental methods of determining molecular weights are of two main types, depending upon whether they are applied to the substance in the vapour state or in solution. Unless we have evidence to the contrary, we then assume that the value determined also applies to the substance as a solid, although this assumption should always be subject to careful scrutiny because association or dissociation may occur. A further difficulty arises, as explained in § 1.15, when a substance is non-volatile and insoluble e.g. silica or carbon. In these cases, which involve macromolecules of indefinite extent, no satisfactory method of determining molecular weight has been devised.

5.2. The Density of a Gas. As has been explained in § 1.10, it follows from Avogadro's hypothesis that the molecular weight of a gaseous substance can readily be determined from its density by direct comparison.

* Hence the alternative name of *relative molecular mass* for molecular weight.

This makes it a matter of considerable importance to distinguish clearly between the various ways in which the density of a gas can be expressed.

The *normal density* is the mass of a specified volume of the gas at s.t.p. The units most commonly used are kilogramme per cubic metre. For example, oxygen has a normal density of 1.4290 kg m^{-3} at s.t.p.

Since the density of a gas is defined as its mass per unit volume, it follows that its density *per unit pressure* is given by the expression $\frac{m}{PV}$, where m is the mass of gas in a volume V at 0°C at a pressure P. Now for an ideal or perfect gas, i.e. one which obeys the ideal gas equation exactly (see § 3.4), this density should have a constant value at constant

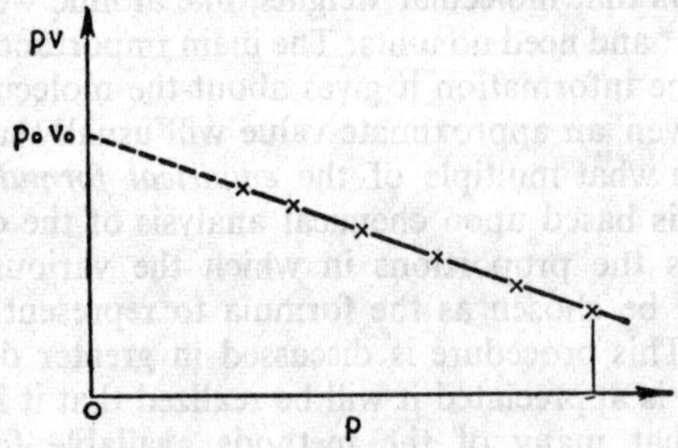

FIG. 5.1 Determination of P_0V_0 by Extrapolation

temperature, since PV is then constant. In practice this is not the case and for all real gases the value of $\frac{m}{PV}$ changes slightly with pressure. The limiting value as the pressure is reduced to zero is known as the *limiting density* of the gas. It cannot be measured directly by experiment so it is determined by extrapolating the straight line graph obtained when the product PV at 0°C is plotted against pressure for low pressures (see Fig. 5.1) and evaluating the expression $\frac{m}{P_0V_0}$. For example, for oxygen the value is 1.4276 kg m^{-3}. The importance of limiting densities is that they are independent of pressure and can therefore be compared without incurring an error due to the deviation of the gases concerned from ideal behaviour at the arbitrary pressure of the atmosphere. The determination of the limiting density of a gas by Regnault's method (§ 5.3) or by some microscale adaptation of it provides one of the few really accurate methods of finding its molecular weight. This is also a useful way of finding the atomic weights of the inert gases, as described in § 2.15.

Just as it is far more convenient in chemistry to refer to the weights of atoms and molecules in terms of *relative* values, so it is usually preferable to express the density of a gas in a relative way, too. The *vapour density* of a gas, which should strictly be called the relative vapour density, is the density of that gas compared with the density of hydrogen measured under the same conditions. By applying Avogadro's hypothesis it has already been established in § 1.10 that this vapour density is equal to one half of the molecular weight, so that its experimental measurement by one of the methods described in the following sections offers a very convenient way of determining the approximate molecular weight of a volatile substance.

5.3. Regnault's Method. A glass globe of known (about 1000 cm³) capacity is completely evacuated and weighed. It is then filled with the gas whose density is required and weighed again. Like all methods involving a direct weighing of gases, this method is open to the objection that even small errors in the weighing tend to give a large percentage error in the final result because the gas weighs so little compared with the apparatus containing it. The following precautions are therefore essential to obtain an accurate result:

1. The globe is counterbalanced by another of the same size, filled with air. Not only does this make a buoyancy correction for the upthrust of the air on the globe unnecessary but it also means that any changes during the experiment in the density of the air or in the amount of moisture adsorbed upon the outer walls of the globe will not matter since both globes will be affected to an equal extent.

2. The whole experiment is often carried out at s.t.p. to avoid the necessity of making corrections based upon the ideal gas laws.

3. The gas must be completely dry.

4. When evacuated the globe tends to contract slightly owing to the effect of the external pressure upon its walls and a small correction must therefore be made for the change in the upthrust it experiences.

Regnault's method, provided these precautions are duly taken, is capable of high accuracy, although it is far too tedious and elaborate for class use. It was whilst using this method for determining the density of nitrogen that Lord Rayleigh discovered the existence of the noble gases (see § 26.3).

Example: Mass of globe evacuated $= m_1$

„ „ „ full of gas $= m_2$

Volume of globe at 0°C $= v$

Then the normal density of the gas $= \dfrac{m_2 - m_1}{v}$

Alternatively, if mass of the same globe full of hydrogen at s.t.p. = m_3.

Then $$\text{v.d.} = \frac{m_2 - m_1}{m_3 - m_1}$$

5.4. Dumas' Method. The substance is vaporized and the mass is found of a given volume of its vapour under known conditions. A clean dry Dumas bulb (see Fig. 5.2) is first weighed open to the air. About 5–10 cm³ of the liquid are then introduced by warming the bulb with one's hands and allowing it to cool again with the open end dipping into the

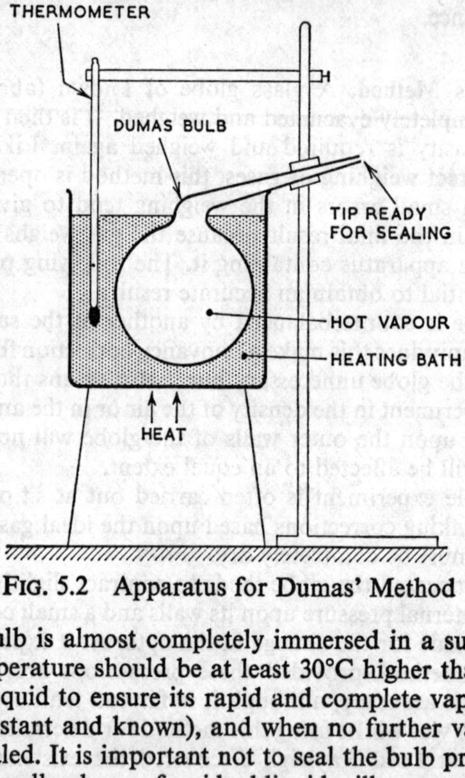

FIG. 5.2 Apparatus for Dumas' Method

liquid. The bulb is almost completely immersed in a suitable heating bath (the temperature should be at least 30°C higher than the boiling point of the liquid to ensure its rapid and complete vaporization and should be constant and known), and when no further vapour escapes the bulb is sealed. It is important not to seal the bulb prematurely because even a small volume of residual liquid will cause a large error in the result. After sealing, the bulb is cooled, dried, and weighed together with any fragment of tip removed in sealing. Now in this weighing the sealed bulb displaces its own volume of air and therefore experiences an appreciable upthrust, which must be taken into account. For this purpose it is necessary to measure the temperature and pressure of the atmosphere at the time of the experiment. Lastly the volume of the bulb is found by breaking the sealed tip under air-free water and weighing the bulb full of water; this weighing need only be accurate to the nearest half gramme because with a total weight of 250–300 gramme the percentage error will be small.

The method is simple and quick and is adaptable to a wide range of temperature, but the results may be in error by anything up to 5%, particularly if the liquid contains any less volatile impurities which tend to remain in the bulb. It also has the disadvantage that a relatively large volume of liquid is required.

Example: Mass of bulb open to the air $= m_1$

„ „ „ full of vapour $= m_2$

„ „ „ „ „ water $= m_3$

Temperature of heating bath $= t_1$

„ „ atmosphere $= t_2$

Pressure „ „ $= P_{atm}$

∴ Volume of bulb at temperature t_2 and pressure $P = m_3 - m_1$ (assuming 1 kg of water occupies 1 dm³)

$$\therefore \text{Volume of air displaced (corrected to s.t.p.)} = (m_3 - m_1) \times \frac{273}{(t_2 + 273)} \times \frac{P}{P_{atm}} = v_1$$

Now 1 dm³ of air at s.t.p. weighs 1.29 g

∴ Mass of air displaced by bulb $= 1.29 \times v_1$ and this is equal to the upthrust on the bulb during the second weighing.

∴ Mass of vapour in bulb = apparent mass + upthrust

$$= (m_2 - m_1) + 1.29 \times v_1$$

Now vol. of vapour at temperature t_1 and pressure $P = m_3 - m_1$

$$\therefore \text{Vol. of vapour at s.t.p.} = (m_3 - m_1) \times \frac{273}{(t_1 + 273)} \times \frac{P}{P_{atm}} = v_2$$

$$\therefore \text{Density of vapour at s.t.p.} = \frac{(m_2 - m_1) + 1.29 \times v_1}{v_2}$$

Now density of hydrogen at s.t.p. $= 0.09$ g dm^{-3}

$$\therefore \text{v.d.} = \frac{(m_2 - m_1) + 1.29 \times v_1}{0.09 \times v_2}$$

5.5. Victor Meyer's Method. In this method the volume occupied by a given mass of vapour under known conditions is determined. The apparatus is shown in Fig. 5.3. The liquid in the outer jacket, which should have a boiling point at least 30°C higher than the substance whose density is being determined, is heated until no more air is displaced. Meanwhile the special small bottle B is weighed empty, filled with the liquid, and weighed again. When the steady state has been reached the bung is momentarily removed and the small loosely-stoppered bottle dropped into the tube. At the same time a graduated tube filled with water is inverted over the delivery tube T to collect the air displaced by the vaporizing liquid. When no more bubbles appear the volume of air displaced is measured (after equalizing the levels inside and outside the tube) and the temperature and pressure of the atmosphere is recorded. It should be noted that it is not necessary in this method to know the temperature of the hot vapour. As the liquid vaporizes it pushes out an equal volume of hot air from the inner container, but this displaced air soon cools down to room temperature. Thus the volume actually measured is that which the vapour would occupy at room temperature (if it did not condense) and this is the only temperature which is relevant in correcting the recorded volume to s.t.p.

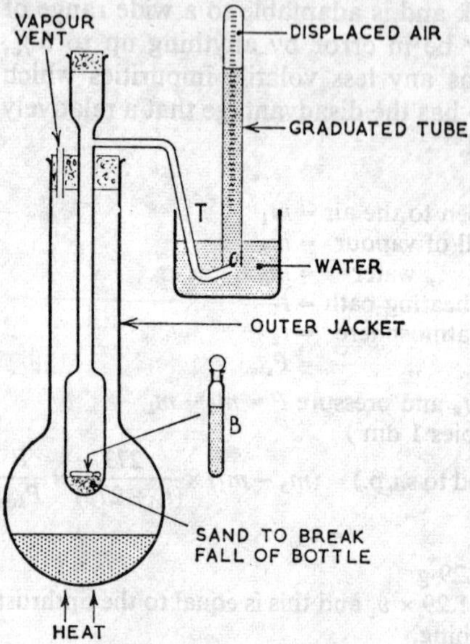

FIG. 5.3 Victor Meyer's Method

This is the best of the simple methods of determining vapour density. It is quick, reasonably accurate (an error of 1–2% is to be expected), and adaptable to a wide range of temperature since a vessel of silica or metal can be used for work at high temperatures. It requires only a small quantity of liquid and little manipulative skill. Its only disadvantages are that it cannot be used for substances which decompose at their normal boiling points (a special method was devised by Hofmann for such cases) or for substances which undergo dissociation on heating.

Example:

Mass of bottle empty $= m_1$

" " " filled with liquid $= m_2$

Vol. of air displaced at temperature t and pressure $P = v$

Let the s.v.p. of water at temperature t be p (obtained from tables)

Then the actual pressure of the air displaced $= (P - p)$ (see § 3.5)

$$\therefore \text{ Vol. of air displaced, corrected to s.t.p.} = v \times \frac{273}{(t+273)} \times \frac{(P-p)}{P_{atm}} = v_1$$

Now mass of liquid taken $= m_2 - m_1$

$\therefore$ Density of vapour at s.t.p. $= \dfrac{m_2 - m_1}{v_1}$

Now density of hydrogen at s.t.p. = 0.09 kg m^{-3}

$$\therefore \text{Vapour density} = \frac{m_2 - m_1}{0.09 \times v_1}$$

5.6. Molecular Weights and Association. Measurements of vapour density reveal that a number of compounds have molecular weights which are higher than expected on valency and other grounds. Such substances are said to be *associated* in the vapour state. At higher temperatures the vapour density usually falls indicating that the association occurs only near the boiling point. Amongst the examples considered in later chapters are aluminium chloride Al_2Cl_6 (§ 19.10), the oxides of phosphorus, arsenic, and antimony e.g. P_4O_6, P_4O_{10}, As_4O_6, Sb_4O_6 (see chapter 23), hydrogen fluoride $(HF)_n$ (§ 25.3), and acetic acid $(CH_3COOH)_2$ (§ 37.2). In these last two cases the association is caused by hydrogen bonding (§ 14.7) which links two or more molecules together into larger units.

5.7. Thermal Dissociation. It is not uncommon to find that the vapour density of a substance decreases with rising temperature and increases again as the temperature is allowed to fall. Such a change in vapour density is a sure sign of *thermal dissociation*, a process whereby on heating a molecule breaks up into smaller molecules which recombine again on cooling. The reversibility of the dissociation is an important feature which distinguishes it from decomposition by heat.

Before considering details of the way in which vapour density varies with temperature during dissociation, it is worth while considering what other experimental evidence of the process is available. Perhaps the most striking is the change in colour which takes place when some substances dissociate. The example of nitrogen tetroxide is described in detail in § 22.12:

$$N_2O_4 \rightleftharpoons 2NO_2 \rightleftharpoons 2NO + O_2$$

Another example is provided by hydrogen iodide, which is colourless just above its boiling point but which turns purple as the temperature is raised owing to the presence of iodine vapour in the mixture:

$$2HI \rightleftharpoons H_2 + I_2$$

Again, phosphorus pentachloride vapour changes from colourless to greenish yellow as the temperature rises, and the pentabromide develops the red-brown colour of free bromine:

$$PCl_5 \rightleftharpoons PCl_3 + Cl_2$$
$$PBr_5 \rightleftharpoons PBr_3 + Br_2$$

Further evidence of dissociation can be obtained by demonstrating the existence of other simpler molecules in the vapour. This is not always easy, but it can be done quite effectively in a number of cases. For example, if ammonium chloride is heated in a tube, as in Fig. 5.4, the litmus papers soon change colour because the less dense ammonia

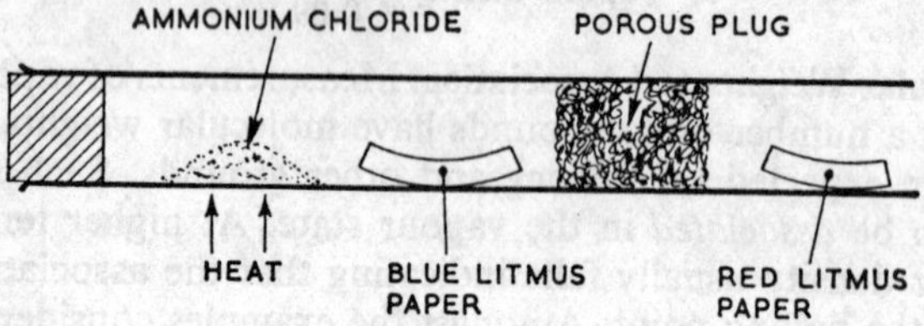

FIG. 5.4 The Dissociation of Ammonium Chloride

molecules diffuse through the porous plug more quickly than those of hydrogen chloride leaving a predominance of the latter in the left hand part of the tube.

$$NH_4Cl \rightleftharpoons NH_3 + HCl.$$

If some gold leaf is inserted into the hot vapour of mercurous chloride it is rapidly amalgamated by the mercury vapour present:

$$Hg_2Cl_2 \rightleftharpoons HgCl_2 + Hg.$$

By cooling a dissociated substance down to room temperature very suddenly it is often possible to isolate samples of the simpler substances before they have the chance to recombine. This technique of 'freezing' an equilibrium mixture is referred to in § 7.8 where it is used as a means of finding the proportion of hydrogen and iodine present in the heated vapour of hydrogen iodide.

5.8. The Degree of Dissociation. If the vapour density of nitrogen tetroxide is measured at various temperatures between its boiling point (22°C) and 200°C, it will be found to behave as in Fig. 5.5. The same kind of change is observed if phosphorus pentachloride is heated to 350°C. In both cases the vapour density falls gradually to one half of its value at the boiling point and then remains constant. This suggests that over the range of temperature considered each of the molecules of these substances dissociates on heating into two smaller molecules. When this splitting is complete the volume occupied at a given pressure is doubled and the density is therefore halved. Now vapour density changes can be used not only to indicate the completion of the dissociation but also to calculate its extent at any intermediate temperature.

The proportion of the molecules which dissociate at any temperature is known as the *degree of dissociation*, α, and is conventionally expressed either as a decimal or as a percentage. Its relationship to vapour density can be made clear by considering one mole of a hypothetical substance AB which dissociates on heating into two simpler molecules A and B:

$$\begin{array}{llll} & AB \rightleftharpoons & A \quad + & B \\ \text{Before dissoc}^n & 1 \text{ mole} & O & O \\ \text{After dissoc}^n & (1-\alpha) \text{ mole} & \alpha \text{ mole} & \alpha \text{ mole} \end{array}$$

Thus the number of molecules has increased as a result of the dissociation in the ratio $\frac{1-\alpha+\alpha+\alpha}{1}$ i.e. $\frac{1+\alpha}{1}$, and by Avogadro's hypothesis this will increase the volume (at constant pressure) in the same proportion. Since, however, the mass of gas concerned is constant, the density of the gas is inversely proportional to its volume, so that

$$\frac{\text{v.d. before dissociation}}{\text{v.d. after dissociation}} = \frac{1+\alpha}{1}$$

In the more general case, when each molecule of AB dissociates into

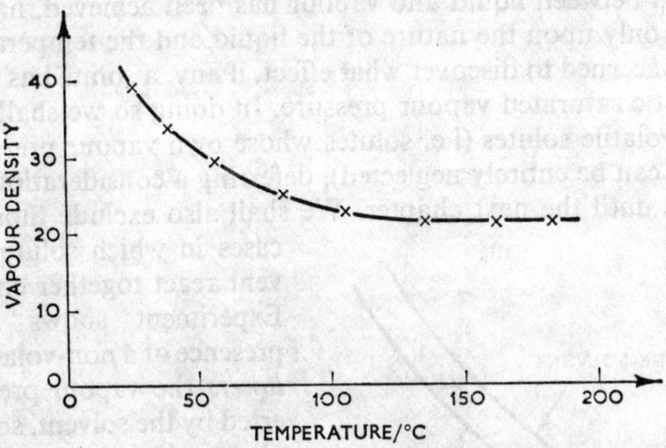

FIG. 5.5 The Dissociation of Nitrogen Tetroxide

n molecules, the relation between vapour density and degree of dissociation becomes

$$\frac{\text{v.d. before dissociation}}{\text{v.d. after dissociation}} = \frac{1+(n-1)\alpha}{1}$$

From this it will be seen that when there is no change in the number of molecules as a result of dissociation and *n* is 1 (as, for example, in the dissociation of hydrogen iodide into hydrogen and iodine) the vapour density does not change with temperature and provides no information upon the extent of dissociation.

The calculation of the degree of dissociation is illustrated by the following example:

Example: The vapour density of phosphorus pentachloride at 220°C and at atmospheric pressure is 64.4. What is its degree of dissociation under these conditions? (P = 31, Cl = 35.5)

$$PCl_5 \rightleftharpoons PCl_3 + Cl_2.$$

Its vapour density before dissociation $= \dfrac{(31 + 5 \times 35.5)}{2} = 104.2.$

Now if α is the degree of dissociation at 220°C and atmospheric pressure

$$\frac{104.2}{64.4} = \frac{1+\alpha}{1}$$

$$\therefore\ \alpha = \frac{104.2 - 64.4}{64.4} = \frac{39.8}{64.4} = 0.618$$

Thus the pentachloride is 61.8% dissociated under these conditions.

5.9. Vapour Pressure of Solutions. It was seen in § 3.16 that every pure liquid exerts a vapour pressure in the space above it which, when equilibrium between liquid and vapour has been achieved, has a value dependent only upon the nature of the liquid and the temperature. We are now concerned to discover what effect, if any, a solute has upon this characteristic saturated vapour pressure. In doing so we shall consider only non-volatile solutes (i.e. solutes whose own vapour pressure is so low that it can be entirely neglected), deferring a consideration of volatile solutes until the next chapter. We shall also exclude those special cases in which solute and solvent react together chemically. Experiment shows that the presence of a non-volatile solute *lowers* the vapour pressure exerted by the solvent, so that any given solution exerts a lower vapour pressure than would the pure solvent at that temperature.

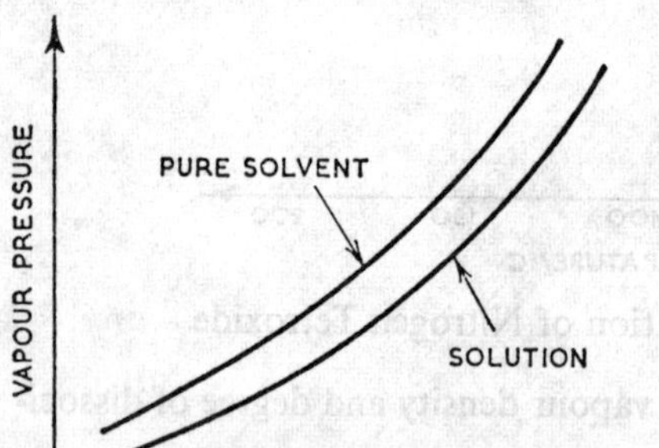

FIG. 5.6 Lowering of the Vapour Pressure

This important effect, which is shown diagrammatically in Fig. 5.6, is explained in terms of the kinetic theory in the following way. The molecules of solute will, of course, be distributed uniformly throughout the solution, so that at any instant some of them will always be found near the surface of the liquid as in Fig. 5.7. Now these solute molecules are non-volatile and show no measurable tendency themselves to escape from the solution at that temperature, but like the solvent molecules all around them they exert a strong inward pull at the surface on those fast-moving solvent molecules about to escape. As a result fewer molecules of solvent escape from the surface per second and since the rate of return of vapour molecules to the liquid is unaltered by the presence

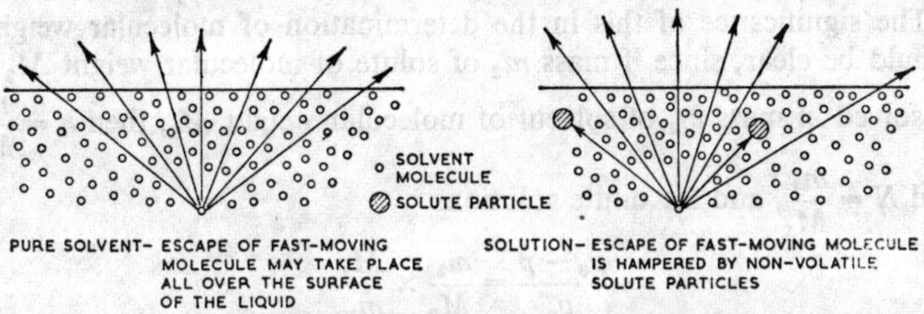

FIG. 5.7 The Effect of a Solute upon Vapour Pressure

of the solute, the pressure set up by the vapour at equilibrium will be less than would otherwise have been the case. The solute molecules, in other words, partially block the escape route of the fast-moving solvent molecules by effectively reducing the area of surface they can penetrate whilst not interfering at all with their rate of return into the solution, so that the equilibrium position between the two opposing processes is altered.

5.10. Raoult's Law. The relationship between the concentration of a non-volatile solute and the lowering of vapour pressure it produces was investigated experimentally by Raoult using several different solvents as well as a variety of solutes. In 1887 he embodied his results in the following very important generalization:

The relative lowering of the vapour pressure of a solution is proportional to the mole fraction of the solute, i.e. to the ratio of the number of molecules of solute to the total number of molecules of solute and solvent in the solution.

The *relative lowering* of vapour pressure is the lowering divided by the vapour pressure of the pure solvent. It is therefore the proportion by which the vapour pressure is lowered; experiment shows that for any given solution it is practically independent of temperature.

If p_0 and p are the vapour pressures of pure solvent and solution respectively and n and N are the respective numbers of molecules of solute and solvent, then Raoult's law may be expressed in the form

$$\frac{p_0 - p}{p_0} = \frac{n}{N + n}$$

Now in a dilute solution n is small compared with N and the expression can therefore be simplified to

$$\frac{p_0 - p}{p_0} = \frac{n}{N}$$

This means that *for dilute solutions the relative lowering of the vapour pressure is proportional to the molecular concentration of the solute.*

The significance of this in the determination of molecular weights should be clear, since if mass m_2 of solute of molecular weight M_2 is dissolved in mass m_1 of solvent of molecular weight M_1, then $n = \dfrac{m_2}{M_2}$ and $N = \dfrac{m_1}{M_1}$, and for dilute solutions

$$\frac{p_0 - p}{p_0} = \frac{m_2}{M_2} \times \frac{M_1}{m_1}$$

Thus if the vapour pressures of solvent and solution of known concentration can be measured, the molecular weight of the solute may readily be found from the expression

$$M_2 = \frac{m_2 M_1}{m_1} \times \frac{p_0}{(p_0 - p)}$$

In fact the measurement of vapour pressure is tedious and generally inaccurate and is rarely used for determining molecular weights because other related experimental methods of much greater accuracy and convenience are available. These other methods, some of which are described in detail in the following pages, are all based indirectly upon the lowering of the vapour pressure by a solute and therefore upon Raoult's law. Now in practice solutions do not obey Raoult's law exactly, particularly when they are concentrated, and slight deviations occur owing to interaction between the molecules of solute and solvent. We define an *ideal solution*, in fact, as one which does conform exactly to Raoult's law over the whole range of concentration, just as we define an ideal gas as one that obeys the ideal gas equation exactly (§ 3.4). Raoult's law and the various experimental methods of determining molecular weight depending upon it should only be applied, therefore, to *dilute solutions in which the solute is neither associated nor dissociated*, and these limitations must be borne constantly in mind.

5.11. Elevation of the Boiling Point. The boiling point of a liquid has been defined in § 3.17 as the temperature at which its vapour pressure reaches 101 325 N m^{-2}. Now since the presence of a non-volatile solute lowers the vapour pressure of a solution throughout the whole range of temperature, it follows that a solution will always boil at a higher temperature than the pure solvent. This is made clear in Fig. 5.8. Provided only dilute solutions are considered, the amount by which the boiling point is raised, known as the elevation of the boiling point, is proportional to the lowering of the vapour pressure (i.e. the various vapour pressure curves in Fig. 5.8. may be regarded as being approximately parallel to each other in the region of the boiling point). Thus if ΔT represents the elevation of the boiling point when the vapour pressure of the solution is denoted by p, then $\Delta T = k(p_0 - p)$. Now we

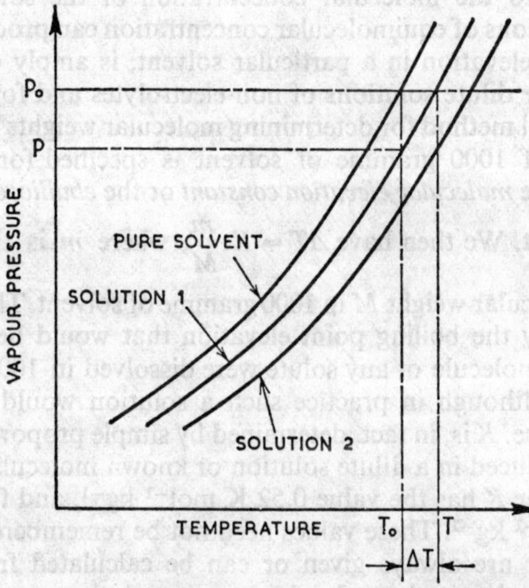

FIG. 5.8 Elevation of the Boiling Point

have already seen from Raoult's law that the relative lowering of the vapour pressure is approximately proportional to the molecular concentration of the solute for a dilute solution, i.e.

$$\frac{p_0 - p}{p_0} = \frac{n}{N}, \quad \text{so} \quad p_0 - p = p_0 . \frac{n}{N}$$

$$\therefore \Delta T = k(p_0 - p) = kp_0 . \frac{n}{N} = k' . \frac{n}{N}$$

(where k and k' are constants, since p_0 is a constant for a given solvent). If mass m_2 of solute of molecular weight M_2 is dissolved in mass m_1 of solvent of molecular weight M_1, then

$$n = \frac{m_2}{M_2} \quad \text{and} \quad N = \frac{m_1}{M_1}$$

$$\therefore \Delta T = k' . \frac{m_2}{M_2} \times \frac{M_1}{m_1}$$

Now for a given quantity of a given solvent, m_1 and M_1 are, of course, constant, so

$$\Delta T = k'' . \frac{m_2}{M_2} \text{ (where } k'' \text{ is another constant)}$$

This result, which predicts that the elevation of the boiling point is proportional to the molecular concentration of the solute and that different solutions of equimolecular concentration can produce the same boiling point elevation in a particular solvent, is amply confirmed by experiment for dilute solutions of non-electrolytes and forms the basis of a very useful method for determining molecular weights of substances in solution. If 1000 gramme of solvent is specified for m_1, then k'' becomes K, the *molecular elevation constant* or the *ebullioscopic constant* of that solvent. We then have $\Delta T = K . \frac{m}{M}$, where m is the weight of solute of molecular weight M in 1000 gramme of solvent. The constant K is theoretically the boiling point elevation that would be produced if one gramme-molecule of any solute were dissolved in 1000 gramme of the solvent, although in practice such a solution would be too concentrated to use. K is, in fact, determined by simple proportion from the elevation produced in a dilute solution of known molecular concentration. For water K has the value 0.52 K mol^{-1} kg^{-1}, and for benzene it is 2.60 K mol^{-1} kg^{-1}. These values need not be remembered because in problems they are always given or can be calculated from the data provided, but the student should be warned that occasionally the constant for 100 gramme of solvent is quoted instead, with a value which is ten times as great as K because it refers to a more concentrated solution. No ambiguity should arise, however, as the relevant mass of solvent should always be clearly stated.

Example: 2.00 g of urea dissolved in 125 g of water causes a boiling point elevation of 0.139 K.

Given that K for 1000 g of water is 0.52 K mol^{-1} kg^{-1}, calculate the molecular weight of urea.

$$\text{Now } \Delta T = K . \frac{m}{M} \text{ where } \begin{cases} \Delta T = \text{b.p. elevation} = 0.139\ \text{K} \\ K = \text{constant for 1000 g} = 0.52\ \text{K mol}^{-1}\ \text{kg}^{-1} \\ m = \text{mass of solute in 1000 g solvent} = 16\ \text{g} \\ M = \text{mol. wt. of solute.} \end{cases}$$

$$\therefore M = \frac{Km}{\Delta T} = \frac{0.52 \times 16}{0.139} = 59.9$$

$\therefore$ Molecular weight of urea is 59.9 ($NH_2.CO.NH_2 \equiv 60$).

5.12. Measurement of Boiling Point Elevation. Although it might seem from the previous section that the molecular weight of a solute can readily be determined by measuring the elevation of boiling point produced in a dilute solution of known concentration, in practice certain difficulties arise which limit the accuracy of the method and make an error of 2–3% inevitable. The chief of these is *superheating*, which has been referred to in § 3.17. Unless special precautions are taken to minimize or prevent it, the temperature at which the liquid boils may rise appreciably above its true boiling point and since the elevation produced is in any case only a small one, the error caused may be considerable. The various experimental methods available differ mainly in the ways in which they attempt to solve this problem.

Another practical difficulty is the great dependence of the boiling point upon pressure; a variation in the pressure of the atmosphere of a mere 40 N m^{-2} during the experiment results in a change of about 0.01 K in the boiling point so that again a large percentage error may be caused. Lastly, the elevation of boiling point, produced will rarely exceed 0·5 K and the temperature must therefore be read accurately to $\frac{1}{100}$K or less if the result is to be of much value. This demands the use of a special type of thermometer, originally devised by Beckmann for the purpose (see Fig. 5.9). Its large bulb and very thin capillary tube give it a high sensitivity and the reservoir of mercury at the top means that it can be adjusted beforehand to record small temperature changes over any selected 5 K range. The fact that the scale is not calibrated to read the actual temperature is no disadvantage, since it is only the difference in the boiling points of the pure solvent and solution which is of interest.

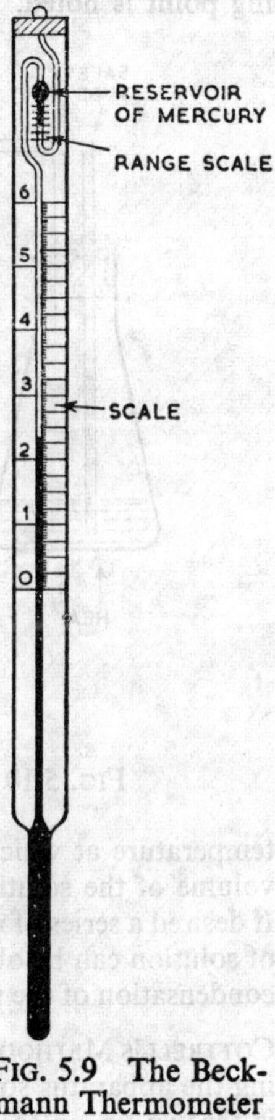

FIG. 5.9 The Beckmann Thermometer

Having reviewed the practical difficulties inherent in this way of determining molecular weights, we are now in a position to consider the principal experimental methods.

LANDSBERGER'S METHOD: The apparatus used is shown in Fig. 5.10. The solution is heated to its boiling point by passing into it the vapour of the pure solvent. In this way superheating is avoided, the source of heating being the latent heat of vaporization given up by the condensed vapour. The uncondensed vapour escapes from a small hole near the top of the inner container and provides a hot jacket around the tube thereby helping to avoid fluctuations in temperature. The heating is first performed using pure solvent in the inner container and the steady boiling point is noted. Then a weighed pellet of solute is added and the

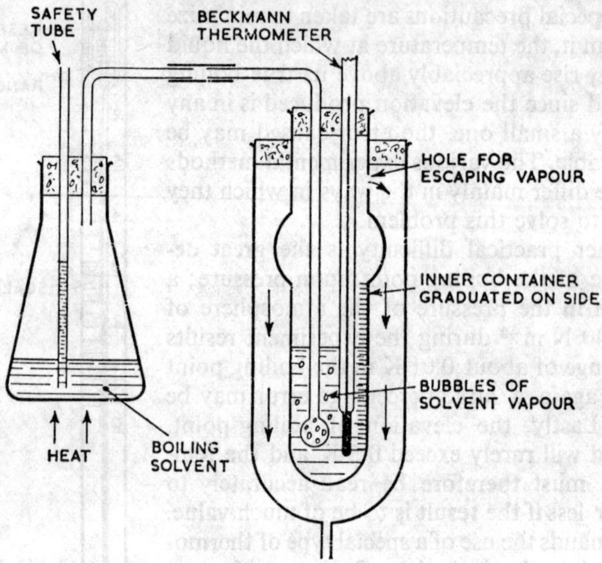

FIG. 5.10 Apparatus for Landsberger's Method

temperature at which the solution boils is recorded together with the volume of the solution (the inner tube is graduated for this purpose). If desired a series of readings of boiling point and corresponding volume of solution can be obtained as the latter becomes steadily diluted by the condensation of the solvent vapour owing to heat losses.

COTTRELL'S METHOD: Superheating is avoided in this method by designing the apparatus so that a steady stream of vapour and boiling liquid is thrown up on to the bulb of the thermometer, which is suspended above the surface of the liquid. This has the further advantage that the temperature recorded is not affected by variations in the hydrostatic pressure of liquid around the bulb. The apparatus used is shown in Fig. 5.11. The procedure is similar to that of the previous method, but in this case the tube is graduated directly in cubic centimetres. Electrical heating can be used to maintain a steady temperature.

Beckmann's Method: This method is now little used, being more susceptible to draughts and superheating than the methods just described. Some glass beads and a piece of platinum wire sealed into the bottom of the inner container are used to facilitate the formation of vapour bubbles in an attempt to minimize superheating.

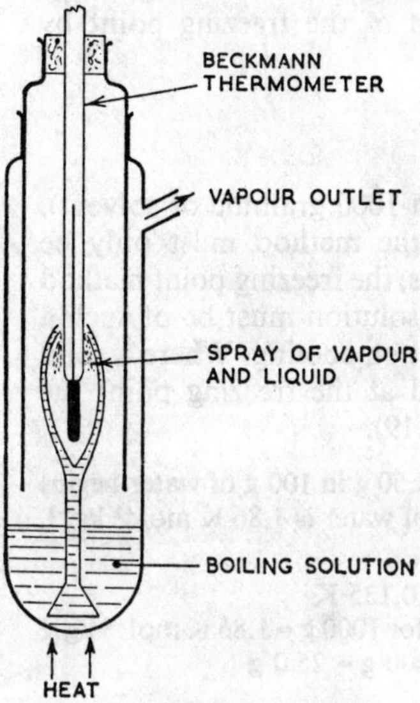

Fig. 5.11 Cottrell's Apparatus

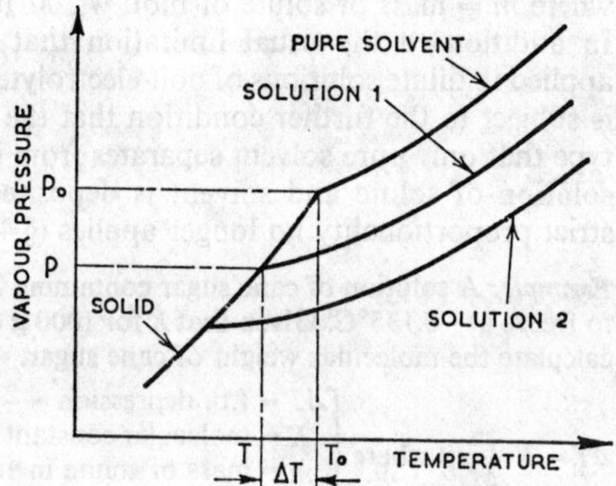

Fig. 5.12 Depression of the Freezing Point

5.13. Depression of the Freezing Point. Fig. 5.12 shows how the lowering of the vapour pressure by a solute depresses the freezing point of the solution below that of the pure solvent. As a result of this lowering the vapour pressure curves of the solution and solid solvent intersect at a lower temperature giving a depression ΔT. Now by similar reasoning to that used in § 5.11, we should expect from Raoult's law that the depression of the freezing point would be proportional to the molecular concentration of the solute. Experiment shows that this is the case for dilute solutions in which the solute is neither dissociated nor associated. Indeed the proportionality between freezing point depression and concentration was discovered as early as 1788 by Blagden and was known as *Blagden's law*. A thorough investigation was made by Raoult from 1878 to 1886 using non-aqueous solvents as well as water; his results confirmed that various solutions of equimolecular concentration produced the same depression in any particular solvent. The depression of freezing point produced when one mole of any solute is dissolved in 1000 gramme of a given solvent is known as the *molecular depression constant* or the *cryoscopic constant*, *K*. It has the value 1.86 K mol^{-1} kg^{-1} for water, and 5.10 K mol^{-1} kg^{-1} for benzene. Once K for any given solvent is known (and it can easily be found by using a solute of known concentration and molecular weight), then other molecular weights can readily be determined from the depression of the freezing point by using the expression

$$\Delta T = K.\frac{m}{M}$$

where m = mass of solute of mol. wt. M in 1000 gramme of solvent). In addition to the usual limitation that the method must only be applied to dilute solutions of non-electrolytes, the freezing point method is subject to the further condition that the solution must be of such a type that only pure solvent separates from it on cooling. Where a solid solution of solute and solvent is deposited at the freezing point the strict proportionality no longer applies (§ 4.19).

Example: A solution of cane sugar containing 2.50 g in 100 g of water begins to freeze at −0.135°C. Given that K for 1000 g of water is 1.86 K mol^{-1} kg^{-1}, calculate the molecular weight of cane sugar.

$$\Delta T = K.\frac{m}{M} \quad \text{where} \begin{cases} \Delta T = \text{f.p. depression} = -\ 0.135\ \text{K} \\ K = \text{molecular constant for 1000 g} = 1.86\ \text{K mol}^{-1}\ \text{kg}^{-1} \\ m = \text{mass of solute in 1000 g} = 25.0\ \text{g} \\ M = \text{required mol. wt.} \end{cases}$$

$$\therefore\ M = \frac{Km}{\Delta T} = \frac{1.86 \times 25}{0.135} = 344$$

∴ Mol. wt. of cane sugar is 344 ($C_{12}H_{22}O_{11}$ = 342).

5.14. Measurement of Freezing Point Depression. The principal drawback of freezing point depression methods is the liability to supercooling, as described in § 4.3. Unless careful precautions are taken, the temperature recorded will be appreciably below the true freezing point of the solution and a considerable error will be incurred. Provided this difficulty is overcome, as it is in the Beckmann method described below, the measurement of freezing point depression is the most accurate and convenient method of determining molecular weights in solution since it is much less susceptible to draughts and variations in atmospheric pressure than the boiling point methods.

BECKMANN'S METHOD: The apparatus used is shown in Fig. 5.13. Supercooling is minimized by arranging the temperature of the cooling bath C to be not more than 5 K below the freezing point of the solvent and by surrounding the inner tube A, into which a known weight of the pure solvent is placed, with an air jacket B, so that the cooling is gentle and slow. When the temperature has fallen to about half a degree below its normal freezing point, the solvent is stirred more vigorously to induce crystallization. 'Seeding' or inoculation with a minute crystal of the solvent is sometimes used. When the solvent begins to freeze the temperature rises sharply to the true freezing point as a result of the release of latent heat of fusion (see Fig. 5.14*a*). This steady temperature is recorded on the Beckmann thermometer, T. The tube A is then removed and warmed slightly to melt the crystals, and the process is repeated until two concordant results have been obtained. A weighed pellet of solute is then introduced through the side arm, D, and allowed to dissolve. The freezing point of the solution is then determined in the same way. The temperature recorded should be the maximum reached when the first crystals of solvent appear because otherwise, as shown in Fig. 5.14*b*, an error may arise from the gradual fall in freezing point with increasing concentration of solution as crystallization occurs. If desired further small additions of solute may be made and the freezing points of the corresponding solutions determined, provided that they are not too concentrated.

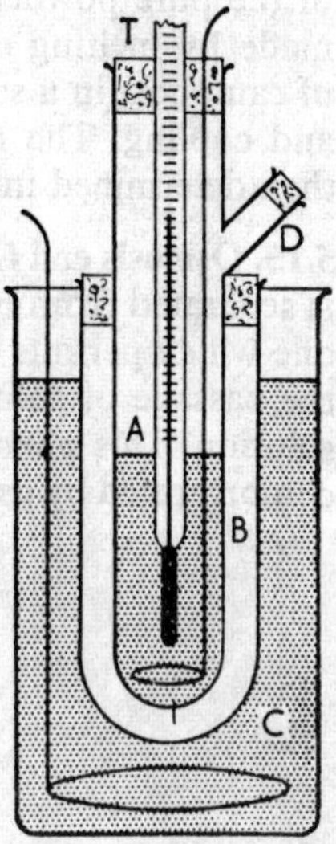

FIG. 5.13 Beckmann's Apparatus

RAST'S METHOD: The molecular depression constant of camphor is unusually high (40 K). This led Rast to devise a special micro-method

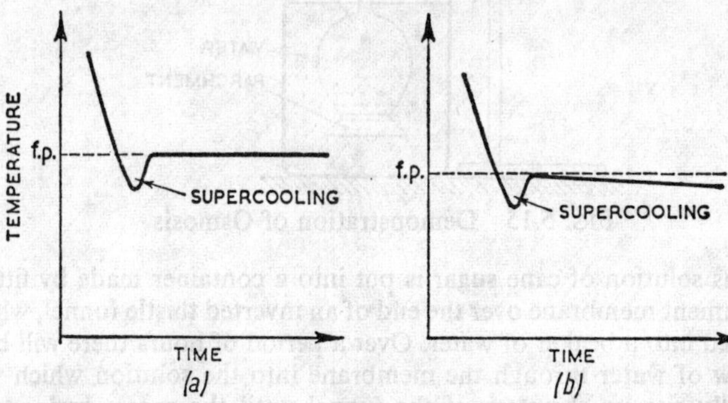

FIG. 5.14 Cooling Curves of (*a*) Pure Solvent and (*b*) Solution

using camphor as the solvent, which has the advantage that the depression produced is large enough to be measured with reasonable accuracy with a normal thermometer. The apparatus used is that for determining the melting point of a solid, as shown in Fig. 32.1. The melting point of the pure powdered camphor is found first. Then a solid solution is made by melting a known amount of solute with ten times its mass of camphor (in a sealed tube to avoid loss of camphor by evaporation) and cooling. The melting point of the powdered solid that results is then determined in the usual way.

5.15. Osmosis and Osmotic Pressure. Experiment shows that if a solution is separated from its pure solvent by a semi-permeable membrane (i.e. one which permits the passage of solvent molecules through it but not the passage of solute), then a net flow of solvent takes place into the solution. This phenomenon, which is known as *osmosis*, can be simply demonstrated by using the apparatus shown in Fig. 5.15. A concentrated

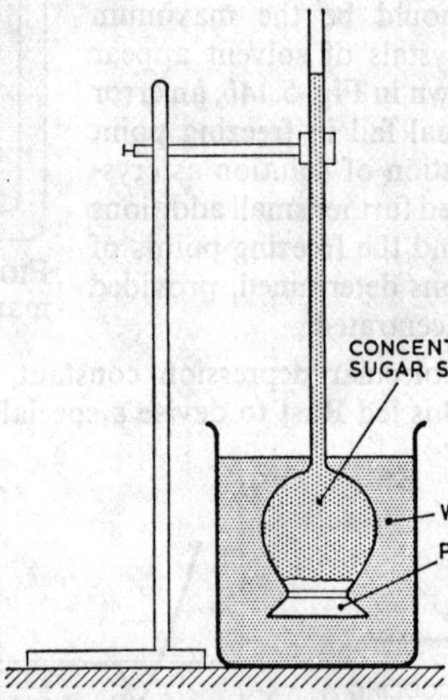

FIG. 5.15 Demonstration of Osmosis

aqueous solution of cane sugar is put into a container made by fitting a parchment membrane over the end of an inverted thistle funnel, which is dipped into a beaker of water. Over a period of hours there will be a net flow of water through the membrane into the solution which will gradually rise up the stem of the funnel until the excess hydrostatic pressure causes water to flow out of the solution as fast as it is flowing in. This excess pressure which must be applied to a solution to prevent a net flow of solvent into it when it is separated from some pure solvent by a perfect semi-permeable membrane is known as the *osmotic pressure* of the solution. Osmosis also takes place between two similar solutions of different strengths when they are separated by a semi-permeable membrane, the solvent always flowing into the solution with the greater molecular concentration. Two solutions are said to be *isotonic* when they have the same osmotic pressure so that no osmosis takes place between them. Clearly solutions which are isotonic with a given solution are also isotonic with each other.

Before going into details concerning the laws of osmosis and the measurement of osmotic pressure, it is vital to get a clear idea of the meaning of these new terms. Osmosis is the fundamental process, and osmotic pressure, far from being the cause of osmosis as some students seem to think, is best defined (as it is here) as the pressure needed to prevent osmosis. When we say that a given solution standing in a beaker has an osmotic pressure of ten times atmospheric at 0°C we do not mean that it is actually exerting a pressure of 1.01×10^6 N m^{-2}—if it were it would soon force its way out of the beaker! What we do mean is that a back pressure of about 10^6 N m^{-2} would be necessary in order to prevent osmosis taking place if that solution were separated by a semi-permeable membrane from pure solvent at 0°C. If this pressure is applied to the solution, then the solvent diffuses through the membrane in both directions at the same rate and so no resultant flow is observed. Thus osmotic pressure is a potential property of a solution which is only of practical significance in a specific context.

5.16. The Laws of Osmosis. The first serious quantitative study of osmosis was made by Pfeffer in 1877. His results were summarized by van't Hoff in 1886 in the following laws:

1. *At constant temperature the osmotic pressure of a solution is proportional to the molecular concentration of the solute*, i.e. $\Pi \propto c$.

2. *For a given solution the osmotic pressure is proportional to the absolute temperature*, i.e. $\Pi \propto T$

where Π is the conventional symbol for osmotic pressure. The first of these laws explains our interest in osmosis in this chapter, since any property of a solution which is proportional to molecular concentration offers a means of determining molecular weights.

On examining these generalizations, van't Hoff saw that they were analogous to the ideal gas laws. For example, if $\Pi \propto c$, then ΠV is a constant, where V is the volume of solution containing one gramme molecule of solute. This corresponds to Boyle's law for gases (§ 3.2). Similarly, the proportionality between osmotic pressure and absolute temperature resembles Charles' law (§ 3.3). Combining these two laws of osmosis we get $\Pi V = kT$, where k is a constant for any one solute. Using Pfeffer's experimental results van't Hoff showed that if Π is measured in newton per metre squared and V in cubic metre, the term k then has the same value for all solutes and is equal to the universal gas constant, R (see § 3.4), i.e. $\Pi V = RT$, where $R = 8.31$ J mol^{-1} K^{-1}. He concluded that *the osmotic pressure of a dilute solution is equal to the pressure which the solute would exert if it were a gas at the same temperature and occupying the same volume as the solution.* Since one mole of any gas occupies 2.24×10^{-2} m^3 at s.t.p. (§ 1.13), it follows that *a solution containing one mole of solute in 22.4* dm^3 *exerts an osmotic pressure of 101 325* N m^{-2} *at 0°C*. This relationship is the best to use for calculating molecular weights by simple proportion from measurements of osmotic pressure, but the usual warning must be given that such calculations should only be applied to dilute solutions in which the solute is neither associated nor dissociated.

Example: What is the molecular weight of cane sugar if a solution containing 75 g dm^{-3} exerts an osmotic pressure of 4.85×10^5 N m^{-2} at 10°C?

75 g in 1 dm^3 exerts an O.P. of 4.85×10^5 N m^{-2} at 283 K

$\therefore\ 75 \times \frac{283}{273}$ g in 1 dm^3 exerts an O.P of 4.85×10^5 N m^{-2} at 273 K

$\therefore\ 75 \times \frac{283}{273} \times 22.4$ g in 22.4 dm^3 exerts an O.P. of 4.85×10^5 N m^{-2} at 273 K

$\therefore\ 75 \times \frac{283}{273} \times 22.4 \times \frac{1.01}{4.85}$ g in 22.4 dm^3 exerts an O.P. of 1.01×10^5 N m^{-2} at 273 K

But 1 mole in 22.4 dm^3 exerts an O.P. of 1.01×10^5 N m^{-2} at 273 K.

$$\therefore\ M = 75 \times \frac{283}{273} \times 22.4 \times \frac{1.01}{4.85} = 363$$

$\therefore$ Molecular weight of cane sugar = 363 ($C_{12}H_{22}O_{11}$ = 342).

5.17. The Measurement of Osmotic Pressure. The simple apparatus shown in Fig. 5.15. is useless for measuring osmotic pressure. For one thing the semi-permeable membrane is imperfect and tends to allow solute molecules to pass through it. Another drawback is that it is difficult to prevent leakage at the edges of the membrane, particularly if the hydrostatic pressure becomes very great. Other objections are the time required to obtain a result and the large and cumbersome nature of the apparatus.

One of the greatest practical difficulties in the study of osmosis is to find really satisfactory semi-permeable membranes. The early workers used membranes made from animal tissues such as bladders for their experiments, but these are not completely semi-permeable and tend to burst when subjected to high pressure. In 1864 Traube discovered that a good semi-permeable membrane could be made artificially from cupric ferrocyanide by adding a solution of a copper salt to potassium ferrocyanide solution:

$$2CuSO_4 + K_4Fe(CN)_6 = Cu_2Fe(CN)_6 \downarrow + 2K_2SO_4$$

By dipping a porous pot containing one of these solutions into a vessel containing the other, Pfeffer precipitated this material in the walls of the pot and obtained a membrane of considerable strength. Using a membrane of this type and a simple mercury manometer to measure pressure, as in Fig. 5.16, he was able to obtain fairly accurate results.

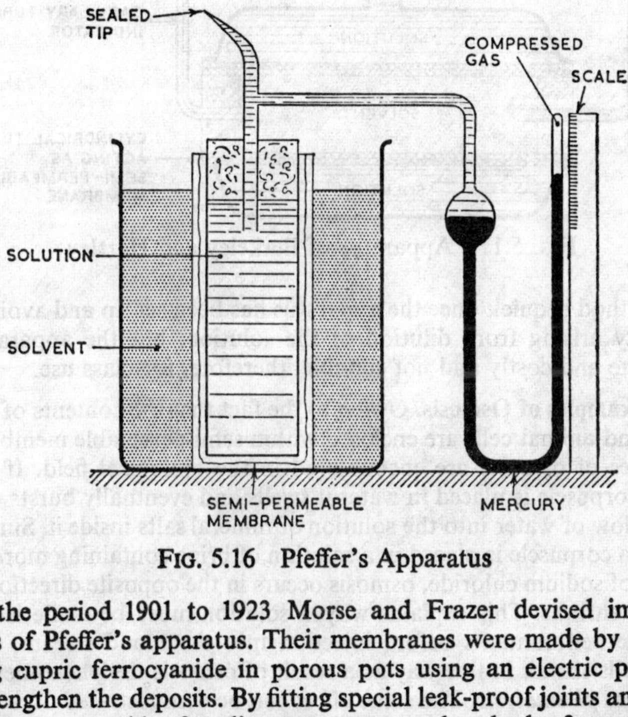

FIG. 5.16 Pfeffer's Apparatus

In the period 1901 to 1923 Morse and Frazer devised improved forms of Pfeffer's apparatus. Their membranes were made by precipitating cupric ferrocyanide in porous pots using an electric potential to strengthen the deposits. By fitting special leak-proof joints and using manometers capable of reading pressures up to hundreds of atmospheres and by controlling the temperature with a thermostat they obtained results of considerable accuracy. The only disadvantages were the slowness of the method and the need to allow for the dilution of the solution which inevitably occurred before the steady state was reached.

An alternative method, based directly upon the definition of osmotic pressure was used by Hartley and Berkeley. In their apparatus, which is shown diagrammatically in Fig. 5.17, a variable external pressure is applied to the solution and the pressure needed to just prevent the process of osmosis occurring is determined. In practice two pressures are measured, one at which solvent just flows into the solution and the other at which the reverse flow is just detectable by means of the capillary tube indicator, and the mean is taken as the osmotic pressure. This

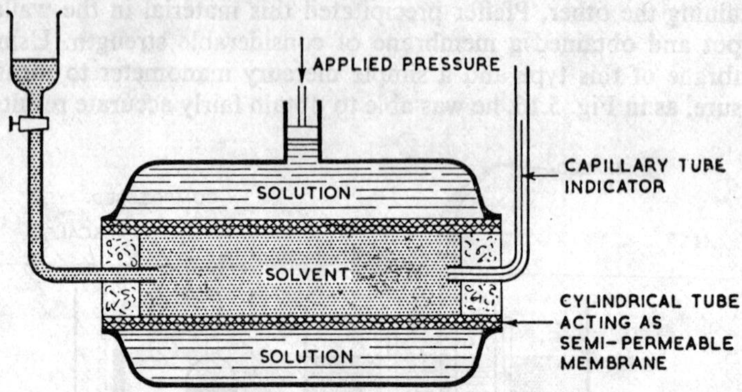

FIG. 5.17 Apparatus of Berkeley and Hartley

null method is quick once the apparatus has been set up and avoids the difficulty arising from dilution of the solution, but the apparatus is elaborate and costly and not suitable, therefore, for class use.

5.18. Examples of Osmosis. Owing to the fact that the contents of many plant and animal cells are enclosed within semi-permeable membranes, examples of osmosis are encountered in the biological field. If a red blood corpuscle is placed in water it swells and eventually bursts owing to the flow of water into the solution of mineral salts inside it. Similarly if such a corpuscle is placed in a solution of brine containing more than 0·91% of sodium chloride, osmosis occurs in the opposite direction and the cell shrinks. This explains why a solution must be made isotonic with the cell contents before injection into the blood stream. When plant cells shrink in this way as a result of osmosis the phenomenon is known as *plasmolysis*. The same effect can be demonstrated on a larger scale by placing in water and in concentrated brine samples of dried fruit (e.g. prunes or currants) or an egg from which the shell has been carefully dissolved away with dilute hydrochloric acid. It is almost certain that osmosis is at least partially responsible for the flow of sap up the stem of a plant or the trunk of a tree and for the absorption of water from the soil by the roots.

If crystals of calcium, chromium, cobalt, copper, iron, magnesium, manganese, and nickel salts are added to a 5% solution of sodium silicate, they begin to dissolve forming a layer of metallic silicate around them by double decomposition. This precipitate acts as a semi-permeable membrane through which water flows to dilute the very concentrated solution of the metallic salt inside until eventually the membrane bursts ejecting the contents in a stream. This in turn forms a layer of membrane around it and the process is repeated. In this way various coloured plant-like growths are obtained which earn for the phenomenon the popular name of a 'chemical garden'.

5.19. Osmotic Pressure and Vapour Pressure. The relationship between the osmotic pressure of a solution and the lowering of its vapour pressure can be established by the following argument.

Imagine the apparatus shown in Fig. 5.18 to be set up and left to attain equilibrium at constant temperature T after evacuation of the bell jar. It may be assumed in this hypothetical case that the membrane is perfectly semi-permeable and that no leakage occurs at its edges. Then equilibrium will be reached when the hydrostatic pressure due to the column of liquid in the stem is equal to the osmotic pressure of the solution Π. Thus if the density of the solution is ρ', then $\Pi = h\rho'$. For a dilute solution the density ρ' will be approximately equal to the density of the pure solvent, ρ, so $\Pi = h\rho$. Now if p is the vapour pressure of the pure solvent at the surface of the liquid in the beaker and p' is the slightly lower vapour pressure at a height h cm above the liquid (which is also the vapour pressure of the solution at the top of the column when equilibrium is established), and d is the average density of the vapour, then
$p - p' = hd$

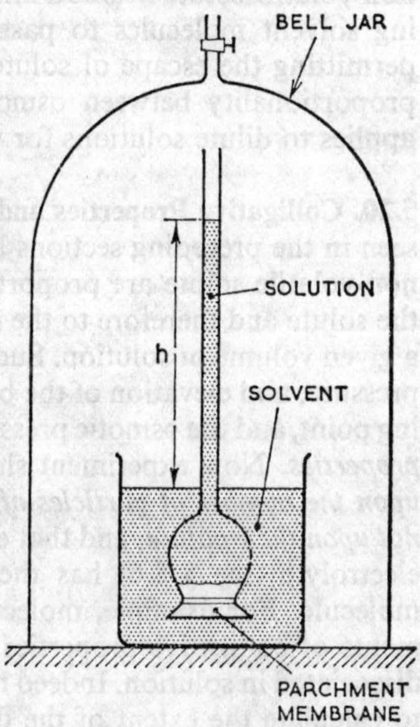

FIG. 5.18 Vapour Pressure and Osmotic Pressure

$$\therefore\ p - p' = \frac{\Pi d}{\rho} \quad \text{(since } h = \frac{\Pi}{\rho} \text{ as shown above).}$$

If the molecular weight of the vapour is denoted by M and its molar volume at temperature T by V, then

$$d = \frac{M}{V} = \frac{Mp}{RT} \quad \text{(since } pV = RT\text{)}$$

$$\therefore p - p' = \frac{\Pi M p}{\rho RT}$$

$$\text{and } \frac{p - p'}{p} = \frac{\Pi M}{\rho RT}$$

Now M, R, T, and p are all constant, so $\frac{p - p'}{p} = k.\Pi$, and *the osmotic pressure is proportional to the relative lowering of the vapour pressure.*

This relationship is hardly surprising, since experiment shows that both properties are proportional to the molecular concentration of the solute. Moreover, in many ways the surface of a solution containing a non-volatile solute acts as a kind of semi-permeable membrane in allowing solvent molecules to pass through it in both directions but not permitting the escape of solute. It should be noted, however, that the proportionality between osmotic pressure and vapour pressure only applies to dilute solutions for which ρ and ρ' may be taken as equal.

5.20. Colligative Properties and Abnormal Molecular Weights. We have seen in the preceding sections how certain properties of a solution of a non-volatile solute are proportional to the molecular concentration of the solute and therefore to the number of molecules of solute present in a given volume of solution. Such properties, e.g. lowering of the vapour pressure, and elevation of the boiling point, and depression of the freezing point, and the osmotic pressure of a solution, are known as *colligative properties.* Now experiment shows that *these properties depend in fact upon the number of particles of solute in a given volume of solution and not upon their nature*, and that each ion formed by the dissociation of an electrolyte (see § 8.6) has the same colligative effect as a complete molecule. For example, molecular weights determined from measurements of colligative properties are abnormally small if the solute is dissociated in solution. Indeed the abnormality can be used in such cases to calculate the extent of the dissociation, as explained in § 8.7. Similarly, colligative measurements reveal that the molecular weights of carboxylic acids such as acetic acid (§ 37.2) when dissolved in benzene are approximately twice as large as those expected from the usual formulae, suggesting that these acids are associated into double molecules in this solvent.

5.21. Other Methods of Determining Molecular Weights.

1. BY DIFFUSION AND EFFUSION MEASUREMENTS: As explained in § § 3.6 and 3.7, gases diffuse and effuse at rates which are inversely proportional to the square roots of their densities and therefore, by Avogadro's hypo-

thesis, to the square roots of their molecular weights. It follows that if the molecular weight of one gas is known, that of the other can be determined. The method is not of much importance, but it has been used for finding the molecular weight of a substance like ozone which is difficult to obtain pure. Worked examples are given in the appropriate sections.

2. BY STEAM DISTILLATION: The principles underlying this method are explained in § 6.13, where a worked example is given. By analysing the distillate obtained from a pair of immiscible liquids, the molecular weight of one can be determined if that of the other is known. The method is of little practical importance.

3. BY GRAVIMETRIC METHODS: Certain special methods can be used for determining the molecular weights of organic acids and amines (see § 32.10). In each case a metallic salt is formed which leaves only the metal when heated strongly.

4. FROM PARTITION COEFFICIENTS: When a solute exists in a different molecular condition in two immiscible solvents, the relative molecular weights can be determined by measuring the partition coefficient of the solute between the two solvents as described in § 6.14. The method, which is of very limited application, can be used to determine the molecular weight of carboxylic acids (e.g. acetic acid and benzoic acid) in benzene.

5.22. Review of Molecular Weight Methods. When a substance is volatile, as the majority of organic compounds are, the molecular weight is best determined from its vapour density, usually by Victor Meyer's method. For substances which are not easily vaporized or which decompose when heated (e.g. urea and sugars), a colligative method is usually chosen. Of these the measurement of freezing point depression is the most accurate and convenient and is generally preferred, but it is unsuitable for substances of very high molecular weight because the depression obtained in a dilute solution is then so small (e.g. an aqueous solution containing 20 kg m^{-3} of solute of molecular weight 10 000 shows a depression of less than 0.004 K). Thus when the molecular weight exceeds about 5000 (as it often does in carbohydrates, proteins, and polymers), it is best determined from measurements of osmotic pressure, particularly since membranes which are semi-permeable to solute molecules of this size are readily obtained. The solution quoted above, for example, containing 20 kg m^{-3} of solute of molecular weight 10 000, will at room temperature exert an osmotic pressure of about 4.7×10^3 N m^{-2}, which can certainly be measured accurately using modern apparatus.

6

Phase Equilibria

6.1. Introduction. The term *phase* is used to describe any part of a system which is itself homogeneous (i.e. uniform throughout so that every sample of it has exactly the same physical properties and chemical composition) and yet is physically distinct from all the other parts of the system. For example, a mixture of a liquid and a gas consists of two different phases, and so does a mixture of a solid and a gas, or a solid and a liquid in equilibrium with each other. On the other hand a mixture of gases constitutes only a single phase no matter how many gases are present. The same is true of a mixture of liquids which dissolve completely in each other, or of an unsaturated solution of a solid in a liquid, since all these examples are homogeneous throughout. Except when a solid solution is formed (§ 4.19) a mixture of solids always consists of as many phases as there are different solids present, and every solid allotropic form is regarded as a separate phase.

In previous chapters we have already considered some simple examples of phase equilibria, such as the equilibrium between a liquid and its saturated vapour (§ 3.16) and that between a saturated solution and excess of solid solute (§ 4.12). In this chapter we shall extend these ideas by making a systematic study of a wide range of phase equilibria, culminating in the important generalization known as *the phase rule.*

6.2. Phase Equilibrium Diagrams. The relation between the solid, liquid, and vapour states of a given substance can be depicted diagrammatically by what is known as the *phase equilibrium diagram* of that substance. Fig. 6.1 is a typical example, in which pressure is plotted against temperature for a single substance. Such a diagram is entirely constructed from experimental data, which it summarizes in a compact form. For example, the curves OA and BO are the saturated vapour pressure curves of the liquid and solid respectively. They intersect at the point O, which represents, therefore, the conditions of temperature and pressure at which the solid and liquid are in equilibrium with each other under their own vapour pressure. This is not exactly the same as the melting point of the solid, which is the temperature at which solid and liquid are in equilibrium under a pressure of 101 325 N m^{-2} (see § 4.3), but the difference is very small. The curve OC shows how this equilibrium point between solid and liquid varies with applied pressure. Since it slopes slightly towards the right in this case, it indicates that the melting point of this substance is raised by the application of pressure.

It is important to appreciate the significance of the various areas, lines, and points in such a phase diagram. As shown in Fig. 6.1, the diagram is divided into three *areas*, labelled solid, liquid, and vapour. Thus at a glance it is possible to see what conditions of temperature and pressure correspond to any one of these physical states of the substance. For example, under the conditions prevailing in the area AOC liquid is

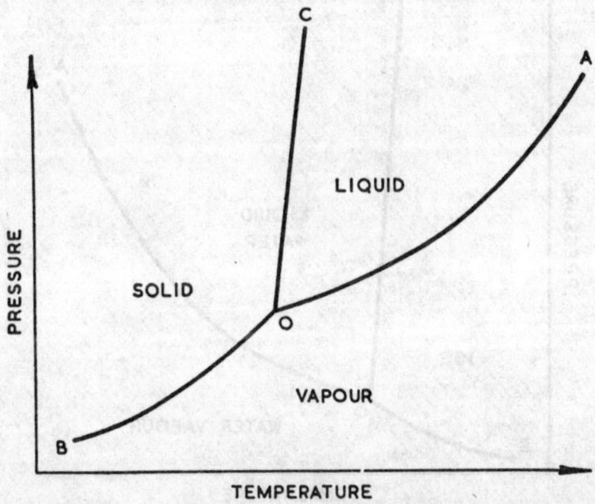

FIG. 6.1 A Typical Phase Equilibrium Diagram

the only stable phase, whilst under the conditions of the area BOA the substance can only exist as a vapour. The *lines* in the diagram represent the conditions under which two of the phases are in equilibrium with each other. At any temperature and corresponding pressure along the line OA, for example, the liquid and vapour can coexist indefinitely, but if either the temperature or the pressure is changed, then one phase disappears. The point O is unique, being the only conditions under which all three phases, solid, liquid, and vapour, are in equilibrium with each other. It is known as the *triple point* for that reason. If the substance is at exactly this temperature and pressure and the conditions are altered, then at least one of the three phases will disappear and probably two.

The line OA ends abruptly at the point A, which corresponds to the critical temperature and pressure of the substance. It is impossible to extend the diagram beyond this point because at higher temperatures and pressures the liquid and vapour are indistinguishable. The practical difficulties of working at very high and at very low temperatures and pressures limit the extent of OB and OC respectively.

6.3. The Phase Diagram of Water. This diagram, which is of particular interest and importance, is shown (not to scale) in Fig. 6.2*. The triple point, O, occurs at 613 N m^{-2} pressure and 0.0075°C, and the critical point, A, at 2.20×10^7 N m^{-2} (217 atm) and 374°C. The line OC inclines very slightly to the left because the melting point of water is lowered by

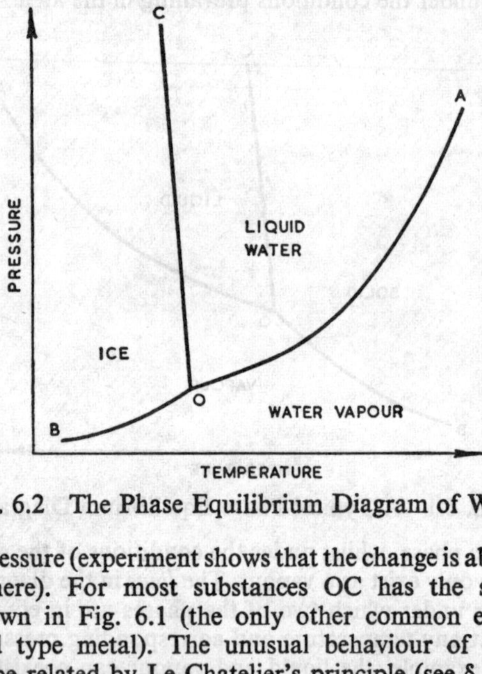

FIG. 6.2 The Phase Equilibrium Diagram of Water

raising the pressure (experiment shows that the change is about 0.0075 K per atmosphere). For most substances OC has the steep positive gradient shown in Fig. 6.1 (the only other common exceptions are bismuth and type metal). The unusual behaviour of water in this respect can be related by Le Chatelier's principle (see § 7.9) to its expansion on freezing, which arises from the very open structure of ice crystals owing to their hydrogen bonding.

6.4. Phase Diagram of Carbon Dioxide. Another unusual equilibrium diagram is that of carbon dioxide (Fig. 6.3), because here the triple point, O, lies above atmospheric pressure. As the diagram shows, liquid carbon dioxide is not stable at any temperature when the external

* The various forms of ice are not shown for the sake of simplicity.

pressure is below 5.16×10^5 N m^{-2}. Thus when solid carbon dioxide is warmed above −78°C at atmospheric pressure it *sublimes* (i.e. it turns

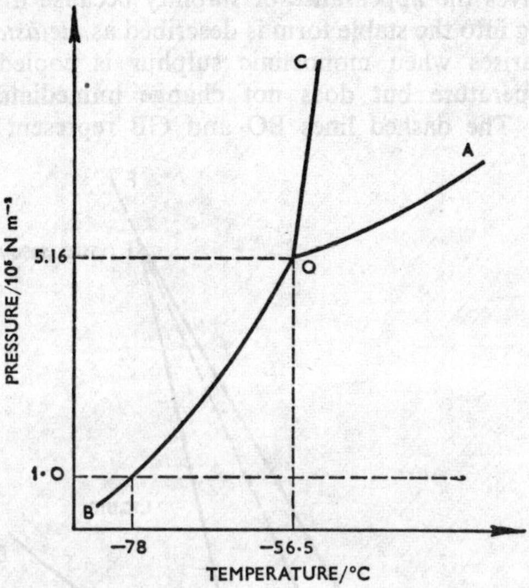

FIG. 6.3 The Phase Equilibrium Diagram of Carbon Dioxide

directly into gas without forming a liquid first), which partly accounts for its use for refrigeration (§ 20.5).

6.5. Types of Allotropy. This phenomenon of allotropy has already been referred to in § 4:10. The three main types are:

1. ENANTIOTROPY: The main characteristics of enantiotropy are:

(*a*) Each allotrope is stable over a definite range of temperature.
(*b*) There is a definite transition temperature at which both solid forms exert the same vapour pressure and can exist in equilibrium with each other.
(*c*) Each allotrope can be converted directly into the other by varying the temperature, i.e. the change from one form into another is reversible.

One of the best examples is provided by sulphur, whose pressure–temperature diagram is shown in Fig. 6.4. The lines AB and BC are the vapour pressure curves of the rhombic and monoclinic allotropes respectively, and B is the transition temperature at which each changes into the other if the temperature is varied slowly. If rhombic sulphur is heated strongly, however, so that its temperature rises rapidly through the transition temperature, it tends to persist for a while even though it is theoretically unstable. Any substance or system which, like this, is unstable but gives the appearance of stability because it is only very slowly changing into the stable form is described as *metastable*. A similar situation arises when monoclinic sulphur is cooled below the transition temperature but does not change immediately into the rhombic form. The dashed lines BO and GB represent the vapour

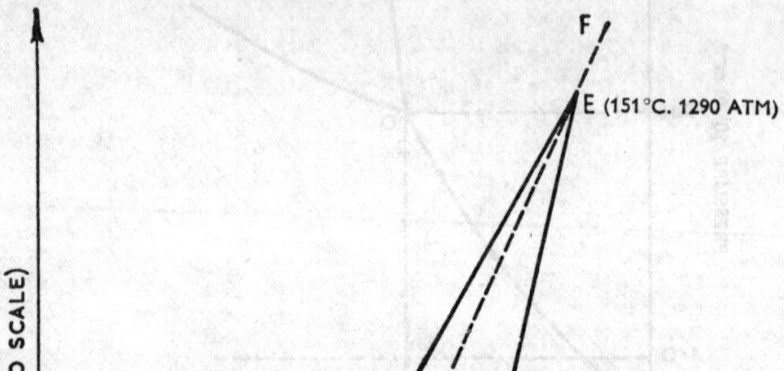

FIG. 6.4 The Phase Equilibrium Diagram of Sulphur

pressure curves of these metastable forms; it will be noticed that they are always higher than the vapour pressure curve of the more stable form.

The line CD is the vapour pressure curve of liquid sulphur and C is therefore the melting point of monoclinic sulphur under its own vapour pressure (viz. 5.3 N m^{-2}). The result of applying a high external pressure is to raise both the transition temperature between the rhombic and monoclinic forms and the melting point of monoclinic sulphur;

these changes are represented by the lines BE and CE, though not to scale since the transition temperature rises by only about 0.045 K for each increase of one atmosphere in pressure. Thus the points B and C and the point E are all triple points where three phases are in equilibrium with each other. At B the rhombic and monoclinic allotropes can coexist with sulphur vapour, at C liquid and vapour sulphur are in equilibrium with the monoclinic form, and at the abnormal conditions represented by E the system can contain rhombic, monoclinic, and liquid sulphur together. A fourth triple point, O, is metastable and represents the special conditions when rhombic sulphur is in equilibrium with the liquid and vapour; it is only achieved in practice by heating rhombic sulphur rapidly to its melting point (114.5°C).

The four areas of the diagram represent the various conditions of temperature and pressure under which rhombic, monoclinic, liquid, and vapour sulphur are stable. For example, the monoclinic form can only exist in a stable condition if the temperature and pressure lie somewhere between the limits set by the points B and E. Since no quadruple point exists, it is impossible for all four phases to be in equilibrium with each other simultaneously.

The transition temperature in enantiotropy is often determined experimentally by observing some property such as colour or volume and finding when it undergoes a sharp change (e.g. by using a dilatometer). Alternatively the variation of solubility or vapour pressure with temperature can be plotted on a graph and the temperature at which the curves of the two forms intersect can be found.

2. MONOTROPY: The chief features of monotropy are:

(*a*) One allotrope is stable and the other metastable at all temperatures up to the melting point.

(*b*) There is no definite transition temperature at which the two solid forms are in equilibrium with each other.

(*c*) The direct change from one allotrope to the other can only be made in one direction, from the metastable to the stable form.

The allotropy of phosphorus is one of the best known examples of this type. A simplified and rather idealized phase diagram is shown in Fig. 6.5. The red allotrope is the stable form at all temperatures up to its melting point, and white phosphorus is always metastable changing only very slowly into the red form at room temperature and rapidly on heating (see § 23.2). AB is the vapour pressure curve of the red form, which melts at B (590°C). Similarly DE is the vapour pressure curve of white phosphorus and E is its melting point (44°C) (note that metastable forms exert a higher vapour pressure at any temperature). BC is the vapour pressure curve of molten red phosphorus; if the liquid is cooled rapidly it follows the path CBE and deposits crystals of the white variety. This illustrates a general tendency whereby the metastable allotrope is usually obtained first when a vapour is condensed or a solution crystallized. The effect on the two melting points of applying a high external pressure is depicted by the lines EF and BF. Thus at the triple point E the solid and liquid states of white phosphorus are in equilibrium with the vapour, and at the triple point B the solid and liquid red forms and phosphorus vapour are in equilibrium. The point F is theoretically the transition temperature between the two forms but

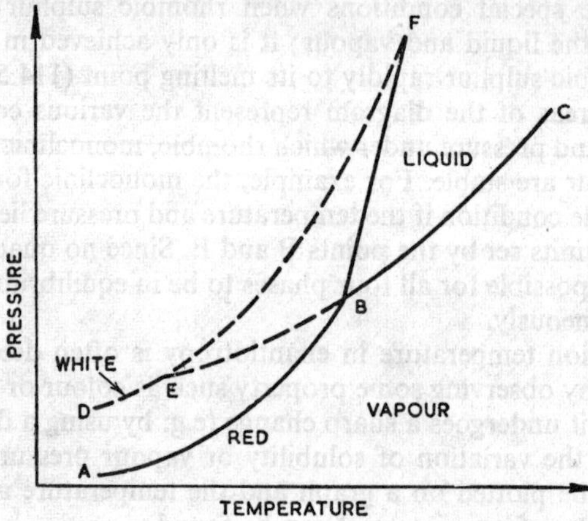

FIG. 6.5 A Simplified Phase Equilibrium Diagram of Phosphorus

it is of little practical importance because it is above the melting point of either form at atmospheric pressure.

DYNAMIC ALLOTROPY: In this type the two allotropes are in dynamic equilibrium with each other over a wide range of temperature, the proportions of each in the gaseous or liquid mixture depending upon the temperature and, sometimes, the pressure. The difference between the allotropes lies not in the way in which the identical particles are packed together in the crystal (as it does in monotropy and enantiotropy) but in the atomicity and shape of the molecules themselves.

For example, the two allotropes oxygen and ozone are continually changing into each other and at any given temperature and pressure there is a fixed proportion of each present in the system when it is at equilibrium (see § 24.3). Liquid sulphur provides another example of the same kind (§ 24.5).

6.6. Solutions of Gases in Liquids. Experiment shows that the mass of a gas which dissolves in a fixed volume of a liquid when the two are in equilibrium depends upon the nature of the gas and the liquid, the temperature, and the pressure. Table 6.1, which gives the volumes of various gases which dissolve in one volume of water at s.t.p., illustrates the wide differences that occur from one gas to another, and Table 6.2

TABLE 6.1. SOLUBILITIES OF VARIOUS GASES IN WATER AT S.T.P.

Gas	NH_3	HCl	SO_2	H_2S	CO_2	O_2	N_2	He
Absorption Coefficient	1300	500	80	4.6	1.7	0.05	0.024	0.009

shows how the solubility of one particular gas (oxygen) varies with the solvent.

With few exceptions the solubility of a gas in a liquid decreases as the temperature is raised, as might be expected from Le Chatelier's principle (§ 7.9), since heat is evolved when gases dissolve and low temperature will therefore favour the exothermic process. For example, the solubility of ammonia in water is 1300 volumes at 0°C, 700 volumes at 20°C, and about 240 volumes at 90°C. This is in direct contrast to the general behaviour of solids, most of which are more soluble at higher

TABLE 6.2. SOLUBILITY OF OXYGEN AT 20°C.

Solvent	*Water*	*Ethyl Alcohol*	*Benzene*	*Acetone*	*Diethyl Ether*
Absorption Coefficient	0.028	0.144	0.163	0.208	0.416

temperature. The effect of temperature can be observed very simply by drawing a tumbler of cold water from a tap and leaving it to stand in a warm room, when dissolved air will be expelled from saturated solution as the temperature of the water rises and will escape as bubbles to the surface. If an aqueous solution of ammonia is boiled for some time in the open air, the dissolved gas is driven off completely and the residual liquid is found to be pure water. This does not show that the solubility of ammonia in water is nil at 100°C, because the gas is not in equilibrium with the solution under these conditions. If the experiment were repeated in such a way that the boiling solution remained in contact with expelled gas at atmospheric pressure, then the concentration of gas in solution would remain at the equilibrium level for that temperature (195 volumes) however long the boiling was continued.

Similarly, if air is bubbled through an aqueous solution of ammonia at room temperature, all the ammonia gas is eventually removed from the solution because the continual flow of air carries away the volatile ammonia and prevents the establishment of equilibrium. These examples emphasize the importance of regarding solutions of gases in liquids in terms of an equilibrium between the two phases.

The solubility of a gas can be expressed in the following ways:

1. As the volume of gas, measured under the conditions of dissolving, which dissolves in unit volume of the solvent. This is known as Ostwald's *coefficient of solubility*, β.

2. As the volume of gas, *reduced to s.t.p.*, which dissolves in unit volume of solvent at any particular temperature under a partial pressure of one atmosphere. This is known as Bunsen's *absorption coefficient*, α. Assuming that the gas laws are obeyed, it follows that $\alpha = \beta \times \dfrac{273}{T}$, where T is the absolute temperature of the experiment.

3. As the mass of gas in gramme which dissolves in 100 gramme of solvent. This is usually expressed simply as a percentage.

4. As the mole fraction of gas in the solution. This method is particularly useful when considering phase equilibria.

6.7. Henry's Law. This states that *at constant temperature the mass of gas which dissolves in a given volume of liquid at equilibrium is proportional to the pressure of the gas.* If the gas obeys Boyle's law, then it follows that the *volume* which dissolves is independent of the pressure. The law does not hold true for aqueous solutions of very soluble gases such as ammonia or hydrogen chloride which not only dissolve physically in water but also react chemically with it:

$$NH_3 + H_2O \rightleftharpoons NH_4^+ + OH^-$$
$$HCl + H_2O \rightleftharpoons H_3O^+ + Cl^-$$

Deviations are also shown by most gases at very high pressures and low temperatures.

As Dalton showed (§ 3.5), Henry's law also applies to individual gases in a mixture provided the pressure of each gas is taken as its partial pressure. For example, the partial pressures of oxygen and nitrogen in the air are approximately in the ratio of 1 : 4. Since oxygen is about twice as soluble in water as nitrogen under the same conditions, the weights of oxygen and nitrogen dissolved in water exposed for a long period to the air are in the ratio 1 : 2.

In the Bosch process for manufacturing hydrogen (§ 16.2) use is made of the effect of pressure upon solubility to remove carbon dioxide from the product by washing; at 50 times atmospheric pressure the dioxide dissolves readily in water and is almost completely removed. Henry's law is also applied in storing acetylene by dissolving it in acetone under a pressure of about 3×10^6 N m^{-2} (30 atm) (§ 34.4). If a diver who is wearing a pressurized suit rises to the surface rapidly, the additional gas (mainly nitrogen) which dissolves in his blood at high pressure tends to escape as bubbles, causing acute pain or even death. For this reason gradual decompression is essential. The use of helium in this connection is referred to in § 26.5.

6.8. Measurement of the Solubility of Gases. Two main methods are in use:

PYKNOMETER METHOD: This is only applicable to gases forming solutions which can be analysed volumetrically, which includes most of the very soluble gases. A pyknometer (Fig. 6.6) is first weighed dry and empty. It is then partially filled with solvent, as shown in the diagram, and the gas is bubbled through it until a saturated solution is obtained. During this part of the experiment the pyknometer must be kept in a thermostat at the desired temperature. The pyknometer is sealed at both ends and reweighed to give the mass of the solution. It is then immersed in a measured excess of a standard solution of a suitable reagent (one that reacts with the dissolved gas) and one end is broken to allow the liquids to mix. The residual reagent is determined by back titration. From the amount of reagent used the mass of gas in the solution can be calculated. Taken from the mass of the solution, this gives the mass of solvent. Hence the solubility can be found.

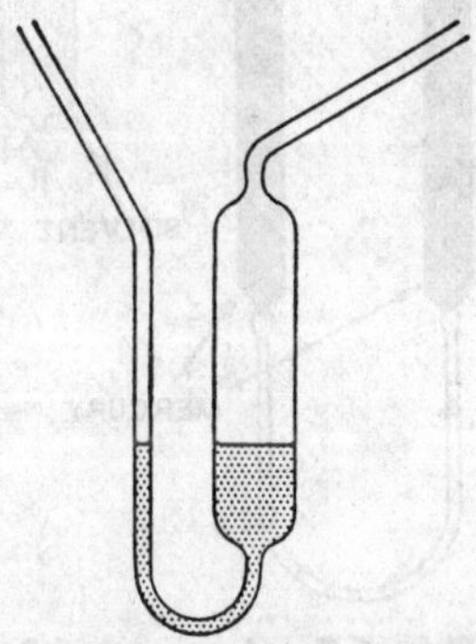

FIG. 6.6 A Pyknometer

OSTWALD'S METHOD: The apparatus used is shown in Fig. 6.7. Before starting the experiment the gas burette A is filled with mercury and the pipette B with solvent, and a current of the chosen gas is passed through the flexible tube C to expel the air from it. Then a suitable volume of the gas is drawn into the burette A through the tap T_1 by lowering the other limb, and after equalizing the mercury levels its volume at atmospheric pressure is measured. The three-way taps T_1 and T_2 are set so that A and B are connected, and a known mass of the liquid in the pipette is then run out through tap T_3, leaving it about half-full, so that the upper part is filled with gas. After closing all the taps the pipette is shaken vigorously for some minutes to ensure thorough mixing of gas and liquid, and A and B are again connected. The shaking process is repeated until no more gas will dissolve, when the volume of gas (at atmospheric pressure) remaining in A is noted. The temperature and pressure of the atmosphere should be recorded; in accurate work the gas pipette B is completely immersed in a thermostat. The solubility is calculated in the following way:

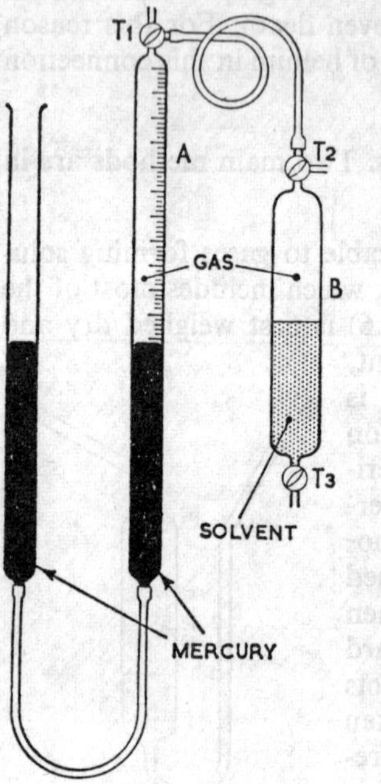

FIG. 6.7 Apparatus for Sparingly Soluble Gases

Example: Total volume of B = v_1

Volume of solvent run out = v_2

Volume of solvent remaining = $v_1 - v_2$

Volume of gas in A originally = v_3

" " " " " finally = v_4

∴ Volume of gas which dissolved = $(v_3 - v_4 - v_2)$

If the temperature is T and the pressure P then the volume of dissolved gas at

$$\text{s.t.p.} = (v_3 - v_4 - v_2) \times \frac{273}{T} \times \frac{P}{P_{atm}}$$

Now this dissolved in $v_1 - v_2$ of solvent.

∴ Absorption coefficient at temperature T and pressure P

$$= \frac{(v_3 - v_4 - v_2)}{(v_1 - v_2)} \times \frac{273}{T} \times \frac{P}{P_{atm}}$$

6.9. Mixtures of Miscible Liquids. Two liquids are said to be *miscible* when they dissolve completely in each other in all proportions to give a homogeneous mixture. For example, when water is added to ethyl alcohol the two liquids mix intimately forming a solution of uniform composition throughout. The vapour pressure above a mixture of two miscible liquids varies with the composition of the mixture in one of three ways:

1. Whatever the composition the vapour pressure always lies between the extreme values of the two pure liquids.
2. At a certain composition the vapour pressure has a maximum value which is greater than that of either pure liquid.
3. At a certain composition the vapour pressure has a minimum value which is less than that of either pure liquid.

Types 2 and 3 give rise to constant boiling point mixtures and are discussed in a later section, only the first type being considered here.

If Raoult's law (§ 5.10) is obeyed exactly by two miscible liquids over the whole range of composition then the vapour pressure exerted above the mixture by each constituent is proportional to its mole fraction

i.e. $$p_A = \frac{n_A}{n_A + n_B}.P \quad \text{and} \quad p_B = \frac{n_B}{n_A + n_B}.P$$

Thus the partial pressure curve of each constituent is a straight line through the origin and the total vapour pressure above the mixture varies regularly with composition, as in Fig. 6.8. Such a mixture is known as an *ideal solution*; the two liquids forming it do not show any heat change when mixed and the final volume of the mixture is exactly equal to the sum of the volumes of the constituents. In practice most real solutions deviate from this ideal behaviour because their molecules

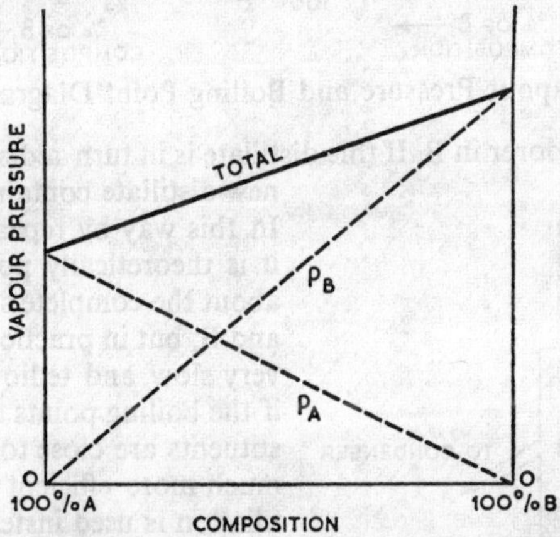

FIG. 6.8 The Vapour Pressure/Composition Curve for an Ideal Solution at Constant Temperature

'interfere' with each other to some extent and exert abnormal attractions for each other, but the concept of an ideal solution is useful, just like that of an ideal gas, because it gives an approximate guide to the behaviour of real substances.

Where deviations from ideal behaviour are only slight and do not lead to a constant boiling point mixture, then vapour pressure and corresponding boiling point curves take the form shown in Fig. 6.9. Mixtures of this type include benzene and toluene, acetone and water, and nitrogen and oxygen in liquid air. At any particular composition of liquid the vapour above the mixture is always richer in the more volatile constituent than is the liquid with which it is in equilibrium. Taking the example shown in Fig. 6.9, if a liquid mixture containing 33% of B is distilled, the vapour which escapes and condenses contains 65% B and 35% A, and the liquid remaining in the flask becomes

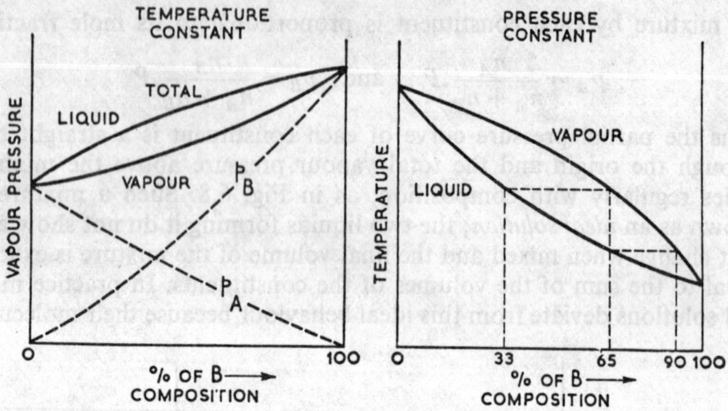

FIG. 6.9 Vapour Pressure and Boiling Point Diagrams (Type 1)

progressively poorer in B. If this distillate is in turn redistilled it gives a new distillate containing 90% of B. In this way by repeated distillation it is theoretically possible to bring about the complete separation of A and B, but in practice this process is very slow and tedious, particularly if the boiling points of the two constituents are close together, and the much more efficient fractional distillation is used instead.

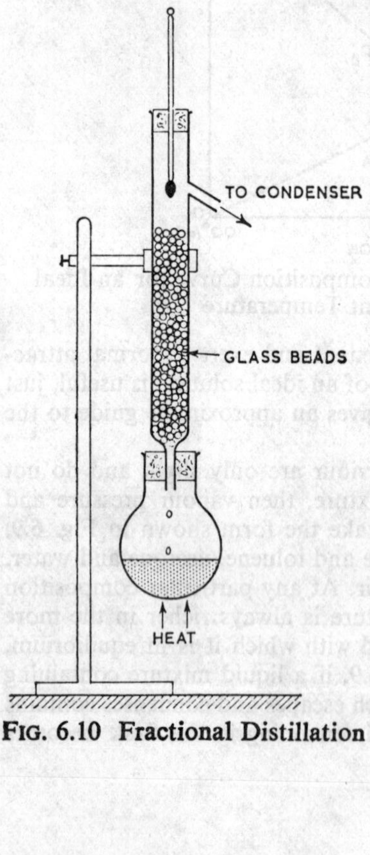

FIG. 6.10 Fractional Distillation

6.10. Fractional Distillation. The apparatus used is shown diagrammatically in Fig. 6.10. The fractionating column is designed to ensure that the ascending vapour is brought into as intimate contact with descending liquid as possible so that an equilibrium between the two exists at every stage. In the laboratory this is achieved by filling the column with glass beads or short pieces of glass tubing orientated at random or by using a column with specially designed bulbous surfaces, but industrial fractionating columns are often extremely complicated structures and may be up to 30 metre tall. Vapour from the flask undergoes repeated condensation and volatilization in the fractionating column as it is successively cooled by the descending liquid and heated by the ascending vapour. The effect is the same as repeated distillation and provided the heating is steady so that a state of equilibrium is established, almost complete separation of the constituents is possible in one operation, even when their boiling points differ by only a few degrees. The more volatile constituent issues from the top of the column whilst the other remains behind in the flask. Fractional distillation is of great importance in the manufacture of oxygen, nitrogen, and the inert gases, and in the refining of petroleum (§ 33.8).

6.11. Constant Boiling Point Mixtures. As explained in § 6.9, in certain cases the vapour pressure above a mixture of miscible liquids does not vary even approximately regularly with composition but reaches a maximum or minimum value at some intermediate composition. Such liquids give rise to *constant boiling point* or *azeotropic mixtures*, which are of two types:

MAXIMUM BOILING POINT MIXTURES: The typical way in which the vapour pressure and the boiling point vary with composition in this type of mixture is shown in Fig. 6.11. On distilling such a mixture the composition of the residue changes gradually and its boiling point rises

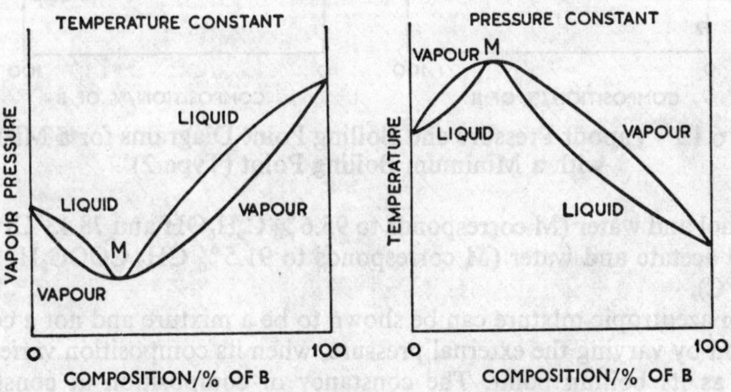

FIG. 6.11 Vapour Pressure and Boiling Point Diagrams for a Mixture with a Maximum Boiling Point (Type 3)

slowly until the liquid remaining in the flask has the composition M, when vapour of this composition distils over unchanged at constant temperature. Thus complete separation of the constituents cannot be achieved by distillation and only that constituent which is present in greater proportion than in the azeotropic mixture can be obtained pure.

Examples of such mixtures include sulphuric acid and water (M corresponds to 98.3% H_2SO_4 and 338°C), nitric acid and water (M corresponds to 68% HNO_3 and 120.5°C) and hydrochloric acid and water (M corresponds to 20.2% HCl and 108.6°C).

MINIMUM BOILING POINT MIXTURES: Typical vapour pressure and boiling point curves for this type of mixture are shown in Fig. 6.12. When such a mixture is fractionally distilled or subjected to repeated redistillation, a distillate of composition M is obtained and this continues to distil over until the whole of one constituent has been vaporized. The pure constituent remaining in the flask then distils over unchanged. Thus it is impossible to separate the two constituents completely by means of distillation. Examples of mixtures of this type include ethyl

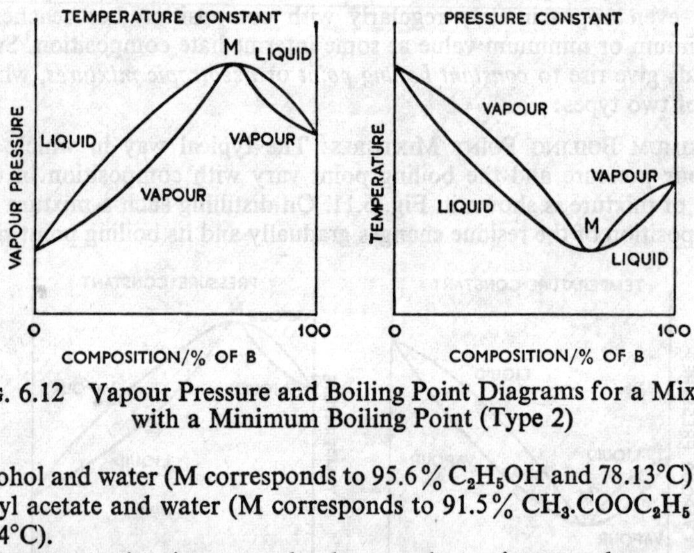

FIG. 6.12 Vapour Pressure and Boiling Point Diagrams for a Mixture with a Minimum Boiling Point (Type 2)

alcohol and water (M corresponds to 95.6% C_2H_5OH and 78.13°C) and ethyl acetate and water (M corresponds to 91.5% $CH_3.COOC_2H_5$ and 70.4°C).

An azeotropic mixture can be shown to be a mixture and not a compound by varying the external pressure, when its composition varies as well as its boiling point. The constancy of composition at constant pressure is often made use of in preparing a standard solution of hydrochloric acid—the maximum boiling point azeotropic mixture eventually obtained by distilling any sample of hydrochloric acid under 101 325 N m^{-2} pressure is merely diluted by the required amount.

6.12. Partially Miscible Liquids. Some liquids are only partially miscible, i.e. they dissolve in each other only in certain proportions to form a homogeneous mixture. For example, if diethyl ether is added to water drop by drop, at first it dissolves forming a solution of uniform composition throughout, but there comes a point when the solution of ether in water is saturated and any further addition results in the formation of a separate layer. The two saturated solutions in equilibrium with each other at this stage are known as *conjugate solutions*. If more ether is added a stage is eventually reached when all the water present dissolves in the large excess of ether and the boundary between the two layers then disappears.

It is found by experiment that the range of composition over which two such liquids are immiscible changes with temperature. The effect is

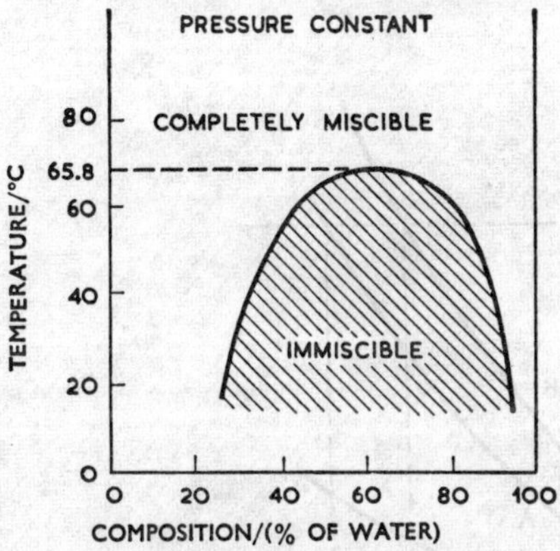

FIG. 6.13 Miscibility of Water and Phenol

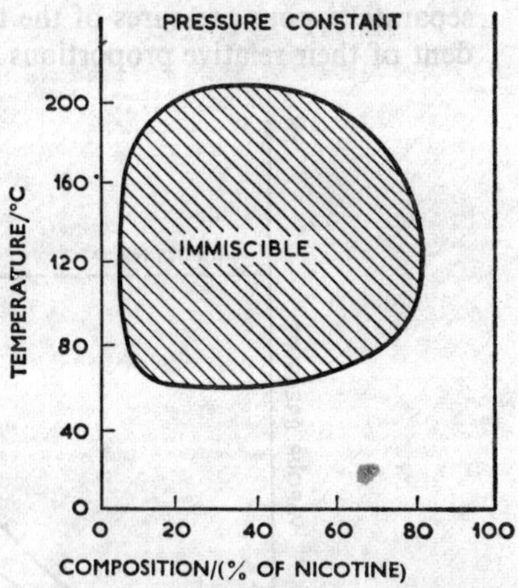

FIG. 6.14 Miscibility of Water and Nicotine

shown for the phenol-water system in Fig. 6.13, where the area corresponding to conditions of immiscibility is shaded. The lowest temperature at which the two liquids are miscible in all proportions is known as the *upper critical solution temperature* or the *consolute temperature* of the system. For phenol and water at atmospheric pressure this temperature is 65.8°C, as shown in the diagram. Experiment shows that it is very sensitive to the presence of impurities, rising several degrees if even small quantities of benzene or naphthalene are present and falling to room temperature on adding about 1% of soap.

Other pairs of liquids are known (e.g. triethylamine and water) which are only completely miscible below a certain temperature and are immiscible in certain proportions above it. The system nicotine-water is unusual in having an upper and a lower critical solution temperature, so that under high pressure its miscibility curve is a closed loop as shown in Fig. 6.14. At the conditions corresponding to the shaded area the two liquids are immiscible and exist as two separate phases, but under all other conditions they form a homogeneous solution.

6.13. Immiscible Liquids and Steam Distillation. Many pairs of liquids are found to be virtually completely immiscible, i.e. they form separate layers when added to each other in almost any proportions. Examples are mercury and water, carbon disulphide and water, and chlorobenzene and water. It is found by experiment that the vapour pressure above a pair of immiscible liquids of this sort is equal to the sum of the separate vapour pressures of the two pure components and is independent of their relative proportions in the mixture. Since such a mixture

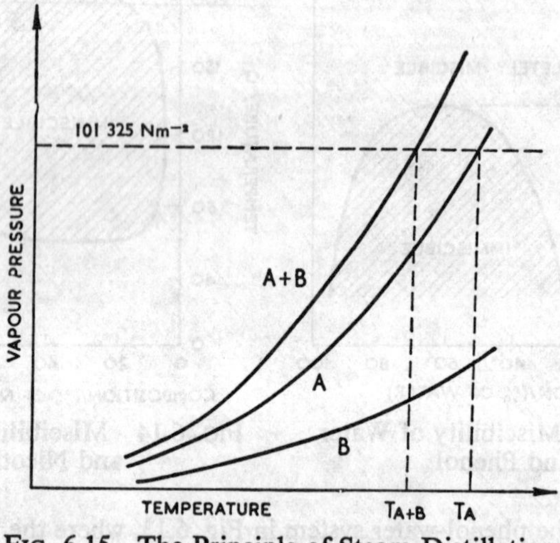

FIG. 6.15 The Principle of Steam Distillation

will boil when the total vapour pressure is equal to the external pressure, it follows that the mixture always boils at a lower temperature than either pure constituent. This point is illustrated in Fig. 6.15.

Advantage is taken of this behaviour of immiscible liquids in the process of *steam distillation*. The apparatus used is shown in Fig. 44.1. Steam from the can bubbles into the liquid mixture in the flask and condenses giving up its latent heat of vaporization and eventually causing the mixture to boil. The vapours of both constituents escape from the mixture in the proportion of their relative vapour pressures at that temperature, so that both are present in the distillate where they form separate layers. Steam distillation makes it possible to recover a liquid like aniline from a mixture containing a large excess of water and solids (see § 44.5), whereas if direct heating were used, local overheating and charring would be inevitable. As the following example shows, steam distillation could be used for determining the molecular weight of a liquid which is immiscible with water, but the method is not of much practical value (§ 5.21).

Example: A mixture of aniline and water boils at 98.5°C, the vapour pressures of the constituents being 5.6×10^3 N m^{-2} and 9.57×10^4 N m^{-2} respectively. The distillate contains 23.2% by weight of aniline. Calculate the molecular weight of aniline, M, given that the molecular weight of water is 18.

Now since the volume of each constituent in the vapour mixture is in proportion to its vapour pressure at the boiling point, it follows that:

$$\frac{\text{mass of aniline in distillate}}{\text{mass of water in distillate}} = \frac{\text{mol. wt. of aniline} \times \text{v.p. of aniline at b.p.}}{\text{mol. wt. of water} \times \text{v.p. of water at b.p.}}$$

$$\therefore \frac{23.2}{76.8} = \frac{M}{18} \times \frac{5.6}{95.7}$$

so

$$M = \frac{23.2 \times 18 \times 95.7}{76.8 \times 5.6} = 93$$

and the molecular weight of aniline is therefore 93.

6.14. The Distribution or Partition Law. If to a mixture of two immiscible liquids a substance which is soluble in both is added in varying amounts, then experiment shows that it distributes itself in such a way that the ratio of its concentrations in the two layers is constant at constant temperature,

i.e.

$$\frac{\text{concn. of solute in liquid A}}{\text{concn. of solute in liquid B}} = \text{constant, } k$$

The generalization, which is subject to certain limiting conditions, is known as the *distribution or partition law*; the constant k is called the *distribution* or *partition coefficient* of that system at that temperature. For example, if iodine is added to a mixture of benzene and water and the mixture is shaken until equilibrium is attained, then the concentration of iodine in the benzene layer is always about 400 times as great as that in the water at room temperature.

The distribution law only holds true if the solute is in the same molecular condition in both solvents. Thus if association takes place in one solvent and not in the other, or if the solute is dissociated to some extent in one solvent only, then the ratio of the concentrations is no longer constant. The quantity of solute added must not be so great that its solubility in either solvent is exceeded and equilibrium must be reached before the concentrations are compared. It is important to notice that the partition coefficient is given by the ratio of the *concentrations* of the solute and not the ratio of the amounts; this latter ratio varies, of course, with the relative volumes of the two liquids.

It should now be seen that Henry's law is really a special case of the distribution law in which the gaseous solute may be regarded as distributing itself between free space on the one hand and the liquid solvent on the other in a constant ratio at constant temperature.

If the solute has a different molecular weight in the two solvents, then a modified distribution law applies. For example, if the molecular weight

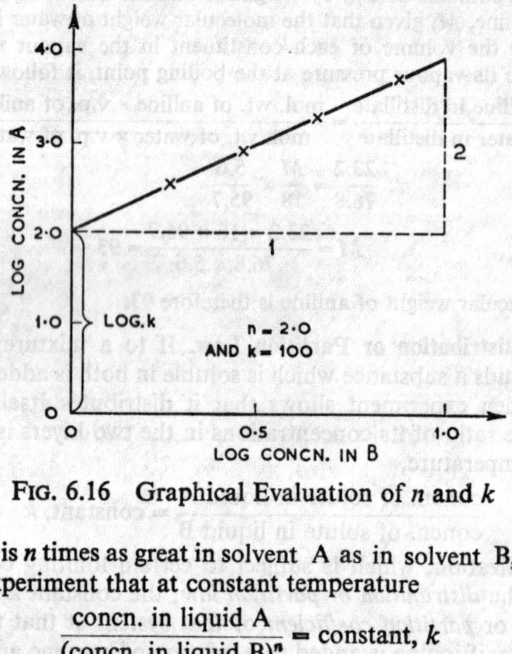

FIG. 6.16 Graphical Evaluation of *n* and *k*

of the solute is *n* times as great in solvent A as in solvent B, then it is found by experiment that at constant temperature

$$\frac{\text{concn. in liquid A}}{(\text{concn. in liquid B})^n} = \text{constant, } k$$

For example, acetic acid and benzoic acid are associated by hydrogen bonding (§ 14.7) into double molecules in benzene but not in water. This follows from the fact that for these systems

$$\frac{\text{concn. of acid in benzene}}{(\text{concn. of acid in water})^2} = \text{constant (at constant temperature).}$$

Thus by evaluating *n* in a given case the molecular weight of the solute in the two solvents can be compared. The best way of determining *n* is to plot a graph of the log of the concentration in liquid A against the log of the concentration in liquid B, as in Fig. 6.16, when the gradient of the straight line obtained will be equal to *n* and the intercept will be equal to log k, since the following relationship applies:

$$\text{log concn. in liquid A} = n \times \text{log concn. in liquid B} + \log k$$

$$\text{i.e.} \quad y = mx + c$$

If n is not an integer it indicates that the association or dissociation is only partial.

The distribution law is often applied industrially and in the laboratory to extract a solute from water. The general procedure is to shake the aqueous solution with another liquid which is immiscible with it but in which the solute is much more soluble, and then to separate the two layers. For example penicillin is extracted from a dilute aqueous solution with chloroform, and aniline is usually reclaimed from a mixture with water by using ether (see § 44.5). The Parkes process for recovering silver from lead is another application of the distribution law (see § 28.14). In all these cases it is more efficient to shake with several small volumes of the extracting solvent in turn, separating each time, than to use the same total volume all at once, as the following calculation shows:

Example: Consider the effect of shaking one dm³ of an aqueous solution containing 11 gramme of solute X with (*a*) 100 cm³ of ether at one time, (*b*) two successive volumes of 50 cm³ of ether, given that the partition coefficient for the system is 100 at that temperature.

(*a*)
$$\frac{\text{concn. of X in ether}}{\text{concn. of X in water}} = \frac{100}{1}, \text{ given}$$

$$\therefore \frac{\text{mass of X in ether layer}}{\text{mass of X in water layer}} = \frac{100}{1} \times \frac{\text{volume of ether}}{\text{volume of water}}$$

$$= \frac{100}{1} \times \frac{100}{1000} = \frac{10}{1}$$

Thus $\frac{10}{11}$ of X is extracted (i.e. 10 gramme) and $\frac{1}{11}$ (i.e. 1 gramme) of X remains in the water layer.

(*b*)
$$\frac{\text{mass of X in ether layer}}{\text{mass of X in water layer}} = \frac{\text{concn. of X in ether}}{\text{concn. of X in water}} \times \frac{\text{volume of ether}}{\text{volume of water}}$$

$$= \frac{100}{1} \times \frac{50}{1000} = \frac{5}{1}$$

Thus $\frac{5}{6}$ of X is extracted each time and $\frac{1}{6}$ remains in the aqueous layer. In the first extraction, $\frac{5}{6} \times 11$ (i.e. 9.167 gramme) are extracted and $\frac{1}{6} \times 11$ (i.e. 1.833 gramme) remain. In the second extraction $\frac{5}{6}$ of the remaining X is extracted (i.e. $\frac{5}{6} \times 1.833 = 1.528$ gramme) whilst $\frac{1}{6}$ (i.e. $\frac{1}{6} \times 1.833 = 0.305$ gramme) remains in the water layer. Thus the mass of X remaining in the water was 1.00 gramme when shaken with a single volume of 100 cm³ of ether and only 0.305 gramme when shaken with two successive portions of 50 cm³.

6.15. Eutectic Mixtures. On slowly cooling a pure liquid a temperature is eventually reached at which solid begins to form. If the cooling is continued the temperature remains constant until all the liquid has solidified, owing to the evolution of latent heat of fusion, and then begins to fall slowly again. Thus if the temperature is plotted against the time, a cooling curve like that in Fig. 4.2 is obtained. In practice supercooling (§ 4.3) often occurs so that the cooling curve shows a dip and hump as shown in the diagram. If the cooling curves of various mixtures of two miscible liquids A and B which do not form solid

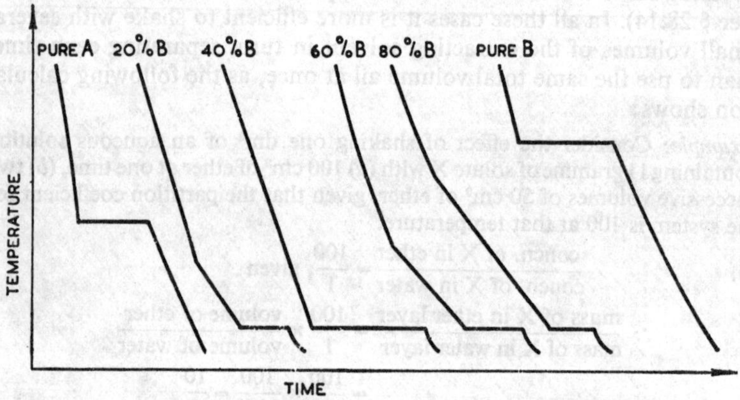

FIG. 6.17 Typical Cooling Curves of A, B, and A + B

solutions in each other are plotted in this way, it is possible to determine the temperature at which solid begins to separate from each mixture by the position of the change in the gradient of each curve. Typical curves are shown in Fig. 6.17; from these it is possible to construct an equilibrium diagram of the type shown in Fig. 6.18 in which freezing points measured at atmospheric pressure are plotted against composition. In practice the lines AE and BE are usually curved (not straight as shown in the figure) because the two constituents of the mixture are slightly soluble in each other in the solid state.

This diagram shows how the presence of each substance in the mixture lowers the freezing point of the other. For example, the point A is the freezing point of pure A and the line AE represents the progressive lowering of that freezing point as the proportion of B increases. Similarly the freezing point of B follows the course BE as the proportion of

A increases. These two lines intersect at the point E which is known as the *eutectic point*. It is the lowest temperature at that pressure at which a mixture of A and B can exist in the liquid state. The mixture corresponding to the point E is known as the *eutectic mixture.*

The lines AE and BE can also be regarded in another way. Consider a mixture M containing a greater proportion of A than the eutectic mixture. When this liquid is cooled it begins to deposit crystals of pure A at temperature T (see Fig. 6.18), so that at this temperature solid A and liquid mixture of composition M are in equilibrium. On further cooling more and more pure A is deposited and the remaining liquid becomes progressively poorer in A (and therefore relatively richer in B)

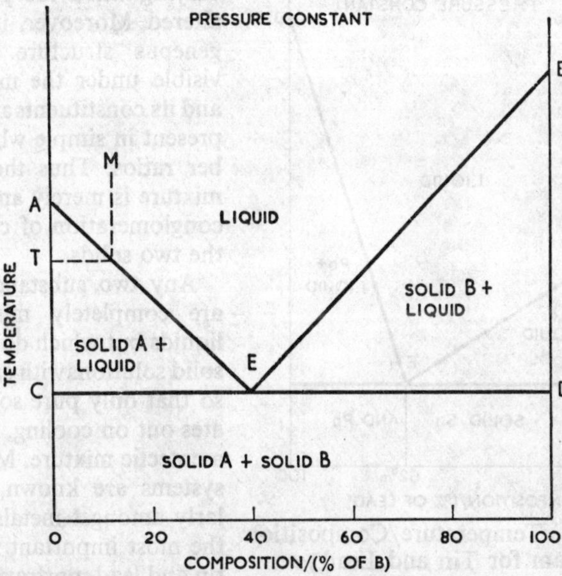

FIG. 6.18 Temperature/Composition Diagram for A and B

until eventually it has the composition of the eutectic mixture, when it all solidifies at the eutectic temperature. Thus AE represents the set of temperatures and corresponding compositions at which the solid A and the liquid phase are in equilibrium. Similarly a mixture richer in B than the eutectic mixture will deposit pure B on cooling until the eutectic composition is reached when it will all solidify. Along BE, therefore, solid B is in equilibrium with the melt. The point E is the only one at that pressure at which solid A, solid B, and the liquid mixture are all three in equilibrium with each other.

If a liquid mixture of the eutectic composition is cooled no solid appears until the eutectic temperature is reached, when the whole mixture freezes sharply depositing both solids simultaneously. This explains why the cooling curve corresponding to 40% of B in Fig. 6.17 has only a single flat portion whereas the cooling curves of the other mixtures show two changes in gradient, one when the first solid begins to separate and the other at the eutectic temperature. Since this eutectic mixture has (at any one pressure) a constant composition and a sharp constant melting point it might, at first sight, be thought to be a compound. That this is not so is clear from the fact that its composition depends upon the pressure, varying when the pressure is altered. Moreover, its heterogeneous structure is often visible under the microscope and its constituents are seldom present in simple whole-number ratios. Thus the eutectic mixture is merely an intimate conglomeration of crystals of the two solids.

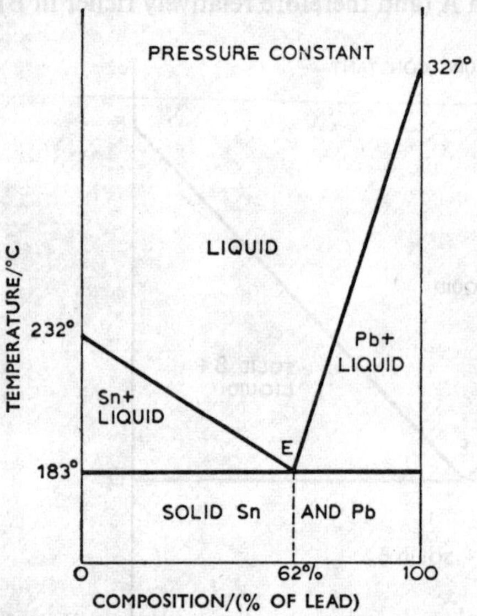

FIG. 6.19 Temperature/Composition Diagram for Tin and Lead

Any two substances which are completely miscible as liquids but which do not form solid solutions with each other, so that only pure solid separates out on cooling, can form a eutectic mixture. Many such systems are known, particularly amongst metals. One of the most important is that of tin and lead, portrayed in Fig. 6.19. The melting points of pure tin and pure lead are 232°C and 327°C respectively, but they form a eutectic mixture with the composition 62% lead and 38% tin melting at only 183°C. It is the low melting points of mixtures of tin and lead which account for the extensive use of these two metals in solders. Other examples are zinc (m.p. 419°C) and cadmium (m.p. 321°C) which form a eutectic mixture melting at 270°C, o-nitrophenol (m.p. 44°C) and p-toluidine (m.p. 43°C) which give one melting at 15.6°C, and naphthalene (m.p. 80°C) and p-nitrotoluene (m.p. 50°C) which form a eutectic mixture melting at 30°C. The eutectic mixture formed by iron and carbon (melting at 1130°C and containing 4.3% of carbon) is of great industrial importance because it largely determines the carbon content of pig iron and the operating temperature of the blast furnace.

Many inorganic salts (e.g. sodium chloride, potassium chloride, potassium iodide) form eutectic mixtures with water. It is difficult to study these systems over the complete range of composition at atmospheric pressure because the water tends to boil, but a typical equilibrium diagram is shown in Fig. 6.20. The line AC represents the freezing points of progressively stronger solutions of the salt in water, pure ice being deposited at these temperatures. The line BC is the solubility curve of the salt in water, i.e. the temperature at which the pure salt is in equilibrium in the solid state with its saturated solution. The two curves intersect at C, the eutectic point, at which the solid salt, ice, and the solution are all three in equilibrium at atmospheric pressure. This is, of course, the lowest temperature at which the salt can exist in solution. In systems involving a salt and water it is often known as the *cryohydric point* and the mixture of that particular composition as a *cryohydrate*.

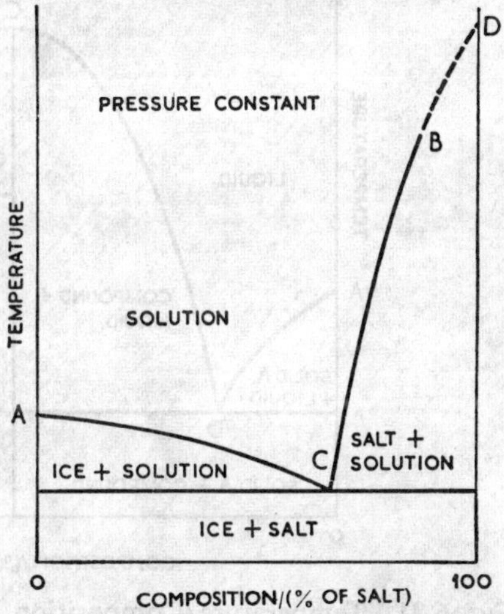

FIG. 6.20 Temperature/Composition Diagram for a Salt and Water

The use of a mixture of salt and ice as a freezing mixture depends upon the above principle. When the salt is added some dissolves in the water giving a system containing solid salt, ice, and salt solution. Since such a system is only stable at the eutectic point, ice will melt and salt will dissolve (both endothermic processes) until the temperature falls to the eutectic temperature, or until one of the solids is all used up. In this way if sodium chloride, the cheapest salt, is used, a cooling mixture of −21°C can be obtained.

Referring again to Fig. 6.20, the point D (where CB extrapolated intersects the axis) represents the melting point of the pure salt, and the dashed curve DB traces the lowering of this melting point brought about by adding water.

In many equilibrium diagrams two eutectic points occur separated by a distinct hump, as shown in Fig. 6.21. This indicates the existence of a compound of the two components A and B; its composition is that of the maximum point, C. Along DC and CE this pure compound separates from the melt, just as pure A separates along AD and pure B along BE. The formation of compounds is very common amongst metals and in inorganic salt/water systems. Ferric chloride, for example, gives an equilibrium diagram with four maxima corresponding to different hydrates.

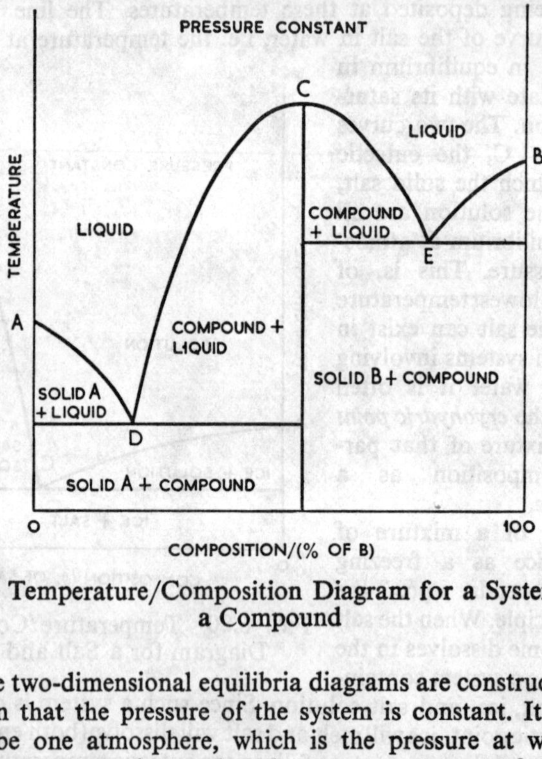

FIG. 6.21 Temperature/Composition Diagram for a System forming a Compound

All these two-dimensional equilibria diagrams are constructed on the assumption that the pressure of the system is constant. It is usually taken to be one atmosphere, which is the pressure at which most measurements are made, so that the vapour pressure of the various solids can be neglected and the diagram then refers to what is known as a *condensed system*. If it is desired to incorporate the effect of pressure upon the various equilibria, it is necessary to construct three-dimensional equilibrium diagrams, which are beyond the scope of this book.

6.16. The Vapour Pressure over Salt Hydrates. The way in which the pressure of water vapour above salt hydrates varies as their composition changes is well illustrated by considering the example of copper sulphate at 25°C. The results are embodied in Fig. 6.22, which is another condensed system diagram based this time upon measurements at constant temperature. Let us start with a concentrated solution of copper sulphate and consider the changes in water vapour pressure as water is progressively removed from the system. At first the vapour pressure falls along the curve AB as the solution becomes more and more concentrated, but when a saturated solution is obtained the vapour pressure remains at a constant value of 3070 N m^{-2} despite further removal of water. At this pressure crystals of the pentahydrate

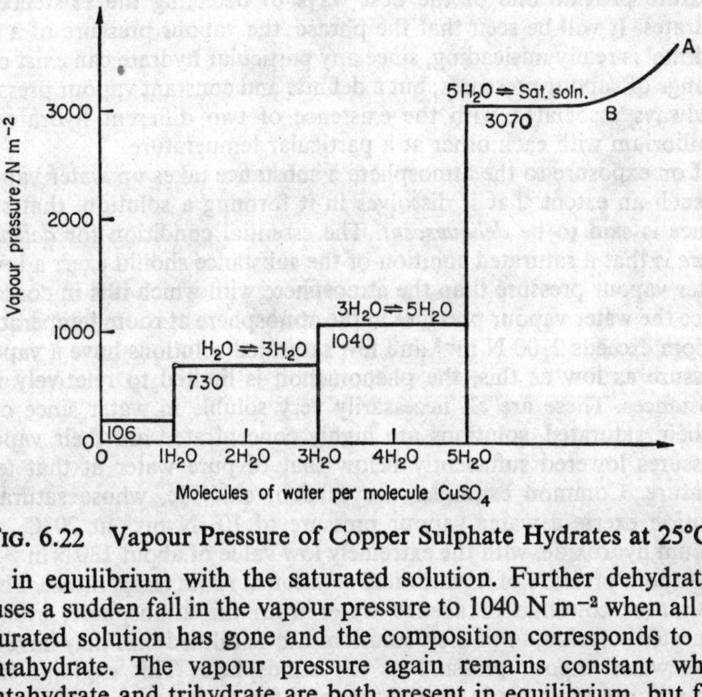

FIG. 6.22 Vapour Pressure of Copper Sulphate Hydrates at 25°C

are in equilibrium with the saturated solution. Further dehydration causes a sudden fall in the vapour pressure to 1040 N m^{-2} when all the saturated solution has gone and the composition corresponds to the pentahydrate. The vapour pressure again remains constant whilst pentahydrate and trihydrate are both present in equilibrium, but falls sharply to 730 N m^{-2} when the last pentahydrate disappears. After a further constancy whilst the trihydrate and monohydrate exist together, the vapour pressure falls again to about 106 N m^{-2} when the composition is that of the monohydrate and stays at this value as long as any water remains in the system.

These changes, all of which are reversible, can be represented by the following equations:

$$\underset{}{CuSO_4 \text{ soln.}} \rightleftharpoons \underset{\text{pentahydrate}}{CuSO_4.5H_2O} + nH_2O$$

$$CuSO_4.5H_2O \rightleftharpoons \underset{\text{trihydrate}}{CuSO_4.3H_2O} + 2H_2O$$

$$CuSO_4.3H_2O \rightleftharpoons \underset{\text{monohydrate}}{CuSO_4.H_2O} + 2H_2O$$

$$CuSO_4.H_2O \rightleftharpoons \underset{\text{anhydrous}}{CuSO_4} + H_2O$$

Similar step-like changes in vapour pressure occur in systems involving other salt hydrates. Indeed these sharp changes in vapour pressure provide one of the best ways of detecting the existence of hydrates. It will be seen that the phrase 'the vapour pressure of a salt hydrate' is really misleading, since any particular hydrate can exist over a range of vapour pressures, but a definite and constant vapour pressure is always associated with the existence of two different hydrates in equilibrium with each other at a particular temperature.

If on exposure to the atmosphere a substance takes up water vapour to such an extent that it dissolves in it forming a solution, that substance is said to be *deliquescent*. The essential condition for *deliquescence* is that a saturated solution of the substance should exert a lower water vapour pressure than the atmosphere with which it is in contact. Since the water vapour pressure in the atmosphere at room temperature seldom exceeds 2100 N m^{-2} and few saturated solutions have a vapour pressure as low as this, the phenomenon is limited to relatively few substances. These are all necessarily very soluble in water since only if their saturated solutions are highly concentrated are their vapour pressures lowered sufficiently below that of pure water at that temperature. Common examples are calcium chloride, whose saturated solution exerts a water vapour pressure of 10^3 N m^{-2} at 20°C, and sodium hydroxide, with the extremely low value of about 130 N m^{-2} for the vapour pressure of its saturated solution at room temperature. These substances continue to absorb water from the atmosphere, even in saturated solution, until their solutions are so diluted that they have the same water vapour pressure as the atmosphere. This explains their frequent use as solids in desiccators and for drying gases and liquids, although two other substances, phosphorus pentoxide and magnesium perchlorate (§ 25.8), which are *hygroscopic* (i.e. absorb moisture from the atmosphere) but not deliquescent, are more efficient.

A substance which, on exposure to the atmosphere, loses water and changes into a lower hydrate or an anhydrous form is said to be *efflorescent*. *Efflorescence* only occurs when the substance exerts a higher water vapour pressure than the atmosphere around it. Well-known examples of efflorescent substances are sodium carbonate decahydrate, $Na_2CO_3.10H_2O$, and sodium sulphate decahydrate,

$Na_2SO_4.10H_2O$, which change into a monohydrate and an anhydrous form respectively on standing in air for some days. For example a mixture of the decahydrate and monohydrate of sodium carbonate exerts a water vapour pressure of about 2400 N m^{-2} at 25°C, when the water vapour pressure in the atmosphere is usually about 2000 N m^{-2}, so efflorescence generally occurs. The process is usually accompanied by crumbling as the original crystal structure changes into that of the lower hydrate.

In any consideration of deliquescence and efflorescence the importance of the humidity of the atmosphere must be stressed since both processes depend upon the relative values of the water vapour pressure in the hydrate system and in the atmosphere. For example, referring again to Fig. 6.22, if the pressure of water vapour in the atmosphere exceeds 3070 N m^{-2} at 25°C (which is very rarely the case), then the pentahydrate of copper sulphate will deliquesce, and if it falls below 1040 N m^{-2} at 25°C (as it does in certain parts of the world), then the pentahydrate will effloresce, but between these two extremes crystals of the pentahydrate are perfectly stable, showing no tendency to lose or gain water from the atmosphere.

6.17. The Phase Rule. The word phase has already been discussed in § 6.1, but before stating this rule two other technical terms must be explained:

1. *The number of components in a system* is the least number of different substances which must be specified to be present in order to describe completely the composition of every phase. For example, in any system consisting of a single substance, only one component is involved even if that substance is present in two or three different phases at the same time. A system containing two different substances which do not react with each other is regarded as containing two components, but a system in which three different compounds are in chemical equilibrium with each other (e.g. $A + B \rightleftharpoons C$) consists of only two components, however, because once any two of the constituents A, B, or C are specified the composition of every phase in the system is settled.

2. *The number of degrees of freedom of a system* is the number of factors such as pressure, temperature, and concentration of the various components which can be varied independently of each other without bringing about a change in the number of phases. For example, in a system consisting of a liquid and its vapour in equilibrium with each other, once one of the two variables temperature and pressure is fixed, the other is also settled. It follows that alteration of either of these factors alone will cause one of the phases to disappear and such a system is said to have only one degree of freedom.

We are now in a position to understand the rule, which was devised by Gibbs in 1876 and can be simply stated in the form:

$$F = C - P + 2$$

where F = the number of degrees of freedom of the system.
C = the number of components in the system at equilibrium.
P = the number of phases present at equilibrium.

As our first example let us consider once again the phase equilibrium diagram shown in Fig. 6.1. This refers to a system containing only one component, so $C = 1$. At the triple point three phases are present, solid, liquid, and vapour, so under these particular conditions of temperature and pressure $P = 3$. Applying the phase rule,

$$F = 1 - 3 + 2 = 0$$

so there are no degrees of freedom in the particular system corresponding to the triple point O. This means that we can alter neither the temperature nor the pressure without also altering the number of phases present, which is confirmed by experiment. Such a system is said to be *invariant*.

A system in which a gas is in equilibrium with its saturated solution in a liquid provides our second example. Here we have two components and two phases, since the solution of the gas in the liquid is itself a single phase, so C and P are both equal to 2. Thus, applying the phase rule:

$$F = 2 - 2 + 2 = 2$$

and this system is *bivariant*, i.e. it has two degrees of freedom the temperature and the pressure. Both of these factors must be specified in order to define the system completely.

For our third example let us consider a mixture of two immiscible liquids A and B. Here there are two components A and B and three phases, liquid A, liquid B, and vapour mixture, all in equilibrium with each other. Thus applying the phase rule to this system, $C = 2$, $P = 3$, and

$$F = 2 - 3 + 2 = 1$$

which means that the system is *univariant*, i.e. it has only one degree of freedom. Thus in this system if we fix the temperature, for example, the vapour pressure exerted above the immiscible liquids and the composition of the vapour phase will both be constant irrespective of the relative amounts of each liquid taken. These predictions are completely fulfilled in practice.

The application of the phase rule to heterogeneous chemical equilibria (i.e. ones in which the reactants and products are not all in the same phase) can be illustrated by considering the thermal dissociation of calcium carbonate:

$$\underset{\text{(solid)}}{CaCO_3} \rightleftharpoons \underset{\text{(solid)}}{CaO} + \underset{\text{(gas)}}{CO_2}.$$

In this system there are two solid phases and one gaseous phase, making a total of three. The number of components, however, is only two, because the nature of the third substance present can always be deduced from a knowledge of the other two. Thus $P = 3$ and $C = 2$ and

$$F = 2 - 3 + 2 = 1$$

so the system has only one degree of freedom. This means that once the temperature of the system is fixed, the equilibrium pressure will be settled also, and the carbon dioxide will exert a characteristic partial pressure at any particular temperature. This prediction, which is fully borne out by experiment, can also be deduced by applying the law of mass action to the equilibrium (see case 4 of § 7.8).

The wide applicability of the phase rule should be apparent from the examples discussed above. Its great importance lies in its ability to predict how many variables will affect a given system in equilibrium. By its help we are able to decide what type of equilibrium can occur in heterogeneous systems and whether it is appropriate in any given case to apply Le Chatelier's principle (§ 7.9) to the system.

7

Kinetics and Chemical Equilibrium

7.1. Introduction. A chemical equation provides information on the ultimate products of a reaction and on the masses obtained from given amounts of reactants, but it conveys nothing about the speed of the reaction. This may vary widely from very slow, as in corrosion, for example, to very fast as in explosive reactions. The study of the rate of a chemical reaction and the factors upon which it depends is called chemical kinetics. These factors are temperature, presence or absence of a catalyst, intimacy of mixing of the reactants, intensity of ultraviolet light, and the concentration of the reactants. Each will be considered in turn.

7.2. Temperature and Rate of Reaction. Reaction rates generally increase with rising temperature, approximately doubling for each rise of 10 K. This means that a reaction proceeds about a million times faster if the temperature is raised by 200 K. This is the principal reason for heating substances undergoing reaction and for operating industrial processes at high temperature. The reverse effect of cooling a reaction in order to slow it down is sometimes useful; examples are given in reaction 2 of § 7.8 and in § 18.4 (process b).

An explanation of the pronounced effect which temperature exerts upon the rate of a reaction is given in § 7.12.

7.3. Catalysis. A substance which alters the speed of a chemical reaction but remains unchanged chemically and in mass at the end of the reaction is called a *catalyst* (positive or negative according to whether it speeds or slows it). Most examples of *catalysis*, the term used to describe the operation of a catalyst, are positive, but negative catalysts include organic compounds added to solutions of hydrogen peroxide to retard their decomposition during storage and lead tetraethyl which is added to petrol to prevent 'knocking' (§ 33.10).

CHARACTERISTICS OF CATALYSTS: Catalytic reactions are very numerous and whilst the catalysts themselves vary widely in type and chemical composition, most have certain general features in common. They are usually specific, i.e. efficient only for one particular reaction. The most striking examples of this are the enzymes, complicated organic substances which catalyse many of the chemical reactions occurring in digestion. Nevertheless certain metals such as platinum and nickel catalyse a wide range of reactions and water vapour appears to act as a catalyst in many gaseous reactions. Again, benzoyl peroxide (§ 47.5) is used as a catalyst in the polymerization of several unsaturated organic compounds, as we shall see in later chapters.

Another characteristic of catalysts is that even minute amounts may enormously increase the rate of reaction. For example, mere traces of copper ions greatly increase the rate at which a bisulphite is oxidized in solution and extremely low concentrations of cobalt ions accelerate the decomposition of hypochlorites. On the other hand, in the Friedel-Crafts reaction (§ 46.2) the anhydrous aluminium chloride catalyst must be present in amounts comparable to the reactants to be effective.

Catalysts do not affect the position of equilibrium in a reversible reaction and hence do not affect the yield of product obtained under given conditions. In the presence of a catalyst the forward and backward reactions are accelerated in the same proportion, so that although the position of equilbrium is unchanged, the state of equilibrium is achieved much more quickly.

Experiments have shown that it is often possible to increase the activity of a catalyst considerably by adding small amounts of another substance, e.g. in the Haber process for synthesizing ammonia (§ 22.4), where molybdenum and the oxides of aluminium and potassium are added to the catalyst of iron. Such substances are called *promoters*. Conversely traces of certain substances destroy the activity of some catalysts; these are known as *catalyst poisons*. For example, arsenic poisons the platinum catalyst used in the contact process for making sulphuric acid (§ 24.11) and consequently all traces of arsenic compounds have to be removed from the reacting gases before use.

Where two substances can react together in more than one way, the choice of catalyst may decide the course of the reaction, presumably by so accelerating one particular reaction that it predominates over the others. For example, carbon monoxide and hydrogen react together at high temperatures and pressures to give methyl alcohol in the presence of zinc oxide catalyst, methane and other volatile hydrocarbons with a catalyst of nickel, and a mixture of higher paraffins with cobalt.

Occasionally one of the products is capable of catalysing the reaction by which it is made. This is called *auto-catalysis*. When an oxalate reacts with acidified potassium permanganate solution, the manganese(II) ions resulting from the reduction of the permanganate catalyse this reaction so that it proceeds at a convenient speed at a lower temperature than the 60°C needed to start it.

Many hydrolyses are catalysed by hydrogen ions or hydroxyl ions in solution, the rate of the reaction being directly proportional to the concentration of the ion concerned. For this reason such hydrolyses are often brought about by solutions of acids or alkalis rather than by water alone. Saponification (§ 38.2) is a good example.

THEORIES OF CATALYSIS: Two main explanations of catalysis have been put forward. *Heterogeneous* catalysis, where reactants and catalyst are in different physical states, as, for example, in the catalysis of gaseous reactions by metals, has been regarded as an example of *adsorption*, i.e. condensation of the reacting gases upon the surface of the metal in a layer perhaps only a few molecules thick (§ 12.9). Presumably the greatly increased concentration of the reactants leads to rapid reaction, the products escaping and being replaced by more reactants. This theory has been elaborated by suggesting that the catalyst has on its surface certain active points where adsorption occurs most readily, and that in this process the molecules of the reactants are activated in some way so that they react together more rapidly (see § 7.12).

The other principal theory of catalysis postulates the continuous formation and decomposition of unstable intermediate compounds. This provides a route whereby the final products are obtained more quickly than by direct combination, presumably because the intermediate reactions are comparatively rapid. The theory finds support from the discovery of traces of intermediate compounds remaining in the mixture at the end of the reaction, e.g. small amounts of purple permanganate have been found when manganese dioxide has been used to catalyse the thermal decomposition of potassium chlorate, and traces of chlorine have been detected in the oxygen produced. The lead chamber process for making sulphuric acid (§ 24.11) is another example where an intermediate compound, nitrosyl sulphuric acid, has been isolated from the reaction mixture. This type of catalysis is discussed in terms of energy of activation in § 7.12.

7.4. Intimacy of Reactants. The rate of a reaction is increased by bringing the reactants into more intimate contact. This is particularly marked in reactions between solids where the state of sub-division and the thoroughness of mixing are important factors. When metals are finely powdered they are often much more reactive owing to their greatly increased surface area, e.g. powdered aluminium, used in the thermite reaction (§ 19.8), is a powerful reducing agent, and finely powdered lead inflames spontaneously in air.

Liquids of different densities often react together only slowly, especially when they are immiscible, unless continually stirred or shaken. Similarly, it is found worth while in some industrial processes to use elaborate devices to ensure thorough mixing of gases with liquids, e.g. in the Solvay process for making sodium carbonate (§ 17.8).

The catalysts used in heterogeneous reactions, such as hydrogenations, are often prepared by chemical means in a special condition of extreme sub-division, so that the largest possible surface area of the metal is exposed to the reacting gases. Similarly, it is the large surface area of metals in the colloidal condition which accounts for their efficiency as catalysts.

7.5. Intensity of Ultra-Violet Light. Certain reactions are greatly accelerated by intense ultra-violet light. For example, hydrogen and chlorine combine only slowly in diffuse light at room temperature, even in the presence of a charcoal catalyst, but react together explosively when exposed to ultra-violet light. Molecules of the reactants absorb the light energy, thereby becoming activated, and react together rapidly in a series of chain reactions. This effect is considered in more detail in § 7.15 and § 7.17.

7.6. Concentration of Reactants. The effect of concentration upon reaction rate is summarized in the *law of mass action* put forward by Guldberg and Waage:

At constant temperature the rate of a reaction is proportional to the active masses of each of the reactants.

The term *active mass* is a special one best interpreted as the concentration of a substance raised to the appropriate power, which is the number of molecules of it written in the equation for that reaction.* The concentration can be expressed in mole dm^{-3} or, where gaseous, as the partial pressure of the substance in the reaction mixture.

Suppose *m* molecules of substance A react with *n* molecules of substance B and let the rate of reaction at any given temperature be *r*, thus:

$$mA + nB \xrightarrow{r}$$

then $r \propto (\text{concentration of A})^m$
and $r \propto (\text{concentration of B})^n$
so $r \propto \{(\text{concentration of A})^m \times (\text{concentration of B})^n\}$
i.e. $r \propto [A]^m[B]^n$, where [X] is a symbol standing for the molar concentration of the substance X
or $r \propto p_A{}^m p_B{}^n$, where p_X is a symbol standing for the partial pressure of X.
$\therefore$ $r = k.[A]^m[B]^n$, where *k* is a constant, called the ***velocity constant*** of that reaction at that temperature. It embodies all the factors, other

* This statement is qualified in § 7.11, where order of reaction is considered.

F

than concentration, which affect reaction rate and it is only constant if these factors are not changed. The velocity constant at any temperature is equal to the rate at which the reaction proceeds when the concentrations of each of the reactants is unity, since under these conditions r is equal to k.

7.7. Reversible Reactions and Chemical Equilibrium. Many reactions are reversible, i.e. they can proceed in both directions under suitable conditions. If substances A and B are brought together they will react to give products C and D, for example, and if we start with C and D under suitable conditions they will react together to form A and B. In practice, if the reactants in a reversible reaction are left in contact an equilibrium will eventually be set up between the two opposing reactions giving a mixture of reactants and products.

It is instructive to consider the steps by which this state of equilibrium is reached. If we start by mixing A and B, then reaction between them will produce quantities of C and D at a rate proportional to the active masses of A and B, which will, at first, be high. Similarly the C and D formed will combine to give A and B at a rate proportional to the active masses of C and D, which will, at first, be very low. So at the beginning the forward reaction will be much faster than the backward reaction and there will be a net shift from the left side to the right side of the equation representing the reaction. As this process continues, the concentrations of A and B will fall, and those of C and D will rise, and gradually the disparity between the rates of the two opposing reactions will diminish, until eventually when equilibrium is reached the reactions proceed in opposite directions at exactly the same speed.

This condition of equilibrium appears, at first sight, to be one of stagnation because no change occurs in the concentrations of the substances in the reaction mixture once it has been reached. But consideration will show that the condition is really a dynamic one, and no net change is noticed because the products of each reaction are being used up as fast as they are being formed.

Now let us apply the law of mass action to the reversible reaction represented by the following equation, assuming that the temperature is constant throughout:

$$m\mathrm{A} + n\mathrm{B} \underset{r_2}{\overset{r_1}{\rightleftharpoons}} p\mathrm{C} + q\mathrm{D}$$

rate of left-to-right reaction, $r_1 = k_1[\mathrm{A}]^m[\mathrm{B}]^n$
rate of right-to-left reaction, $r_2 = k_2[\mathrm{C}]^p[\mathrm{D}]^q$

where k_1 and k_2 are the velocity constants of the forward and backward reactions respectively.

At equilibrium, $r_1 = r_2$

$$\therefore \ k_1[A]^m[B]^n = k_2[C]^p[D]^q$$

$$\therefore \ \frac{[C]^p[D]^q}{[A]^m[B]^n} = \frac{k_1}{k_2} = K$$

where K is a constant called the *equilibrium constant* of that reaction at that temperature.

It is important to notice that K is given by the ratio of the two velocity constants k_1 and k_2. Since these usually alter to a different degree when temperature is varied, it follows that K will vary with temperature also, but the addition of a catalyst to the system will affect the values of k_1 and k_2 in the same proportion and so not alter the value of K. This agrees with experimental observation.

To avoid confusion it is conventional to write the expression for K with the concentrations of the products (the substances on the right-hand side of the equation as written) in the numerator and with the concentrations of the reactants in the denominator. If this convention is ignored the value of K obtained is the reciprocal of the accepted value.

7.8. Application of the Law of Mass Action to Chemical Equilibria. This is illustrated by the following cases:

(1) REACTANTS AND PRODUCTS IN THE LIQUID STATE:

$$\underset{\text{ethyl alcohol}}{C_2H_5OH} + \underset{\text{acetic acid}}{CH_3.COOH} \rightleftharpoons \underset{\text{ethyl acetate}}{CH_3.COOC_2H_5} + \underset{\text{water}}{H_2O}$$

Let a mole of the alcohol react with b mole of the acid in volume V, and let x mole of ethyl acetate be present when the state of equilibrium is reached.

	C_2H_5OH +	$CH_3.COOH$ $\rightleftharpoons$	$CH_3.COOC_2H_5$ +	H_2O
Initial amounts (mole):	a	b	0	0
Equilibrium amounts (mole):	$a-x$	$b-x$	x	x
„ concn. (mole dm^{-3}):	$\frac{(a-x)}{V}$	$\frac{(b-x)}{V}$	$\frac{x}{V}$	$\frac{x}{V}$

Now $$\frac{[CH_3.COOC_2H_5].[H_2O]}{[C_2H_5OH].[CH_3COOH]} = K$$

Taking the molar concentrations at equilibrium as the active masses, we get:

$$\frac{\frac{x}{V} \times \frac{x}{V}}{\frac{(a-x)}{V} \times \frac{(b-x)}{V}} = K = \frac{x^2}{(a-x)(b-x)}$$

If a, b, and x are given, then K can be calculated for that system at that temperature. Once we know K we can calculate the concentrations present in the equilibrium mixture at that temperature starting from any given amounts of reactants and products.

Example: One mole of ethyl alcohol was added to one mole of acetic acid and kept at 25°C; at equilibrium there were $\frac{1}{3}$ mole of each of these reactants remaining and $\frac{2}{3}$ mole of each of the products. How many gramme of ethyl acetate will be present in the equilibrium mixture when 138 gramme of ethyl alcohol are added to 120 gramme of acetic acid and left to reach equilibrium at 25°C?

Now $$K = \frac{\frac{2}{3} \times \frac{2}{3}}{(1 - \frac{2}{3})(1 - \frac{2}{3})} = 4$$

Mol. wt. of $C_2H_5OH = 46$ $\quad \therefore a = 3$ mole

Mol. wt. of $CH_3COOH = 60$ $\quad \therefore b = 2$ mole

$$K = \frac{x^2}{(a - x)(b - x)} \quad \therefore 4 = \frac{x^2}{(3 - x)(2 - x)}$$

$\therefore 4(6 - 5x + x^2) = x^2$

$\therefore 3x^2 - 20x + 24 = 0$

Solving this quadratic equation we get: $x = 5.1$ or $x = 1.57$.
The higher value is rejected as it is clearly impossible to produce over 5 mole of ethyl acetate from the quantities of reactants provided.

$\therefore$ There are 1.57 mole of ethyl acetate present at equilibrium. Mol. wt. of $CH_3.COOC_2H_5 = 88$.

$\therefore$ There are 1.57 × 88, i.e. 138.2 gramme of ethyl acetate in the equilibrium mixture.

This example demonstrates several points often overlooked by students approaching this type of problem for the first time. Molar concentrations are used in the expression for K, so data quoted in gramme has to be converted to mole first. Two values of the unknown are provided by the quadratic equation, but usually one of these can be rejected outright as being incompatible with the original data. The final result should always be stated in words and should provide the information required by the question, not just left in the form $x = 1.57$.

(2) Reactants and Products are Gaseous:

	H_2 hydrogen	+	I_2 iodine	$\rightleftharpoons$	2HI hydrogen iodide
Initial amounts (mole)	a		b		0
Equilibrium amounts (mole):	$(a - x)$		$(b - x)$		$2x$
Equilibrium concn. (mole dm^{-3}):	$\frac{(a - x)}{V}$		$\frac{(b - x)}{V}$		$\frac{2x}{V}$

Applying the law of mass action:

$$\frac{[HI]^2}{[H_2][I_2]} = K_c^* = \frac{\left(\frac{2x}{V}\right)^2}{\frac{(a-x)}{V} \times \frac{(b-x)}{V}} = \frac{4x^2}{(a-x)(b-x)}$$

It will be noticed that in this example, as in the previous one, the factor V cancelled and did not appear in the final expression for K. It must not be assumed that this happens in every case, but it does occur whenever the total number of molecules of reactants and products are equal, as in these two equations. In such reactions the position of equilibrium is independent of pressure and the composition of the equilibrium mixture is not affected by changes in pressure, a fact confirmed by experimental observation.

Alternatively, using partial pressures in this example instead of molar concentrations, we get:

	H_2	+	I_2	$\rightleftharpoons$	2HI
Initial amounts (mole):	a		b		0
Equilibrium amounts (mole):	$a-x$		$b-x$		$2x$

Now the partial pressure of a substance, X, is given by the expression:

$$p_x = \frac{\text{number of mole of X in system}}{\text{total number of mole in system}} \times P$$

where P is the external or total pressure of the system.

$$\therefore p_{HI} = \frac{2x}{(a+b)} \times P \quad p_{H_2} = \frac{(a-x)}{(a+b)} \times P \quad p_{I_2} = \frac{(b-x)}{(a+b)} \times P$$

because the total number of mole in the system

$$= a - x + b - x + 2x$$
$$= a + b$$

Applying the law of mass action:

$$K_p = \frac{(p_{HI})^2}{p_{H_2} \times p_{I_2}} = \frac{\left(\frac{2xP}{(a+b)}\right)^2}{\frac{(a-x)P}{(a+b)} \times \frac{(b-x)P}{(a+b)}} = \frac{4x^2}{(a-x)(b-x)}$$

The best way of verifying the law of mass action experimentally is to bring together varied but known concentrations of reactants at a given temperature, determine the composition of the equilibrium mixture in

* In applying the law of mass action to gaseous equilibria two values of the equilibrium constant may be obtained, depending on whether molar concentrations or partial pressures are taken as the active masses. These two constants are usually distinguished by using the symbols K_c and K_p respectively.

each case, and show the constancy in the value of K. This particular reaction is well suited to this treatment because the amount of iodine present can be determined volumetrically. The system is cooled suddenly to room temperature, which so lowers the rates of the forward and backward reactions that it prevents any substantial readjustment of the position of equilibrium before the analysis is complete.

(3) Another Gaseous Reaction:

$$\underset{\text{nitrogen}}{N_2} + \underset{\text{hydrogen}}{3H_2} \rightleftharpoons \underset{\text{ammonia}}{2NH_3}$$

Applying the law of mass action to this equilibrium we get:

$$\frac{[NH_3]^2}{[N_2][H_2]^3} = K$$

Suppose that 3 mole of hydrogen are mixed with 1 mole of nitrogen in volume V, and that x mole of nitrogen are used up at equilibrium.

	N_2	+	$3H_2$	$\rightleftharpoons$	$2NH_3$
Initial amounts (mole):	1		3		0
Equilibrium amounts (mole):	$1 - x$		$3 - 3x$		$2x$
Equilibrium concns. (mole dm^{-3}):	$\frac{(1-x)}{V}$		$\frac{(3-3x)}{V}$		$\frac{2x}{V}$

$$\therefore K_c = \frac{\left(\frac{2x}{V}\right)^2}{\left(\frac{1-x}{V}\right)\left(\frac{3-3x}{V}\right)^3} = \frac{4x^2V^2}{27(1-x)^4} = \frac{4x^2}{27(1-x)^4P^2}$$

where P = pressure of the system at equilibrium.

$$\therefore \frac{4x^2}{27(1-x)^4P^2} \text{ is constant at constant temperature.}$$

If x is small, which, in fact, it is under normal experimental conditions, then $(1 - x)^4$ is approximately equal to 1.

$$\therefore \frac{4x^2}{27P^2} \text{ is approximately constant}$$

$\therefore x^2$ is approximately proportional to P^2
$\therefore x$ „ „ „ P.

So we conclude that the yield of ammonia is approximately proportional to the applied pressure, which result is confirmed by experiment.

(4) A Thermal Dissociation: As explained in §§ 5.7 and 22.12, nitrogen

tetroxide when heated undergoes reversible dissociation into nitrogen dioxide in accordance with the equation:

$$N_2O_4 \rightleftharpoons 2NO_2$$

Suppose there are originally *a* mole of the tetroxide in volume *V*, and let *x* be the number of mole which are dissociated at any chosen temperature.

	N_2O_4	NO_2
Initial amounts (mole):	a	0
Equilibrium amounts (mole):	$a-x$	$2x$
Equilibrium concns. (mole dm^{-3}):	$\frac{a-x}{V}$	$\frac{2x}{V}$

Applying the law of mass action to this equilibrium we get:

$$\frac{[NO_2]^2}{[N_2O_4]} = K_c$$

$$\therefore \frac{\left(\frac{2x}{V}\right)^2}{\left(\frac{a-x}{V}\right)} = K_c = \frac{4x^2}{(a-x)V}$$

Thus at a constant temperature the extent of the dissociation *x* varies with the volume *V* and therefore with the pressure of the system. Since the term *V* appears in the denominator of the expression for the equilibrium constant, we conclude that the greater the pressure the smaller will be the dissociation, so a low pressure will encourage the dissociation and a high pressure reduce it.

(5) A HETEROGENEOUS EQUILIBRIUM: Each reaction considered so far has been an example of *homogeneous* equilibrium in which the reactants and products are all gaseous or all in the liquid state at the equilibrium temperature. In the following reaction, which is a *heterogeneous* equilibrium, the substances involved are not all in the same phase or physical state:

$CaCO_3$	$\rightleftharpoons$	CaO	+	CO_2
calcium carbonate (solid)		calcium oxide (solid)		carbon dioxide (gas)

Strictly speaking the law of mass action should not be applied to heterogeneous equilibria, but this difficulty is overcome by considering only the vapour phase here. The solids present do exert very small but finite vapour pressures, and these are taken as their active masses in the expression for *K*. Now the active mass of a solid at constant temperature must be constant because the vapour pressure of that solid is constant, being independent of the amount of solid present (§ 4.4).

Applying the law of mass action to this equilibrium and using partial pressures, we get:

$$\frac{p_{CaO} \times p_{CO_2}}{p_{CaCO_3}} = K_p$$

But, as explained above, p_{CaO} and p_{CaCO_3} are themselves constant at constant temperature.

$$\text{So } p_{CO_2} = K_p \times \frac{\text{a constant}}{\text{a constant}} = \text{a constant.}$$

This means that the partial pressure of carbon dioxide present at equilibrium is constant at constant temperature and does not depend upon the amounts of the two solids present. This remarkable prediction is fully confirmed by experiment, the characteristic equilibrium pressure

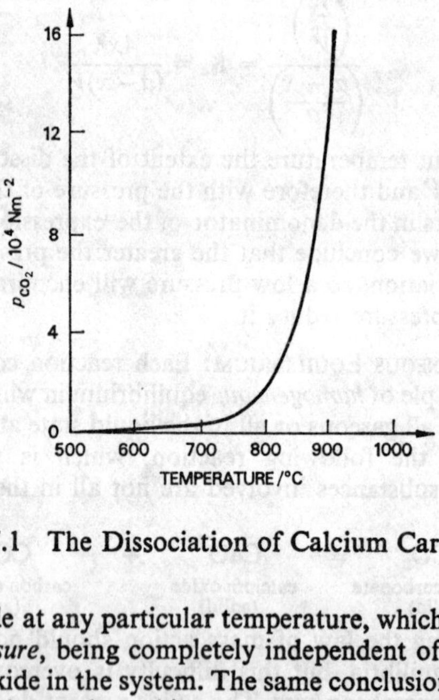

FIG. 7.1 The Dissociation of Calcium Carbonate

of carbon dioxide at any particular temperature, which is known as its *dissociation pressure*, being completely independent of the amounts of carbonate and oxide in the system. The same conclusion can be reached by applying the phase rule to this system, as demonstrated at the end of § 6.17. Fig. 7.1 shows how the extent of the dissociation varies with the temperature; above 900°C the partial pressure of carbon dioxide exceeds 101 325 N m^{-2} and the gas escapes rapidly into the atmosphere. The industrial importance of this reaction is described in § 18.14.

7.9. Le Chatelier's Principle. This was originally stated thus: *If a system is in equilibrium and a constraint be applied, the system will respond in such a way as to tend to annul the constraint.*

In this form the principle can be applied generally not only to chemical equilibria, but also to physical equilibria such as those existing between an undissolved solid and its saturated solution, or between a liquid and its vapour. Examples of this sort will be found in the appropriate chapters although, as emphasized in § 11.7, great care must be taken to ensure that the system to which the principle is applied really is in equilibrium. Here we are concerned with the application of the principle to chemical equilibria only, for which its meaning will be clearer if it is reworded thus:

If a system is in equilibrium and one of the factors pressure, temperature, or the concentration of a component, is altered, then the system responds in such a way as to oppose, or tend to oppose, the change that has been made.

Let us apply this principle to a hypothetical example:

$$\text{A} + 3\text{B} \underset{\text{endothermic}}{\overset{\text{exothermic}}{\rightleftharpoons}} 2\text{C} + \text{D}$$

	1 molecule	3 molecules	2 molecules	1 molecule
∴	1 volume	3 volumes	2 volumes	1 volume
	4 volumes		⇌ 3 volumes	

CHANGE OF PRESSURE: If, in a system in equilibrium, the pressure is raised, then Le Chatelier's principle predicts that the system will so respond to this external change as to oppose its effect, i.e. the position of equilibrium will alter by promoting that reaction which results in lower pressure because it proceeds with a reduction in volume. So in this case a higher pressure will cause a shift in the position of equilibrium from left to right and lead to a higher proportion of C and D in the reaction mixture. Lowering the pressure of the system will have the opposite effect.

CHANGE IN TEMPERATURE: By similar reasoning, if the system is in equilibrium and the temperature is raised, this change will be opposed by promoting the reaction which proceeds with the absorption of heat, i.e. the endothermic reaction, causing a shift in the position of equilibrium to the left here and increasing the proportions of A and B. Conversely, lowering the temperature of the system will give a higher yield of C and D.

CHANGES IN CONCENTRATION: Partial or complete removal of D from the equilibrium mixture will upset the system in equilibrium. Le Chatelier's principle predicts that the consequence of such a change will be a shift in the position of equilibrium to the right with the production of more C and D, thereby opposing the reduction in the concentration of D that has been made. Care is needed in applying the principle to predict the effect of changes in concentration because removal or addition of one component of the system may alter the concentrations of the other substances present, producing contrary effects upon the equilibrium, and Le Chatelier's principle, being only a qualitative guide, cannot be used to decide which of these will predominate.

7.10. Applications of Le Chatelier's Principle. From the above treatment it should be clear that the application of Le Chatelier's principle to chemical equilibria leads to these two generalizations:

(i) When an increase in pressure is applied to a system in equilibrium it causes the equilibrium to be displaced in that direction which leads to a contraction in volume.

(ii) When the temperature is raised in a system in equilibrium, it causes the equilibrium to be displaced in that direction which leads to an absorption of heat, i.e. it favours the endothermic reaction.

The value of Le Chatelier's principle in predicting the conditions for maximum yield of a particular product can be demonstrated by applying these general conclusions to some important reversible reactions:

$$(1) \qquad N_2 \quad + \quad 3H_2 \quad \underset{\text{endothermic}}{\overset{\text{exothermic}}{\rightleftharpoons}} \quad 2NH_3$$

	1 molecule	3 molecules		2 molecules
∴	1 volume	3 volumes		2 volumes
	4 volumes		⇌	2 volumes

High pressure will favour the left-to-right reaction because it results in a decrease in volume. High temperature will favour the endothermic reaction, the dissociation of ammonia into its elements. We conclude that the conditions necessary for the highest possible yield of ammonia are high pressure and low temperature.

$$(2) \qquad 2SO_2 \quad + \quad O_2 \quad \underset{\text{endothermic}}{\overset{\text{exothermic}}{\rightleftharpoons}} \quad 2SO_3$$

2 molecules	1 molecule		2 molecules
2 volumes	1 volume		2 volumes
3 volumes		⇌	2 volumes

Again application of a higher pressure will drive the equilibrium to the right because there is a reduction in volume as this reaction proceeds. Higher temperature will promote the endothermic right-to-left reaction, as in (1). If a large excess of oxygen is added to the system in equilibrium this will drive the equilibrium further to the right. Thus the optimum conditions for a high yield of sulphur trioxide from a given amount of sulphur dioxide would appear to be high pressure, low temperature, and a large excess of air or oxygen.

(3) $N_2 + O_2 \underset{\text{exothermic}}{\overset{\text{endothermic}}{\rightleftharpoons}} 2NO$

1 molecule	1 molecule		2 molecules
1 volume	1 volume		2 volumes
2 volumes		⇌	2 volumes

Altering the pressure will not affect the position of equilibrium because there is no change in volume when the reactions take place. Raising the temperature will displace the equilibrium in the direction of the endothermic reaction, i.e. to the right. Thus the yield of nitric oxide is not affected by pressure and is highest at high temperature.

(4) $CO + H_2O \underset{\text{endothermic}}{\overset{\text{exothermic}}{\rightleftharpoons}} CO_2 + H_2$

1 molecule	1 molecule		1 volume	1 molecule
1 volume	1 volume		1 volume	1 volume
2 volumes		⇌	2 volumes	

Again, changing the pressure has no effect upon the position of equilibrium. Raising the temperature of the system will favour the endothermic right-to-left reaction, so the highest yield of hydrogen will be obtained by using as low a temperature as possible.

It must be stressed that Le Chatelier's principle, valuable as it is as a means of predicting the conditions which provide the highest yield of a desired product, is not concerned at all with the factors governing the rate at which that yield is obtained. For instance, when ammonia is synthesized by the Haber process the operating temperature is about 500°C, despite the prediction in (1) above that a low temperature is desirable for the highest possible yield. This is because at lower temperatures nitrogen and hydrogen combine too slowly, even in the presence of a catalyst. The process is operated, therefore, at the lowest temperature consistent with a convenient rate of reaction, the chosen temperature representing a compromise between the conflicting demands of high yield and high rate. The same considerations apply to the oxidation of sulphur dioxide (reaction 2 above), the operating temperature in the contact process being about 500°C (see § 24.11), and to the Bosch process (reaction 4 above), which is normally operated at 450°C (see § 16.2).

7.11. Order of Reaction. The rate of a reaction can be found experimentally by determining the concentration of one of the reactants or products at suitable intervals and calculating the rate of change. This may be done by removing portions from the reaction mixture from time to time and analysing them volumetrically, usually by back-titration methods to prevent further reaction occurring during the estimation. It is often preferable, however, to use a physical method such as observation of changes in volume, pressure, refractive index, colour, or optical rotation (§ 48.2), since there is then no danger of the measurements interfering with the speed or course of the reaction.

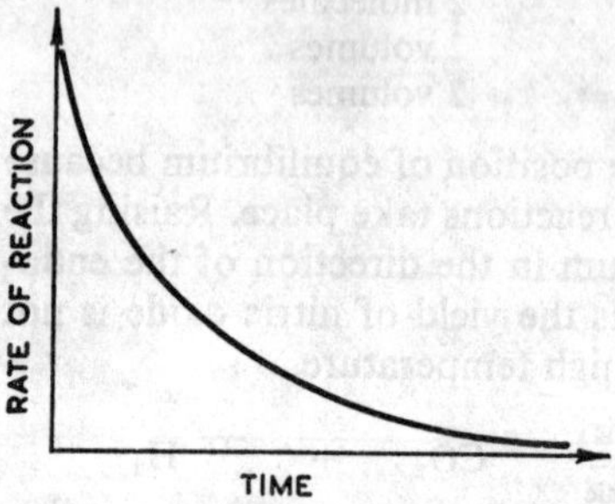

FIG. 7.2 Variation of Reaction Rate with Time

As a result of these experiments it is found that a reaction tends to get slower and slower as the reactants are used up. Fig. 7.2 shows a typical fall in reaction rate with time. It is also found that reactions differ in the way in which their rates depend upon the concentration of reactants, this difference being expressed in terms of *order of reaction.*

7.12. First Order Reactions. These are reactions where experimental investigation shows that the rate is proportional to the concentration of only one substance. Most of them are reactions in which a substance undergoes chemical decomposition or radioactive decay, as in the following examples:

$$\underset{\text{tertiary butyl alcohol}}{C_4H_9OH} = \underset{\text{cyclobutane}}{C_4H_8} + H_2O$$

$$\underset{\text{nitrogen pentoxide}}{2N_2O_5} = 4NO_2 + O_2$$

In some reactions between two substances one of the reactants is present in such great excess that its concentration remains virtually unchanged throughout the reaction, so that the rate varies only with the concentration of the other reactant and the reaction is, in practice, of the first order. The hydrolysis of cane sugar (§ 42.4) is a good example:

$$\underset{\text{sucrose}}{C_{12}H_{22}O_{11}} + H_2O = \underset{\text{glucose}}{C_6H_{12}O_6} + \underset{\text{fructose}}{C_6H_{12}O_6}$$

Let a represent the initial concentration of reactant and $(a - x)$ the concentration of reactant remaining at time t. Then in a first order reaction the rate of reaction $\frac{dx}{dt}$ will be proportional to the concentration of reactant at time t, i.e. $(a - x)$, and we can write

$$\text{rate of reaction} = \frac{dx}{dt} = k(a - x) \quad \text{(where } k \text{ is a constant.)}$$

Thus $$\frac{dx}{(a-x)} = k\,dt$$

and on integrating, we get

$$-\log_e (a-x) = kt + \text{integration constant}$$

Now $x = 0$ when $t = 0$, so substituting these values in the equation reveals the integration constant to be $-\log_e a$.

$$\therefore \log_e a - \log_e (a-x) = kt$$

$$\therefore kt = \log_e \frac{a}{(a-x)} = 2{\cdot}303 \log_{10} \frac{a}{(a-x)}$$

Thus in a first order reaction if $(a - x)$, the concentration of reactant at time t, is determined at intervals and a graph of $\log_e \frac{a}{a-x}$ is plotted against time, we obtain a straight line through the origin as in Fig. 7.3(a), with a gradient equal to k, the reaction rate constant.

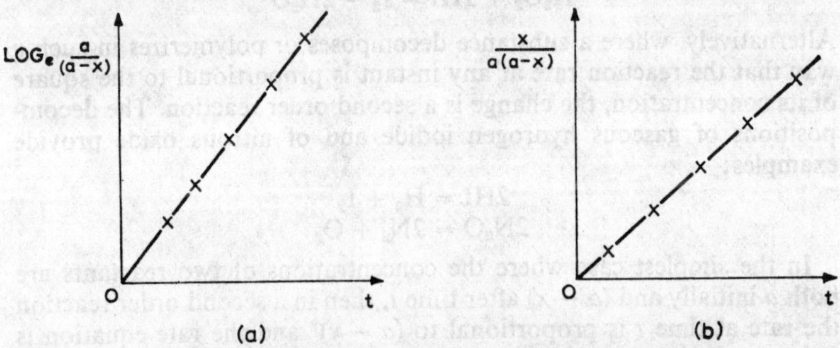

FIG. 7.3 Graphs distinguishing the Order of Reaction (a) First Order Reaction; (b) Second Order Reaction. In each case the gradient is equal to the rate constant, k

We can obtain an expression for the half-life, i.e. the time taken for 50% of the reactant to be consumed, by substituting $x = \frac{a}{2}$ in the equation for a first order reaction.

$$kt_{\frac{1}{2}} = \log_e \frac{a}{a/2} = \log_e 2$$

$$\therefore t_{\frac{1}{2}} = \frac{\log_e 2}{k} = \frac{0{\cdot}693}{k}$$

Thus in a reaction of the first order the half-life is completely independent of the initial concentration of the reactants, and the reaction takes exactly the same length of time to reach halfway to completion whatever the concentration originally taken. This is a matter which can easily be put to the test experimentally and can therefore be used as a means of recognizing first order reactions.

7.13. Second Order Reactions. In these reactions the rate is proportional to the concentrations of two substances or to the square of the concentration of one reactant. For example, if in a reaction between two substances it is found by experiment that the rate of reaction at any instant is proportional to the product of their concentrations, then that reaction is of the second order. The hydrolysis of ethyl acetate (§ 38.2) with sodium hydroxide solution is of this kind:

$$CH_3.COOC_2H_5 + NaOH = CH_3.COONa + C_2H_5OH$$

and the reaction between hydrogen peroxide solution and hydriodic acid is another example:

$$H_2O_2 + 2HI = I_2 + 2H_2O$$

Alternatively, where a substance decomposes or polymerizes in such a way that the reaction rate at any instant is proportional to the square of its concentration, the change is a second order reaction. The decompositions of gaseous hydrogen iodide and of nitrous oxide provide examples:

$$2HI = H_2 + I_2$$
$$2N_2O = 2N_2 + O_2$$

In the simplest case where the concentrations of two reactants are both a initially and $(a - x)$ after time t, then in a second order reaction the rate at time t is proportional to $(a - x)^2$ and the rate equation is written

$$\text{rate of reaction} = \frac{dx}{dt} = k.(a - x)^2$$

$$\therefore \frac{dx}{(a - x)^2} = k dt$$

which on integration gives

$$\frac{1}{(a - x)} = kt + \text{integration constant}$$

Now $x = 0$ when $t = 0$, so the integration constant is $\frac{1}{a}$ and

$$\frac{1}{(a - x)} = kt + \frac{1}{a}$$

or
$$\frac{x}{a(a-x)} = kt$$

Thus when $\frac{x}{a(a-x)}$ is plotted against time, a second order reaction of this simple type gives a straight line through the origin, as in Fig. 7.3(b), with a gradient equal to the rate constant k.

Again, an expression for the half-life can be obtained by substituting $x = \frac{a}{2}$ when $t = t_{\frac{1}{2}}$:

$$\frac{a/2}{a(a-a/2)} = kt$$

$$\therefore t_{\frac{1}{2}} = \frac{1}{ak}$$

It is thus a characteristic of second order reactions in which the initial concentrations of the reactants are equal that the half-life is inversely proportional to that initial concentration, which enables us to distinguish such reactions experimentally from those of the first order.

7.14. Molecularity. Many chemical reactions, although normally represented by a single chemical equation, take place in fact in a number of consecutive stages or steps, some fast and some relatively slow. Very often the intermediate products formed during these individual stages are free radicals which are too unstable to be isolated, but a great deal of experimental evidence has been accumulated in recent years on the complicated mechanisms of such reactions. Just as a chain is only as strong as its weakest link, so the overall rate of a reaction is limited by its slowest stage, which is known as the rate-determining stage for that reason. *The molecularity of a reaction is defined as the number of molecules taking part in a single-step reaction or in the rate-determining stage of a composite reaction.* Reactions with a molecularity of one and two are known as *unimolecular* and *bimolecular* respectively.

The thermal decomposition of nitrogen dioxide, for example, is bimolecular

$$2NO_2 = 2NO + O_2$$

whereas the decomposition of nitrous oxide which is normally represented by the equation

$$2N_2O = 2N_2 + O_2$$

has a unimolecular rate-determining step as its first stage:

$$N_2O = N_2 + \bar{O}$$

Unlike order of reaction, which can be determined without any knowledge of mechanism merely from experimental measurements of the rate of a reaction, molecularity is a theoretical quantity which can only be deduced when details of the rate-determining stage of a reaction are known. When a reaction is a simple one proceeding in only one stage, then the order of reaction and molecularity are usually the same; a good example is the decomposition of hydrogen iodide which is both bimolecular (because two HI molecules are involved in the reaction) and of the second order (because the rate of the reaction is proportional to the square of the hydrogen iodide concentration):

$$2HI = H_2 + I_2$$

A warning must be given about chemical equations. Most of them are composite ones embodying the various stages of a reaction in one overall relationship. They indicate the reactants and final products and the relative quantities of each of them involved in any reaction but they convey no information about the mechanism or actual course of the reaction. The student will soon be convinced that this is true if he considers the following equations and reflects for a moment on the remoteness of the chance that these large numbers of reactant molecules might collide together simultaneously in order to react.

$$3Cu + 8HNO_3 = 3Cu(NO_3)_2 + 2NO\uparrow + 4H_2O$$

$$2KMnO_4 + 10FeSO_4 + 8H_2SO_4$$
$$= K_2SO_4 + 2MnSO_4 + 5Fe_2(SO_4)_3 + 8H_2O$$

$$4P + 3NaOH + 3H_2O = 3NaH_2PO_2 + PH_3\uparrow$$

$$4HN_3 + 5O_2 = 4NO + 6H_2O$$

7.15. Chain Reactions. When methane and chlorine react together under suitable conditions (§ 33.2) research has shown that the first step consists of reaction between a methane molecule and an individual atom of chlorine which is present in the mixture:

$$CH_4 + Cl\cdot = CH_3\cdot + HCl$$

The methyl radical so formed then reacts with a chlorine molecule producing methyl chloride and another atom of chlorine thus:

$$CH_3\cdot + Cl_2 = CH_3Cl + Cl\cdot$$

This series of reactions then repeats itself over and over again, each cycle regenerating an active chlorine atom.

A reaction of this kind involving a complete cycle of steps which is capable of unlimited repetition is called a *chain reaction*. In such reactions free radicals or individual atoms are usually formed, as in the above example; they are very active chemically and react rapidly causing the chain to spread through the mixture of gases. Sometimes several million cycles are repeated before such a chain of reactions is terminated. Termination usually happens when the radicals or atoms which initiate the cycle are removed from the mixture by combination with each other thus:

$$CH_3\cdot + CH_3\cdot = CH_3.CH_3$$

or $$Cl\cdot + Cl\cdot = Cl_2$$

or $$Cl\cdot + CH_3\cdot = CH_3Cl$$

The reaction of hydrogen and chlorine is another good example of a chain reaction. The stages of the cycle are believed to be:

$$H_2 + Cl\cdot = HCl + H\cdot$$
$$H\cdot + Cl_2 = HCl + Cl\cdot$$

In both of these examples, the mechanism suggested for the reaction presupposes the existence in the gaseous mixture of highly reactive individual atoms of chlorine. These atoms, which are responsible for initiating the chain reactions, are produced when a chlorine molecule undergoes dissociation

$$Cl_2 = Cl\cdot + Cl\cdot$$

which it does when heated to a high temperature or exposed to ultraviolet light.-It follows that although these reactions are infinitely slow at room temperature in the dark, once the conditions of light or temperature needed to initiate them are provided, they proceed at great speed and can be explosive in nature.

Many polymerizations are chain reactions, some of them being of considerable industrial importance (§§ 49.2 and 49.3). In reactions of this kind a substance such as benzoyl peroxide (§ 47.5) is frequently used as a catalyst to initiate the reactions by producing a supply of free radicals. In the same way, chain reactions are notoriously sensitive to substances which act as negative catalysts or inhibitors; these are usually impurities or specially chosen substances capable of uniting with and removing free radicals from the reaction mixture so that the propagating cycle is terminated (see § 33.10).

7.16. Energy of Activation. As explained in § 7.2, the rate of most chemical reactions is approximately doubled for each 10 K rise in temperature. This is an experimental fact which demands an explanation. Now calculations based upon the kinetic theory show that the number of collisions per second between molecules of a gas only increases a few per cent when the temperature of the system rises by 10 K, so clearly the sharp increase in reaction rate cannot be explained in that way. Moreover, these calculations indicate that only a small proportion of the molecular collisions do, in fact, result in reaction and that in some cases the reaction would be many millions of times faster if every collision was fruitful. This led Arrhenius to propose that two colliding molecules must possess a certain minimum amount of energy, known as the *energy of activation*, before they could react together. Now it can be calculated that the fraction of the molecules in a gaseous system possessing this critical energy at any instant is approximately $e^{-E/RT}$, where E is the energy of activation of the reaction, R is the universal gas constant, and T is the absolute temperature. Thus in a reaction with a high energy of activation the proportion of collisions leading to reaction would be very small and that reaction would be slow. This is generally true of reactions between covalent molecules (§ 14.4), such as those

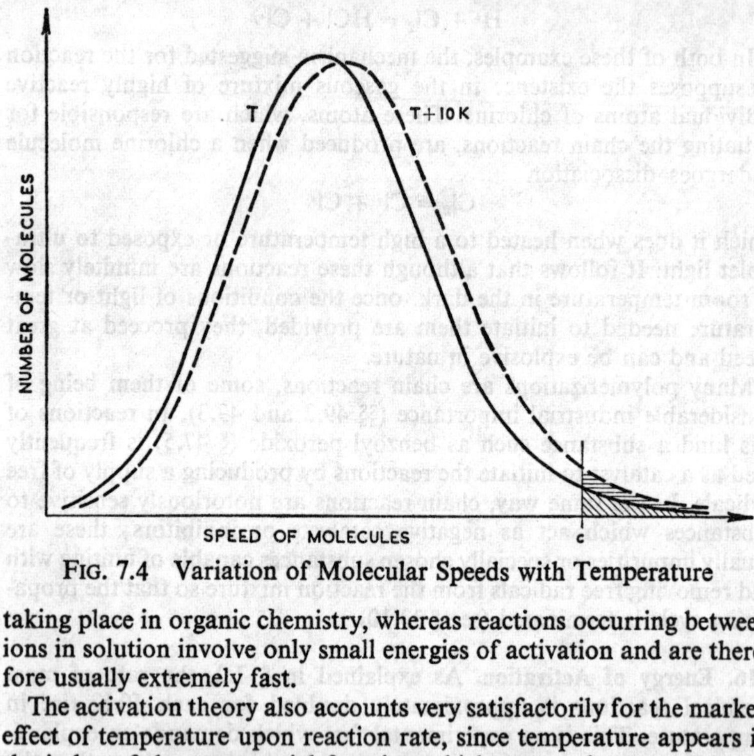

FIG. 7.4 Variation of Molecular Speeds with Temperature

taking place in organic chemistry, whereas reactions occurring between ions in solution involve only small energies of activation and are therefore usually extremely fast.

The activation theory also accounts very satisfactorily for the marked effect of temperature upon reaction rate, since temperature appears in the index of the exponential function which means that even a small rise in temperature causes a rapid increase in the proportion of activated molecules. This point can be made clear by considering briefly the way in which the kinetic energy of a gas is believed to be distributed between its molecules. It can be shown that provided the number of molecules considered is extremely large, which of course is always the case in practice, their velocities at any instant will be as in Fig. 7.4. As a result of random collisions some molecules will be virtually stationary whilst others will be moving at enormous speeds, but the bulk of the molecules will have a velocity between these two extremes. The smooth curve shows the velocities at temperature T and the dashed curve the different distribution when the temperature is 10 K higher. It will be seen that the whole curve has been displaced a little to the right and its shape slightly altered by the rise in temperature, but the important thing to notice is that the proportion of molecules with sufficient energy to react (those with a velocity greater than c) has approximately doubled, as shown by the shaded areas.

The energy of activation idea also helps to explain the working of some types of catalyst. It is suggested that catalysis often brings about an increase in the rate of reaction by making it possible to reach the product by an alternative route each stage of which has a lower energy of activation than the uncatalysed reaction and is therefore relatively fast. It is rather like a person trying to get from one point to another in a mountainous area who finds it quicker to climb over a number of small peaks and follow a roundabout route than to scale the much larger mountain which lies directly between the two points (see Fig. 11.6).

The relationship between the energy of activation and the heat of reaction is discussed in § 11.9.

7.17. Photochemistry. In §§ 7.5 and 7.15 reference has been made to the part that light can play in initiating certain chain reactions. The study of the effect of light upon chemicals and chemical reactions is known as *photochemistry*. When a molecule absorbs a quantum of light energy it is temporarily raised to an excited state. Whilst in this condition it has the energy to participate in chemical changes from which it would otherwise be precluded by high activation energy. For example, absorption of ultra-violet light by a chlorine molecule gives it such an enhanced energy content that rupture of the Cl—Cl bonding and dissociation into single atoms usually results:

$$Cl_2 = Cl\cdot + Cl\cdot$$

Ultra-violet light tends to have a much greater photochemical effect than visible light because of its shorter wavelength, the energy contained in a quantum of light being proportional to the frequency.

Photochemical activation is common in reactions between the halogens and aliphatic and aromatic hydrocarbons; irradiation with ultra-violet light is frequently used as an alternative to heating to accelerate a reaction which would otherwise proceed only very slowly (see § 43.3, reaction 6, and § 46.2).

Many compounds of silver and gold, particularly the halides, are very sensitive to light and photochemical activation, a fact of great importance as it forms the basis of photographic processes (§ 28.18). The tendency of certain dyes to fade on exposure to intense light probably arises from the absorption of light energy by the molecules of dyestuff which whilst so activated undergo changes to colourless compounds. But by far the most important photochemical effect of all is the complicated series of chemical changes known as *photosynthesis* in which plants by absorbing light energy are able to convert carbon dioxide and water into sugars and other complicated molecules which they require as food (§ 20.4).

Photochemical activation has a negative aspect as well. One way of inhibiting and slowing a reaction which is not desired is to exclude strong light so that activation is less likely to occur. For this reason certain chemicals are stored in dark glass bottles e.g. hydrogen peroxide solution (§ 16.6), silver nitrate (§ 28.19), and chloroform (§ 40.8), so that decomposition or oxidation of the reagent during storage is reduced to a minimum.

8

Electrochemistry I—The Ionic Theory

8.1. Introduction. Electrochemistry is the study of the various chemical effects of electricity. With the discovery by Volta in 1800 of a way of obtaining electricity by chemical means, this branch of chemistry soon became a subject of great experimental and theoretical interest, and it has remained so to this day. An *electrolyte* is a substance which conducts electricity when dissolved in water* or when molten; other substances are known as *non-electrolytes.* For example, all acids, bases, and

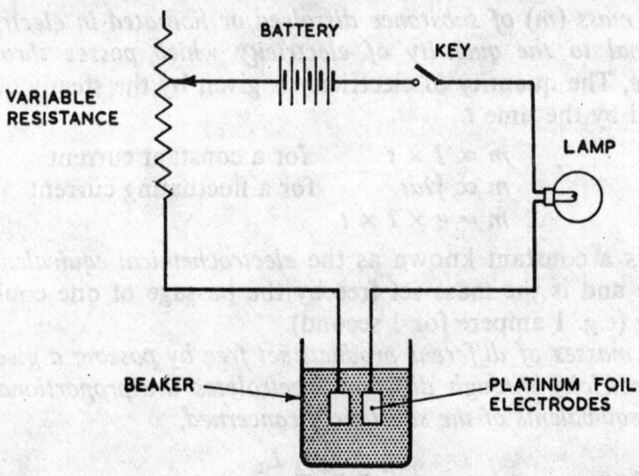

FIG. 8.1 Apparatus to Demonstrate Electrolytic Conduction

salts are electrolytes, but substances like sugar and alcohol are not. This distinction can easily be demonstrated by setting up the apparatus shown in Fig. 8.1. When a potential difference is applied to the electrodes, no detectable current flows if the beaker contains pure water or a solution of sugar or alcohol, but as soon as an electrolyte such as dilute sulphuric acid or sodium chloride is added the lamp lights,

* The term electrolyte is sometimes applied to the solution conducting the current and not just to the solute as here.

showing that the solution is now conducting. Similarly, when a potential difference is applied to two carbon electrodes dipping into a U-tube containing molten lead bromide, a current flows through the circuit and the lamp lights, showing the molten salt to be a conductor, and after a short while globules of molten lead appear at the cathode and bromine vapour is given off at the anode.

The conduction of electricity by an electrolyte, which is known as *electrolysis*, differs from metallic conduction in two main respects. Firstly electrolysis is invariably accompanied by chemical changes resulting, for example, in the liberation of a gas, the deposition of a metal, or the dissolution of an electrode, whereas only physical changes occur when an electric current passes through a piece of metal. The second difference concerns the way in which conductivity varies with temperature; an electrolyte conducts electricity more readily at higher temperatures whereas in metals the reverse is the case.

8.2. Faraday's Laws. In 1833 Faraday summarized his experimental results in his laws of electrolysis:

1. *The mass (m) of substance dissolved or liberated in electrolysis is proportional to the quantity of electricity which passes through the electrolyte.* The quantity of electricity is given by the steady current I multiplied by the time t.

i.e. $m \propto I \times t$ for a constant current

or $m \propto \int I dt$ for a fluctuating current

$$\therefore m = e \times I \times t$$

where e is a constant known as the *electrochemical equivalent* of the substance and is the mass set free by the passage of one coulomb of electricity (e.g. 1 ampere for 1 second).

2. *The masses of different products set free by passing a given quantity of electricity through different electrolytes are proportional to the chemical equivalents of the substances concerned,*

i.e. $$\frac{m_1}{m_2} = \frac{e_1}{e_2} = \frac{E_1}{E_2}$$

where m_1 and m_2 are the masses liberated by the same quantity of electricity from substances with chemical equivalents E_1 and E_2 respectively. It follows from this law that E/e is a constant and that the same quantity of electricity will liberate one gram-equivalent of any product, i.e. 1.008 g of hydrogen, or 107.88 g of silver, or 31.785 g of copper. This quantity, which is 96 487 coulomb mol^{-1},* is known as the *Faraday constant*, and is denoted by the symbol F.

* This figure is usually expressed to only three significant figures, i.e. as 96 500 C mol^{-1}.

3. *The products of electrolysis appear only at the electrodes*, i.e. the wires or plates leading the current into and out of the electrolyte. Faraday called the positive electrode the *anode* and the other the *cathode*, terms which have grown into general use.

These laws are obeyed exactly over very wide ranges of temperature, concentration, and current. Advantage was taken of their exactness to define the unit of current, the ampere, in terms of its electrolytic effect, but with the advent of SI units this definition has now been superseded (see Appendix II). A silver coulometer which involves a silver anode, a platinum cathode, and a solution of silver nitrate as electrolyte, is used for measuring quantities of electricity very accurately.

Example: **Calculate the mass of copper deposited on the cathode when a current of 0.200 ampere is passed through a solution of copper sulphate for ten minutes. The equivalent weight of copper is 31.78.**

Now 96 500 coulomb deposit one gramme-equivalent of copper.

∴ 96 500 „ „ 31.78 g of copper.

$$\therefore 0.200 \times 10 \times 60 \text{ coulomb deposit } \frac{31.78 \times 0.200 \times 10 \times 60}{96\,500} \text{ g of copper}$$

$$= 0.0395 \text{ g of copper}$$

Since the charge on the electron, e, has been determined accurately by Millikan's experiment, which is described in most textbooks of physics, the Avogadro constant (§ 1.12) can be found by dividing the Faraday constant F by the electronic charge:

$$F = 96\,487 \text{ C mol}^{-1}$$
$$e = 1.602 \times 10^{-19} \text{ C}$$
$$\therefore N_A = \frac{F}{e} = \frac{96\,487}{1.602 \times 10^{-19}} = 6.023 \times 10^{23} \text{ mol}^{-1}$$

This method is generally regarded as one of the most reliable and accurate of the many methods available for determining N_A.

Faraday's second law was used historically for comparing two equivalent weights as described in § 2.3. Although this was an accurate method it was only of limited application.

8.3. Onm's Law. Experiment shows that *at constant temperature the current which flows through a conductor (electrolytic or metallic) is proportional to the potential difference applied across its ends.* The ratio of current to applied voltage is known as the *resistance* of the conductor. The unit of resistance is the ohm; a conductor has a resistance of one ohm when it allows a current of one ampere to pass through it when subjected to a potential difference of one volt. At constant temperature the resistance of a given conductor is directly proportional to its length and inversely proportional to its cross-sectional area

i.e. $R = \rho.\frac{l}{A}$ where R = resistance
ρ = resistivity (rho)
l = length
A = area of cross-section.

Thus the resistivity of a conductor, which is a constant for any given substance, is the resistance between two opposite faces of a unit cube of that substance since l/A is then equal to unity.

8.4. Conductivity. In electrochemistry we are rarely concerned with the resistivity of an electrolyte, preferring to use the *conductivity** instead. This is the reciprocal of the resistivity and is defined as the conductance

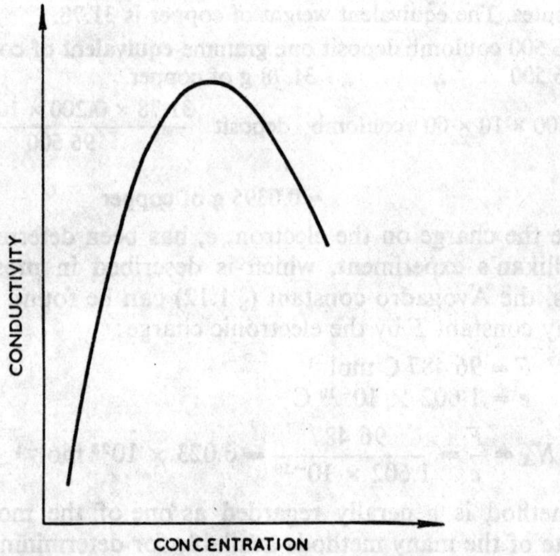

FIG. 8.2 Variation of Conductivity with Concentration

of a one centimetre cube of the substance or solution. The greater the conductivity the more readily does the electrolyte allow the passage of an electric current through it. Conductivity κ (kappa), is usually expressed in reciprocal ohm per centimetre (Ω^{-1} cm^{-1}), one Ω^{-1} corresponding to the flow of a current of one ampere when the potential difference applied to the conductor is one volt, i.e. $\kappa = 1/\rho$. The con-

* It has been agreed that the adjective '*specific*' shall always mean '*per unit mass*'. Thus the former terms 'specific resistance' and 'specific conductance' are no longer appropriate and have been superseded by resistivity and conductivity respectively.

ductivity of an electrolyte depends not only upon its nature but also upon the concentration of the solution. The typical way in which it varies with concentraion is shown in Fig. 8.2.

MEASUREMENT OF CONDUCTIVITY: This is normally carried out by the Kohlrausch method which uses the Wheatstone bridge circuit shown in Fig. 8.3 to compare the resistance of a cell containing the

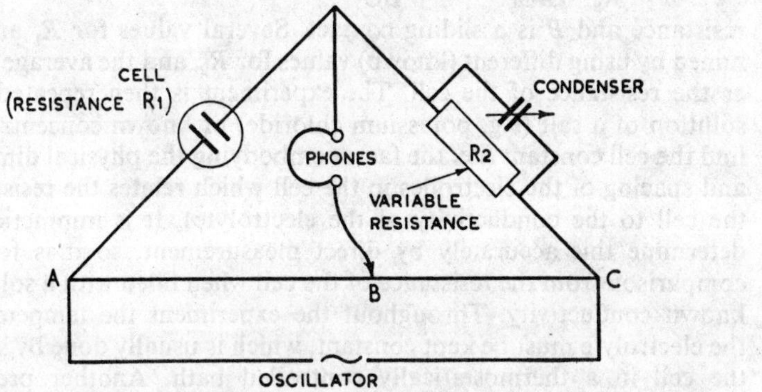

FIG. 8.3 Measurement of Conductivity

electrolyte with a known but variable resistance R_2. The cell used is shown in Fig. 8.4. It is made of silica or specially resistant glass and is fitted with two platinum electrodes (usually coated with platinum black)

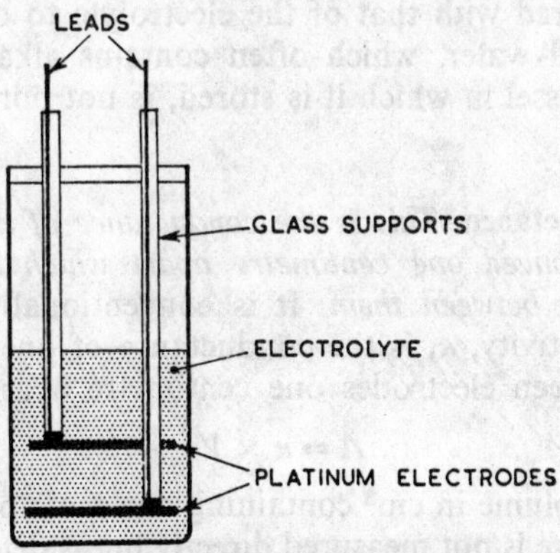

FIG. 8.4 Conductance Cell

which are fixed rigidly in position so that their spacing apart is constant. Alternating current from an induction coil or valve oscillator is used (about 1000 Hz) to avoid polarization effects (i.e. changes in the resistance of the cell due to the formation of bubbles of gas on the surfaces of the electrodes as a result of electrolysis). The null position is indicated by minimum sound in the headphones, which are connected to a valve amplifier to increase their sensitivity. The sharpness of the minimum is usually increased by connecting a small variable condenser in parallel with the post office box R_2 to balance the capacity of the cell. At the minimum $\frac{R_1}{R_2} = \frac{AB}{BC}$, so $R_1 = \frac{AB}{BC} \times R_2$ where AC is a wire of uniform resistance and B is a sliding contact. Several values for R_1 are determined by using different (known) values for R_2, and the average is taken as the resistance of the cell. The experiment is then repeated with a solution of a salt (e.g. potassium chloride) of known concentration to find the cell constant (i.e. the factor embodying the physical dimensions and spacing of the electrodes in the cell which relates the resistance of the cell to the conductivity of the electrolyte). It is impracticable to determine this accurately by direct measurement, so it is found by comparison from the resistance of the cell when filled with a solution of known conductivity. Throughout the experiment the temperature of the electrolyte must be kept constant, which is usually done by standing the cell in a thermostatically controlled bath. Another precaution which is essential if an accurate result is to be obtained is to use *conductivity water* for making the solution of the electrolyte. This is water which has been specially purified by deionization or by repeated distillation in tin vessels until its conductivity has a constant and extremely low value (about $5 \times 10^{-7}\ \Omega^{-1}\ cm^{-1}$ at 18°C) which is negligible compared with that of the electrolyte to be dissolved in it. Ordinary distilled water, which often contains alkali dissolved in it from the glass vessel in which it is stored, is not pure enough for this purpose.

8.5 Molar Conductance. *This is the conductance of a solution between two electrodes spaced one centimetre apart which enclose one mole of the electrolyte between them.* It is conventionally denoted by Λ. Since the conductivity, κ, is the conductance of one cubic centimetre of solution between electrodes one centimetre apart, it follows that

$$\Lambda = \kappa \times V$$

where V = the volume in cm^3 containing one mole of electrolyte. The molar conductance is not measured directly but is determined from the conductivity by using this relationship. A physical picture, such as that shown in Fig. 8.5, may help to make clear the distinction between the two.

The molar conductance of a solution of any given electrolyte depends upon the temperature and the concentration. Experiment shows that at room temperature there is an increase of about 2% in conductance for each 1 K rise in temperature. It is significant that this figure agrees with the decrease in viscosity of the solvent with rising temperature, suggesting that there may be a connection between the two properties.

The two general ways in which the molar conductance varies with concentration are shown in Fig. 8.6 where it is plotted against the dilution, i.e. the volume of solution in dm^3 which contains one mole of solute. Curve A is typical of most salts, and of hydrochloric, sulphuric, and nitric acids, and also of sodium and potassium hydroxides. In these cases the conductance is always high and increases only slightly on dilution, soon reaching a maximum value known as the *molar conductance at infinite dilution* and denoted by Λ_∞. Substances whose molar conductance varies in this way are known as *strong electrolytes*. Curve B is typical of most organic acids, and of ammonium hydroxide, and of certain mercury salts. In concentrated solutions the molar conductance is very low but it increases steadily with dilution. Substances of this type are known as *weak electrolytes*. No sharp distinction exists between the two types and some electrolytes of intermediate character are known.

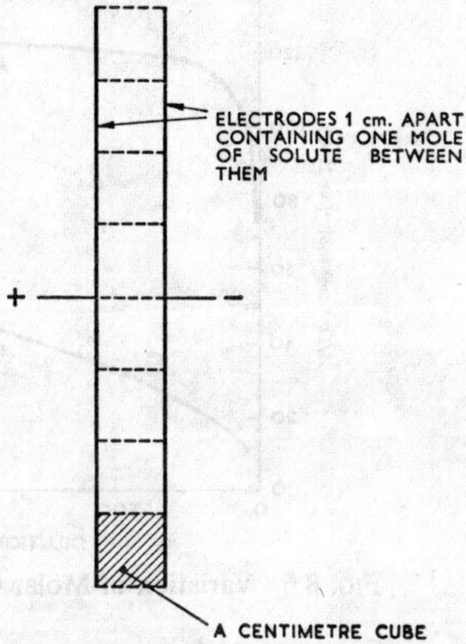

FIG. 8.5 Molar Conductance

The importance of molar conductance is that its variation with concentration faithfully portrays the variation in the conducting power of the electrolyte since the quantity of electrolyte between the electrodes in Fig. 8.5 remains the same whatever the dilution. This is because on dilution the volume of solution taken changes so that one mole of electrolyte is always present. This is not so with conductivity where the variation with concentration represents the resultant of two different changes—one in the conducting ability of the electrolyte and the other in the quantity of electrolyte contained in the cubic centimetre of solution between the two electrodes. Thus changes caused by dilution in the effectiveness of an electrolyte as a conductor of electricity are obscured in the case of conductivity, κ.

For a strong electrolyte the molar conductance at infinite dilution is best determined by plotting the molar conductance against the

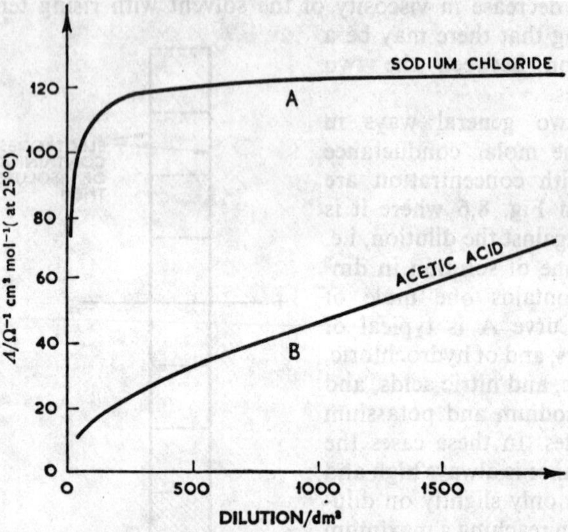

FIG. 8.6 Variation of Molar Conductance with Dilution

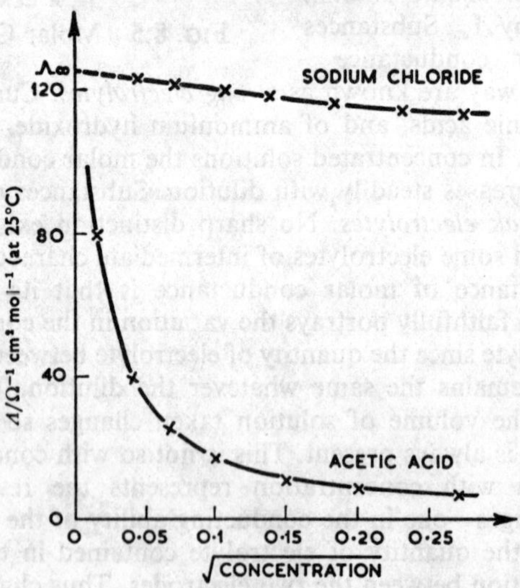

FIG. 8.7 Variation of Molar Conductance with the Square Root of Concentration

square root of the concentration. For very dilute solutions this gives a straight line graph (see Fig. 8.7) which can then be extrapolated to cut the axis, the intercept being taken as Λ_∞. Weak electrolytes also have a characteristic value for the molar conductance at infinite dilution, but in their case it cannot be determined directly because, as Fig. 8.7 shows, the graph is not a straight line. It is calculated therefore by applying Kohlrausch's law of independent migration of ions as described in § 8.8.

8.6. The Simple Ionic Theory. During the nineteenth century numerous attempts were made to explain the experimental facts described in the preceding sections. The most notable of these was the theory of electrolytic dissociation, generally known simply as the ionic theory, which was put forward by Arrhenius between 1883 and 1887. Because it successfully related so many hitherto disconnected facts and because it has proved so fruitful over the years in advancing our knowledge and understanding of electrochemistry, this theory deserves special consideration.

The theory postulates that in solution all electrolytes are spontaneously dissociated to some extent into positive and negative ions. These *ions* are atoms or groups of atoms carrying a net charge which is numerically equal to the valency number of the element or radical; they have completely different properties from the atoms from which they are derived. The dissociation of an electrolyte into ions in solution is regarded as a reversible process, the extent of dissociation depending upon the nature of the electrolyte, the concentration of the solution, and the temperature. Arrhenius explained the variation in the molar conductance with dilution by proposing that in a solution of a strong electrolyte the proportion dissociated is high at all concentrations and increases when the solution is diluted, soon reaching a maximum value when dissociation is complete and all the electrolyte exists as separate ions. He suggested that for a weak electrolyte the proportion which is dissociated in a solution of ordinary concentration is very small but that this proportion increases steadily with dilution so that the electrolyte is completely dissociated when the solution is infinitely dilute.

Electrolytic conduction is explained as being due to the bodily movement of these dissociated ions through the solution towards the electrode of opposite sign under the influence of the electric field. At the electrodes a process of discharge occurs giving rise to a metallic deposit, the liberation of a gas, or some other chemical change. According to the Arrhenius theory the amount of product liberated at the electrodes during electrolysis is proportional to the number of ions discharged per second and the time, i.e. to the total quantity of electricity which passes through the electrolyte, which explains Faraday's first law. Similarly, the masses of different products liberated by a given quantity of electricity are related to their chemical equivalents because the masses of the different ions per unit of charge are proportional to their equivalent weights.

One of the principal difficulties of earlier ionic theories was to explain why electrolytic conduction obeyed Ohm's law. Arrhenius did this by proposing that an electrolyte partially dissociates into ions *as soon as it dissolves*. Thus ions are present in the solution all the time, even before any potential difference is applied to the electrodes, and no electrical energy is consumed in ionizing the solute. During electrolysis, therefore, the dissociated ions move towards the appropriate electrodes and are discharged at a rate which is proportional to the attractive force acting upon them as a result of the applied potential difference.

The Arrhenius theory also explains why solutions of electrolytes exert higher osmotic pressures or show larger depressions of freezing point or elevations of boiling point than might be expected from their molecular weights as calculated from the atomic weights of their constituent elements. Because electrolytes are partially dissociated into ions the number of particles of solute in solution is always greater than would be the case if dissociation did not occur, and since these colligative properties depend for their magnitude on the number of particles and not upon their nature (§ 5.20) the effects are abnormally large for solutions of electrolytes. Indeed these very abnormalities were used by Arrhenius and van't Hoff to calculate the extent of the dissociation, as described in the next section.

One of the achievements of the Arrhenius theory is the simple explanation it offers of the common properties shown by similar compounds. For example, all solutions of cupric salts give a black precipitate with hydrogen sulphide, a brown precipitate with potassium hexacyanoferrate(II) solution, a light blue precipitate with sodium hydroxide solution, and a deep blue solution with an excess of ammonium hydroxide solution. They also deposit metallic copper on the cathode when electrolysed. These are all characteristic reactions of the copper(II) ion, Cu^{2+}, and every solution containing it will behave in this way. Similarly all solutions of metallic chlorides give a white precipitate with silver nitrate solution and with mercurous nitrate solution and with lead acetate solution in the cold, since these are the reactions of the Cl^- ion. Thus a dilute solution of cupric chloride reacts just as if it contains independent collections of cupric and chloride ions. Similarly Arrhenius realized that the properties of all electrolytes, at least as far as their dilute solutions are concerned, are the sum of the properties of the dissociated ions they contain. Kohlrausch had already shown that this was true of conductance in his law of independent migration of ions (§ 8.8).

It is shown in § 11.6 that the heat of neutralization of all strong acids and bases is approximately the same, amounting to about 57 300 joule per mole. According to Arrhenius the salts formed by such neutralizations are strong electrolytes which remain almost completely dissociated into ions in solution so the process is effectively the combination of hydrogen and hydroxyl ions to form undissociated water molecules:

e.g. $$NaOH + HCl = NaCl + H_2O$$

becomes $$Na^+ + OH^- + H^+ + Cl^- = Na^+ + Cl^- + H_2O$$

or $$OH^- + H^+ = H_2O$$

Now since these strong acids and bases are almost completely dissociated into ions in solution the same number of ions are present from each mole and hence the same quantity of heat is evolved each time. Support for this view comes from a study of the equilibrium:

$$H^+ + OH^- \rightleftharpoons H_2O$$

which indicates from the change in the position of equilibrium with temperature that the heat evolved when a mole of water is formed from its ions is, in fact, about 57 300 joule.

The simple ionic theory can be used in like manner to explain many experimental observations including the hydrolysis of salts, the common-ion effect, the precipitation of insoluble compounds by double decomposition, the action of indicators, the effectiveness of buffer solutions, the solubility of salts of weak acids in solutions of strong acids, and so on. All these topics are discussed in the next chapter and explanations of them are given in terms of the ionic theory. Indirectly they provide impressive confirmation of the existence of ions, since it is very unlikely that any other idea could offer such a satisfactory explanation of such diverse phenomena.

The most powerful evidence of all for the ionic theory was not available at the time the theory was put forward, but only became known in the twentieth century as a result of X-ray analysis of solids. As described in § 4.6, many salts are now believed to be composed solely of ions in the solid state, their crystalline structures consisting of ionic lattices and not molecules. When such solids dissolve in water many of these constituent ions are separated by the solvent, which therefore plays an active part in the process of dissolving, and the dissociated ions then move about freely in the solution. Thus a theory of electrolytic conduction no longer has to explain the presence of ions in the solution of an electrolyte and it would be difficult to imagine any modern theory which did not take their presence for granted. The origin of these ions has also been satisfactorily explained in this century by the various theories of bonding, particularly the electronic theory described in chapter 14, so that many of the difficulties encountered by Arrhenius in justifying the existence of ions to his contemporaries do not now arise.

8.7. The Degree of Dissociation. The proportion of an electrolyte which is dissociated into ions in any given solution is known as the degree of dissociation, α. It is expressed either as a decimal or as a percentage. Arrhenius proposed that since all electrolytes are completely dissociated in an infinitely dilute solution, the degree of dissociation at any particular dilution is given by the ratio of its equivalent conductance at that dilution to its equivalent conductance at infinite dilution, i.e. $\alpha = \Lambda/\Lambda_{\infty}$.

Now about that time van't Hoff, who had been investigating the abnormally large colligative effects exhibited by solutions of electrolytes, suggested the use of a term i, known as the *van't Hoff factor*, to relate the observed colligative value to the value calculated on the basis of no dissociation.

$$\text{Thus} \quad i = \frac{\text{observed osmotic pressure}}{\text{calculated osmotic pressure}}$$

$$= \frac{\text{observed freezing point depression}}{\text{calculated freezing point depression}}$$

$$= \frac{\text{observed boiling point elevation}}{\text{calculated boiling point elevation}}$$

and $\Pi V = iRT$ for a given solution.

Let us consider an electrolyte each molecule of which dissociates into n ions in solution and let its degree of dissociation at a particular concentration be α (expressed as a decimal). Then if 1 gramme-molecule of the electrolyte is present in that solution, α gramme-molecules will dissociate forming $n\alpha$ gramme-ions and leaving $(1 - \alpha)$ gramme-molecules undissociated. Thus the total of solute particles in solution, gramme-molecules and gramme-ions together, is $1 - \alpha + n\alpha$, i.e. $1 + (n - 1)\alpha$. This compares with only 1 gramme-molecule when no dissociation occurs.

$$\therefore i = \frac{1 + (n-1)\alpha}{1} \quad \text{and} \quad \alpha = \frac{(i-1)}{(n-1)}.$$

Arrhenius used this deduction to calculate the degree of dissociation of various electrolytes at particular concentrations from the colligative properties of their solutions and compared the values of α so obtained with those found from measurements of equivalent conductance. In most cases there was remarkably close agreement between the two values of α, which did much to strengthen confidence in the ionic theory. It is now known, however, that in the case of strong electrolytes this agreement was largely fortuitous and was not as significant as it appeared.

Example: Calculate the degree of dissociation of sodium chloride from the following data which were obtained with a solution of the same concentration:

(*a*) The molar conductance at 25°C is 96.2 Ω^{-1} cm^2 mol^{-1}, whereas the corresponding value at infinite dilution is 126.5 Ω^{-1} cm^2 mol^{-1}.

(*b*) The freezing point of the solution (containing 12.5 g in 1000 g of water) is − 0.70°C. The cryoscopic constant for 1000 g is 1.86 K and the calculated molecular weight, assuming no dissociation into ions, is 58.5.

(*a*) $$\alpha = \frac{\Lambda}{\Lambda_\infty} = \frac{96.2}{126.5} = 0.761 \text{ or } 76.1\%$$

(*b*) $$\Delta T = \frac{Km}{M} \quad \text{(see § 5.13)}$$

where ΔT = freezing point depression = 0.70 K
K = cryoscopic constant for 1000 g = 1.86 K
m = mass of solute in 1000 g solvent = 12.5 g
M = apparent mol. wt. of solute

$$\therefore M = \frac{Km}{\Delta T} = \frac{1.86 \times 12.5}{0.70} = 33.2$$

Now $$i = \frac{\text{calculated mol. wt.}}{\text{apparent mol. wt.}} = \frac{58.5}{33.2}$$

and $$\alpha = \frac{i-1}{n-1} \quad \text{where } n \text{ is 2 here } (NaCl \rightleftharpoons Na^+ + Cl^-)$$

$$\therefore \alpha = \frac{58.5 - 33.2}{33.2} = \frac{25.3}{33.2} = 0.762 \text{ or } 76.2\%$$

It would appear from this data, therefore, that sodium chloride is about 76% dissociated into ions in a solution of this concentration. As we shall see in later sections, this conclusion is no longer accepted, a different interpretation being put upon the result.

8.8. Kohlrausch's Law of Independent Migration of Ions. In 1876 Kohlrausch noticed that pairs of sodium and potassium salts with the same anion showed a constant difference (21 Ω^{-1} cm^2 mol^{-1}) in molar conductance at infinite dilution (see Table 8.1). Similar results were observed with other pairs of salts which had a cation or anion in common. Kohlrausch concluded that *at infinite dilution each ion shows a characteristic conductance which is independent of the other ions present in the solution.* Thus the molar conductance at infinite dilution for any electrolyte is equal to the sum of the molar conductances of its cation and anion, since each ion contributes a definite amount to the total conductance of the electrolyte:

i.e. $$\Lambda\infty = \Lambda_+ + \Lambda_-$$

where Λ_- and Λ_+ are the *ionic conductances* at infinite dilution of the cation and anion respectively.

TABLE 8.1. MOLAR CONDUCTANCES AT INFINITE DILUTION

Salt	$\Lambda\infty$ *at* 18°C $\Omega^{-1}cm^2mol^{-1}$	
KCl	130.1	difference = 21.1 Ω^{-1} cm² mol⁻¹
NaCl	109.0	
KNO_3	126.3	difference = 21.0 Ω^{-1} cm² mol⁻¹
$NaNO_3$	105.3	

The importance of this law is that it can be used to calculate molar conductances at infinite dilution (such as those of weak electrolytes which cannot be determined by extrapolation), provided the values for three suitable strong electrolytes are known. The calculation is illustrated by the following example:

Example: Calculate the molar conductance at infinite dilution of acetic acid at 25°C, given that at that temperature the molar conductances at infinite dilution of hydrochloric acid, sodium chloride, and sodium acetate are 426, 126, and 91 Ω^{-1} cm² mol⁻¹ respectively.

Now $$\Lambda_{CH_3COOH} = \Lambda_{CH_3COO^-} + \Lambda_{H^+}$$

$$\Lambda_{HCl} = \Lambda_{H^+} + \Lambda_{Cl^-}$$

$$\Lambda_{NaCl} = \Lambda_{Na^+} + \Lambda_{Cl^-}$$

$$\Lambda_{CH_3COONa} = \Lambda_{CH_3COO^-} + \Lambda_{Na^+}$$

$$\therefore \Lambda_{CH_3COOH} = \Lambda_{HCl} + \Lambda_{CH_3COONa} - \Lambda_{NaCl}$$

$$= 426 + 91 - 126$$

$$= 391\ \Omega^{-1}\ cm^2\ mol^{-1}$$

Now the fraction of the total current which is carried through the solution by each type of ion depends upon the relative velocities of the cation and anion. This fraction, which is known as the ***transport number*** of the ion, is given for a simple binary electrolyte by the expression:

$$t_+ = \frac{\text{current carried by anion}}{\text{total current}} = \frac{v_+}{v_+ + v_-}$$

where v_+ and v_- are the absolute velocities (see below) of the anion and cation respectively. The transport number can be measured experimentally by various methods, details of which are beyond the scope of this book. From a determination of the transport number of an ion, its ionic conductance at infinite dilution can easily be calculated since $t_+ = \Lambda_+/\Lambda\infty$ and therefore $\Lambda_+ = t_+ \times \Lambda\infty$. It follows that a list of ionic conductances at infinite dilution can be drawn up for the various ions from which that of compounds can be calculated by straightforward addition. Table 8.2 gives some typical values at 25°C. The conductance

TABLE 8.2. IONIC CONDUCTANCES AT INFINITE DILUTION

Cation	*$\Lambda\infty$ at 25°C* Ω^{-1} cm^2 mol^{-1}	*Anion*	*$\Lambda\infty$ at 25°C* Ω^{-1} cm^2 mol^{-1}
H^+	349.8	OH^-	198
K^+	73.5	$\frac{1}{2}SO_4^{2-}$*	79.7
Ag^+	61.9	Cl^-	76.3
$\frac{1}{2}Mg^{2+}$*	53.2	NO_3^-	71.4
Na^+	50.1	CH_3COO^-	40.9

* The $\frac{1}{2}$ is necessary here because one gramme-equivalent is concerned, not one gramme-ion.

of an ion depends upon the speed at which it moves through the solution. The main retarding force which limits the movement of ions when a potential difference is applied is the frictional drag experienced by the ion and its attendant solvent molecules owing to the viscosity of the solvent. Thus the speed of an ion depends upon a number of factors including the nature of the solvent, the temperature, the potential gradient applied, the charge on the ion, the size of the ion, and the degree to which it is solvated. The velocity of an ion under a potential gradient of one volt per centimetre is known as the *ionic mobility* of that ion in that solvent at that temperature. For most ions in aqueous solution its value lies between 3×10^{-6} metre per second and 8×10^{-6} metre per second at room temperature, which corresponds to a speed of about 2 cm per hour, but H^+ and OH^- ions move considerably faster. Absolute ionic velocities can be determined experimentally, or they can be calculated by dividing the molar conductance of an ion at infinite dilution by 96 500.

The relatively high conductance of the H^+ ion is a point of special interest. Not only does it explain the high conductance of strong acids, which yield these ions on dissociation, but it suggests that this ion moves through the solution in a different way from the other ions. It is thought that the H^+ ion, which is only a proton when not hydrated, is able to transfer from one water molecule to another during its passage through the solution instead of clinging only to the same water molecules all the time. The effect of this is much the same as that of a child who whilst taking part in a race by paddling rafts, jumps from time to time from one raft to another ahead of it, using the rafts almost as stepping stones. Naturally his progress will be much faster as a result than it would have been by staying on the same raft all the time. A similar mechanism is believed to take place, but to a lesser extent, with the OH^- ion. The conductometric titration of acids and alkalis is based upon the exceptional conductances of the H^+ and OH^- ions (see § 9.15).

The solubility of a sparingly soluble salt such as silver chloride or barium sulphate can be determined from the conductance of its saturated solution provided the ionic conductance of its component ions are known, as the following calculation illustrates:

Example: The conductivity of a saturated solution of silver chloride at 25°C is found to be $1.50 \times 10^{-6}\ \Omega^{-1}\ cm^{-1}$, after allowing for the conductivity of water. What is the solubility of silver chloride in gramme per dm³ if

$$\Lambda_{Ag^+} = 62 \quad \text{and} \quad \Lambda_{Cl^-} = 76?$$

Now a sparingly soluble salt like this can be assumed to be completely dissociated into ions even in saturated solution. Consequently, by Kohlrausch's law, its equivalent conductivity is equal to the sum of the ionic conductances,

i.e. $\Lambda_\infty = \Lambda_{Ag^+} + \Lambda_{Cl^-}$

$= 62 + 76 = 138$

But $\Lambda_\infty = \kappa . V,$

where κ = conductivity of saturated solution

V = the volume *in cm³* containing 1 gramme-equivalent

$$\therefore \text{ Concentration in gramme-equivalents per dm}^3 = \frac{1000}{V}$$

$$= \frac{1000\kappa}{\Lambda_\infty}$$

$$= \frac{1000 \times 1.50 \times 10^{-6}}{138}$$

$$\therefore \text{ Solubility in gramme per dm}^3 = \frac{1000 \times 1.50 \times 10^{-6} \times 143.5}{138}$$

$$= 1.56 \times 10^{-3}$$

8.9 Modern Ionic Theories. In his theory of electrolytic dissociation Arrhenius proposed that the degree of dissociation of an electrolyte at any dilution was equal to the ratio of its molar conductance at that dilution to its molar conductance at infinite dilution, i.e. $\alpha = \Lambda/\Lambda\infty$. When the degree of dissociation of a strong electrolyte is calculated in this way, the electrolyte is found not to obey Ostwald's dilution law, as explained in the next chapter. Several other pieces of experimental evidence weigh against the Arrhenius theory as it applies to strong electrolytes. Studies of absorption spectra show no evidence of undissociated molecules in solutions of salts, and the various heats of neutralization of strong acids and strong bases are so concordant that they suggest, contrary to the Arrhenius theory, that all these electrolytes are dissociated to exactly the same extent in solution and that this extent does not depend upon their concentrations. Moreover, measurements show that transport numbers vary with the concentration of the solution, which would certainly not be expected from the Arrhenius theory

These findings have necessitated an important modification of the original Arrhenius theory as far as *strong* electrolytes are concerned. In relating the degree of dissociation to the ratio of the molar conductances, Arrhenius made the assumption that the velocities of the ions did not change on dilution, so that the number of ions present in the solution was the only variable. This view is no longer accepted and *strong electrolytes are now regarded as being completely dissociated into ions in aqueous solution at all dilutions*. The variation in their conductance and colligative effects on dilution is attributed solely, therefore, to changes in the velocities of their ions as a result of interionic attraction, and the ratio $\Lambda/\Lambda\infty$, now known as the *conductance ratio* or the *apparent degree of dissociation*, is really a measure of the way in which the velocities of ions from strong electrolytes change with concentration.

The modern view, devised by Debye and Hückel in 1923 and developed further by Onsager in 1927, is that in a concentrated solution of a strong electrolyte each ion is surrounded in its immediate vicinity by a cloud or atmosphere of ions which are mostly of opposite charge. In the absence of a potential gradient in the solution this arrangement of ions is symmetrical, but once a potential difference is applied to the electrodes asymmetry arises, as in Fig. 8.8, because the positive and negative ions are attracted in opposite directions. Thus as each ion moves through the solution its ionic atmosphere exerts a drag upon it owing to the excess of ions of opposite charge in its wake and this has the effect of slowing it down appreciably. The situation is rather like a man trying to force his way through a dense crowd, many members of which are moving in the opposite direction. The analogy is all the more realistic if we imagine each person in the crowd to be carrying a couple of large cases (adhering molecules of solvent) which continually get caught in those of other people. Just as the denser the crowd the more the man's progress is impeded, so the more concentrated the solution the greater is the retarding effect of interionic attraction.

It is possible that the attraction between oppositely charged ions is so great in concentrated non-aqueous solutions of strong electrolytes, particularly when the ions are small and highly charged, that a small proportion of the ions may momentarily form pairs or groups. In such cases the electrolyte is said to be completely *ionized* in solution but not completely *dissociated*. This is not, of course, the same as Arrhenius proposed in his original theory, since he imagined an equilibrium between un-ionized molecules and dissociated ions, whereas in this modern theory the solute exists as ions under all conditions.

In their original form these modern theories were highly mathematical but experimental support for them is obtained by applying a very large potential gradient indeed to the solution, when the conductance is appreciably increased. Presumably the ions move so fast under these conditions that they shed their solvating molecules and their retarding atmosphere of oppositely charged ions entirely and stream towards the attracting electrode. Similarly, if an alternating potential

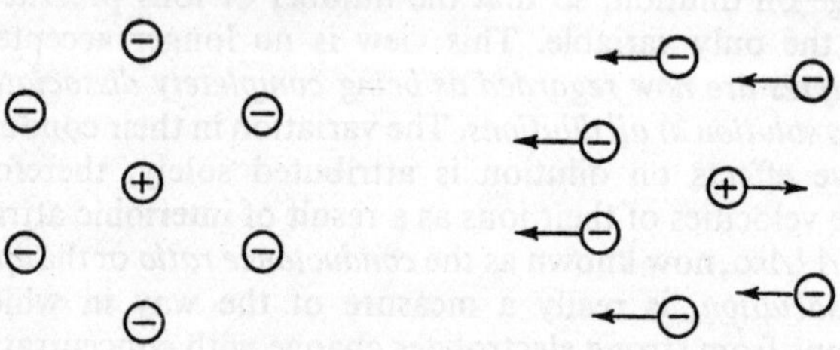

NO ELECTRIC FIELD, SYMMETRICAL (a) UNDER POTENTIAL GRADIENT, ASYMMETRICAL (b)

FIG. 8.8 Diagram to Illustrate Ionic Atmosphere

difference of very high frequency is applied, the conductance is also raised; in this case the ions change direction so frequently per second that the ionic atmosphere gets no time to build up asymmetrically around them and impede their movements.

For weak electrolytes interionic attraction is slight, since the concentration of ions in solution is never large, and the velocity of the ions does not change very much, therefore, with dilution. Consequently only a small error is incurred in taking the conductance ratio as the degree of dissociation for weak electrolytes and in general the Arrhenius theory accounts satisfactorily for their behaviour. A clear distinction must be drawn here between the terms *strong* and *weak* as applied to electrolytes and the terms *concentrated* and *dilute* as applied to solutions. An electrolyte is strong or weak depending upon the extent to which it is dissociated into ions in, say, a molar solution, and the way in which its molar conductance changes with dilution, whereas whether a solution is concentrated or dilute depends upon how much solute it contains in unit volume. For example, sulphuric acid is always a strong acid whether dilute or concentrated, and acetic acid is a weak acid even in concentrated solution.

8.10. The Hydroxonium Ion. In his original theory Arrhenius proposed that when acids dissociated in solution they yielded hydrogen ions, H^+, thus:

$$HCl \rightleftharpoons H^+ + Cl^-$$
$$CH_3.COOH \rightleftharpoons H^+ + CH_3COO^-$$

With the growth of our knowledge of atomic structure we now realize that it is highly improbable that the hydrogen ion, which is merely a proton and therefore only about $\frac{1}{10000}$ the size of other ions, would exist independently in solution. Any ion carrying such a high density of charge is likely to be unstable and will tend to form a hydrate by combining loosely with one or more water molecules. Reference has already been made to this tendency in explaining the abnormally high equivalent conductance of the H^+ ion in solution. Thus the modern view is that solutions of acids contain not just the simple hydrogen ion, but rather the *hydroxonium ion*, H_3O^+ (also known as the oxonium ion or the hydronium ion), and that the dissociation of acids in solution is best represented by equilibria of the type:

$$HCl + H_2O \rightleftharpoons H_3O^+ + Cl^-$$
$$CH_3COOH + H_2O \rightleftharpoons H_3O^+ + CH_3COO^-$$

One advantage of this is that it recognizes the active part played by the solvent in the ionization of these acids. Indeed in the completely anhydrous form hydrogen chloride and acetic acid are covalent substances which do not conduct electricity; they do not turn litmus red, corrode metals, or attack carbonates releasing carbon dioxide. The same is true of a solution of hydrogen chloride in benzene or toluene. It is only in the presence of water that these substances show their usual acidic properties. In the same way the dissociation of water may be expressed in terms of the equilibrium:

$$H_2O + H_2O \rightleftharpoons H_3O^+ + OH^-$$

Although the existence of the hydroxonium ion in aqueous solutions of acids is now universally accepted, for the sake of simplicity it is still common practice to use the symbol H^+ in describing ionic equilibria, as is done in the following chapters. The hydrogen ion is not the only one, of course, to be hydrated in this way and experimental evidence provided by a study of transport numbers suggests that most other ions are also linked loosely to one or more water molecules in aqueous solution.

8.11. Modern Ideas on Acids and Bases. Another consequence of this modern view of dissociation is a wider conception of the terms acid and base. In 1923 Brönsted and Lowry put forward independently the definition of *an acid as any substance which donates protons* (*i.e. hydrogen ions*) *to other substances*, so that acid salts and the ammonium ion are included with the usual range of compounds. Similarly, *a base is defined as any substance which acts as a proton acceptor*; this includes substances such as water, amines, and ammonia which contain a lone pair of electrons (see § 14.4) and all acid radicals, as well as the usual compounds which yield hydroxyl ions. Thus acids and bases are related in the following way:

$$\text{acid} \rightleftharpoons H^+ + \text{base}$$

It will be seen, therefore, that water can act both as a base and as an acid, and does in fact do both simultaneously during its dissociation:

$$H_2O \rightleftharpoons H^+ + OH^-$$
$$H^+ + H_2O \rightleftharpoons H_3O^+$$

These definitions are preferred to the narrower ones formerly in use (acids and bases are substances yielding H^+ and OH^- ions respectively in aqueous solution) because they are also applicable to non-aqueous solvents and because they account for the way in which the strength of an acid varies with the nature of the solvent.

The views of Brönsted and Lowry have been extended further by Lewis, who has defined a base as a substance possessing a lone pair of electrons capable of being used to link it to another atom, and an acid as a substance willing to accept these electrons from a base. Details of this viewpoint lie beyond the scope of this book, but even on the briefest consideration it should be clear that by defining acids as electron acceptors rather than as proton donors, Lewis includes certain substances not usually regarded as acids at all.

9

Electrochemistry II—Ionic Equilibria

9.1. Introduction. In studying ionic equilibria we shall indicate when one side of the equation greatly preponderates over the other by using special arrows as in the following examples:

$$NaCl \rightleftharpoons Na^+ + Cl^-$$

(virtually all the sodium chloride is dissociated into separate ions)

$$H_2O \rightleftharpoons H^+ + OH^-$$

(most water molecules are undissociated, with only a small proportion of ions). Where no great disparity exists in the position of equilibrium, or where the disparity is unpredictable or immaterial, the usual reversible arrows will be used.

9.2. Ostwald's Dilution Law. Let us consider a solution containing one mole of a typical simple binary weak electrolyte AB (i.e. one in which each molecule dissociates into two ions in solution) in volume V of solution. Let α be the degree of dissociation under these conditions. Then provided the solution is kept at constant temperature we can apply the law of mass action (§ 7.6) to the equilibrium between AB and its ions by taking the active mass of each ion to be its ionic concentration:

	AB	$\rightleftharpoons$	A^+	+	B^-
Amount present at equilibrium	$(1-\alpha)$ mole		α mole		α mole
Concentration " "	$\frac{(1-\alpha)}{V}$ mol dm^{-3}		$\frac{\alpha}{V}$ mol dm^{-3}		$\frac{\alpha}{V}$ mol dm^{-3}

Applying the law of mass action:

$$\frac{[A^+][B^-]}{[AB]} = \text{constant, at constant temperature.}$$

But $$[A^+] = \frac{\alpha}{V} \text{ mol dm}^{-3} = [B^-]$$

and $$[AB] = \frac{(1-\alpha)}{V} \text{ mol dm}^{-3}$$

$$\therefore \frac{\frac{\alpha}{V} \times \frac{\alpha}{V}}{\frac{(1-\alpha)}{V}} = \frac{\alpha^2}{(1-\alpha)V} = \text{constant, } K \text{ (at constant temperature)}$$

This is a statement in symbols of *Ostwald's dilution law* first put forward in 1888. The constant K is known as the *dissociation constant* of the weak electrolyte at that temperature.

The law is easily verified by determining the degree of dissociation, α, at various dilutions and substituting corresponding values of V and α in the expression $\frac{\alpha^2}{(1-\alpha)V}$. It is found that for weak electrolytes K is fairly constant, but that for strong electrolytes there is a wide variation in its value with changing concentration. In fact a weak electrolyte is often defined as one which obeys Ostwald's law and it was the inapplicability of the law to strong electrolytes which first raised doubts about the Arrhenius theory. The term dissociation constant clearly has no meaning, therefore, when applied to strong electrolytes, which explains why when we meet it K always has a very low value, usually less than 10^{-2} mol dm^{-3}. Since K varies with temperature the latter should always be stated when K is quoted unless it is 25°C, the arbitrary standard.

9.3. The Strength of Acids and Bases. The dissociation constant, K, is a very important index of the strength of an acid or base and is generally preferred to the degree of dissociation, α, for this purpose because unlike the latter it does not vary with concentration. Thus the relative strengths of weak electrolytes can best be determined by comparing their dissociation constants at the same temperature as follows:

Acetic acid: $CH_3COOH \rightleftharpoons CH_3COO^- + H^+$

$$\frac{[CH_3.COO^-][H^+]}{[CH_3COOH]} = K_{CH_3COOH} = 1.85 \times 10^{-5} \text{ mol dm}^{-3} \text{ at } 25°C$$

Prussic acid: $HCN \rightleftharpoons H^+ + CN^-$

$$\frac{[H^+][CN^-]}{[HCN]} = K_{HCN} = 7.1 \times 10^{-10} \text{ mol dm}^{-3} \text{ at } 25°C$$

Ammonium Hydroxide: $NH_4OH \rightleftharpoons NH_4^+ + OH^-$

$$\frac{[NH_4^+][OH^-]}{[NH_4OH]} = K_{NH_4OH} = 1.80 \times 10^{-5} \text{ mol dm}^{-3} \text{ at } 25°C$$

Thus acetic acid and ammonium hydroxide are electrolytes of closely comparable strength and both are much stronger than prussic acid, which is a very weak electrolyte.

The degree of dissociation, α, at any concentration can readily be calculated from the dissociation constant in the following way:

Example: Calculate the degree of dissociation and the concentration of H^+ ions in a decimolar solution of acetic acid at 25°C, given that its dissociation constant at that temperature is 1.85×10^{-5} mol dm^{-3}.

$$CH_3.COOH \rightleftharpoons CH_3.COO^- + H^+$$

Amounts: $(1-\alpha)$ mole $\quad \alpha$ mole $\quad \alpha$ mole

Concns: $\dfrac{(1-\alpha)}{V}$ mol dm^{-3} $\quad \dfrac{\alpha}{V}$ mol dm^{-3} $\quad \dfrac{\alpha}{V}$ mol dm^{-3}

From Ostwald's dilution law $\dfrac{\alpha^2}{(1-\alpha)V} = K_{CH_3COOH} = 1.85 \times 10^{-5}$ mol dm^{-3}

$$\therefore \alpha^2 = 1.85 \times 10^{-5}(1-\alpha)V.$$

Now V is the volume (in dm^3) containing one mole of acid, and is 10 here, since the solution is decimolar.

$$\therefore \alpha^2 = 1.85 \times 10^{-5} \times (1-\alpha) \times 10$$
$$= 1.85 \times 10^{-4} \times (1-\alpha)$$
$$\therefore \alpha^2 + 1.85 \times 10^{-4}\alpha - 1.85 \times 10^{-4} = 0.$$

Solving this quadratic equation we get:

$$\alpha = 1.35 \times 10^{-2} = 0.0135.$$

Thus the degree of dissociation in this solution is 0.0135 or 1.35%.

Now $\quad [H^+] = \dfrac{\alpha}{V} = \dfrac{1.35 \times 10^{-2}}{10} = 1.35 \times 10^{-3}$ mol dm^{-3}

Thus the concentration of hydrogen ions in the solution is

1.35×10^{-3} mol dm^{-3}.*

In calculations of this sort, where α is known to be small, it is often permissible to approximate by calling $(1-\alpha)$ unity. In this case the error in the value of α by doing so would be less than 1%, since α^2 then equals 1.85×10^{-4} and $\alpha \simeq \sqrt{1.85 \times 10^{-4}} \simeq 1.36 \times 10^{-2}$. When this approximation is made, Ostwald's dilution law becomes $\dfrac{\alpha^2}{V} \simeq K$ so $\alpha^2 \simeq KV$ and $\alpha \simeq \sqrt{KV}$. Thus when $V = 1$, $\alpha \simeq \sqrt{K}$, i.e. the degree of dissociation of a weak electrolyte in normal solution is approximately equal to the square root of its dissociation constant. It also means that the degree of dissociation of a weak electrolyte is proportional to the square root of the dilution.

The dissociation constant of a weak electrolyte remains unchanged

* In this calculation we have ignored the H^+ ions from the dissociation of water because, as will be clear later, in this case they are negligible compared with the concentration produced by dissociation of the acid.

not only on dilution but also on addition of other electrolytes to the solution, even when the same ion is common to both equilibria. This is important as the basis of the 'common ion effect' described in § 9.7.

According to the modern view outlined in the previous chapter, in aqueous solution strong electrolytes are completely dissociated into ions at all concentrations. It follows, therefore, that the concentration of H^+ ions in a solution of a strong acid is effectively the same as the concentration of the acid. For example, in decimolar hydrochloric acid, which by definition contains 0.1 mole of acid per dm^3, the concentration of hydrogen ions will be 0.1 mole per dm^3. Frequent use of this fact will be made later in this chapter.

Polybasic acids (i.e. acids containing more than one hydrogen atom in each molecule replaceable by a metal) dissociate into ions in solution in more than one stage and have, therefore, more than one dissociation constant.

e.g.

$$H_2S \rightleftharpoons H^+ + HS^-$$

$$\therefore \frac{[H^+][HS^-]}{[H_2S]} = K_1$$

$$HS^- \rightleftharpoons H^+ + S^{2-}$$

$$\therefore \frac{[H^+][S^{2-}]}{[HS^-]} = K_2$$

Multiplying these two expressions together we get:

$$\frac{[H^+]^2[S^{2-}]}{[H_2S]} = K_1 \times K_2 = K_{H_2S}$$

The approximate values of K_1 and K_2 are 10^{-7} and 10^{-15} respectively, so the dissociation constant K_{H_2S} has a value of about 10^{-22} at 25°C. Similarly, tribasic orthophosphoric acid dissociates in three stages as follows:

$$H_3PO_4 \rightleftharpoons H^+ + H_2PO_4^- \rightleftharpoons 2H^+ + HPO_4^{2-} \rightleftharpoons 3H^+ + PO_4^{3-}$$

giving rise to three dissociation constants $K_1(10^{-2})$, $K_2(10^{-7})$, and $K_3(10^{-12})$. Thus a solution of this acid contains all three types of orthophosphate ion in equilibrium with H^+ ions and a relatively large concentration of undissociated molecules.

9.4. The Dissociation of Water. If water is purified by repeated distillation or by using a mixture of ion-exchange resins (§ 16.5), its molar conductance falls to a steady, very low value. That this slight conductance remains, no matter how many times the water is distilled or treated with resins, shows that however pure it is water always contains some ions, which are believed to be formed as a result of its own dissociation:

$$H_2O + H_2O \rightleftharpoons H_3O^+ + OH^-$$

or, more simply:

$$H_2O \rightleftharpoons H^+ + OH^-$$

Applying the law of mass action to this equilibrium at constant temperature and taking the concentrations of the ions expressed in mole per dm³ as their active masses, we get:

$$\frac{[H^+].[OH^-]}{[H_2O]} = \text{constant}, K$$

$$\therefore\ [H^+].[OH^-] = K \times [H_2O]$$

Now $[H_2O]$, which is the molar concentration of undissociated water molecules in water, is effectively constant since water is an extremely weak electrolyte only very slightly dissociated into ions. For example, even if the dissociation of water is increased a thousandfold, there will be negligible *percentage* change in the value of $[H_2O]$ and we can therefore regard it as constant under all conditions, i.e. $[H_2O]$ = constant.

$$\therefore\ [H^+].[OH^-] = K \times \text{constant} = K_W \text{ (at constant temperature)}$$

where the constant K_W is known as the *ionic product for water.*

Since each molecule of water gives rise to one hydrogen ion and one hydroxyl ion when it dissociates, the concentration of these two ions must necessarily be the same in pure water. Experiment shows that this concentration is about 10^{-7} mol dm^{-3} at 25°C, i.e. $[H^+] = [OH^-] = 10^{-7}$ mol dm^{-3}. K_W has the value of approximately 1×10^{-14} mol^2 dm^{-6} at 25°C, therefore, and increases to about 50×10^{-14} mol^2 dm^{-6} at 100°C. From this change in K_W we can deduce that the heat of the reaction $H^+ + OH^- \rightleftharpoons H_2O$ is about 57 300 joule (see § 8.6). A solution in which there is a preponderance of hydrogen ions over hydroxyl ions, i.e. one in which the hydrogen ion concentration exceeds 10^{-7} mol dm^{-3} at 25°C, is said to be *acidic,* whilst a solution in which the hydrogen ion concentration is less than 10^{-7} mol dm^{-3} is called an *alkaline* solution. Only at *neutrality* are the concentrations of H^+ and OH^- ions exactly equal.

It is important to realize that however much the concentration of H^+ and OH^- ions is altered by the addition of acids or alkalis, the ionic product remains constant at constant temperature. For example, the H^+ ion concentration may vary very widely from 1 mol dm^{-3} in a molar solution of hydrochloric acid to 10^{-14} mol dm^{-3} in a molar solution of sodium hydroxide, yet the product $[H^+].[OH^-]$ always has the value 10^{-14} mol^2 dm^{-6} at 25°C. Thus it is the product of the two ionic concentrations which is constant, not the concentrations of the individual ions.

9.5. Hydrogen Ion Concentration and pH. In describing the acidity or alkalinity of a solution it is adequate to quote the hydrogen ion concentration alone, since the hydroxyl ion concentration can always be deduced from the relationship $[OH^-] = \frac{10^{-14}}{[H^+]}$ at 25°C. Now in view of the wide variation which is possible in the H^+ ion concentration, it is convenient to express it on a logarithmic scale, and the one suggested by Sorensen in 1909 is now in general use. *The negative logarithm to the base 10 of the H^+ ion concentration (in* mol dm^{-3}*) is called the pH of a solution*, i.e. $pH = -\log_{10} [H^+] = \log_{10} \frac{1}{[H^+]}$ and $[H^+] = 10^{-pH}$

The sign is reversed in order to obtain a scale on which all the numbers normally encountered are positive.

Example: The concentration of H^+ ions in two solutions are *(a)* 1×10^{-3} mol dm^{-3} and *(b)* 5×10^{-10} mol dm^3. What is the pH of each solution?

$$(a)\ [H^+] = 10^{-3} \text{ mol dm}^{-3}$$
$$\therefore pH = -\log_{10} (10^{-3}) = -(-3) = 3$$
$$(b)\ [H^+] = 5 \times 10^{-10} \text{ mol dm}^{-3}$$
$$\therefore pH = -\log_{10} (5 \times 10^{-10})$$
$$= -\log_{10} (10^{0.70} \times 10^{-10})$$
$$= -\log_{10} (10^{-9.30})$$
$$= -(-9.30) = 9.3$$

Thus the solutions have a pH of 3.0 and 9.3 respectively.

The reverse calculation is sometimes needed:

Example: A solution has a pH of 4.1. What is its H^+ ion concentration?

Now $\quad 4.1 = -\log_{10} [H^+]$

so $\quad \log_{10} [H^+] = -4.1$

$$= -5.0 + 0.9 = \bar{5}.9$$

$\therefore [H^+] = \text{antilog. } \bar{5}.9 = 8 \times 10^{-5}$ mol dm^{-3}

The relationship between the acidity or alkalinity of a solution and its pH is shown diagrammatically in Fig. 9.1. An acidic solution, it should be noted, has a high concentration of H^+ ions but a low pH value, i.e. its pH is less than 7. It must also be remembered that the pH scale is a logarithmic one and that a change of only one unit corresponds to a tenfold change in the hydrogen ion concentration.

The hydrogen ion concentration of a solution may be found in various ways. For a weak electrolyte it can be calculated from the degree of dissociation as determined from the conductivity ratio or the van't Hoff factor. The hydrogen ion concentration can also be deduced from its catalytic effect upon the hydrolysis of esters (§ 38.2) or, more simply, by a comparative colorimetric method as explained in § 9.14. Alternatively the pH can be found by measuring the potential difference

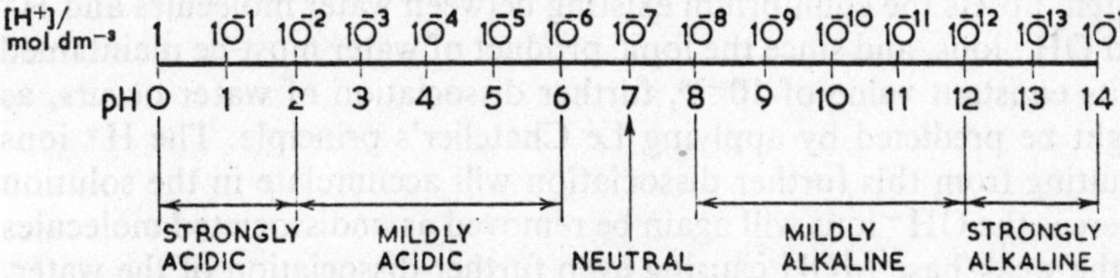

FIG. 9.1 The pH Scale

developed when a glass electrode and a standard electrode (§ 10.3) are in contact with the solution (this is the principle of the *pH meter*, which is popular in industry because it gives direct readings of pH quickly and accurately and can be used by unskilled labour for routine measurements).

9.6. The Hydrolysis of Salts. An aqueous solution of a salt of a strong acid and a strong base (e.g. sodium chloride or potassium nitrate) has a pH of about 7 at room temperature and is therefore neutral. However, salts formed from a weak acid and a strong base (e.g. sodium phosphate or potassium acetate) give an alkaline solution when dissolved in water, and those obtained from a strong acid and a weak base (e.g. ammonium chloride) give acidic solutions. These facts can readily be explained by examining the ionic equilibria involved in a typical case.

Let us consider a salt MA which dissociates completely in aqueous solution as follows:

$$MA \rightarrow M^+ + A^-$$

The water present is also dissociated, but only, of course, to a very small extent:

$$H_2O \rightleftharpoons H^+ + OH^-$$

Thus the solution of a salt contains four types of ion and these will give rise to two further equilibria:

$$M^+ + OH^- \rightleftharpoons MOH$$

and $$H^+ + A^- \rightleftharpoons HA$$

The extent to which these equilibria proceed to the right depends upon the strength of MOH and HA as electrolytes. If MA is the salt of a strong acid and a weak base, for example, then few, if any, of the H^+ and A^- ions will unite since the acid HA will be completely dissociated in solution, but most of the M^+ and OH^- ions present will combine to form undissociated MOH until the expression $\frac{[M^+][OH^-]}{[MOH]}$ has the low value of the dissociation constant K_{MOH}. This removal of OH^- ions from the system upsets the equilibrium existing between water molecules and H^+ and OH^- ions, and since the ionic product of water must be maintained at its constant value of 10^{-14}, further dissociation of water occurs, as might be predicted by applying Le Chatelier's principle. The H^+ ions resulting from this further dissociation will accumulate in the solution whereas the OH^- ions will again be removed as undissociated molecules of the weak base MOH causing even further dissociation of the water. Thus the result of this interaction between the two equilibria will be to produce a large preponderance of H^+ ions in the solution, making the solution acidic. In a similar way, if MA is the salt of a weak acid and a strong base then the concentration of OH^- ions in its solution will be much higher than the concentration of H^+ ions and its solution will be alkaline. Sodium carbonate, which is often titrated against acids in volumetric analysis, is a good example of this.

It is the *relative* strength of MOH and HA which determines whether the solution will show any marked acidity or alkalinity. Salts formed from acids and bases which are equally strong (e.g. sodium chloride) or equally weak (e.g. ammonium acetate) will be approximately neutral since their solutions will contain equal concentrations of H^+ and OH^- ions. The phenomenon is known as *hydrolysis* because it involves the interaction of the ions of the salt with those of water. Effectively the process is the opposite of neutralization and can be represented by the general equation:

$$\text{salt} + \text{water} \rightleftharpoons \text{acid} + \text{base}.$$

For a strong acid and a weak base the overall change is:

$$M^+ + A^- + H_2O \rightleftharpoons MOH + H^+ + A^-$$

Hence, applying the law of mass action to this equilibrium and assuming that the ionic concentrations can be taken as the active masses:

$$\frac{[MOH][H^+][A^-]}{[M^+][A^-][H_2O]} = K_H = \frac{K_W}{K_{MOH}}$$

where K_H is the *hydrolysis constant* of the salt
K_W is the ionic product of water
K_{MOH} is the dissociation constant of the weak base.

Similarly the equation for the hydrolysis of a salt of a weak acid HA and a strong base MOH is:

$$M^+ + A^- + H_2O \rightleftharpoons M^+ + OH^- + HA$$

and the hydrolysis constant is given by:

$$\frac{[M^+][OH^-][HA]}{[M^+][A^-][H_2O]} = K_H = \frac{K_W}{K_{HA}}$$

Two other aspects of the hydrolysis of salts must be mentioned. Certain salts (e.g. the trichlorides of bismuth and antimony) undergo hydrolysis to give a precipitate of an oxysalt when their solution in hydrochloric acid is diluted:

$$BiCl_3 + H_2O \rightleftharpoons BiOCl\downarrow + 2HCl$$

(bismuth oxychloride or bismuthyl chloride)

These hydrolyses are reversible and the precipitate redissolves on adding concentrated hydrochloric acid. Secondly, some salts of weak acids and weak bases are so completely and rapidly hydrolysed that they cannot exist in aqueous solution. For example, aluminium sulphide gives an immediate precipitate of aluminium hydroxide when added to water, and magnesium nitride and calcium phosphide are instantly hydrolysed to their respective hydride and hydroxide:

$$Al_2S_3 + 6H_2O = 2Al(OH)_3\downarrow + 3H_2S\uparrow$$
$$Mg_3N_2 + 6H_2O = 3Mg(OH)_2\downarrow + 2NH_3\uparrow$$
$$Ca_3P_2 + 6H_2O = 3Ca(OH)_2\downarrow + 2PH_3\uparrow$$

9.7. The Common Ion Effect. If a strong electrolyte such as a salt is added to a solution of a weak electrolyte so chosen because one of the ions into which it dissociates in solution is the same as one of the ions from the salt, the degree of dissociation of the weak electrolyte is decreased as a result of the *common ion effect*. The strong electrolyte is, of course, completely dissociated into its ions in solution, so that a high concentration of these ions is produced when it dissolves. This affects the equilibrium between the undissociated molecules of the weak electrolyte and its component ions, causing some of the latter to combine so that their concentrations are restored to a level in keeping with the dissociation constant of the weak electrolyte. This effect is well illustrated by considering a saturated solution of hydrogen sulphide. Ignoring the two-stage dissociation of hydrosulphuric acid, we can express its partial dissociation by the equation:

$$H_2S \rightleftharpoons 2H^+ + S^{2-}$$

so that at constant temperature $\frac{[H^+]^2.[S^{2-}]}{[H_2S]} = K_{H_2S}$

If a strong acid such as hydrochloric acid is now added to the solution, the concentration of H^+ ions is enormously increased thus:

$$HCl \rightarrow H^+ + Cl^-$$

As a consequence H^+ and S^{2-} ions unite to form undissociated H_2S molecules until the expression $\frac{[H^+]^2.[S^{2-}]}{[H_2S]}$ is again equal to K_{H_2S} at that temperature. Thus the effect of adding the hydrochloric acid is to suppress the dissociation of the hydrosulphuric acid and substantially lower the concentration of free S^{2-} ions in solution. Since H_2S is a weak electrolyte only slightly dissociated into ions in saturated solution, the value of $[H_2S]$ is relatively large and shows little percentage change, so the product $[H^+]^2.[S^{2-}]$ is approximately constant. Thus the sulphide ion concentration is approximately inversely proportional to the square of the hydrogen ion concentration and a tenfold increase in the latter reduces the former to about a hundredth of its original value.

Another example is provided by the addition of ammonium chloride (or ammonium nitrate) to a solution of ammonium hydroxide, which has the effect of partially suppressing its dissociation and lowering the concentration of OH^- ions in solution. The common ion here is the NH_4^+ ion, as the following equations indicate:

$$NH_4Cl \rightarrow NH_4^+ + Cl^-$$
$$NH_4OH \rightleftharpoons NH_4^+ + OH^-$$

Nor is the common ion effect limited to cases in which the dissociation of the weak electrolyte is reduced. For example, if ammonium hydroxide solution is added to a saturated solution of hydrosulphuric acid, H_2S, the removal of H^+ ions from the system by combining with OH^- ions from the ammonium hydroxide causes the acid to dissociate more extensively. Here OH^- ions are common to the dissociations of both the ammonium hydroxide and the water and the H^+ ions to the dissociations of the water and the hydrosulphuric acid.

As explained in a later section, advantage is taken of these examples to separate the metals into various groups in qualitative analysis. Thought will show that no new principle is involved in the common ion effect, however, and that the results it achieves can be successfully predicted and explained by applying Le Chatelier's principle to the various ionic equilibria concerned.

One example of the common ion effect which is frequently quoted is the addition of hydrogen chloride gas to a nearly saturated solution of sodium chloride, when precipitation of the salt occurs (see § 17.10). The explanation which is usually given is that on dissolving in the water the hydrogen chloride dissociates strongly into H^+ and Cl^- ions, and the high concentration of the common chloride ion causes the solubility of the sodium chloride to be exceeded. There is undoubtedly some truth in this, but the common ion effect does not wholly account for the precipitation since a precipitate of sodium chloride is also obtained if hydrogen bromide gas is bubbled into the solution. It seems likely that when these very soluble gases dissolve and dissociate, substantial numbers of water molecules are used for hydrating the resulting ions and this increases the effective concentration of the sodium chloride causing its precipitation.

9.8. Solubility Product. We can determine the conditions for precipitation by applying the law of mass action to the equilibrium that exists between a sparingly soluble salt in the solid state and its ions in solution. Let us consider a saturated solution of the binary electrolyte MA in contact with an excess of undissolved solid at constant temperature. The following equilibrium is set up, since such a salt is a strong electrolyte completely dissociated into its ions in solution:

$$\text{MA solid} \rightleftharpoons M^+ + A^-$$

Applying the law of mass action to this and taking the concentrations of the individual ions as their active masses, we get:

$$\frac{[M^+][A^-]}{[\text{MA solid}]} = \text{constant}$$

Now the active mass of any solid is a constant at constant temperature, so [MA solid] is constant.

$\therefore$ $[M^+][A^-] = \text{constant} = S.P._{M.A.}$, the *solubility product* of the salt MA at that temperature.

For example, the solubility product of silver bromide, $S.P._{AgBr}$, is $[Ag^+][Br^-]$ and has the value 4×10^{-13} mol dm^{-3}. Similarly, for an electrolyte M_pA_q which dissociates thus:

$$M_pA_q \rightleftharpoons pM^{m+} + qA^{n-}$$

the solubility product is the product $[M^{m+}]^p[A^{n-}]^q$, in which the ionic concentrations are raised to the appropriate power. Thus the expression for the solubility product of calcium phosphate, which has the formula $Ca_3(PO_4)_2$, is $[Ca^{2+}]^3.[PO_4{}^{3-}]^2$.

The importance of solubility product is that it indicates the maximum possible value of the ionic product in a solution when the ionic concentrations are expressed in mol dm^{-3}; in an unsaturated solution, of course, the ionic product will have a lower value. The concept of solubility product can only be applied to sparingly soluble electrolytes because only in those cases are the ionic concentrations sufficiently low to be taken as their active masses,* and it is altogether wrong, therefore, to refer to the solubility product of soluble salts such as sodium chloride. A consequence of this limitation is that the numerical value of

* The concept of ionic activities, devised to take account of inter-ionic attraction, is beyond the scope of this book.

solubility products is always very small, rarely exceeding 10^{-2} and sometimes being as low as 10^{-50} for salts of extremely low solubility. Since solubility product varies with temperature, the latter should always be quoted if it is other than the arbitrarily selected standard of 25°C.

The solubility product of a salt can be calculated from its solubility in the following way:

Example: What is the solubility product of silver chloride if its saturated solution contains 1.60×10^{-3} g dm^{-3} at 20°C? (Ag = 108, Cl = 35.5).

$$\text{Solubility of silver chloride} = \frac{1.60 \times 10^{-3}}{143.5} \text{ mol dm}^{-3}$$

$$= 1.1 \times 10^{-5} \text{ mol dm}^{-3}$$

Now $\quad AgCl \text{ solid} \rightleftharpoons Ag^+ + Cl^-$

Thus the concentration of Ag^+ and Cl^- ions in the saturated solution is the same and when expressed in mol dm^{-3} is equal to the solubility of the salt in mol dm^{-3}, since silver chloride is a strong electrolyte and is completely dissociated into its ions in solution.

$$\therefore [Ag^+] = [Cl^-] = 1.1 \times 10^{-5} \text{ mol dm}^{-3}$$

$$\therefore S.P._{AgCl} = [Ag^+].[Cl^-] = 1.1 \times 10^{-5} \times 1.1 \times 10^{-5} \text{ mol}^2 \text{ dm}^{-6}$$
$$= 1.2 \times 10^{-10} \text{ mol}^2 \text{ dm}^{-6}$$

So the solubility product of silver chloride at 20°C is 1.2×10^{-10} mol^2 dm^{-6}.

Solubility products are usually determined experimentally in this way by first finding the solubility of the sparingly soluble salt either by volumetric analysis or from its molar conductance (§ 8.8).

It is important to realize that although the solubility product of a salt is constant at constant temperature, the concentration of the individual ions in its saturated solution may vary over a very wide range. Naturally these ions are present in equivalent concentrations when the saturated solution is prepared by simply dissolving the salt in the solvent, but there may be a big difference in their concentrations when the solution is prepared by double decomposition or by mixing two solutions with a common ion. In such circumstances the solubility product can be used to determine when precipitation will begin and what concentration of residual ions will remain in solution. These important applications of solubility product are illustrated in the next two sections.

9.9. Application of Solubility Product to Qualitative Analysis.

GROUP* I: The solubility products of silver chloride, mercurous chloride, and lead chloride at 15°C are 1.2×10^{-10} mol^2 dm^{-6}, 2.5×10^{-18}

* The word *group* in this section refers to a collection of elements which are precipitated by the same reagent in qualitative analysis and not, of course, to the periodic classification.

$mol^2\ dm^{-6}$, and $2.0 \times 10^{-4}\ mol^3\ dm^{-9}$ respectively. Thus if any of these three metals are present in the solution when the hydrochloric acid is added they will be precipitated as their chlorides because these solubility products will be exceeded. For example, if a silver salt is present and the concentration of Cl^- ions in the solution after precipitation is 0.5 $mol\ dm^{-3}$ owing to the excess of hydrochloric acid, if follows that the concentration of Ag^+ ions remaining in the solution cannot exceed the very low figure of $2.4 \times 10^{-10}\ mol\ dm^{-3}$, since $[Ag^+][Cl^-] = 1.2 \times 10^{-10}$ at room temperature.

The solubility product of lead chloride increases markedly with temperature, which explains why it is always necessary to cool a group I solution thoroughly before proceeding to group II, lest the bulk of any lead present should escape precipitation in group I because the solubility product of its chloride is not exceeded. Even at room temperature precipitation of lead chloride is not complete and enough lead ions usually remain in solution to give a precipitate of lead sulphide in group II. Another interesting feature of lead is its ability to form a soluble complex ion (§ 9.16) with Cl^- ions when the concentration of the latter is very high:

$$Pb^{2+} + 4Cl^- \rightleftharpoons [PbCl_4]^{2-}.$$

This explains why a group I solution which has been made by dissolving the solid to be analysed in concentrated hydrochloric acid should always be diluted, since the formation of the complex ion so reduces the concentration of the simple lead ion that again the solubility product of lead chloride is often not exceeded.

Group II: The precipitating reagent here is hydrogen sulphide in the presence of hydrochloric acid. As has already been explained in § 9.7, the hydrochloric acid suppresses the dissociation of hydrosulphuric acid by the common ion effect, lowering the concentration of S^{2-} ions from about $10^{-15}\ mol\ dm^{-3}$ (which is its value in a saturated solution at room temperature) to about $10^{-25}\ mol\ dm^{-3}$. Since the concentration of metal ions in the group solution will be of the order of $10^{-1}\ mol\ dm^{-3}$, this means that only the least soluble sulphides with solubility products below about 10^{-26} will be precipitated and the others will remain in solution. Thus sulphides such as bismuth (10^{-72}). mercuric mercury (10^{-54}), copper (10^{-45}), cadmium (10^{-29}), lead (10^{-29}), and tin (10^{-28}) appear in this group, whereas those of nickel (10^{-24}), zinc (10^{-23}), cobalt (10^{-22}), and manganese (10^{-15}) do not. This enables us to limit group II to some seven or eight metals and so facilitates their identification.

If the concentration of hydrochloric acid is made too high, then the dissociation of hydrosulphuric acid may be suppressed to such an extent that even lead, cadmium, and stannous sulphides, whose solubility products are near the cut-off value, may not be precipitated in this group. To guard against this danger, after treatment with hydrogen sulphide the group solution should always be diluted and tested again with hydrogen sulphide before proceeding to group III.

GROUP III: The precipitating reagent in this group is ammonium hydroxide in the presence of a high concentration of ammonium chloride. As explained in § 9.7, the dissociation of the ammonium hydroxide is suppressed by the common ion effect exerted by the ammonium chloride, so that the solubility products of some of the more soluble hydroxides are not exceeded and their precipitation does not occur in this group. The solubility products of the hydroxides of chromium, aluminium, and ferric iron are so low (10^{-30}, 10^{-33}, and 10^{-38} respectively), that these three are completely precipitated as their hydroxides. The following calculation shows that this is not the whole truth, however, and that some other factor must be at work as well to prevent the precipitation of nickel, cobalt, and zinc as their hydroxides.

Let us assume that the ammonium hydroxide in the group solution is decinormal:

$$NH_4OH \rightleftharpoons NH_4^+ + OH^-$$

Then, since it is a weak electrolyte and its degree of dissociation α is small, *when no ammonium chloride is added,*

$$\alpha^2/V = K_{NH_4OH} = 1.8 \times 10^{-5}, \text{ and } V = 10 \text{ dm}^3$$
$$\therefore \alpha^2 = 1.8 \times 10^{-4} \quad \text{and} \quad \alpha = 1.3 \times 10^{-2}$$

$$\therefore [OH^-] = \frac{\alpha}{V} = 1.3 \times 10^{-3} \text{ mol dm}^{-3}$$

On adding the ammonium chloride the concentration of the NH_4^+ ion is raised to about 1.5M, since this salt is a strong electrolyte completely dissociated into ions in solution:

$$NH_4Cl \rightarrow NH_4^+ + Cl^-$$

$\therefore$ $[NH_4^+] \simeq 1.5$, since the NH_4^+ ions from the ammonium hydroxide can be neglected compared with those from the strong electrolyte ammonium chloride.

But $$\frac{[NH_4^+][OH^-]}{[NH_4OH]} = K_{NH_4OH} = 1.8 \times 10^{-5} \text{ mol dm}^{-3}$$

$$\therefore \frac{1.5 \times [OH^-]}{0.1} = 1.8 \times 10^{-5}, \text{ and } [OH^-] = 1.2 \times 10^{-6} \text{ mol dm}^{-3}$$

Thus the effect on the OH^- ion concentration of adding the ammonium chloride is to reduce it about a thousandfold from about 10^{-3} to 10^{-6} mole per cubic decimetre.

Now for a trivalent metal the solubility product is given by

$$[M^{3+}][OH^-]^3 = S.P._{M(OH)_3}.$$

Assuming that the concentration of the metal ion $[M^{3+}]$ is originally 10^{-1} mol dm^{-3}, the ionic product will be $10^{-1} \times (10^{-3})^3 = 10^{-10}$ without ammonium chloride and $10^{-1} \times (10^{-6})^3 = 10^{-19}$ in the presence of ammonium chloride, so in either case the hydroxides of aluminium, chromium, and ferric iron will be precipitated.

For a divalent metal, however, the solubility product is given by $[M^{2+}][OH^-]^2 = S.P._{M(OH)_2}$. Again assuming that $[M^{2+}]$ is originally 10^{-1} mol dm^{-3}, the ionic product will be about $10^{-1} \times (10^{-3})^2 = 10^{-7}$ in the absence of ammonium chloride and about $10^{-1} \times (10^{-6})^2 = 10^{-13}$ in its presence. Now the solubility products of nickel, cobalt, zinc, and manganese hydroxides are about 10^{-19}, 10^{-18}, 10^{-17}, and 10^{-14} respectively, so according to the above calculation, all these divalent metals would be expected to be precipitated as their hydroxides in this group, even in the presence of the ammonium chloride, since all their solubility products would be exceeded.

The main explanation of this conflict between experimental fact (that only the hydroxides of aluminium, chromium, and ferric iron are precipitated in this group) and theoretical prediction based upon the solubility product concept almost certainly lies in the formation of soluble complex ions with ammonia molecules, which abound in ammonium hydroxide solution.* Three of the metals concerned, nickel, cobalt, and zinc, are known to form such complex ions very freely, e.g. $[Ni(NH_3)_6]^{2+}$, $[Co(NH_3)_6]^{3+}$, and $[Zn(NH_3)_2]^{2+}$. This would remove many of the simple metallic ions from the solution and prevent the solubility products of their hydroxides from being exceeded in the presence of ammonium chloride. Manganese, however, is not much affected in this way, and being near the border line of the solubility product cut-off it tends to be precipitated slowly as its hydroxide in group III if insufficient ammonium chloride is added. This is particularly liable to happen if one of the group III metals is present as well as manganese; the phenomenon, which is known as co-precipitation, is best avoided by using ample ammonium chloride and by filtering the group III solution promptly.

GROUP IV: In this group the precipitating reagent is either hydrogen sulphide in the presence of ammonium hydroxide solution or ammonium sulphide solution. In either case the concentration of S^{2-} ions in

* The concentration of free ammonia molecules in ammonia solution far exceeds the concentration of ammonium hydroxide molecules, although the fact is rarely recognized in equations which should strictly be written:

$$NH_3.H_2O \rightleftharpoons NH_4^+ + OH^-$$

solution is very high (as explained in § 9.7) and those metals with insoluble sulphides which escaped precipitation in group II (i.e. nickel, cobalt, zinc, and manganese) will be precipitated here.

GROUP V: The precipitating reagent is ammonium carbonate which gives a high concentration of CO_3^{2-} ions in solution and brings about the complete precipitation of calcium, strontium, and barium carbonates which have solubility products of 10^{-8}, 10^{-9}, and 8×10^{-9} respectively.

9.10. Application of Solubility Product to Volumetric Analysis. The concept of solubility product can be applied to precipitation reactions in volumetric analysis enabling us to calculate the ionic concentrations at the end-point. A good example is afforded by the titration of a neutral chloride solution with silver nitrate solution by Mohr's method, using potassium chromate as the indicator:

$$\text{e.g. } AgNO_3 + NaCl = AgCl\downarrow + NaNO_3$$

Now $$AgCl \text{ solid} \rightleftharpoons Ag^+ + Cl^-$$

$$\therefore [Ag^+][Cl^-] = S.P._{AgCl} = 1.2 \times 10^{-10} \text{ mol}^2 \text{ dm}^{-6}$$

The desired end-point of this titration is when neither silver nor chloride ions are in excess in the solution, i.e. when

$$[Ag^+] = [Cl^-] = \sqrt{1.2 \times 10^{-5}} = 1.1 \times 10^{-5} \text{ mol dm}^{-3}$$

We must aim to use, therefore, that concentration of chromate indicator which just gives a precipitate of silver chromate when the concentration of Ag^+ ions in solution is 1.1×10^{-5} mol dm^{-3}.

Now $$Ag_2CrO_4 \rightleftharpoons 2Ag^+ + CrO_4^{2-}$$

$$\therefore [Ag^+]^2[CrO_4^{2-}] = S.P._{Ag_2CrO_4} = 2.4 \times 10^{-12} \text{ mol}^3 \text{ dm}^{-9}$$

$$\therefore [CrO_4^{2-}] = \frac{2.4 \times 10^{-12}}{[Ag^+]^2} = \frac{2.4 \times 10^{-12}}{1.2 \times 10^{-10}} \text{ mol dm}^{-3}$$

$$= 2.0 \times 10^{-2} \text{ mol dm}^{-3}$$

Thus the concentration of chromate ion should be about M/100 at the end-point, which means that it ought to be about M/50 at the start of the titration. This can be achieved by adding about 0.5 cm³ of bench potassium chromate solution to each 25 cm³ of halide solution in the conical flask.

This case illustrates very well that one substance may be less soluble than another and yet have a higher solubility product. As the example in § 9.8 shows, the solubility of silver chloride is 1.1×10^{-5} mol dm^{-3} or 1.60×10^{-3} g dm^{-3} at 20°C. The solubility of silver chromate can be calculated from its solubility product as follows:

$$[Ag^+]^2[CrO_4^{2-}] = S.P._{Ag_2CrO_4} = 2.4 \times 10^{-12} \text{ mol}^3 \text{ dm}^{-9}$$

Now in its saturated solution $[Ag^+] = 2[CrO_4^{2-}]$

$$\therefore\ 4[CrO_4^{2-}]^3 = 2.4 \times 10^{-12}\ \text{mol}^3\ \text{dm}^{-9}$$

$$\text{and } [CrO_4^{2-}] = \sqrt[3]{\frac{2.4}{4}} \times 10^{-4} = \sqrt[3]{0.6} \times 10^{-4} = 0.84 \times 10^{-4}\text{mol dm}^{-3}$$

Thus the solubility of silver chromate is 8.4×10^{-5} mole per cubic decimetre or $8.4 \times 10^{-5} \times 332 = 2.8 \times 10^{-2}$ g dm^{-3} at 20°C. The greater solubility of silver chromate is explained by the occurrence of the term $[Ag^+]^2$ in the expression for its solubility product.

9.11. Solubility of Salts of Weak Acids in Strong Acids. It is an experimental fact that salts of weak acids are much more soluble in a solution of a strong acid than they are in water. For example, calcium orthophosphate is only sparingly soluble in water at room temperature, yet it dissolves freely in dilute hydrochloric acid. This phenomenon can be explained by considering the ionic equilibria existing in the solution.

When solid calcium phosphate is added to water and left to achieve equilibrium a saturated solution is formed in which the concentrations of calcium and phosphate ions are very low and are governed by the relationship:

$$Ca_3(PO_4)_2\ \text{solid} \rightleftharpoons 3\ Ca^{2+} + 2\ PO_4^{3-}$$

where $$[Ca^{2+}]^3[PO_4^{3-}]^2 = S.P._{Ca_3(PO_4)_2} = 10^{-20}$$

The water is also dissociated, of course, but only to a very small extent, so that the concentration of H^+ and OH^- ions in solution is also very low:

$$H_2O \rightleftharpoons H^+ + OH^-$$

Under these circumstances very little combination of H^+ and PO_4^{3-} ions takes place, despite the fact that orthophosphoric acid is a weak acid, because the expression $\frac{[H^+]^3[PO_4^{3-}]}{[H_3PO_4]}$ soon attains the value corresponding to $K_{H_3PO_4}$:

$$H_3PO_4 \rightleftharpoons 3H^+ + PO_4^{3-}$$

Now when hydrochloric acid is added it dissociates completely into ions, so the concentration of H^+ ions in the solution is raised to a relatively very high level:

$$HCl \rightarrow H^+ + Cl^-$$

This causes hydrogen ions and phosphate ions to unite in large numbers to form undissociated molecules of orthophosphoric acid, so that eventually the expression $\frac{[H^+]^3[PO_4^{3-}]}{[H_3PO_4]}$ is restored to its original low value of $K_{H_3PO_4}$. In the process the PO_4^{3-} ion concentration is reduced to such a low level that the ionic product of calcium phosphate falls well below its solubility product causing more of the solid calcium phosphate to dissolve. The interaction between the various equilibria continues until a high concentration of orthophosphoric acid molecules exists in solution and much of the solid calcium phosphate has gone into solution. Thus the overall effect of the hydrochloric acid is to suppress the already small dissociation of the weak orthophosphoric acid by means of the common ion effect and thereby artificially to lower the concentration of phosphate ions in the solution so that more of the salt dissolves. A similar explanation can be provided for the solubility of salts of other weak acids, e.g. oxalates and tartrates.

9.12. Indicators. An indicator for acid–alkali titrations is a substance which shows different colours when the pH of its environment is altered. It is usually a weak organic acid or base capable of undergoing a tautomeric change (Table 48.1) within its molecule, which accounts for the colour change, but for our purposes it will suffice to regard it simply as a substance which dissociates into an ion of different colour thus:

$$\underset{\text{colour 1}}{HA} \rightleftharpoons H^+ + \underset{\text{colour 2}}{A^-}$$

Since the dissociation is only partial and is reversible, its extent will depend very much upon the hydrogen ion concentration of the solution to which it is added. In a strongly acidic solution, as Le Chatelier's principle predicts, the dissociation would be suppressed and colour 1 would predominate, whereas in a solution with a low concentration of H^+ ions the dissociation would be largely complete and the colour of the anion, colour 2, would be seen. In practice, although particles of both colours will always be present in the solution, only one colour will be noticeable, the predominant one, unless the ratio of the concentrations of the two types of particle is less than 10 : 1, when some intermediate colour will be observed. This means that as far as the human eye is concerned the colour change of an indicator takes place gradually over a band of about two units of pH, only being complete when there is a hundredfold change in H^+ ion concentration. It follows that if an indicator is to give a precise end-point by changing colour sharply on adding a single drop of acid or alkali, it is necessary that the pH of the solution should change suddenly by at least two units.

Now although indicators are weak electrolytes, they vary greatly in their dissociation constants, which means that they change colour over widely different bands of pH. This point is illustrated in Fig. 9.2, which shows the ranges of some well-known indicators. The H^+ ion concentration of the centre-point of each range is easily calculated from the value of the dissociation constant, since $\frac{[H^+][A^-]}{[HA]} = K_{Ind.}$ and so $[H^+] = \frac{K_{Ind.}[HA]}{[A^-]}$. Now at the centre-point the two types of coloured particle will be present in equal concentrations, i.e. $[HA] = [A^-]$, and $[H^+]$ will therefore be equal to $K_{Ind.}$. This explains why air-free distilled water is alkaline to methyl orange ($K_{Ind.} = 2 \times 10^{-4}$ mol dm^{-3}, so central pH is 3.7, neutral to litmus ($K_{Ind.} = 1 \times 10^{-7}$ mol dm^{-3}, so central pH is 7.0), and acidic to phenolphthalein ($K_{Ind.} = 2 \times 10^{-10}$ mol dm^{-3}, so central pH is 9.7). At first sight the fact that various

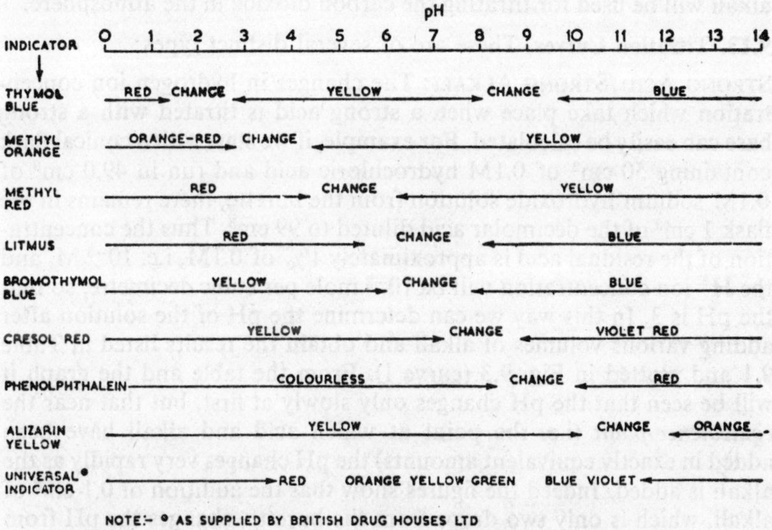

FIG. 9.2 Colours and Ranges of Common Indicators

indicators change colour at different H^+ ion concentrations might seem to be a disadvantage in judging endpoints since they would appear to disagree in their verdicts, but in practice, as we shall see in later sections, this difference between them is of great value and importance.

When a carbonate is titrated with an acid, the solution becomes saturated with carbon dioxide and its pH changes to about 4.4. Although this is a big enough change to interfere when methyl red is the indicator, it does not affect methyl orange, which is why the latter is preferred for titrations involving carbonates. If methyl red is used for such titrations it is necessary to boil off the carbon dioxide from the solution before judging the end-point.

When water is stored in an open vessel an equilibrium is soon set up with the gases in the atmosphere, and the carbon dioxide which dissolves changes the pH to about 5.7. This is important when phenolphthalein is being used as the indicator and alkali is being added from the burette, since owing to the action of carbon dioxide upon pH the pink-purple colour which marks the end-point disappears again after a short period If a further drop of alkali is added the pink colour reappears, since the pH is then temporarily restored to about 10, only to fade away again slowly on standing in air as the pH of the solution falls. The end-point in such a titration is clearly the moment when the indicator colour persists on shaking for, say, five seconds, since otherwise the alkali will be used for titrating the carbon dioxide in the atmosphere.

9.13. Titration Curves. These are of several distinct types:

STRONG ACID/STRONG ALKALI: The changes in hydrogen ion concentration which take place when a strong acid is titrated with a strong base can easily be calculated. For example, if we start with a conical flask containing 50 cm^3 of 0.1M hydrochloric acid and run in 49.0 cm^3 of 0.1M sodium hydroxide solution from the burette, there remains in the flask 1 cm^3 of the decimolar acid diluted to 99 cm^3. Thus the concentration of the residual acid is approximately 1% of 0.1M, i.e. 10^{-3} M, and the H^+ ion concentration will be 10^{-3} mole per cubic decimetre, so that the pH is 3. In this way we can determine the pH of the solution after adding various volumes of alkali and obtain the results listed in Table 9.1 and plotted in Fig. 9.3 (curve I). From the table and the graph it will be seen that the pH changes only slowly at first, but that near the *equivalence-point* (i.e. the point at which acid and alkali have been added in exactly equivalent amounts) the pH changes very rapidly as the alkali is added. Indeed the figures show that the addition of 0.1 cm^3 of alkali, which is only two drops from the burette, changes the pH from 4.3 to 9.7, producing a half-millionfold change in hydrogen ion concentration. If normal solutions are used the effect is even more striking, the hydrogen ion concentration being made over ten million times as small by the addition of only a few drops of alkali near the equivalence point. These calculated figures can be confirmed experimentally by using a pH meter.

Now the aim of any titration is to determine the equivalence-point, i.e. to find what volumes of the two solutions are chemically equivalent to each other. Ideally the end-point, which is the point at which the titration is stopped, should coincide exactly with this equivalence-point, and to achieve this the indicator used must show a sharp colour change at the right pH. From titration curve I it should be clear that any indicator which changes colour between pH 4 and pH 10 will be satisfactory when titrating a strong acid with a strong base in view of the sudden and very wide swing in pH at the equivalence-point. Thus methyl orange and phenolphthalein will both change colour at about the same time in this type of titration and either may be used with confidence.

WEAK ACID/STRONG ALKALI: When a 0.1M solution of acetic acid (a typical weak acid) is titrated with a 0.1M solution of sodium hydroxide, the change in pH is as shown in curve II of Fig. 9.3. There is very little

TABLE 9.1. TITRATION OF 50 CM³ OF 0.1M HCl WITH STRONG ALKALI

Volume of alkali added / (cm³ 0.1M NaOH)	$[H^+]$ / mol dm^{-3}	pH
0	10^{-1}	1.0
25.0	3.3×10^{-2}	1.5
45.0	5×10^{-3}	2.3
49.0	1×10^{-3}	3.0
49.5	5×10^{-4}	3.3
49.9	1×10^{-4}	4.0
49.95	5×10^{-5}	4.3
50.00	1×10^{-7}	7.0
50.05	2×10^{-10}	9.7
50.1	1×10^{-10}	10.0
50.5	2×10^{-11}	10.7
51.0	1×10^{-11}	11.0
55.0	2×10^{-12}	11.7

variation at first as the alkali is added, the pH being considerably higher for a given volume of alkali than in curve I because the salt formed is extensively hydrolysed, but near the equivalence-point there is again a rapid swing in pH, although the change is not as great as before. These results are calculated by making certain assumptions which are only approximately true. For example, the presence of acetate ions in solution is regarded as being solely due to the complete dissociation of sodium acetate, so their concentration is taken as the concentration of alkali added. Again the concentration of undissociated acetic acid molecules present is calculated from the amount of residual acid, since it is a weak acid only slightly dissociated in solution.

Then since
$$\frac{[H^+].[CH_3COO^-]}{[CH_3COOH]} = K_{CH_3COOH} = 1.8 \times 10^{-5} \text{ mol dm}^{-3}$$

$$[H^+] = 1.8 \times 10^{-5} \times \frac{[CH_3.COOH]}{[CH_3.COO^-]} \text{ mol dm}^{-3}$$

$$\simeq 1.8 \times 10^{-5} \times \frac{[\text{residual acid}]}{[\text{alkali added}]} \text{mol dm}^{-3}$$

Thus when the amount of alkali added is enough to neutralize exactly half the acid originally taken, so that the concentrations of acetic acid molecules and acetate ions are the same, the hydrogen ion concentration

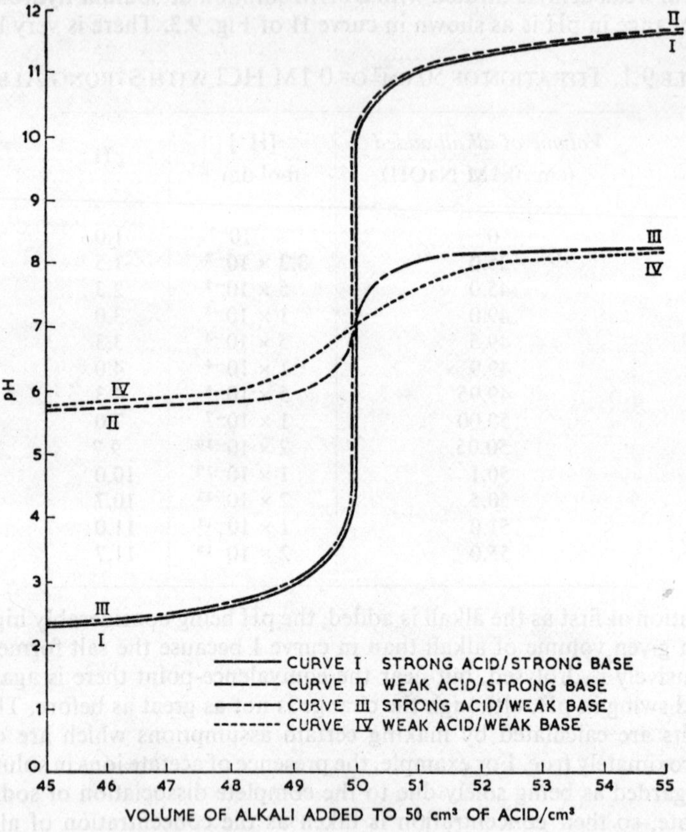

FIG. 9.3 Titration Curves

is numerically equal to the dissociation constant of the acid, i.e. 1.8×10^{-5} mol dm^{-3}, and the pH is then 4.74.

In this titration the best indicator is one which changes colour when the pH is about 9, since it is then that the equivalence-point is reached and the pH changes most rapidly with added alkali (see curve II in Fig. 9.3). Thus phenolphthalein, which changes colour in this region is a very suitable indicator, whereas the end-point indicated by methyl orange (at pH 3.7, say) would be seriously different from the equivalence-point and would not, in any case, be precise, because at that stage the addition of a few drops of alkali produces only a small change in pH.

STRONG ACID/WEAK ALKALI: Using a similar procedure it is possible to construct the *titration curve*, as it is called, for the neutralization of a strong acid by a weak base, e.g. hydrochloric acid and ammonium hydroxide solution. This is shown as curve III in Fig. 9.3. From this curve it is clear that the equivalence-point is reached when the pH is about 4 or 5 and a suitable indicator must change colour in that region. Moreover, it is then that the pH is most responsive to addition of alkali, so a sharp and precise end-point is possible. Methyl orange and methyl red are both good indicators, therefore, for this type of titration, whereas phenolphthalein is useless and will give a large error.

WEAK ACID/WEAK ALKALI: Curve IV in Fig. 9.3 shows the changes in pH which occur when a weak acid such as acetic acid is titrated against a weak base such as ammonium hydroxide. The extensive

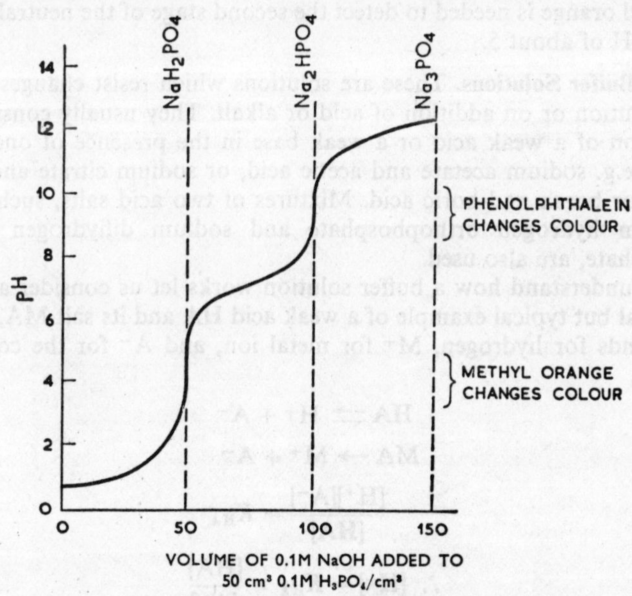

FIG. 9.4 Titration Curve of a Polybasic Acid

hydrolysis of the salt formed in this type of titration prevents any sudden swing in pH near the equivalence-point. No indicator will be satisfactory, therefore, since none will give a sharp and reliable end-point. For this reason this type of titration must be avoided even if it means titrating the weak acid and the weak base against some strong electrolytes whose relative concentrations are known.

POLYBASIC ACIDS: Another type of titration curve which is of particular interest is the one obtained when an alkali is added to a polybasic acid. Fig. 9.4 shows the changes in pH which occur when orthophosphoric acid, H_3PO_4, is titrated with sodium hydroxide. This titration curve shows two rapid swings in pH, neither of them very sharp, corresponding to the first and second stages of the acid's dissociation. A third inflection of the curve, this time very indistinct, occurs in strongly alkaline solution. This particular diagram explains why methyl orange is suitable for detecting the first end-point and phenolphthalein the second, and why no indicator is satisfactory for judging the position of the third dissociation (see § 23.10).

A curve with two inflections is obtained when sodium carbonate is titrated with a strong acid. When the carbonate has been changed to bicarbonate the pH is about 9, so phenolphthalein responds, but methyl orange is needed to detect the second stage of the neutralization at a pH of about 5.

9.14. Buffer Solutions. These are solutions which resist changes in pH on dilution or on addition of acid or alkali. They usually consist of a solution of a weak acid or a weak base in the presence of one of its salts, e.g. sodium acetate and acetic acid, or sodium citrate and citric acid, or borax and boric acid. Mixtures of two acid salts, such as disodium hydrogen orthophosphate and sodium dihydrogen orthophosphate, are also used.

To understand how a buffer solution works let us consider a hypothetical but typical example of a weak acid HA and its salt MA, where H stands for hydrogen, M^+ for metal ion, and A^- for the common anion.

Then $$HA \rightleftharpoons H^+ + A^-$$

and $$MA \rightarrow M^+ + A^-$$

Now $$\frac{[H^+][A^-]}{[HA]} = K_{HA}$$

$$\therefore\ [H^+] = K_{HA} \times \frac{[HA]}{[A^-]}$$

where [HA] is effectively the concentration of weak acid taken, since it will be only very slightly dissociated in the presence of its salt, and where $[A^-]$ is effectively the concentration of the salt because this is completely dissociated into its ions.

$$\therefore [H^+] \simeq K_{HA} \times \frac{[acid]}{[salt]}.$$

This explains why the hydrogen ion concentration of a buffer solution is not much affected by dilution, since the relative concentration of the acid and salt will remain approximately the same.

If an acid is added to such a buffer solution most of the added H^+ ions will combine with the A^- ions present to form undissociated molecules of the weak acid HA in order to keep the expression $\frac{[H^+][A^-]}{[HA]}$ at its constant value, K_{HA}, and the pH will be little altered as a result. If an alkali is added to the buffer solution, most of the added OH^- ions are removed from solution by combination with H^+ ions to form undissociated molecules of water, further H^+ ions being furnished by the dissociation of HA, so again the pH changes only slightly. Thus the stability of the pH can be attributed to the high concentration of A^- ions, which act as a trap for added H^+ ions, and the large concentration of HA molecules, which provide a reservoir of H^+ ions for combining with added OH^- ions.

The chief use of buffer solutions in the laboratory is for making solutions of known and constant pH. Such solutions cannot be made accurately by adding acids and alkalis to each other in calculated quantities because of the large swing in pH which occurs near the equivalence-point. Nor would they be easily preserved when made because of the sensitivity of their pH to dissolved gases (e.g. carbon dioxide) or alkali from glass vessels. By a suitable choice of substances it is possible to prepare buffer solutions of any pH between 1 and 13. These solutions of known pH can then be used to calibrate indicators or to determine the pH of a solution colorimetrically. Special mixtures of indicators (known as *universal indicators*) are available, which show various characteristic colours for each pH over a wide range (see Fig. 9.2). One of these can be used to find the pH of a solution approximately, and then one particular indicator can be applied to the solution and the colour obtained can be carefully compared by means of a special comparator with the colours shown by that indicator in a series of buffer solutions in which the pH differs by only 0.1.

It is important to be able to calculate the pH of any particular buffer solution and also the composition of the mixture needed to make a buffer solution with a given pH. We have shown that

$$[H^+] \simeq K_{HA} \times \frac{[acid]}{[salt]}$$

Taking logarithms of both sides we get

$$\log_{10}[H^+] = \log_{10} K_{HA} + \log_{10} \frac{[\text{acid}]}{[\text{salt}]}$$

Multiplying each side by -1 gives

$$-\log_{10} [H^+] = -\log_{10} K_{HA} - \log_{10} \frac{[\text{acid}]}{[\text{salt}]}$$

Now $\quad -\log_{10} [H^+] = \text{pH}$ (see § 9.5)

$$\therefore \text{pH} = -\log_{10} K_{HA} - \log_{10} \frac{[\text{acid}]}{[\text{salt}]}$$

It follows from this equation that the pH of a given buffer solution depends upon the *ratio* of the concentrations of the acid and salt , and not upon their actual values. In the special case when acid and salt are present in equimolar concentrations,the hydrogen ion concentration is equal to K_{HA}. Clearly the magnitude of this dissociation constant is the predominant influence upon the pH of a buffer solution, although the relative concentrations of acid and salt taken determine the exact value within a narrow range.

Example: Calculate the pH of a buffer solution which is 0.1 molar with respect to the boric acid ($K = 6.3 \times 10^{-10}$ mol dm^{-3}) and 0.2 molar with respect to borax.

$$\text{pH} = -\log_{10} K - \log_{10} \frac{[\text{acid}]}{[\text{salt}]}$$

$$= -\log_{10} 6.3 \times 10^{-10} - \log_{10} \frac{0.1}{0.2}$$

$$= 9.2 - \bar{1}.7$$

$$= 9.2 + 0.3$$

$$= 9.5$$

Example: In what proportions must 0.1M solutions of acetic acid and sodium acetate be mixed in order to obtain a buffer solution of pH 4.3? ($K_{HAc} = 1.8 \times 10^{-5}$ mol dm^{-3}).

$$\log_{10} \frac{[\text{acid}]}{[\text{salt}]} = -\log_{10} K_{HAc} - \text{pH}$$

$$= -\log_{10} 1.8 \times 10^{-5} - 4.3$$

$$= 4.75 - 4.3$$

$$= 0.45$$

$$\therefore \frac{[\text{acid}]}{[\text{salt}]} = \frac{2.8}{1}$$

Since both constituents are of equimolar concentration, the buffer solution must be made from 2.8 volumes of the acid to 1 volume of the salt.

Buffer solutions are also of great importance in medicine and biochemistry, where pH values are often critical and have to be maintained at a steady value. For example, the blood contains dissolved carbonic acid and bicarbonates and also phosphates which maintain its pH at about 7.4, and a change of ± 0.5 would probably be fatal. Injections into the bloodstream, therefore are usually buffered so as not to upset this delicate balance. Similarly, many fermentation processes and enzyme reactions depend critically upon the pH, which must vary only within narrow limits. Buffer solutions are widely used in electroplating and also in processed foods and drinks to prevent excessive acidity.

9.15. Electrometric Titrations. These are of two types:

1. CONDUCTOMETRIC TITRATION: This involves measuring the conductivity of the solution at various stages of the titration and plotting a graph as in Fig. 9.5. As explained in § 8.8, hydrogen ions and hydroxyl ions have a much greater conductivity than other ions, so that during the titration of a strong acid with a strong base the conductivity of the solution in the conical flask will reach a minimum near the equivalence-point, when the total concentration of these highly conducting ions is least. This minimum in the conductivity is taken as the end-point, as in Fig. 9.5(*a*). A different graph, like that shown in Fig. 9.5(*b*) is obtained

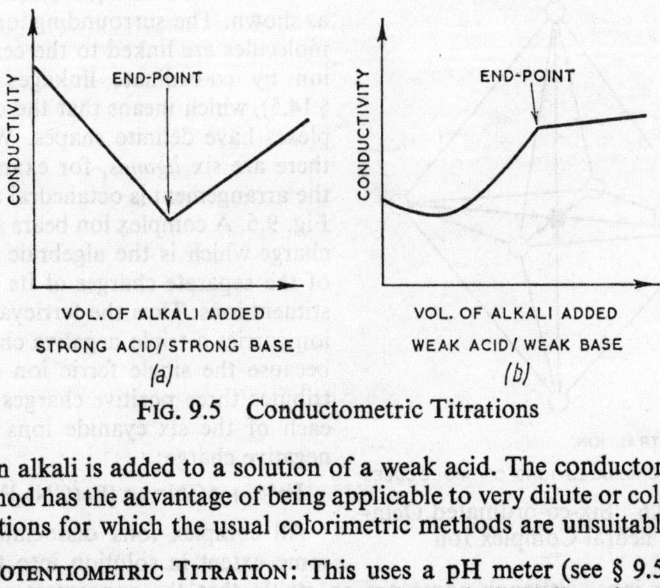

FIG. 9.5 Conductometric Titrations

when alkali is added to a solution of a weak acid. The conductometric method has the advantage of being applicable to very dilute or coloured solutions for which the usual colorimetric methods are unsuitable.

2. POTENTIOMETRIC TITRATION: This uses a pH meter (see § 9.5), i.e. a system involving a calomel reference electrode (§ 10.3) and a special glass electrode which develops a potential directly proportional to the

pH of the solution in which it is immersed. From the readings of potential difference as measured with a potentiometer a titration curve like those in Fig. 9.3 can be plotted. The end-point is taken as the point at which the pH changes most rapidly with added titrant. The method is accurate, even when the solutions are dilute, and can be applied to coloured or opaque solutions. It has the further advantages of needing only small quantities of titrant, of being applicable to many types of titration including redox (§ 15.4) and precipitation reactions, and of being adaptable for automatic use.

9.16. Complex Ions. Many cases are known where an ion, particularly an ion of a transition metal (see § 27.2) joins with two or more ions of opposite charge or with two or more neutral molecules to form a larger ion which has completely different properties from the simple ion(s) it contains. For example, copper(II) forms the following ions:

$$Cu^{2+} + 4NH_3 \rightleftharpoons [Cu(NH_3)_4]^{2+}$$
$$Cu^{2+} + 4CN^- \rightleftharpoons [Cu(CN)_4]^{2-}$$
$$Cu^{2+} + 4H_2O \rightleftharpoons [Cu(H_2O)_4]^{2+}$$
$$Cu^{2+} + 4Cl^- \rightleftharpoons [CuCl_4]^{2-}$$

Larger ions such as these are known as *complex ions*; it is conventional to enclose them in square brackets as shown. The surrounding ions or molecules are linked to the central ion by co-ordinate linkages (see § 14.5), which means that the complexes have definite shapes. When there are six *ligands*, for example, the arrangement is octahedral as in Fig. 9.6. A complex ion bears a net charge which is the algebraic sum of the separate charges of its constituent ions. Thus the ferricyanide ion carries a triple negative charge because the single ferric ion contributes three positive charges and each of the six cyanide ions one negative charge:

$$Fe^{3+} + 6CN^- \rightleftharpoons [Fe(CN)_6]^{3-}$$

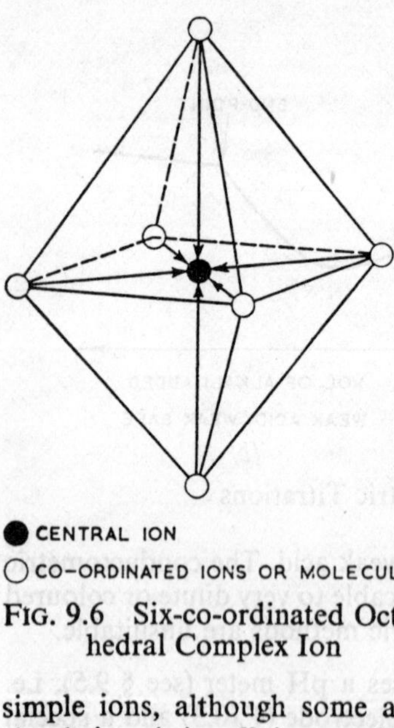

FIG. 9.6 Six-co-ordinated Octahedral Complex Ion

All complex ions dissociate to some extent in solution into their simple ions, although some are so stable that the concentration of simple ions is insufficient to show the usual reactions. For example the hexacyanoferrate(II) ion, formed by adding an excess of potassium cyanide solution to a iron(II) salt, is very stable and when crystals of potassium hexacyanoferrate(II) are dissolved in water the solution gives no precipitate of iron(II) hydroxide with sodium hydroxide solution because the concentration of Fe^{2+} ions is too low to cause the solubility product of the hydroxide to be exceeded. Nor does it give a precipitate of silver cyanide with silver nitrate solution, showing that the concentration of simple cyanide ions must also be extremely low:

$$Fe^{2+} + 6CN^- \rightleftharpoons [Fe(CN)_6]^{4-}.$$

On the other hand, a solution of potassium ferrocyanide gives a gelatinous brown precipitate of cupric ferrocyanide with copper sulphate solution which is not given by solutions containing simple iron(II) or cyanide ions:

$$2\,CuSO_4 + K_4Fe(CN)_6 = Cu_2Fe(CN)_6\downarrow + 2K_2SO_4$$

These differences between the properties of complex ions and their simple ions distinguishes a complex ion from a double salt such as ferrous ammonium sulphate, $FeSO_4.(NH_4)_2SO_4.6H_2O$, which behaves in solution as a mere mixture of the simple ions of which it is composed.

The law of mass action can be applied to the equilibrium existing between a complex ion and its simple constituent ions at constant temperature. If we consider the complex ion formed between zinc ions and ammonia molecules we get:

$$Zn^{2+} + 2NH_3 \rightleftharpoons [Zn(NH_3)_2]^{2+}$$

$$\therefore \frac{[Zn(NH_3)_2]^{2+}}{[Zn^{2+}][NH_3]^2} = K_{stab.}$$

where $K_{stab.}$ is known as the *stability constant* of the ion.

Detection of Complex Ions: The presence of complex ions can be detected in various ways:

1. Metals which form complex anions, e.g. $[Ag(CN)_2]^-$, move towards the anode in electrolysis showing that the particle of which they are a part bears a negative charge overall.

2. When complex ions are formed their solutions show abnormally low colligative effects and lower conductivity than expected because they contain fewer particles of solute than would otherwise be present. For example, when a little mercuric iodide is added to potassium iodide solution the freezing point depression is smaller than before because the simple ions have united as follows:

$$2K^+ + 2I^- + HgI_2 \rightleftharpoons 2K^+ + [HgI_4]^{2-}$$

When the two salts are present in the molecular proportion of two of potassium iodide to one of mercuric iodide, the depression is only three-quarters of its original value since there are only three ions in solution for every four that existed before.

3. When a salt which is only sparingly soluble in water dissolves freely in an aqueous solution, it is usually a sign that a soluble complex ion is formed. For example, silver chloride dissolves readily in ammonium hydroxide solution, which contains a high concentration of ammonia molecules:

$$Ag^+ + Cl^- + 2NH_3 \rightleftharpoons \underset{\text{Soluble complex salt}}{[Ag(NH_3)_2]^+Cl^-}$$

4. Changes in colour often occur when a complex ion is formed. The familiar deep blue colour of the complex ion $[Cu(NH_3)_4]^{2+}$ is a good example.

5. X-ray analysis often reveals the presence of complex ions in the crystal.

USES OF COMPLEX IONS: The soluble complex ion formed between silver ions and thiosulphate ions is used in photography as a means of 'fixing' the negative, i.e. removing the unchanged silver halide after exposure (see § 24.14). The complex ion which copper forms with ammonia, known as Schweitzer's solution, dissolves cellulose and is used in the manufacture of rayon (see § 28.9). Sodium or potassium argento-cyanide* (§ 28.14) and auricyanide* (§ 28.21) are used for extracting silver and gold from their ores and for electroplating (see § 28.15). In analysis complex ions are used for separating precipitates, e.g. silver chloride from mercurous chloride by adding ammonia solution, or detecting one metal of a group in the presence of another, e.g. cadmium in the presence of copper by forming the complex cyanide (see § 28.7). The ferrocyanide* and ferricyanide* ions are used for detecting ferric and ferrous ions respectively (§ 30.18), and the cobaltinitrite* ion $[Co(NO_2)_6]^{3-}$ for precipitating potassium from solution in group 6 (§ 27.15). The complex ion formed from ethylene diamine tetra-acetic acid (EDTA) is used to remove hardness from water (§ 16.5).

9.17. Ion Exchange Processes. The earliest examples of ion exchange substances were some naturally occurring silicates such as zeolites which were found to have the property of removing certain metallic ions from aqueous solution and retaining them in the solid material, releasing other loosely-held ions into solution in their place. From these a complex aluminium silicate known as *permutit* was developed which has been extensively used for softening water because it substituted Na^+ ions for the Ca^{2+} and Mg^{2+} ions present in hard water (§ 16.5) and could be regenerated with concentrated brine.

* For alternative nomenclature, see Appendix I, page 850.

More recently whole ranges of synthetic resins have been made with very valuable ion exchange properties. These resins consist of solid cross-linked polymers (§ 49.2) of high molecular weight derived from certain organic substances such as phenols, formaldehyde, and styrene, to which are attached specific ionizable groups. If these groups are strongly acidic ones such as sulphonic acid, $—SO_3H$, or carboxyl, —COOH, then the resins are used for exchanging cations, but if basic groups such as amino or quaternary ammonium groups are present then they act as anion exchangers. The exchange of ions is a physical and completely reversible process, and like permutit these polymeric resins can be regenerated when exhausted and used over and over again.

Ion exchange processes are of great chemical and industrial importance. They are used for extracting and purifying uranium from its ores (§ 27.20), for separating nuclear fission products from each other, for the purification of water, for separating the lanthanides or rare earth elements, for recovering precious metals from solutions, and for treating and disposing of very poisonous or dangerously radioactive waste. Many clays in soil also act as ion exchangers, retaining ammonium ions from added fertilizer or manure. Ion exchange chromatography (§ 12.12) has been developed as a technique for separating substances in mixtures and has been very successfully applied to amino-acids (§ 41.12) by using anion exchange resins, and to the rare earths (§ 13.8) and the trans-uranic elements (§ 31.7) using cation exchange resins.

10

The Electrochemical Series

10.1. Introduction. One of the most important properties of an element is the comparative readiness with which it gives up or accepts electrons and forms a positively or negatively charged ion. This chapter is primarily concerned with the various ways in which this tendency is expressed.

10.2. Electrode Potential. If a piece of metal is placed in a solution containing its ions (such as a solution of one of its salts) then a potential difference known as the *electrode potential* of the metal will be set up between the metal and the solution. For example, if a rod of zinc is stood in a beaker containing a solution of zinc sulphate, then the zinc will become negatively charged with respect to the solution and will maintain a constant potential difference provided the conditions are not changed. On the other hand when copper is placed in contact with copper sulphate solution the metal assumes a different charge with respect to the solution and its electrode potential is smaller.

The electrode potential can be explained in terms of an equilibrium between the atoms of the metal and its ions in solution. It is believed that when the metal is placed in the solution some of its atoms tend to give up electrons to the piece of metal and go into solution as positively charged ions. At the same time some of the metallic ions already in solution tend to take up electrons from the piece of metal and deposit themselves as neutral atoms. Whichever tendency is greater in a given case determines whether an element becomes negatively or positively charged compared with the solution, but when equilibrium is eventually attained the two opposing processes continue at the same rate and there is no further change in potential difference. A rod of zinc, for example, will bear an accumulation of negative charges owing to the net ionization of some of its atoms, and this will attract an atmosphere of positively charged zinc ions around the electrode to form a double layer as shown in Fig. 10.1. The equilibrium can be represented, therefore, as:

$$Zn \rightleftharpoons Zn^{2+} + 2e$$

Since an electrode potential depends not only upon the nature of the metal but also upon the concentration of its ions in solution and upon the temperature, it is necessary to specify both of these variables before comparing the values for different elements. The potential set up when an electrode is in contact with a molar solution of its ions at 25°C is arbitrarily chosen as the basis for comparison and is known as the *standard electrode potential* of the element. It is not possible to measure the absolute value of an electrode potential directly because in doing so another electrode must necessarily be introduced to complete the circuit and it will have an electrode potential of its own. Thus a potentiometer

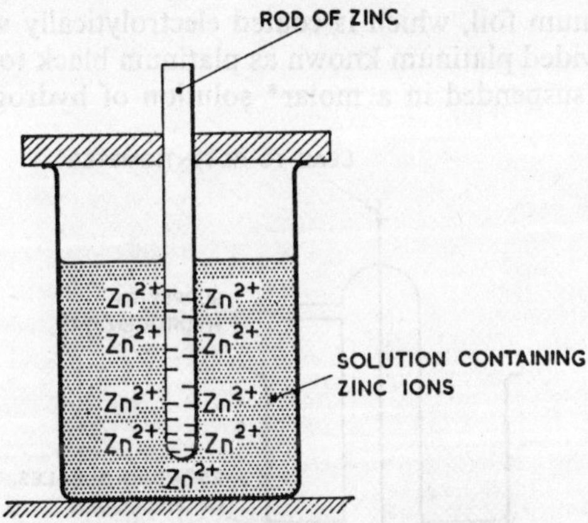

FIG. 10.1 Equilibrium between Zinc and its Ions in Solution

will record only the difference between the values of the two electrode potentials. This difficulty has been overcome by adopting the standard electrode potential of one particular element, hydrogen, as an arbitrary standard and giving it the value zero. The standard electrode potential of any other element is then expressed in terms of the potential difference (in volt) which exists between its electrode and a normal hydrogen electrode, which is described in the next section. This potential difference is known as the standard electrode potential of the element *on the hydrogen scale*. Typical values for some well-known metals are listed in Table 10.1.*

10.3. Measurement of Electrode Potential. The normal hydrogen electrode which is used as a standard is shown in Fig. 10.2. It consists of a

* Many books, particularly those originating in America, use the opposite sign convention, but the one adopted here is that recommended by the International Union of Pure and Applied Chemistry.

TABLE 10.1. STANDARD ELECTRODE POTENTIALS ON THE HYDROGEN SCALE

Element	E_0/V	*Element*	E_0/V
K	−2.92	Sn	−0.14
Ca	−2.87	Pb	−0.13
Na	−2.71	H	0.00
Mg	−2.38	Cu	+0.34
Al	−1.67	Hg	+0.80
Zn	−0.76	Ag	+0.80
Fe	−0.44	Au	+1.55

piece of platinum foil, which is coated electrolytically with a form of very finely-divided platinum known as platinum black to give it a large surface area, suspended in a molar* solution of hydrogen ions. Pure

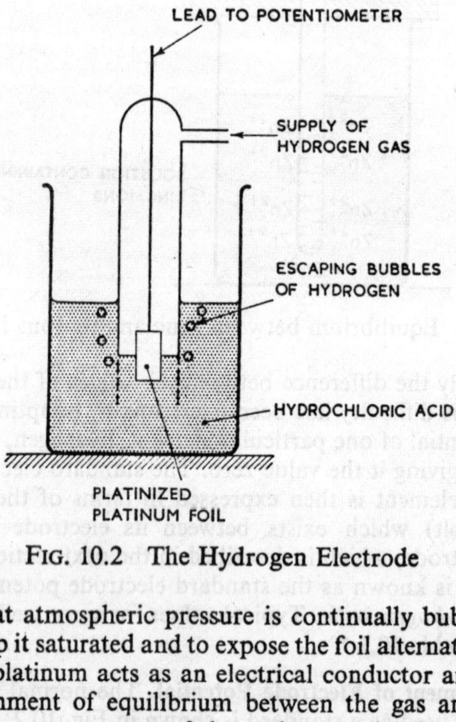

FIG. 10.2 The Hydrogen Electrode

hydrogen gas at atmospheric pressure is continually bubbled into the solution to keep it saturated and to expose the foil alternately to gas and solution. The platinum acts as an electrical conductor and also facilitates the attainment of equilibrium between the gas and its ions in solution. In any measurement of electrode potential the two electrodes concerned must be joined electrolytically. A salt bridge containing a

* In practice a 1.18M solution of hydrochloric acid is used so that the activity (i.e. the effective concentration) of the hydrogen ions in solution is exactly unity.

concentrated solution of potassium chloride is commonly used to provide a highly conducting path between the two, as in Fig. 10.3. The potential difference is measured by means of a potentiometer or valve

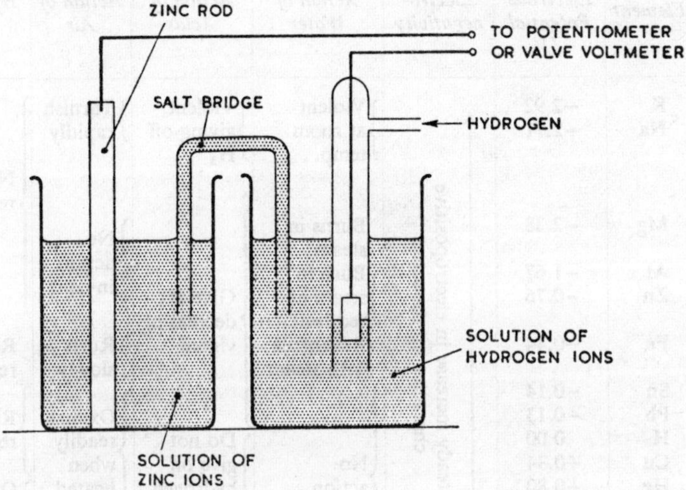

FIG. 10.3 Measurement of Electrode Potential

voltmeter, using a standard cell such as a Weston cell (§ 10.6) for comparison.

The standard electrode potentials of most non-metallic elements have also been determined experimentally. For example, a normal chlorine electrode can be constructed resembling the hydrogen electrode. The electrode potentials of very reactive elements which combine with water (e.g. alkali metals, fluorine, etc.) can be determined indirectly.

Although the normal hydrogen electrode is the ultimate standard in measurements of electrode potential, it is not very convenient to use in practice and a calomel electrode (Fig. 10.4) is generally

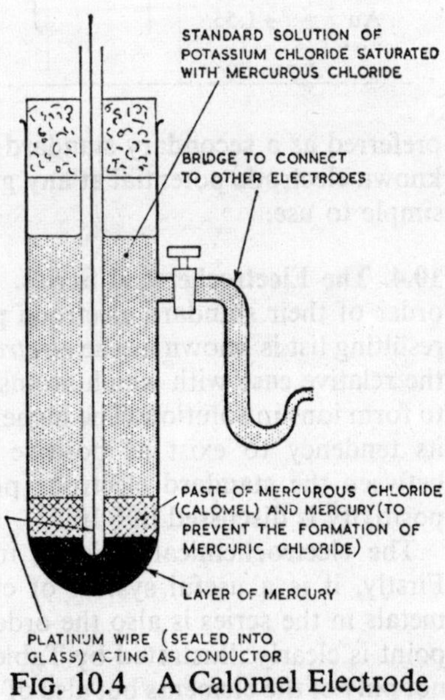

FIG. 10.4 A Calomel Electrode

TABLE 10.2. THE ELECTROCHEMICAL

Element	*Standard Electrode Potential /V*	*Electro-negativity*	*Action of Water*	*Action of Acids*	*Action of Air*	*Action of Hydrogen on Heated Oxide*
K	−2.92	↑ Steady increase in electropositive character up the series	Violent at room temp.	Violent, giving off H_2	Tarnish rapidly	No reduction
Na	−2.71					
Mg	−2.38		Burns in steam	Give H_2 decreasing vigour	No action in cold	
Al	−1.67		Burn in steam at red heat			
Zn	−0.76					
Fe	−0.44		Reversible with steam		Rusts slowly	Reversible reduction
Sn	−0.14		No action		Oxidize readily when heated	Rapid reduction
Pb	−0.13					
H	0.00			Do not give off hydrogen		
Cu	+0.34					
Hg	+0.80					Oxide decomposes when heated alone
Ag	+0.80			Attacked only by certain acids	No action	
Pt	+1.25					
Au	+1.55					

preferred as a secondary standard. This has a constant and accurately known electrode potential at any given temperature and is reliable and simple to use.

10.4. The Electrochemical Series. When elements are arranged in the order of their standard electrode potentials on the hydrogen scale the resulting list is known as the *electrochemical series.* This series indicates the relative ease with which atoms of the elements surrender electrons to form ions in solution. The higher up the series a metal is the stronger its tendency to exist as positive ions in solution. The relationship between the standard electrode potential of a metal and its electropositivity is discussed in § 10.7.

The electrochemical series is important for a variety of reasons. Firstly, it is a useful system of classification, since the order of the metals in the series is also the order of their reactivity in general. This point is clearly illustrated by Table 10.2 which refers to the properties not only of the elements but also of some of their important compounds.

SERIES AND CHEMICAL PROPERTIES

Action of Carbon on Heated Oxide	*Nature of Hydroxide*	*Action of Heat on Nitrates*	*Solubility of Sulphides*	*Natural Occurrence as:*	*Element*
Carbide formed at high temp.	Stable to heat and strongly basic	Give nitrite + $O_2\uparrow$	Very soluble	Chlorides	Potassium
					Sodium
	Gives oxides on heating Weakly basic or amphoteric	Give oxide + NO_2 + $O_2\uparrow$	Hydrolysed, by water	Carbonate Oxide	Magnesium
			Pptd from alkaline soln.		Aluminium
Reduced to metal				Oxides and Sulphides	Zinc
					Iron
			Pptd even from acid soln.		Tin
					Lead
					Hydrogen
					Copper
Oxide decomposes when heated alone	Unstable at room temp.	Give metal + $NO_2\uparrow$ + $O_2\uparrow$		Sulphides	Mercury
					Silver
				Free metal	Platinum
					Gold

A few exceptions (e.g. aluminium and magnesium) arise because some metals are coated with an impervious layer of insoluble oxide which inhibits further attack by the atmosphere.

Secondly, the series shows at a glance which elements will displace each other from a solution of their salts. For example, if a piece of iron is placed in a solution of copper sulphate, some of the iron goes into solution as ions and metallic copper is deposited upon the surface of the iron because iron is higher up the electrochemical series and has a greater tendency than copper to exist as positive ions in solution:

$$Fe + CuSO_4 = FeSO_4 + Cu\downarrow$$

i.e. $$Fe + Cu^{2+} \rightarrow Fe^{2+} + Cu$$

Similar displacements occur when a zinc-copper couple is prepared by immersing granulated zinc in copper sulphate solution for a few minutes, and when zinc dust is added to argentocyanide solution in the recovery of silver from its ores (§ 28.14). The same idea can be used to account for the ability of those metals above hydrogen in the series to displace hydrogen ions as hydrogen gas from acids or even, in the case of metals at the top of the series, from water or steam, whereas metals such as copper and silver which are below hydrogen in the series, never cause the liberation of hydrogen. It also explains why chlorine will replace bromine or iodine from solutions of the ions of these elements, as, for example, in the extraction of bromine from sea-water (§ 25.9):

$$Cl_2 + MgBr_2 = MgCl_2 + Br_2$$

i.e. $$Cl_2 + 2Br^- \rightarrow 2Cl^- + Br_2$$

Thirdly, the series explains why metals such as zinc, magnesium, and aluminium provide valuable sacrificial protection of iron against corrosion. These metals, which are all above iron in the series, dissolve preferentially on exposure, transferring electrons to the iron with which they are in contact in the process. This tends to prevent the ionization of the iron, so that it remains intact, whereas a metal like tin which stands below iron has less tendency to dissolve than the exposed iron and may even accelerate its corrosion by setting up a galvanic cell in the opposite direction (see Fig. 29.2).

Fourthly, the electrochemical series has a quantitative significance in that the e.m.f. of any cell formed from two metallic electrodes is given by the algebraic difference between their electrode potentials. For example, a Daniell cell (see Fig. 10.5) consists of copper and zinc electrodes in solutions of their respective ions and therefore has an e.m.f. of 0.34 – (–0.76), i.e. 1.10 volt.

One of the most important applications of the electrochemical series is its use to predict or explain the products obtained during the electrolysis of aqueous solutions. In general, the metal with the lowest electrode potential will be discharged first, since it has the least tendency to persist in solution in the ionic condition, and only when practically all of its ions have been removed from the solution will any of the metals higher up the series be deposited. This selective or preferential discharge of ions is an important feature in various industrial processes, particularly in the electrolytic refining of copper (see § 28.2) and in electroplating. It is also the basis of a method of analysis known as polarography, as described in the next section.

A word of warning is necessary here, however, lest the student should regard the electrochemical series as an infallible guide to the course of an electrolysis. Certain other factors also govern the order of discharge such as the concentration of the various competing ions, the nature of the electrode (this affects the overpotential, as described in the next section), the current density (i.e. the current flowing over unit surface area of electrode), and the temperature. Only when all these factors have been considered is it possible to decide with any confidence what products are likely to be obtained under a given set of conditions. For example, if a cathode is immersed in a solution containing similar concentrations of zinc sulphate and copper sulphate and a gradually increasing potential is applied, only copper will be deposited at first, followed eventually, when the concentration of copper(II) ions has fallen to a very low value, by the deposition of zinc. On the other hand if an excess of potassium cyanide is added to the solution before apply-

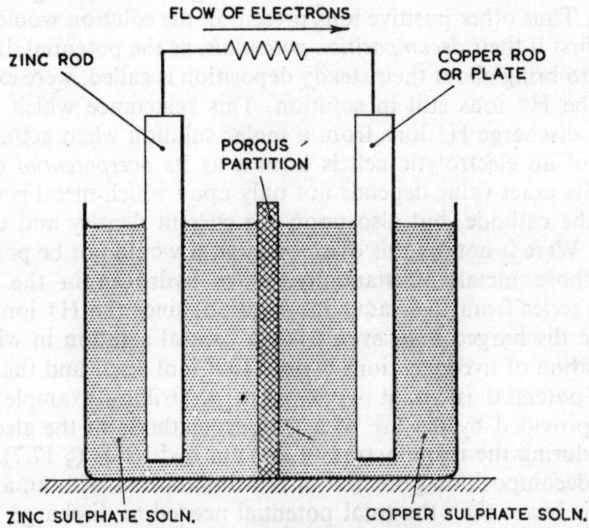

FIG. 10.5 The Daniell Cell

ing a potential, both zinc and copper will combine with the cyanide to form complex ions of the type $[Cu(CN)_4]^{2-}$ and $[Zn(CN)_4]^{2-}$. In both cases, therefore, the concentration of simple metallic ions in the solution will be reduced to a low level, but the effect is much more pronounced with copper owing to the greater stability of its complex ion. As a result both copper and zinc will be deposited simultaneously as brass plating on the cathode when a suitable potential is applied.

10.5. Overpotential. Suppose that an electrode of platinized platinum (i.e. one of platinum covered with platinum black) is placed in a solution containing a variety of positive ions in equivalent concentrations and a steadily increasing negative potential is applied, it is found that the various ions are discharged in the order of the electrochemical series, those at the bottom of the series which have the least tendency to exist as positive ions (e.g. silver and copper) first, then hydrogen, and then the metals higher up the series. If a cathode of some other material is used, however, the order of discharge is unlikely to correspond with that of the electrochemical series in that hydrogen ions may be discharged only at higher voltages than their position in the series would lead one to expect.

For example, for a current density of 10^3 ampere per metre squared cathodes of bismuth, graphite, mercury, lead, silver, zinc, tin, or iron would all require an additional potential of about a volt on the cathode before the H^+ ions in solution were discharged and hydrogen gas was liberated. Thus other positive ions present in the solution would be discharged first if their *decomposition potentials*, as the potential difference required to bring about their steady deposition is called, were exceeded, leaving the H^+ ions still in solution. This reluctance which a metal shows to discharge H^+ ions from a molar solution when acting as the cathode of an electrolytic cell is known as its *overpotential* or overvoltage. Its exact value depends not only upon which metal is used for making the cathode, but also upon the current density and the temperature. Were it not for this overpotential, it would not be possible to deposit those metals substantially above hydrogen in the electrochemical series from their aqueous solution, since the H^+ ions would always be discharged first, even from a neutral solution in which the concentration of hydrogen ions is only 10^{-7} mol dm^{-3} and the decomposition potential is about -0.41 volt. A striking example of this effect is provided by the use of a mercury cathode in the electrolysis of brine during the manufacture of sodium hydroxide (§ 17.7). In this case the decomposition potential of the sodium ions with an amalgam cathode is lower than the total potential needed to discharge H^+ ions from solution, so sodium ions are discharged preferentially. In a second cell nearby the sodium amalgam is then exposed to water in the presence of iron, which has a lower overpotential than mercury and which brings about the exchange of Na^+ ions for H^+ ions.

The high overpotential of mercury towards hydrogen ions is made use of in an important method of analysis known as *polarography.* A mercury cathode with a continually renewed surface is obtained by allowing the metal to drop slowly from a fine jet. When the applied potential is gradually raised the ions of metals present in the solution are discharged at the cathode in turn and are recognized from their characteristic decomposition potentials. The size of the peak current which flows at any given decomposition potential is proportional to the concentration of that particular ion in the solution, so the method can easily be adapted to give quantitative results.

The overpotential shown by various metals towards hydrogen is slightly greater for ions of deuterium than for ions of the much commoner isotope protium. As a result protium ions, $^1H^+$, are discharged preferentially during electrolysis and the proportion of deuterium remaining in the electrolyte steadily increases. Advantage is taken of this enrichment process in the manufacture of deuterium oxide ('heavy water') by electrolysis (see § 16.3).

Overpotential is by no means confined to hydrogen, and is shown towards oxygen particularly, which explains why the halogens are often liberated at the anode instead of oxygen during the electrolysis of aqueous solutions of halides and halogen acids. Attempts have been made to explain the phenomenon of overpotential in terms of *polarization*, i.e. interference in the normal processes of discharge caused by some slow stage such as bubble formation on the surface of the electrode or an accumulation of uncombined atoms, but details of these theories are beyond the scope of this book.

10.6. Cells. A cell is a means of producing electrical energy by chemical means. The earliest type was that invented by Volta in 1800. It consisted of two plates, one of zinc and the other of silver, separated by sheets of paper soaked in brine, the e.m.f. being the difference in the electrode potentials of the two metals. A number of simple cells of this kind were connected in series (forming a *pile*) to give a larger e.m.f. Primitive though this source of power was, it and cells like it using zinc and copper enabled experimenters like Davy to carry out numerous electrochemical experiments including the first isolation of the six strongly electropositive metals sodium, potassium, calcium, strontium, barium, and magnesium.

The construction of the *Daniell cell* has already been shown in Fig. 10.5. The e.m.f. can be made as large as possible by using a dilute solution of zinc sulphate and a saturated solution of copper sulphate, so that there is the greatest possible disparity in the electrode potentials of the two metals.

The *Weston cell* is illustrated in Fig. 10.6. Its importance lies in the constancy and reproducibility of its e.m.f., which makes it an excellent standard for comparison purposes in, for example, potentiometers. At 20°C the e.m.f. is 1.018 64 V.

As explained in § 10.2, the electrode potential of any particular metal depends upon the concentration of its ions in the solution with which it is in contact. Consequently if two electrodes of the *same* metal are placed in solutions of different concentration linked by a salt bridge, a potential difference will be developed between them equal to the difference in their electrode potentials. This is the principle of the so-called *concentration cell*, which derives its electrical energy from the transfer of ions from a solution of one concentration to another.

A *dry cell* usually consists of a graphite cathode surrounded by a moist paste of ammonium chloride, zinc chloride, carbon black, starch, and manganese dioxide (to prevent polarization), the whole being enclosed in a zinc case which acts as the anode (see Fig. 10.7). The reactions which occur during the working of the cell, although in fact very complicated, can be represented simply as follows:

At the anode zinc goes into solution in the electrolyte leaving an accumulation of electrons on the metal:

$$Zn \rightarrow Zn^{2+} + 2e$$

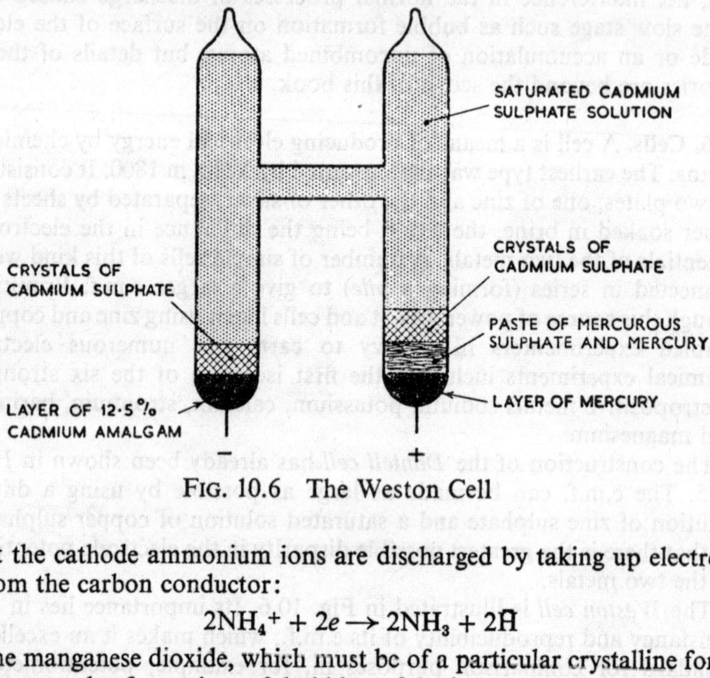

FIG. 10.6 The Weston Cell

At the cathode ammonium ions are discharged by taking up electrons from the carbon conductor:

$$2NH_4^+ + 2e \rightarrow 2NH_3 + 2\bar{H}$$

The manganese dioxide, which must be of a particular crystalline form, prevents the formation of bubbles of hydrogen on the cathode by oxidizing the hydrogen atoms to water:

$$2MnO_2 + 2\bar{H} \rightarrow Mn_2O_3 + H_2O$$

The ammonia diffuses through the electrolyte and combines with the zinc chloride to form a compound of the type $Zn(NH_3)_2Cl_2$. Although the e.m.f. of the single cell is only about 1.5 volt, a larger e.m.f. can be obtained by connecting a number of cells in series, when the arrangement is known as a *battery*. Its greatest advantages over other cells are its convenience, compactness, and portability, but it is non-reversible and has to be discarded after use. The *lead accumulator* (see § 21.22) is reversible and easily recharged, and its principal use, therefore, is (as its name suggests) for storing electric power rather than for generating it.

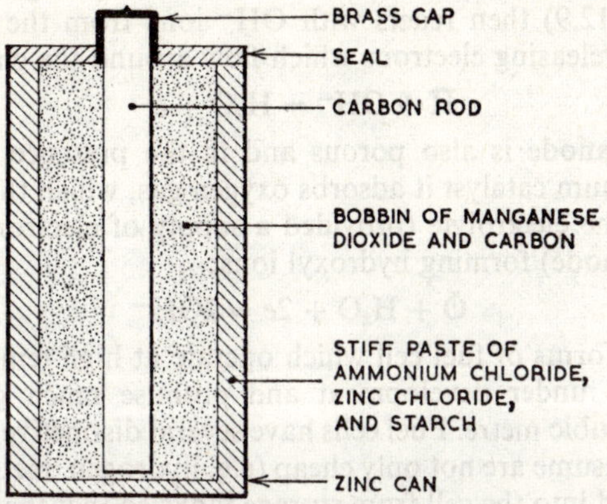

FIG. 10.7 A Typical Dry Cell

FUEL CELLS: In recent years much research has been done with the aim of finding a means of converting chemical energy continuously and directly into electricity. Devices for achieving this, which are known as

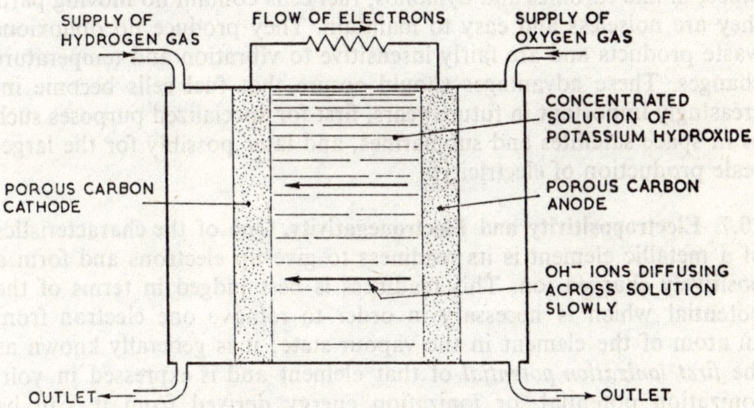

FIG. 10.8 A Simple Fuel Cell

fuel cells, are still only in the development stage, but the principles of their operation can easily be understood by considering the simplest type, one consuming hydrogen and oxygen at room temperature. Such a cell is shown diagrammatically in Fig. 10.8. Hydrogen gas diffuses through the porous cathode, which usually consists of carbon impregnated with finely-divided platinum to catalyse the chemical changes and prevent overpotential effects. Some of the hydrogen gas which is adsorbed (§ 12.9) then reacts with OH^- ions from the electrolyte to form water, releasing electrons which flow around the circuit:

$$\bar{H} + OH^- = H_2O + e$$

The carbon anode is also porous and in the presence of the finely-divided platinum catalyst it adsorbs oxygen gas, which then reacts with water from the electrolyte (provided a supply of electrons is available from the cathode) forming hydroxyl ions:

$$\bar{O} + H_2O + 2e = 2OH^-$$

Improved forms of fuel cell which operate at high temperatures and pressures are under development and promise much greater power outputs per cubic metre. Fuel cells have several distinctive features. The fuels they consume are not only cheap (e.g. hydrogen, methane, oxygen) but can be fed into the cell from storage tanks so that the cell can operate continuously over long periods without recharging. The efficiency of such cells is likely to be very high, approaching 70–75% compared with a maximum efficiency of only about 40–45% for conventional sources of electric power. Moreover, when fully developed they are expected to be much lighter and more compact than accumulators. Since, unlike turbines and dynamos, fuel cells contain no moving parts they are noiseless and easy to maintain. They produce no obnoxious waste products and are fairly insensitive to vibration and temperature changes. These advantages should ensure that fuel cells become increasingly important in future years, first for specialized purposes such as in space satellites and submarines, and later possibly for the large-scale production of electricity.

10.7. Electropositivity and Electronegativity. One of the characteristics of a metallic element is its readiness to give up electrons and form a positively charged ion. This readiness is best judged in terms of the potential which is necessary in order to remove one electron from an atom of the element in the vapour state; it is generally known as the *first ionization potential* of that element and is expressed in volt. Ionization potential, or ionization energy derived from it, can be determined experimentally for each element from spectroscopic measurements. Values compiled in this way for the first sixty elements are listed in Table 13.8 and commented upon in § 13.11.

Those metals (e.g. sodium and potassium) with the lowest ionization potentials and therefore the greatest tendency to form positive ions are referred to as being strongly *electropositive* whilst metals with the least tendency to lose electrons during chemical combination (e.g. silver, gold, and mercury) are described as weakly electropositive. Although the electropositivity of a metal is very closely related to its standard electrode potential, when the metals are arranged in a list with the most electropositive ones first (see Table 10.3) the order is not exactly the same as in the electrochemical series. For example, lithium stands above sodium and potassium in the electrochemical series but is less electropositive than these elements. The reason for this discrepancy is that

TABLE 10.3. METALS ARRANGED IN ORDER OF ELECTROPOSITIVITY

Cs	Rb	K	Na	Ba	Li	Ca	Mg	Al	Zn	Sn

standard electrode potential is a measure of the readiness of the metal in its standard state to change not only into an ion, but into an ion *in solution*, so that the total energy change involved is the sum of several distinct stages thus:

$$\text{molecule} \xrightarrow[\text{atomization}]{\text{energy of}} \text{single atom} \xrightarrow[\text{energy}]{\text{ionization}} \text{ion} \xrightarrow[\text{hydration}]{\text{energy of}} \text{ion in solution}$$

Hence although the ionization energy (and therefore the electropositivity) largely determines the standard electrode potential of an element, its energy of atomization and the energy of hydration of its ions in solution are contributory factors and these may vary widely from one metal to another. In the case of lithium, for example, its high energy of hydration which arises from the small size of its ion (see Table 13.7) accounts for its high position in the electrochemical series.

The usefulness of electropositivity as an idea has led to the development of a similar concept which can be applied to non-metallic elements, which form negatively charged ions by gaining electrons. Such elements are said to be *electronegative*, electronegativity being a measure of their appetite for electrons and of their readiness to accept them and change into negative ions. For example, fluorine, which is the most electronegative element of all, has the greatest affinity for electrons, whilst oxygen, nitrogen, and chlorine are also strongly electronegative, i.e. electron-attracting.

Attempts have been made to correlate the contrary concepts of electropositivity and electronegativity and to express them numerically in a way which would make it possible to compare the values for various elements. The suggestion was made by Pauling that all elements, metals and non-metals alike, should be regarded as being electronegative in different degrees, electronegativity being defined here in the strict sense as the power of the atom to attract electrons to itself whilst in chemical combination. He suggested that their character in this respect should be expressed on a single scale ranging from 4.0 for fluorine to about 0.7

for caesium, which is with francium the most weakly electronegative (i.e. most strongly electropositive) alkali metal. This has the advantage of dispensing with positive and negative signs by arbitrarily choosing a zero of electronegativity which is less than that of any known element. Some typical values, which are derived from thermochemical and spectroscopic measurements, are given in Table 10.4. As explained in § 14.7, these electronegativity values can be used for predicting the kind

TABLE 10.4. PAULING'S ELECTRONEGATIVITY VALUES

						H
						2.1
Li	Be	B	C	N	O	F
1.0	1.5	2.0	2.5	3.0	3.5	4.0
Na	Mg	Al	Si	P	S	Cl
0.9	1.2	1.5	1.8	2.1	2.5	3.0
K	Ca		Ge	As	Se	Br
0.8	1.0		1.7	2.0	2.4	2.8
Rb	Sr		Sn	Sb	Te	I
0.8	1.0		1.7	1.8	2.1	2.4
Cs	Ba					
0.7	0.9					

of bonding likely to be formed between two particular atoms and for explaining the degree of polarity displayed by a given covalent bond. The word electropositive as applied to metals, however, is a familiar and valuable one which is not lightly discarded and the term will be used extensively in this book in its broader sense to convey a general readiness to give up electrons and form positive ions.

The various factors which affect the electronegativity of an element are discussed in § 13.9. The same section includes a brief account of the way in which electronegativity varies from one element to another. Further references to its variation within a group of the periodic classification will be found in the summaries provided at the end of the chapters dealing with inorganic chemistry.

10.8. Electropositivity and the Extraction of Metals. The electropositivity (or electronegativity in the Pauling sense) of a metal is a very important factor in determining the method used for extracting it from its ores. To start with, as Table 10.2 clearly shows, the type of compound which occurs naturally is related to the electropositivity of the metal. Moreover, the least electropositive metals (e.g. platinum, gold, silver, mercury, and copper) often occur naturally as elements. Such metals have little affinity for electronegative elements and the reduction of their ores is usually a simple process because their compounds are easily decomposed to the metal when heated either alone or with a reducing agent. The more electropositive metals such as lead, tin, nickel, cobalt, cadmium, iron, and zinc are readily extracted by roasting their oxides with carbon or carbon monoxide, although the reaction between zinc oxide and carbon monoxide is reversible and incomplete and so carbon must be used in that case (§ 29.2). Ores of the more strongly electropositive metals are not easily reduced when heated with carbon, except at very high temperatures when carbides tend to form, so an alternative method has to be adopted. The oxides of manganese and chromium can be reduced by the Goldschmidt process (§ 19.8), and compounds of titanium, magnesium, and aluminium, can be reduced to the metal by heating them with even more electropositive elements such as alkali or alkaline earth metals, although cheaper methods are usually preferred for their manufacture. The extraction of aluminium presents special problems, as explained in § 19.8, but in general the most highly electropositive metals form ionic chlorides which conduct electricity when molten and which deposit the metal at the cathode when electrolysed. They are best made, therefore, by electrolysis, using electrical energy to separate the metals from the electronegative elements to which they are so firmly bound; these processes are similar to ones used by Davy for isolating the metals over 150 years ago.

10.9. The Cell Convention. A conventional way has been devised of representing a cell in symbols. The two half-cell reactions are written in a single line joined by a vertical dotted line which stands for the porous diagram separating the two sections of the cell. For example, the Daniell cell is written:

$$Zn\ (s)\ |\ Zn^{2+}(aq)\ |\ Cu^{2+}(aq)\ |\ Cu\ (s);\ E = +\ 1.10\ \text{volt}$$

The continuous vertical lines indicate the division between the metal electrodes and ions in solution. The e.m.f. of the cell expressed in volt is written as positive if the current flow is from the right side to the left (i.e., the electron flow is from the left hand electrode to the one on the right). Thus the sign given to the cell e.m.f. indicates the polarity of the right hand electrode.

11

Thermochemistry and Energetics

11.1. Introduction. Thermochemistry is concerned with the heat changes which take place during chemical reactions. Every chemical change is accompanied by the evolution or absorption of heat, although in many cases this is not obvious because the amount is so small that no noticeable change in temperature occurs. Reactions which take in heat as they proceed are said to be *endothermic*, and those which give out heat *exothermic*.

These heat changes are given the general name of heats of reaction, but it is often found more convenient to classify them and refer to them as heats of formation, combustion, neutralization, and solution, according to the type of change involved. *In every case the convention is followed that heat evolved is negative in sign and heat absorbed is positive.* This practice, which is now generally adopted throughout the world, has the advantage that an exothermic reaction is depicted as a negative heat change corresponding to the fall in the heat content of the system; heat is given out and the products contain that much less energy than the reactants did. Similarly, an endothermic reaction, which is regarded as involving a positive heat change, results in a gain in heat energy for the system. Students should beware, however, since many old textbooks, particularly those published in Europe, still employ the opposite convention, regarding heat evolved as positive. To prevent any confusion arising it is always a good plan to state clearly, as we have done here, which convention is being used.

Two other conventions generally adopted in thermochemistry also need explanation. In SI the unit of heat and energy is the *joule* (J),* which is the energy used when a force of one newton acts over a distance of one metre. However, since the heat changes associated with chemical reactions are usually large, amounting to many thousands of joule, and are rarely known to the nearest joule, it is preferable to express them in *kilojoule*. This larger unit (which, as the name implies, is equal to 1000 joule) is distinguished from the latter by using the abbreviation

* Before SI the *calorie* was widely used; 1 calorie = 4.184 joule.

'kJ'. Secondly, when it is desired to convey information about the heat change which accompanies a reaction, this is usually done by placing the symbol ΔH (which denotes the heat of reaction *at constant pressure*) at the end of the equation as in the examples quoted in this chapter. The Greek letter Δ (pronounced 'delta') conventionally stands for the change in a quantity and H for the heat content or *enthalpy*.

In a reversible reaction the heat change of the forward (i.e. left-to-right) reaction is equal in magnitude but opposite in sign to the heat change of the backward or right-to-left reaction. This statement, which is known as the *law of Lavoisier and Laplace* after the scientists who discovered it in 1780, can be deduced from a much wider generalization, the *law of conservation of energy*, since otherwise it would be possible to obtain an unlimited amount of energy merely by repeatedly reversing a reaction in which more heat was given out than taken in. In reversible thermochemical equations the sign of the heat change is always understood to refer to the forward reaction.

11.2. Hess's Law. In 1840 Hess summarized his experimental findings in his *law of constant heat summation* which states that *the heat change in a given reaction depends only upon the initial and final states of the system and is independent of the path followed (provided that heat is the only form of energy to enter or leave the system)*. This may be illustrated by supposing that a substance A can be converted into another substance D either directly, when the heat of reaction is w joule, or indirectly by way of substances B and C when the heats of reaction of the three stages are x, y, and z joule respectively, thus:

$$\mathrm{A} \rightarrow \mathrm{D}; \Delta H = -w \text{ J}$$
$$\mathrm{A} \rightarrow \mathrm{B}; \Delta H = -x \text{ J}$$
$$\mathrm{B} \rightarrow \mathrm{C}; \Delta H = -y \text{ J}$$
$$\mathrm{C} \rightarrow \mathrm{D}; \Delta H = -z \text{ J}$$

Then according to Hess's law the heat of reaction in going from A to D is the same whether the change is achieved directly or through a series of separate stages, and $w = x + y + z$.

Hess's law is important because it enables us to add and subtract heat changes algebraically and so calculate heats of reaction which it is impossible to determine directly by experiment. This process is demonstrated in the worked examples given later in the chapter. Like the law of Lavoisier and Laplace, Hess's law can be deduced theoretically from the law of conservation of energy.

We are now in a position to consider the definitions of the various terms used in thermochemistry and to illustrate each of them with an example.

11.3. Heat of Reaction. *This is the amount of heat absorbed or evolved when the quantities of reactants stated in the equation, expressed in mole, react together.* For example, in the synthesis of ammonia from its elements, which proceeds according to the equation:

$$N_2 + 3H_2 \rightleftharpoons 2NH_3; \Delta H = -92 \text{ kJ mol}^{-1}$$

the heat of reaction is 92 kilojoule because that amount of heat is liberated when 28 gramme of nitrogen combine with 6 gramme of hydrogen to form 34 gramme of ammonia. The experimental measurement of heat of reaction is described in § 11.8, and its significance in terms of chemical affinity and energetics is discussed in § 11.9.

The heat of a reaction depends not only upon the weight of reagents used, but also upon the temperature and pressure of the reacting system and the physical states of the reactants and products, since latent heats of fusion and evaporation may be involved as well. To overcome this difficulty heats of reaction are usually quoted for 25°C and one atmosphere pressure and where any doubt or ambiguity exists the phases of the substances involved in the reaction are indicated by inserting into the thermochemical equation the symbols g, l, s, and aq, standing for gas, liquid, solid, or dilute aqueous solution respectively. For example, the following equations may be used to represent the reaction of hydrogen with iodine at 25°C and 375°C respectively:

$$H_2\,(g) + I_2\,(s) = 2HI\,(g); \Delta H = +\,11.3 \text{ kJ mol}^{-1}$$
$$H_2\,(g) + I_2\,(g) = 2HI\,(g); \Delta H = -\,53.6 \text{ kJ mol}^{-1}$$

The heat of sublimation of iodine accounts for about 38 kJ^{-1} mol of the difference, the rest being due to the change in the heat of this reaction with temperature.

The value of a given heat of reaction also depends upon whether the measurement is made at constant volume or at constant pressure. For reactions involving only solids and liquids there is hardly any difference between the two, since the change in volume in such cases is usually negligible, but where the reactants or products are gases, the heat of reaction at constant pressure, ΔH, may differ appreciably from ΔU, the heat of reaction at constant volume. The relation between these two is like that between the two specific heat capacities of a gas, C_p and C_v, (see § 3.11), since $\Delta H = \Delta U + P\Delta V$, where P is the constant pressure of the system and ΔV is the change in volume that occurs at pressure P. Thus $\Delta H = \Delta U +$ external work done by the system at constant pressure. Since most experiments are carried out at atmospheric pressure, the heat change measured is usually ΔH, but when a bomb calorimeter is used the volume of the system is constant and the result obtained, ΔU, must then be corrected slightly if the heat of reaction at constant pressure is required.

11.4. Heat of Combustion. *This is the amount of heat evolved when one mole of a given substance is burned in an excess of oxygen.* For example, the heat of combustion of methane is 890 kilojoule because this amount of heat is liberated when 16 gramme of it are burned completely in oxygen:

$$CH_4\,(g) + 2O_2\,(g) = CO_2\,(g) + 2H_2O\,(l);\ \Delta H = -890\ \text{kJ mol}^{-1}$$

The reference to combustion in excess of oxygen is necessary to exclude the possibility of only partial combustion to, for example, carbon monoxide, when the heat evolved would be appreciably less.

The heat of combustion of a substance is important for several reasons. Firstly, it is readily determined experimentally by using a bomb calorimeter as described in § 11.8. This determination is applied industrially to suitable substances such as hydrocarbons in order to find their heating value as fuels; its use as a means of calculating heats of formation is referred to in the next section. Secondly, by comparing the heats of combustion of two different allotropes of an element it is possible to estimate the difference in their energy content. For example, when diamond is burned in excess of oxygen slightly more heat is evolved than from an equal mass of graphite, showing diamond to be the allotrope with the higher intrinsic energy.

The experimental figures are:

$$C\ (\text{diamond}) + O_2\,(g) = CO_2\,(g);\ \Delta H = -395.4\ \text{kJ mol}^{-1}$$
$$C\ (\text{graphite}) + O_2\,(g) = CO_2\,(g);\ \Delta H = -393.5\ \text{kJ mol}^{-1}$$

Applying Hess's law (§ 11.2) to these results, we conclude that 1900 joule would be absorbed if 12 gramme of graphite were changed completely into diamond. Similar energy differences exist between other pairs of allotropes, the more stable form of an element always possessing less energy.

11.5. Heat of Formation. *This is the amount of heat absorbed or evolved when one mole of any given compound is made from its constituent elements in their normal states.* The reference to normal states is necessary because the heat of formation depends to some extent upon the physical conditions of the elements and their allotropic form. For example, the heat of formation of carbon dioxide is −406 kilojoule because that is the quantity of heat which is given out when 44 gramme of it are formed from oxygen and graphite (the stable allotrope of carbon) at 25°C and 101 325 N m^{-2} pressure:

$$C + O_2 = CO_2;\ \Delta H = -406\ \text{kJ mol}^{-1}$$

Since the heat of formation is defined in terms of one mole of a compound, thermochemical equations are often written for convenience with fractional coefficients before the molecules of reactants thus:

$$H_2\,(g) + \tfrac{1}{2}O_2\,(g) = H_2O(l);\, \Delta H = -284 \text{ kJ mol}^{-1}$$

(This means that 284 kilojoule of heat are evolved when 18 gramme of water are formed from 2 gramme of hydrogen and 16 gramme of oxygen at 25°C and atmospheric pressure. Although the use of fractions in this way may seem rather odd at first sight, it is perfectly acceptable. Where, however, it is desired to eliminate fractions by multiplying both sides of the equation by an appropriate factor, the value of ΔH must be increased appropriately. Thus the above equation becomes:

$$2H_2\,(g) + O_2\,(g) = 2H_2O\,(l);\, \Delta H = -568 \text{ kJ mol}^{-1}$$

and the heat of reaction is twice as great in this case as the heat of formation.

The heat of formation of a compound is a useful guide to its stability and reactivity. Most compounds are formed exothermically, but some such as acetylene, carbon disulphide, ozone, the oxides of nitrogen, and hydrogen peroxide, have positive heats of formation:

e.g. $$N_2 + O_2 \rightleftharpoons 2NO;\, \Delta H = +181 \text{ kJ mol}^{-1}$$

In general endothermic compounds (i.e. ones whose formation from their elements involves the absorption of heat) are unstable and highly reactive; when heated many of them decompose readily into simpler substances with the release of energy. The heat of formation of a binary compound is also closely related to the relative electronegativity (§ 10.7) of its constituent elements. Generally speaking, the greater the difference in electronegativity, the more exothermic the heat of formation and the more stable the compound formed. This explains why heats of formation have been used as a means of comparing and estimating the electronegativities of various elements.

Heats of formation are rarely measured directly, since direct synthesis of a compound from its elements is usually impracticable, but the heat of formation is generally calculated from the heat of combustion of the compound, as shown in the following examples:

Example I: Calculate the heat of formation of carbon monoxide,* given that its heat of combustion is −283 kilojoule per mole and that the heat of formation of carbon dioxide is −394 kilojoule per mole.

Let the heat of formation of carbon monoxide be $-Q$ kJ mol^{-1}.

Then

$$C + \tfrac{1}{2}O_2 = CO;\, \Delta H = -Q \text{ kJ mol}^{-1} \quad (1)$$

$$CO + \tfrac{1}{2}O_2 = CO_2;\, \Delta H = -283 \text{ kJ mol}^{-1} \quad (2)$$

$$C + O_2 = CO_2;\, \Delta H = -394 \text{ kJ mol}^{-1} \quad (3)$$

* This cannot be measured directly because some carbon dioxide is always formed as well when carbon is burnt in a limited supplv of air.

Now by Hess's law, (1) + (2) = (3)

$$\therefore\ Q + 283 = 394$$

$$\therefore\ Q = 111\ \text{kJ mol}^{-1},\ \text{so that}\ \Delta H = -\ 111\ \text{kJ mol}^{-1}$$

i.e. 111 kilojoule of heat would be evolved in forming one mole of carbon monoxide from its elements.

Example II: Calculate the heat of formation of methane, given that the heats of combustion of carbon, hydrogen, and methane are 393 kJ mol^{-1}, 285 kJ mol^{-1}, and 887 kJ mol^{-1} respectively.*

$$C + O_2 = CO_2; \quad \Delta H = -393\ \text{kJ mol}^{-1} \quad (1)$$

$$H_2 + \tfrac{1}{2}O_2 = H_2O; \quad \Delta H = -285\ \text{kJ mol}^{-1} \quad (2)$$

$$CH_4 + 2O_2 = CO_2 + 2H_2O; \quad \Delta H = -887\ \text{kJ mol}^{-1} \quad (3)$$

Multiplying (2) by 2:

$$\therefore\ 2H_2 + O_2 = 2H_2O; \quad \Delta H = -570\ \text{kJ mol}^{-1} \quad (4)$$

Adding (1) and (4):

$$C + 2H_2 + 2O_2 = CO_2 + 2H_2O; \quad \Delta H = -963\ \text{kJ mol}^{-1} \quad (5)$$

Subtracting (3) from (5):

$$C + 2H_2 = CH_4; \quad \Delta H = -76\ \text{kJ mol}^{-1}$$

Thus the heat of formation of methane is -76 kilojoule per mole.

Example III: Given that the heats of combustion of acetylene, ethylene, and hydrogen are -1310 kJ mol^{-1}, -1393 kJ mol^{-1}, and -285 kJ mol^{-1} respectively, calculate the heat change for the reaction: $C_2H_2 + H_2 = C_2H_4$.

$$C_2H_4 + 3O_2 = 2CO_2 + 2H_2O; \quad \Delta H = -1393\ \text{kJ mol}^{-1} \quad (1)$$

$$C_2H_2 + 2\tfrac{1}{2}O_2 = 2CO_2 + H_2O; \quad \Delta H = -1310\ \text{kJ mol}^{-1} \quad (2)$$

$$H_2 + \tfrac{1}{2}O_2 = H_2O; \quad \Delta H = -285\ \text{kJ mol}^{-1} \quad (3)$$

Adding (2) and (3):

$$C_2H_2 + H_2 + 3O_2 = 2CO_2 + 2H_2O; \quad \Delta H = -1595\ \text{kJ mol}^{-1} \quad (4)$$

Subtracting (1) from (4):

$$C_2H_2 + H_2 = C_2H_4; \quad \Delta H = -202\ \text{kJ mol}^{-1}$$

Hence 202 kilojoule are given out when one mole of acetylene is converted into ethylene.

11.6. Heat of Neutralization. *This is the amount of heat evolved when one mole of H^+ ions from an acid reacts with one mole of OH^- ions from an alkali.* Thus the heat of neutralization of sodium hydroxide by hydrochloric acid is 57.36 kilojoule because this amount of heat is given out when 40 gramme of sodium hydroxide react with 36.5 gramme of

* The negative signs of heats of combustion are often omitted by examiners from examination questions as being understood, since heats of combustion are always necessarily exothermic (see definition given in § 11.4).

hydrochloric acid in dilute solution:

$NaOH (aq) + HCl (aq) = NaCl (aq) + H_2O; \Delta H = -57.36$ kJ mol^{-1}

Other examples are:

$KOH (aq) + \frac{1}{2}H_2SO_4 (aq)$
$= \frac{1}{2}K_2SO_4 (aq) + H_2O; \Delta H = -57.94$ kJ mol^{-1}

$NaOH (aq) + HNO_3 (aq)$
$= NaNO_3 (aq) + H_2O; \Delta H = -57.32$ kJ mol^{-1}

$\frac{1}{2}Ca(OH)_2 (aq) + HCl (aq)$
$= \frac{1}{2}CaCl_2 (aq) + H_2O; \Delta H = -57.40$ kJ mol^{-1}

$\frac{1}{2}Ca(OH)_2 (aq) + HNO_3 (aq)$
$= \frac{1}{2}Ca(NO_3)_2 (aq) + H_2O; \Delta H = -57.36$ kJ mol^{-1}

$\frac{1}{2}Ba(OH)_2 (aq) + HCl (aq)$
$= \frac{1}{2}Ba(Cl_2)_2 (aq) + H_2O; \Delta H = -57.36$ kJ mol^{-1}

$NaOH (aq) + \frac{1}{2}H_2SO_4 (aq)$
$= \frac{1}{2}Na_2SO_4 (aq) + H_2O; \Delta H = -59.50$ kJ mol^{-1}

These figures demonstrate that in the neutralization of a strong acid by a strong base the amount of heat liberated is approximately the same in every case. The evidence which this provides in support of the ionic theory is referred to in § 8.6, where it is explained that the fundamental change common to all neutralizations is believed to be $H^+ + OH^- = H_2O$. On the other hand when weak acids or weak bases are involved the heat of neutralization is less, as the following examples show:

$NH_4OH (aq) + HCl (aq) = NH_4Cl (aq) + H_2O; \Delta H = -51.5$ kJ mol^{-1}
$NH_4OH (aq) + HCN (aq) = NH_4CN (aq) + H_2O; \Delta H = -5.4$ kJ mol^{-1}

The explanation is that these weak electrolytes are only partially dissociated into ions in solution so that when neutralization takes place some energy is used in drawing apart the undissociated ions before they can react. Thus the heat of neutralization observed in such cases is only the net heat change and is less than 57.4 kJ mol^{-1} by the amount of energy needed to bring about complete dissociation of the electrolytes into ions.

The basicity of an acid (i.e. the number of hydrogen atoms in one molecule which are replaceable by a metal) can be determined from measurements of the heat evolved during the neutralization of the acid, provided its molecular weight is known. For example, if 50 cm³ portions of a molar solution of a dibasic acid are mixed in three calorimeters with 50 cm³, 100 cm³, and 150 cm³ of M alkali solution respectively and the heat change is determined each time, it is found that altogether about twice as much heat is liberated in the second case as in the first, but that with 150 cm³ of alkali the heat change does not increase further. Hence complete neutralization of 50 cm³ of a molar solution of an acid can be accomplished by using 100 cm^3 of a molar solution of alkali, showing the acid to be dibasic.

11.7. Heat of Solution. *This is the amount of heat absorbed or evolved when one mole of a given substance is dissolved in so much water that further dilution results in no detectable heat change.* For example, the heat of solution of anhydrous sodium carbonate is −23 kilojoule because that amount of heat is given out when 106 gramme of it are dissolved in a large volume of water:

$$Na_2CO_3\,(s) + nH_2O = Na_2CO_3\,(aq);\ \Delta H = -23\ kJ\ mol^{-1}$$

Some substances dissolve very exothermically (e.g. sodium hydroxide or sulphur trioxide), whilst others (e.g. sodium nitrite) absorb heat when they dissolve. The reason for these wide differences in behaviour is that the heat of solution is only the resultant* of a number of simultaneous and relatively large heat changes. For example, when anhydrous calcium chloride is added to water a certain amount of energy is required to separate the ions from each other in solution (an endothermic change) but a much larger amount of heat is evolved as a result of the hydration of the ions (highly exothermic change). Again, some heat is absorbed during the dilution of the concentrated solution of calcium chloride, so that the heat of solution of the anhydrous salt is the net effect of these various changes and is, in fact, exothermic. On the other hand, crystals of the hexahydrate $CaCl_2.6H_2O$ dissolve endothermically because the ions in this form are already extensively hydrated and the heat liberated on this account does not predominate over the endothermic changes. This example emphasizes how important it is to specify clearly the degree of hydration of a salt when quoting its heat of solution.

The heat of solution of a substance can be related to the way in which its solubility varies with temperature by applying Le Chatelier's principle (§ 7.9) to the physical equilibrium existing between a saturated solution and excess of undissolved solid. Where a solid dissolves endothermically in such circumstances one would expect an increase in temperature to cause more of it to dissolve, since in that way heat is taken in and the rise in the temperature of the system is opposed. Thus such a substance would, like most substances, be more soluble at higher temperatures and less soluble at lower temperatures. A word of warning is needed here, however, to guard against the common error when applying the principle in this way of using the heat change that occurs when a solid is dissolved in pure water rather than in its already nearly saturated solution as a guide to its solubility curve. Sodium hydroxide, for example, gives out a great deal of heat when added to pure water and dissolves so exothermically that the water sometimes boils, but it

* See Fig. 11.8 and the last paragraph of § 11.12.

dissolves endothermically in its nearly saturated solution. It is this second heat change that matters when applying Le Chatelier's principle, and when it is used the increase in solubility of sodium hydroxide with rising temperature can be correctly predicted.

11.8. Measurement of Heat Changes. The method used depends upon the type of change. Where a large heat change such as a heat of combustion is being measured, or where the reactants or products include gases, bomb calorimetry is employed, but heats of solution or neutralization are best determined by variations of the standard method of mixtures. An example of each method is described below:

THE BOMB CALORIMETER: This consists of a strong cylindrical steel vessel (see Fig. 11.1) lined with enamel to prevent corrosion. If it is desired to determine the heat of combustion of carbon, for example, a

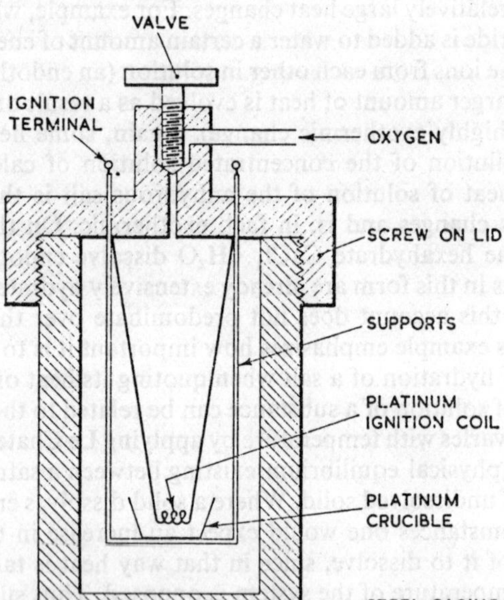

FIG. 11.1 A Bomb Calorimeter

known mass (about a gramme) of the element is placed in a platinum crucible inside the bomb, the lid is screwed on tightly, and oxygen is pumped in through a valve until the pressure inside is about 2×10^6 $N\ m^{-2}$ (20 atm). The bomb is then immersed in a known mass of water in a calorimeter as in Fig. 11.2 and left to attain a steady temperature, which is carefully measured with a Beckmann thermometer (§ 5.12) to the nearest $\frac{1}{100}$° C. This is necessary because the combustion may produce a rise in temperature of only a few degrees in the water. The carbon is ignited electrically by passing a current through the platinum coil and the temperature of the water, which is stirred continuously, is recorded at the end of each minute. Readings are continued until the temperature has shown a fall for five consecutive minutes, to enable a cooling correction to be made (see Fig. 11.3). Allowance must be made for the thermal capacities of the two calorimeters together with the

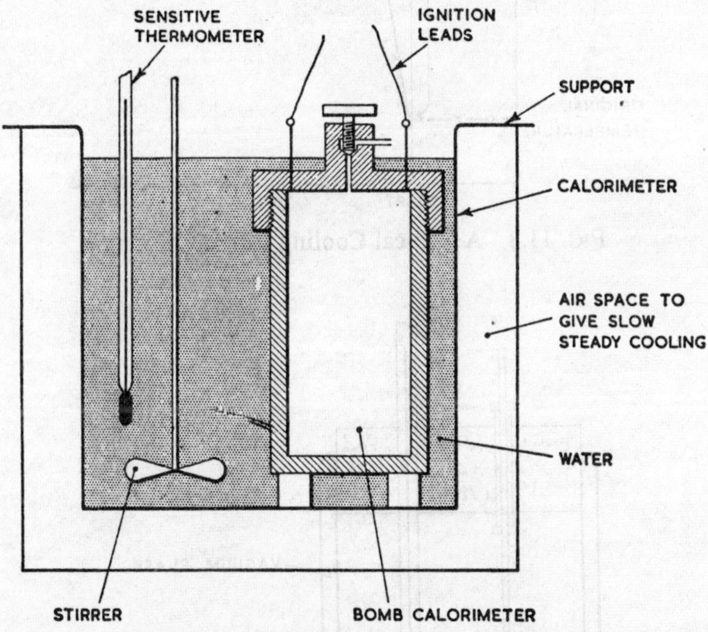

FIG. 11.2 Apparatus for Determining the Heat of Combustion

stirrer and thermometer, and for the heat generated by the ignition current. This gives the heat evolved from the combustion of a known mass of carbon, from which the heat of combustion can easily be calculated.

METHOD OF MIXTURES: The apparatus used is shown in Fig. 11.4. The first step is to determine the water equivalent of the calorimeter with stirrer and thermometer by finding the temperature rise which occurs when a known mass of hot water of known temperature is added. If, for example, the heat of neutralization of a M solution of an acid by a M solution of an alkali is required, then equal volumes of each solution

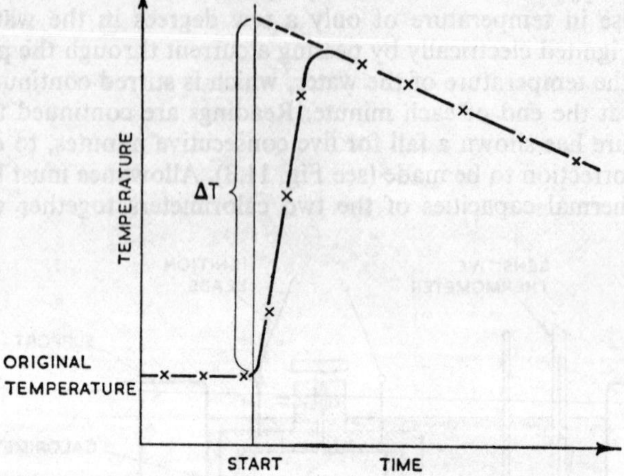

FIG. 11.3 A Typical Cooling Curve Correction

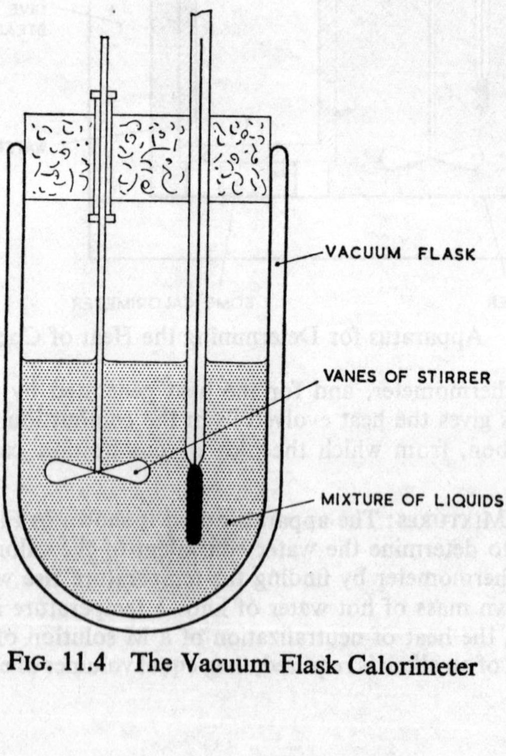

FIG. 11.4 The Vacuum Flask Calorimeter

are taken and their initial temperatures are carefully measured. The two solutions are then mixed together in the calorimeter and stirred continuously, the temperature of the mixture being recorded at half-minute intervals until a definite decline is apparent. The readings may be plotted on a graph and a cooling correction applied in the usual way, although when a vacuum flask is used as the calorimeter this correction is hardly necessary because the heat losses are so small. From the results the heat evolved from molar quantities of acid and base can be calculated.

11.9. Energetics. The natural tendency of any system when left to itself is to assume the lowest possible energy state; this is seen in all branches of science and forms the basis of one of its greatest generalizations, the second law of thermodynamics. The way in which water flows to the lowest possible level and heat flows from a hotter body to a cooler one, and a gas diffuses throughout a vessel, and an electric charge flows from one point to another of lower potential, are common examples of this universal tendency. It is not surprising, therefore, to find that in all chemical reactions which proceed spontaneously there is a loss of energy in the system. In most cases, but by no means in all, there is also a net evolution of heat, i.e. the reaction is exothermic overall.

This raises the problem of why all such reactions do not proceed with vigour and speed. Why, for example, should a lump of coal not burst into flames when exposed to the air, since its combustion to carbon dioxide is a highly exothermic change? Similarly, why is it possible for a mixture of hydrogen and oxygen to remain unchanged in a gas jar apparently indefinitely when the combination of the two gases to form water would result in a system of much lower energy? The accepted explanation is that before two substances can combine together their molecules must possess at the moment of impact at least a certain amount of energy, known as the *energy of activation* of that reaction (see § 7.16). Without this energy the collision is unfruitful and no reaction occurs. Thus if the lump of coal is first heated until it is nearly red-hot it will burn spontaneously in air, and if the mixture of hydrogen and oxygen is ignited or sparked, water will be formed. Since both reactions are highly exothermic, the heat released by the initial changes will then serve to activate other molecules in the vicinity and so maintain the reaction.

The energy of activation probably represents the energy needed to pull apart some of the atoms in the original molecules so that they are in a suitable condition to react together. The very high speed of reactions between ions in solution, which indicates their low activation energies, provides evidence in favour of this suggestion, since in these ionic reactions many of the particles are already separated as a result of electrolytic dissociation. Similarly, investigation of certain very fast reactions such as those between methane and chlorine or between hydrogen and chlorine has revealed the presence of an abundance of free radicals in these systems during the reactions, which explains their low energies of activation and their extremely high speeds.

The energy changes occurring during an exothermic reaction are illustrated diagrammatically in Fig. 11.5. The level A represents the average energy of the reactants, B the energy needed before reaction occurs, and C the average energy of the products. The difference between

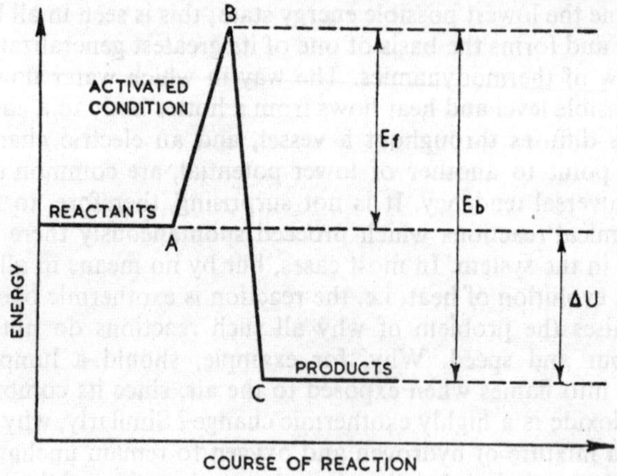

FIG. 11.5 Relationship between Heat of Reaction and Energy of Activation

A and B is the energy of activation of the forward reaction E_f, and the difference between B and C is the energy of activation of the backward reaction, E_b. From the diagram it will be seen that *the heat of the reaction at constant volume, ΔU, is equal to the difference between these two energies of activation.* Heat of reaction is, in fact, the resultant of the various energy changes occurring during the reaction, changes arising from the making and breaking of chemical bonds and from the difference in the thermal capacities of reactants and products, and being only a net energy change it bears no direct relationship to the energy of activation.

The effect of using a catalyst to accelerate a reaction is shown diagrammatically in Fig. 11.6; as explained in § 7.16 the energies of activation of the intermediate stages are usually much smaller than that of the original reaction, so it proceeds comparatively rapidly although the overall energy change is the same.

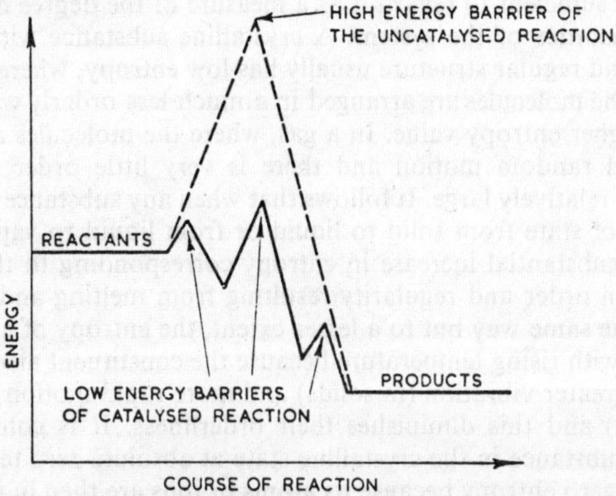

FIG. 11.6 The Effect of a Catalyst upon Energy of Activation

11.10. Entropy and Free Energy. From the fact that the majority of spontaneous reactions (i.e. reactions which under suitable conditions are capable of proceeding of their own accord) are exothermic and result in a decrease in the heat content or enthalpy of the system, it might be concluded that this was the only factor which determined whether any particular reaction is spontaneous or not. In fact, some spontaneous reactions take place endothermically so that heat is absorbed as they proceed and there is a net gain in enthalpy, as the following examples illustrate:

$$N_2 + O_2 \rightleftharpoons 2NO;\ \Delta H = +181\ \text{kJ mol}^{-1}$$
$$3O_2 \rightleftharpoons 2O_3;\ \Delta H = +284\ \text{kJ mol}^{-1}$$
$$C_{(s)} + H_2O_{(g)} = CO_{(g)} + H_{2(g)};\ \Delta H = +131\ \text{kJ mol}^{-1}$$

The readiness with which some salts dissolve in water with strong absorption of heat also demonstrates the limitation of exothermicity as a criterion for spontaneous changes. For example, when crystals of ammonium nitrate, ammonium chloride, or sodium nitrite are dissolved in water, so much heat is absorbed that their solutions are noticeably cooled. At first sight all these examples seem contrary to the general tendency to lose energy which was described in the previous section as a characteristic of all spontaneous changes. The explanation of this paradox lies in the realization that the change in enthalpy *alone* is completely unreliable as a predictor of spontaneous reactions and that some other factor must exist which also exerts an important influence.

This other factor is known as the *entropy* of the system. Although entropy is really a sophisticated thermodynamic concept, for our purposes it is sufficient to regard it as a measure of the degree of randomness or disorder of the system. A crystalline substance with a highly ordered and regular structure usually has low entropy, whereas a liquid in which the molecules are arranged in a much less orderly way tends to have a higher entropy value. In a gas, where the molecules are in continual and random motion and there is very little order at all, the entropy is relatively large. It follows that when any substance undergoes a change of state from solid to liquid or from liquid to vapour, there will be a substantial increase in entropy corresponding to the marked decrease in order and regularity resulting from melting and vaporization. In the same way but to a lesser extent, the entropy of a substance increases with rising temperature because the constituent atoms or ions undergo greater vibration (in solids) and more rapid motion (in liquids and gases) and this diminishes their orderliness. It is convenient to regard a substance in the crystalline state at absolute zero temperature as having zero entropy because its atoms or ions are then in rigid order and perfect regularity. Its entropy value at any other temperature is known as its molar entropy, S_m, and is expressed in entropy units of joule per kelvin or JK^{-1}, the standard value, S°_m, being the entropy value at 25°C and 101 325 N m^{-2} pressure.

Many spontaneous physical and chemical changes involve an increase in entropy. For example, when two pure gases are mixed and allowed to diffuse into each other there is a gain in entropy corresponding to the decrease in order. The molecules become randomly mixed instead of being separated into two distinct samples. Similarly, when a solid dissolves in a liquid the resulting solution has less regularity than the original crystal because the particles become scattered throughout the solution and mix homogeneously with the solvent. Diffusion of liquids, evaporation of a liquid, osmosis, sublimation, enantiotropic transition, thermal dissociation and thermal decomposition are other examples where spontaneous changes are accompanied by gains in entropy. Because of this natural tendency in spontaneous physical and chemical changes towards greater disorder and randomness, it is a fundamental belief of all scientists, based upon a great range of experience, that the entropy of the universe as a whole is continually increasing.

The thing which really matters in deciding the feasibility of a chemical reaction is the change which occurs in the *free energy* of the system. At any temperature T, this is given by the expression

$$\Delta G = \Delta H - T\Delta S$$

where ΔG = the change in free energy of the system

ΔH = the change in enthalpy of the system
T = the absolute temperature, K, of the system
ΔS = the change in entropy of the system

Thus the change in free energy incorporates changes in both the enthalpy and the entropy of the system and is the resultant of them both. In general terms, changes in *enthalpy* during a reaction arise from the making and breaking of atomic and ionic bonds and represent the net changes in energy involved in changing the distribution of the bonding electrons, whereas changes in *entropy* are concerned with the organization and states of the constituent atoms and ions in the reactants and products and the relative amounts of energy concerned with molecular translation, vibration, and rotation.

Only reactions which bring about a decrease in the free energy of the system can proceed spontaneously, since this fall in free energy is the essential driving force of the reaction. If a reaction is exothermic and also results in greater disorder, then the changes in enthalpy and entropy combine to ensure that there is a large decrease in free energy and such a reaction is fully capable of proceeding of its own accord. On the other hand, reactions which would result in absorption of heat and a loss in entropy are not spontaneous because in such reactions the free energy change ΔG would be positive. When a reaction is endothermic however, or when it involves a decrease in entropy, then it will only be spontaneous if the resultant of the two opposing influences leads to a net decrease in free energy. In such cases the sign of ΔG at any temperature will depend upon which term is greater and therefore represents a balance of opposing influences. At normal temperatures the term involving entropy is not usually large, amounting to only a few kilojoule per mole, but at higher temperatures the product $T\Delta S$ will tend to be much more significant and may even become predominant.

The reaction used industrially for making water gas (§ 20.15) provides a good example of the balance between enthalpy and entropy:

$$\text{C (graphite)} + H_2O(g) = CO(g) + H_2(g);\ \Delta H = +\ 131\ \text{kJ mol}^{-1}.$$

This reaction is strongly endothermic, and although there is a gain in entropy in forming the water gas from the reactants (ΔS is positive), the subtracted value of $T.\Delta S$ does not nearly compensate at room temperature for the large gain in enthalpy, so that at room temperature ΔG is positive and the reaction is not spontaneous. However, at temperatures in excess of 700°C the entropy term $T.\Delta S$ becomes so substantial that it exceeds the positive value of ΔH. It follows that at these high temperatures ΔG is negative corresponding to a decrease in free energy and the reaction is therefore spontaneous.

Another example of industrial importance is the synthesis of ammonia (§ 22.4):

$$N_2(g) + 3H_2(g) \rightleftharpoons 2NH_3(g); \Delta H = -92 \text{ kJ mol}^{-1}$$

Here the formation of ammonia results in a decrease in entropy, so ΔS is negative. At room temperature the enthalpy term predominates over the value of $T\Delta S$ and there is a fall in free energy ΔG, so the reaction is spontaneous and a good yield of ammonia can theoretically be expected. It must be stressed here that considerations of free energy act as a guide to the position of equilibrium only and give no information at all on the rate of reaction, which in this case is infinitely slow at room temperature. Above 600°C the entropy term becomes so significant that it outweighs the change in enthalpy and ΔG becomes positive, so the reaction is not spontaneous above this temperature, a conclusion in keeping with experimental findings.

It should be noted that ΔG is given by the difference between ΔH, the total energy transferred to or from the environment during the reaction to keep the temperature constant, and $T\Delta S$, the energy associated with the internal changes within the system. Thus the change in free energy in a closed system is in effect a measure of the maximum amount of energy which is available to do work outside the system, although in practice in any particular reaction it may not be possible to harness all this free energy to do useful work.

The change in free energy, ΔG, for a particular reaction will vary with temperature because the components from which it is derived are dependent upon temperature. For purposes of comparison it is usual to consider the standard free energy change, ΔG°, which is the change in free-energy which occurs with molar quantities of reactants and products at 25°C and at 101 325 N m^{-2} pressure. For a *reversible* reaction at constant temperature the standard free energy change ΔG° is related to the equilibrium constant K_p at an absolute temperature T by the expression:

$$\Delta G^\circ = -RT \log_e K_p = -2{\cdot}303RT \log_{10} K_p$$

which is known as the *van't Hoff Isotherm.* Its importance lies in its use for calculating the equilibrium constant at any given temperature from a knowledge of ΔG° or alternatively in determining ΔG° from experimentally measured values of K_p. Thus in a reversible reaction ΔG° and K_p both indicate *how far* reaction will proceed spontaneously and they define the position of equilibrium which will eventually be attained at that temperature. The greater the negative value of ΔG°, the farther the reaction will go towards completion until eventually a state is reached where no further fall in free energy is possible either by a decrease in enthalpy or by an increase in entropy. At this point ΔG°

is zero and the system is in equilibrium under those conditions. It is important to realize, however, that ΔG° and the equilibrium constant give no information at all on the *rate* of the reaction which may be very slow indeed, depending upon the conditions.

11.11. Bond Dissociation Energy. As explained in the previous section, the change in enthalpy ΔH represents the net energy change per mole which results from the breaking and making of chemical bonds during a reaction. By determining experimentally the value of ΔH in a series of carefully chosen reactions involving covalent bonds and by using spectroscopic measurements of heats of dissociation of diatomic molecules, it is possible to ascribe definite amounts of energy to the breaking or making of specific covalent bonds. The energy required to break a particular covalent bond under standard conditions (25°C, 101 325 N m^{-2} pressure, for one mole of substance) and separate the two atoms from each other completely is known as its *bond dissociation energy*. A list of bond dissociation energies has been compiled, some of which are given in Table 11.1.

TABLE 11.1. BOND DISSOCIATION ENERGIES/kJ mol^{-1}

C—C	343	C—O	339
C═C	611	O—H	460
C≡C	803	N≡N	941
C—H	414	H—H	435

The main value of these bond energies is that they can be added together to give an approximate value of an unknown heat of formation or to calculate a heat of reaction, as the following examples show:

Example I: Calculate the heat of formation of ethane given that the bond dissociation energies of C—C and C—H bonds are 343 kJ mol^{-1} and 414 kJ mol^{-1} respectively, the heat of dissociation of hydrogen is 435 kJ mol^{-1}, and the heat of vaporization of graphite is 711 kJ mol^{-1}.

$$2C\text{ (graphite)} + 3H_2\text{ (g)} = C_2H_6\text{ (g)};\ \Delta H = ?$$

Now $C\text{ (graphite)} \longrightarrow C\text{ (g)};\ \Delta H = +711\text{ kJ mol}^{-1}$

and $H_2\text{ (g)} \longrightarrow 2H;\ \Delta H = +435\text{ kJ mol}^{-1}$

$\therefore$ $2C\text{ (graphite)} \longrightarrow 2C\text{ (g)};\ \Delta H = +1422\text{ kJ mol}^{-1}$ (1)

and $3H_2\text{ (g)} \longrightarrow 6H;\ \Delta H = +1305\text{ kJ mol}^{-1}$ (2)

Thus the total energy required to break bonds in forming a molecule of ethane from graphite and hydrogen is the sum of (1) and (2), i.e. + 2727 kJ mol^{-1}.

Now the bond energy of C—C is −343 kJ mol^{-1} (3)

and „ „ „ C—H „ −414 kJ mol^{-1}

so the energy for 6 (C—H) bonds is 6 × −414 = −2484 kJ mol^{-1}. (4)

Thus the total energy released by making bonds in a molecule of ethane is the sum of (3) and (4), i.e. -2827 kJ mol^{-1}

$$\therefore \text{ the net energy change is } +2727 - 2827 = -100 \text{ kJ mol}^{-1}$$

Hence the heat of formation of ethane is -100 kJ mol^{-1}, i.e. 100 kilojoule of heat are evolved when one mole of ethane is formed from its elements in their standard states.

Example II: From the bond dissociation energies given in Table 11.1, calculate the heat of the reaction:

$$C_2H_2 + H_2 = C_2H_4$$

H—C≡C—H + H—H = 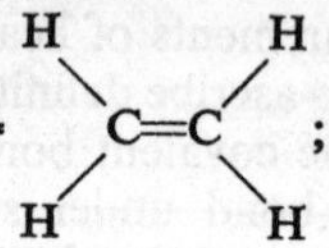 ; $\Delta H = ?$

Breaking one C≡C bond requires +803 kJ mol^{-1}
„ two C—H „ „ +828 kJ mol^{-1}
„ one H—H „ „ +435 kJ mol^{-1}

So the total energy required to break bonds is +2067 kJ mol^{-1}.

Making one C=C bond releases -611 kJ mol^{-1}
„ four C—H „ „ -1656 kJ mol^{-1}

So the total energy released in making bonds is -2267 kJ mol^{-1}.
Thus the net energy change, ΔH, is -200 kJ mol^{-1}, i.e. 200 kilojoule of heat would be evolved when one mole of ethylene was made by hydrogenation of acetylene.

In calculations of this kind it is assumed that the energy associated with any particular covalent bond is always the same irrespective of the molecule in which it occurs. In practice this is usually true, at least approximately, but sometimes there is a significant difference in bond energy values between different molecules because of a degree of polarity in the residual part of the molecule. For example, the bond dissociation energy in methane, CH_4, is 435 kilojoule for the first C—H bond broken, 433 kilojoule for the second and third, and 339 kilojoule for the fourth. Similar variations in the bond dissociation energy of O—H are found when the water molecule is disrupted. In such cases it is usual in theoretical calculations to accept the *average* value of bond dissociation energy as the energy needed to break or make the bond in question.

Occasionally heats of formation calculated from the generally accepted values for the bond dissociation energies show a marked discrepancy from the values found experimentally. Where the actual heat of formation of a compound is appreciably higher than the value expected from bond energy considerations, then the explanation is usually that the compound concerned has a resonance structure (§ 14.11). Typical examples are the oxides of nitrogen and carbon and aromatic compounds such as benzene and toluene.

11.12. Lattice Energy and the Born–Haber Cycle. The energy liberated when one mole of an ionic crystal lattice is formed from its constituent ions in the gaseous state is known as the *lattice energy* of that compound. For example, the lattice energy of sodium chloride is the energy involved in the change:

$$Na^+(g) + Cl^-(g) \longrightarrow Na^+Cl^-(s); \quad \Delta H = \text{lattice energy}$$

Lattice energy is in effect potential energy stored in the ionic crystal lattice. It is the resultant of all the electrostatic attractions and repulsions between the ions in the three-dimensional structure. It cannot be determined directly by experiment, but it can be calculated theoretically by regarding a lattice simply as a collection of spherical ions, each with its charge evenly distributed, spaced apart at distances corresponding to the ionic radii as determined by X-ray analysis (see §§ 4.5 and 13.9).

Alternatively, lattice energies can be obtained indirectly by applying Hess's law to a cycle of energy changes each stage of which can be determined experimentally. Such a cycle, known as the *Born–Haber cycle* after the two chemists who devised it in 1919, traces the energy changes involved in converting the elements in their standard atomic states into an ionic lattice by two alternative routes. By adding the energy changes of the various individual stages and comparing the totals for the two parts of the cycle it is possible to deduce a value for the lattice energy. The process is illustrated for sodium chloride in Fig. 11.7. Equating the two overall energy changes for the reaction

$$Na\,(s) + \tfrac{1}{2}\,Cl_2\,(g) = Na^+\,Cl^-(s)$$

gives

$$109 + 121 + 494 - 364 + \text{lattice energy} = -410$$

Hence the lattice energy of sodium chloride = 770 kJ mol^{-1}.

Now the value of the lattice energy of sodium chloride obtained by theoretical calculation is 766 kJ mol^{-1}. The very close agreement between this and the value determined by applying the Born–Haber cycle is typical of the halides of the alkali metals in general and lends powerful support to the concept of the ionic lattice for these compounds. Where there is a marked discrepancy between the two values, as in the silver halides and zinc sulphide, it suggests that in these cases the structure is not a purely ionic one and that the bonding between the elements is partially covalent in character (§ 14.7).

It should be noted that the predominant influences upon the value

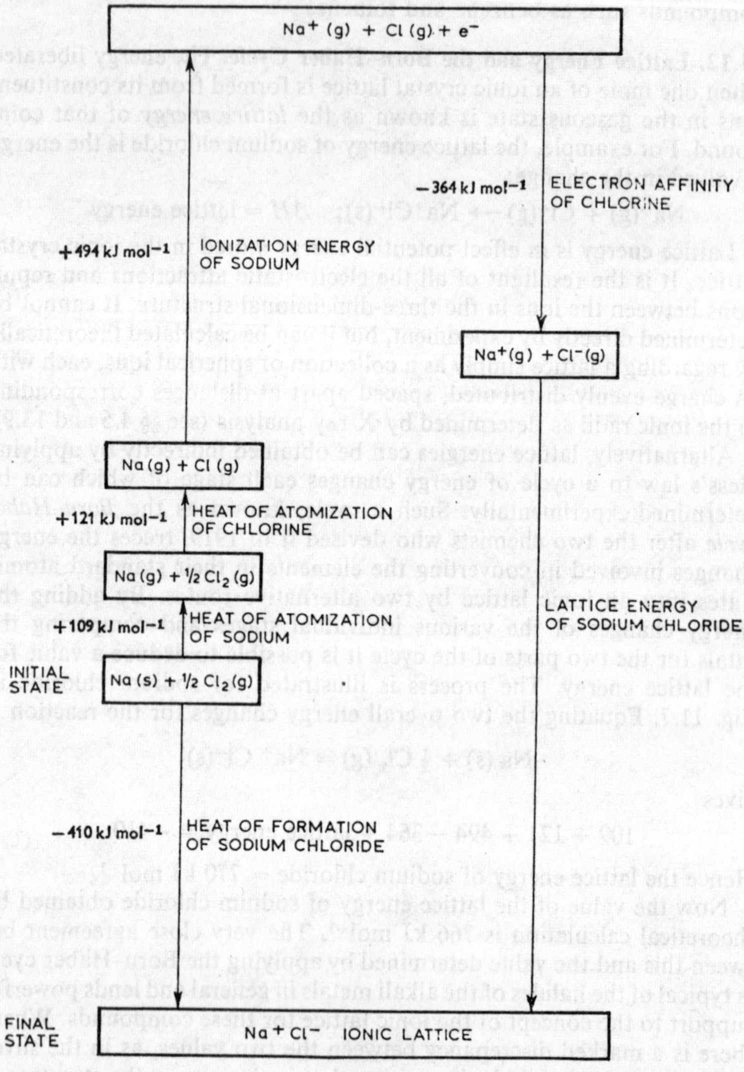

Fig. 11.7 The Born–Haber Cycle for Sodium Chloride

of the heat of formation of an ionic lattice from its elements are the two largest items, the ionization energy and the lattice energy. These are, of course, of opposite sign, since ionization of a metal is always an endothermic change and ionic lattice formation is always an exothermic process. It follows that in general an ionic compound will only be formed where the lattice energy exceeds the ionization energy of the metal, so that the energy released during lattice formation is enough to compensate for the energy required to bring about ionization. By using the simple ionic model as a theoretical basis we can calculate the lattice energies we should expect for various hypothetical ionic compounds

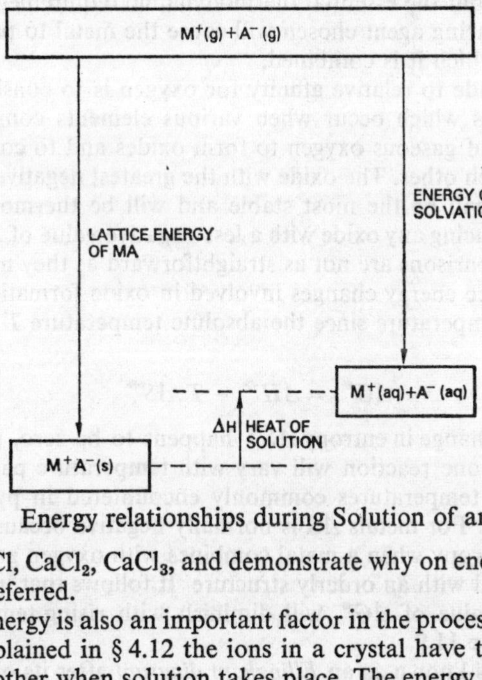

FIG. 11.8 Energy relationships during Solution of an Ionic Solid

such as CaCl, $CaCl_2$, $CaCl_3$, and demonstrate why on energetic grounds $CaCl_2$ is preferred.

Lattice energy is also an important factor in the process of dissolving, since as explained in § 4.12 the ions in a crystal have to be separated from each other when solution takes place. The energy needed for this comes mainly from the *energy of solvation* (called the *energy of hydration* where water is the solvent). It will be seen from Fig. 11.8 that the heat of solution (§ 11.7) is the difference between the lattice energy and the energy of solvation, and although it can be either positive or negative depending upon their relative magnitudes, it is always much smaller than either component.

11.13. Free Energy and the Extraction of Metals. Many metals occur as minerals in the form of the oxide, because of the preponderance of oxygen in the earth's crust (about 50% by mass) and its high electronegativity. Other metals which occur naturally as sulphides are usually converted into oxides as the first step in their extraction. It follows that the common feature of many metal extraction processes is the heating of the oxide with some element or compound which has a greater affinity for oxygen than the metal has, so that reduction occurs. The choice of reducing agent is often complicated by the necessity of avoiding reaction between the metal and the reducing agent, so that, for example, carbon cannot be used to reduce aluminium, uranium, or titanium oxides because these metals form carbides at the operating temperatures, but the essential thermodynamic requirement is that the particular reducing agent chosen will cause the metal to part from the oxygen with which it is combined.

The best guide to relative affinity for oxygen is to consider the free energy changes which occur when various elements combine with a fixed amount of gaseous oxygen to form oxides and to compare these values with each other. The oxide with the greatest negative free energy of formation will be the most stable and will be thermodynamically capable of reducing any oxide with a less negative value of $\Delta G^{\ominus}$. However such comparisons are not as straightforward as they might appear because the free energy changes involved in oxide formation can vary widely with temperature since the absolute temperature T is a term in the expression

$$\Delta G^{\ominus} = \Delta H^{\ominus} - T.\Delta S^{\ominus}$$

So unless the change in entropy $\Delta S^{\ominus}$ happens to be zero, the value of $\Delta G^{\ominus}$ for any one reaction will vary with temperature particularly at the very high temperatures commonly encountered in pyrometallurgical processes. For metals ΔS is normally negative because there is a decrease in entropy when a metal combines with oxygen gas to form a crystalline solid with an orderly structure. It follows that in such cases the negative value of $\Delta G^{\ominus}$ will diminish with rising temperature as shown in Figure 11.9.

This figure is known as an *Ellingham diagram* after its originator; it shows the variation in standard free energy changes with temperature in a number of oxidation reactions for one mole of oxygen at a pressure of 101 325 N m^{-2}. It will be seen from the diagram that $\Delta G^{\ominus}$ tends to vary linearly with temperature, although there are changes in gradient at the melting point and, more obviously, the boiling point of each metal. These abrupt changes in slope result from the increased randomness of the reactants, particularly at the boiling point of the metal, where its entropy increases by the molar latent heat of evaporation divided by the absolute temperature.

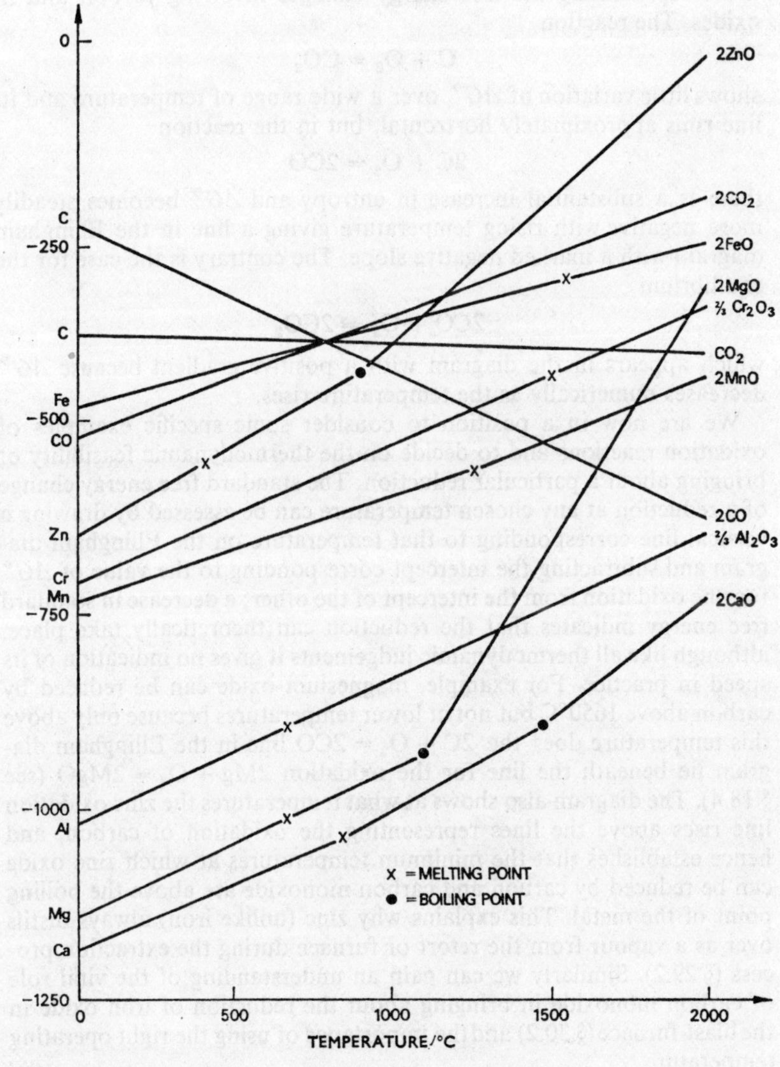

FIG. 11.9 Ellingham Diagram for Oxides

For reduction purposes the most interesting lines in Fig. 11.9 are those representing the free energy changes involving carbon and its oxides. The reaction

$$C + O_2 = CO_2$$

shows little variation of ΔG° over a wide range of temperature and its line runs approximately horizontal, but in the reaction

$$2C + O_2 = 2CO$$

there is a substantial increase in entropy and ΔG° becomes steadily more negative with rising temperature giving a line in the Ellingham diagram with a marked negative slope. The contrary is the case for the equilibrium

$$2CO + O_2 \rightleftharpoons 2CO_2$$

which appears in the diagram with a positive gradient because ΔG° decreases numerically as the temperature rises.

We are now in a position to consider some specific examples of oxidation reactions and to decide on the thermodynamic feasibility of bringing about a particular reduction. The standard free energy change of a reduction at any chosen temperature can be assessed by drawing a vertical line corresponding to that temperature on the Ellingham diagram and subtracting the intercept corresponding to the value of ΔG° for one oxidation from the intercept of the other; a decrease in standard free energy indicates that the reduction can theoretically take place, although like all thermodynamic judgements it gives no indication of its speed in practice. For example, magnesium oxide can be reduced by carbon above 1650°C but not at lower temperatures because only above this temperature does the $2C + O_2 = 2CO$ line in the Ellingham diagram lie beneath the line for the oxidation $2Mg + O_2 = 2MgO$ (see § 18.4). The diagram also shows at what temperatures the zinc oxidation line rises above the lines representing the oxidation of carbon, and hence establishes that the minimum temperatures at which zinc oxide can be reduced by carbon and carbon monoxide are above the boiling point of the metal. This explains why zinc (unlike iron) always distils over as a vapour from the retort or furnace during the extraction process (§ 29.2). Similarly we can gain an understanding of the vital role of carbon monoxide in bringing about the reduction of iron oxide in the blast-furnace (§ 30.2) and the importance of using the right operating temperature.

The Ellingham diagram is not limited, however, to explaining reductions by carbon and its oxide. Fig. 11.9 also demonstrates why aluminium can be used to reduce the oxides of manganese (§ 27.10) and chromium (§ 27.7) to their respective metals in the Goldschmidt process; in general any element can replace any other from its oxide if its standard free energy change during oxidation $\Delta G^{\ominus}$ has a higher negative value, i.e. if its line is beneath the other in the Ellingham diagram at that temperature. Although Fig. 11.9 suggests that calcium oxide cannot be reduced by aluminium, this is in fact the method now used for extracting calcium from quicklime. This reduction is possible thermodynamically *in vacuo* because at very low pressure the calcium oxidation line has a much steeper slope and rises at high temperatures above the line for the oxidation of aluminium giving a net fall in the free energy of the system when the reduction occurs.

12

The Colloidal State

12.1. Introduction. A colloid is not a particular type of substance, as was first thought, but a substance in a particular state of sub-division. In fact most substances can be obtained in a colloidal condition if sufficient trouble is taken, although some substances assume it more readily than others. A substance is colloidal when it is dispersed or scattered as very fine particles, droplets, or bubbles in another substance of different phase. It is the presence of two distinct phases in a colloidal mixture which distinguishes it from a true solution. In a colloidal system the

TABLE 12.1. TYPES OF COLLOIDAL SYSTEM

Disperse Phase	*Dispersion Medium*	*Name*	*Examples*
Solid	Solid	Solid sol	Coloured glasses
Liquid	Solid	Gel	Jellies
Gas	Solid	Solid foam	Pumice stone
Solid	Liquid	Sol	Metals, metallic salts
Liquid	Liquid	Emulsion	Milk, oil/water mixtures
Gas	Liquid	Foam	Whipped cream, lather
Solid	Gas	Aerosol	Smokes, clouds
Liquid	Gas	Aerosol	Fogs, mists, sprays

finely-divided material is known as the *disperse phase* and the continuous medium as the *dispersion medium*. The principal types of colloidal dispersion, with examples of each, are listed in Table 12.1.

In this chapter we shall consider sols, gels, and emulsions, dealing in detail with only the first of these as it is the most important type from our point of view.

12.2. Colloidal Sols. In most ordinary solutions the particles of solute are less than 10^{-9} m in diameter, but in a colloidal sol the particles range from about 10^{-9} m upwards to about $2 \cdot \times 10^{-7}$ m in diameter. They are too small, therefore, to be seen under an ordinary microscope (the lower limit, dictated by the wavelength of light, is about 2×10^{-7} m) or to be separated by ordinary filtration. Particles in a colloidal sol, then, are intermediate in size between those in a true solution and those making up a coarse suspension. No sharp limits exist, in fact, between these three, the characteristic properties changing only gradually as the particle size increases. It is nevertheless convenient to study colloidal sols apart from the others because of their distinctive optical, kinetic, and electrical properties. The three main types are *lyophobic sols*, *lyophilic sols*, and *colloidal electrolytes*.

12.3. Properties of Lyophobic Sols. These are sols in which little attraction exists between the colloidal particles and the dispersion medium, e.g. metals, and metallic sulphides, chlorides, and hydroxides. The dispersed phase consists of polymolecular aggregates, each particle containing enough small molecules to bring it within the colloidal range. Sols of this type display the following characteristics:

Optical Properties: If a beam of light is passed through a lyophobic sol, some of the light is scattered by the colloidal particles so that if they are viewed under a microscope directed at right angles to the beam they appear to an observer as bright spots of light against a dark background.

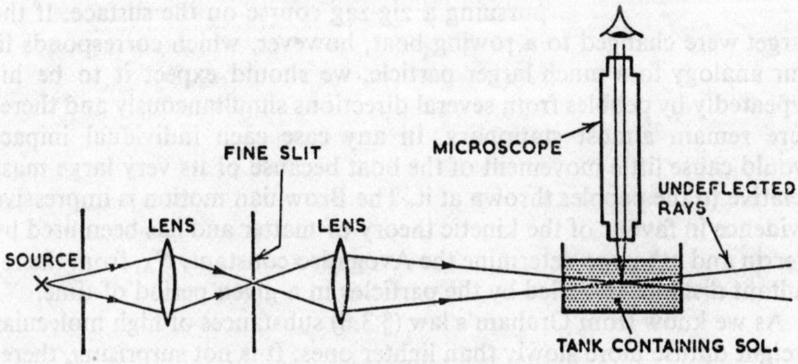

Fig. 12.1 The Ultramicroscope

The actual particles themselves are not visible, but only the light scattered by them, and the sol looks perfectly clear when viewed in the ordinary way. This scattering of light is known as the *Tyndall effect* after its discoverer, and a microscope arranged in this way (see Fig. 12.1) is called an *ultramicroscope*. The Tyndall effect is also shown by other colloidal systems, as for example when a searchlight, car headlamp, or cinema projector shines a powerful beam through fog or smoke-filled air. The importance of this property of sols is that it clearly demonstrates the presence of distinct particles and reveals their movements, which would not be visible in any other way. The colour of the sol depends upon its particle size, to which it is an approximate guide.

KINETIC PROPERTIES: When colloidal particles are observed under an ultramicroscope they are found to be in continual, rapid, and random motion in all directions owing to bombardment by molecules of the dispersing medium. The motion is known as *Brownian movement* (see Fig. 12.2) because it was discovered by the botanist Brown in 1827 whilst he was observing grains of pollen suspended in water. Molecular collisions are experienced, of course, by all particles suspended in a medium whatever their dimensions, but only when the particles are of colloidal size, or even smaller, is any net movement visible as a result. Larger particles, which are hit by many molecules from all directions at once, show no such motion because the effects of the simultaneous collisions cancel each other. The situation is rather like that of an empty petrol can floating in a swimming bath. If pelted with pebbles from all sides it would tend to change its direction abruptly each time it was hit, pursuing a zig-zag course on the surface. If the target were changed to a rowing boat, however, which corresponds in our analogy to a much larger particle, we should expect it to be hit repeatedly by pebbles from several directions simultaneously and therefore remain almost stationary. In any case each individual impact would cause little movement of the boat because of its very large mass relative to the pebbles thrown at it. The Brownian motion is impressive evidence in favour of the kinetic theory of matter and has been used by Perrin and others to determine the Avogadro constant, N_A, from the resultant distance travelled by the particles in a given period of time.

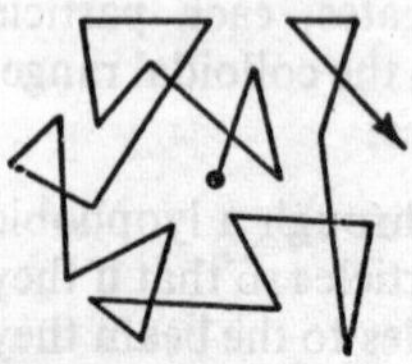

FIG. 12.2 Brownian Movement

As we know from Graham's law (§ 3.6) substances of high molecular weight diffuse more slowly than lighter ones. It is not surprising, therefore, to find that ordinary molecules or ions diffuse much more rapidly through a membrane of parchment, collodion, or cellophane than relatively heavy colloidal particles. We can take advantage of this difference to purify sols from true solutes, the process being known as *dialysis*. Fig. 12.3 shows a typical *dialyser*, as the apparatus is called, in which the dispersing medium is being continually renewed.

It is important at this stage to distinguish clearly between the various processes of separation which are in practical use. When a liquid is filtered in the ordinary way, only those particles with a diameter greater than approximately 10^{-6} metre are retained on the filter paper, any particles of colloid or true solute present passing through into the filtrate with the solvent. In *ultrafiltration*, in which a liquid is forced under pressure through a membrane made by specially treating filter paper with collodion to reduce the size of its pores, larger colloidal particles are retained as well as those of coarse suspension size, whilst any particles in true solution collect in the filtrate with the solvent. Dialysis, on the other hand, is a selective diffusion process in which the membrane used is permeable to molecules and ions (and also therefore to solvent) but not to particles of colloidal size. This is in contrast to osmosis (§ 5.15), where the membrane is permeable only to molecules of the solvent, allowing no particles of the solute at all to pass through it. In osmosis the solvent diffuses through the membrane in both

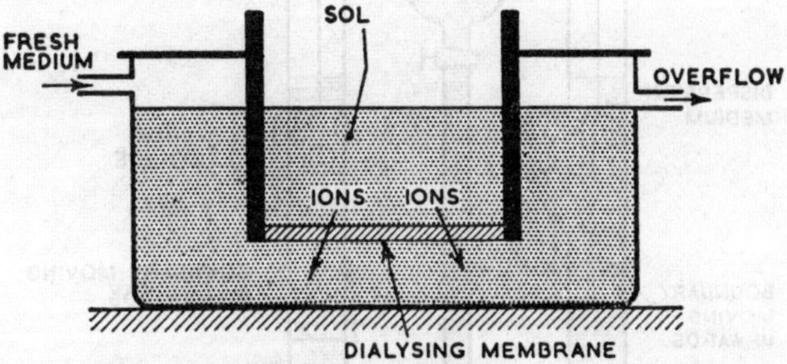

FIG. 12.3 A Dialyser

directions at different rates, leading, unless an external pressure is applied, to dilution of the solution. When a colloidal solution is separated from pure dispersing medium by a semi-permeable membrane of this kind, osmosis does occur, but the osmotic pressure is much less than for a true solution of similar concentration because it depends upon the number of particles in solution, and for a given mass of solute this number is much smaller for colloidal particles owing to their greater individual mass. Of the various colligative properties only osmosis is used for determining the mass of colloidal particles since the depressions of the freezing point and the elevations of the boiling point produced by colloidal sols are too small to be measured.

ELECTRICAL PROPERTIES: If an electric field is applied to a lyophobic sol, its colloidal particles migrate towards one of the electrodes, showing that they must be electrically charged. This process, which is known as *electrophoresis*, or sometimes as *cataphoresis*, may be demonstrated by using the apparatus shown in Fig. 12.4. The sol, which usually has a distinctive colour, is placed in the U-tube with a layer of dispersion medium above it in each limb. On applying a potential of about 200 volt (d.c.) to the electrodes, the boundaries of the sol are seen to move in one direction at a rate of about 5×10^{-5} m s^{-1}. The sign of the charge on the particles can be deduced from the direction of migration; most metals, sulphur, and metallic sulphides are negatively charged, whereas particles of metallic hydroxides usually carry a positive charge. Electrophoresis is employed industrially for depositing substances such as rubber latex, ceramics, or metals by making the article to be plated one of the electrodes. The advantages of the process over electrolysis are its much greater speed and its applicability to a wider range of materials.

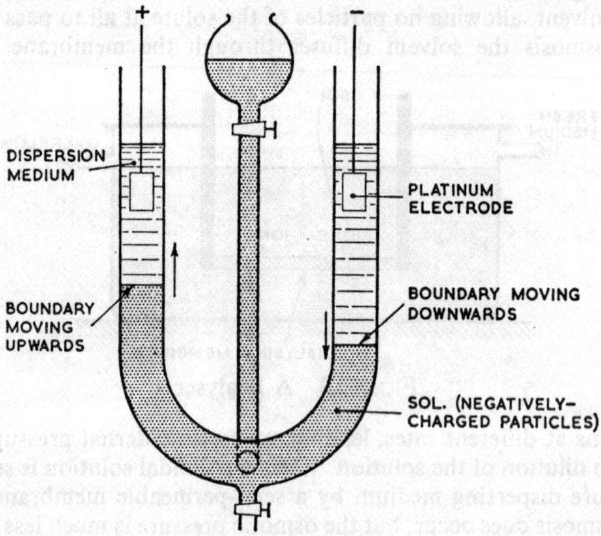

FIG. 12.4 Electrophoresis

The charge borne by colloidal particles results from the adsorption of ions onto their surfaces. For example, a sol of ferric hydroxide readily adsorbs ferric ions leaving an excess of free hydroxyl ions in the solution. Since the colloidal system as a whole is electrically neutral it follows that the dispersion medium must bear an equal and opposite charge to the particles. This can be demonstrated by the process of *electro-osmosis* using the apparatus shown in Fig. 12.5. The sol is placed in the inner compartment which is separated from the electrodes by two dialysing membranes. On applying a suitable potential difference to the electrodes, a flow of medium takes place which shows itself by a gradual change in levels in the two outer compartments.

COAGULATION AND SEDIMENTATION: The presence of adsorbed charges on colloidal particles in a lyophobic sol explains why they do not normally coagulate into larger particles and undergo sedimentation (i.e. gradually fall to the bottom of the vessel under the influence of gravity) as coarse suspensions do. Since all particles in a particular sol bear the same charge, the mutual repulsion of like charges keeps them apart. As long as they remain at a size within the colloidal range they are unlikely to settle as a precipitate because they are being continually buffeted about in all directions as a result of the Brownian movement. It follows, however, that if the charges on the particles are neutralized in any way, then the usual forces of attraction acting between the molecules will lead to rapid coagulation and precipitation. This happens, for example, if arsenious sulphide sol, which contains negatively charged

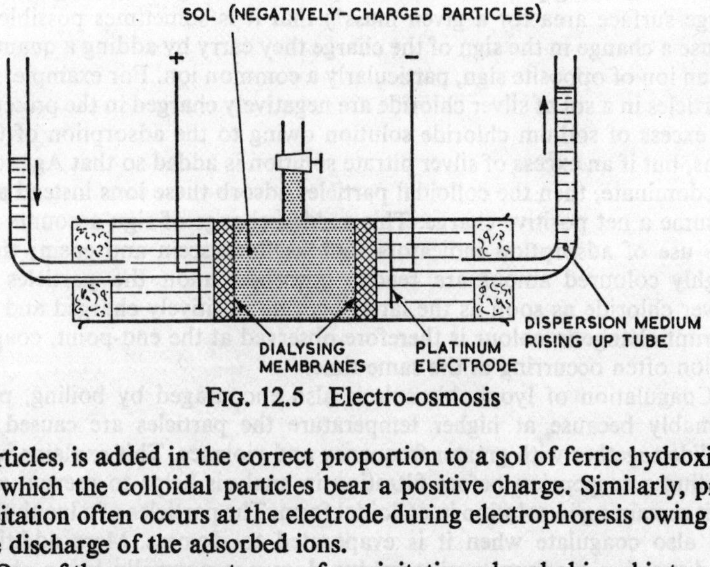

FIG. 12.5 Electro-osmosis

particles, is added in the correct proportion to a sol of ferric hydroxide in which the colloidal particles bear a positive charge. Similarly, precipitation often occurs at the electrode during electrophoresis owing to the discharge of the adsorbed ions.

One of the commonest ways of precipitating a lyophobic sol is to add a solution of an electrolyte. The amount which must be added depends critically upon the valency of the appropriate ion. It is found that in general trivalent ions are much more effective in promoting coagulation than divalent ones, which in turn are between fifty and one hundred times as effective as any univalent ion. This relationship between precipitating power and valency, which is known as the *Schulze-Hardy rule*, explains the use of aluminium sulphate (§ 19.12) and other salts containing the Al^{3+} ion for water purification and sewage treatment and in a styptic pencil to stop bleeding. The polyvalent cation rapidly neutralizes the negative charges on the colloidal particles, making it possible for them to coagulate and be precipitated. The formation of a delta at the mouth of a river is also partly caused by the precipitating action of the Na^+ ions in salt water on the colloidal particles of clay.

Lyophobic sols are sometimes stabilized by adding to them a quantity of lyophilic sol such as gelatin. This *protective colloid*, as it is called, has the effect of making the original sol much less susceptible to coagulation by electrolytes. It is probable that the charged particles of the lyophobic sol tend to be adsorbed onto the surface of the highly solvated lyophilic particles and are thereby prevented from coalescing in the usual way. For example, gelatin is often added to ice-cream during manufacture to prevent the colloidal particles of ice from coagulating and giving it a gritty structure.

The adsorbing power of colloidal particles is so great (owing to their large surface area for a given mass*) that it is sometimes possible to cause a change in the sign of the charge they carry by adding a quantity of an ion of opposite sign, particularly a common ion. For example, the particles in a sol of silver chloride are negatively charged in the presence of excess of sodium chloride solution owing to the adsorption of Cl^- ions, but if an excess of silver nitrate solution is added so that Ag^+ ions predominate, then the colloidal particles adsorb these ions instead and assume a net positive charge. This sudden change of sign accounts for the use of adsorption indicators such as fluorescein and eosin; their highly coloured anions are readily adsorbed upon the particles of silver chloride as soon as the latter become positively charged and an abrupt change in colour is therefore observed at the end-point, coagulation often occurring at the same time.

Coagulation of lyophobic sols is also encouraged by boiling, presumably because at higher temperature the particles are caused to collide together with greater frequency and violence. This explains why boiling a suspension before filtration in analysis helps to prevent any of it remaining in solution in colloidal form. The particles of a lyophobic sol also coagulate when it is evaporated to dryness. Mere addition of dispersion medium to the residue does not normally bring about the formation of a sol again, in contrast to the usual behaviour of lyophilic sols.

12.4. Properties of Lyophilic Sols. In this type the colloidal particles show considerable affinity for the dispersion medium and are often extensively solvated. The particles themselves consist of single

* For example, when a lump of substance is sub-divided into cubes with sides of 10^{-8} m length (and many colloidal particles in a sol are smaller than this) its surface area is increased a millionfold.

molecules which are so large that they come within the colloidal range. Many of them occur naturally and are of great biological importance, e.g. starch, dextrin, agar, gelatin, albumen, casein, and other proteins. Their characteristics are different from those of lyophobic sols in many respects. For example, owing to the similarity between the refractive index of a particle of lyophilic sol and that of the dispersion medium very little light is scattered by the Tyndall effect and consequently the particles are not easily detected under

TABLE 12.2. A COMPARISON OF LYOPHOBIC AND LYOPHILIC SOLS

Lyophobic Sols	*Lyophilic Sols*
1. The particles are readily detected by means of the ultramicroscope.	1. The particles are not easily detected by the ultramicroscope.
2. The surface tension is similar to that of the pure medium.	2. The surface tension is generally lower than that of the pure medium.
3. The viscosity is similar to that of the pure medium.	3. The sol is usually much more viscous than the pure medium.
4. The particles migrate in one characteristic direction during electrophoresis.	4. During electrophoresis the particles may migrate in either direction or even not migrate at all, depending upon the pH of the medium.
5. Even small concentrations of electrolytes can cause precipitation.	5. Small concentrations of electrolytes have little effect.
6. On cooling or evaporation they give solids which do not readily form sols again.	6. On cooling or evaporation they give gels which readily form sols again on adding medium.
7. The sols show only slight colligative properties.	7. The sols show only slight colligative properties.
8. The sols can be purified by dialysis.	8. The sols can be purified by dialysis.

the ultramicroscope. Because of their extensive solvation lyophilic sols are not readily precipitated on adding electrolytes and it is often possible to neutralize the charges on their colloidal particles completely without bringing about coagulation. However if high concentrations of suitable electrolytes are used, coagulation does occur owing to desolvation of the disperse phase; this process is commonly known as 'salting-out'. The charge borne by the colloidal particles depends very much upon their environment, particularly upon the pH of the medium, since the particles readily combine with or adsorb H^+ and OH^- ions. Unlike lyophobic sols, this charge is often due to partial electrolytic dissociation of the molecule rather than to the adsorption of ions from solution. The pH at which the particles bear no net charge is known as the *iso-electric point*. It follows that a given lyophilic sol may migrate in either direction during electrophoresis or even, at the iso-electric point, not move at all, depending upon the conditions. One important application of this is the use of electrophoresis for distinguishing between and separating proteins in a mixture on the basis of their different rates of migration in media of different pH. Details of the apparatus will be found in § 12.16.

In lyophilic sols the surface tension is usually lower and the viscosity normally very much higher than in the pure dispersion medium, so that many of these sols are notably viscous liquids (e.g. glue, gelatin, and starch). They also differ from lyophobic sols in that the gels formed from them by cooling or evaporation readily yield sols again on adding a fresh quantity of medium. Lyophilic sols resemble lyophobic ones in having very slow rates of diffusion. Indeed it was this property which first attracted Graham's attention to them and enabled him to separate them from true solutes by means of dialysis.

The properties of lyophobic and lyophilic sols are compared in Table 12.2.

12.5. Colloidal Electrolytes. This type of colloid, which includes soaps (§ 38.5), synthetic detergents, and many dyestuffs and proteins, is composed of ions of a special sort, each having a long-chain hydrocarbon structure attached to a polar grouping at one end. These ions are arranged in the sol to form highly charged aggregates of colloidal size known as *micelles*, as shown diagrammatically in Fig. 12.6, giving an unusual combination of properties. Like lyophobic and lyophilic sols they have a low osmotic pressure, diffuse only slowly, and often show the Tyndall effect, but unlike them colloidal electrolytes are highly conducting. They are also good emulsifying agents because non-polar materials (e.g. oils, fats, and greases) dissolve in the hydrocarbon part of the micelles whilst water-soluble substances are adsorbed by the ions at the fringe. It is this and their low surface tension which gives colloidal electrolytes their excellent

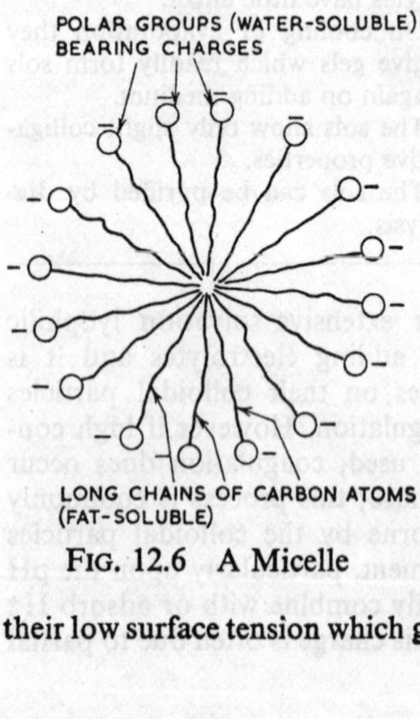

FIG. 12.6 A Micelle

detergent and wetting properties leading to their extensive use as cleansing agents.

12.6. Preparation of Colloidal Sols. The methods available are of two main types—disintegration methods (1–3) in which, as the name suggests, the substance is broken down into particles of colloidal size, and condensation methods (4–7) in which the colloidal particles are made from smaller units by some process of controlled aggregation:

1. Many lyophilic sols may be made directly by warming the substance (usually a gel) with a suitable dispersion medium, e.g. gelatin and warm water.

2. One of the best methods of preparing lyophobic sols is by *peptization*, i.e. the addition of a solution containing a small concentration of ions (usually common ions) to a fine suspension or precipitate of the substance. The particles adsorb the ions readily and are dispersed into the medium as a result of the repulsion of their similar charges. For example, sols are obtained when hydrogen sulphide is passed into a cold aqueous suspension of arsenic trisulphide As_2S_3, and when a little dilute hydrochloric acid or dilute silver nitrate solution is added to a precipitate of silver chloride, and when cold sodium hydroxide solution is added to a precipitate of aluminium hydroxide. The practical implications of peptization should be obvious as far as the washing of precipitates in analysis is concerned.

3. By using a *colloid mill* in which the substance is ground between two metal discs rotating rapidly in opposite directions. A protective colloid is usually employed to act as a stabilizer.

4. Metal sols are best made by reducing solutions of their salts in the dispersion medium. Reducing agents such as hydrogen, carbon monoxide, formaldehyde, hydrazine, and tannic acid are used to avoid ionic by-products which might cause coagulation. A protective colloid such as gelatin helps to stabilize the metal sol. For example, a gold sol is obtained when chloroauric acid (§ 28.22) is treated with formaldehyde, hydrazine, or tannic acid:

$$HAuCl_4 + 3H = Au + 4HCl$$

5. Various chemical methods such as hydrolysis, oxidation, and double decomposition are widely used. The general principle of all of them is to produce by some means a highly supersaturated solution of the substance from which it rapidly separates and condenses into particles of colloidal size. Thus metallic hydroxide sols are often prepared by the hydrolysis of metallic salts, e.g. by adding a concentrated solution of ferric chloride to boiling water:

$$FeCl_3 + 3H_2O \rightleftharpoons Fe(OH)_3 + 3HCl$$

Sols of sulphur are obtained by mixing solutions of hydrogen sulphide and sulphur dioxide, or by adding acids to solutions of thiosulphates. Similarly, a sol of hydrated silica may be prepared by adding dilute hydrochloric acid to sodium silicate solution. In all of these cases the sol should be at least partially freed of ions by dialysis if it is to be kept for any length of time.

6. A substance is often thrown out of solution in particles of colloidal size by suddenly changing the solvent, e.g. sols are obtained when concentrated alcoholic solutions of sulphur or phosphorus are poured into a large excess of water.

7. By using *Bredig's arc*, in which two metal electrodes are allowed to strike an electric arc whilst immersed in a cooled dispersion medium in the presence of traces of alkali. The metal is probably vaporized by the heat of the spark, condensing into colloidal particles.

12.7. Emulsions. When two immiscible liquids are so thoroughly mixed together that one is dispersed in the other in globules of colloidal size, the product is called an *emulsion*. Unless an emulsifying agent is present the droplets of the disperse phase tend to coalesce causing the emulsion to separate on standing into two distinct layers. For example, in milk, in which colloidal globules of oil are suspended in water, the protein casein acts as an emulsifier. Emulsions resemble lyophobic sols in many ways despite the fact that the droplets are often considerably larger than the lyophobic particles. They show the Tyndall effect and Brownian motion, they usually migrate towards the anode during electrophoresis, and they are readily separated by adding solutions containing polyvalent positive ions. Emulsions are of great importance in everyday life; examples include butter, margarine, mayonnaise, hair-cream, many cosmetic creams and ointments, and certain paints, varnishes, and liquid polishes.

12.8. Gels. A gel is a colloidal dispersion of a liquid in a solid. The solid material usually has a very open structure rather like a honeycomb or sponge with the molecules of liquid trapped in the spaces of the network. Those gels like gelatin which retain a high proportion of liquid are soft and elastic, but gels in which the dehydration is almost complete, such as silica gel, are hard and rigid. Gels of the first type are formed when a hot concentrated lyophilic sol is cooled; its viscosity increases gradually until the whole mass sets to a semi-solid (table jelly is a good example). When it is warmed such a gel reverts to the original sol. For such gels it is equally possible to hold the converse view—that a gel is in fact a dispersion of fibrous solid particles in a continuous liquid medium, the proportions being such that the whole system acquires a semi-rigidity. This explains why the process of *gelation* (gel formation) is so easily reversed by adding more liquid or by heating, and also accounts for the readiness with which diffusion occurs in this type of gel. In non-elastic gels such as silica gel, however, the solid is best regarded as the continuous medium, the minute spaces in its sponge-like structure being filled with varying amounts of water or air according to its degree of hydration.

One of the characteristic properties of gels is their readiness to absorb large quantities of liquid, often swelling considerably in the process. Indeed silica gel is so porous when dehydrated that it is often used as a drying agent for gases.

12.9. Adsorption. Although adsorption is one of the most prominent characteristics of substances in the colloidal state, it should not be concluded that only colloids exhibit the phenomenon. In fact many substances are known to possess this power of attracting and retaining molecules or ions upon their surfaces. The readiness with which activated charcoal adsorbs various gases, particularly at low temperatures, is referred to in § 20.3. The process is reversible, the gas being completely released again on strong heating, which suggests that the mechanism is generally a physical one possibly involving attachment of the gas molecules to the surface of the solid by relatively weak van der Waals bonds (§ 4.9). On the other hand the adsorptive powers of certain transition metals (e.g. nickel, cobalt, copper, iron, and platinum) which are extensively used as catalysts in industrial oxidations and hydrogenations (see § 27.2) are probably of a chemical nature since they are often irreversible, highly specific, and encouraged by heating. On the surface of any solid there will be a great number of unsatisfied valency forces where the exposed ions or molecules in the lattice are incompletely co-ordinated, and it is likely that this unsaturation at the surface is mainly responsible for the ability of certain substances to adsorb gases or take up solutes from solution. Aluminium hydroxide, for example, adsorbs dyes very readily, which accounts for its use as a mordant (§ 19.10)and for its detection by means of aluminon (§ 19.15). Adsorption of solutes by solids is also very important in tanning, in the commercial extraction of vitamins, in sugar refining (§ 42.4), and in water purification. Advantage is taken of adsorption to judge end-points in volumetric analysis by means of adsorption indicators or the starch-iodine coloration.

12.10. Flotation. The ores of many metals (e.g. copper, zinc, nickel, and lead) occur mixed with a large excess of rock and clay known as *gangue* and the first stage of their extraction consists of separating the mineral from the bulk of worthless material. This is usually done by crushing the mixture to a fine powder and adding it to a tank of water to which certain substances which form colloidal particles in solution have been added in small concentration. These *flotation agents*, as they are called, are of two main kinds, frothers (e.g. pine oil, creosote, and soaps) which by lowering the surface tension of the liquid facilitate the formation of a stable froth, and collectors (e.g. xanthates and dithiophosphates) which selectively adsorb the fine particles of mineral and attach them to passing air bubbles. On agitating the liquid by means of compressed air, therefore, particles of the mineral tend to collect in the froth at the surface, leaving the unwanted material to settle to the bottom of the tank. This flotation method of concentrating ores is widely used because when only the concentrate has to be transported and reduced there is a substantial saving in freight and fuel costs.

12.11. Chromatography. This is the general name now given to those processes whereby the separation of a mixture of substances can be brought about by taking advantage of the different degrees to which they undergo adsorption, partition, or ion-exchange when passing over material of a different phase. These processes may involve the flow of a solution of the mixture through a column of adsorbent in powder form as in the various types of *column chromatography*, or through discs, strips, or sheets of special adsorbent paper as in *paper chromatography*, or through a film of adsorbent solid as in *thin-layer chromatography*. The same principle applies in *gas chromatography*, where the solvent is replaced by a stream of gas which carries the vapours from a volatile mixture at differing rates though a column packed with adsorbent material impregnated with a liquid.

The name chromatography was devised because in its original applications it involved mixtures of coloured substances which when separated gave distinct rings or zones of colour known as *chromatograms*. Nowadays chromatography is applied to colourless as well as coloured substances and the process is no longer solely dependent upon the visual observation of separated colours.

The great importance of chromatography lies in its valuable advantages over other analytical methods. It can be used successfully to separate mixtures of substances which are very similar physically and chemically (e.g. amino-acids, hydrocarbons, sugars, phenols, rare earth metals) and which necessitate lengthy and tedious separation processes when the usual physical and chemical methods are employed. It can be adapted to apply to very small quantities of material, making it possible to detect the presence of only a microgramme of a substance. It not only provides a means of identifying the components present in a mixture but it can also be used for quantitative analysis; gas chromatography in particular permits a rapid and accurate assessment of the proportions of each constituent. Chromatography has been used to purify products industrially and as a means of checking purity.

12.12. Column Chromatography.

ADSORPTION TYPE: If a solution of a mixture in a suitable solvent is allowed to flow down a vertical column filled with highly adsorbing material, the most readily adsorbed solutes tend to be retained near the top and the others penetrate various distances down the column depending upon the degree to which they are adsorbed and how much pure solvent is added at the top afterwards. If the mixture consists of highly coloured substances, the result is a series of coloured bands or zones spread down the column as in Fig. 12.7. The commonest adsorbents are alumina, chalk, charcoal, magnesia, starch, and dried silica gel, all of which are available in special grades with the optimum particle size to give uniform packing and easy percolation. Typical solvents include petrol, ether, benzene, acetone, chloroform, and various alcohols. Separation of the components of the mixture can be achieved either by providing a continuous flow of solvent which will eventually wash each constituent in turn out of the base of the column (a process known as *elution*), or by extruding the column of adsorbent and extracting the adsorbed substances from the mechanically separated zones of the chromatogram. These methods have been widely used for detecting and separating plant pigments and similar organic mixtures.

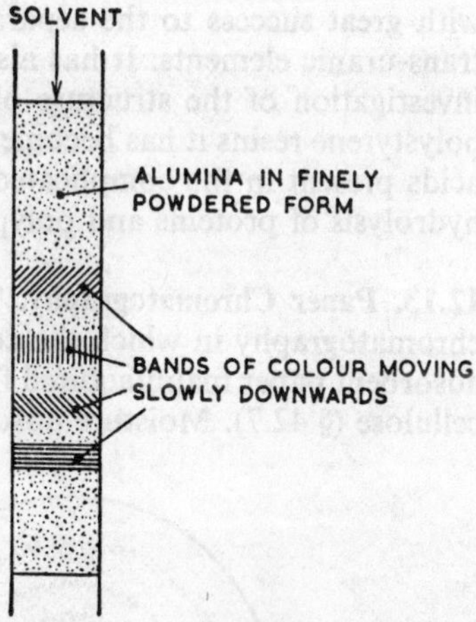

FIG. 12.7 Column Chromatography

PARTITION TYPE: In this type of chromatography the partition principle described in § 6.14 is at work. Separation is achieved because the solutes in the mixture distribute themselves in different degrees between the solvent flowing down the column and the water which is adsorbed on the solid material used for packing the column. This consists of finely-powdered cellulose or silica gel or a natural porous material such as kieselguhr, so treated as to contain a high proportion of adsorbed moisture. The apparatus used and the technique for effecting a separation are similar to those of the adsorption type. The method has proved useful for the isolation and detection of amino-acids, sugars, and other water-soluble compounds.

ION EXCHANGE TYPE: Columns packed with synthetic ion exchange resins made from cross-linked polystyrene (§ 9.17) have been used to separate particular substances from solutions of mixtures. The principle underlying the method is that certain ions undergo exchange more readily than others and are preferentially retained in the resin, whilst the remaining ions in the mixture are carried down the column by an excess of solvent. The trapped ions can afterwards be released from the resin by changing the pH or by washing with a solution in which they form complex ions (e.g. citrates) and can then be collected as a separate fraction from the base of the column. The method has been applied with great success to the separation of the rare earth metals and the trans-uranic elements. It has also made a valuable contribution to the investigation of the structure of proteins, since by using sulphonated polystyrene resins it has become possible to identify the various amino-acids present in the complicated mixture which often results from the hydrolysis of proteins and polypeptides (§ 41.16).

12.13. Paper Chromatography. This is essentially a form of partition chromatography in which the column is replaced by a sheet of special adsorbent paper manufactured from cotton fibre and composed of pure cellulose (§ 42.7). Moisture adsorbed on the cellulose fibres acts as the

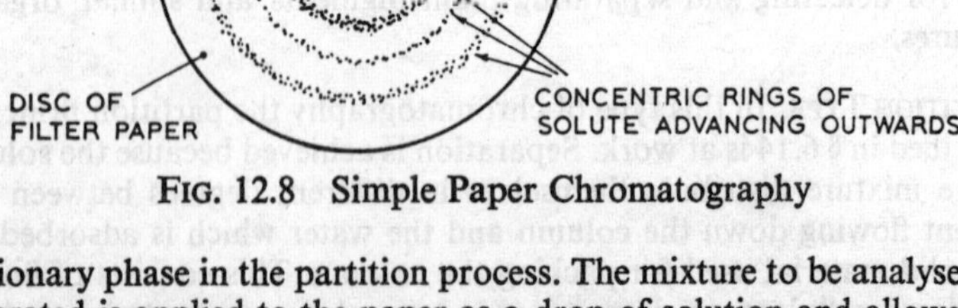

FIG. 12.8 Simple Paper Chromatography

stationary phase in the partition process. The mixture to be analysed or separated is applied to the paper as a drop of solution and allowed to dry. The paper is then brought into contact with a suitable solvent which spreads through the paper by capillary action, the various solutes advancing at different rates behind the front of the liquid to produce the familiar chromatogram. The solvent is often a mixture of several liquids, of which *n*-butanol, ethanol, acetic acid, hydrochloric acid, water, ammonium hydroxide, and pyridine are the commonest, and is chosen for its effectiveness in bringing about the separation of the particular mixture in hand, the selection often being a matter of trial and error.

The simplest form of paper chromatography merely uses a disc of filter paper with the drop of solution at the centre; as solvent spreads outwards a series of concentric rings are produced on the paper as in Fig. 12.8. Alternatively, strips or sheets of paper can be suspended in a tank as shown in Fig. 12.9 and the solvent allowed to ascend or descend

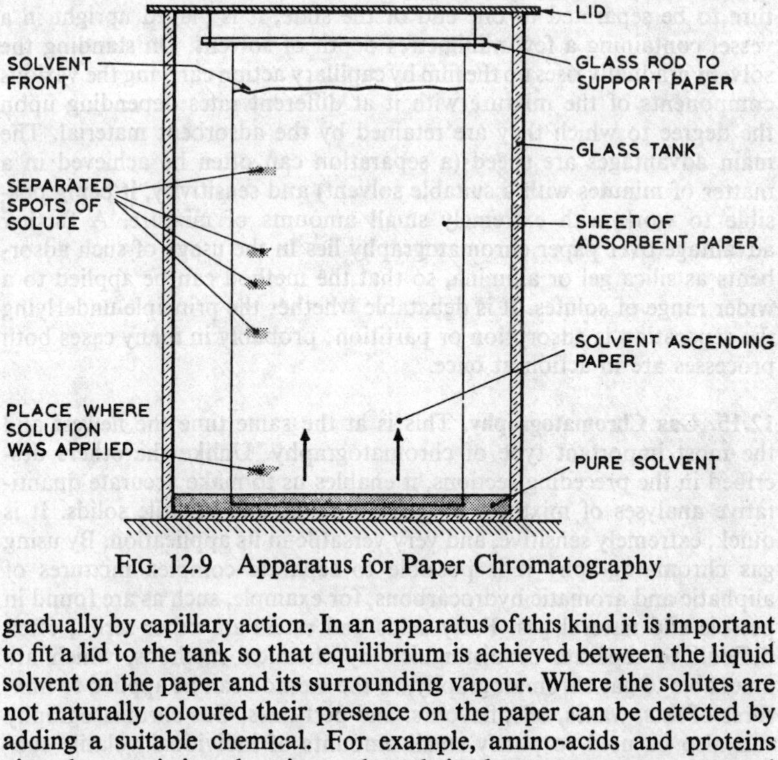

FIG. 12.9 Apparatus for Paper Chromatography

gradually by capillary action. In an apparatus of this kind it is important to fit a lid to the tank so that equilibrium is achieved between the liquid solvent on the paper and its surrounding vapour. Where the solutes are not naturally coloured their presence on the paper can be detected by adding a suitable chemical. For example, amino-acids and proteins give characteristic colorations when their chromatograms are treated with the reagent ninhydrin, which is widely used for this purpose.

When a single flow of solvent is incapable of effecting complete separation of a mixture, two-dimensional paper chromatography can be employed. A chromatogram prepared from a spot of solution applied to one corner of a sheet of paper is dried and then turned through 90°

K

and subjected to a second process with a different solvent. In this way those spots which may not have separated with the first solvent are carried along by the second to differing extents and can be distinguished on the paper. This technique has proved very useful in analysing mixtures of amino-acids.

12.14. Thin Layer Chromatography, T.L.C. In this type a thin layer of adsorbent supported on a glass plate is used in place of a column or sheet of paper. The simplest apparatus consists of a microscope slide coated with a thin film of adsorbent which has been thoroughly dried by heating. After applying a drop of the solution containing the mixture to be separated to one end of the slide, it is placed upright in a vessel containing a few millimetres depth of solvent. On standing the solvent gradually rises up the film by capillary action carrying the various components of the mixture with it at different rates depending upon the degree to which they are retained by the adsorbent material. The main advantages are speed (a separation can often be achieved in a matter of minutes with a suitable solvent) and sensitivity, it being possible to work with extremely small amounts of mixture. A further advantage over paper chromatography lies in the usage of such adsorbents as silica gel or alumina, so that the method can be applied to a wider range of solutes. It is debatable whether the principle underlying the separation is adsorption or partition; probably in many cases both processes are in action at once.

12.15. Gas Chromatography. This is at the same time the newest and the most important type of chromatography. Unlike the others described in the preceding sections, it enables us to make accurate quantitative analyses of mixtures of gases, liquids, and volatile solids. It is quick, extremely sensitive, and very versatile in its application. By using gas chromatography it is possible to separate complex mixtures of aliphatic and aromatic hydrocarbons, for example, such as are found in petrols and natural gas. Fatty acids and esters are other series, often difficult to separate or distinguish by chemical methods, which are amenable to gas chromatography, so the method can be applied to mixtures of fats, waxes, oils, flavours, and perfumes. Gas chromatography can also be used to purify small amounts of individual volatile substances and as a test of the purity of a substance, since the presence of even a trace of impurity can readily be detected.

The apparatus used in gas chromatography is shown diagrammatically in Fig. 12.10. A small quantity of the volatile mixture is injected into a steady stream of an unreactive gas which acts as carrier. On passing through the column, the vapours of constituents of the mixture undergo adsorption or partition to differing extents thereby suffering delays which are characteristic of those particular substances. Their emergence from the end of the column one by one at distinct intervals is detected by a special device and is recorded automatically on a chromatogram.

Gas chromatography is of two types depending upon either the partition principle or direct adsorption. The *partition* kind, which is often referred to as *gas-liquid chromatography* or G.L.C., is much the more important of the two and can be used for analysing mixtures of liquids and volatile solids. In this type the column is packed with granules of inert supporting material (e.g. 'Celite', firebrick, or glass beads) coated with a non-volatile liquid such as silicone oil, polyethylene glycols, or detergents. The carrier gases most frequently used are nitrogen, hydrogen, helium, carbon dioxide, and argon. Capillary columns in which

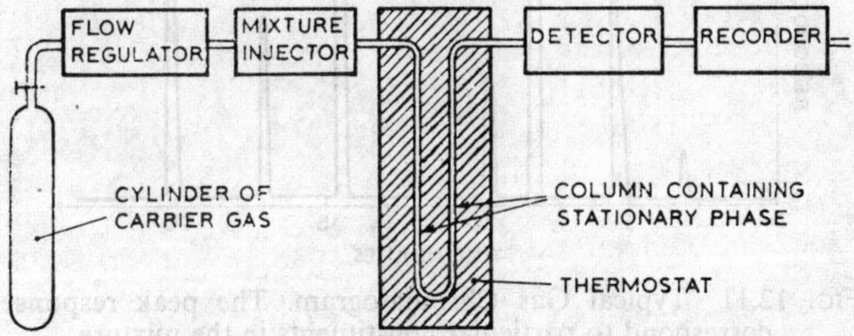

FIG. 12.10 Apparatus for Analysis by Gas Chromatography

the walls are coated with liquid are sometimes used because they give a much faster separation. They are usually very long tubes of well under a millimetre in diameter and can be made of glass, nylon, or metal. Only very small quantities of mixture (a microgramme is usually enough) are needed to perform an analysis.

In *adsorption* or *gas-solid chromatography* the column, which consists of a glass, metal, or plastic tube between 1 mm and 10 mm in diameter and 1 metre and 20 metre in length, contains a finely powdered adsorbent such as charcoal, silica gel, alumina, or linde molecular sieve 5A, the latter being particularly effective in separating the permanent gases hydrogen, oxygen, nitrogen, methane, and carbon monoxide. The adsorbents must be freed from adsorbed water by heating before use and the carrier gas must be thoroughly dried.

The detector used in gas chromatography is usually some form of thermal conductivity gauge known as a catharometer. Basically it consists of a fine wire heated by an electric current, the temperature and therefore the resistance of which varies with the composition of the gas flowing over it. In recent years various flame ionization detectors have been devised which are capable of much greater sensitivity than catharometers and have greatly increased the usefulness of gas chromatography. The type of chromatogram obtained from the recorder consists of a number of sharp peaks, each one corresponding to a substance in the original mixture, as shown in Fig. 12.11. These peaks can

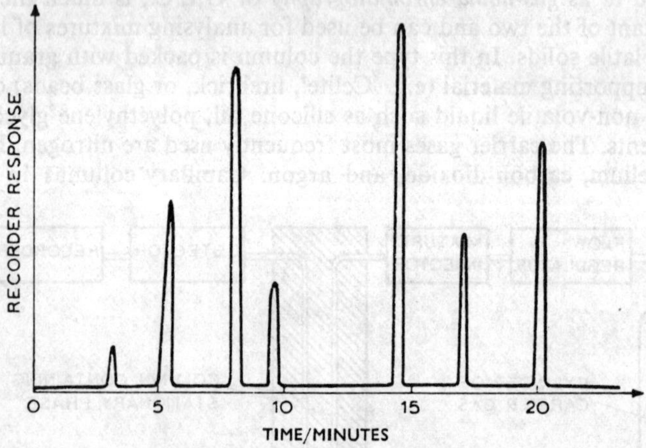

FIG. 12.11 Typical Gas Chromatogram. The peak responses correspond to particular constituents in the mixture

be identified by performing calibration experiments with known substances or by using physical methods such as spectroscopy on collected samples of the vapours.

12.16. Paper Electrophoresis. In recent years a technique has been devised for carrying out electrophoresis (§ 12.3) in such a way that the electrolyte is supported upon a medium of adsorbent paper such as filter paper. The process has proved very valuable for separating mixtures of serum proteins, amino-acids, nucleic acids, and the collections of pigments in inks, paints, and dyes.

The apparatus used is shown in Fig. 12.12. A strip of adsorbent paper soaked in a buffer solution (§ 9.14), which acts as a conductor, is draped from a supporting rod so that each end dips into a vessel containing some of the buffer solution and an electrode. A drop of the solution containing the mixture to be separated is added to the paper at its central point and a d.c. potential is applied. Some ions move towards the anode at speeds depending upon the ionic charge and the size of the solvated ion, others move towards the cathode at different speeds, whilst covalent molecules and electrolytes at their isoelectric point (§ 12.4) remain stationary. After a suitable interval the applied potential is switched off and the strip of paper dried rapidly to prevent

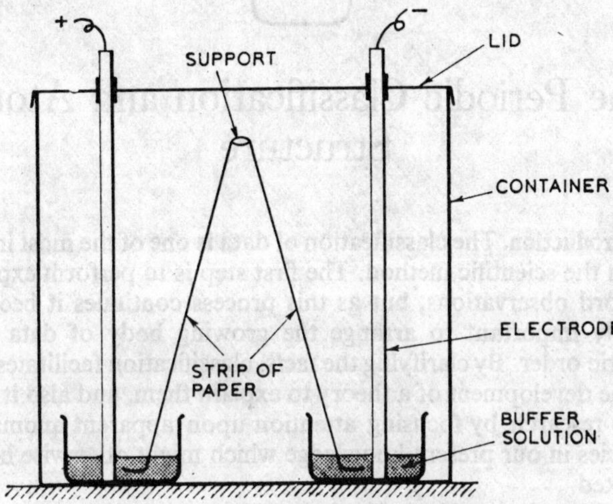

FIG. 12.12 Apparatus for Paper Electrophoresis

diffusion of ions. The positions of the various constituents of the mixture can be detected on the resulting *electrophoretogram* by staining, as in paper chromatography. The separation process can be elaborated by taking advantage of the different rates of migration in media of different pH by using a series of buffer solutions in turn. In an alternative form of apparatus the strip of paper is laid between two sheets of glass to minimize evaporation of the solution.

13

The Periodic Classification and Atomic Structure

13.1. Introduction. The classification of data is one of the most important stages in the scientific method. The first step is to perform experiments and record observations, but as this process continues it becomes increasingly important to arrange the growing body of data in some systematic order. By clarifying the facts, classification facilitates the next stage, the development of a theory to explain them, and also it acts as a guide to research by focusing attention upon apparent anomalies and deficiencies in our present knowledge which might otherwise have been overlooked.

During the first half of the nineteenth century many new elements were discovered and much information on their atomic weights, their physical and chemical properties, and the chemistry of their compounds was collected. It was only natural, therefore, that attempts should have been made to classify this data. For example, in 1829 Döbereiner drew attention to the existence of *triads*, i.e. groups of three very similar elements in which the atomic weight of the middle member was approximately half-way between those of the other two (e.g. Cl,Br,I and Ca,Sr,Ba). Other attempts to relate the properties of the elements to their atomic weights followed this very fragmentary classification, but none met with much success before 1860 mainly because of the unreliability of atomic weight values up to that time. It was not until Cannizzaro had established the truth of Avogadro's hypothesis and had used it in 1859 to find molecular and atomic weights (§ 1.11), and until Stas had determined certain atomic weights accurately in 1860, that real progress could be made.

One of the earliest classifications was that put forward by Newlands in 1865. He arranged the known elements in order of increasing atomic weight and showed that similar physical and chemical properties were to be found in every eighth element in the series. This led him to propose a 'law of octaves', but his suggestion was received with ridicule. The first comprehensive and successful classifications were those produced independently by Mendeléeff and Lothar Meyer in 1869.

13.2. Mendeléeff's Classification. Mendeléeff presented his classification with such vigour and confidence that it soon won universal acceptance and became the prototype of all subsequent attempts to classify the elements. Starting with hydrogen he placed all the then known elements in order of ascending atomic weight and demonstrated that when this was done there was a periodicity in the properties of the elements (i.e. a recurrence at regular intervals of elements with similar properties). He summarized this in the statement that *the properties of the elements were a periodic function of their atomic weights.* To emphasize this periodicity Mendeléeff arranged the elements in a table in which similar elements were placed near each other. Table 13.1 shows a modern form of Mendeléeff's classification, which has been adapted to include the noble gases and other elements not known in his time. It consists of seven horizontal series, called *periods*, so arranged that similar elements are placed beneath each other in nine vertical columns, called *groups.* It is conventional to use arabic numerals for the periods and roman for the groups.

Elements in the same group usually have the same valency. They also have similar physical and chemical properties and they form similar compounds. Any variation in their properties tends to be a regular one down the group. For example, group VII contains the halogens, fluorine to astatine. These five elements closely resemble each other chemically, and so do their corresponding compounds. Where differences exist, as in their physical properties, the changes are regular and gradual as Table 13.2 shows. Group I contains the alkali metals, which are also notable for their similarity. Again there are marked trends in physical and chemical properties in passing from lithium to francium, and again these follow a regular pattern.

The elements in the same period are characterized not by their similarity but by the smooth and regular way in which their valencies and properties change in proceeding from one to another. These trends across a period are illustrated for elements of the third period in Table 13.3.

In constructing his table Mendeléeff encountered a number of difficulties. One of the foremost of these was that the periods were not all of the same length, the later ones containing a much larger number of elements than the earlier periods. He overcame this by 'telescoping' these longer periods, i.e. doubling them back so that, for example, in the fourth period copper was allocated to group I with potassium, and zinc to group II with calcium, and so on. In this way the elements at the

TABLE 13.1. A MODERN FORM OF MENDELEEFF'S CLASSIFICATION

Group	I		II		III		IV		V		VI		VII		VIII	O
Sub-group	A	B	A	B	A	B	A	B	A	B	A	B	A	B		
1st Period														H		(He)
2nd Period	Li		Be			B		C		N		O		F		(Ne)
3rd Period	Na		Mg			Al		Si		P		S		Cl		(Ar)
4th Period	K	Cu	Ca	Zn	(Sc)	(Ga)	Ti	(Ge)	V	As	Cr	Se	Mn	Br	Fe Co Ni	(Kr)
5th Period	Rb	Ag	Sr	Cd	Y	In	Zr	Sn	Nb	Sb	Mo	Te	(Tc)	I	Ru Rh Pd	(Xe)
6th Period	Cs	Au	Ba	Hg	La+ rare earths	Tl	(Hf)	Pb	Ta	Bi	W	(Po)	(Re)	(At)	Os Ir Pt	(Rn)
7th Period	(Fr)		(Ra)		(Ac)		Th		(Pa)		U					

Elements with symbols in brackets were undiscovered when Mendeléeff drew up his classification.

end of these longer periods such as arsenic, selenium, and bromine, fell into the appropriate groups, and the total number of groups in the table was kept small. Even so Mendeléeff found difficulty in fitting some elements into any existing group, e.g. the elements iron, cobalt, and nickel lying between manganese and copper in the fourth period. These three elements, and subsequent triads like them, he called 'transition elements',* and he assigned them to a single new group, VIII.

A consequence of this telescopic arrangement was the creation of *sub-groups* or families within each of the existing groups, to separate the obvious occupants of that group from those placed in it for convenience. It was a weakness in the classification that, apart from valency, the elements of one sub-group often showed very little resemblance to those of the other, e.g. the metals copper, silver, and gold in sub-group IB

TABLE 13.2. TRENDS IN PHYSICAL PROPERTIES IN GROUP VIIB

Property / *Element*	*m.p.* °C	*b.p.* °C	*crit. temp.* °C	*density (of solid)* g cm^{-3}	*latent ht* kJ mol^{-1}	*colour of vapour*
Flourine	−220	−188	−129	1.3	6.7	pale greenish-yellow
Chlorine	−102	−34	145	1.9	18.4	deeper greenish-yellow
Bromine	−7	59	310	3.4	31.0	dark red
Iodine	114	184	550	4.9	43.5	violet

have little in common with the alkali metals of sub-group IA, and manganese in group VIIA is very different from the halogens in group VIIB.

Another difficulty which Mendeléeff surmounted was the obvious misplacings which occurred when the known elements, arranged in order of ascending atomic weight, were allocated to each group in turn. He maintained, for example, that in the fourth period there was an element missing from the known series between calcium and titanium, and two others between zinc and arsenic, and he insisted upon leaving spaces in his table for these unknown elements. Not only did he predict the existence of elements not then discovered, but by a process of comparison with neighbouring elements and skilful interpolation in the trends apparent in the periods and groups, he boldly predicted the

* This modern meaning of the term transition element is explained in § 13.8.

properties of these elements and their principal compounds. His predictions were impressively confirmed in later years when the missing elements (scandium, gallium, and germanium) were discovered, and by his spectacular success in this matter Mendeléeff did much to strengthen confidence in his classification.

The atomic weights of a few elements were seriously in error and Mendeléeff used his classification to correct these. For example, beryllium, which had an equivalent weight of 4.54, was considered to be

TABLE 13.3. TRENDS IN PROPERTIES ACROSS THE THIRD PERIOD

Elements	Na	Mg	Al	Si	P	S	Cl	A
Hydrides	NaH	MgH_2	AlH_3	SiH_4 etc.	PH_3	H_2S	HCl	—
Principal valency	1	2	3	4	3	2	1	0
Oxides	Na_2O	MgO	Al_2O_3	SiO_2	P_2O_3 P_2O_5	SO_2 SO_3	Cl_2O etc.	—
Character of oxide	strongly basic	basic	amphoteric	weakly acidic	acidic	strongly acidic		—
m.p./°C	98	650	660	1410	600	119	−102	−189
b.p./°C	883	1100	2400	3300	c.300	444	−34	−186
ϱ/g cm^{-3}	0.97	1.7	2.7	2.4	2.3	2.1	1.6	1.4
Electronegativity	steady increase in electronegativity →							—

trivalent because of its likeness to aluminium and because Dulong and Petit's Rule (§ 2.12), to which it is now known to be an exception, gave a rough atomic weight of 13.9. Mendeléeff showed that there was no space for it in his table between carbon (12) and nitrogen (14), nor would such a placing match its chemical properties, but that a gap existed between lithium and boron which suited it well. In this way he demonstrated that its valency was two not three and its correct atomic weight 2 × 4.54, i.e. 9.08. Other examples of erroneous atomic weights corrected in this way are given in § 2.14.

13.3. Lothar Meyer's Classification. In 1869 Lothar Meyer devised a method of classification based upon atomic *volume* as well as atomic weight. He constructed a diagram like that shown in Fig. 13.1, in which atomic volume (obtained by dividing atomic weight by the density in

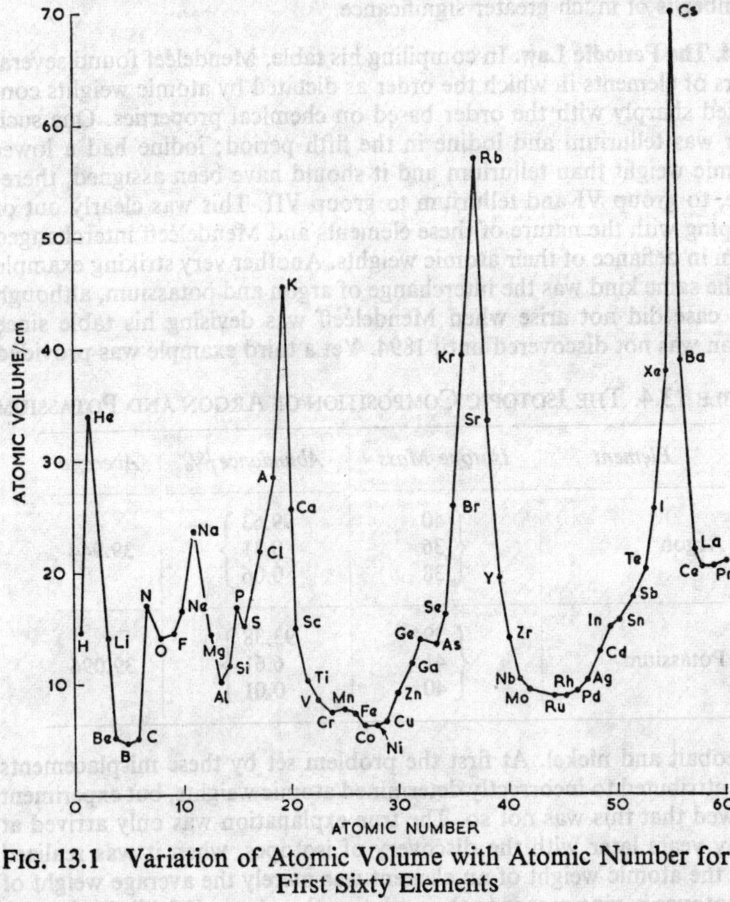

FIG. 13.1 Variation of Atomic Volume with Atomic Number for the First Sixty Elements

g cm^{-3} of the solid element) was plotted against atomic weight, and he showed that similar elements were situated in similar positions on the curve, e.g. all the alkali metals were situated on the crests. This quantity atomic volume is the volume in cubic centimetre of one mole of the element in the solid state. It is effectively a measure of the space occupied by the crystal divided by the total number of atoms in that crystal. Hence it includes not only the volume of the atom but also the space between adjacent atoms, and it depends, therefore, on the way in which the atoms are packed together in the crystal. In recent times it has become possible by physical methods to estimate the actual radius of the atom, and the curve obtained by plotting this against atomic number is of much greater significance.

13.4. The Periodic Law. In compiling his table, Mendeléef found several pairs of elements in which the order as dictated by atomic weights conflicted sharply with the order based on chemical properties. One such pair was tellurium and iodine in the fifth period; iodine had a lower atomic weight than tellurium and it should have been assigned, therefore, to group VI and tellurium to group VII. This was clearly out of keeping with the nature of these elements and Mendeléeff interchanged them in defiance of their atomic weights. Another very striking example of the same kind was the interchange of argon and potassium, although this case did not arise when Mendeléeff was devising his table since argon was not discovered until 1894. Yet a third example was provided

TABLE 13.4. THE ISOTOPIC COMPOSITION OF ARGON AND POTASSIUM

Element	*Isotope Mass*	*Abundance/%*	*Average*
Argon	40 36 38	99.63 0.31 0.06	39.944
Potassium	39 41 40	93.38 6.61 0.01	39.096

by cobalt and nickel. At first the problem set by these misplacements was attributed to incorrectly determined atomic weights, but experiment showed that this was not so. The true explanation was only arrived at many years later with the discovery of isotopes, when it was realized that the atomic weight of an element was merely the average weight of its isotopes having regard to their relative abundance (§ 2.18). In the case of argon and potassium, for example, the atomic weights are not in the expected order because the commonest isotope of potassium is its lightest, whereas the heaviest of the stable argon isotopes is the most abundant. As a result the average weight of potassium atoms is less than the average weight of argon atoms, as shown in Table 13.4.

In modern classifications this problem is overcome by arranging the elements in order of increasing atomic *number* instead of increasing atomic *weight*. As explained in § 1.14 the atomic number is a much more fundamental quantity; it uniquely characterizes an element and when it is chosen as the basis of the classification every element falls into its appropriate place. Indeed, atomic number is sometimes defined in terms of the periodic classification as the number corresponding to the position of the element in the table, hydrogen having an atomic number of one, helium two, lithium three, and so on. The periodic law can then be restated in the form: *the properties of the elements are a periodic function of their atomic numbers.* Two points emerge from this view, firstly that since elements with every atomic number from 1 to 103 are now known, there can be no more missing elements or missing groups (after the discovery of the inert gases, doubts arose lest there might be other groups of elements in existence which had not yet been discovered), and secondly that the position of each element in the table can be unambiguously settled by determining its atomic number by Moseley's method, as described in the next section. This last point has been important in classifying the rare earth elements referred to later.

13.5. Moseley's Law. When X-rays are examined by passing them through a crystal, as in Fig. 4.4, it is found that they consist not only

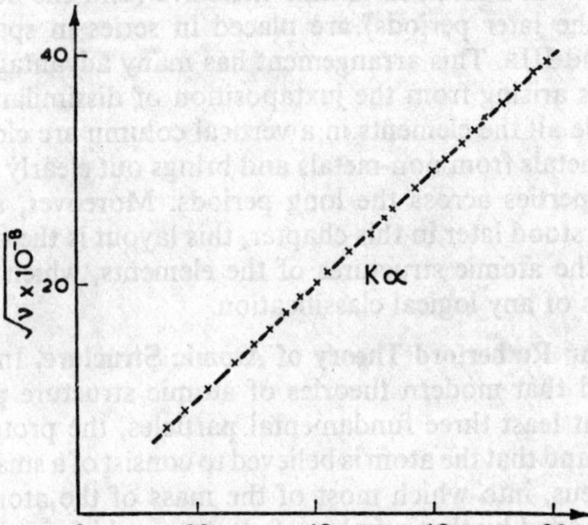

FIG. 13.2 Moseley's Law

of a continuous band of scattered radiation but also of certain prominent rays of definite wavelength which are characteristic of the particular element used for making the anode. When the latter is composed of one of the lighter metals two distinct spectral lines are obtained, known as the K series. Heavier elements give X-ray spectra containing additional lines known as the L and M series respectively. In 1914 Moseley examined the X-ray spectra of nearly forty different elements, using a crystal of potassium ferrocyanide to diffract the radiation produced on to a photographic plate. He found that the wavelength of the characteristic X-rays decreased as elements of higher atomic weight were used as anodes. When he plotted a graph of the square root of the frequency (ν) against the atomic number (Z) he obtained approximately straight lines for each series, as shown in Fig. 13.2 for the K series. This result, which can be expressed in the form $\nu = a(Z - b)^2$ where a and b are constants for the lines of any one series, is known as *Moseley's law*. It states that *the square root of the characteristic frequency of X-radiation from an element is a linear function of its atomic number*. Its great importance lies in the reality it gives to the concept of atomic number and in the method it provides for determining it experimentally.

13.6. Other Classifications. Numerous attempts have been made to improve upon Mendeléeff's classification, and the best of these for our purposes is the long or extended one shown in Table 13.5. It is to this classification that reference will be made in the following chapters. In it the elements from scandium to zinc inclusive (and the corresponding elements in the later periods) are placed in series in spaces between groups IIA and IIIB. This arrangement has many advantages. It avoids the difficulties arising from the juxtaposition of dissimilar sub-groups, as in this table all the elements in a vertical column are closely related. It separates metals from non-metals and brings out clearly the trends in chemical properties across the long periods. Moreover, and this will only be understood later in this chapter, this layout is the one in closest accord with the atomic structures of the elements, which must be the ultimate basis of any logical classification.

13.7. The Bohr Rutherford Theory of Atomic Structure. In chapter 1 it was explained that modern theories of atomic structure postulate the existence of at least three fundamental particles, the proton, neutron, and electron, and that the atom is believed to consist of a small positively-charged nucleus, into which most of the mass of the atom is concentrated, surrounded by that number of electrons which just balances the positive charge on the nucleus. The distribution of the electrons in space was not discussed in that chapter, but must be considered now if the periodicity described above is to be explained.

When an element is heated to a very high temperature in a flame or by means of an electric arc, it emits light of certain definite wavelengths which are characteristic of that element. For example, if the spectrum of

Table 13.5. The Long or Extended Form of the Periodic Classification

IA	IIA	IIIA	IVA	VA	VIA	VIIA	VIII			IB	IIB	IIIB	IVB	VB	VIB	VIIB	O
H 1																H 1	He 2
Li 3	Be 4											B 5	C 6	N 7	O 8	F 9	Ne 10
Na 11	Mg 12											Al 13	Si 14	P 15	S 16	Cl 17	Ar 18
K 19	Ca 20	Sc 21	Ti 22	V 23	Cr 24	Mn 25	Fe 26	Co 27	Ni 28	Cu 29	Zn 30	Ga 31	Ge 32	As 33	Se 34	Br 35	Kr 36
Rb 37	Sr 38	Y 39	Zr 40	Nb 41	Mo 42	Tc 43	Ru 44	Rh 45	Pd 46	Ag 47	Cd 48	In 49	Sn 50	Sb 51	Te 52	I 53	Xe 54
Cs 55	Ba 56	La 57	Hf 72	Ta 73	W 74	Re 75	Os 76	Ir 77	Pt 78	Au 79	Hg 80	Tl 81	Pb 82	Bi 83	Po 84	At 85	Rn 86
Fr 87	Ra 88	Ac 89															

Ce 58	Pr 59	Nd 60	Pm 61	Sm 62	Eu 63	Gd 64	Tb 65	Dy 66	Ho 67	Er 68	Tm 69	Yb 70	Lu 71
Th 90	Pa 91	U 92	Np 93	Pu 94	Am 95	Cm 96	Bk 97	Cf 98	Es 99	Fm 100	Md 101	No 102	Lw 103

hydrogen is examined with a spectroscope it is found to contain four prominent lines in the visible region, as shown in Fig. 13.3. These atomic or line spectra, as they are called, led Bohr to propose in 1913 that the electrons in an atom are not randomly distributed but are arranged in a series of shells or orbits which are situated at various distances from the nucleus corresponding to differing energy levels. As long as an electron remains within a particular orbit its energy is constant, but a move to a more distant shell is always accompanied by the

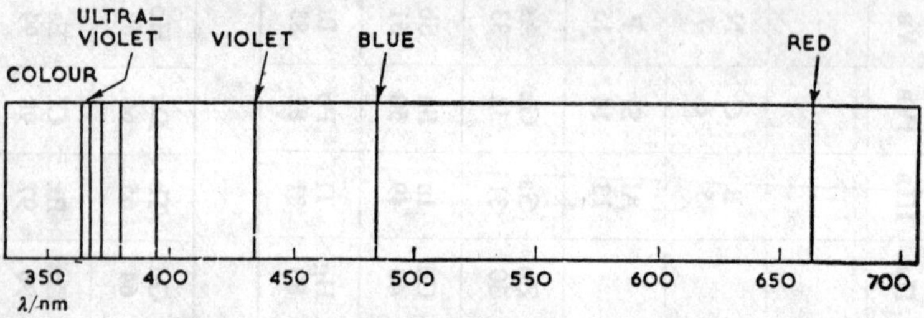

FIG. 13.3 The Spectrum of Hydrogen

absorption of a definite amount of energy and a jump to another shell closer to the nucleus inevitably results in the emission of radiation of definite wavelength. By calculating the various energy levels or shells that may exist in an atom Bohr was able to explain successfully the line spectrum of hydrogen. He further proposed that the innermost shell of any atom contains a maximum of two electrons, that the second shell can accommodate up to eight, and that if the atom contains more than ten electrons, some occur in a third shell further from the nucleus, and so on.

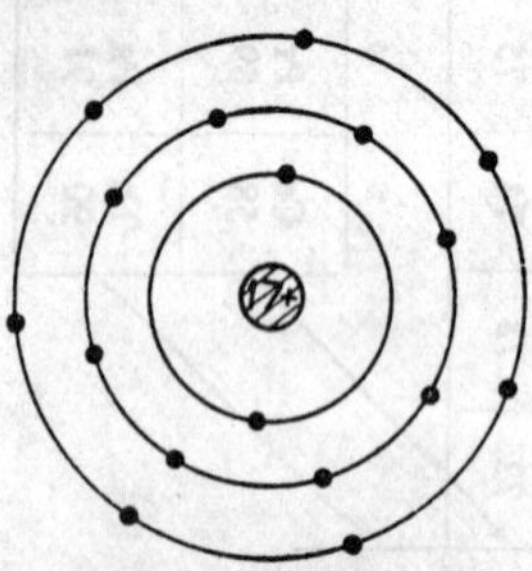

FIG. 13.4 The Electronic Configuration of Chlorine (2,8,7)

Table 13.6 shows the electronic configurations of the first twenty elements of the periodic table. The numbers beneath the symbols indicate the number of electrons in each shell. For example, chlorine (2,8,7) has two electrons in its innermost shell, eight in the second shell, and seven in its outermost shell as shown diagrammatically in Fig. 13.4. From this table it is clear that when the elements are arranged in the order of their atomic numbers, there is a periodicity in the number of electrons in the outermost shells of their atoms. This is believed to be the cause of the repetition at regular intervals in the properties of the elements which underlies the periodic classification. For example, all the atoms of the alkali metals have one electron in their outermost shells, and it is this common feature which accounts for their similar chemical behaviour. Similarly, all the halogens are alike in possessing seven electrons in the outermost shells of their atoms.

We conclude that it is the electrons, especially the electrons in the outermost shell of an atom, which play a decisive part in determining the chemical properties of an element. The nucleus, and the electrons in any

TABLE 13.6. THE ELECTRONIC CONFIGURATIONS OF THE FIRST TWENTY ELEMENTS

						H 1	He 2
Li 2,1	Be 2,2	B 2,3	C 2,4	N 2,5	O 2,6	F 2,7	Ne 2,8
Na 2,8,1	Mg 2,8,2	Al 2,8,3	Si 2,8,4	P 2,8,5	S 2,8,6	Cl 2,8,7	Ar 2,8,8
K 2,8,8,1	Ca 2,8,8,2						

shell which is full, have much less effect upon chemical properties, although the nuclear charge and the number of filled electron shells does influence the size of the atom and the ease with which it loses or gains electrons in exchanges with other atoms.

13.8. The Transition Metals. The electronic configurations of only the first twenty elements were shown in Table 13.6, because after calcium an interesting change occurs which deserves special consideration. The twenty-first element, scandium, has one more electron than calcium to balance the additional proton in its nucleus, and we should expect to find this additional electron in the fourth shell, giving it a configuration of 2,8,8,3. However, spectroscopy shows scandium's configuration to be 2,8,9,2 and that of the next element, titanium, to be 2,8,10,2. It appears that such structures are more stable than those in which the additional electrons occur in the outermost shell, presumably because they possess less energy. The eight atoms which follow scandium and titanium also have their extra electrons in the third shell, so that the number rises to eighteen, which is the maximum that shell can accommodate. After that successive elements have additional electrons in the outermost shell, so that the fourth period ends with the inert gas krypton (2,8,18,8), in which the outermost shell appears filled. The series of ten elements from scandium to zinc is known as the *first series of transition metals.* These elements are considered in greater detail in Chapter 27.

A similar thing happens in the fifth period, where a series of ten elements from yttrium to cadmium occurs in which the penultimate shell is built up from eight to eighteen electrons, forming a *second series of transition metals.* This period ends with the inert gas xenon, of configuration 2,8,18,18,8. In the sixth period a further complication occurs. Not only is there a *third series of ten transition elements* in which the number of electrons in the penultimate shell is increased to eighteen, but within this series is another of fourteen elements, known as the *rare earths* or *lanthanons,* in which the number of electrons in the fourth shell is increased from eighteen to thirty-two. These elements, which are all extremely similar in their chemical character, are conventionally placed in the same space in the periodic table, corresponding to the atomic number 57, which is that of lanthanum, despite the fact that their atomic numbers cover the range from 57 to 71 (see Table 13.5).

The seventh period starts in the normal way with the two elements francium and radium, and then comes actinium, the first member of another transition series in which the additional electrons appear in the penultimate shell and the fifth shell. This period is incomplete because the later members of it are unstable, undergoing radioactive decay into simpler smaller atoms (see § 31.6). Until 1940 uranium was the heaviest atom known, but since then a number of trans-uranic elements have been made artificially and their properties examined (§ 31.7). From this work it is clear that after actinium there begins a new series, the *actinons* which corresponds to the lanthanon series in the previous period.

13.9. Atomic and Ionic Radii. In § 4.11 we saw how it is possible to determine the radii of atoms and ions by means of X-ray and electron diffraction measurements. Naturally the sizes are only approximate because the apparent radius of any given atom tends to vary slightly with the type of bonding involved, but the results are of such great interest that they deserve special consideration at this stage. They are collected together diagrammatically in Table 13.7, arranged in the order of the extended form of the periodic classification.*

Examination of this table shows several well-defined trends. Firstly, the size of atoms of elements in the same period decreases steadily as

* Credit for this ingenious idea is due to Dr. J. A. Campbell of the Harvey Mudd College, Claremont, California, and I am grateful to him for permission to use it here. See *J. Chem. Education,* **23,** 525 (1946).

the atomic number rises. Now it is to be expected that the larger nuclear charge would exert a progressively stronger attraction upon the electrons around it and would tend to pull them in more closely in the heavier atoms, but what is notable is that this attraction by the nucleus increases more rapidly across a period than the forces of mutual repulsion amongst the electrons. As a result not only does the radius of the atom decrease with rising atomic number, but the retentive force exerted by the nucleus upon the electrons at the periphery of the atom also increases markedly across a period so that the elements become progressively less electropositive (or more electronegative) as we go from left to right.

Another obvious trend is the increase in atomic radius within a given group as the atomic number rises. For example, the alkali metals increase from 0.133 nm (lithium) to 0.27 nm (francium), and the halogens show an equally striking change from 0.072 nm (fluorine) to 0.14 nm (astatine). Clearly the increase in the attractive force exerted by the nucleus as its positive charge gets greater is more than counterbalanced by the considerable additional space taken up by the extra shell of electrons added in each period, although this effect becomes less pronounced amongst the heavier elements. It is not surprising, therefore, to find that the elements of any one group become increasingly electropositive (or decreasingly electronegative) with rising atomic number.

This and the trend referred to in the previous paragraph explain why fluorine is the most electronegative element and oxygen the next and why francium and caesium are the most electropositive of all. Nuclear charge and atomic size are not, however, the only factors governing electronegativity, which is also affected by the way in which the electrons in the atom are arranged. Presumably the inner electrons, especially those in shells which are completely filled, exercise some sort of screening effect and so reduce the attraction which the positive charge on the nucleus exerts upon the electrons at the periphery of the atom. This may explain why zinc is so much more electropositive than copper and cadmium than silver, and why the inert gases are not electronegative like the halogens.

It is also possible to draw some important conclusions about the size of ions from the Campbell Table. Clearly a *cation* (a positively charged ion which would be attracted to the cathode during electrolysis) is always considerably smaller than the atom from which it is made, and in many cases, especially amongst the lighter metals, it is less than half the size of the original atom. The opposite effect is apparent for *anions* (negatively charged ions which would move to the anode during electrolysis), for they are always much larger than the atoms from which they are derived. These effects follow naturally from the processes of

TABLE 13.7. ATOMIC AND IONIC RADII

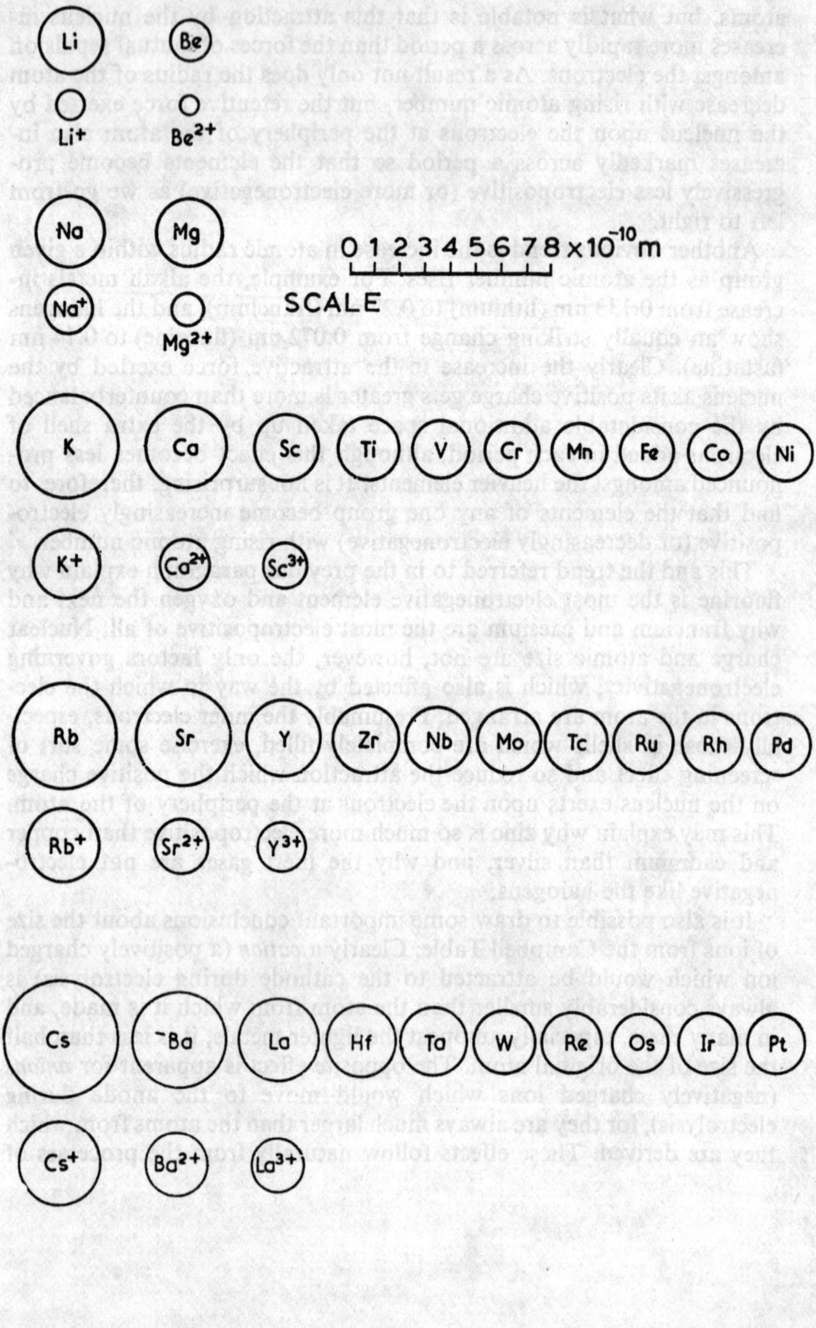

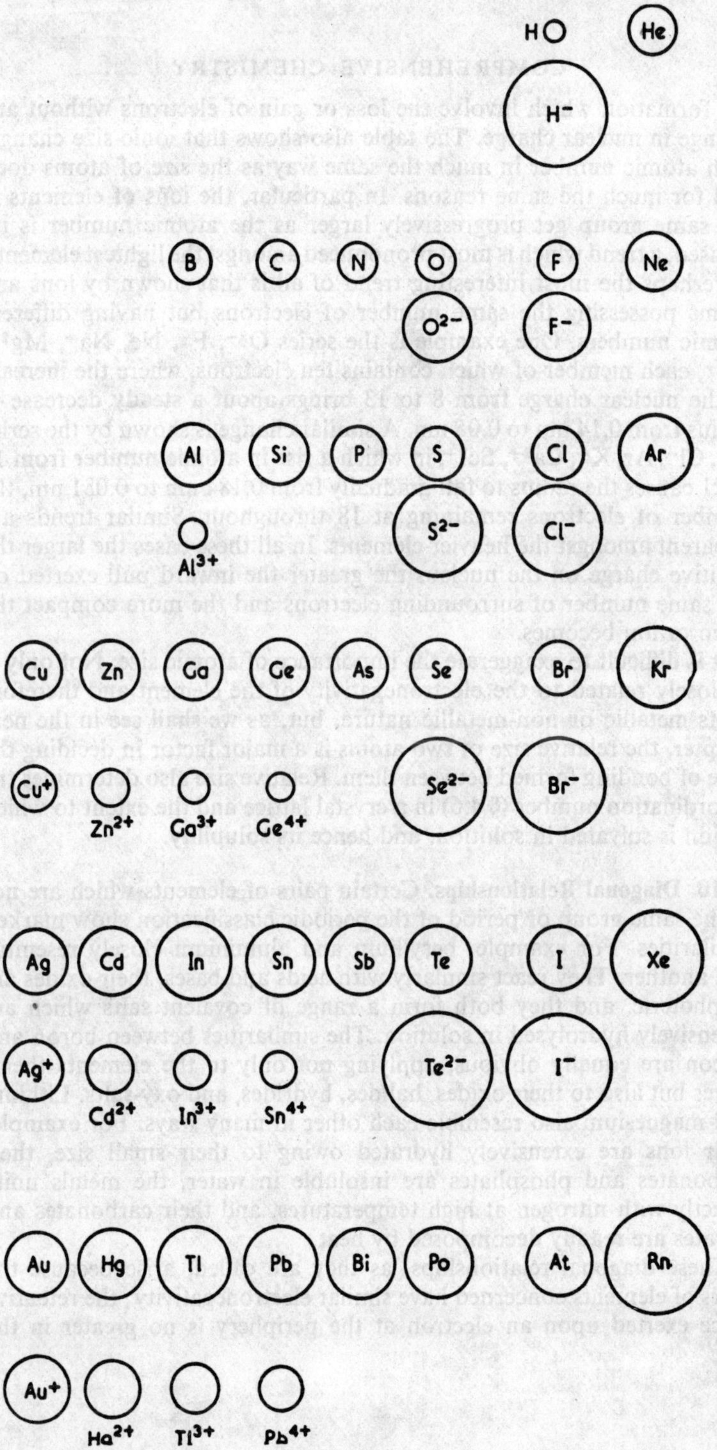
H
He
H⁻
B
C
N
O
F
Ne
O²⁻
F⁻
Al
Si
P
S
Cl
Ar
S²⁻
Cl⁻
Al³⁺
Cu
Zn
Ga
Ge
As
Se
Br
Kr
Cu⁺
Zn²⁺
Ga³⁺
Ge⁴⁺
Se²⁻
Br⁻
Ag
Cd
In
Sn
Sb
Te
I
Xe
Ag⁺
Cd²⁺
In³⁺
Sn⁴⁺
Te²⁻
I⁻
Au
Hg
Tl
Pb
Bi
Po
At
Rn
Au⁺
Hg²⁺
Tl³⁺
Pb⁴⁺

ion formation which involve the loss or gain of electrons without any change in nuclear charge. The table also shows that ionic size changes with atomic number in much the same way as the size of atoms does, and for much the same reasons. In particular, the ions of elements in the same group get progressively larger as the atomic number is increased, a trend which is most pronounced amongst the lightest elements.

Perhaps the most interesting trend of all is that shown by ions and atoms possessing the same number of electrons but having different atomic numbers. One example is the series O^{2-}, F^-, Ne, Na^+, Mg^{2+}, Al^{3+}, each member of which contains ten electrons, where the increase in the nuclear charge from 8 to 13 brings about a steady decrease in radius from 0.14 nm to 0.05 nm. A similar change is shown by the series S^{2-}, Cl^-, Ar, K^+, Ca^{2+}, Sc^{3+}, in which a rise in atomic number from 16 to 21 causes the radius to fall gradually from 0.184 nm to 0.081 nm, the number of electrons remaining at 18 throughout. Similar trends are apparent amongst the heavier elements. In all these cases the larger the positive charge on the nucleus the greater the inward pull exerted on the same number of surrounding electrons and the more compact the atom or ion becomes.

It is difficult to exaggerate the importance of atomic size. Not only is it closely related to the electronegativity of the element and therefore to its metallic or non-metallic nature, but, as we shall see in the next chapter, the relative size of two atoms is a major factor in deciding the type of bonding formed between them. Relative size also determines the co-ordination number (§ 4.6) in a crystal lattice and the extent to which an ion is solvated in solution, and hence its solubility.

13.10. Diagonal Relationships. Certain pairs of elements which are not in the same group or period of the periodic classification show marked similarities. For example, beryllium and aluminium closely resemble one another. They react similarly with acids and bases, their oxides are amphoteric, and they both form a range of covalent salts which are extensively hydrolysed in solution. The similarities between boron and silicon are equally obvious, applying not only to the elements themselves but also to their oxides, halides, hydrides, and oxy-salts. Lithium and magnesium also resemble each other in many ways. For example, their ions are extensively hydrated owing to their small size, their carbonates and phosphates are insoluble in water, the metals unite directly with nitrogen at high temperatures, and their carbonates and nitrates are readily decomposed by heat.

These diagonal relationships, as they are called, arise because the pairs of elements concerned have similar electronegativity; the retentive force exerted upon an electron at the periphery is no greater in the heavier atom of the pair, despite its larger nuclear charge, because of the mutual repulsion amongst its more numerous electrons.

13.11. Ionization Energy. As explained in § 10.7 the amount of energy which is required to remove one electron from the atom of an element in the vapour state is an important quantity which is characteristic of that element and has a substantial influence on its chemical properties and behaviour. It is known as the *first ionization energy* of the element and is usually expressed in kilojoule per mole.

TABLE 13.8. IONIZATION ENERGIES OF THE FIRST SIXTY ELEMENTS

Element	*Atomic Number*	*Ionization Energy* ΔU/kJ mol^{-1}	*Element*	*Atomic Number*	*Ionization Energy* ΔU/kJ mol^{-1}
H	1	1310	Ga	31	577
He	2	2372	Ge	32	761
Li	3	519	As	33	946
Be	4	900	Se	34	941
B	5	799	Br	35	1142
C	6	1088	Kr	36	1351
N	7	1402	Rb	37	402
O	8	1314	Sr	38	548
F	9	1682	Y	39	615
Ne	10	2079	Zr	40	661
Na	11	494	Nb	41	665
Mg	12	736	Mo	42	686
Al	13	577	Tc	43	699
Si	14	787	Ru	44	711
P	15	1017	Rh	45	720
S	16	1000	Pd	46	803
Cl	17	1255	Ag	47	732
Ar	18	1519	Cd	48	866
K	19	418	In	49	556
Ca	20	586	Sn	50	707
Sc	21	632	Sb	51	833
Ti	22	657	Te	52	870
V	23	648	I	53	1008
Cr	24	653	Xe	54	1171
Mn	25	716	Cs	55	372
Fe	26	761	Ba	56	502
Co	27	757	La	57	536
Ni	28	736	Ce	58	669
Cu	29	745	Pr	59	561
Zn	30	908	Nd	60	611

Table 13.8 lists the ionization energies* of the first sixty elements of the periodic classification as determined experimentally from spectroscopic measurements. From this table it is clear that ionization energies differ greatly from one element to another and that there is a periodicity in this variation corresponding closely to the periodic law. This becomes even clearer when these ionization energies are plotted against atomic

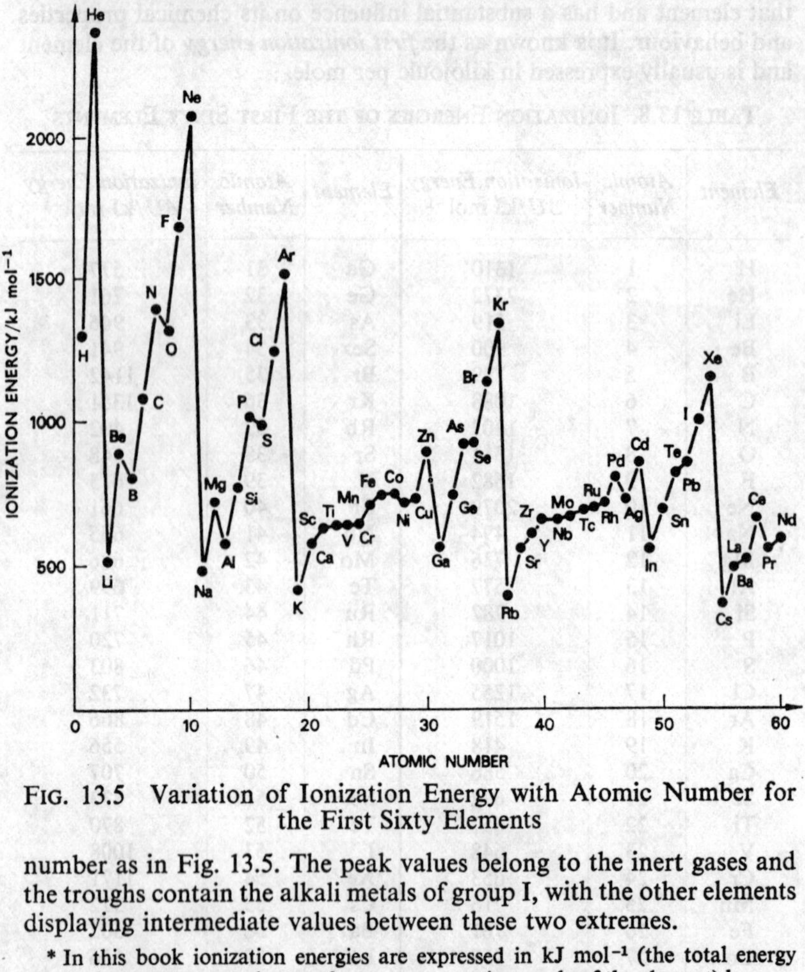

FIG. 13.5 Variation of Ionization Energy with Atomic Number for the First Sixty Elements

number as in Fig. 13.5. The peak values belong to the inert gases and the troughs contain the alkali metals of group I, with the other elements displaying intermediate values between these two extremes.

* In this book ionization energies are expressed in kJ mol^{-1} (the total energy required to remove one electron from every atom in a mole of the element) because these units facilitate direct comparison between ionization energies and the energy changes which occur during chemical reactions. Some chemists, however, still use the non-SI molar electron-volt, $N_A \times e \times V$, a unit with the value 96 487 J mol^{-1}, for this purpose.

This means that it is relatively easy to remove an electron from an atom of lithium, sodium, or potassium and form a positively-charged ion, whereas the energy required to do the same in a halogen atom or in a noble gas would be so prohibitive that we should expect it to happen but rarely. It is this ease with which some elements part with electrons which distinguishes them as metals from the other elements and which has a decisive influence not only on their chemical properties but also (as we shall see in the next chapter) on the way in which they combine to form compounds.

This periodicity in ease of ionization is directly related to the periodicity which exists in electronic configuration. The elements most willing to give up electrons are those with one, two, or three in the outermost shell of their atoms, whereas those with the highest energies of ionization have outer shells which are full or nearly full.

If a second electron is to be removed from an atom, an even greater amount of energy is required because of the electrostatic attraction between the negatively-charged electron and the already positively-charged ion. It follows that for any particular element the second ionization potential is always greater than the first, and the third higher than the second, and so on. A comparison of the successive ionization energies for any one element can be very revealing because it often indicates that one, two, or three electrons, as the case may be, are relatively much easier to remove from the atom than the rest. For example, in sodium the first, second, and third ionization energies are 494, 4562, and 6905 kilojoule per mole respectively, showing that the first electron is the only one that is loosely bound. Similarly, in magnesium the successive ionization energies are 737, 1444, and 7742 kilojoule per mole, implying that two (and only two) electrons in each magnesium atom are held relatively loosely. Aluminium, on the other hand, has values of 577, 1816, 2745, and 11 592 kilojoule per mole for its ionization energies, so that in this case there are three electrons in each atom which are held much less firmly than the rest. The significance of these facts will be appreciated in the next chapter where the electronic theory of valency is considered.

The magnitude of the ionization energy of an element depends upon a number of conflicting factors and is the resultant of them all. As explained in § 13.9, the distance of the electron from the nucleus of the atom is a major influence, and so are the atomic number and the degree to which the other electrons in the atom provide screening or shielding from the strong attractive force of the positive charge on the nucleus. A further factor is the extent to which that particular electron is able to penetrate this screening, which varies from one electron to another depending, as we shall see in § 13.13 upon the shape of its orbit.

It is important to stress that ionization energy applies to the element *in the vapour state*: i.e. $M(g) \longrightarrow M^+(g) + e$. This must be clearly distinguished from the energy required to change an atom into an ion *in solution*, because that involves the energy of solvation which may be substantial in an ion possessing a high charge and small size. It must also be distinguished from *ionization potential*, which is the potential difference *in volt* which has to be applied to bring about the removal of one electron from an atom of the element.

The energy change in kilojoule when one mole of an electronegative element in the gaseous state acquires electrons to form negative ions is known as the *electron affinity* of the element. For example, the electron affinity of chlorine is 368 kJ mol^{-1} because this is the amount of energy liberated in the change:

$$Cl\,(g) + e \longrightarrow Cl^-\,(g)$$

Energy changes resulting in ion formation by a *gain* of electrons are known as electron affinities, and energy changes resulting in ion formation by a *loss* of electrons as ionization energies.

13.12. Atomic Spectra. We have seen in § 13.7 how when any element is heated to a very high temperature in a flame or arc or when an electric discharge is passed through a gaseous element at low pressure, it emits radiation of certain definite wavelengths which are characteristic of that particular element. The spectrum obtained in this way consists of a large number of fine lines spaced apart, each line corresponding to one of these characteristic wavelengths, stretching from the ultra-violet through the visible region into the infra-red. A long and intensive study of the spectrum of hydrogen established that it consists of a number of series of spectral lines which are related to each other, these series being named after their discoverers as the Balmer, Paschen, Lyman, Brackett, and Pfund series. It was eventually realized that in all these series the wavelength λ was given by the expression

$$\frac{1}{\lambda} = R_\infty\left(\frac{1}{m^2} - \frac{1}{n^2}\right)$$

where R_∞ is a constant called *Rydberg's constant*, and where m and n are integers, m determining which series is being considered and n corresponding to each of the individual lines in that particular series. For example, in the Lyman series, which occurs in the ultra-violet region, m is unity, so

$$\frac{1}{\lambda} = R_\infty\left(\frac{1}{1^2} - \frac{1}{n^2}\right)$$

and *n*, known as the principal quantum number, consists of integers 2, 3, 4, etc.

A big step forward was taken when Bohr interpreted these spectra in terms of the quantum theory and related the discrete lines in the spectrum of hydrogen to the various energy levels in the atom which could be occupied by electrons. The whole Lyman series could be explained, for example, in terms of an electron assuming each one of a number of excited energy states in turn and reverting each time to its stablest position of lowest energy, known as the ground state, with the emission of radiation of characteristic wavelength. Clearly a limit is eventually reached when the energy given to excite the electron is sufficient not only to raise it to the highest energy level but to remove it from the attraction of the nucleus altogether; the energy needed to do this is, of course, the ionization energy of hydrogen, as described in the previous section.

13.13. Atomic Orbitals. Although the theory of atomic structure put forward by Bohr is adequate for many purposes, in certain respects it has been found wanting. In view of its great simplicity this is hardly surprising, the remarkable thing being what a vast range of phenomena can be satisfactorily explained by making a few simple assumptions about the structure of the atom. Before referring to more recent ideas it is important to emphasize that the diagram in Fig. 13.4 is, after all, only a conventional representation of a three dimensional model, and that this model, like any other, is itself bound to be misleading in certain respects, suggesting as it does that electrons are solid particles which can be represented by dots moving about on the surfaces of concentric spheres.

Despite the success of the Bohr theory in explaining the spectrum of hydrogen, it proved unsatisfactory when applied to the spectra of more complicated atoms. It also failed to provide any explanation of a number of chemical phenomena, and a demand arose for a more sophisticated theory of atomic structure. At the same time there was a growing realization that electrons could not be regarded solely in terms of particles carrying a negative charge. Most of our modern views on atomic structure are derived from mathematical reasoning known as wave mechanics which does not readily lend itself to representation by physical models. The modern tendency is to regard electrons rather as clouds of negative charge with some of the properties of waves as well as particles. This view was reinforced by Heisenberg, in 1927 when he suggested in his *uncertainty principle* that it was impossible to specify precisely the simultaneous position and velocity of any given electron.

The first refinement was the realization that the electrons in any one shell of an atom are not all exactly equivalent. This developed into the belief that the electrons in a particular shell are often distributed between several sub-levels which differ in their energy and distance from the nucleus, and that these sub-levels may be further divided into a number of different *orbitals* with distinctive shapes and symmetries. These orbitals are not paths along which the electrons move, but complicated three dimensional figures which indicate density of charge and so help us to estimate the probability of finding an electron at any particular position at any instant. In his *exclusion principle* Pauli postulated that each orbital can be occupied by up to two electrons which can be distinguished from each other by possessing opposite *spins*.

The number of possible orbitals in any given shell is n^2, where n is the principal quantum number of the shell. Thus the first K shell, where n is one, contains one orbital and a maximum of two electrons. The second L shell contains four orbitals and up to eight electrons, and the third M shell has nine orbitals and a maximum of eighteen electrons.

In every shell there is one orbital which is spherically symmetrical, known as the '*s*' orbital. In the K shell this is the only orbital available and is denoted by 1*s*, the number standing for the shell's principal quantum number, n, and the latter indicating the type of orbital* (see Fig. 13.6). In the L shell there are two sub-levels, one consisting of an *s* orbital which is spherically symmetrical and known as 2*s*, and the other at a slightly higher energy level involving three '*p*' orbitals shaped rather like dumbbells. These are usually distinguished as $2p_x$, $2p_y$, and $2p_z$ because in their symmetry they can be regarded as being orientated along three orthogonal axes x, y, and z, as in Fig. 13.7. In the third M shell two electrons are found in the lowest energy sub-level, a 3*s* orbital, six more electrons in the slightly higher three 3*p* orbitals, and ten more electrons can be accommodated, if necessary, in what are known as five 3*d* orbitals at a slightly higher energy sub-level still. The fourth N shell is made up of one 4*s* orbital, three 4*p* orbitals, five 4*d* orbitals, and seven 4*f* orbitals, the *d* and *f* orbitals also possessing distinctive symmetries and shapes.

The relative energies of the *s*, *p*, *d*, and *f* orbitals for the lowest principal quantum numbers are shown diagrammatically in Fig. 13.8. Table 13.9 shows how the electrons in the first eighteen elements are arranged in the various *s* and *p* orbitals.† In every case the configuration adopted is the one involving the least energy and therefore the greatest stability for the atoms concerned. It will be seen from this Table that

* The orbitals were originally designated *s*, *p*, *d*, and *f* after the sharp, principal, diffuse, and fundamental spectral lines.

† The electronic configurations of the transition metals are shown in Table 27.3.

two electrons occupying the same orbital always have opposite spin in accord with Pauli's exclusion principle, and that where sufficient orbitals of the same energy sub-level are available, electrons tend to occupy these singly rather than share an orbital. This rule, first formulated by Hund, is well illustrated by the nitrogen atom, where owing to mutual

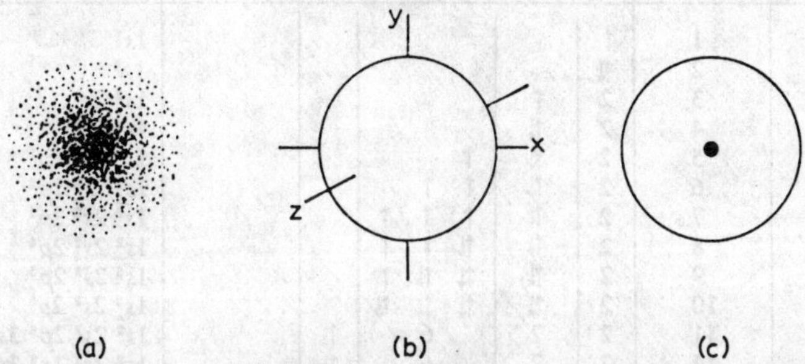

FIG. 13.6 Three representations of a 1*s* Orbital, (a) as a section through a charge cloud, dots showing the electron density, (b) as a spherical boundary surface containing the electron charge, (c) as a circle centred on the nucleus.

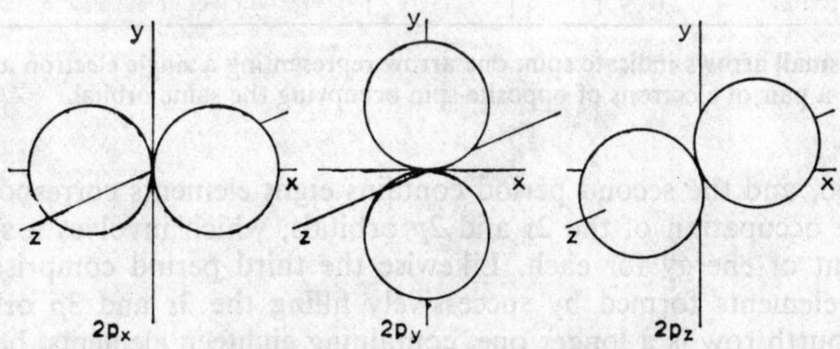

FIG. 13.7 Representations of the three 2*p* Orbitals

repulsion the three electrons in the 2*p* sub-level occupy three separate orbitals and are unpaired and have parallel spin.

Fig. 13.8 enables us to see how the periodicity in the chemistry of the elements depends upon their extranuclear structure and can be used to explain the basis of the periodic classification. It is instructive to imagine that we are building up an atom by the progressive addition of electrons. Each row of the periodic table corresponds to the filling of a number of orbitals with approximately the same energy requirement. Thus the first period is complete with only two elements in it when the 1*s* orbital

TABLE 13.9. THE ELECTRONIC CONFIGURATIONS BY ORBITALS OF THE FIRST EIGHTEEN ELEMENTS

Element	*Atomic Number*	*Orbitals* 1*s*	2*s*	2*p*	3*s*	3*p*	*Electronic Configuration*
H	1	↑					$1s^1$
He	2	↑↓					$1s^2$
Li	3	2	↑				$1s^2\ 2s^1$
Be	4	2	↑↓				$1s^2\ 2s^2$
B	5	2	↑↓	↑			$1s^2\ 2s^2\ 2p^1$
C	6	2	↑↓	↑ ↑			$1s^2\ 2s^2\ 2p^2$
N	7	2	↑↓	↑ ↑ ↑			$1s^2\ 2s^2\ 2p^3$
O	8	2	↑↓	↑↓ ↑ ↑			$1s^2\ 2s^2\ 2p^4$
F	9	2	↑↓	↑↓ ↑↓ ↑			$1s^2\ 2s^2\ 2p^5$
Ne	10	2	↑↓	↑↓ ↑↓ ↑↓			$1s^2\ 2s^2\ 2p^6$
Na	11	2	2	6	↑		$1s^2\ 2s^2\ 2p^6\ 3s^1$
Mg	12	2	2	6	↑↓		$1s^2\ 2s^2\ 2p^6\ 3s^2$
Al	13	2	2	6	↑↓	↑	$1s^2\ 2s^2\ 2p^6\ 3s^2\ 3p^1$
Si	14	2	2	6	↑↓	↑ ↑	$1s^2\ 2s^2\ 2p^6\ 3s^2\ 3p^2$
P	15	2	2	6	↑↓	↑ ↑ ↑	$1s^2\ 2s^2\ 2p^6\ 3s^2\ 3p^3$
S	16	2	2	6	↑↓	↑↓ ↑ ↑	$1s^2\ 2s^2\ 2p^6\ 3s^2\ 3p^4$
Cl	17	2	2	6	↑↓	↑↓ ↑↓ ↑	$1s^2\ 2s^2\ 2p^6\ 3s^2\ 3p^5$
Ar	18	2	2	6	↑↓	↑↓ ↑↓ ↑↓	$1s^2\ 2s^2\ 2p^6\ 3s^2\ 3p^6$

The small arrows indicate spin, one arrow representing a single electron and two arrows a pair of electrons of opposite spin occupying the same orbital.

is filled, and the second period contains eight elements corresponding to the occupation of the 2*s* and 2*p* orbitals, which involves a similar amount of energy for each. Likewise the third period comprises the eight elements formed by successively filling the 3*s* and 3*p* orbitals. The fourth row is a longer one, containing eighteen elements, because as is clear from Fig. 13.8, the 4*s*, 3*d*, and 4*p* orbitals all require approximately the same energy. Similarly, the fifth period corresponds to the filling of the 5*s*, 4*d*, and 5*p* orbitals, making eighteen elements in all, and the sixth period to the occupation of the 6*s*, 4*f*, 5*d*, and 6*p* orbitals by thirty-two electrons all with about the same energy. This figure also explains certain features of the transition metals (see chapter 27) such as their general similarity and their variable valency, since clearly there are only very small energy differences between these particular configurations. It also highlights the fact that in the atoms of the noble gases all the orbitals that are occupied are filled to capacity.

It must be emphasized that the geometrical shapes given to atomic orbitals are only designed to express in visual form the most likely

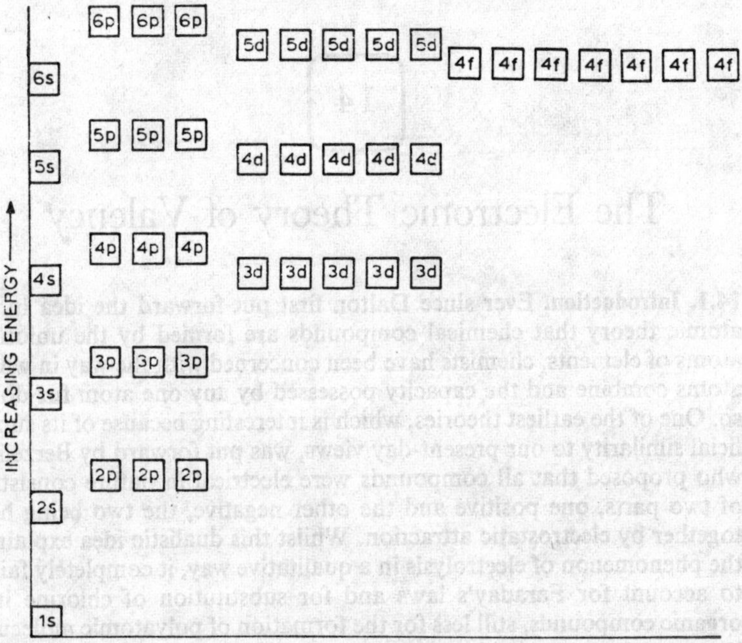

FIG. 13.8 Energy Levels of Atomic Orbitals for Elements of Low Atomic Number

For elements of high atomic number the relative positions of the 3*d* and 4*s* orbitals and the 4*d* and 5*s* orbitals in this diagram are interchanged.

disposition of the electrons within the atom as calculated by means of wave mechanics. The different shapes are primarily intended to convey the presence or absence of symmetry in the spatial distribution of the electrons. For example, in an *s* orbital there is at any instant an equal probability of finding an electron in every direction at any given distance from the nucleus. This should be contrasted with a *p* orbital where there is a very pronounced directional character; the probability of finding an electron is greatest along the axis of the dumbbell and is zero in the plane through the centre of the nucleus at right angles to this axis. When atoms combine together the orbitals involved in the bonding are believed to be modified into hybrid orbitals as described in § 14.11.

14

The Electronic Theory of Valency

14.1. Introduction. Ever since Dalton first put forward the idea in his atomic theory that chemical compounds are formed by the union of atoms of elements, chemists have been concerned with the way in which atoms combine and the capacity possessed by any one atom for doing so. One of the earliest theories, which is interesting because of its superficial similarity to our present-day views, was put forward by Berzelius, who proposed that all compounds were electrical in nature consisting of two parts, one positive and the other negative, the two being held together by electrostatic attraction. Whilst this dualistic idea explained the phenomenon of electrolysis in a qualitative way, it completely failed to account for Faraday's laws and for substitution of chlorine into organic compounds, still less for the formation of polyatomic molecules by many elements, and it was eventually abandoned. The middle of the nineteenth century saw a rapid advance in organic chemistry in particular, leading to the realization by Frankland and Kekulé that the combining capacities of the elements carbon, hydrogen, oxygen, and nitrogen were constant and could be represented by simple whole numbers. So was born the concept of quantivalence or valency number, which has already been defined in § 2.9. By representing these valencies by short lines or links attached in appropriate numbers to the symbol for an atom, graphic formulae (§ 32.3) were constructed in great numbers to explain the constitution and reactions of organic compounds. Eventually the phenomenon of stereoisomerism (§ 48.2) led van't Hoff and Le Bel to propose independently that the valency bonds of the carbon atom were directed tetrahedrally in space, since this was the only arrangement which would explain the observed facts. The Arrhenius ionic theory and the exciting new discoveries in physics made towards the end of the nineteenth century revived an interest in an electrical theory of valency, an interest that was heightened by Werner's work on inorganic co-ordination compounds.

However, it was not until the acceptance of the Bohr–Rutherford theory of atomic structure that a satisfactory explanation of chemical bonding was possible. In the *electronic theory of valency*, put forward independently in 1916 by Kossel and Lewis, and subsequently elaborated by Langmuir, Sidgwick, and others, chemical combination was brilliantly accounted for in terms of the sharing or transfer of electrons between the atoms. It is with an outline of this theory and of its application to simple examples that we shall be concerned in this chapter, leaving an account of the newer approaches of wave-mechanics, known as the valence-bond and the molecular-orbitals methods, to more advanced books.

14.2. Valency and Electronic Configuration. One of the starting points of the electronic theory of valency is the observation that the valency number of an element is closely related to the number of electrons in the outermost shell of its atom. For example, as Table 14.1 demonstrates, the valency of each element in the second period is either equal to the number of outermost electrons or is equal to the shell maximum (eight) minus this number. The same kind of relationship holds true for elements in the other periods. It will also be noticed that atoms of the noble gases (§ 26.1) all possess outermost electron shells containing the maximum number of electrons, from which it is concluded that this

TABLE 14.1. RELATIONSHIP BETWEEN VALENCY AND ELECTRONIC CONFIGURATION IN THE SECOND PERIOD

Element	Li	Be	B	C	N	O	F	Ne
Electronic Configuration	2,1	2,2	2,3	2,4	2,5	2,6	2,7	2,8
Valency Number	1	2	3	4	3	2	1	0

common feature is the cause of their great chemical stability and small reactivity. Atoms with these particular electronic configurations appear to be saturated in the sense that they have no capacity for combining with other atoms; also they have exceptionally high values for their first ionization potentials. It is this association of filled electron shells with stability which gives rise to the fundamental postulate of the electronic theory of valency that *every atom shows a tendency in chemical combination so to adjust the number of electrons in its outermost shell that it adopts the stable electronic configuration of one of the noble gases.* For example, when an atom of an alkali metal combines with another atom it gives up the single electron in its outermost shell and acquires the same configuration as the nearest noble gas. Again, a halogen atom always tends to gain an electron from any atom with which it combines, thereby bringing the electrons in its outermost shell up to the maximum of eight.

According to the electronic theory there are several different ways in which atoms can combine, giving rise to various types of chemical linkage which can be distinguished by the characteristic properties they confer upon the molecules containing them. In the following sections we shall consider each type of bond in turn and see how its formation affects the electronic configuration of an atom.

14.3. The Electrovalent Bond. In this type one atom gives up one or more of its electrons to the atom(s) with which it combines. When this happens the atom which loses electrons becomes positively charged because it now contains fewer electrons than protons, and the atom which accepts the donated electron(s) becomes negatively charged because it then has an excess of electrons over protons. *The bond is therefore an electrostatic one between oppositely charged ions which have been created by the transfer of one or more electrons from one atom to the other*. For this reason it is often called the *ionic bond*.

A good example of an electrovalent bond is provided by sodium chloride. When sodium and chlorine combine together the following processes occur:

$$\underset{(2,8,1)}{Na} \rightarrow \underset{(2,8)}{Na^+} + e$$

and

$$\underset{(2,8,7)}{Cl} + e \rightarrow \underset{(2,8,8)}{Cl^-}$$

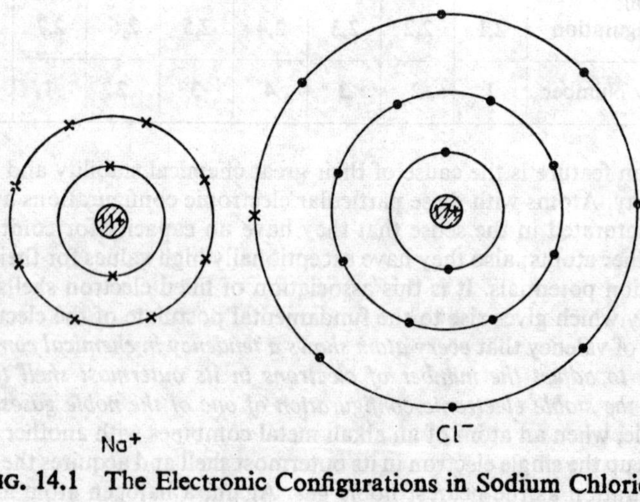

FIG. 14.1 The Electronic Configurations in Sodium Chloride

These changes are shown diagrammatically in Fig. 14.1, where electrons originating in the sodium atom are shown as crosses and those in the chlorine atom as dots.* Except in spin, all electrons are identical, of course, regardless of their atom of origin, but the distinction is made in these diagrams in order to make the transfer of the electron obvious. Thus we believe that sodium chloride is really Na^+Cl^- and that its crystals consist entirely of sodium and chloride ions, as described in § 4.6.

When magnesium burns in oxygen it forms magnesium oxide, each magnesium atom giving up two electrons to an oxygen atom as follows:

$$\underset{(2,8,2)}{Mg} \rightarrow \underset{(2,8)}{Mg^{2+}} + 2e$$

and

$$\underset{(2,6)}{O} + 2e \rightarrow \underset{(2,8)}{O^{2-}}$$

As shown in Fig. 14.2, the solid oxide consists of an aggregate of Mg^{2+} and O^{2-} ions, all with the same electronic configuration, that of neon (2,8). These ions are not, of course, atoms of neon because their nuclei still contain either twelve or eight protons, whereas any atom of neon

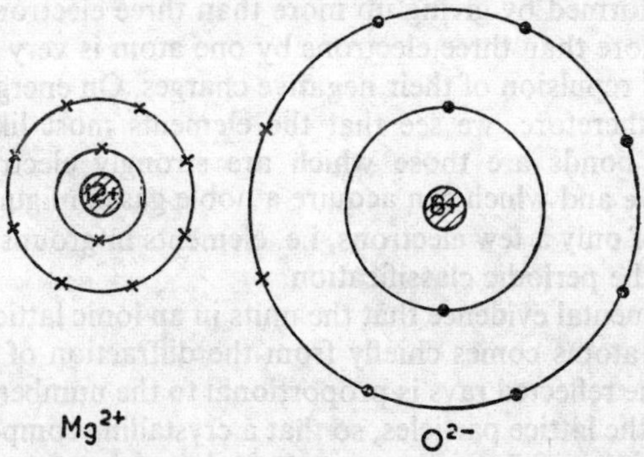

FIG. 14.2 The Electronic Configurations in Magnesium Oxide

necessarily has ten protons in its nucleus. This emphasizes two important facts about atomic number—that it alone characterizes an element, not the number of electrons which happen to surround the nucleus at any particular instant, and that it remains unchanged during chemical combination, being entirely unaffected by the transfer of electrons from one atom to another.

Calcium fluoride is another example of an electrovalent compound.

* In this, as in all electronic configuration diagrams in this chapter, atoms and ions are drawn approximately to the scale: 1 cm ≡ 8×10^{-11} m, using the atomic and ionic radii shown in Table 13.7.

In this case the two outermost electrons in the calcium atom are transferred to two different fluorine atoms as follows:

$$\underset{(2,8,8,2)}{Ca} \rightarrow \underset{(2,8,8)}{Ca^{2+}} + 2e$$

and

$$\underset{(2,7)}{2F} + 2e \rightarrow \underset{(2,8)}{2F^-}$$

Thus this compound consists of Ca^{2+} and F^- ions, containing twice as many of the latter as the former, which is why we normally write its formula as CaF_2.

It is important to note that in all these examples the ions formed possess the stable electronic configurations of the noble gases and that once these configurations have been achieved there is no tendency whatsoever to lose or gain any further electrons. Thus the valency number of any atom forming electrovalent bonds is that number of electrons it must lose or gain in order to acquire the configuration of the nearest noble gas. Now an appreciable amount of energy is needed to remove electrons from even an electropositive atom and consequently few ions are formed by giving up more than three electrons. Similarly, the gain of more than three electrons by one atom is very rare because of the mutual repulsion of their negative charges. On energy considerations alone, therefore, we see that the elements most likely to form electrovalent bonds are those which are strongly electropositive or electronegative and which can acquire a noble gas configuration by the loss or gain of only a few electrons, i.e. elements in groups IA, IIA, VIB and VIIB of the periodic classification.

The experimental evidence that the units in an ionic lattice are indeed ions and not atoms comes chiefly from the diffraction of X-rays. The intensity of the reflected rays is proportional to the number of electrons possessed by the lattice particles, so that a crystalline compound gives a series of reflections of maximum and minimum intensity for various values of θ, the angle of incidence, in accord with Bragg's law (§ 4.5). Now in some compounds these minima are found to be points of complete extinction, so that in these cases the reflected X-rays must not only be in anti-phase but must also be of equal intensity so that they cancel each other completely. We conclude that the units in these lattices (e.g. potassium chloride or sodium fluoride) must contain the same number of electrons, which can only be the case if the elements exist in the ionic condition, i.e. as K^+ (2,8,8) and Cl^- (2,8,8) or as Na^+ (2,8) and F^- (2,8).

The electrovalent bond confers certain characteristic properties upon the compounds containing it. These properties are described in detail in § 4.6 in an account of the ionic type of lattice, so it will suffice here merely to illustrate them by the typical example of sodium chloride. This compound has a high melting point and boiling point and is soluble in water. It conducts electricity readily when molten or when dissolved in water. In aqueous solution it reacts rapidly with other electrovalent substances displaying all the usual properties of sodium and chloride ions.

14.4. The Covalent Bond. *The essential feature of this type is the sharing of pairs of electrons between two atoms, each contributing an equal number of electrons to the bond.* These shared electrons may be regarded as circulating around each nucleus in turn in perhaps figure-of-eight fashion, or they may be looked upon as a cloud of negative charge situated somewhere in the region between the two nuclei and under the partial control of both. By sharing electrons in this way each atom effectively increases the number of electrons in its outermost shell bringing it near to the stable configuration of one of the noble gases.

In methane CH_4, for example, each hydrogen atom is linked to the

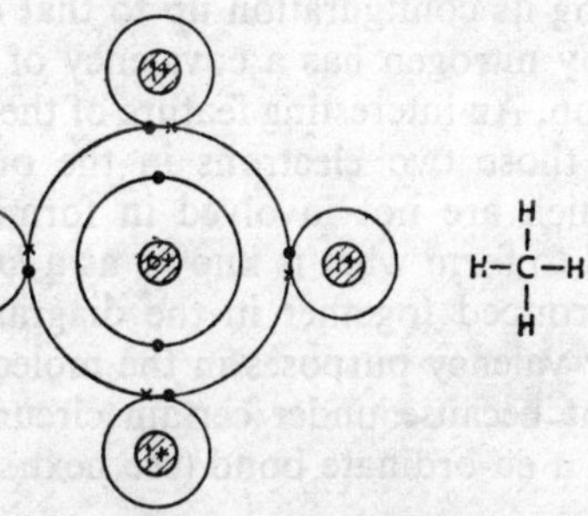

FIG. 14.3 The Electronic Configurations in Methane

carbon atom by a pair of shared electrons as in Fig. 14.3, where electrons originating in a hydrogen atom are shown as crosses and those from the carbon atom as dots. In this way each hydrogen atom gains control of an additional electron and assumes the stable configuration of helium, whilst the carbon atom builds up the number of electrons in its outermost shell from four to eight as in neon. In the diagram these four covalent bonds are shown arranged at right angles to each other in one plane, but this is done only for convenience and the bonds are in fact directed tetrahedrally in space owing to their strong mutual repulsion. The convention of portraying covalent bonds by short lines or strokes joining the atoms has already been referred to at the beginning of this chapter. This practice is used extensively in organic chemistry for writing graphic formulae (§ 32.3), each line standing for a *pair* of shared electrons.

Ammonia, NH_3, provides another familiar example of covalent bonding, as shown in Fig. 14.4. Since the outermost shell of the nitrogen

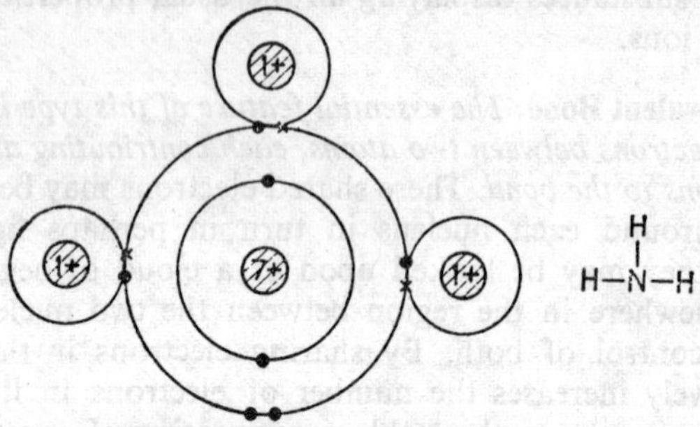

FIG. 14.4 The Electronic Configurations in Ammonia

atom already contains five electrons, only three bonds are necessary in this case to bring its configuration up to that of the inert gas neon, which explains why nitrogen has a covalency of only three compared with four for carbon. An interesting feature of the ammonia molecule is what happens to those two electrons in the outermost shell of the nitrogen atom which are not involved in forming a covalent bond. These are believed to form what is known as a *lone pair* (which is why they are shown grouped together in the diagram) and although not actually in use for valency purposes in the molecule of ammonia they are very important because under certain circumstances they can be employed to form a co-ordinate bond (see next section) and thus they

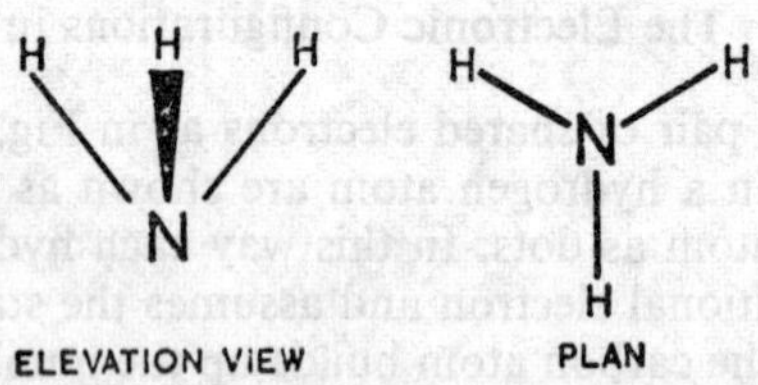

FIG. 14.5 The Shape of the Ammonia Molecule

represent a latent or potential valency. Again the molecule of ammonia has been depicted for convenience as if it were planar in the electron diagram, whereas in fact it has the shape shown in Fig. 14.5.

The molecules of most non-metallic elements contain covalent bonds. For example, Fig. 14.6 shows the electronic configurations in molecules of hydrogen, fluorine, and chlorine, and demonstrates how it is possible, according to the electronic theory, for two atoms of the same electronegativity to join stably together. This is the problem which all the nineteenth-century theories of valency failed to solve, since clearly no stable bond of an electrostatic kind could be expected to exist between identical atoms bearing like charges.

A covalent bond may involve the sharing of one, two, or even three pairs of electrons. This is well illustrated by the three hydrocarbons

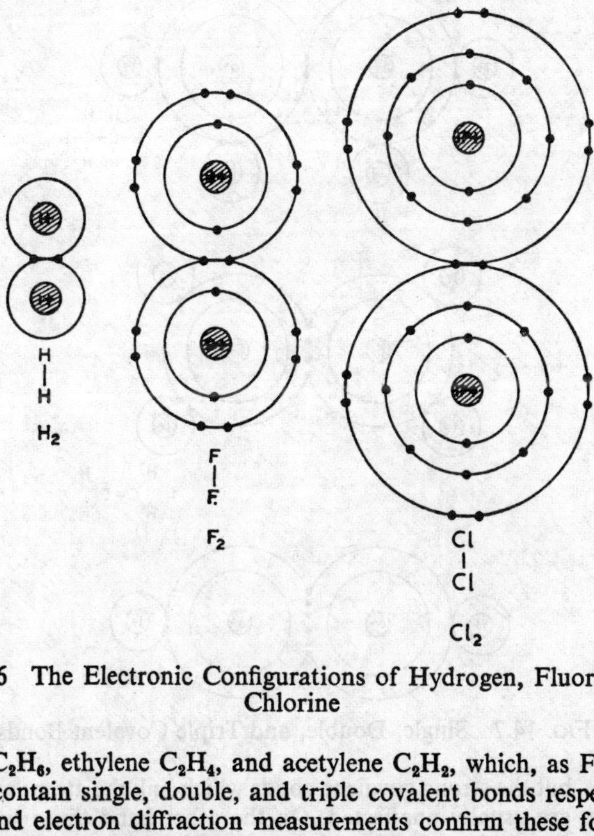

FIG. 14.6 The Electronic Configurations of Hydrogen, Fluorine and Chlorine

ethane C_2H_6, ethylene C_2H_4, and acetylene C_2H_2, which, as Fig. 14.7 shows, contain single, double, and triple covalent bonds respectively. X-ray and electron diffraction measurements confirm these formulae since they show that the carbon atoms are 0.154 nm apart in compounds containing C—C, 0.134 nm apart in compounds containing C=C, and 0.120 nm apart in C≡C compounds. Presumably these differences in interatomic distances arise from distortion of the electron orbitals in the doubly and triply bound carbon atoms.

Covalent compounds, unless their molecular weight is high, are volatile, i.e. they are gases at room temperature or low boiling point liquids. This is because these substances usually consist of small separate molecules held together in a lattice by weak forces of the van der Waals type (§ 4.9). When such substances are heated the cohesive forces between the molecules break readily causing the substances to melt and

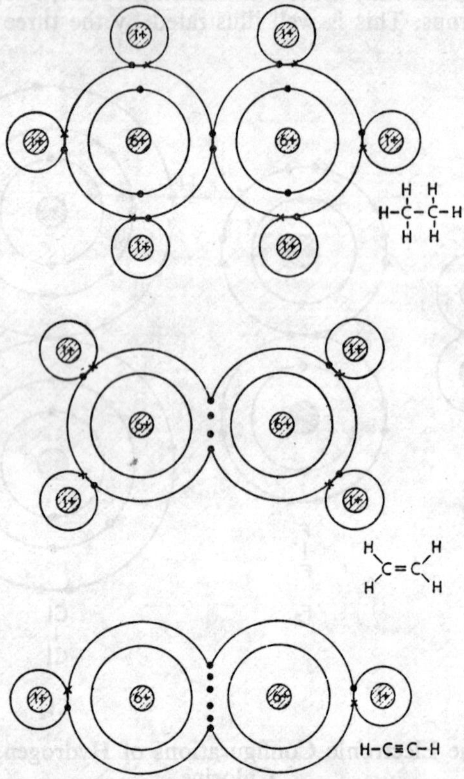

FIG. 14.7 Single, Double, and Triple Covalent Bonds

vaporize, but the strong covalent bonds which link the atoms within the molecule are usually unaffected. On the other hand those substances such as diamond and silicon carbide which form covalent giant molecules (§ 4.8) are hard and non-volatile. Other characteristics of covalency are non-conduction of electricity (because there are no ions available to carry the current), sparing solubility in water (unless the substance and water react together chemically), high solubility in non-polar solvents such as benzene, and rates of reaction which are often slow compared with those of ions in aqueous solution. The fact that covalent bonds are directed in space has already been referred to in considering the examples of methane and ammonia. It is this important property which gives rise to stereoisomerism (§ 48.1) and which determines the shape of covalent molecules (see § 14.8). All these characteristics are illustrated admirably by methane, phosphorus trichloride, and carbon tetrachloride, which are typical covalent compounds.

Almost all the elements except the noble gases and the highly electropositive metals of groups IA and IIA can form covalent bonds. For elements like carbon, silicon, and hydrogen, which rarely combine electrovalently, covalency is the principal type of linkage, and most organic compounds are completely covalent. Covalent bonds are also usually formed when two electronegative elements combine together and when weakly electropositive metals such as gold, silver, and mercury combine with weakly electronegative atoms or groups.

It should not be thought that an atom always acquires the electronic configuration of a noble gas when it forms covalent bonds. For example, in boron trichloride, BCl_3, the boron atom has only six electrons in its outermost shell and not the eight found in neon. However, as one might expect, it is then a very unsaturated atom readily gaining control over two additional electrons by accepting a co-ordinate bond (see below), as in $BCl_3.NH_3$ and the fluoborate ion $[BF_4]^-$. Another familiar exception is phosphorus pentachloride, whose formula is conventionally written as PCl_5. Since the phosphorus atom has five electrons in its outermost shell to start with and since it gains control over five more in forming covalent bonds with the chlorine atoms, its final configuration is 2,8,10, i.e. two more electrons than argon. Such a molecule certainly exists in the vapour, although it tends to dissociate into the more stable trichloride and chlorine, but in the solid state the pentachloride is an ionic compound with the structure $[PCl_4]^+[PCl_6]^-$, as explained in § 23.7. Yet another example is provided by the stable compound sulphur hexafluoride, SF_6, which clearly demonstrates the ability of elements in the third period, unlike those in the second, to raise the number of their outermost electrons to twelve in some of their compounds.

14.5. The Co-ordinate Bond. This type resembles the covalent linkage in that pairs of electrons are shared between two atoms, but the essential difference is that *in a co-ordinate bond one atom contributes both the electrons which make up the bond.* This one-sided origin of the electrons causes a pronounced polarity in the linkage, which accounts for the alternative names of *semi-polar bond* and *dative covalency.* It is conventional to denote the co-ordinate bond by an arrow pointing away from the atom which provides the electrons, as shown in the examples quoted below.

Many complex ions (§ 9.16) contain co-ordinate bonds. For example, when the simple cupric ion Cu^{2+} combines with a molecule of ammonia a co-ordinate bond is formed by the nitrogen atom, which uses its lone pair of electrons for the purpose as shown in Fig. 14.8. In practice the

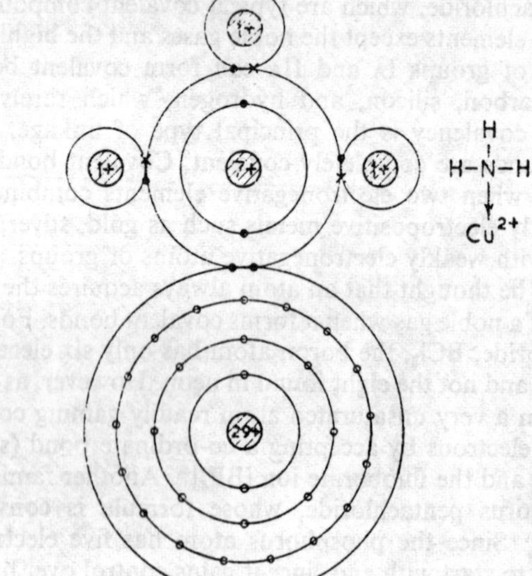

FIG. 14.8 A Typical Co-ordinate Bond

tendency is for four ammonia molecules to link to the cupric ion in this way forming the tetramminocopper(II) ion which confers a deep blue colour to Schweitzer's reagent (§ 28.9). Since each co-ordinate bond provides the copper(II) ion with two additional electrons, the number in its outermost shell is raised to eight:

$$Cu^{2+} + 4NH_3 \rightleftharpoons \left[\begin{array}{c} NH_3 \\ \downarrow \\ H_3N \rightarrow Cu \leftarrow NH_3 \\ \uparrow \\ NH_3 \end{array} \right]^{2+}$$

Other groups which readily form co-ordinate bonds with ions of the transition metals include water, the chloride ion, and the cyanide ion. All these *ligands*, as they are called, contain lone pairs of electrons which can be shared with the central ion. For example, six CN^- ions are linked co-ordinately around each ferrous or ferric ion in the ferro- and ferricyanides, also known as hexacyanoferrate II and III ions respectively:

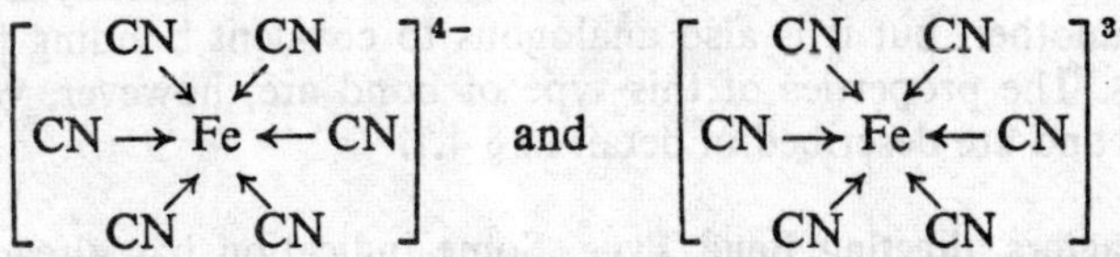

One of the simplest and commonest examples of the co-ordinate bond is the ammonium ion NH_4^+, which may be regarded as being formed by the sharing of a lone pair from the nitrogen atom with a proton thus:

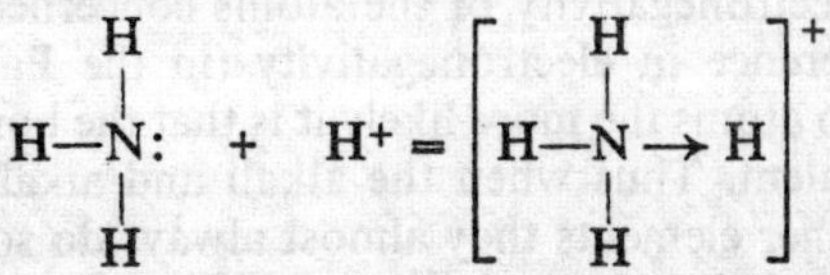

In practice, however, we rarely make any distinction between the four linkages of the nitrogen atom when we write the formula for the ammonium ion, since once they are formed all four bonds are, of course, identical.

Another compound containing the co-ordinate bond is phosphoryl chloride, $POCl_3$ (sometimes known by its older name of phosphorus oxychloride), which has the formula

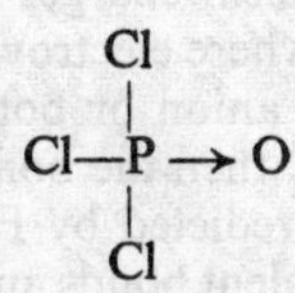

Here each chlorine atom is linked to the phosphorus atom by an ordinary single covalent bond as in phosphorus trichloride, i.e. by a pair of shared electrons, one originating from the chlorine and the other from the phosphorus. On the other hand the linkage between the phosphorus and oxygen atoms is a co-ordinate bond since both electrons making up the bond have come originally from the outermost shell of the phosphorus atom where they existed as a lone pair. In this way the number of electrons in the outer shell of the oxygen atom is raised to the maximum of eight as found in the noble gas neon.

As might be expected from their similar nature, co-ordinate bonds closely resemble covalent bonds in properties, although compounds containing a co-ordinate bond are usually less volatile than purely covalent compounds.

14.6. The Metallic Bond. This is the type formed when atoms are linked together in a metallic lattice. The nature of the bond is not clear; to some extent it is electrical in nature, consisting of positive ions embedded in an atmosphere of loosely held electrons which move readily from one ion to another, but it is also analogous to covalent bonding in some respects. The properties of this type of bond are, however, very distinctive and are described in detail in § 4.7.

14.7. Factors affecting Bond Type. Some indication has already been given in the preceding sections of which elements commonly form a particular type of bond, but it is now possible to consider in a more general way the factors governing the choice. One of the chief of these is the relative electronegativity of the atoms concerned. In general the greater the difference in electronegativity (in the Pauling sense—see § 10.7) of the two atoms the more likely it is that the bond between them will be electrovalent. Thus when the alkali and alkaline earth metals combine with other elements they almost always do so electrovalently, and ionic bonds are also formed when weakly electronegative elements such as the transition metals and aluminium combine with very strongly electronegative elements such as fluorine or oxygen. On the other hand when elements of about the same electronegativity combine together they usually do so covalently by sharing pairs of electrons. The part played by atomic size in determining the electronegativity has been discussed in § 13.9; the relative sizes of the atoms taking part in a bond and the magnitude of the nuclear charges have a big influence upon the type of bond. For example, where electrovalency would result in a very small cation or a very large anion or both, particularly if the charge on the cation would be large, then the bonding between the two atoms is likely to be covalent, as predicted by Fajans in 1923. This explains why boron forms no electrovalent bonds and why the iodides, bromides, and chlorides of beryllium and aluminium are covalent in character.

It would be wrong to conclude from the descriptions given so far that there is in practice any sharp distinction between electrovalency and covalency. Indeed the modern view is that a bond between atoms of two different elements is rarely entirely electrovalent or completely covalent but is usually intermediate between these two extreme types, with the characteristics of one of them predominating. When two identical atoms are linked together covalently it is reasonable to imagine the pair of electrons which composes the bond to be situated midway between the two nuclei, shared equally between the two, so that the bond is purely covalent. When a covalent bond exists between two atoms of different character, however, one being more electronegative than the other, then the electrons which form the linkage will be displaced towards the nucleus of the more electronegative atom owing to its greater affinity for electrons. This displacement may cause an asymmetry in the distribution of charge within the molecule giving rise to a pronounced polarity, of which the dipole moment (§ 4.9) is the best measure. The greater the disparity in electronegativity of the two atoms the more polar the covalent bond becomes, so that in extreme cases the electrons concerned are transferred completely to one atom, forming an electrovalent bond. Hydrogen chloride is a good example of a covalent molecule exhibiting appreciable polarity. When it dissolves in water it immediately ionizes and dissociates, behaving in aqueous solution as a strong acid.

A case of special interest arises when hydrogen combines with certain strongly electronegative elements such as fluorine, oxygen, and nitrogen. The polarity in the covalent bond between the two atoms is then so marked that it leads adjacent molecules to associate into pairs or even chains (see Fig. 14.9), each positive pole aligning itself towards the

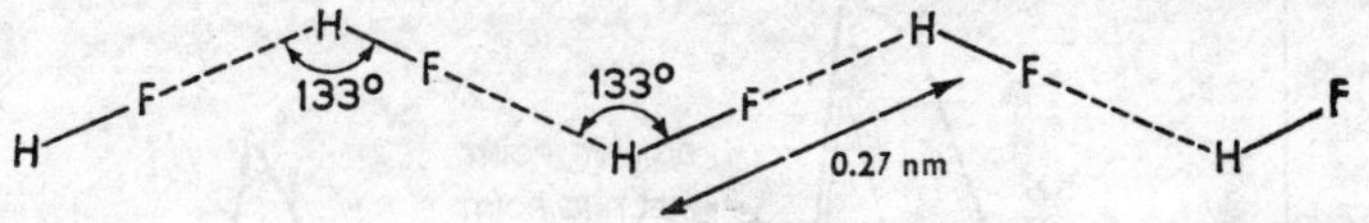

FIG. 14.9 The Association of Hydrogen Fluoride by Hydrogen Bonding

negative one in its neighbour. When this happens the molecules are said to be linked by *hydrogen bonds*. These hydrogen bonds, which arise from the very small size of the hydrogen atom and its ability to approach closely to the electronegative elements, are much weaker than covalent or electrovalent bonds and are easily broken by thermal agitation, so the degree of association tends to decrease with rising temperature. The important effects which hydrogen bonds have on the physical properties of a compound are clearly seen from the examples of hydrogen fluoride (§ 25.3) and water (§ 16.4), and are strikingly illustrated in Fig. 3.14.

It is worth emphasizing that the various bonds in a molecule are not necessarily all of the same kind. For example, in the compound ammonium chloride there is an electrovalent bond between the ammonium and chloride ions, and covalent or co-ordinate linkages between the nitrogen and hydrogen atoms as shown in the following formula:

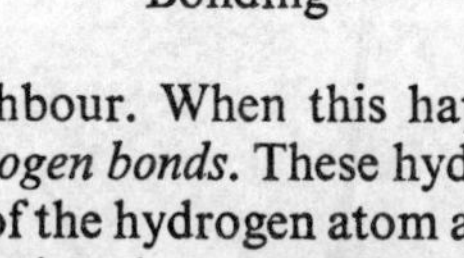

This diversity of bonding is reflected in the physical properties of the compound. The electrovalency explains why it is a solid at room temperature which dissolves readily in water to give a conducting solution with all the characteristic properties of ammonium and chloride ions, whilst the covalent and co-ordinate bonds help to account for its ready sublimation and its tendency to undergo thermal dissociation.

14.8. Periodicity of Physical Properties. The way in which the melting points and boiling points of the first 18 elements (hydrogen to argon) vary with atomic number is shown in Fig. 14.10. The marked periodicity, with high peaks in the middle of each period, arises directly from the differences in the molecular structures of these elements. Some, such as hydrogen, the noble gases, nitrogen, oxygen, fluorine, and chlorine, exist as small discrete molecules with only van der Waals forces (§ 4.9)

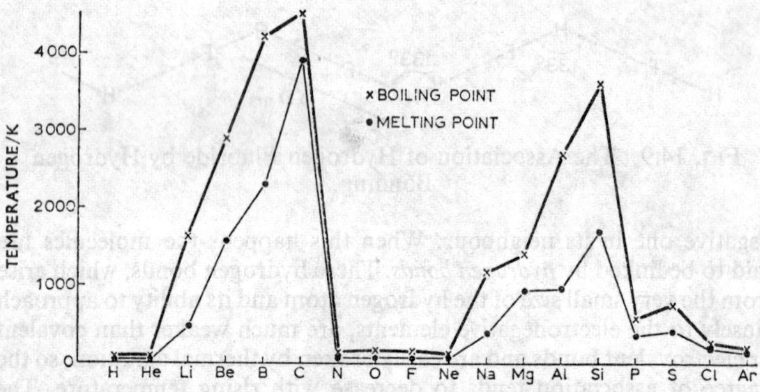

Fig. 14.10 Variation of Volatility with Atomic Number

acting between them. These weak cohesive forces are easily overcome by thermal agitation; such elements have very low melting points and boiling points and are sufficiently volatile to exist as gases at room temperature. In contrast to this, other elements form metallic lattices (§ 4.7) or homopolar lattices (§ 4.8) in which firm bonds between adjacent atoms have to be broken before melting or vaporization can take place. This requires considerable amounts of thermal energy and explains their relatively high melting and boiling points. The effect is greatest for boron, carbon and silicon which form giant crystal lattices in which each atom is joined to several other surrounding atoms by strong covalent bonds.

The latent heats of fusion and vaporization show a similar periodicity, with distinct peaks in the middle of each period and very low values in between. Again the variation reflects the relative amounts of energy required to disrupt the crystal lattices and separate the atoms, as explained in § 4.3.

14.9. Molecular Shape. From chapter 4 it should be clear that metallic and ionic 'molecules', and also covalent molecules of the macro type, all lack definite shape or extent. Small covalent molecules of the kind that make up a molecular lattice do exhibit a variety of shapes, however, owing to the directed nature of the covalent bond. This property arises because the orbitals of the electrons participating in a covalent bond are localized in particular directions causing the bonds to lie at definite

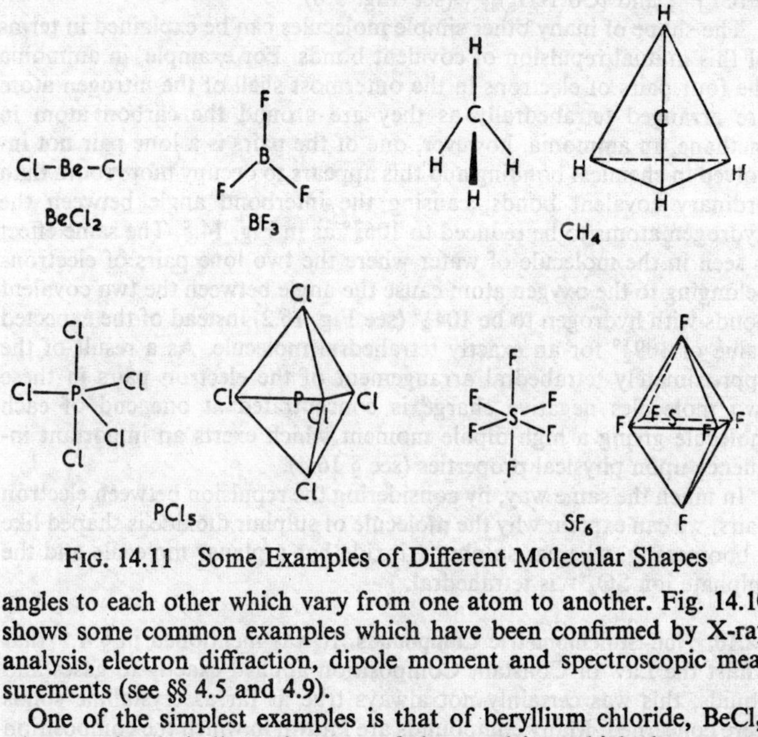

FIG. 14.11 Some Examples of Different Molecular Shapes

angles to each other which vary from one atom to another. Fig. 14.10 shows some common examples which have been confirmed by X-ray analysis, electron diffraction, dipole moment and spectroscopic measurements (see §§ 4.5 and 4.9).

One of the simplest examples is that of beryllium chloride, $BeCl_2$, whose molecule is linear because of the repulsion which the two covalent bonds exert upon each other. Similarly, in boron trifluoride, BF_3, the three bonds all lie in the same plane at an angle of 120° to each other owing to their mutual repulsion. When an atom forms four single covalent bonds, as carbon does in methane or carbon tetrachloride, they are directed tetrahedrally for the same reason, although where double or trible bonds are formed a planar or linear molecule is often produced as in carbon dioxide, carbon disulphide, and acetylene. This tetrahedral arrangement of the single carbon bonds had already been established by van't Hoff and Le Bel in 1874 by purely stereochemical reasoning (see § 48.2). In phosphorus pentachloride, where five such bonds are formed, they are arranged in a trigonal bypyramid as shown in the diagram, with one bond above, one below, and three directed at 120° to each other in the equatorial plane. The shape of sulphur hexafluoride, SF_6, is octahedral with the six fluorine atoms arranged symmetrically around the sulphur atom. The same octahedral arrangement is found in six-co-ordinated complexes such as $[Fe(CN)_6]^{3-}$, $[PtCl_6]^{2-}$, and $[Co(NH_3)_6]^{3+}$ (see Fig. 9.6).

The shape of many other simple molecules can be explained in terms of this mutual repulsion of covalent bonds. For example, in ammonia the four pairs of electrons in the outermost shell of the nitrogen atom are arranged tetrahedrally as they are around the carbon atom in methane. In ammonia, however, one of the pairs is a lone pair not involved in chemical bonding and this appears to occupy more room than ordinary covalent bonds, causing the interbond angle between the hydrogen atoms to be reduced to 106¾° as in Fig. 14.5. The same effect is seen in the molecule of water where the two lone pairs of electrons belonging to the oxygen atom cause the angle between the two covalent bonds with hydrogen to be 104½° (see Fig. 16.2) instead of the expected value of 109½° for an exactly tetrahedral molecule. As a result of the approximately tetrahedral arrangement of the electron pairs in these two molecules negative charge is concentrated at one end of each molecule giving a high dipole moment which exerts an important influence upon physical properties (see § 16.4).

In much the same way, by considering the repulsion between electron pairs, we can explain why the molecule of sulphur dioxide is shaped like a boomerang, whereas sulphur trioxide has a planar molecule and the sulphate ion SO_4^{2-} is tetrahedral.

14.10. Non-Stoichiometric Compounds. It was mentioned in § 1.3 that whilst the Law of Constant Composition applied exactly to gases and liquids, this was certainly not always true as far as crystalline solids were concerned. Many compounds are known in which the composition varies over a range and does not show exact constancy. Such substances are called non-stoichiometric compounds, or, alternatively, are referred to as Berthollide or non-Daltonian structures. For example, the composition of ferrous oxide never corresponds exactly with the formula FeO, but normally lies in the range $Fe_{0.86}O$ to $Fe_{0.96}O$. Similarly in ferrous sulphide the proportion of iron varies from sample to sample but is usually appreciably less than that implied by the conventional formula FeS. These deviations from stoichiometry arise because the compounds are ionic solids in which the predominant factor in deciding the composition is a balance of charges on the ions. Some ferrous ions are missing from the crystal lattices leaving gaps or 'holes', but the overall electrical balance is maintained by some of the remaining ferrous Fe^{2+} ions increasing their charges and assuming the ferric Fe^{3+} state. It will be seen that to compensate electrically for the loss of each ferrous ion from the lattice, two ferric ions have to be formed.

Other common examples of non-stoichiometric compounds are to be found in the hydrides, nitrides, and carbides of the transition metals (§ 27.1). These are of a different type and are known as *interstitial structures* because the atoms of non-metals tend to lodge in the interstices or empty spaces in between the larger metal atoms. The carbon steels are important examples of structures of this kind.

It is now clear that whereas the laws of chemical combination listed in the first chapter apply to compounds with a covalent molecular structure, they are not necessarily valid for structures of an ionic or metallic nature where the concept of a molecule has little meaning.

14.11. Resonance. The idea of *resonance*, which is sometimes known by the alternative name of *mesomerism*, results from the application of quantum mechanics to the electronic theory of valency. It can be used to explain those many cases where no single structure of the conventional type satisfactorily accounts for all the reactions of a compound. Such a substance is said to be a *resonance hybrid* of two or more forms each of which is responsible for some of its properties but which differ from each other only in the way the valency electrons are distributed between the various nuclei. Unlike tautomerism, which is a type of isomerism in which a compound exists as a mixture of two forms in dynamic equilibrium (see Table 48.1), when resonance occurs the hybrid is a new and distinct structure possessing some of the properties of each *canonical form*, as the extreme forms are called, but being intermediate in character between them. The nature of a resonance hybrid is very well illustrated by the following analogy, reprinted with permission from G. W. Wheland, *Resonance in Organic Chemistry*, John Wiley & Sons, Inc., New York, 1955:

'A mule is a hybrid between a horse and a donkey. This does not mean that some mules are horses and the rest are donkeys, nor does it mean that a given mule is a horse part of the time and a donkey the rest of the time. Instead it means that a mule is a new kind of animal, neither horse nor donkey, but intermediate between the two and partaking to some extent of the character of each.'

Resonance can be recognized in several ways. The resonance compound is always more stable than any of the canonical forms, i.e. its heat of formation is greater and its heat of combustion is less than expected by comparison with normal compounds. This is an important feature of resonance structures and helps to explain their stability, e.g. benzene and acetic acid. Another characteristic of resonance is that the various bond lengths in the molecule, as measured by the diffraction of X-rays or electrons, are intermediate between the extreme values expected in the canonical forms. For example, the carbonate ion, CO_3^{2-}, is regarded as a resonance hybrid of the three planar forms

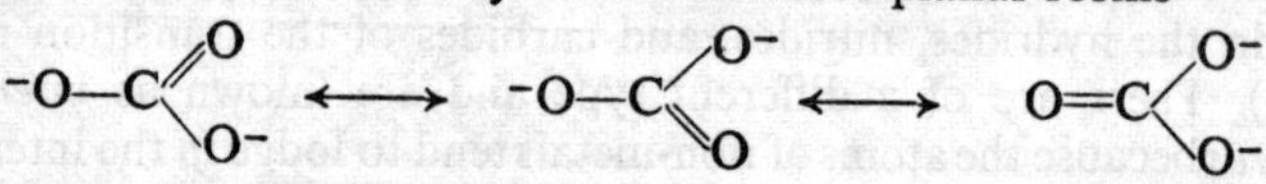

in which the distance between the carbon atom and each of the oxygen atoms in the molecule is exactly the same, 0.131 nm,* compared with the usual values of 0.143 nm for C—O and 0.124 nm for C=O. Again, in benzene the distance between adjacent carbon atoms is 0.139 nm compared with bond lengths of 0.154 nm for C—C and 0.134 nm for C=C (see § 43.3). A further example is provided by urea (§ 41.9) which is believed to be a hybrid of the forms

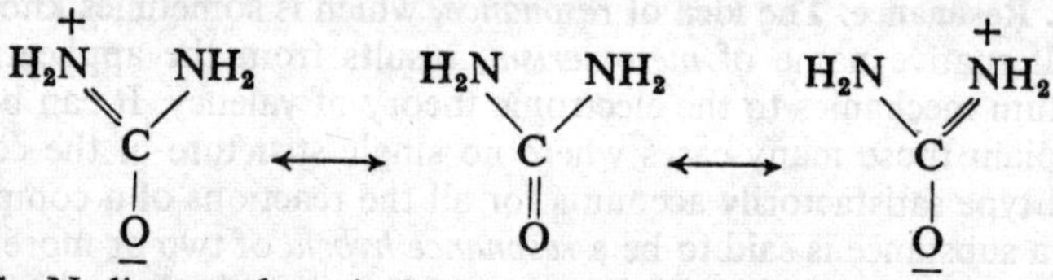

The C—N distance here is 0.137 nm, instead of the usual 0.147 nm found in other compounds. Yet another indication of resonance is that the dipole moment of the substance may differ from the value calculated for the canonical forms.

Other examples of resonance include the oxides of nitrogen (see chapter 22), carbon monoxide (§ 20.7), carbon dioxide (§ 20.5), carboxylic acids (§ 37.2), esters (§ 38.1), amides (§ 41.8), and many aromatic compounds.

14.12. Hybrid Orbitals. When atoms combine together covalently as described in § 14.4, two general principles apply. The first is that chemical combination usually involves the pairing in one orbital of two previously unpaired electrons of appropriate energy and opposite spin.

* Bond lengths are conventionally expressed in nanometre, 1 nm being 10^{-9} metre.

Thus, for example, in forming ammonia the nitrogen atom links to three hydrogen atoms by pairing each of its unpaired $2p$ electrons with an electron from a hydrogen atom. Secondly, when several paired electron orbitals are formed in this way by covalent bonding they tend, owing to strong mutual repulsion between the electrons in adjacent orbitals, to orientate themselves in space so that they are spread as far apart as possible. A number of examples have been quoted in § 14.8 to illustrate this tendency and the effect that it has upon molecular shape.

When a covalent bond is formed we can represent the linkage between the atoms as an overlap of some of their orbitals, so that the electrons actually involved in the bond are influenced by both atomic nuclei and their charge density distribution is altered accordingly. It is not surprising, therefore, that in the process of bond formation the original orbitals are modified and become *hybrid orbitals*. These do not resemble the orbitals described in § 13.13 but are a kind of mixture embodying some features of each type in their mathematical derivation. One of the best examples is provided by the carbon atom which in the process of chemical combination promotes one of the electrons in its $2s$ orbital to the vacant $2p$ orbital and then involves all four electrons, one from the s orbital and the other three from p orbitals in four covalent bonds. Such a hybrid orbital is known as a sp^3 orbital; in it the bonds are directed tetrahedrally in space as they are in methane, carbon tetrachloride, and diamond.

Another example of hybridization occurs in boron trichloride, BCl_3, where the boron atom forms a sp^2 orbital which gives a planar molecule with bonds at 120° to each other. The molecule of beryllium chloride in the vapour state provides a third example; this time a sp hybrid orbital is formed in which the bonds are orientated in diametrically opposite directions giving a linear molecule.

Thus the concept of hybridization can be used to explain why the covalencies of these elements are higher than would be expected from their electronic configurations and why the bonds formed by any one element are equivalent in every way. It also offers an explanation of the spatial orientation of the bonds and hence of the shape of the molecules which result.

14.13. Variable Valency. Many elements show variable valency, i.e. they exert different combining capacity under different conditions. This phenomenon, which is one of the distinctive characteristics of the transition metals, is explained in terms of electronic configurations in chapter 27, where numerous examples are given. It is also considered in chapter 15, where the concepts of oxidation and reduction are discussed in terms of the transfer of electrons and in Appendix I in terms of nomenclature.

15

Oxidation and Reduction

15.1. Introduction. As described in § 24.2, oxygen is a highly reactive element which, under appropriate conditions, combines directly with every other element except the noble gases. In view of its reactivity and its ready availability from the atmosphere, its reactions with other substances, elements and compounds, figured prominently in the experiments of the early chemists. It is not surprising, therefore, that in a desire to classify the reactions they observed and to emphasize their common features, these chemists should have grouped together all the reactions in which substances combined with oxygen and applied the general term *oxidation* to them:

e.g.
$$2Mg + O_2 = 2MgO$$
$$6PbO + O_2 = 2Pb_3O_4$$
$$2H_2 + O_2 = 2H_2O$$
$$2C_6H_5.CHO + O_2 = 2C_6H_5.COOH \text{ (see § 47.3)}$$
$$4HI + O_2 = 2I_2 + 2H_2O \text{ (see § 25.14)}$$
$$2CH_3.CH_2OH + O_2 = 2CH_3.CHO + 2H_2O \text{ (see § 36.2)}$$

Most of these changes take place with elemental oxygen, but the last three are very slow under these conditions and are best effected indirectly using potassium permanganate solution, chromic acid, or hydrogen peroxide solution.

In the same way, the term *reduction* was originally used to describe those reactions in which oxygen was removed from a compound, as in the following examples:

$$PbO + H_2 = Pb + H_2O$$
$$Fe_2O_3 + 3CO = 2Fe + 3CO_2$$
$$Cr_2O_3 + 2Al = 2Cr + Al_2O_3$$

In due course these narrow definitions of oxidation and reduction gave way to more comprehensive ones embodying a much wider range of reactions, as described in the following sections.

15.2. Oxidation. As more reactions were studied it was seen that many of them were fundamentally of the same type as oxidation, involving the combination of a metal or hydrogen with an electronegative element or radical so that the proportion of the latter in the substance was increased. The term oxidation was extended, therefore, to include not only combination of a substance with oxygen or the removal of hydrogen from a compound but also reactions in which oxygen atoms take no part such as combination with fluorine, chlorine, bromine, iodine, sulphur, and other electronegative elements:

$$2Al + 3Cl_2 = Al_2Cl_6$$
$$H_2 + I_2 = 2HI$$
$$H_2S + Br_2 = 2HBr + S$$

A further important extension of the concept of oxidation was its application to cases where the valency of a metal was increased as result of the reaction.* For example, ferrous chloride reacts with chlorine to form ferric chloride:

$$2FeCl_2 + Cl_2 = 2FeCl_3$$

and ferrous sulphate is converted into ferric sulphate when treated with potassium permanganate solution or when boiled with potassium chlorate solution:

$$2FeSO_4 + H_2SO_4 + \bar{O} = Fe_2(SO_4)_3 + H_2O$$

The fundamental change in both of these reactions is from ferrous iron Fe^{2+} to ferric iron Fe^{3+}, and since this corresponds exactly to the change from ferrous oxide to ferric oxide, it is clearly an example of oxidation:

$$2FeO + \bar{O} = Fe_2O_3$$

Similarly, changes such as the conversion of cuprous copper into cupric and stannous tin into stannic and plumbous lead into plumbic are all regarded as oxidations:

$$Cu_2Cl_2 + 2HCl + \bar{O} = 2CuCl_2 + H_2O$$
$$SnCl_2 + 2HgCl_2 = SnCl_4 + Hg_2Cl_2$$
$$PbCl_2 + Cl_2 = PbCl_4$$

In all these changes the positive charge on a cation is increased or the negative charge on an anion is decreased, e.g. $Fe^{2+} \rightarrow Fe^{3+}$, or $I^- \rightarrow I$, etc. This led eventually to an even broader view of oxidation as any process which involves the loss of electrons. Thus an atom or radical which gives up electrons to another is said to be *oxidized*, and

* The transition metals (chapter 27) readily show variable valency and afford numerous examples of such changes.

any substance which habitually gives up oxygen or increases the valency of a metal by accepting electrons is termed an *oxidizing agent.* We arrive, therefore, at a full definition of oxidation as *any change in which the proportion of oxygen or other electronegative element in a substance is increased or in which there is a loss of electrons.*

15.3. Reduction. Just as the idea of oxidation has undergone a tremendous growth, as outlined in the preceding section, so there has been a corresponding development in the meaning of the term reduction over the years. The original concept was first broadened to include not only the removal of oxygen from a compound or the addition of hydrogen to form water but also the removal of other chemically similar electronegative elements such as the halogens and sulphur. Later, those reactions in which the valency exerted by a metal was reduced such as the change from ferric sulphate to ferrous sulphate brought about by nascent hydrogen (§ 16.2) were also regarded as reductions. The final stage in this development was to view reduction as *any process which involves the gain of electrons by a substance.* Those substances which possess the characteristic property of giving up electrons to others or taking up oxygen from them are called *reducing agents*, and those which accept electrons are said to be *reduced.*

In developing these concepts of oxidation and reduction we have greatly increased the number of reactions to which the terms can be applied, but at the same time we have sacrificed some of the precision of the original ideas. This illustrates the difficulty of defining a term satisfactorily in science; a compromise often has to be sought between a definite but very narrow and rigid meaning and something much more comprehensive but noticeably less precise.

15.4. Redox Reactions. We have seen how oxidation is essentially a process involving the surrender of electrons and reduction one in which electrons are accepted. In view of the complementary nature of these changes it is hardly surprising to find that in fact they never occur singly but that every oxidation is necessarily accompanied by a reduction and vice versa. We therefore arrive at the conclusion that reduction and oxidation reactions are not separate processes but rather different aspects of the same fundamental change, the transfer of electrons from one atom or ion to another. This view is reflected in the name reduction–oxidation reactions, usually abbreviated to the shorter *redox reactions*, now given to changes of this kind.

The idea of electron transfer, which forms the basis of all redox reactions, can be made clearer by considering some typical examples. In each case the changes are expressed ionically as well as in terms of a conventional equation:

1. $2HgCl_2 + SnCl_2 = Hg_2Cl_2 + SnCl_4$

mercuric chloride, stannous chloride, mercurous chloride, stannic chloride

oxidation: $Sn^{2+} \rightarrow Sn^{4+} + 2e$

reduction: $2Hg^{2+} + 2e \rightarrow Hg_2{}^{2+}$

2. $CuSO_4 + Zn = ZnSO_4 + Cu$

copper sulphate, zinc, zinc sulphate, copper

oxidation: $Zn \rightarrow Zn^{2+} + 2e$

reduction: $Cu^{2+} + 2e \rightarrow Cu$

3. $KIO_3 + 5KI + 6HCl = 3I_2 + 3H_2O + 6KCl$

pot. iodate, pot. iodide, iodine

oxidation: $5I^- \rightarrow 2\frac{1}{2}I_2 + 5e$

reduction: $IO_3{}^- + 6H^+ + 5e \rightarrow 3H_2O + \frac{1}{2}I_2$

4. $10FeSO_4 + 2KMnO_4 + 8H_2SO_4$
$= 5Fe_2(SO_4)_3 + K_2SO_4 + 2MnSO_4 + 8H_2O$

ferrous sulphate, potassium permanganate, ferric sulphate, manganous sulphate

oxidation: $10Fe^{2+} \rightarrow 5Fe_2{}^{3+} + 10e$

reduction: $2MnO_4{}^- + 16H^+ + 10e \rightarrow 2Mn^{2+} + 8H_2O$

5. $K_2Cr_2O_7 + 6FeSO_4 + 7H_2SO_4$
$= 3Fe_2(SO_4)_3 + K_2SO_4 + Cr_2(SO_4)_3 + 7H_2O$

potassium dichromate, ferrous sulphate, ferric sulphate, chromic sulphate

oxidation $6Fe^{2+} \rightarrow 6Fe^{3+} + 6e$

reduction: $Cr_2O_7{}^{2-} + 14H^+ + 6e \rightarrow 2Cr^{3+} + 7H_2O$

15.5. The Redox Series. Although all substances which show a general tendency to give up electrons are regarded as reducing agents, the tendency is much greater in some substances than in others. Similarly, the affinity for electrons varies widely from one oxidizing agent to another. Whilst the differences can be indicated in a general way by the use of such adjectives as strong and weak, the need still arises to devise some means of comparing quantitatively the relative strengths of different oxidizing and reducing agents. This can be done by assigning a numerical value to the tendency of a substance to lose electrons. For an element the standard electrode potential (§ 10.2) serves very well for this purpose, when it is referred to as its *standard oxidation potential.* The elements can then be arranged in a series (see Table 15.1) according to their oxidation potentials. This series, known as the *redox series*, indicates the relative willingness of a substance to lose or gain electrons. Those substances at the head of the series, such as the alkali and alkaline earth metals, lose their electrons most readily and are therefore the strongest reducing agents, whereas those substances at the end of the series such as fluorine or oxygen have the greatest affinity for electrons and are therefore the strongest oxidizing agents.

TABLE 15.1. STANDARD REDOX POTENTIALS

Reaction Equation	$E^{\ominus}$ /V*
$Li \rightleftharpoons Li^+ + e$	−3.03
$K \rightleftharpoons K^+ + e$	−2.92
$Ca \rightleftharpoons Ca^{2+} + 2e$	−2.87
$Na \rightleftharpoons Na^+ + e$	−2.71
$Mg \rightleftharpoons Mg^{2+} + 2e$	−2.38
$Al \rightleftharpoons Al^{3+} + 3e$	−1.67
$Zn \rightleftharpoons Zn^{2+} + 2e$	−0.76
$Fe \rightleftharpoons Fe^{2+} + 2e$	−0.44
$\frac{1}{2}H_2 \rightleftharpoons H^+ + e$ (for pH = 7)	−0.41
$Sn \rightleftharpoons Sn^{2+} + 2e$	−0.14
$Pb \rightleftharpoons Pb^{2+} + 2e$	−0.13
$\frac{1}{2}H_2 \rightleftharpoons H^+ + e$ (arbitrary zero)	0.00
$Sn^{2+} \rightleftharpoons Sn^{4+} + 2e$	+0.14
$Cu^+ \rightleftharpoons Cu^{2+} + e$	+0.16
$Cu \rightleftharpoons Cu^{2+} + 2e$	+0.34
$I^- \rightleftharpoons \frac{1}{2}I_2 + e$	+0.54
$Fe^{2+} \rightleftharpoons Fe^{3+} + e$	+0.77
$Hg_2^{2+} \rightleftharpoons 2Hg^{2+} + 2e$	+0.91
$Br^- \rightleftharpoons \frac{1}{2}Br_2 + e$	+1.07
$Cl^- \rightleftharpoons \frac{1}{2}Cl_2 + e$	+1.36
$Mn^{2+} + 4H_2O \rightleftharpoons MnO_4^- + 8H^+ + 5e$	+1.52
$Pb^{2+} \rightleftharpoons Pb^{4+} + 2e$	+1.80
$F^- \rightleftharpoons \frac{1}{2}F_2 + e$	+2.84

* The sign convention used here is the one recommended by the International Union of Pure and Applied Chemistry (1953). Many American authors use the opposite convention.

The value of the redox series is increased by including in it not only systems involving an element and its ions but also systems in which two different ions of one element are in equilibrium with each other. For example, iron may exist either as ferrous ions, Fe^{2+}, or as ferric ions, Fe^{3+}, and it is possible by dipping a platinum electrode into a mixture containing both ions in molar† concentration at 25°C and measuring

† Strictly the activity (i.e. the effective concentration having regard for interionic attraction) should be unity here, but the distinction is small and is ignored in this chapter.

the potential difference set up between the electrode and a standard hydrogen electrode (§ 10.3) to determine the standard redox potential of the system Fe^{3+}/Fe^{2+} ions. The redox potential of the system Sn^{4+}/Sn^{2+} can be determined in a similar way.

One use of oxidation potentials is for predicting which redox reactions are theoretically possible. For example, since the change

$$Fe \rightleftharpoons Fe^{2+} + 2e$$

has an oxidation potential of −0.44 volt, we can successfully predict that elemental iron will reduce hydrogen ions to hydrogen gas, e.g. it will liberate hydrogen from a normal solution of hydrochloric acid. Care is needed, however, in applying our knowledge of oxidation potentials in this way because we can only predict what is energetically possible and not what actually takes place, and many reactions which on energy grounds are perfectly feasible are too slow to be of practical value.

15.6. Electrolytic Oxidation and Reduction. Since the term oxidation is applied to any process in which an atom or radical loses electrons, the changes taking place at an anode during electrolysis (e.g. the discharge of anions or the dissolution of the electrode) may be regarded as oxidations. Similarly, the gain of electrons which occurs at the cathode during electrolysis comes within the broad definition of reduction. It should be noted that although the oxidation and reduction are effected at two different places in the solution, one process still cannot occur without the other and they inevitably take place simultaneously and to exactly the same extent, since the same number of electrons are involved at each electrode. For example, when molten sodium chloride is electrolysed, chlorine ions give up electrons at the anode and are oxidized to chlorine atoms, whilst at the cathode sodium ions accept electrons and are reduced to atoms of sodium. Thus electrolysis can be used to extract both of these elements from the melt (see § 17.4). Electrolytic oxidation is such a powerful method that it can even be used to oxidize fluoride ions to fluorine, a change which cannot be brought about by any other means (§ 25.2).

When oxygen is liberated at the anode during an electrolysis it tends to oxidize the material of the electrode. For example, during the manufacture of aluminium the carbon anode is gradually burnt away as described in § 19.8. Another example arises when an aluminium anode is used for electrolysing chromic acid solution, the oxygen evolved converting the surface of the aluminium into a strongly adhering layer of oxide which greatly increases the resistance of the metal to corrosion. This process, which is known as *anodizing* the aluminium, is widely used industrially. It has the further advantage that the oxide layer readily adsorbs dyes giving the metal an attractive coloured appearance.

Many other examples of anodic oxidation and cathodic reduction are given in this book. For example, hydroxylamine is best prepared by reducing nitric acid electrolytically, using a lead or mercury cathode (see § 22.9), and hydrogen peroxide is usually made industrially by the electrolysis of concentrated sulphuric acid under special conditions (§ 16.6). Many of the highly electropositive metals are extracted electrolytically from their compounds; details of these processes are described in the appropriate chapters.

15.7. Equivalent Weights of Oxidizing and Reducing Agents. In §§ 2.5 and 2.6 the equivalent weights of oxidizing and reducing agents were defined in terms of the number of parts by mass of them which yielded or combined with 8 parts by mass of available oxygen respectively. The definition of the equivalent weight of an oxidizing agent can now be extended to all cases involving electron transfer by defining it as the molecular weight* divided by the number of electrons gained by each molecule of oxidizing agent during the oxidation. For example, permanganate in acid solution oxidizes in accordance with the ionic equation:

$$MnO_4^- + 8H^+ + 5e = Mn^{2+} + 4H_2O$$

Hence, the equivalent weight of the permanganate is one-fifth of its molecular weight under these conditions. However, in neutral solution permanganate oxidizes according to the equation:

$$MnO_4^- + 2H_2O + 3e = MnO_2 + 4OH^-$$

so that in this case the equivalent weight is one-third of the molecular weight. Another example is provided by the oxidizing action of dichromate in acid solution, when the ionic equation is:

$$Cr_2O_7^{2-} + 14H^+ + 6e = 2Cr^{3+} + 7H_2O$$

Here the equivalent weight is clearly one-sixth of its molecular weight, since each gramme molecule gains six electrons during the oxidation.

Similarly, the equivalent weight of a reducing agent can be defined electronically as the molecular weight divided by the number of electrons lost from each of its molecules during the reaction. For example, the ionic equation for the reducing action of oxalic acid is:

$$C_2O_4^{2-} - 2e = 2CO_2$$

from which the equivalent weight as a reducer is seen to be one-half of

* It is customary in this context to use the term 'molecular weight' when referring to salts, although strictly speaking the phrase 'empirical formula weight' would be more accurate.

its molecular weight. Similarly, when thiosulphate is oxidized by iodine the ionic equation has the form:

$$2S_2O_3^{2-} - 2e = S_4O_6^{2-}$$

so that the equivalent weight of the thiosulphate as a reducing agent is equal to its molecular weight.

15.8 Oxidation Number. It is highly desirable to have available some means whereby the oxidation state of an element in any given situation can be expressed numerically, particularly in redox reactions and cases where an element is capable of exhibiting variable valency. *Oxidation number* is a concept for doing this which has gained widespread recognition because of its usefulness and general applicability.

The oxidation number of an element in any particular molecule or ion can be determined by applying a set of rules as follows:

1. The oxidation number of an element in the free state is zero.
2. The algebraic sum of the oxidation numbers of all the atoms in a molecule is zero. In a compound, the more electronegative elements will have negative oxidation numbers and the more electropositive ones positive numbers. This is true not only of ionic compounds but also of molecules in which the bonding is covalent, when each atom present is given an oxidation number equal to the charge it could be imagined to have if the substance were composed of ions.
3. The algebraic sum of the oxidation numbers of all the atoms in an ion is equal to the overall charge on that ion. In a monatomic ion the element has an oxidation number equal to the ionic charge.
4. When co-ordination bonds are formed in a complex ion, the oxidation numbers of the elements involved are not affected.

Examples: Atoms in molecules of oxygen, hydrogen, nitrogen, sulphur, chlorine, or crystalline metals all have an oxidation number of zero.

In KCl, with structure K^+Cl^-, the oxidation number of potassium is +1 and that of chlorine −1. In Mg Br_2, composed of Mg^{2+} and Br^- ions, magnesium has an oxidation number of +2 and each bromine −1, giving zero overall.

In the nitrate ion, NO_3^-, each oxygen atom has an oxidation number of −2 and the nitrogen therefore has an oxidation number of +5 to make the algebraic sum of oxidation numbers equal to the single negative charge on the ion.

In the sulphate ion, SO_4^{2-}, each oxygen contributes −2 giving a total of −8, so the sulphur atom has an oxidation number of +6 to make the algebraic sum equal to the charge on the ion. In carbon tetrachloride, CCl_4, each chlorine atom is regarded as having an oxidation number of −1 and the carbon atom a value of +4, even though the molecule is covalent and contains no ions.

It follows from these rules that some elements have fixed oxidation numbers in all their compounds. For example, all the alkali metals have an oxidation number of +1, and the oxidation numbers of the alkaline earth metals are invariably +2. Similarly, hydrogen has an oxidation number of +1 in all of its compounds except the metallic hydrides (when it is −1), and oxygen has the value −2 in all its compounds except peroxides and F_2O.

Many elements have several different oxidation numbers depending upon the molecule or ion considered. For example, sulphur has the oxidation number −2 in H_2S and sulphides, +4 in SO_2 and sulphites, +6 in SO_3, SF_6, H_2SO_4, and sulphates, and +7 in $H_2S_2O_8$ and persulphates. In such cases it is helpful to construct a special chart showing the oxidation numbers assigned to the element in its various compounds. This is done for nitrogen in Table 22.1, for sulphur in Table 24.2, for chlorine in Table 25.4, and for the transition metals vanadium, chromium, and manganese in Table 27.2.

The concept of oxidation number provides us with a further definition of oxidation as *any change which brings about an algebraic increase in the oxidation number of an element*. For example, when methane undergoes successive substitution with chlorine atoms to form CH_3Cl, CH_2Cl_2, $CHCl_3$, and CCl_4 (§ 33.2), the oxidation number of the carbon atom changes from −4 to −2, 0, +2, and +4 respectively as oxidation occurs. Similarly, a change which results in a decrease in oxidation number is termed a reduction. It follows from the rules determining oxidation numbers that in any redox reaction the total increase in oxidation numbers must always exactly equal the total decrease, and this principle can be used as a helpful guide in constructing balanced equations for complicated redox reactions.

Another important application of oxidation numbers is in chemical nomenclature. As explained in Appendix I, the Stock system now adopted internationally uses the concept of oxidation number to distinguish between the different compounds of a metal which exhibits variable valency by inserting into the name the oxidation state of the metal written in Roman numerals in parentheses. Thus the two chlorides of copper are known as copper(I) chloride and copper(II) chloride instead of cuprous and cupric respectively. This use of oxidation numbers is particularly valuable for those transition metals such as vanadium, chromium, and manganese which can exist in several different oxidation states in their various compounds.

16

Hydrogen, Water, and Hydrogen Peroxide

16.1. Introduction. In many respects hydrogen is an element of quite exceptional interest and importance. Its atom has the simplest structure of all, consisting in its commonest isotope of only a single proton and electron, which accounts for its ability to form both covalent and electrovalent bonds. Because of this relative simplicity the study of the spectrum of hydrogen (see Fig. 13.3) has made a unique contribution to our understanding of the structure of atoms. Hydrogen is not easy to place in the periodic classification; it shows certain similarities to the alkali metals of Group IA and to the halogens of Group VIIB, yet it is neither strongly electropositive nor strongly electronegative. Its use as an arbitrary standard in the measurement of electrode potentials is explained in § 10.2, and its former use as a standard for equivalent and atomic weights is referred to in §§ 2.1 and 2.8. Of all real gases, hydrogen most closely resembles a perfect or ideal gas in its behaviour, and this has led to its adoption as a standard in gas thermometry. Its oxide water is the most important of all chemical compounds, being essential to every form of plant and animal life. Lastly, the fusion of atoms of hydrogen into atoms of other elements (see § 31.11) has proved to be the most energetic nuclear change possible and is the one which seems chiefly responsible for the immense energies of the stars.

16.2. Hydrogen, H. Electronic configuration: 1.

OCCURRENCE: Hydrogen occurs naturally combined with oxygen, carbon, and other elements in water, petroleum, natural gas, and organic matter, but the earth's atmosphere contains less than one part in a million of hydrogen gas.

LABORATORY PREPARATION: The following methods are available:

1. By the action of acids on metals: Zinc, aluminium, magnesium, and iron are commonly used with dilute hydrochloric or sulphuric acid, or tin with concentrated hydrochloric acid. No heating is necessary, so a Kipp generator (§ 24.6) can be employed. The product is pure enough for most purposes, but traces of arsine, phosphine, and hydrogen sulphide present can be removed by passing it through acidified potassium permanganate solution and the gas can be dried with anhydrous calcium chloride or phosphorus pentoxide.

e.g. $$Sn + 2HCl = SnCl_2 + H_2\uparrow$$
$$Zn + H_2SO_4 = ZnSO_4 + H_2\uparrow$$

2. By the action of alkalis on metals: Certain metals liberate pure hydrogen from concentrated solutions of caustic alkalis, especially on heating:

e.g. $$2Al + 2NaOH + 2H_2O = 2NaAlO_2 + 3H_2\uparrow$$
(sodium aluminate soln.)
$$Zn + 2NaOH = Na_2ZnO_2 + H_2\uparrow$$
(sodium zincate soln.)

3. By electrolysis: Platinum electrodes are used to electrolyse a dilute aqueous solution of sulphuric acid or an alkali; the hydrogen is liberated at the cathode. If very pure hydrogen is required, this is the best method, since the product contains only traces of oxygen (diffused from the anode region) and moisture. The former can be removed by passing the gas over platinum black which catalyses the combination of hydrogen and oxygen to form water, and the water vapour can be removed by passing it first over anhydrous calcium chloride then over phosphorus pentoxide.

The changes occurring during the electrolysis can be represented in the following way:

At the cathode: hydrogen ions (the only cations present in dilute sulphuric acid and the least electropositive ones in solutions of strong alkalis) are discharged, the overpotential (§ 10.5) being negligible with a platinum electrode, and hydrogen gas is liberated:

$$H^+ + e \rightarrow H$$
then $$2H \rightarrow H_2\uparrow$$

At the anode: hydroxyl ions, being less electronegative than HSO_4^- and SO_4^{2-} ions, are discharged, the overpotential again being negligible:

$$4OH^- \rightarrow 4e + O_2\uparrow + 2H_2O$$

The balanced changes are therefore:

$$4H^+ + 4OH^- \rightarrow 2H_2\uparrow + O_2\uparrow + 2H_2O$$
and $$2H_2O \rightarrow 2H^+ + 2OH^-$$

giving an overall equation:

$$2H^+ + 2OH^- \rightarrow 2H_2\uparrow + O_2\uparrow$$

It will be observed that these changes are exactly the reverse of those taking place in the simple fuel cell described in § 10.6.

4. By the action of water on hydrides and metals: This method is not important because of its high cost:

e.g.
$$Ca + 2H_2O = Ca(OH)_2 + H_2\uparrow$$
$$CaH_2 + 2H_2O = Ca(OH)_2 + 2H_2\uparrow$$

MANUFACTURE: Several methods are in use:

1. From water gas by the Bosch process: When steam is passed over white hot coke a mixture of hydrogen and carbon monoxide called water gas results (§ 20.15):

$$C + H_2O = CO + H_2$$

This water gas is then mixed with more steam and passed over a catalyst of iron, activated by traces of chromic oxide which acts as a promoter, at 450°C, when a further change produces more hydrogen:

$$CO + H_2O \rightleftharpoons CO_2 + H_2$$

This reaction is exothermic when proceeding from left to right, and so by Le Chatelier's principle (§ 7.10) the lower the temperature the higher will be the yield of hydrogen. The operating temperature, therefore, is the lowest consistent with a convenient rate of reaction. The carbon dioxide is removed by washing with water under pressure (see Henry's law, § 6.7). Any unchanged carbon monoxide is absorbed in ammoniacal cuprous chloride solution.

2. From methane: This is the principal method in countries possessing an abundant supply of natural gas (§ 33.7). Methane is mixed with steam and a limited supply of oxygen (not air, to avoid the entry of nitrogen), and passed over a nickel catalyst at 900°C, when the following reactions take place:

$$CH_4 + H_2O = CO + 3H_2 \text{ (endothermic reaction)}$$
$$2CH_4 + O_2 = 2CO + 4H_2 \text{ (exothermic reaction).}$$

By carrying out these two changes simultaneously the necessity for external heating is avoided. The product is mixed with more steam and passed over an iron catalyst at 450°C (as in the Bosch process), when the carbon monoxide present is converted into additional hydrogen and carbon dioxide, which is removed by washing under pressure.

3. From petrol: As described in § 33.11 and § 43.3, petrol is often subjected to various catalytic reforming processes in order to improve its anti-knock properties and to meet the demand for benzene and toluene. Large quantities of hydrogen are obtained as by-products.

4. From Naphtha: Large amounts of hydrogen are now made by the catalytic steam-reforming of naphtha, as described in § 20.16, by subjecting the 'lean gas' to the second stage of the Bosch process.

5. By electrolysis: Hydrogen of high purity is obtained as a by-product in the manufacture of sodium hydroxide by electrolysing brine (§ 17.7).

6. By the action of steam on iron at about 700°C:

$$3Fe + 4H_2O \rightleftharpoons Fe_3O_4 + 4H_2$$

Although theoretically reversible, in practice this reaction proceeds continuously from left to right because the hydrogen is blown out of contact with the iron by the current of steam. The iron oxide is periodically reduced back to the metal by passing water gas over it.

PROPERTIES OF HYDROGEN: Colourless, odourless gas, b.p. = −253°C, m.p. = −257°C, slightly soluble in water (about 2% at room temperature). It is the lightest gas known with a density of 0.0899 kg m^{-3} at s.t.p. Liquid and solid hydrogen are non-metallic in properties.

REACTIONS OF HYDROGEN: Hydrogen always has a valency of one. It combines with other elements to form two main types of hydrides, but in either case the hydrogen atom gains control of an extra electron thereby acquiring the stable electronic configuration of the inert gas helium. For example, it combines directly on heating with the highly electropositive alkali and alkaline earth metals forming solid electrovalent hydrides containing the H^- ion:

e.g. $$Ca + H_2 = CaH_2$$

These hydrides yield hydrogen on treatment with cold water.

Hydrogen also combines directly on heating with most non-metallic elements forming volatile covalent hydrides, particularly if a suitable catalyst is present. Many of these hydrides are important and familiar substances; their chemistry is considered in detail in later chapters. For example, with fluorine it combines explosively, even at low temperature:

$$H_2 + F_2 = H_2F_2$$

With chlorine the reaction is explosive in u.v. light, but in diffuse light it proceeds steadily at room temperature in the presence of a charcoal catalyst:

$$H_2 + Cl_2 = 2HCl$$

The reactions with the less electronegative bromine and iodine are reversible and slow, needing a platinum catalyst and a temperature of about 200°C:

$$H_2 + Br_2 \rightleftharpoons 2HBr$$
$$H_2 + I_2 \rightleftharpoons 2HI$$

Mixtures of hydrogen and oxygen explode violently on ignition, but the two gases react steadily when passed over a platinum catalyst, or when hydrogen burns at a jet in air or oxygen, evolving much heat:

$$2H_2 + O_2 = 2H_2O$$

Hydrogen combines with nitrogen extremely slowly unless an iron catalyst and a temperature of 400-500°C is used:

$$3H_2 + N_2 \rightleftharpoons 2NH_3$$

This reaction, which is of great industrial importance, is discussed in sections 7.8, 7.10, 11.10, and 22.4.

Hydrogen is absorbed on heating with certain transition metals such as palladium, platinum, and nickel. The phenomenon is called *occlusion* and is of use in purifying the gas since the metallic hydrides formed break up again on reheating, releasing the absorbed gas.

Because of its affinity for oxygen, hydrogen is a strong reducing agent; many metallic oxides are reduced to the metal on heating in hydrogen, but not the oxides of the very electropositive metals, e.g. sodium, potassium, calcium, and magnesium:

e.g. $$CuO + H_2 = Cu + H_2O$$

Unsaturated organic compounds react readily with hydrogen in the presence of a nickel catalyst at about 150°C; such reactions are called *hydrogenations*. Some, such as the manufacture of margarine from oils and petrol from coal, are of considerable commercial importance.

USES OF HYDROGEN: Great quantities are used to manufacture ammonia (§ 22.4), methyl alcohol (§ 35.3), hydrochloric acid (§ 25.6), margarine (§ 38.4), and synthetic petrol (§ 33.11). The oxyhydrogen flame is used for welding and cutting steel. Some hydrogen is used to fill balloons, but non-inflammable helium is now preferred. In the laboratory hydrogen is a valuable reducing agent. It has been found experimentally that hydrogen is much more active at the moment of its generation, when it is called *nascent* hydrogen. Presumably it is then in the form of single atoms of the element instead of the diatomic molecules found in hydrogen gas. The commonest sources of nascent hydrogen are zinc and dilute hydrochloric acid, tin and concentrated hydrochloric acid, aluminium–mercury or zinc–copper couples, sodium and alcohol, sodium amalgam and water, sodium borohydride (§ 19.7), and lithium aluminium hydride (§ 19.14).

16.3. Isotopes of Hydrogen. Three isotopes are known; their atomic structures are shown diagrammatically in Fig. 16.1. *Protium* ^1_1H, vastly predominates in a sample of ordinary hydrogen. *Deuterium*, ^2_1H, which is usually denoted by the symbol D, is present in ordinary hydrogen to the extent of one part in about six thousand. It can be separated from protium by gaseous diffusion (§ 3.6) or it can be made by any of the usual methods from its oxide D_2O, which is commonly known as *heavy water*. The latter is best obtained by electrolysis of dilute aqueous solutions of acids or alkalis when the ions of protium present are

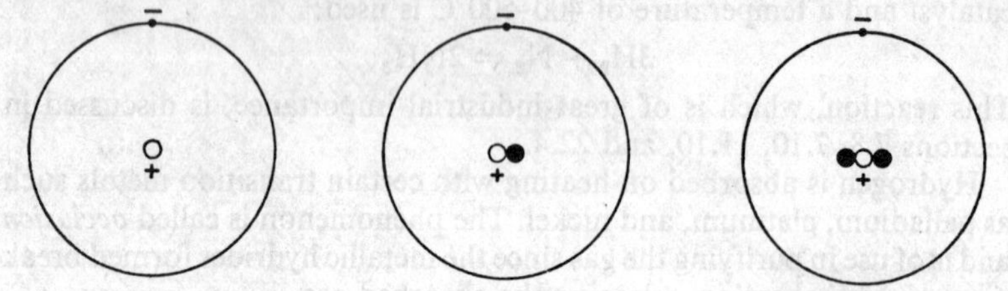

FIG. 16.1 The Isotopes of Hydrogen

preferentially discharged at the platinum cathode (see § 10.5), leaving the electrolyte richer in deuterium oxide than it was before. By repeatedly adding fresh water and electrolysing, it is eventually possible to produce deuterium oxide of any desired concentration, although the process is very costly. Heavy water is very important as a moderator in atomic reactors (§ 31.11) and it is widely used for this purpose in atomic-powered submarines and surface vessels. Its physical properties are compared with those of water in Table 16.1. The third isotope of hydro-

TABLE 16.1. COMPARISON OF WATER AND DEUTERIUM OXIDE

Property	*Water*	*Deuterium oxide*
Melting point/°C	0	3.8
Boiling point/°C	100	101.4
Density/g cm^{-3}	1.00	1.11
Max.^{m} density at	4°C	11.6 °C

gen, which is known as *tritium*, ^3_1H, is present only in traces in ordinary hydrogen, but it can be made artificially. It is unstable, undergoing radioactive decay, with a half-life of about 12.5 years (§ 31.6). The isotopes of hydrogen differ noticeably in physical properties owing to the large percentage difference in their masses, but they are very similar chemically, deuterium being slightly less reactive than hydrogen in many chemical changes.

16.4. Water, H_2O. This, the most important compound of hydrogen, can be shown to be its oxide either by direct synthesis from the elements or by electrolytic decomposition into hydrogen and oxygen. Its composition has been established quantitatively with great accuracy by a number of classical experiments, details of which will be found in reference books.

Physical measurements have established that the water molecule is shaped like an arrow-head, as in Fig. 16.2 with a definite angle, 104°40′, between the covalent bonds. As a result of its shape and the asymmetrical distribution of electrons between its atoms owing to the big difference in their electronegativities (see § 10.7), the molecule is highly polar. This polarity explains why water is such an excellent solvent for electrovalent substances and why water molecules tend to cling readily

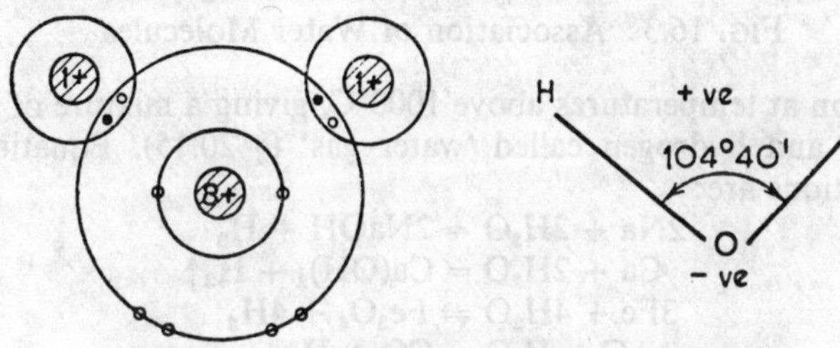

FIG. 16.2 The Shape of the Water Molecule

to ions, forming hydrates in aqueous solution and in crystals. It also explains the formation of hydrogen bonds (§ 14.7) between water molecules, leading to polymeric molecules of formula $(H_2O)_n$ where n varies from about eight at 0°C to a value near unity at 100°C (see Fig. 16.3). Thus the effective molecular weight of water in the liquid state is often much higher than 18 which explains the abnormally high values for its melting and boiling points when compared with hydrogen sulphide.

When associated in this way into a network of molecules linked by hydrogen bonds, water occupies a greater volume than it does as single molecules, which explains both the unusual manner in which the density of water varies with rising temperatures, showing a maximum at 4°C, and also the expansion of water when it changes into crystals of ice, where hydrogen bonding is predominant.

REACTIONS OF WATER: At very high temperatures water undergoes thermal dissociation into its elements, thus:

$$2H_2O \rightleftharpoons 2H_2 + O_2$$

but this dissociation is not appreciable below 2000° C, when it occurs to the extent of about 1%.

Electropositive metals react with water releasing hydrogen. For example, water reacts with the alkali metals violently, even in the cold; with calcium steadily at room temperature and vigorously when warmed; and with magnesium only above 100°C. Being less electropositive, iron reacts with steam at 700°C, and then only reversibly. Water also reacts

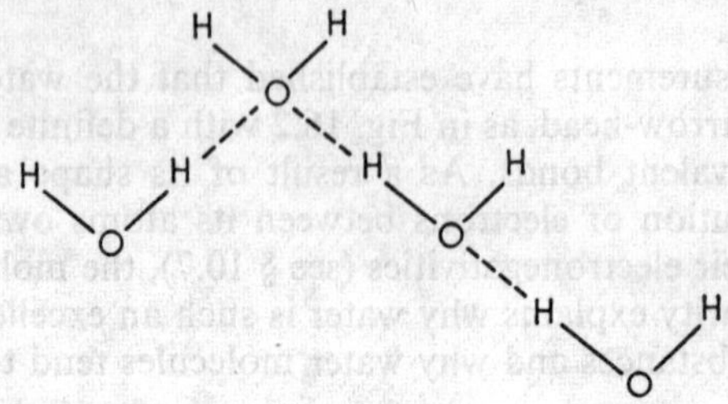

FIG. 16.3 Association of Water Molecules

with carbon at temperatures above 1000°C, giving a mixture of carbon monoxide and hydrogen called 'water gas' (§ 20.15). Equations for these reactions are:

$$2Na + 2H_2O = 2NaOH + H_2$$
$$Ca + 2H_2O = Ca(OH)_2 + H_2\uparrow$$
$$3Fe + 4H_2O \rightleftharpoons Fe_3O_4 + 4H_2$$
$$C + H_2O = CO + H_2$$

Reactions of water with compounds are so numerous and widespread that the general term *hydrolysis* is used to describe them. Many organic compounds such as esters, cyanides, halides, acid chlorides, and proteins combine with water giving important products; examples of these hydrolyses will be found in the appropriate chapters. In some cases these reactions are slow, and alkalis and acids are used as catalysts. Many inorganic compounds react with cold water readily, as the following examples show, liberating hydrides:

$$PCl_5 + 4H_2O = H_3PO_4 + 5HCl\uparrow$$
$$PCl_3 + 3H_2O = H_3PO_3 + 3HCl\uparrow$$
$$Mg_3N_2 + 6H_2O = 3Mg(OH)_2 + 2NH_3\uparrow$$
$$Ca_3P_2 + 6H_2O = 3Ca(OH)_2 + 2PH_3\uparrow$$
$$CaC_2 + 2H_2O = Ca(OH)_2 + C_2H_2\uparrow$$
$$Al_2S_3 + 6H_2O = 2Al(OH)_3\downarrow + 3H_2S\uparrow$$

Salts of weak acids and bases undergo hydrolysis in aqueous solution, giving alkaline and acidic solutions respectively. Examples of this are discussed and explained in chapter 9, where the electrolytic dissociation of water into ions is considered in detail.

Experiments have shown that substances which normally react together readily, and even violently, such as hot sodium and chlorine or hydrogen and chlorine, often fail to combine if they are previously intensively dried over a long period, This has led to the suggestion that water, present in the form of its vapour, catalyses these reactions.

16.5. Hardness of Water. Hard water is water which does not readily form a lather when shaken with soap solution. The hardness of a particular sample of water is a quantitative measure of its reluctance to form a lather under prescribed conditions.

Soaps are sodium and potassium salts of certain organic acids, such as stearic, oleic and palmitic acids (§ 38.5); when they dissolve in water they lower its surface tension and facilitate cleansing. Hardness is caused by dissolved metallic ions which react with the soap converting it to an insoluble curd or scum. Calcium and magnesium salts, particularly their sulphates and bicarbonates, are the commonest substances responsible.

e.g. $$\underset{\text{sodium stearate (a soap)}}{2C_{17}H_{35}COONa} + \underset{\text{(present in hard water)}}{CaSO_4} = Na_2SO_4 + \underset{\text{calcium stearate (scum)}}{(C_{17}H_{35}COO)_2Ca\downarrow}$$

This reaction wastes soap which would otherwise be available in solution for cleansing and it produces a dirty scum which clings to the article being washed. Hard water has the further disadvantage that it often deposits a scale or fur on boiling which blocks the circulation of water in pipes and wastes fuel by lining the inside of the boiler with a bad conductor of heat. One advantage of detergents over soaps is that, unlike the latter, they do not form insoluble products with the calcium and magnesium ions present in hard water and consequently their efficiency is not impaired.

If a sample of water containing the bicarbonates of calcium and magnesium is boiled, these salts are decomposed to the carbonates, which being insoluble are precipitated. Hardness removed in this way is referred to as *temporary*, whereas hardness not removable by boiling, such as that caused by dissolved sulphates, is referred to as *permanent*.

REMOVAL OF HARDNESS: Hardness can be removed in a number of ways, the choice depending upon convenience and relative cost, but in all the methods the underlying principle is the same, that of removing the calcium and magnesium ions from solution and thus preventing them from reacting with the soap. The following are the principal methods employed:

1. Distillation: water is vaporized and condensed again, leaving the non-volatile mineral salts in the boiler. This removes all the hardness, but is expensive.

2. *Addition of sodium carbonate:* this removes all the hardness by precipitating calcium and magnesium ions as their insoluble carbonates. Many soap powders and bath salts contain washing soda for this purpose

$$CaSO_4 + Na_2CO_3 = Na_2SO_4 + CaCO_3\downarrow$$

3. *Addition of calcium hydroxide in calculated amount:* this converts bicarbonates in the water to carbonates, which are precipitated. It is a cheap method, but it only removes temporary hardness and care must be taken not to add excess of lime which would itself cause additional hardness.

$$Ca(HCO_3)_2 + Ca(OH)_2 = 2CaCO_3\downarrow + 2H_2O.$$

4. *Permutit method:* water is passed over permutit, a complex aluminium silicate material containing sodium ions which it readily exchanges for the calcium and magnesium ions in the hard water. Periodically the silicate has to be reactivated by soaking it in concentrated brine, when the trapped calcium and magnesium ions are replaced by a fresh supply of sodium ions.

e.g. $\underset{\text{(in solution) causes hardness}}{Ca^{2+}} + \underset{\text{(from permutit)}}{2Na^+} \rightleftharpoons \underset{\text{(in solution) does not cause hardness}}{2Na^+} + \underset{\text{(trapped in permutit)}}{Ca^{2+}}$

5. *Ion exchange resins:* in recent years complex organic compounds with the property of removing dissolved ions from a sample of water have been synthesized from phenols, formaldehyde, and certain amines.* Some of these resins remove cations by exchanging them for hydrogen ions, whilst others remove dissolved anions. If both types are present then the final product is completely free of dissolved ions and not only shows no hardness but is comparable with distilled water in its purity.

6. *E.D.T.A. method:* When the sodium salt of ethylene diamine tetra-acetic acid (E.D.T.A.) is added to hard water, its complex anion forms very stable complexes with any calcium and magnesium ions present, thereby preventing them from reacting with soap to produce a scum:

$$\text{e.g.} \left[\begin{array}{l} CH_2.N.(CH_2.COO)_2 \\ | \\ CH_2.N.(CH_2.COO)_2 \end{array}\right]^{4-} + Ca^{2+} = \left[\begin{array}{ll} CH_2.N.(CH_2.COO)_2 & \\ | & Ca \\ CH_2.N.(CH_2.COO)_2 & \end{array}\right]^{2-}$$

This method of softening water is used industrially but the product is not potable.

7. *Phosphate method:* when sodium dihydrogen phosphate is heated under suitable conditions it is converted into a highly polymerized metaphosphate $(NaPO_3)_n$, where *n* is of the order of 150. Such a com-

* Further details are given in § 9.17.

pound, sold commercially as 'Calgon', has the property of removing calcium and magnesium ions from solution, probably by adsorption upon its colloidal anions.

ESTIMATION OF HARDNESS: This is best performed by titrating a given volume of the water with standard soap solution from a burette until a lather is obtained which persists for at least one minute after shaking. In comparing the hardness of two samples of water, allowance should be made for the amount of soap required to provide a lather by titrating an equal volume of distilled water with the soap solution and subtracting this volume of soap solution from the titres before comparing them.

16.6. Hydrogen Peroxide, H_2O_2.

LABORATORY PREPARATION: By the action of a dilute acid upon a metallic peroxide:

e.g. $$Na_2O_2 + H_2SO_4 = Na_2SO_4 + H_2O_2$$

The reactants should be kept ice-cold because hydrogen peroxide decomposes on warming. If barium peroxide and a calculated amount of dilute sulphuric acid are used, then the by-product is insoluble and can be removed by filtering, leaving a 10%–20% solution of hydrogen peroxide which can be concentrated by distilling under reduced pressure:

$$BaO_2 + H_2SO_4 = BaSO_4\downarrow + H_2O_2$$

Unfortunately commercial barium peroxide, made by heating barium oxide in air to 500°C, is unsuitable for this reaction because a coating of insoluble barium sulphate forms on its particles, stopping the reaction. It must first be converted to the hydrated form by dissolving it in dilute hydrochloric acid and treating the product with barium hydroxide solution, when the following changes occur:

$$BaO_2 + 2HCl = BaCl_2 + H_2O_2$$
$$Ba(OH)_2 + H_2O_2 + 6H_2O = BaO_2.8H_2O\downarrow$$

This hydrated peroxide can be filtered off and treated with dilute sulphuric acid, as described above.

MANUFACTURE: Two processes are in use:

1. *Electrolytic Method:* By electrolysing concentrated sulphuric acid at 5°C, using platinum electrodes and a high current density. Under these conditions persulphuric acid is formed at the anode and this slowly hydrolyses, reforming sulphuric acid and giving hydrogen peroxide solution. The latter is concentrated by distillation under reduced pressure. The changes are:

$$2H_2SO_4 \rightleftharpoons \underset{\text{to anode}}{2HSO_4^-} + \underset{\text{to cathode}}{2H^+}$$

At the anode: $\begin{cases} 2HSO_4^- \rightarrow H_2S_2O_8 + 2e \\ \text{followed by} \\ H_2S_2O_8 + 2H_2O = 2H_2SO_4 + H_2O_2. \end{cases}$

At the cathode: $2H^+ + 2e \rightarrow 2H \rightarrow H_2\uparrow$ (by-product).

2. *Autoxidation Method:* A derivative of the organic compound anthraquinone is reduced with gaseous hydrogen by using finely-divided palladium as a catalyst. When air is bubbled through the hydroxyl compound produced in this way, it is oxidized back again to the original substance (which can be used over again), leaving a dilute solution of hydrogen peroxide as by-product. This solution is then concentrated by vacuum distillation.

PROPERTIES OF HYDROGEN PEROXIDE: Anhydrous hydrogen peroxide is a colourless liquid, ρ 1.45 g cm^{-3}, boiling at 70°C under 4×10^3 N m^{-2} pressure, and freezing at −1°C. It is rarely encountered in the laboratory because it is too dangerous to handle, being unstable and highly reactive, and dilute solutions of hydrogen peroxide are commonly used.

REACTIONS OF HYDROGEN PEROXIDE: Hydrogen peroxide, even in dilute solution, decomposes readily on warming to give water and oxygen:

$$2H_2O_2 = 2H_2O + O_2\uparrow$$

This decomposition, which is exothermic, is accelerated by alkalis, rough surfaces, finely divided metals, and by certain solids, particularly manganese dioxide. It is retarded by the presence of many organic substances such as glycerol, urea, and alcohol, which act as negative catalysts, and small amounts of these substances are often added to commercial solutions to preserve them. For the same reason concentrated solutions of hydrogen peroxide are sometimes stored in bottles lined with a layer of paraffin wax.

Hydrogen peroxide is a useful oxidizing agent. Essentially it reacts in this way:

$$H_2O_2 = H_2O + \bar{O} \text{ (where } \bar{O} \text{ is an atom of available oxygen).}$$

Examples of its oxidizing action are:

(*a*) it liberates iodine from acidified potassium iodide solution:

$$H_2O_2 + 2KI + 2HCl = I_2 + 2KCl + 2H_2O$$

(*b*) it converts ferrous salts to ferric salts:

$$2FeSO_4 + H_2O_2 + H_2SO_4 = Fe_2(SO_4)_3 + 2H_2O$$

(*c*) it converts sulphites to sulphates:

$$Na_2SO_3 + H_2O_2 = Na_2SO_4 + H_2O$$

(*d*) it converts black lead sulphide to white lead sulphate:

$$PbS + 4H_2O_2 = PbSO_4 + 4H_2O.$$

This last reaction has been used to restore oil paintings which have darkened by exposure to hydrogen sulphide in the atmosphere, the original lead pigments having changed to lead sulphide.

When in contact with substances which readily lose oxygen, hydrogen peroxide acts as a reducing agent. The fundamental reaction here is:

$$H_2O_2 + \bar{O} = H_2O + O_2\uparrow$$

Examples of its reducing action are:

(*a*) it evolves oxygen with acidified potassium permanganate solution:

$$2KMnO_4 + 3H_2SO_4 + 5H_2O_2 = K_2SO_4 + 2MnSO_4 + 8H_2O + 5O_2\uparrow$$

(*b*) it reduces silver oxide to the metal:

$$Ag_2O + H_2O_2 = 2Ag + H_2O + O_2\uparrow$$

Hydrogen peroxide is a very weak acid; its solution turns litmus red and liberates carbon dioxide from sodium carbonate:

$$Na_2CO_3 + H_2O_2 = Na_2O_2 + H_2O + CO_2\uparrow$$

DETECTION OF HYDROGEN PEROXIDE: Two specific tests are used in addition to its general oxidizing action:

1. A little ether is added to some potassium dichromate solution which has been acidified with dilute sulphuric acid. When hydrogen peroxide solution is added a deep blue coloration appears in the upper ether layer due to unstable perchromic acid, which slowly decomposes to oxygen and green chromic sulphate solution. This is a sensitive test.

2. If hydrogen peroxide solution is added to a very dilute solution of titanium sulphate, an intense yellow-orange colour results.

ESTIMATION OF HYDROGEN PEROXIDE: In view of its liability to decompose in storage, it is important to be able to estimate a solution of hydrogen peroxide before use. Two methods are available:

1. By titration against standard potassium permanganate solution in the presence of dilute sulphuric acid. No indicator is needed because the permanganate solution is run into the peroxide solution until a pink colour just persists on shaking.

2. By adding a known volume of hydrogen peroxide solution to excess of acidified potassium iodide solution and determining the iodine liberated by using a standard solution of sodium thiosulphate.

This method is preferable where an organic preservative has been added to the peroxide solution.

STRENGTH OF HYDROGEN PEROXIDE SOLUTIONS: This is usually expressed as the number of volumes of oxygen gas at s.t.p. which could be obtained by the complete decomposition of the hydrogen peroxide in unit volume of the solution. Thus a '10-volume solution', which is a strength frequently encountered in the laboratory, would yield ten times its own volume of oxygen gas when completely decomposed. An example will show how to calculate the strength of such a solution in terms of percentage by mass:

$$\underset{2(2\ +\ 32)}{2H_2O_2} = 2H_2O + O_2$$

$$68 \text{ gramme} \equiv 1 \text{ mole of oxygen.}$$

Now 1 mole of any gas occupies 22.4 dm^3 at s.t.p. So 68 gramme of hydrogen peroxide are equivalent to 22.4 dm^3 of oxygen at s.t.p. But 22.4 dm^3 of oxygen can be liberated from 2.24 dm^3 of a 10-volume solution, by definition. ∴ 2.24 dm^3 of such a solution contain 68 gramme of hydrogen peroxide, i.e. a 10-volume solution contains

$$\frac{68}{2.24} = 30.4 \text{ gramme per cubic decimetre of hydrogen peroxide.}$$

Thus a 10-volume solution is about 3% hydrogen peroxide by mass.

USES OF HYDROGEN PEROXIDE: Its dilute solutions are extensively used as a bleaching agent, particularly for delicate materials such as wool, silk, ivory, teeth and hair, for which chlorine is unsuitable, and also as a disinfectant. It is used in concentrated solution or in the anhydrous form, as a fuel in submarines and as a propellent in rockets and guided missiles. Use is made either of its highly exothermic decomposition or of its violent oxidizing action on certain organic compounds. Hydrogen peroxide is also employed in the manufacture of organic peroxy compounds, and in the laboratory as an oxidizing agent.

STRUCTURE OF HYDROGEN PEROXIDE: The molecule has the structure H—O—O—H, since it gives rise to crystalline derivatives in which X-ray analysis reveals the O—O linkage. Physical measurements, such as dipole moment and infra-red absorption spectrum, confirm this and show that the shape of the molecule resembles that of a partly open book, the O—O bond coinciding with the binding and the O—H bonds lying in the planes of the pages, as shown in Fig. 16.4. All true peroxides contain the O—O bond and give hydrogen peroxide solution on treatment with dilute acids. Substances which do not do this, such as lead dioxide and nitrogen tetroxide, should not be called peroxides at all.

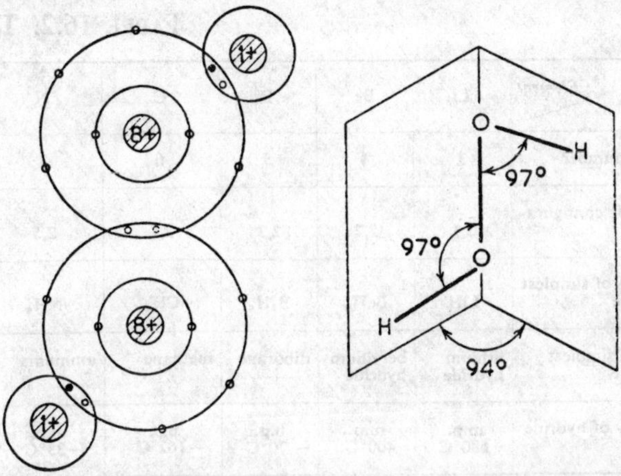

FIG. 16.4 The Structure of Hydrogen Peroxide

16.7. Hydrides. Many elements form compounds with hydrogen known as *hydrides*. There are five main types, depending upon their structure.

The alkali and alkaline earth metals form *ionic hydrides* which contain H^- ions; these are crystalline solids which react readily with water giving gaseous hydrogen and which conduct electricity when molten, releasing hydrogen gas at the anode.

e.g. $$CaH_2 + 2H_2O = Ca(OH)_2 + H_2 \uparrow$$

Most non-metallic elements form *covalent hydrides*; these are volatile compounds, gaseous at room temperature except where they are extensively associated by hydrogen bonding as in hydrogen fluoride and water (see § 14.7). In dissolving in water some of these hydrides react to give basic (e.g. ammonia) or acidic (e.g. hydrogen chloride or hydrogen bromide) solutions which conduct electricity and are ionic in character.

e.g. $$HCl + H_2O \rightleftharpoons H_3O^+ + Cl^-$$

Many transition metals absorb hydrogen readily when heated in the gas forming *interstitial hydrides,* so called because the hydrogen is believed to occupy the interstices or gaps in between the relatively large atoms of metal in the lattice. Such hydrides are usually non-stoichiometric compounds (§ 14.10).

A few elements such as boron and aluminium can form *electron-deficient hydrides* in which there are special linkages of a bridging kind to supplement the covalent bonds. Such hydrides react with water giving gaseous hydrogen and are strong reducing agents.

TABLE 16.2. THE HYDRIDES

Property \ Element	Li	Be	B	C	N	O	F
Atomic number	3	4	5	6	7	8	9
Electronic configuration	2,1	2,2	2,3	2,4	2,5	2,6	2,7
Formula of simplest hydride	LiH	BeH_2	B_2H_6	CH_4	NH_3	H_2O	HF
Name of simplest hydride	lithium hydride	beryllium hydride	diborane	methane	ammonia	water	hydrog fluorid
Volatility of hydride	m.p. 680°C	m.p. 400°C	b.p. − 70°C	b.p. −162°C	b.p. −33°C	b.p. 100 °C	b.p. 19°C
Structure of hydride	ionic $Li^+ H^-$	ionic $Be^{2+} 2H^-$	covalent	covalent H | H—C—H | H	covalent H H H \ | / N	covalent H H \ / O	covale H—
Nature of molecule	electrolyte when molten, giving $H_2\uparrow$ at anode	electrolyte when molten, giving $H_2\uparrow$ at anode	electron-deficient	tetrahedral bonding; no polarity	pyramidal bonding; polar molecule forming hydrogen bonds	arrowhead bonding; very polar molecule forming hydrogen bonds	highly polar molecu forming hydrog bonds readily
Reaction of hydride with water	reacts readily, giving $H_2\uparrow$	reacts readily, giving $H_2\uparrow$	reacts readily, giving $H_2\uparrow$	slightly soluble, giving neutral solution	very soluble, giving basic solution	—	very soluble giving acidic solutior
References	§ 17.2 § 17.3			§ 33.2	§ 22.4	§ 16.4	§ 25.3

Boron and aluminium also form *complex hydrides* which are crystalline solids containing the anions BH_4^- and AlH_4^- linked to positive metallic ions. The two commonest, sodium borohydride $NaBH_4$ and lithium aluminium hydride $LiAlH_4$, are important reducing agents and are described in §§ 19.7 and 19.14 respectively.

A general comparison of the simplest hydrides of the fifteen elements lithium—chlorine appears in Table 16.2, which clearly illustrates the periodic nature of their properties.

HE ELEMENTS Li—Cl

Ne	Na	Mg	Al	Si	P	S	Cl
10	11	12	13	14	15	16	17
2,8	2,8,1	2,8,2	2,8,3	2,8,4	2,8,5	2,8,6	2,8,7
—	NaH	MgH_2	$(AlH_3)n$	SiH_4	PH_3	H_2S	HCl
—	sodium hydride	magnesium hydride	aluminium hydride	silane	phosphine	hydrogen sulphide	hydrogen chloride
—	m.p. 600°C	m.p. 470°C	decomposes when heated	b.p. −112°C	b.p. −87°C	b.p. −60°C	b.p. −85°C
—	ionic $Na^+ H^-$	ionic $Mg^{2+} 2H^-$	covalent	covalent H | H—Si—H | H	covalent H H H \ | / P	covalent H H \ / S	covalent H—Cl
—	electrolyte when molten, giving H_2↑ at anode	electrolyte when molten, giving H_2↑ at anode	electron-deficient	tetrahedral bonding; no polarity	pyramidal bonding; polar molecular	arrowhead bonding; polar molecular	highly polar molecule forming hydrogen bonds
—	reacts readily, giving H_2↑	reacts readily, giving H_2↑	reacts readily, giving H_2↑	soluble	soluble giving neutral solution	slightly soluble, giving weakly acidic solution	very soluble, giving strongly acidic solution
	§ 17.4		§ 19.13	§ 20.18	§ 23.3	§ 24.6	§ 25.6

17

The Alkali Metals

17.1. Introduction. The six elements lithium, sodium, potassium, rubidium, caesium and francium, known collectively as the alkali metals, occur together in Group IA of the Periodic Classification. Only sodium and potassium will be studied in detail, as these are by far the most important and are in many respects typical of the others.

17.2. Lithium, Li. Electronic configuration: 2, 1.

OCCURRENCE: The element does not occur in the free state, but it is found in certain minerals of which the complex silicates lepidolite and spodumene and the fluophosphate amblygonite are the most important.

EXTRACTION: By electrolysis of a molten mixture of lithium chloride and potassium chloride at 400°C using a graphite anode and a steel cathode. The cell resembles that used for making sodium, which is shown in Fig. 17.1.

PROPERTIES OF LITHIUM: It is a soft, lustrous, white metal which tarnishes rapidly in damp air. It melts at 180°C and boils at 1330°C. Its density is 0.53 g cm^{-3}.

REACTIONS OF LITHIUM: When melted in air or oxygen it catches fire and burns with a white flame forming the oxide with only a trace of peroxide. The metal decomposes water readily releasing hydrogen:

$$2Li + 2H_2O = 2LiOH + H_2\uparrow$$

Lithium is less electropositive than sodium or potassium although it stands higher in the electrochemical series (see § 10.7). It combines readily with hydrogen above 450°C forming a stable hydride and with nitrogen forming a nitride:

$$2Li + H_2 = 2LiH$$
$$6Li + N_2 = 2Li_3N$$

The molten metal attacks glass readily and lithium must not be heated, therefore, in glass vessels.

USES OF LITHIUM: The metal is increasingly used in lead alloys for making bearings and in aluminium alloys; even small proportions of lithium have a marked hardening effect. It is also used for removing traces of dissolved gases from molten copper and aluminium and for making lithium compounds especially the hydride.

17.3. Compounds of Lithium. In general these resemble the corresponding sodium and potassium compounds, although a few notable differences exist. For example, lithium chloride is extremely deliquescent and more soluble in water than sodium or potassium chloride. Again, the fluoride, carbonate, phosphate, and oxalate of lithium are only sparingly soluble in water and are precipitated from solutions in which they are formed:

$$LiCl + KF = LiF\downarrow + KCl$$
$$2LiCl + Na_2CO_3 = Li_2CO_3\downarrow + 2NaCl$$
$$3LiCl + Na_3PO_4 = Li_3PO_4\downarrow + 3NaCl$$

Unlike the hydroxides, carbonates, and nitrates of the other alkal metals, those of lithium are decomposed to the oxide on moderate heating in hydrogen:

$$2LiOH = Li_2O + H_2O$$
$$Li_2CO_3 = Li_2O + CO_2\uparrow$$

Lithium oxide dissolves slowly in water giving the hydroxide, which dissociates in solution thus:

$$LiOH \rightleftharpoons Li^+ + OH^-$$

Lithium hydride is a stable white solid, m.p. 680°C, which releases hydrogen at the *anode* when melted and electrolysed:

$$LiH \rightleftharpoons Li^+ + H^-$$

It reacts vigorously with water liberating hydrogen:

$$LiH + H_2O = LiOH + H_2\uparrow$$

If lithium hydride is mixed with anhydrous aluminium chloride in dry ethereal solution, the important reducing agent lithium aluminium hydride is formed (see § 19.13):

$$4LiH + AlCl_3 = LiAlH_4 + 3LiCl\downarrow$$

USES OF LITHIUM COMPOUNDS: The chloride is used as a dehumidifier in air conditioning and the fluoride as a flux in welding. Other lithium salts are used in making ceramics, fireworks, and lubricants, and for treating gout.

17.4. Sodium, Na. Electronic Configuration 2, 8, 1.

OCCURRENCE: The metal does not occur naturally, being much too reactive, but it is abundant in compounds, chiefly as sodium chloride,

NaCl, but also as sodium sesqui-carbonate, $Na_2CO_3.NaHCO_3.2H_2O$, sodium nitrate $NaNO_3$ (§ 17.11), sodium sulphate $Na_2SO_4.10H_2O$, sodium borate $Na_2B_4O_7.10H_2O$ (§ 19.4), sodium iodate $NaIO_3$ (§ 25.13), cryolite $3NaF.AlF_3$ (§19.8), and in complex silicates.

EXTRACTION: Sodium is so electropositive that the chemical reduction of its compounds is not practicable, and an electrolytic method involving fused salts is necessary. The cell used is shown diagrammatically in Fig. 17.1. The electrodes are shaped to bring large areas close

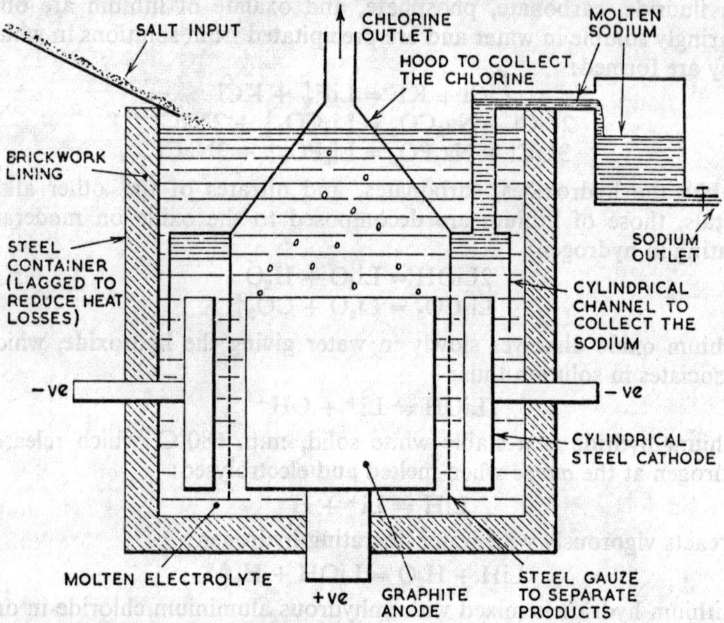

FIG. 17.1 The Downs Cell for Making Sodium

together, thereby lowering the resistance of the cell. The sodium chloride electrolyte is kept molten by the heat from the current.

At the cathode: $Na^+ + e \rightarrow Na$

Molten sodium, guided by the gauze, rises through the electrolyte and collects in the container.

At the anode: $Cl^- \rightarrow Cl + e$

$$2Cl \rightarrow Cl_2\uparrow$$

The electrolysis of molten sodium chloride presents considerable technical difficulties involving corrosion, evaporation of the sodium (its vapour pressure is nearly half atmospheric at 820°C), and rapid diffusion of the molten metal into the electrolyte. These difficulties have been overcome by adding substances such as calcium chloride to the electrolyte to lower its melting point (§ 5.13) so that the process can be operated at about 600°C. As a result this process has completely superseded the older one of electrolysing molten sodium hydroxide.

PROPERTIES OF SODIUM: It is a soft, silvery metal with m.p. 97.8°C, b.p. 880°C, and ρ 0.97 g cm^{-3}. It is lustrous when freshly cut, but it tarnishes rapidly on exposure to air. To prevent its oxidation in this way, sodium is usually stored under paraffin or solvent naphtha. It is a good conductor of heat and electricity. Its vapour is monatomic.

REACTIONS OF SODIUM: It is highly electropositive and very reactive. When heated in air it burns with a brilliant yellow flame forming a mixture of oxide and peroxide, in which the latter predominates:

$$4Na + O_2 = 2Na_2O$$
$$2Na + O_2 = Na_2O_2$$

Sodium combines directly when heated with halogens, sulphur, and phosphorus. With hydrogen above 300°C it forms a solid hydride with the structure Na^+H^-; this gives off hydrogen on contact with cold water:

$$2Na + H_2 = 2NaH$$
$$NaH + H_2O = NaOH + H_2\uparrow$$

When added to water, sodium reacts vigorously, liberating hydrogen. The heat of the reaction melts the metal to a globule which moves about on the surface:

$$2Na + 2H_2O = 2NaOH + H_2\uparrow$$

Sodium decomposes acids explosively, but with anhydrous ethyl alcohol it reacts slowly at room temperature, releasing a steady stream of hydrogen and forming a solution of sodium ethoxide or ethylate:

$$2Na + 2C_2H_5OH = 2C_2H_5ONa + H_2\uparrow$$

Sodium reduces many metallic halides and oxides to the metals on heating:

e.g.
$$AlCl_3 + 3Na = Al + 3NaCl$$
$$TiCl_4 + 4Na = Ti + 4NaCl$$

If a stream of dry ammonia gas is passed over sodium at 300–400°C, sodamide results:

$$2Na + 2NH_3 = 2NaNH_2 + H_2$$

With mercury sodium forms an amalgam, which is liquid if the proportion of sodium is below 1 %. This amalgam reacts quietly with cold water liberating hydrogen.

USES OF SODIUM: It is used to manufacture sodium peroxide, sodium cyanide, and the alloy, Na_4Pb, used to make lead tetraethyl, the petrol additive (§ 40.5). Liquid sodium is being used as a coolant in nuclear reactors to remove heat from the core (§ 31.11); because of its low melting point, high thermal conductivity and high specific heat capacity. Sodium and alcohol, and sodium amalgam and water are useful reducing agents, particularly in organic chemistry. Small amounts of sodium are used in vapour discharge lamps for street lighting, and for organic syntheses, and in the extraction of titanium (§ 27.3).

17.5 Sodium Monoxide, Na_2O. Made by heating sodium peroxide gently with excess sodium, or by burning sodium in a limited supply of air:

$$Na_2O_2 + 2Na = 2Na_2O$$

It is a colourless solid with strongly basic properties. It reacts violently with water giving sodium hydroxide solution and explosively with acids forming salts:

$$Na_2O + H_2O = 2NaOH$$

17.6. Sodium Peroxide, Na_2O_2. Made by burning sodium in excess of dry air or oxygen. It is a pale yellow solid which is slowly converted into sodium hydroxide and sodium carbonate on exposure to air, and for this reason samples should be stored in air-tight containers.

It reacts readily with excess of cold water, forming a solution which contains sodium hydroxide and hydrogen peroxide, and is used for bleaching:

$$Na_2O_2 + 2H_2O = 2NaOH + H_2O_2$$

This reaction is exothermic and if the temperature is allowed to rise either by the addition of large amounts of the peroxide or by external heating, the hydrogen peroxide decomposes releasing oxygen; the overall reaction then becomes:

$$2Na_2O_2 + 2H_2O = 4NaOH + O_2\uparrow$$

With dilute acids in the cold, sodium peroxide gives solutions of hydrogen peroxide:

e.g. $$Na_2O_2 + H_2SO_4 = Na_2SO_4 + H_2O_2$$

Sodium peroxide reacts with carbon dioxide, forming sodium carbonate:

$$2Na_2O_2 + 2CO_2 = 2Na_2CO_3 + O_2\uparrow$$

Use is made of this reaction in air purifiers in, for example, submarines and portable breathing apparatus.

Sodium peroxide is a powerful oxidizing agent. It reacts vigorously with many organic compounds, often igniting them, so care should be taken when using sodium peroxide not to leave any in contact with filter-paper, cotton-wool, wood or similar materials. Sodium peroxide oxidizes most metals on heating, often doing so explosively if the metals are powdered. It also converts the oxides of transition metals into the corresponding oxy-salts, e.g. chromic oxide into sodium chromate.

17.7. Sodium Hydroxide (Caustic Soda), NaOH.

MANUFACTURE FROM SODIUM CARBONATE: In this method, called Gossage's process, slaked lime and dilute sodium carbonate solution (made cheaply by the Solvay process, see § 17.8) are heated together in large tanks, when the following equilibrium is set up:

$$Ca(OH)_2 + Na_2CO_3 \rightleftharpoons CaCO_3\downarrow + 2NaOH$$

Although theoretically reversible, in practice most of the sodium carbonate is converted into sodium hydroxide because the calcium carbonate, being insoluble, is precipitated from the reaction mixture as fast as it is formed, thereby preventing the back reaction. The sodium hydroxide solution is decanted from the sludge and concentrated or evaporated to dryness. It contains a small proportion of unchanged sodium carbonate, but this is not objectionable for most commercial uses.

FIG. 17.2 Simple Brine Cell

MANUFACTURE FROM SODIUM CHLORIDE: When a solution of sodium chloride is electrolysed using the cell shown in Fig. 17.2, the following changes occur:

$$NaCl \rightleftharpoons Na^+ + Cl^-$$
$$H_2O \rightleftharpoons H^+ + OH^-$$

At the anode, Cl^- ions are discharged and chlorine gas is liberated. At the cathode, H^+ ions are discharged, being less electropositive than Na^+ ions, and hydrogen gas is liberated. More water molecules then dissociate to maintain the ionic product $[H^+][OH^-]$ at its constant value (§ 9.4). Thus the effect of continued electrolysis is the removal of Cl^- and H^+ ions and the gradual accumulation of Na^+ and OH^- ions, producing sodium hydroxide solution. After a while, however, the chlorine gas which dissolves in the brine around the anode diffuses through the solution and reacts with the sodium hydroxide produced forming sodium chloride and either sodium hypochlorite or chlorate, according to the temperature (§ 25.5). For this reason a simple cell of this type is not suitable for making sodium hydroxide and two special types of cell are used in which the products are kept apart:

(*a*) *Mercury Cathode Cell:* See Fig. 17.3. Chlorine gas is liberated at the graphite anode, as before, and the sale of this by-product reduces the cost of the process. At the cathode, which consists of a thin layer of mercury flowing across the sloping floor of the cell, H^+ ions are not dis-

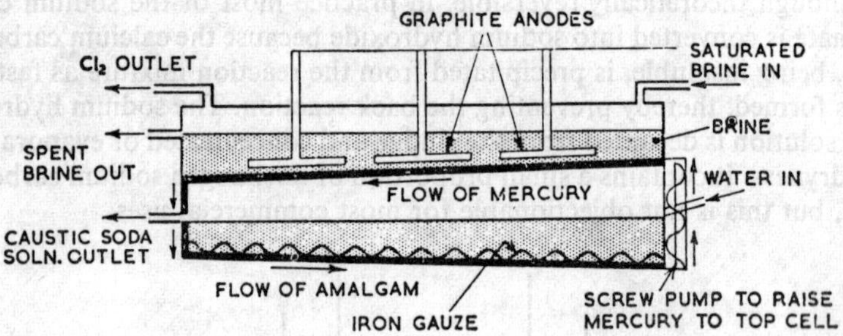

FIG. 17.3 The Mercury Cathode Cell for Making Sodium Hydroxide

charged, as in the simple cell, because mercury shows great reluctance to discharge H^+ ions from a solution—it is said to have a high overpotential (§ 10.5). Instead Na^+ ions are discharged and form a liquid amalgam which flows out of the cell into a second tank full of water. Here, in the presence of iron, which does not have a high overpotential, H^+ ions are discharged, giving hydrogen gas as a useful by-product, and sodium goes into solution as ions:

$$2H^+ + 2Na/Hg = 2Na^+ + 2Hg + H_2\uparrow$$

As a result Na^+ and OH^- ions accumulate in the tank; the sodium hydroxide solution can be concentrated or, if required, evaporated to dryness. The product is more expensive than that of other methods, mainly due to unavoidable losses of the costly mercury, but it is purer, being completely free of carbonate and chloride.

(*b*) *Diaphragm Cell:* See Fig. 17.4. As before, chlorine liberated at the anode is a useful by-product. The positive ions in the electrolyte are drawn through the diaphragm to the steel cathode, where H^+ ions are discharged forming hydrogen gas, another by-product. More water molecules dissociate, providing an accumulation of Na^+ and OH^- ions. The solution slowly percolates through the porous diaphragm and drips to the bottom of the container, from which it is drawn off periodically. It contains about 10% caustic soda and about 20% sodium

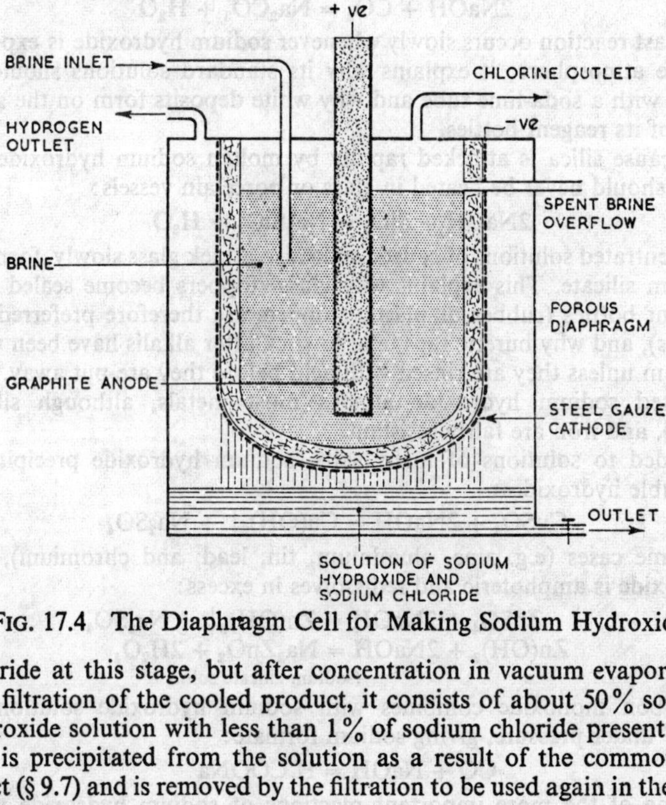

FIG. 17.4 The Diaphragm Cell for Making Sodium Hydroxide

chloride at this stage, but after concentration in vacuum evaporators and filtration of the cooled product, it consists of about 50% sodium hydroxide solution with less than 1% of sodium chloride present. The salt is precipitated from the solution as a result of the common-ion effect (§ 9.7) and is removed by the filtration to be used again in the cell.

PROPERTIES OF SODIUM HYDROXIDE: It is a white, deliquescent solid, m.p. 318°C. It is usually supplied to laboratories in the form of pellets or sticks, but to most industrial users as a 50% solution. It is very soluble in water, dissolving exothermically, to give a strongly alkaline solution with caustic properties (it is dangerous to handle the solid or its concentrated solutions).

REACTIONS OF SODIUM HYDROXIDE: It is a strong electrolyte completely dissociated into its ions in solution:

$$NaOH \rightleftharpoons Na^+ + OH^-$$

It neutralizes all acids, however weak, and converts amphoteric and acidic oxides to sodium salts:

e.g. $$2NaOH + Al_2O_3 = 2NaAlO_2 + H_2O$$

(sodium aluminate soln.)

$$2NaOH + SO_2 = Na_2SO_3 + H_2O$$

$$2NaOH + CO_2 = Na_2CO_3 + H_2O$$

The last reaction occurs slowly whenever sodium hydroxide is exposed to the atmosphere; it explains why its standard solutions should be fitted with a soda-lime tube and why white deposits form on the stoppers of its reagent bottles.

Because silica is attacked rapidly by molten sodium hydroxide the solid should never be heated in glass or porcelain vessels:

$$2NaOH + SiO_2 = Na_2SiO_3 + H_2O$$

Concentrated solutions of caustic soda also attack glass slowly, forming sodium silicate. This explains why glass stoppers become sealed into reagent bottles (rubber or plastic stoppers are therefore preferred for alkalis), and why burette taps tend to stick after alkalis have been used in them unless they are rinsed with acid before they are put away.

Fused sodium hydroxide attacks most metals, although silver, nickel, and iron are fairly resistant.

Added to solutions of many salts, sodium hydroxide precipitates insoluble hydroxides:

e.g. $$CuSO_4 + 2NaOH = Cu(OH)_2\downarrow + Na_2SO_4$$

In some cases (e.g. zinc, aluminium, tin, lead, and chromium), the hydroxide is amphoteric and redissolves in excess:

e.g. $$ZnSO_4 + 2NaOH = Zn(OH)_2\downarrow + Na_2SO_4$$

$$Zn(OH)_2 + 2NaOH = Na_2ZnO_2 + 2H_2O$$

(sodium zincate soln.)

Carbon monoxide combines with sodium hydroxide solution at 150°C under pressure, giving sodium formate:

$$CO + NaOH = H.COONa$$

Some of the more important reactions of sodium hydroxide with elements are listed below; details are given in the appropriate chapters:

$$4P + 3NaOH + 3H_2O = 3NaH_2PO_2 + PH_3\uparrow$$

$$Cl_2 + 2NaOH = NaCl + NaClO + H_2O \text{ (in the cold)}$$

$$3Cl_2 + 6NaOH = 5NaCl + NaClO_3 + 3H_2O \text{ (when heated)}$$

$$4S + 6NaOH = 2Na_2S + Na_2S_2O_3 + 3H_2O$$

$$Zn + 2NaOH = Na_2ZnO_2 + H_2\uparrow$$

$$2Al + 2NaOH + 2H_2O = 2NaAlO_2 + 3H_2\uparrow$$

Heated with ammonium salts, caustic soda solution liberates ammonia, and similarly gives phosphine with phosphonium iodide:

e.g. $$NH_4Cl + NaOH = NaCl + NH_3\uparrow + H_2O.$$

USES OF SODIUM HYDROXIDE: It is a chemical of great industrial importance, with many uses. Huge amounts are used in the manufacture of soap (§ 38.5), in paper-making, in the purification of bauxite before extracting aluminium (§ 19.8), and in the manufacture of rayon (artificial silk) from cellulose (§ 42.7). The textile industry also uses sodium hydroxide for bleaching, and for treating cotton to improve its appearance and to facilitate dyeing. It is also used to make sodium formate and other sodium compounds, and in the refining of petroleum and oils. In the laboratory it is used as a strong alkali and as an analytical reagent, but soda-lime, made by slaking quicklime with sodium hydroxide solution, is often preferred for decarboxylation (§ 33.2) or for absorbing carbon dioxide, because it is less easily fused and is not deliquescent.

17.8. Sodium Carbonate, Na_2CO_3.

MANUFACTURE BY THE SOLVAY OR AMMONIA-SODA PROCESS: This is carried out in a tall tower, shown diagrammatically in Fig. 17.5, down which flows brine saturated with ammonia. Carbon dioxide gas is forced up the tower, which is constructed to ensure thorough mixing of gas and liquid. The following reactions occur:

$$NH_3 + CO_2 + H_2O \rightleftharpoons NH_4HCO_3$$
$$NH_4HCO_3 + NaCl \rightleftharpoons NH_4Cl + NaHCO_3\downarrow$$

Although reversible, the second reaction proceeds continually to the right because the sodium hydrogen carbonate, being almost insoluble in cold brine by the common-ion effect (§ 9.7) is precipitated and thus removed from the equilibrium. The reactions are exothermic, so cooling is necessary at the base of the tower. The suspended hydrogen carbonate is filtered off, washed, and heated strongly:

$$2NaHCO_3 = Na_2CO_3 + H_2O + CO_2\uparrow$$

The carbon dioxide so released is used again in the tower, supplemented by a fresh supply obtained by heating limestone to about 1000°C. The quicklime produced is slaked and heated with the ammonium chloride solution to recover ammonia, which is used again:

$$CaCO_3 = CaO + CO_2\uparrow$$
$$CaO + H_2O = Ca(OH)_2$$
$$2NH_4Cl + Ca(OH)_2 = CaCl_2 + 2NH_3\uparrow + 2H_2O.$$

The raw materials (brine, coal, limestone, and a little ammonia) are

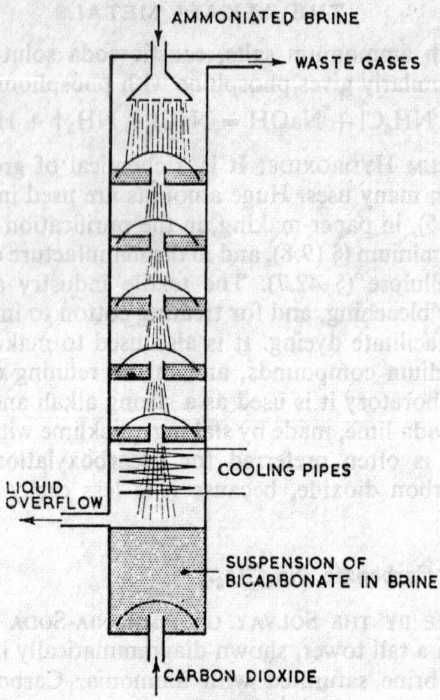

Fig. 17.5 The Solvay Tower

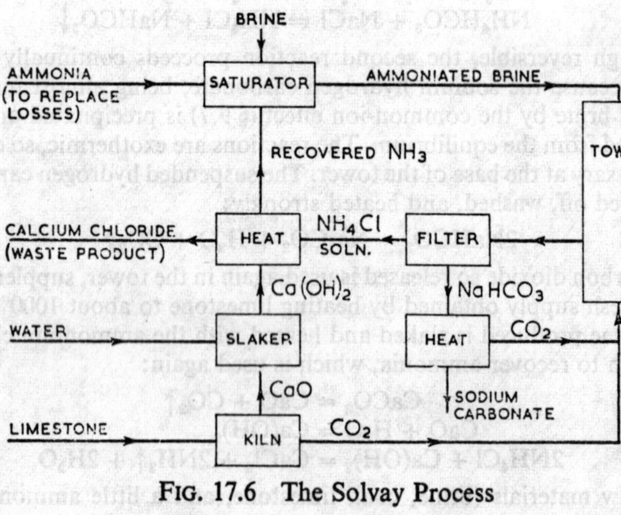

Fig. 17.6 The Solvay Process

plentiful and cheap, and calcium chloride is the only wasted product. The whole process is portrayed diagrammatically in Fig. 17.6.

PROPERTIES OF SODIUM CARBONATE: It exists in several forms, as shown in Table 17.1.

All these forms have the same chemical properties. They dissolve in water giving a solution which is alkaline by hydrolysis, as explained in

TABLE 17.1. FORMS OF SODIUM CARBONATE

Formula	Na_2CO_3	$Na_2CO_3.H_2O$	$Na_2CO_3.7H_2O$	$Na_2CO_3.10H_2O$
Chemical name	anhydrous	monohydrate	heptahydrate	decahydrate
Common name	soda ash	crystal carbonate	—	washing soda
Appearance	white powder	colourless glassy crystals		
Stability in air	hygroscopic	stable	effloresces	effloresces
Preparation	by heating hydrates or bicarbonate strongly	crystallizes from solns above 35.3°C	crystallizes from solns. between 32°C and 35.3°C	crystallizes from solns. below 32° C

§ 9.6. This solution precipitates carbonates, basic carbonates or hydroxides from solutions of most metallic salts, and gives carbon dioxide with acids.

Anhydrous sodium carbonate melts at 850°C without undergoing any appreciable decomposition. A 50% mixture with potassium carbonate melts at about 700°C and is used in analysis, under the name of 'fusion mixture', to convert very insoluble salts into carbonates and oxides which can be dissolved in acids.

USES OF SODIUM CARBONATE: Large amounts are needed for the manufacture of glass, sodium hydroxide (by the Gossage process), borax, sodium silicate (§ 20.20), sodium phosphate, and paints. It is also used in paper-making, and as a water-softener (§ 16.5). In the laboratory it is used to standardize acids and as an analytical reagent.

17.9. Sodium Hydrogen Carbonate, Sodium Bicarbonate, $NaHCO_3$. Manufactured by the Solvay process, described above. It can be prepared by saturating cold sodium carbonate solution with carbon dioxide.

It is a white solid, only sparingly soluble in water at room temperature. Its solution is mildly alkaline by hydrolysis, appearing alkaline to methyl orange but acidic to phenolphthalein. If sodium hydrogen carbonate or its solution is heated, it decomposes thus:

$$2NaHCO_3 = Na_2CO_3 + CO_2\uparrow + H_2O$$

If equimolecular solutions of sodium carbonate and bicarbonate are mixed and allowed to crystallize above 35°C, white crystals of sodium sesqui-carbonate, $Na_2CO_3.NaHCO_3.2H_2O$ appear. This compound is a useful volumetric standard, being neither deliquescent nor efflorescent. Its solution shows the reactions of both constituents.

Sodium bicarbonate is used in baking powder, and in proprietary digestive powders to correct excessive acidity of the stomach.

17.10. Sodium Chloride, NaCl. It occurs naturally as rock-salt, and as a saturated (i.e. 25%) solution called brine, and in sea-water, which contains about 2·5% salt. Although some mines are still worked for rock-salt, the bulk of the sodium chloride used industrially is obtained from brine, which is either supplied direct as the liquid, or is evaporated first to the solid. Pure sodium chloride can be obtained as a precipitate by bubbling hydrogen chloride through a saturated impure solution, making use of the common-ion effect (§ 9.7), but recrystallization cannot be used because the solubility of sodium chloride increases only slightly with rising temperature (see Fig. 17.8 and §4.16).

It is a white crystalline solid, melting at 800°C and boiling at 1420°C. X-ray analysis shows that it consists of a three-dimensional network of Na^+ and Cl^- ions (§ 4.6). It is soluble in water, giving a conducting solution which displays all the usual reactions of chlorides (§ 25.7).

As shown in Table 17.2, sodium chloride, particularly as brine, is used in great quantities in the manufacture of other sodium compounds and as the raw material from which sodium and chlorine are made industrially. It is used to glaze earthenware, on which it forms an impervious coating of sodium silicate during firing. It is also used to melt ice and snow on roads (§ 6.15), and to separate soap during its manufacture (§ 38.5). Pure sodium chloride is used for seasoning and preserving food; it is an essential constituent of our diet.

17.11. Sodium Nitrate, $NaNO_3$. It occurs naturally in huge deposits in rainless areas of Chile, mixed with sand, clay, and other sodium compounds. It is often called 'Chile saltpetre' for this reason. The nitrate is extracted by treating the 'caliche', as the mixture is called, with hot water, from which sodium nitrate of not less than 95% purity crystallizes on cooling. The mother liquor contains sodium iodate. the principal source of iodine (§25.13).

Sodium nitrate is a white solid, m.p. 311°C; it is hygroscopic and very soluble in water. Heated above 400°C, it decomposes into sodium nitrite and oxygen:

$$2NaNO_3 = 2NaNO_2 + O_2\uparrow$$

It is used as a nitrogenous fertilizer, and for making potassium nitrate (§ 17.20), and sodium nitrite (§ 22.16).

Table 17.2. Products made from Sodium Chloride

SODIUM CHLORIDE

Process	Product	Uses
Solvay process (§ 17.8)	sodium hydrogen carbonate, $NaHCO_3$	→ baking powders, health salts
↓ hea[t] (from sodium hydrogen carbonate)	sodium carbonate, Na_2CO_3	→ glass, borax, paints, paper, soap powders, textiles, sodium compds
electrolysis of soln. (§17·7); also ↓ $Ca(OH)_2$, Gossage process (§17.7) (from sodium carbonate)	sodium hydroxide, NaOH	→ soap, paper, rayon, textiles, sodium formate, pure alumina
electrolysis of soln. (§17.7)	chlorine, Cl_2	→ solvents, petrol additives, plastics, drugs, bleaches, insecticides, anaesthetics, refrigerants, bromine
↓ from chlorine; also from hot conc. H_2SO_4	hydrochloric acid, HCl	→ chlorides, pickling metals
hot conc. H_2SO_4	sodium sulphate, Na_2SO_4	→ glass, paper, health salts
electrolysis of cold soln.	sodium hypochlorite, NaOCl	→ antiseptics, bleaches
electrolysis of hot soln.	sodium chlorate, $NaClO_3$	→ weed-killer
electrolysis of melt	sodium, Na	→ sodium peroxide, sodium cyanide, and Na_4Pb.

17.12. Sodium Sulphate, Na_2SO_4. Made by heating sodium chloride strongly with concentrated sulphuric acid:

$$2NaCl + H_2SO_4 = Na_2SO_4 + 2HCl\uparrow$$

The hydrogen chloride is a valuable by-product.

It is a white solid, m.p. 844°C, soluble in water, giving a solution which readily shows supersaturation. Above 32.38°C it deposits the anhydrous salt from solution, but below this transition temperature it crystallizes as the decahydrate, $Na_2SO_4.10H_2O$, known as Glauber's salt. The latter effloresces slowly in air to the anhydrous form.

The solubility curve of sodium sulphate is of interest because of the sharp discontinuity at the transition temperature, T (see Fig. 17.7),

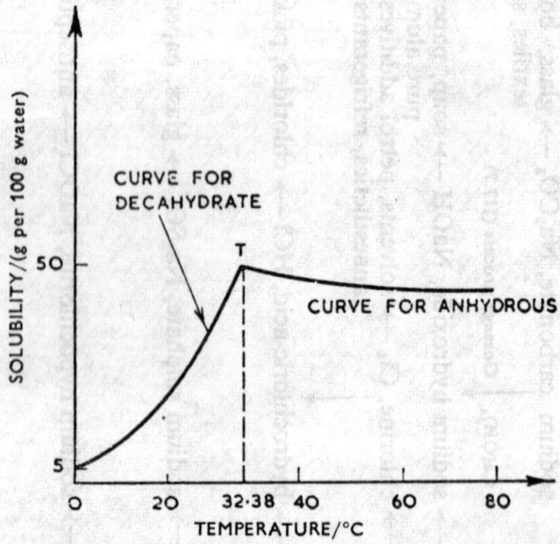

FIG. 17.7 Solubility Curve of Sodium Sulphate

where the solubility curves of the two different forms intersect.

Anhydrous sodium sulphate is used in large amounts to make glass and paper. Glauber's salt has a laxative action and is a constituent of many proprietary health salts.

17.13. Sodium Cyanide, NaCN. This is made from hydrocyanic acid (§ 33.9) or by heating sodium in dry ammonia, and pouring the molten sodamide onto red-hot charcoal, when the following reactions occur:

$$2NaNH_2 + C = Na_2NCN + 2H_2$$

$$\underset{\text{(sodium cyanamide)}}{Na_2NCN} + C = 2NaCN$$

Sodium cyanide has the ionic structure, Na^+CN^-. It is a white solid, soluble in water, giving a solution which is alkaline by hydrolysis. When this solution is warmed or acidified, very poisonous hydrogen cyanide, HCN, is released (§ 20.11).

Large amounts of sodium cyanide are used in extracting silver and gold from their ores (§ 28.14 and § 28.21) and in electroplating processes (§ 28.15).

17.14. Detection of Sodium. Compounds of sodium give an intense yellow flame coloration. Since they are nearly all soluble in water it is not easy to detect them by precipitation reactions, but the following tests work well with concentrated solutions, provided all other ions except potassium are absent:

(*a*) zinc uranyl acetate dissolved in acetic acid gives a yellow precipitate.

(*b*) potassium antimonate gives a white precipitate on standing.

17.15. Potassium, K. Electronic configuration: 2, 8, 8, 1.

Occurrence: Potassium is far too reactive to occur naturally as the metal, but it is widely distributed in minerals, the most important deposits of which occur in places where inland seas have undergone natural evaporation, e.g. at Stassfurt in Germany, and in the Dead Sea area, and in parts of U.S.A. The main commercial source is carnallite, $KCl.MgCl_2.6H_2O$, but kainite, $KCl.MgSO_4.3H_2O$, and schönite, $K_2SO_4.MgSO_4.6H_2O$, and sylvine, KCl, are sometimes used. Potassium is present in most clays and silicate rocks, and also in sea-water (about 0.05%), and wood ash (about 30% of which is potassium carbonate).

Extraction: By electrolysis of molten potassium cyanide in a cell similar to the one used for sodium.

Properties of Potassium: It is very similar to sodium physically and chemically, but being more electropositive, it is even more reactive. When added to water the metal melts and ignites, burning with a purple flame. Potassium is a mixture of three isotopes, the least abundant of which, ^{40}K, is radioactive with a half-life of about 10^9 years (§ 31.6).

17.16. Compounds of Potassium. The metal itself has few uses, but potassium compounds are of great importance in agriculture as fertilizers. Plants take up potassium ions from the soil during growth and these have to be replaced if soil fertility is to be maintained.

In general potassium compounds closely resemble those of sodium, but they tend to be less soluble in water, less commonly hydrated as crystals, and less frequently hygroscopic. In the laboratory, potassium compounds are only preferred to their cheaper sodium counterparts when they offer some specific advantage of this sort.

17.17. Potassium Hydroxide, (Caustic Potash), KOH. Made by methods similar to those used to make sodium hydroxide, which it closely resembles. It is more soluble in alcohol than is sodium hydroxide, and its alcoholic solution is an important reagent in organic chemistry. Aqueous potassium hydroxide solution is used as an absorbent of carbon dioxide because potassium carbonate is more soluble in water than sodium carbonate and is therefore less likely to be deposited. The main industrial use of potassium hydroxide is in the manufacture of soap (§ 38.5) and other potassium compounds.

17.18. Potassium Carbonate, K_2CO_3. It cannot be made by the Solvay method because potassium bicarbonate, being much more soluble than sodium bicarbonate, would not be precipitated. Instead it is manufactured by passing carbon dioxide into a suspension of hydrated magnesium carbonate in potassium chloride solution, when a sparingly soluble double salt is precipitated:

$$2KCl + CO_2 + 3MgCO_3.3H_2O = 2KHCO_3.MgCO_3.4H_2O\downarrow + MgCl_2$$

The precipitate is filtered off and decomposed by boiling with water, giving a precipitate of magnesium carbonate which can be hydrated and used again, and leaving a solution from which potassium carbonate is crystallized.

Potassium carbonate is a white, hygroscopic solid of m.p. 900°C. It is very soluble in water, giving an alkaline solution by hydrolysis. It can exist as tri-, di-, and monohydrates, all of which lose their water of crystallization on heating above 130°C. In its reactions it closely resembles sodium carbonate. It is used to make hard glass, toilet soap, and various potassium compounds, and in the laboratory as a drying agent.

17.19. Potassium Chloride, KCl. It occurs naturally, mainly as carnallite, $KCl.MgCl_2.6H_2O$, from which it is extracted by fractional crystallization. It is a white solid, m.p. 770°C, soluble in water, with reactions which are very similar to those of sodium chloride. It is the source of most other potassium compounds, but its chief use is as a fertilizer, huge quantities being required for this purpose.

17.20. Potassium Nitrate, (Saltpetre or Nitre), KNO_3. Made by adding potassium chloride to a hot saturated solution of sodium nitrate:

$$NaNO_3 + KCl \rightleftharpoons KNO_3 + NaCl$$

Because sodium chloride is the least soluble of these substances in hot water (see Fig. 17.8), much of it is precipitated from solution and filtered off while hot, causing the position of equilibrium to move to the right. On cooling the mixture, most of the potassium nitrate formed crystallizes out, because it is much less soluble in cold water than in hot (the solubility in 100 g of water is 13 g at 0°C, and 246 g at 100°C).

It is a white solid, m.p. 340°C, which, like sodium nitrate, decomposes to the nitrite on strong heating. It is used as a fertilizer, and for

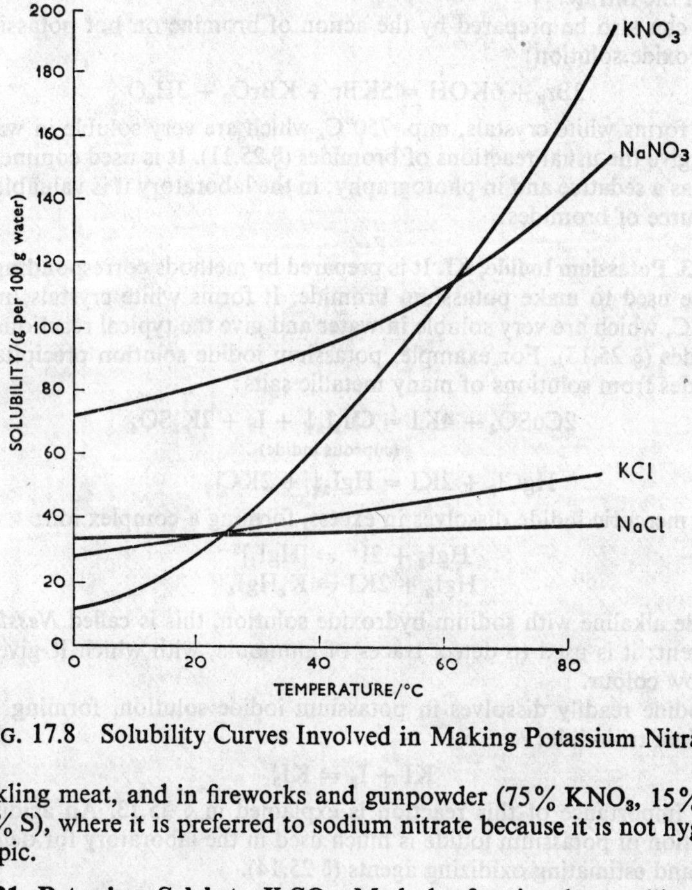

FIG. 17.8 Solubility Curves Involved in Making Potassium Nitrate

pickling meat, and in fireworks and gunpowder (75% KNO_3, 15% C, 10% S), where it is preferred to sodium nitrate because it is not hygroscopic.

17.21. Potassium Sulphate, K_2SO_4. Made by fractional crystallization from solutions of naturally-occurring double salts. It is a white solid, m.p. 1060°C, soluble in water, from which it crystallizes in the anhydrous form. It is used as a fertilizer, and in making alums (§ 19.13), and in the manufacture of hard glass.

17.22. Potassium Bromide, KBr. It is manufactured by treating iron filings with bromine and water and boiling the product with potassium carbonate:

$$3Fe + 4Br_2 = Fe_3Br_8$$
$$Fe_3Br_8 + 4K_2CO_3 = Fe_3O_4\downarrow + 8KBr + 4CO_2\uparrow$$

The iron oxide is filtered off and the potassium bromide is crystallized from the filtrate.

It can also be prepared by the action of bromine on hot potassium hydroxide solution:

$$3Br_2 + 6KOH = 5KBr + KBrO_3 + 3H_2O$$

It forms white crystals, m.p. 750°C, which are very soluble in water and give the usual reactions of bromides (§ 25.11). It is used commercially as a sedative and in photography; in the laboratory it is valuable as a source of bromides.

17.23. Potassium Iodide, KI. It is prepared by methods corresponding to those used to make potassium bromide. It forms white crystals, m.p. 680°C, which are very soluble in water and give the typical reactions of iodides (§ 25.15). For example, potassium iodide solution precipitates iodides from solutions of many metallic salts:

$$2CuSO_4 + 4KI = \underset{\text{(cuprous iodide)}}{Cu_2I_2\downarrow} + I_2 + 2K_2SO_4$$

$$HgCl_2 + 2KI = HgI_2\downarrow + 2KCl$$

The mercuric iodide dissolves in excess, forming a complex ion:

$$HgI_2 + 2I^- \rightleftharpoons [HgI_4]^{2-}$$

or

$$HgI_2 + 2KI \rightleftharpoons K_2HgI_4$$

Made alkaline with sodium hydroxide solution, this is called *Nessler's* reagent; it is used to detect traces of ammonia, with which it gives a yellow colour.

Iodine readily dissolves in potassium iodide solution, forming potassium tri-iodide:

$$KI + I_2 \rightleftharpoons KI_3$$

The importance of this reaction is explained in § 25.13. An acidified solution of potassium iodide is much used in the laboratory for detecting and estimating oxidizing agents (§ 25.14).

17.24. Potassium Permanganate, $KMnO_4$. This compound is prepared by strongly heating potassium hydroxide and manganese dioxide with an oxidizing agent such as potassium chlorate:

$$MnO_2 + 2KOH + \bar{O} = K_2MnO_4 + H_2O$$

The product, which is a dark green mass, is cooled, powdered, and dissolved in water and saturated with carbon dioxide. This converts the manganate to permanganate, which is filtered free of suspended manganese dioxide, using a glass wool filter, and concentrated until crystallization occurs:

$$3K_2MnO_4 + 2CO_2 = 2K_2CO_3 + 2KMnO_4 + MnO_2\downarrow$$

Potassium permanganate forms purple-black crystals, which are moderately soluble in water (about 6 gramme dissolve in 100 gramme of water at room temperature), giving a purple solution which decomposes when boiled:

$$2KMnO_4 = K_2MnO_4 + MnO_2\downarrow + O_2\uparrow$$

It is a very powerful oxidizing agent. The solid converts concentrated hydrochloric acid into chlorine in the cold (§ 25.5), and explodes when heated with sulphur, phosphorus, or organic matter. The solution, in the presence of dilute sulphuric acid, reacts with reducing agents in this way:

$$2KMnO_4 + 3H_2SO_4 = K_2SO_4 + 2MnSO_4 + 3H_2O + 5\bar{O}$$

This reaction takes place quantitatively with ferrous salts, oxalates, nitrites, and hydrogen peroxide, for example, and it is used to estimate solutions of these substances volumetrically. No indicator is necessary in these titrations when the products are colourless, because the presence of even a slight excess of permanganate solution causes a pale pink colour to persist on shaking, and this is taken to be the end point. Examples of the full equations are:

$$2KMnO_4 + 10FeSO_4 + 8H_2SO_4 = K_2SO_4 + 2MnSO_4 + 5Fe_2(SO_4)_3 + 8H_2O$$

$$2KMnO_4 + 5H_2C_2O_4 + 3H_2SO_4 = K_2SO_4 + 2MnSO_4 + 10CO_2\uparrow + 8H_2O$$

$$2KMnO_4 + 5H_2O_2 + 3H_2SO_4 = K_2SO_4 + 2MnSO_4 + 5O_2\uparrow + 8H_2O$$

The calculation of the equivalent weight of potassium permanganate as an oxidizing agent is discussed in § 2.5 and again in § 15.7.

In alkaline or neutral solution, potassium permanganate behaves differently with reducing agents, giving first green potassium manganate solution and finally a brown precipitate of manganese dioxide, thus:

$$2KMnO_4 + H_2O = 2KOH + 2MnO_2\downarrow + 3\bar{O}.$$

This reaction is used to detect unsaturation in organic compounds (§ 34.6).

17.25. Potassium Dichromate, $K_2Cr_2O_7$. It is made by strongly heating chromite (§ 27.7) with sodium carbonate and lime in air:

$$4FeCr_2O_4 + 8Na_2CO_3 + 7O_2 = 8Na_2CrO_4 + 2Fe_2O_3 + 8CO_2$$

The chromate is dissolved in water and treated with concentrated sulphuric acid:

$$2Na_2CrO_4 + H_2SO_4 = Na_2SO_4 + Na_2Cr_2O_7 + H_2O$$

On addition of potassium chloride, the less soluble potassium dichromate crystallizes out. The crystals are orange-red, anhydrous, non-hygroscopic, and since they are much less soluble in cold water than in hot, they are readily purified by recrystallization (§ 4.16).

Potassium dichromate is a powerful oxidizing agent. In acid solution it reacts with reducing agents thus:

$$K_2Cr_2O_7 + 4H_2SO_4 = K_2SO_4 + Cr_2(SO_4)_3 + 4H_2O + 3\bar{O}$$

Important examples are:

$$K_2Cr_2O_7 + 6FeSO_4 + 7H_2SO_4$$
$$= K_2SO_4 + Cr_2(SO_4)_3 + 3Fe_2(SO_4)_3 + 7H_2O$$
$$K_2Cr_2O_7 + 6KI + 7H_2SO_4 = 4K_2SO_4 + Cr_2(SO_4)_3 + 3I_2 + 7H_2O$$
$$K_2Cr_2O_7 + 3SO_2 + H_2SO_4 = K_2SO_4 + Cr_2(SO_4)_3 + H_2O$$

The change from the orange of dichromate to the green of chromic sulphate is used as a test for reducing gases such as sulphur dioxide and hydrogen sulphide.

Potassium dichromate is used as a primary standard in volumetric analysis and as an oxidizing agent in organic chemistry. Industrially it is used for oxidations and for making chromium compounds, including chrome alum.

17.26. Detection of Potassium. Its compounds give a violet flame coloration, which is visible through blue glass even in the presence of sodium compounds. Concentrated solutions of potassium compounds give:

(1) a yellow precipitate with sodium cobaltinitrite solution.

(2) a white precipitate with perchloric acid.

(3) a white precipitate with tartaric acid, especially on standing (see § 48.4).

These precipitates can be used to detect potassium provided the ions of ammonium and metals other than sodium and potassium have been removed.

17.27. Summary. There are certain features of the alkali metals, common to them all, which make them such a clearly defined Group:

(*a*) In physical properties the metals are very similar, but they differ from most other metals in being soft, with low densities and low melting points. They emit electrons when illuminated by ultra-violet light.

(*b*) They are the most electropositive elements (§ 10.7), being very reactive chemically and showing a high affinity for electronegative elements, with which they combine exothermically to form very stable compounds. Because of this the alkali metals are strong reducing agents.

(*c*) Their oxides are strongly basic, dissolving in water to form alkalis, from which property the metals derived their name.

(*d*) With the exception of lithium, their carbonates, nitrates, and hydroxides are not decomposed to the oxide when heated, and their hydrogen carbonates exist as solids at room temperature.

(*e*) Their compounds are nearly all soluble in water, colourless (except where the anion is coloured), and non-volatile, with high melting points and boiling points. They conduct electricity readily when fused and in aqueous solution.

(*f*) They form crystalline hydrides from which hydrogen is liberated at the anode during electrolysis of the melt.

(*g*) Their salts with weak acids are extensively hydrolysed, giving alkaline solutions.

(*h*) They are always univalent.

These characteristics can be explained in terms of atomic structure and the electronic theory of valency. All the alkali metal atoms have a single electron in their outermost shells, and they form compounds by giving up this electron to an electron-accepting atom, leaving an ion with a single positive charge and the stable configuration of an inert gas. Thus all their compounds (except a few of those of lithium) are ionic and display the characteristics of electrovalency (§ 14.3).

Their reactivity and electropositivity arise directly from the ease with which these outermost electrons are surrendered. This tendency is greatest in the largest atoms, where the filled inner shells of electrons have a screening effect upon the attraction which the positively charged nucleus exerts upon the distant outermost electron.

It will be seen that lithium and its salts differ from the other alkali metals and their corresponding salts in a number of ways. In most of these (e.g. in the relative thermal instability of its peroxide, hydroxide, and carbonate and in the sparing solubility of some of its salts) lithium shows a strong resemblance to magnesium, providing a good example of the diagonal relationship referred to in § 13.10.

The alkali metals are compared with the alkaline earth metals in Table 18.2 at the end of the next chapter.

18

The Alkaline Earth Metals

18.1. Introduction. Group IIA of the Periodic Classification contains the elements beryllium, magnesium, calcium, strontium, barium, and radium, known as the alkaline earth metals, although originally the name was limited to the series from calcium to radium. Magnesium and calcium are by far the commonest and most important chemically, so only these will be studied in detail, but the summary at the end of the chapter deals with the Group as a whole, emphasizing the similarities and differences which exist between its members.

18.2. Beryllium, Be. Electronic configuration: 2, 2.

OCCURRENCE: The only important ore is beryl, a complex alumino-silicate of beryllium containing about 5% of the metal.

EXTRACTION: By electrolysis of a molten mixture of beryllium and sodium chlorides at 350° C, using a nickel cathode and a graphite anode. The anhydrous beryllium chloride is made by heating the oxide with carbon to 800°C in chlorine. The metal is refined by melting it in an atmosphere of hydrogen or by distillation in a vacuum.

PROPERTIES OF BERYLLIUM: It is a hard greyish metal, m.p. 1280°C, ρ 1.8 g cm^{-3}. When very pure it is malleable and ductile, but the impure metal is brittle at room temperature. Beryllium and its compounds are very poisonous when inhaled into the lungs as dust or vapour.

REACTIONS OF BERYLLIUM: It burns in oxygen with a brilliant flame forming the oxide, but it only combines with nitrogen above 500°C and it does not decompose steam even at red heat. It dissolves in dilute acids and in concentrated solutions of caustic alkalis liberating hydrogen:

$$Be + 2HCl = BeCl_2 + H_2\uparrow$$
$$Be + H_2SO_4 = BeSO_4 + H_2\uparrow$$
$$Be + 2NaOH = \underset{\text{(sodium beryllate soln.)}}{Na_2BeO_2} + H_2\uparrow$$

USES OF BERYLLIUM: Its principal use is in copper alloys which are valued for their hardness, high electrical conductivity, and resistance to corrosion and are widely used for making instrument springs and electrical contacts and switches. It is also used for deoxidizing molten metals (e.g. copper and aluminium), and for making the 'windows' of X-ray tubes. Because of its very low absorption of neutrons and its high melting point it at one time seemed likely to be very suitable for making fuel cans for atomic reactors (§ 31.11), but in 1962 it was decided to abandon the idea owing to severe metallurgical difficulties. If these are eventually overcome the metal is likely to find increasing use in the aircraft and missile fields owing to its high strength to weight ratio.

18.3. Compounds of Beryllium. In many ways these resemble the corresponding compounds of aluminium, providing yet another example of the diagonal relationship referred to in § 13.10. For example, the anhydrous chloride is a deliquescent white solid, m.p. 405°C, which fumes in moist air and which dissolves readily in water giving an acidic solution by hydrolysis. It is soluble in organic solvents and is a poor conductor of electricity when molten, suggesting covalency. The oxide is a white solid formed by strongly heating the hydroxide, carbonate, nitrate, or sulphate. Its very high melting point (above 2500°C) and its stability lead to its use as a refractory and for making sparking plug insulators where its high thermal conductivity and low electrical conductivity are an advantage. The hydroxide is amphoteric; when freshly prepared it dissolves readily in acids and alkalis and also in ammonium carbonate solution (which distinguishes it from aluminium hydroxide):

$$Be(OH)_2 + H_2SO_4 = BeSO_4 + 2H_2O$$
$$Be(OH)_2 + 2NaOH = Na_2BeO_2 + 2H_2O$$

Beryllium salts are usually hydrated (e.g. $BeCl_2.4H_2O$ and $BeSO_4.4H_2O$) and soluble in water. Beryllium forms a large number of complexes with organic compounds.

18.4. Magnesium, Mg. Electronic configuration: 2,8,2.

OCCURRENCE: The metal is too reactive to occur naturally, but it is found in the following minerals: dolomite $CaCO_3.MgCO_3$, magnesite $MgCO_3$, kieserite $MgSO_4.H_2O$, carnallite $KCl.MgCl_2.6H_2O$, and in many silicates, including talc and asbestos. Magnesium is also present in sea water as the chloride, sulphate, and bromide (although these salts are present in low concentration, each cubic mile of ocean is estimated to contain over five million tons of magnesium)*, and in certain spring waters as the sulphate and bicarbonate. It is a constituent of chlorophyll, the green colouring matter of plants.

* In SI units, about 1.5 kg m^{-3}.

EXTRACTION: Two methods are in use; both start from magnesium oxide, made from dolomite and sea water in the following way. Dolomite is first heated strongly, when it decomposes to the oxides of calcium and magnesium. These are slaked and added to sea water in large tanks, bringing about the precipitation of dissolved magnesium salts as the insoluble hydroxide:

$$CaCO_3.MgCO_3 = CaO + MgO + 2CO_2\uparrow$$

$$CaO + MgO + 2H_2O = Ca(OH)_2 + Mg(OH)_2\downarrow$$

e.g. $\underset{\text{(in sea water)}}{MgCl_2} + Ca(OH)_2 = CaCl_2 + Mg(OH)_2\downarrow$

The sludge of magnesium hydroxide is filtered off and heated strongly to the oxide:

$$Mg(OH)_2 = MgO + H_2O$$

(*a*) *Electrolytic Reduction Process:* Magnesium oxide is mixed with powdered coke and heated strongly in a stream of chlorine, when it is converted to the anhydrous chloride:

$$MgO + C + Cl_2 = MgCl_2 + CO\uparrow$$

The molten chloride is then electrolysed, using the cell shown diagrammatically in Fig. 18.1. Amounts of sodium or potassium chloride are

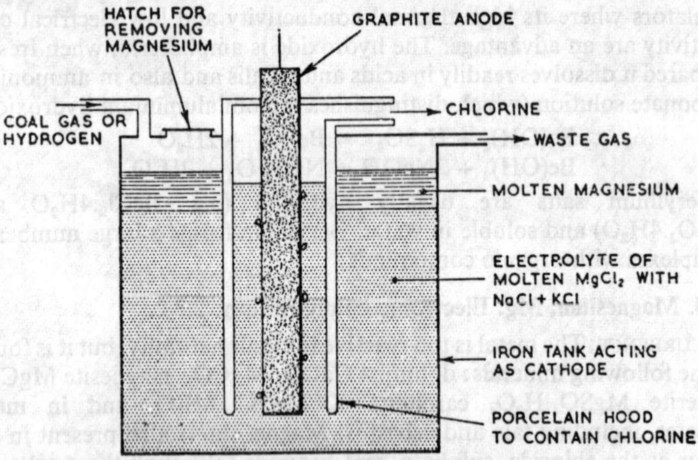

FIG. 18.1 Manufacture of Magnesium

added to lower the melting point of the electrolyte, which is kept at about 700°C. At anode: $Cl^- \rightarrow Cl + e \qquad 2Cl \rightarrow Cl_2\uparrow$

At cathode: $Mg^{2+} + 2e \rightarrow Mg$

At the anode, chlorine ions are discharged and the liberated chlorine, which is prevented from spreading through the cell by the porcelain head, is used to make more magnesium chloride.

At the cathode magnesium ions are discharged and the molten metal rises to the surface of the electrolyte, to be removed periodically by ladle. It is protected from reaction with oxygen and nitrogen by an atmosphere of hydrogen or coal gas. The product is at least 99.9% pure.

(*b*) *Carbon Reduction Process:* Magnesium oxide is heated to about 2000°C with carbon in an electric furnace, when the following equilibrium is set up:

$$MgO + C \rightleftharpoons Mg + CO$$

On cooling the system slowly the position of equilibrium moves to the left, and only at very high temperatures is there a high proportion of metallic magnesium in the reaction mixture. This difficulty is overcome by cooling the system rapidly with hydrogen, which lowers the temperature from 2000°C to about 200°C so suddenly that most of the magnesium vapour condenses to dust before it has time to revert to the oxide. The product is purified by distillation *in vacuo*.

PROPERTIES OF MAGNESIUM: Lustrous white metal, m.p. 650°C, b.p. 1100°C, ρ 1.74 g cm^{-3}. It is a good conductor of heat and electricity and is easily machined.

REACTIONS OF MAGNESIUM: It is highly electropositive, displaying a strong affinity for oxygen and other electronegative elements. For example, when heated above its melting point in air it ignites, burning with a brilliant white light to the oxide and nitride:

$$2Mg + O_2 = 2MgO$$
$$3Mg + N_2 = Mg_3N_2$$

The metal also combines on heating with sulphur, chlorine, and phosphorus, but not with hydrogen. Once ignited, magnesium will continue to burn in steam, sulphur dioxide, carbon monoxide and dioxide, and the oxides of nitrogen, reducing these gases in every case. The powdered metal is a powerful reducing agent, exploding when heated with many oxysalts. Silica is reduced on heating with magnesium, and so are the halides of metals such as titanium and uranium which are less electropositive than magnesium:

$$2Mg + SiO_2 = 2MgO + Si$$
$$2Mg + TiCl_4 = 2MgCl_2 + Ti$$
$$2Mg + UF_4 = 2MgF_2 + U$$

On exposure to moist air magnesium forms a protective film of oxide, which prevents further oxidation. With boiling water or steam it reacts, liberating hydrogen:

$$Mg + H_2O = MgO + H_2\uparrow$$

Magnesium dissolves readily in dilute hydrochloric acid and dilute sulphuric acid giving hydrogen:

$$Mg + H_2SO_4 = MgSO_4 + H_2\uparrow$$

With very dilute nitric acid, some hydrogen is released, but more concentrated nitric acid gives oxides of nitrogen. Magnesium will displace less electropositive metals from solutions of their salts, going into solution as Mg^{2+} ions.

USES OF MAGNESIUM: Its principal use is in alloys (with aluminium, zinc and other metals) providing strong light materials for the construction of aircraft and cars. Its low absorption of neutrons and its excellent machinability have led to its widespread use in nuclear reactors for making the cans which contain the uranium fuel, but its low melting point and its inflammability are disadvantageous here. Magnesium is used in flares, photographic flashlight powders, fireworks and incendiary bombs, and as a deoxidant in metallurgy. Another growing use is for the sacrificial protection of pipelines, piers and ships against corrosion; ingots of magnesium connected to the steel structures corrode preferentially, being more electropositive (§ 10.4), and their periodical replacement is cheaper and more convenient than painting the whole structure at frequent intervals. Magnesium is also used in increasing amounts in the extraction of the important metals titanium (§ 27.3) and uranium (§ 27.20).

18.5. Magnesium Oxide (Magnesia), MgO. Made by heating the nitrate, carbonate or hydroxide, or by burning the metal in oxygen. It is a white powder, only slightly soluble in water; its solution is mildly alkaline and is used to counter gastric acidity. Magnesium oxide has a very high melting point (over 2800°C) and is extremely stable, showing no sign of decomposing at this temperature. For these reasons it is extensively used as a refractory lining for furnaces. It reacts with acids giving the corresponding magnesium salts. As described on the preceding page, its reduction with carbon is reversible and is complete only at very high temperatures.

18.6. Magnesium Hydroxide, $Mg(OH)_2$. It is precipitated when excess caustic alkali solution is added to a solution of a magnesium salt:

$$MgCl_2 + 2NaOH = Mg(OH)_2\downarrow + 2NaCl$$

On heating it loses water, yielding the oxide:

$$Mg(OH)_2 = MgO + H_2O\uparrow$$

It is a white solid, only slightly soluble in water, giving an alkaline solution. Magnesium hydroxide dissolves in ammonium chloride solution forming a complex ion, which partly explains why it is not precipitated with the hydroxides of aluminium, chromium and iron in qualitative analysis (§ 9.9).

18.7. Magnesium Carbonate, $MgCO_3$. Occurs naturally as magnesite and dolomite. It is prepared by mixing solutions of sodium hydrogen carbonate and a magnesium salt (if sodium carbonate solution is used a hydrated basic carbonate is precipitated instead). It is a white solid, almost insoluble in water but readily soluble in acids:

$$MgCO_3 + H_2SO_4 = MgSO_4 + CO_2\uparrow + H_2O$$

It goes into solution as the hydrogen carbonate when brought into contact with water containing dissolved carbon dioxide, causing temporary hardness of water (§ 16.5):

$$MgCO_3 + CO_2 + H_2O \rightleftharpoons Mg(HCO_3)_2$$

When heated it decomposes thus:

$$MgCO_3 = MgO + CO_2$$

Magnesium carbonate is used in toothpastes, as a filler for paper, rubber and paints, and in mineral form as the source of magnesium and many of its compounds.

18.8. Magnesium Chloride, $MgCl_2$. Present in sea water and in the mineral carnallite. If dilute hydrochloric acid is added to magnesium carbonate and the solution concentrated, the hydrate $MgCl_2.6H_2O$ crystallizes out; it is deliquescent and very soluble in water. Anhydrous magnesium chloride can be made by heating magnesium oxide strongly with carbon and chlorine, or by passing chlorine over the heated metal, but it cannot be made by heating the hydrate in air because the salt is hydrolysed by its own water of crystallization to an oxychloride:

$$2\ MgCl_2.6H_2O = MgO.MgCl_2 + 2HCl\uparrow + 11H_2O\uparrow$$

Anhydrous magnesium chloride is used in the electrolytic extraction of magnesium and in making special cements with magnesia.

18.9. Magnesium Sulphate, $MgSO_4$. Occurs naturally as kieserite and is present in sea water and in some springs. It is one of the salts responsible for permanent hardness of water (§ 16.5). Magnesium sulphate is made commercially by fractional crystallization from minerals or by treating magnesium carbonate with dilute sulphuric acid. Various hydrates exist, of which the heptahydrate, $MgSO_4.7H_2O$, known as Epsom salt, crystallizes at room temperature. All are soluble in water and give the anhydrous salt on heating to 200°C. Magnesium sulphate is used as a laxative (it is a constituent of many 'health salts'), as a filler for paper, and in the treatment of textile fibres to render them non-inflammable. It is also used in dyeing and tanning processes.

18.10. Magnesium Nitride, Mg_3N_2. This is formed when magnesium is heated strongly in nitrogen or ammonia, or in a limited supply of air:

$$3Mg + N_2 = Mg_3N_2$$
$$3Mg + 2NH_3 = Mg_3N_2 + 3H_2$$

It is a colourless solid which reacts with water, even in the cold, giving ammonia: $Mg_3N_2 + 6H_2O = 3Mg(OH)_2 + 2NH_3\uparrow$

18.11. Detection of Magnesium. If disodium hydrogen phosphate solution is added to a solution of a magnesium salt in the presence of excess ammonium hydroxide and ammonium chloride, a white precipitate of magnesium ammonium phosphate is obtained. To avoid interference when using this test in qualitative analysis, all metals other than sodium, potassium, and magnesium must first be removed from solution.

e.g. $MgSO_4 + Na_2HPO_4 + NH_4OH = MgNH_4PO_4\downarrow + Na_2SO_4 + H_2O$

As a confirmatory test a little *magneson* reagent (p-nitrobenzene-azo-resorcinol) should be added to the suspected magnesium salt solution and the latter made alkaline with caustic soda solution. A blue precipitate results if magnesium is present. Again care is needed to eliminate interfering metals before applying this test.

18.12. Calcium, Ca. Electronic configuration: 2,8,8,2.

OCCURRENCE: The metal is too reactive to occur naturally, but it is abundant in the combined state as the following minerals: Limestone, chalk, calcite, aragonite, marble and coral $CaCO_3$, dolomite $CaCO_3.MgCO_3$, gypsum $CaSO_4.2H_2O$, anhydrite $CaSO_4$, apatite $CaF_2.3Ca_3(PO_4)_2$, fluorspar CaF_2, and in many silicates. Bones and teeth consist largely of calcium compounds.

EXTRACTION: Calcium is now made by mixing very pure lime with aluminium powder and heating strongly in a steel retort from which the air has been evacuated. The oxide is reduced by the Goldschmidt process (§ 19.8), the principles of which are explained in §11.13:

$$3CaO + 2Al = Al_2O_3 + 3Ca\uparrow$$

The calcium distils out of the retort and is condensed in vacuo.

For many years calcium was made by the electrolysis of molten anhydrous calcium chloride to which some calcium fluoride had been added to lower the melting point (§ 5.13).

At the anode: $Cl^- \rightarrow Cl + e$ $2Cl^- \rightarrow Cl_2\uparrow$
At the cathode: $Ca^{2+} + 2e \rightarrow Ca$
The type of cell that was used is shown in Fig. 18.2.

PROPERTIES OF CALCIUM: It is a moderately hard grey metal, m.p. 851° C, b.p. 1480°C, ρ 1.55 g cm^{-3}.

REACTIONS OF CALCIUM: It is a highly electropositive, reactive metal. It tarnishes rapidly on exposure to the air, forming a layer of oxide and

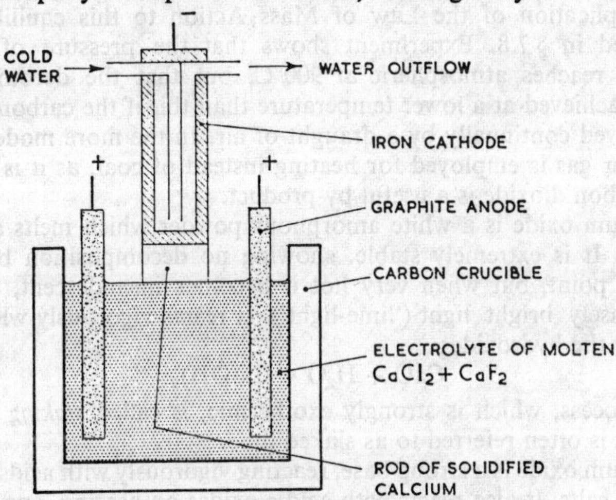

FIG. 18.2 Manufacture of Calcium

hydroxide, and on heating it combines directly with nitrogen, chlorine, sulphur, and hydrogen:

e.g. $$3Ca + N_2 = Ca_3N_2$$
$$Ca + Cl_2 = CaCl_2$$

Calcium reacts readily with cold water and vigorously with hot water and acids:

$$Ca + 2H_2O = Ca(OH)_2 + H_2\uparrow$$

If ammonia is passed over heated calcium, it reacts thus:

$$3Ca + 2NH_3 = Ca_3N_2 + 3H_2$$

USES OF CALCIUM: It is used to make calcium hydride and uranium (§ 27.20), as a chemical means of removing the last traces of gas from vacuum tubes, and as a deoxidant in metal castings.

18.13. Calcium Hydride, CaH_2. Made by passing hydrogen over the heated metal:

$$Ca + H_2 = CaH_2$$

It is a white solid, with the structure $Ca^{2+}2H^-$. It reacts readily with cold water, liberating hydrogen, and its main use is therefore as a portable (but expensive) source of the gas:

$$CaH_2 + 2H_2O = Ca(OH)_2 + 2H_2\uparrow$$

18.14. Calcium Oxide (Quicklime), CaO. Made industrially by heating limestone in vertical or rotary kilns to 800–1000°C:

$$CaCO_3 \rightleftharpoons CaO + CO_2\uparrow$$

The application of the Law of Mass Action to this equilibrium is discussed in § 7.8. Experiment shows that the pressure of carbon dioxide reaches atmospheric at 900°C, but that the decomposition can be achieved at a lower temperature than this if the carbon dioxide is removed continually by a draught of air. In the more modern kilns producer gas is employed for heating instead of coal, as it is cleaner. The carbon dioxide is a useful by-product.

Calcium oxide is a white amorphous powder which melts at about 2600°C. It is extremely stable, showing no decomposition below its melting point, but when very hot it becomes incandescent, emitting an intensely bright light ('lime-light'). It reacts vigorously with water forming the hydroxide:

$$CaO + H_2O = Ca(OH)_2$$

This process, which is strongly exothermic, is called *slaking* and the product is often referred to as slaked lime.

Calcium oxide is a strong base, reacting vigorously with acids to form calcium salts. It also reacts with acidic oxides on heating, one of these reactions, that with silica, being of great industrial importance in the manufacture of iron (see § 30.2):

$$CaO + SiO_2 = CaSiO_3 \text{ (calcium silicate).}$$

When heated to about 2000°C with carbon, calcium carbide is formed as described in § 18.22:

$$CaO + 3C = CaC_2 + CO$$

USES OF CALCIUM OXIDE: Its principal uses are in making slaked lime and calcium carbide and in smelting processes to bring about the removal of silicaceous material as a fusible silicate slag. Calcium oxide is also used as a refractory lining in furnaces, where its infusibility and basic character are valuable assets, in glass making and as a fertilizer. In the laboratory quicklime is used for drying ammonia gas and for making soda-lime.

18.15. Calcium Hydroxide (Slaked Lime), Ca(OH)₂. Made industrially by adding water to calcium oxide:

$$CaO + H_2O = Ca(OH)_2$$

The lumps of quicklime crumble to a fine powder during the slaking. The product is a white solid, only slightly soluble in water, the solubility decreasing with rising temperature. The solution, which is alkaline, is called *lime water*. An aqueous suspension of slaked lime is known as *milk of lime* from its appearance.

Slaked lime reacts with acids and acidic oxides forming calcium salts. If carbon dioxide is bubbled through lime water, it turns milky owing to precipitation of calcium carbonate as fine particles (test for the gas):

$$Ca(OH)_2 + CO_2 = CaCO_3\downarrow + H_2O$$

On passing more carbon dioxide the precipitate gradually disappears as the carbonate is converted to the soluble hydrogen carbonate:

$$CaCO_3 + CO_2 + H_2O \rightleftharpoons Ca(HCO_3)_2$$

Calcium hydroxide reacts with chlorine in several ways, depending upon the conditions. Below 35°C chlorine converts slaked lime into bleaching powder (§ 18.23), but at higher temperatures calcium chloride and oxygen are formed:

$$2Ca(OH)_2 + 2Cl_2 = 2CaCl_2 + 2H_2O + O_2\uparrow$$

Milk of lime gives the hypochlorite in the cold when saturated with chlorine, and the chlorate when hot, calcium chloride being formed in both cases:

$$2Ca(OH)_2 + 2Cl_2 = CaCl_2 + Ca(OCl)_2 + 2H_2O$$
$$6Ca(OH)_2 + 6Cl_2 = 5CaCl_2 + Ca(ClO_3)_2 + 6H_2O$$

Calcium hydroxide releases ammonia when heated with ammonium salts:

e.g. $$2NH_4Cl + Ca(OH)_2 = CaCl_2 + 2NH_3\uparrow + 2H_2O$$

Added to water containing hydrogen carbonates it brings about the precipitation of calcium carbonate (see § 16.5 for application of this):

e.g. $$Ca(HCO_3)_2 + Ca(OH)_2 = 2CaCO_3\downarrow + 2H_2O$$

USES OF CALCIUM HYDROXIDE: It is used to make builders' mortar (§ 18.24) and plaster, to prevent acidity of the soil, to manufacture bleaching powder, to make caustic soda by the Gossage process, and for softening water. It is also used as a cheap alkali for neutralizing acids, for liberating ammonia from ammonium compounds, for the purification of coal gas and sugar, and for glass making. In the laboratory, lime water is used to detect carbon dioxide.

18.16. Calcium Carbonate, $CaCO_3$. Occurs abundantly as minerals such as limestone, chalk, marble and coral. It is *dimorphous*, i.e. it can exist in two different crystalline forms, called calcite and aragonite. A white solid, it is practically insoluble in pure water, but it dissolves slowly in the presence of dissolved carbon dioxide giving calcium hydrogen carbonate solution which accounts for the hardness of water in limestone areas (§ 16.5):

$$CaCO_3 + CO_2 + H_2O \rightleftharpoons Ca(HCO_3)_2$$

It dissolves in acids readily, giving off carbon dioxide:

$$CaCO_3 + 2HCl = CaCl_2 + H_2O + CO_2\uparrow$$

Its most important property is its thermal dissociation into calcium oxide and carbon dioxide, to which reference has been made in § 18.14.

USES OF CALCIUM CARBONATE: It is used in great quantities to make quicklime, for neutralizing soil acidity in agriculture, in the manufacture of cement (§ 18.25), and as a raw material in the Solvay process (§ 17.8) and in the smelting of iron (§ 30.2). In the pure, finely-powdered form it is used in toothpastes and cosmetics.

18.17. Calcium Hydrogen Carbonate, (Calcium Bicarbonate), $Ca(HCO_3)_2$. No solid exists at room temperature, but its solution can be made by passing carbon dioxide into a suspension of calcium carbonate. This solution undergoes decomposition, slowly in the cold and rapidly on warming or boiling, producing a precipitate of the carbonate:

$$Ca(HCO_3)_2 \rightleftharpoons CaCO_3\downarrow + CO_2\uparrow + H_2O$$

This change explains the removal of temporary hardness by boiling (§ 16.5) and also the formation of stalactites and stalagmites in caves (see Fig. 18.3).

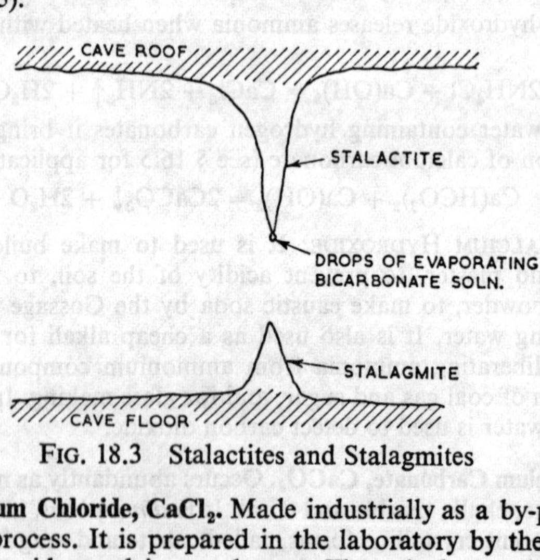

FIG. 18.3 Stalactites and Stalagmites

18.18. Calcium Chloride, $CaCl_2$. Made industrially as a by-product of the Solvay process. It is prepared in the laboratory by the action of hydrochloric acid on calcium carbonate. The anhydrous salt is a white solid, m.p. 780°C; it is very soluble in water and extremely deliquescent. It forms a hydrate, $CaCl_2.6H_2O$, which melts at 30°C and changes to a dihydrate at about 200°C and which loses all of its water of crystallization above 260°C, undergoing partial hydrolysis to the oxychloride. Calcium chloride combines with ammonia, amines, and alcohols forming compounds of the type $CaCl_2.8NH_3$ and $CaCl_2.2C_2H_5OH$, and consequently it must not be used to dry these compounds.

USES OF CALCIUM CHLORIDE: Advantage is taken of its deliquescence to lay dust on roads and to keep cotton thread moist during spinning. The anhydrous salt is used for the production of metallic calcium and as a drying agent in the laboratory.

18.19. Calcium Fluoride, CaF_2. This occurs naturally, particularly in the U.S.A., as fluorspar or fluorite. When pure it is a white solid melting at about 1400°C. Unlike the other halides of calcium it is practically insoluble in water and dilute acids. When heated with concentrated sulphuric acid it gives anhydrous hydrogen fluoride:

$$CaF_2 + H_2SO_4 = CaSO_4 + 2HF\uparrow$$

Its main use is as a flux in the smelting of metals and the manufacture of vitreous enamels, but large amounts are also needed for making hydrogen fluoride (§ 25.3) and hence fluorine (§ 25.2).

18.20. Calcium Sulphate, $CaSO_4$. It occurs naturally as anhydrite and gypsum, and as a cause of permanent hardness in spring-water. The dihydrate, being only sparingly soluble in water, is obtained as a white precipitate when dilute sulphuric acid or a soluble sulphate is added to a solution of a calcium salt below 66°C. When heated just above 100°C it loses part of its water of crystallization, leaving a hemihydrate $CaSO_4.\frac{1}{2}H_2O$ (or $2CaSO_4.H_2O$), commonly called 'Plaster of Paris'. This material reverts to the dihydrate when mixed with water, setting rapidly to a hard mass and undergoing slight expansion in the process. When heated above 180°C the dihydrate changes into the anhydrous form, which does not hydrate readily on the addition of water.

Calcium sulphate is used in making plaster for building, and as a filler for paper, and in the manufacture of Plaster of Paris, used for taking casts and for immobilizing broken limbs. Two recent uses are important. Anhydrite is now used on a large scale for making sulphuric acid and ammonium sulphate, and gypsum is used to treat soil after it has been flooded by sea water to render it fertile and workable again.

18.21. Calcium Phosphate, $Ca_3(PO_4)_2$. This compound occurs naturally as rock phosphate and as apatite. It is also present in bones and bone ash. It is only sparingly soluble in water, but it dissolves readily in dilute acids (§ 9.11). Its chief uses are in the manufacture of superphosphate fertilizer (§ 23.12) and as the source of elemental phosphorus (§ 23.2).

18.22. Calcium Carbide, CaC_2. Made by heating quicklime with powdered coke to about 2000°C in an electric furnace (see Fig. 18.4):

$$CaO + 3C = CaC_2 + CO\uparrow$$

The molten carbide is run off periodically, cooled, and broken into

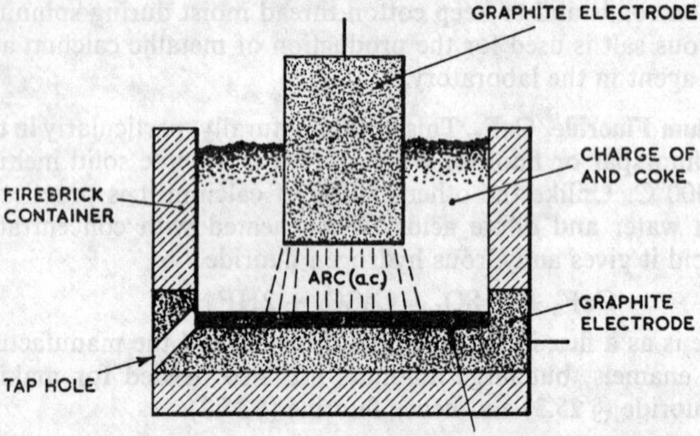

FIG. 18.4 Manufacture of Calcium Carbide

lumps. When pure it is white, but the commercial product is usually grey, or even black, owing to impurities.

It reacts readily with cold water producing the gas acetylene (§ 34.4), which is its main industrial use:

$$CaC_2 + 2H_2O = Ca(OH)_2 + C_2H_2\uparrow$$

When heated strongly in nitrogen it is converted to calcium cyanamide, which is used as a fertilizer:

$$CaC_2 + N_2 = CaNCN + C$$

18.23. Bleaching Powder. When chlorine reacts with calcium hydroxide below 35°C, the product is a white, amorphous solid called bleaching powder. Its exact chemical composition is not known, but it behaves chemically as if it contained calcium hypochlorite, $Ca(OCl)_2$, and basic calcium chloride, $CaCl_2.Ca(OH)_2.H_2O$. The manufacture is carried out in towers designed to ensure that the upflowing chlorine mixes intimately with the finely-powdered slaked lime. The reaction is exothermic, and cooling is necessary to prevent the formation of calcium chlorate.

Bleaching powder reacts with carbon dioxide from the atmosphere producing hypochlorous acid, which accounts for its oxidizing and bleaching actions. With dilute acids it gives chlorine, even in the cold, and this 'available chlorine', as it is called, is usually expressed as a percentage by weight of the solid, reaching about 36% in freshly made material. It can be estimated by adding a known weight of the bleaching powder to a solution containing excess of acidified potassium iodide, and titrating the liberated iodine with a standard solution of sodium thiosulphate using starch as indicator.

When some bleaching powder is warmed with a dilute solution of a cobalt salt, oxygen is evolved rapidly. This is a catalytic reaction, even traces of the cobalt compound bringing about the decomposition.

Bleaching powder is not completely soluble in water, and an aqueous suspension is normally used for bleaching wood pulp, linen and cotton, and also as a disinfectant ('chloride of lime'). Its main advantage over gaseous chlorine arises from its easier storage and safer transport.

18.24. Mortar. This important building material is made by mixing slaked lime with sand and water. It gradually loses moisture on standing, setting to a hard mass which binds together the bricks. The sand renders the mass porous and prevents cracking during this period of setting. In course of time carbon dioxide from the atmosphere converts the lime into calcium carbonate, providing a hard crust:

$$Ca(OH)_2 + CO_2 = CaCO_3 + H_2O$$

18.25. Cement. This is the general name given to various mixtures of silicates, aluminates, and ferrites, formed from limestone and clay. The raw materials are mixed with water to form a sludge or slurry, which is then heated strongly in long rotary kilns, using producer gas or powdered coal as fuels. The dehydrated product is cooled and ground to a fine powder. In use, this powdered cement is mixed with water and sand, when complicated hydration reactions occur, resulting in the formation of a hard mass. The rate of setting and the hardness of the product depend upon the ingredients used. If gravel is mixed with the cement before setting, the product is called *concrete*, a material much used in constructional work. Its strength can be greatly increased by embedding in it steel rods or netting, when it is said to be 'reinforced'.

18.26. Detection of Calcium. Solutions of calcium compounds give:

(1) a white precipitate with ammonium carbonate solution, in the presence of ammonium chloride solution (given also by barium and strontium salts).

(2) a white precipitate with ammonium oxalate solution (given also by barium and strontium salts). This precipitate is insoluble in acetic acid, but it dissolves in dilute hydrochloric acid.

(3) a yellow solution with potassium chromate solution, but no precipitate, which distinguishes calcium salts from those of barium.

(4) no precipitate with saturated calcium sulphate solution, even on standing, which distinguishes calcium salts from those of barium and strontium.

(5) a brick-red flame coloration.

18.27. Strontium, Sr. Electronic configuration: 2,8,18,8,2.

It occurs in the minerals, celestine, $SrSO_4$, and strontianite, $SrCO_3$, neither of which is common. Like calcium, it is extracted by aluminothermic reduction of its oxide. Strontium closely resembles calcium physically and chemically, but tends to be more reactive.

18.28. Strontium Compounds. These are similar to the corresponding compounds of calcium. The chloride, sulphate, nitrate, and chromate of strontium are less soluble than their calcium counterparts, and strontium hydroxide is more strongly basic than calcium hydroxide, but other differences are few. The main use of strontium compounds is in fireworks and flares because of the crimson colour they impart to flames, but strontium hydroxide is also used extensively in sugar-refining (§ 42.4).

18.29. Detection of Strontium. Solutions of strontium compounds give:

(1) white precipitates with ammonium carbonate solution and ammonium oxalate solution, as do calcium and barium compounds.

(2) a white precipitate, especially on standing for a while, with a saturated solution of calcium sulphate; (barium compounds give an immediate precipitate and calcium compounds none at all).

(3) a yellow precipitate, soluble in acetic acid, with a concentrated solution of potassium chromate; (barium chromate is insoluble in acetic acid).

(4) a crimson flame coloration.

18.30. Barium, Ba. Electronic configuration: 2,8,18,18,8,2.

It occurs naturally as barytes, $BaSO_4$, and witherite, $BaCO_3$. The metal is extracted by heating the oxide strongly with powdered aluminium *in vacuo*. Barium resembles calcium and strontium physically and chemically, but is more reactive than either. For example, it inflames spontaneously in moist air, and it reacts vigorously with even cold water.

18.31. Barium Oxide, BaO. When heated to about 500° C in air, it combines with oxygen to form the peroxide, which breaks up again at 800°C:

$$2BaO + O_2 \rightleftharpoons 2BaO_2$$

This was the basis of the now obsolete Brin's process for obtaining oxygen from the air.

When barium oxide is slaked, the hydroxide produced, which is often referred to as 'baryta water', is a stronger base than the hydroxides of calcium and strontium. Barium oxide is used to coat the cathodes of radio valves because it markedly increases the emission of electrons, and for manufacturing barium peroxide (§ 16.6).

18.32. Barium Sulphate, $BaSO_4$. It occurs naturally as barytes. It is a dense, white, very insoluble compound, which is precipitated when dilute sulphuric acid or a soluble sulphate is added to a solution of a barium salt:

$$BaCl_2 + Na_2SO_4 = BaSO_4\downarrow + 2NaCl$$

It is used as a filler for paper (especially wallpaper), rubber, soap, and linoleum, because it gives opacity and body to these materials. A similar use is as a pigment in paints, usually mixed with zinc sulphide. Because it is also opaque to X-rays, barium sulphate is given to patients to swallow before radiography of their digestive organs.

18.33. Detection of Barium. Solutions of barium salts give:

(1) white precipitates with ammonium carbonate solution and ammonium oxalate solution.

(2) an immediate white precipitate with dilute sulphuric acid.

(3) a yellow precipitate with potassium chromate solution, insoluble in acetic acid.

(4) a yellow-green flame coloration.

18.34. Summary. This study of the alkaline earth metals reveals their similarity physically and chemically. They are all divalent and strongly electropositive, standing high in the electrochemical series (§ 10.4) and showing a strong affinity for electronegative elements, with which they combine exothermically to form stable compounds. For example, their oxides have very high heats of formation, are not easily reduced by chemical means, and dissolve in water giving basic hydroxides. All their compounds are colourless, except where the anion is coloured, and at room temperature they are not appreciably hydrolysed in solution. Most of their salts are soluble in water, but their carbonates, phosphates, fluorides, and oxalates generally are not. Although the metals themselves show individual differences, all are grey and moderately hard, with high melting and boiling points and low densities.

Despite these similarities, the alkaline earth metals show certain well defined differences, as indicated in Table 18.1. The most important is a marked increase in electropositivity with increasing atomic weight.

TABLE 18.1. DIFFERENCES BETWEEN THE GROUP IIA METALS

Feature	Be	Mg	Ca	Sr	Ba
Melting point/°C	1280	650	851	750	710
Hardness	Hard	Moderately hard			Fairly soft
Ignition in air	Ignites only above melting points		Ignites on heating		Inflames spontaneously
Action with water	Slowly with steam	Very slow in cold Steadily with steam	Steadily with cold water	Readily with cold water	Vigorously in the cold
With hydrogen	Hydride not formed directly, even on heating		Combine directly on heating		Forms hydride without heating
Oxide character	Amphoteric	Mildly basic	Strongly basic		Very strongly basic
Solubility of hydroxide	Very insoluble	Sparingly soluble	Slightly soluble		Soluble
Stability of peroxide	Peroxide is unknown	Unstable	Decomposes on heating		Stable up to 800°C
Stability of carbonate	Decomposes on gently heating	Decomposes at 550°C	Decomposes at 900°C	Decomposes at 1100°C	Decomposes only above 1400°C
Solubility of sulphate	Soluble	Soluble	Slightly soluble	Sparingly soluble	Very insoluble
Solubility of chromate	Soluble	Soluble	Soluble	Sparingly soluble	Insoluble
Chloride melting point/°C	405	715	780	870	960
Solubility of fluoride	Soluble	Sparingly soluble	Insoluble	Insoluble	Insoluble

The chemistry of the alkaline earth metals can be explained in terms of their atomic structures. Each atom has two electrons in its outermost shell and these are weakly bound compared with the others. In chemical combination these two electrons are given to electron-accepting atoms of electronegative elements, leaving stable positively charged ions with inert gas configurations. This tendency is greatest in the largest atoms in the Group, where the attractive force of the nucleus is least. Thus, with the exception of beryllium, which differs from the other members of the Group in this respect because of the small size of its atom, all the compounds of these metals are ionic and display the characteristic properties of electrovalency.

The alkaline earth metals are compared with the alkali metals in Table 18.2.

TABLE 18.2. COMPARISON OF THE METALS OF GROUPS IA AND IIA

Alkali Metals	*Alkaline Earth Metals*
1. Soft.	Moderately hard.
2. Low melting points (28°C — 180°C).	Much higher melting points (700°C — 1280°C).
3. Reactive and highly electropositive.	Reactive and highly electropositive.
4. Hydroxides very strongly basic.*	Hydroxides less strongly basic.
5. Carbonates not decomposed below melting point.*	Carbonates decomposed to oxides on strong heating.
6. Heated nitrates give nitrite + $O_2\uparrow$.	Heated nitrates give oxide, $NO_2\uparrow + O_2\uparrow$
7. Hydroxides stable to heat.*	Hydroxides give oxides on heating.
8. Hydrogen carbonates stable solids at room temperature.*	Hydrogen carbonates known only in solution.
9. Nearly all compounds soluble in water.*	Many compounds soluble in water, but not carbonates, phosphates, sulphates, and fluorides.
10. Colourless ions.	Colourless ions.
11. Compounds non-volatile.	Compounds non-volatile.
12. Molten compounds conduct electricity.	Molten compounds conduct electricity (except Be).
13. Characteristic flame colorations.	Characteristic flame colorations.
14. Crystalline hydrides containing H^-	Crystalline hydrides containing H^-.
15. Salts with weak acids hydrolysed in solution.	Salts with weak acids not hydrolysed at room temperature.
16. Invariably univalent.	Invariably divalent.
17. Form peroxides on heating.	Only barium forms a peroxide directly.
18. Do not form nitrides directly.*	Form stable nitrides on heating in N_2.
19. Do not form carbides directly.*	Form stable carbides on heating with carbon.

* Except lithium.

19

Boron and Aluminium

19.1. Introduction. Of the elements in Group IIIB of the Periodic Classification, only the first two, boron and aluminium, are sufficiently important to be studied here.

19.2. Boron, B. Electric configuration: 2,3.

OCCURRENCE: None occurs free in nature, but deposits of borax, $Na_2B_4O_7.10H_2O$, and kernite, $Na_2B_4O_7.4H_2O$ exist, and boric acid, H_3BO_3, is sometimes found in the waters of hot springs.

EXTRACTION: Boric oxide is reduced by heating it strongly with a highly electropositive metal, such as magnesium or sodium, in the absence of air:

$$B_2O_3 + 3Mg = 3MgO + 2B$$

When the product is boiled with dilute hydrochloric acid and filtered, the other substances all dissolve leaving the powdered element.

Alternatively, if boron tribromide is reduced by heating to about 1300°C in hydrogen, a crystalline form of boron is obtained:

$$2BBr_3 + 3H_2 = 2B + 6HBr$$

PROPERTIES OF BORON: The amorphous form is a dark brown, very hard powder, melting at about 2000°C. Crystalline boron is black and very hard.

REACTIONS OF BORON: The crystalline material is extremely inert, but amorphous boron combines with fluorine at room temperature, and with chlorine, bromine, and sulphur on heating. Heated above 700°C in air it ignites forming the oxide and nitride:

$$4B + 3O_2 = 2B_2O_3$$
$$2B + N_2 = 2BN$$

When heated with concentrated sulphuric acid or concentrated nitric acid, boron is oxidized to boric acid. It reacts with fused alkalis forming metaborates:

$$2B + 2NaOH + 2H_2O = 2NaBO_2 + 3H_2\downarrow$$

USES OF BORON: The element is used to make boron steels, which are renowned for their hardness and which are widely used as control rods in atomic reactors (§ 31.11) because boron absorbs neutrons so readily. Boron and borides are used as abrasives. Boron is an essential element in plant metabolism and fertile soil must contain traces of boron compounds; a deficiency leads to disease and low yields. Boron compounds are becoming increasingly important as rocket fuels because of their very high energy/weight ratio.

19.3. Boric Acid, (Orthoboric Acid), H_3BO_3. Made by treating a hot solution of a borate with hydrochloric acid, when crystals of boric acid separate on cooling. Being much more soluble in hot water than in cold, it is easily purified by recrystallization. It forms colourless crystals which undergo progressive dehydration on heating:

At 100°C: $H_3BO_3 = H_2O + HBO_2$ (metaboric acid).
Above 200°C: $2HBO_2 = B_2O_3 + H_2O$

Boric oxide is a colourless, glassy solid which shows amphoteric properties and dissolves readily in water to form a solution of boric acid.

Boric acid is a very weak acid ($K = 5.8 \times 10^{-10}$ mol dm^{-3}), and its salts with the alkali metals are therefore extensively hydrolysed in solution. It is used medicinally as an antiseptic (boracic acid powder) and as a preservative of food in some countries.

19.4. Sodium Tetraborate, (Borax), $Na_2B_4O_7.10H_2O$. Some occurs naturally, but most borax is made from the mineral kernite, found in large deposits in California, by dissolving it in hot water and crystallizing. Above 60°C, the pentahydrate separates, but below this temperature the solution deposits the decahydrate. The anhydrous salt is obtained by heating these hydrates to about 400°C.

When heated strongly borax swells to a white frothy mass, and then fuses, giving a colourless, glassy solid on cooling:

$$Na_2B_4O_7 = 2NaBO_2 + B_2O_3$$

This product readily combines with metallic oxides on heating, giving borates, some of which (e.g. Cr, Co, Cu, Mn) display characteristic colours in the borax bead test in qualitative analysis.

Borax dissolves in water giving a strongly alkaline solution by hydrolysis:

$$Na_2B_4O_7 + 7H_2O \rightleftharpoons \underset{\text{(very weak acid)}}{4H_3BO_3} + \underset{\text{(very strong base)}}{2NaOH}$$

The boric acid is too weak to change the colour of methyl orange, so with this indicator borax can be used to standardize hydrochloric acid, when the following relationship applies:

$$Na_2B_4O_7 \equiv 2NaOH \equiv 2HCl.$$

The addition of glycerol or mannitol to boric acid converts it into a much stronger, monobasic, complex acid which can be titrated against solutions of alkalis using phenolphthalein as indicator.

USES OF BORAX: The principal use is for vitreous enamelling baths and domestic appliances; the glazed surface is resistant to heat, stains and scratches. It is also used for glazing tiles and pottery, as a flux in soldering, and to produce a good finish in laundering. Borax is used to make borosilicate glass, e.g. 'Pyrex', which is valuable in the home and the laboratory because of its very low coefficient of thermal expansion. Another use of borax is to make peroxyborates (e.g. $NaBO_2.H_2O_2.3H_2O$), which are important cleansing and bleaching agents, present in proprietary washing powders and in numerous pharmaceutical products. In the laboratory borax is used for standardizing acids, for making buffer solutions (§ 9.14), and in the borax bead test in analysis.

19.5 Boron Trifluoride, BF_3. When boric oxide is heated with calcium fluoride and concentrated sulphuric acid, a colourless gas, boron trifluoride, is obtained:

$$B_2O_3 + 3CaF_2 + 3H_2SO_4 = 2BF_3\uparrow + 3CaSO_4 + 3H_2O$$

The gas, which boils at −100°C, fumes in moist air and reacts readily with water giving boric acid and fluoroboric acid solution:

$$4BF_3 + 3H_2O = H_3BO_3 + 3HBF_4$$

Many compounds containing a lone pair of electrons will combine with boron trifluoride by forming a co-ordinate bond (see § 14.5) so increasing the number of electrons in the outermost shell of the boron atom to eight. For example, with ammonia and with diethyl ether it forms the following compounds:

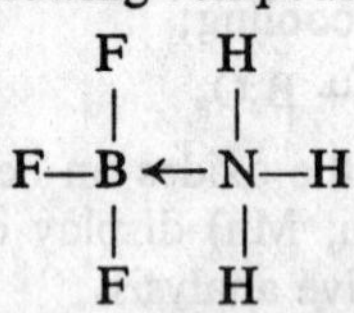

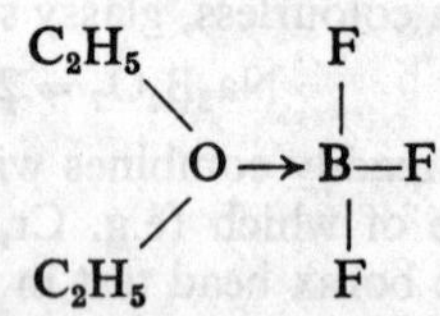

This willingness to accept electrons explains the importance of boron trifluoride as a catalyst in organic chemistry where it is extensively used in polymerization and reforming processes (§ 33.11).

19.6. Detection of Borates. When a borate is mixed in an evaporating basin with concentrated sulphuric acid and alcohol and ignited, the mixture burns with a green-edged flame owing to the presence of volatile ethyl borate. To avoid possible confusion, copper and barium salts must be shown to be absent.

19.7. Sodium Tetrahydridoborate (Sodium Borohydride) $NaBH_4$. This is made by heating trimethyl borate (prepared by esterifying boric acid with methyl alcohol) with powdered sodium hydride at 250°C and extracting the cooled product with liquid ammonia which dissolves the borohydride but not the sodium methoxide:

$$4NaH + B(OCH_3)_3 = NaBH_4 + 3CH_3ONa$$

It forms white non-volatile crystals, stable in dry air below 300°C, which are soluble in water and alcohol but insoluble in ether. Ionic in structure, these crystals contain the tetrahedral BH_4^- ion which is isoelectronic with CH_4 and NH_4^+.

Sodium borohydride is a valuable mild reducing agent. It is rapidly hydrolysed by acidic solutions yielding hydrogen:

$$NaBH_4 + 4H_2O = NaOH + H_3BO_3 + 4H_2\uparrow$$

It reduces iron (III) solutions to iron(II) and can be used to prepare inorganic hydrides. In organic chemistry it is used to reduce the carbonyl group $>C{=}O$ in aldehydes, ketones, and acid chlorides (§§ 35.2 and 35.4), but it does not reduce alkenes, carboxylic acids, or amides. Its main advantage over the more powerful lithium aluminium hydride (§ 19.14) is that sodium borohydride can be used in aqueous and alcoholic solution. The mechanism of its nucleophilic reducing action is discussed in § 50.7.

19.8. Aluminium, Al. Electronic configuration: 2,8,3.

OCCURRENCE: Although aluminium does not occur naturally as the metal, it is abundant as silicates in rocks and clays. The principal source of aluminium is the mineral bauxite, $Al_2O_3.H_2O$ and $Al_2O_3.3H_2O$, huge deposits of which are found in various parts of the world. Another important mineral is cryolite, Na_3AlF_6, found in Greenland.

EXTRACTION: Before studying the details of the modern process it is instructive to consider the peculiar difficulties associated with the extraction of this, the commonest metal in the earth's crust. No economic process has yet been devised for extracting aluminium from rocks and clays, owing to their high silica content. Aluminium oxide is not reduced by heating it in hydrogen; strongly heating it with carbon only produces the carbide, Al_4C_3. The anhydrous chloride cannot be electrolysed because it is covalent and virtually non-conducting, and the oxide does not melt below 2000°C, making its direct electrolysis impracticable. The problem has been solved by electrolysing a solution of aluminium oxide in molten cryolite. The extraction proceeds in two stages:

(*a*) *Purification of Bauxite:* Because it is difficult and expensive to remove impurities from the metal after electrolysis, the crude bauxite has to be purified thoroughly first. In the Baeyer process this is achieved by dissolving the powdered mineral in a hot concentrated solution of sodium hydroxide under pressure, when sodium aluminate solution forms:

$$2NaOH + Al_2O_3 = 2NaAlO_2 + H_2O$$

Iron oxide, titanium dioxide, and most of the silicates present do not dissolve in the alkali and are filtered off as a sludge. The filtrate is diluted and 'seeded' with aluminium hydroxide crystals, and stirred for several days, when hydrolysis of the aluminate occurs thus:

$$NaAlO_2 + 2H_2O = Al(OH)_3\downarrow + NaOH$$

The precipitated hydroxide is filtered off, washed, dried, and heated strongly to the pure oxide, whilst the sodium hydroxide solution is concentrated and used again.

(*b*) *The Electrolysis:* The type of cell used is shown in Fig. 19.1. The purified oxide is added periodically to the melt, which is maintained at

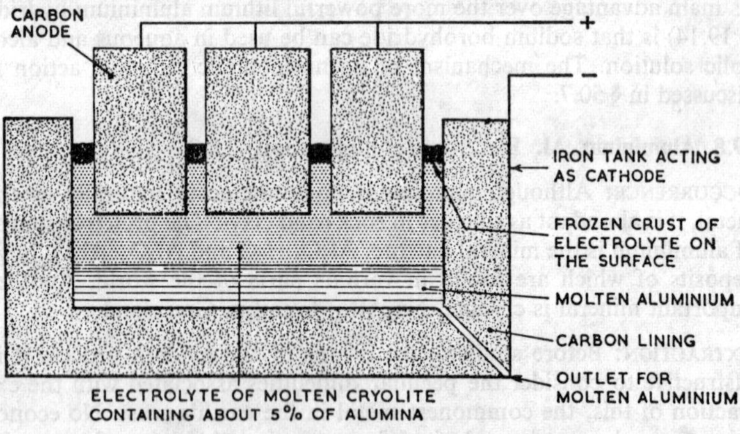

FIG. 19.1 Manufacture of Aluminium

about 950°C by the heating effect of the huge current. Aluminium ions are discharged at the cathode and the molten metal, which collects at the base of the cell, is drawn off from time to time. The cryolite, which is made synthetically, is only used up in small amounts, and the overall change is:

$$Al_2O_3 \rightleftharpoons 2Al^{3+} + 3O^{2-}$$

The oxygen liberated at the anode attacks and gradually burns away the carbon blocks, which have to be continually replaced, adding substantially to the cost of the process. The product is about 99.5% pure, containing traces of iron and silica. For each kg of aluminium produced about 150 g of sodium hydroxide, 500 g of carbon and 6×10^7 J of electricity are consumed, which explains why aluminium production is restricted to countries having abundant electric power.

Properties of Aluminium: It is a highly lustrous, white metal, melting at 660°C and boiling at about 2400°C. Aluminium is malleable and ductile. It is a very good conductor of heat and electricity. An important property is its low density, 2.7 g cm^{-3}, which is only about a third of that of iron.

Reactions of Aluminium: Although highly electropositive, it is resistant to corrosion. It is not attacked by water or steam, and it is converted into the oxide only when heated above 800°C. This anomalous behaviour is explained by the existence of a thin coherent layer of oxide, which forms on the surface of the metal as soon as it is exposed to moist air, and which prevents further reaction. This also explains why nitric acid, even when concentrated, has no action on the metal. If this protective film is removed by rubbing the surface with mercury or mercuric chloride solution, corrosion takes place rapidly and the metal is soon covered with a mass of white aluminium hydroxide. This is the basis of the reducing action of the aluminium-mercury couple.

When heated aluminium combines directly with halogens, oxygen, sulphur, nitrogen, phosphorus, and carbon, all the reactions being exothermic. The metal dissolves readily in warm hydrochloric acid, releasing hydrogen, and in hot concentrated sulphuric acid yielding sulphur dioxide:

$$2Al + 6HCl = Al_2Cl_6 + 3H_2\uparrow$$
$$2Al + 6H_2SO_4 = Al_2(SO_4)_3 + 3SO_2\uparrow + 6H_2O$$

Aluminium is attacked by solutions of sodium and potassium hydroxide, particularly when they are warm and concentrated, releasing hydrogen and forming an aluminate solution:

$$2Al + 2NaOH + 2H_2O = 2NaAlO_2 + 3H_2\uparrow$$

If powdered aluminium is mixed with the oxide of most of the less electropositive metals and the mixture is ignited with a fuse of burning magnesium, a vigorous and highly exothermic reaction takes place, and the oxide is reduced to the metal. This is the basis of the *Goldschmidt process* for preparing metals such as calcium, strontium, barium, manganese, chromium, and molybdenum from their oxides. A mixture of powdered aluminium and ferric oxide is known as *thermite*; it is used in incendiary bombs and for welding steel:

$$Fe_2O_3 + 2Al = Al_2O_3 + 2Fe$$

The theoretical explanation of the ability of aluminium to reduce other metallic oxides in this way is given in terms of free energy in § 11.13.

USES OF ALUMINIUM: As might be expected from its physical properties these are very numerous. It is used for making cooking utensils, brewing vats, and food processing vessels since aluminium compounds are non-poisonous, and for making electrical conductors for transmission lines, where the saving in weight compared with copper permits wider spacing of pylons. In the form of foil, aluminium is widely used for packaging and for capping milk-bottles. When powdered and suspended in oil it gives a useful paint, used for mirrors on account of its high reflectivity. Large amounts of aluminium are converted into alloys, such as those containing a few per cent of magnesium and copper, which are harder and stronger than pure aluminium but possess the same valuable properties of lightness and corrosion resistance, and are useful for constructing aircraft, cars, and the superstructure of ships. Aluminium is also used in buildings and constructional work, for the thermite reaction, and for removing dissolved oxygen and nitrogen from molten metals to prevent blow-holes in castings.

By making an aluminium article the anode of a cell in which dilute chromic or sulphuric acid is the electrolyte, it is possible to produce a thicker and harder film of aluminium oxide on the surface of the metal than exists naturally. This process, called *anodizing*, gives increased resistance to corrosion and by absorbing dyes can provide decorative finishes to a wide range of aluminium articles (see § 15.6).

19.9. Aluminium Oxide, (Alumina), Al_2O_3. It occurs naturally not only as bauxite, but also as corundum, ruby, emerald, topaz, sapphire, and amethyst, the characteristic colours of these forms being caused by impurities. It can be made by heating the hydroxide strongly:

$$2Al(OH)_3 = Al_2O_3 + 3H_2O\uparrow$$

It is a white crystalline powder, renowned for its hardness and infusibility. Besides being the source of aluminium metal, it is used as an abrasive, for making refractories and ceramics, and for adsorption in chromatography (§ 12.12).

19.10. Aluminium Hydroxide, $Al(OH)_3$. Made by adding ammonium hydroxide solution to a solution of an aluminium salt:

$$Al_2(SO_4)_3 + 6NH_4OH = 2Al(OH)_3\downarrow + 3(NH_4)_2SO_4$$

It is thrown down as a white gelatinous precipitate which becomes crystalline on warming.

Aluminium hydroxide is amphoteric, dissolving in acids to give salts containing the hydrated Al^{3+} ion and in caustic alkalis to form aluminates containing the hydrated AlO_2^- ion:

$$Al^{3+} + 3OH^- \rightleftharpoons Al(OH)_3 \rightleftharpoons AlO_2^- + H^+ + H_2O$$

e.g.
$$Al(OH)_3 + 3HCl = AlCl_3 + 3H_2O$$
$$Al(OH)_3 + NaOH = NaAlO_2 + 2H_2O$$

Thus when caustic alkalis are added to solutions of aluminium salts, the precipitated hydroxide redissolves in excess.

The extensive use of aluminium hydroxide as a mordant arises from its marked capacity, when freshly prepared, for adsorbing dyes (§ 12.9). The usual procedure is to precipitate the hydroxide on the fibres of the cloth and then dip the latter into the dyestuff, when some of the dye will be adsorbed and will cling permanently to the fibres.

19.11. Aluminium Chloride, $AlCl_3$ or Al_2Cl_6. When aluminium is dissolved in hydrochloric acid and the solution is concentrated, colourless crystals of the hexahydrate, are deposited:

$$2Al + 6HCl = Al_2Cl_6 + 3H_2\uparrow$$

The anhydrous chloride cannot be made by heating the hydrate in air, because hydrolysis occurs, so it is usually made by passing dry chlorine or hydrogen chloride over heated aluminium, or by strongly heating alumina and carbon in an atmosphere of chlorine:

$$2Al + 3Cl_2 = Al_2Cl_6$$
$$Al_2O_3 + 3C + 3Cl_2 = Al_2Cl_6 + 3CO.$$

Anhydrous aluminium chloride is an extremely deliquescent, crystalline solid which sublimes at 183°C and dissolves freely in organic liquids. It fumes in moist air, owing to hydrolysis.

It is evident from its properties that aluminium chloride has a covalent structure. If the molecular formula is $AlCl_3$, then there are only six electrons present in the outermost shell of each atom of aluminium; it is believed that these simple molecules are linked together in pairs by co-ordinate bonds, as shown below, so that each atom of aluminium then has eight electrons in its outermost shell. Vapour density measurements, which confirm this belief, reveal that dissociation into $AlCl_3$ molecules starts above 400°C and is complete above about 800°C:

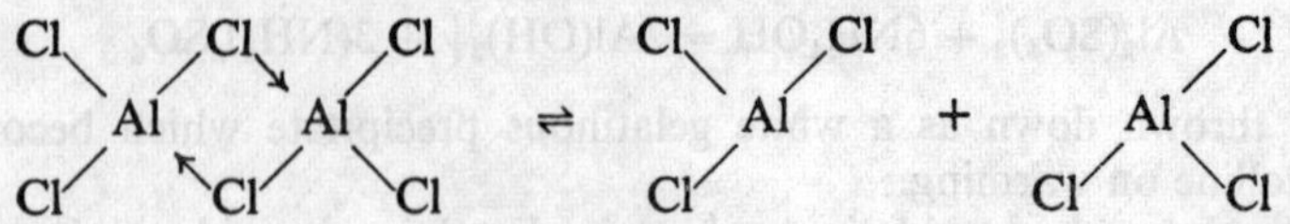

The anhydrous salt is an important catalyst, used in the Friedel–Crafts reaction (§ 46.2) and in the 'cracking' of petroleum (§ 33.11).

Anhydrous aluminium chloride reacts vigorously with water. The covalent chloride is rapidly converted into a hydrated aluminium cation and a chloride anion, the energy of hydration of the small highly-charged aluminium ion being sufficient to separate these ions completely in solution:

$$AlCl_3 + 6H_2O = [Al(H_2O)_6]^{3+} + 3Cl^-$$

This hexa-aquo complex ion in which the six water molecules are disposed octahedrally about the central atom, linked to it by their lone pairs of electrons, is present in crystals of the hexahydrate and in solutions of aluminium salts. It is acidic because it dissociates readily in solution yielding hydroxonium ions by a series of changes in which the co-ordinated water molecules successively give up hydrogen ions thus:

$$[Al(H_2O)_6]^{3+} + H_2O \rightleftharpoons [Al(H_2O)_5.OH]^{2+} + H_3O^+$$
$$[Al(H_2O)_5.OH]^{2+} + H_2O \rightleftharpoons [Al(H_2O)_4.(OH)_2]^+ + H_3O^+$$

If OH^- ions are added to the solution the process goes a stage further and hydrated aluminium hydroxide is precipitated as a gelatinous mass:

$$[Al(H_2O)_4.(OH)_2]^+ + OH^- \rightleftharpoons [Al(H_2O)_3.(OH)_3]\downarrow + H_2O$$

An excess of alkali causes this sequence to continue with the formation of the soluble anions $[Al(OH)_4.(H_2O)_2]^-$, $[Al(OH)_5.H_2O]^{2-}$, and $[Al(OH)_6]^{3-}$, known as *aluminates*, which explains the amphoteric character of aluminium hydroxide.

A similar explanation, i.e. partial dissociation of aquo complexes in solution, accounts for the acidity displayed by salts of other metals such as iron, chromium, copper, and zinc.

19.12. Aluminium Sulphate, $Al_2(SO_4)_3$. Made by dissolving aluminium, or its oxide or hydroxide, in hot concentrated sulphuric acid and concentrating the solution, when colourless crystals of $Al_2(SO_4)_3.18H_2O$ separate on cooling. Dissolved in water they give an acidic solution by hydrolysis.

USES OF ALUMINIUM SULPHATE: It is much used for 'sizing' paper, i.e. it is added to the pulp to give it body and strength. It is also used for tanning leather, for waterproofing cloth, and as a mordant for cotton, which is dipped in turn in solutions of the sulphate and sodium hydroxide in order to precipitate aluminium hydroxide on the fibres before dyeing. Solutions of aluminium sulphate are used extensively in sewage treatment and water purification to bring about the coagulation of negatively-charged colloidal particles of organic matter (§ 12.3).

19.13. The Alums. This series of double sulphates has the general formula:

$$M_2SO_4 . N_2(SO_4)_3 . 24H_2O,$$

where M is a univalent ion such as Na^+, K^+, or NH_4^+, and N is a tri-valent ion such as Fe^{3+}, Al^{3+}, or Cr^{3+}.

e.g.	$K_2SO_4 . Al_2(SO_4)_3 . 24H_2O$	potash alum
	$(NH_4)_2SO_4 . Al_2(SO_4)_3 . 24H_2O$	ammonium alum
	$K_2SO_4 . Cr_2(SO_4)_3 . 24H_2O$	chrome alum
	$(NH_4)_2SO_4 . Fe_2(SO_4)_3 . 24H_2O$	ferric alum.

They are formed by mixing hot concentrated solutions of their component sulphates and cooling, when the alum crystals are deposited because of their low solubility in cold water. They are easily purified by recrystallization because their solubility increases markedly with rising temperature.

The following special methods of preparation are sometimes used: *Potash Alum:* Enough aluminium foil is added to some caustic potash solution to leave some undissolved at the end of the reaction. The solution of potassium aluminate is filtered, and dilute sulphuric acid is added until the solution is acidic (four gram-equivalents of acid are required for every gram-equivalent of alkali taken). The solution is then concentrated until it deposits crystals of the alum on cooling. These are filtered off, washed with a little cold water and dried; they can be recrystallized if necessary. Equations for the reactions are:

$$2Al + 2KOH + 2H_2O = 2KAlO_2 + 3H_2\uparrow$$

$$2KAlO_2 + 4H_2SO_4 = \underbrace{K_2SO_4 + Al_2(SO_4)_3}_{\text{alum}} + 4H_2O$$

Chrome Alum: A steady stream of sulphur dioxide is passed through a solution of potassium dichromate, acidified with sulphuric acid, until the solution turns green. Deep purple crystals of chrome alum separate from the concentrated solution on cooling. The equation for the reduction is:

$$K_2Cr_2O_7 + 3SO_2 + H_2SO_4 = \underbrace{K_2SO_4 + Cr_2(SO_4)_3}_{\text{alum}} + H_2O$$

The crystals are filtered off, washed, and dried.

The alums are isomorphous with each other (§ 2.13), readily forming overgrowths; their crystals are usually octahedral in shape. It is important to realize that the alums are double salts and not complex salts. In solution they behave simply as a mixture of their component sulphates and give the reactions of their individual cations.

19.14. Lithium Aluminium Hydride, Lithium Tetrahydridoaluminate, $LiAlH_4$. Made by adding lithium hydride to anhydrous aluminium chloride in ethereal solution, filtering, and evaporating the solvent:

$$4LiH + AlCl_3 = LiAlH_4 + 3LiCl\downarrow$$

It is a white solid which reacts vigorously with water, releasing nascent hydrogen:

$$LiAlH_4 + 4H_2O = LiOH + Al(OH)_3 + 4H_2\uparrow$$

This leads to its use in preparing inorganic hydrides and for reducing transition metals to lower oxidation states.

Lithium aluminium hydride is a powerful reducing agent widely used in organic chemistry because of its rapid action, its high yield, and its convenience. In ethereal solution it reduces the carbonyl group $>C{=}O$ in aldehydes, ketones, carboxylic acids, esters, amides, and acid chlorides and the cyanide group $-C{\equiv}N$ in nitriles. It does not reduce the ethylenic bond $>C{=}C<$ and so can be used for selective reductions in unsaturated compounds. The mechanism of its nucleophilic reducing action is discussed in § 50.7.

19.15. Detection of Aluminium. Solutions of aluminium salts give a white gelatinous precipitate of aluminium hydroxide on addition of alkalis. If sodium or potassium hydroxide solution is used, the precipitate redissolves in excess. A specific test for aluminium depends upon the outstanding adsorptive power of freshly precipitated aluminium hydroxide; ammonium hydroxide solution is added to a solution of an aluminium salt in the presence of a little *aluminon* (sodium alizarin sulphonate) and on adding a little acetic acid a red precipitate is obtained.

19.16. Summary. The two elements studied in this chapter are so dissimilar that at first sight it is not easy to justify their inclusion in the same Group. Boron is a non-metal with a very high melting point and poor electrical conductivity. Its oxide is weakly acidic and very soluble in water. In all its compounds boron is covalent, and the B^{3+} ion does not exist. On the other hand aluminium is a strongly electropositive metal with fairly low melting point and very good electrical conductivity.

Its oxide is amphoteric and insoluble in water. Aluminium forms both electrovalent and covalent compounds. For example, the oxide, fluoride, and sulphate are ionic, containing the Al^{3+} ion, but the chloride is predominantly covalent. Aluminium is also capable of forming complex anions; these include the AlO_2^- ion in aluminates, the AlH_4^- ion in lithium aluminium hydride, and the AlF_6^{3-} ion in cryolite. Its salts with strong acids are colourless, soluble in water, frequently hydrated, and hydrolysed in solution, and its salts with weak acids, where they exist at all, are instantly hydrolysed by cold water to the hydroxide. The reactivity of aluminium is, however, much less than would be expected of so electropositive a metal because a protective film of oxide covers its surface.

The atoms of both boron and aluminium are alike in having three electrons in their outermost shells. The differences between these two elements arise principally from the big difference in the size of their atoms. In boron the atom is small and compact, and the nuclear charge exerts a powerful attractive force upon the outermost electrons, preventing their escape. This ensures a complete absence of metallic character in the element and makes electrovalency by the loss of electrons impossible. In the larger aluminium atom the outermost electrons are not nearly so strongly held, and the binding force is insufficient to prevent their escape to strongly electron-accepting groups with the formation of the hexa-aquo ion $[Al(H_2O)_6]^{3+}$ in its hydrated salts.

There are marked similarities between beryllium and aluminium and between boron and silicon. These 'diagonal relationships' are explained in § 13.10.

20

Carbon and Silicon

20.1. Introduction. Of the elements in Group IVB of the Periodic Classification, viz. carbon, silicon, germanium, tin and lead, we shall study the chemistry of only the first two here, leaving the others to be considered in the next chapter.

20.2. Carbon, C. Electronic configuration: 2,4.

OCCURRENCE: The element occurs naturally as diamond and graphite, and in an impure form as coal. Combined with other elements carbon occurs in petroleum and natural gases, in minerals such as limestone and dolomite, and as carbon dioxide in the atmosphere. It is an essential constituent of all forms of plant and animal life.

REACTIONS OF CARBON: It is an infusible, unreactive substance, insoluble in water, alkalis, or acids. When heated with metallic elements it forms carbides, many of which are hard infusible materials used as abrasives. It also combines with oxygen, fluorine and sulphur when heated. It is a good reducing agent, converting many metallic oxides to the metals on heating (§ 10.8) and reducing steam to hydrogen at high temperatures (§ 20.15).

20.3. Allotropy of Carbon. Two crystalline allotropes are known, diamond and graphite. In addition other forms of carbon exist, known collectively as 'amorphous carbon', but recent work has shown that these varieties often consist of minute particles of graphite orientated at random.

DIAMOND: This allotrope is found as small crystals sparsely embedded in silicate rock in various parts of the world, particularly in Africa. When pure it is colourless and transparent, with a density of 3.51 g cm^{-3}. Diamond is the hardest naturally-occurring substance known, which accounts for its uses in rock-drilling and glass-cutting, for making bearings, as dies for drawing wire, and as an abrasive generally. Its high refractive index and dispersive power give it a sparkling brilliance which makes it highly prized as jewellery; colourless stones are broken or cut in such a way as to enhance this effect.

Diamond is chemically inert. It burns to carbon dioxide on heating above 800°C in oxygen and combines with fluorine at about 700°C forming carbon tetrafluoride, but it is unaffected by acids and alkalis. There have been many attempts to make artificial diamonds, but only in recent years have they met with success. When a real diamond is examined with X-rays, they reveal its structure to be a three-dimensional network of covalently linked equidistant carbon atoms arranged tetrahedrally in space as in Fig. 20.1. Thus a crystal of diamond is effectively one giant molecule, consisting of billions of carbon atoms linked rigidly together, which explains its great hardness. Melting a substance with this structure necessarily involves breaking a vast number of strong covalent bonds; this requires a great amount of energy and explains the infusibility of diamond.

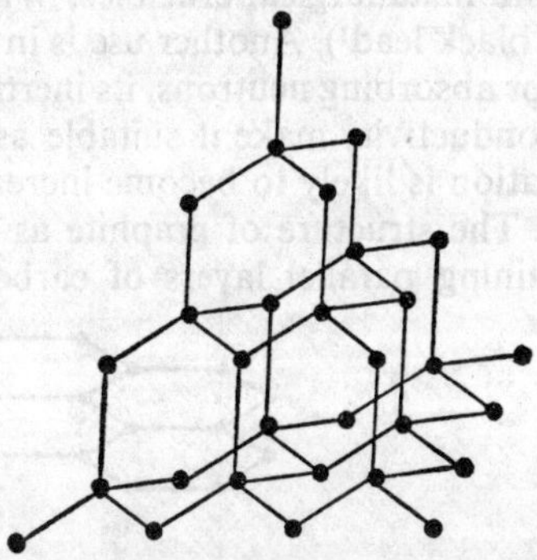

FIG. 20.1 The Structure of Diamond

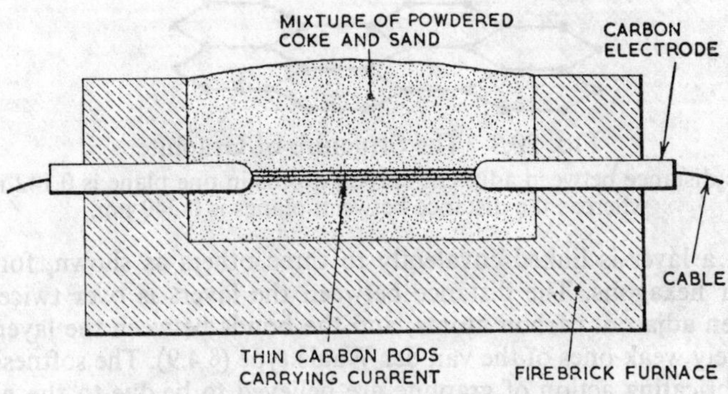

FIG. 20.2 Manufacture of Graphite

GRAPHITE: Deposits of this allotrope occur in many localities, but the supply from natural sources is inadequate and has to be supplemented by manufactured material. In the Acheson process it is made by heating powdered coke with sand in an electric furnace for about twenty hours (see Fig. 20.2). The process is operated where hydro-electric power is plentiful, e.g. at Niagara, U.S.A.

Graphite is an opaque solid with a grey metallic lustre and a soft greasy feel. It is less dense than diamond (its density is 2.3 g cm^{-3}), from which it also differs in being a good conductor of heat and electricity. Graphite is used as a lubricant, particularly where high temperatures make oil unsuitable, and for making electrodes, dynamo brushes, and metallurgical crucibles. When mixed with clay it is used in pencils ('black lead'). Another use is in nuclear reactors, where its low capacity for absorbing neutrons, its inertness, its infusibility and its good thermal conductivity make it suitable as a moderator (see § 31.11). This application is likely to become increasingly important in future years.

The structure of graphite as revealed by X-ray analysis is one containing parallel layers of carbon atoms, as in Fig. 20.3. Each atom

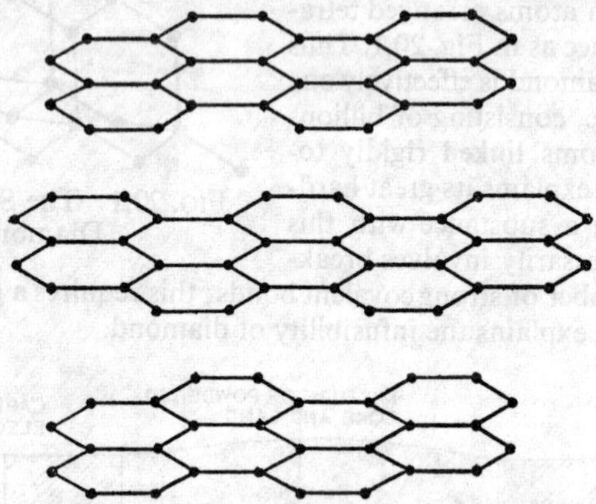

FIG. 20.3 The Structure of Graphite

The distance between adjacent carbon atoms in one plane is 0.142 nm whereas the distance between planes is 0.340 nm.

within a layer is linked covalently to three others as shown, forming regular hexagons. The distance between the layers is over twice that between adjacent carbon atoms, and the bonds between the layers are relatively weak ones of the van der Waals type (§ 4.9). The softness and the lubricating action of graphite are believed to be due to the ability of adjacent layers to slide over each other when under stress.

AMORPHOUS CARBON: The principal varieties are:

Coke: This is formed when coal is carbonized, i.e. heated strongly in the absence of air (§ 20.13). The chief uses of coke are as a fuel, as a reducing agent in smelting metals, for the production of gaseous fuels such as producer gas and water gas, and for the manufacture of calcium carbide, graphite, silicon carbide and carbon disulphide, as shown in Table 20.1.

TABLE 20.1. PRODUCTS MADE FROM COKE

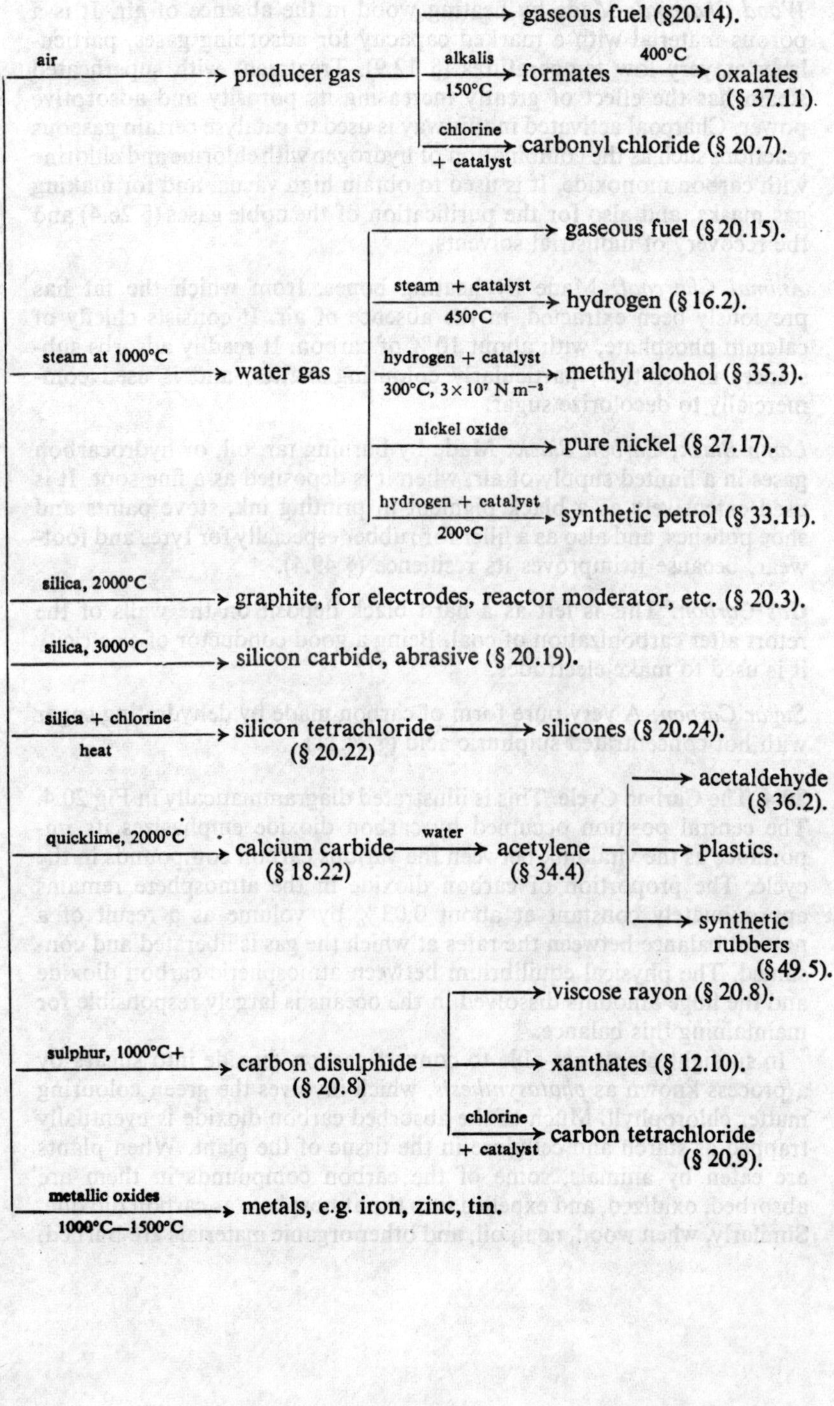

Wood Charcoal: Made by heating wood in the absence of air. It is a porous material with a marked capacity for adsorbing gases, particularly at very low temperatures (§ 12.9). Treatment with superheated steam has the effect of greatly increasing its porosity and adsorptive power. Charcoal activated in this way is used to catalyse certain gaseous reactions such as the combination of hydrogen with chlorine and chlorine with carbon monoxide. It is used to obtain high vacua, and for making gas masks, and also for the purification of the noble gases (§ 26.4) and the recovery of industrial solvents.

Animal Charcoal: Made by heating bones, from which the fat has previously been extracted, in the absence of air. It consists chiefly of calcium phosphate, with about 10% of carbon. It readily adsorbs substances in solution, particularly colouring matter, and is used commercially to decolorize sugar.

Lamp Black, Carbon Black: Made by burning tar, oil, or hydrocarbon gases in a limited supply of air, when it is deposited as a fine soot. It is used extensively as a black pigment in printing ink, stove paints and shoe polishes, and also as a filler for rubber especially for tyres and footwear, because it improves its resilience (§ 49.4).

Gas Carbon: This is left as a hard black deposit on the walls of the retort after carbonization of coal. Being a good conductor of electricity it is used to make electrodes.

Sugar Carbon: A very pure form of carbon made by dehydrating sugar with hot concentrated sulphuric acid (§ 24.11).

20.4. The Carbon Cycle. This is illustrated diagrammatically in Fig 20.4. The central position occupied by carbon dioxide emphasizes its importance as the vital link between the various carbon compounds in the cycle. The proportion of carbon dioxide in the atmosphere remains approximately constant at about 0.03% by volume as a result of a natural balance between the rates at which the gas is liberated and consumed. The physical equilibrium between atmospheric carbon dioxide and the huge amounts dissolved in the oceans is largely responsible for maintaining this balance.

In sunlight plants are able to convert carbon dioxide into sugars by a process known as *photosynthesis*, which involves the green colouring matter chlorophyll. Much of the absorbed carbon dioxide is eventually trapped as starch and cellulose in the tissue of the plant. When plants are eaten by animals, some of the carbon compounds in them are absorbed, oxidized, and expelled into the atmosphere as carbon dioxide. Similarly, when wood, coal, oil, and other organic materials are burned, much of their carbon content is oxidized to carbon dioxide and returned to the atmosphere, and the natural decay of plant and animal remains leads to the release of further carbon dioxide.

The carbon dioxide in the atmosphere contains a very small but definite proportion of the radioactive carbon isotope, $^{14}_{6}C$, produced from nitrogen by the action of cosmic rays. This isotope, which has a half-life of about five thousand years, takes part in the carbon cycle, so

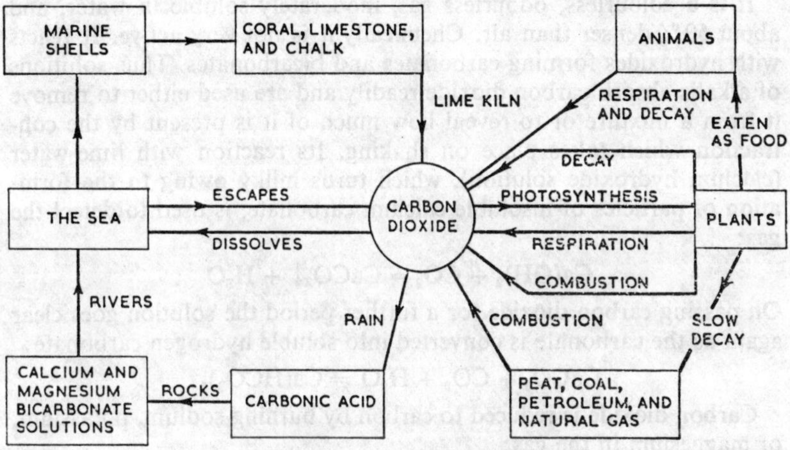

FIG. 20.4 The Carbon Cycle

that all living matter contains a constant proportion of the isotope as a result of natural exchange processes. When a plant or animal dies these exchanges cease and its radioactivity declines. Hence by measuring the amount of radioactivity of any organic material it is possible to estimate the time which has elapsed since it ceased to participate in the carbon cycle exchanges. This method of dating has proved of great value in archeology during recent years; it can also be used to distinguish genuine antiquities from frauds.

20.5. Carbon Dioxide, CO_2. Its occurrence in the atmosphere and the central part it plays in the carbon cycle have been described in the preceding section. It is formed when carbon or carbon compounds are burned in a plentiful supply of air or oxygen, and when carbonates and bicarbonates are treated with acids or decomposed by heating. In the laboratory it is usually prepared by the action of dilute hydrochloric acid on marble, a Kipp generator (§ 24.6) being used when large quantities are required:

$$CaCO_3 + 2HCl = CaCl_2 + H_2O + CO_2\uparrow$$

It is dried with anhydrous calcium chloride or phosphorus pentoxide, and collected by upward displacement of air.

Carbon dioxide is obtained industrially as a by-product of various fermentation processes, such as brewing (§ 35.2), or from lime kilns (§ 18.14), when it is separated from the other flue gases by absorbing it in cold potassium carbonate solution and expelling it again by boiling:

$$K_2CO_3 + CO_2 + H_2O \rightleftharpoons 2KHCO_3.$$

It is a colourless, odourless gas, moderately soluble in water, and about 50% denser than air. Chemically it is not very active. It reacts with hydroxides forming carbonates and bicarbonates. Thus solutions of alkalis absorb carbon dioxide readily and are used either to remove it from a mixture or to reveal how much of it is present by the contraction which takes place on shaking. Its reaction with lime-water (calcium hydroxide solution), which turns milky owing to the formation of particles of insoluble calcium carbonate, is used to detect the gas:

$$Ca(OH)_2 + CO_2 = CaCO_3\downarrow + H_2O$$

On passing carbon dioxide for a further period the solution goes clear again as the carbonate is converted into soluble hydrogen carbonate:

$$CaCO_3 + CO_2 + H_2O \rightleftharpoons Ca(HCO_3)_2$$

Carbon dioxide is reduced to carbon by burning sodium, potassium, or magnesium in the gas:

$$CO_2 + 2Mg = 2MgO + C$$

Passed over red-hot carbon it is reduced to carbon monoxide; this reaction which is reversible, is of great importance in the blast furnace and in the manufacture of gaseous fuels:

$$CO_2 + C \rightleftharpoons 2CO$$

USES OF CARBON DIOXIDE: Dissolved in water under pressure it is sold as 'soda-water' and mineral water drinks because of its refreshing taste. Cooled to −78°C it solidifies to a white solid, often referred to as 'dry-ice', which is used for refrigeration; on warming it sublimes to the vapour, leaving no liquid (see § 6.4 for the explanation of this). Its other advantages over ice as a coolant for food and ice cream include its lower temperature, its greater efficiency volume for volume (because of its higher specific gravity and its higher latent heat), and its germicidal action. Contrary to expectations it does not evaporate away rapidly because an insulating layer of dense vapour tends to surround the lump of solid.Carbon dioxide is used to put out fires, and many extinguishers are devices for producing a supply of carbon dioxide foam to cool the burning material and prevent the access of oxygen.

A lot of carbon dioxide is used in the Solvay process (§ 17.8) and for making urea industrially (§ 41.9). Another important use of the gas is in gas-cooled nuclear reactors (§ 31.11) to convey the heat liberated in the pile to the steam-raising plant. Carbon dioxide is chosen as coolant because it possesses the most suitable physical properties and because it causes negligible corrosion of the metal surfaces at high temperatures.

COMPOSITION OF CARBON DIOXIDE: Since the gas is formed when pure carbon is burned in oxygen, it must contain only these two elements. If its formula is C_xO_y, the values of x and y can be found by using the apparatus shown in Fig. 20.5. The volume of oxygen taken, which must

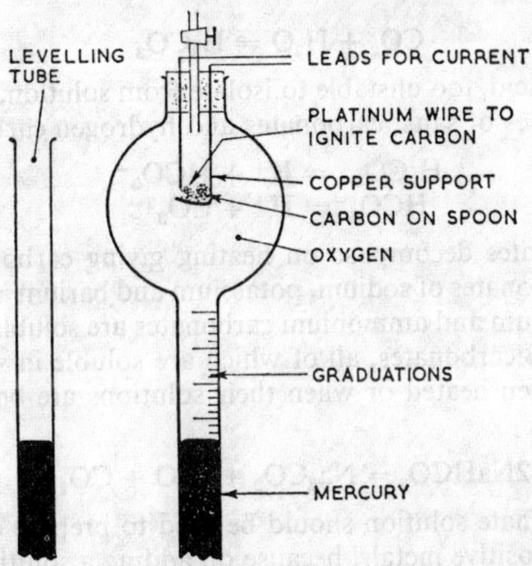

FIG. 20.5 Apparatus for Volume Composition of Carbon Dioxide

be in excess, is measured at room temperature and the carbon then heated electrically so that it burns to carbon dioxide. When the gases in the eudiometer have cooled to room temperature again it is found that there is no change in volume as a result of the combustion.

i.e. carbon + 1 volume of oxygen = 1 volume of C_xO_y

By Avogadro's hypothesis, since the volumes were measured under the same physical conditions,

carbon + 1 molecule of oxygen = 1 molecule of C_xO_y

i.e. carbon + 2 atoms ,, ,, = C_xO_y (since oxygen is diatomic)

$\therefore y = 2.$

Now v.d. is 22, so mol. wt. = 44 and $x = 1$

$\therefore$ Formula of carbon dioxide is CO_2.

STRUCTURE OF CARBON DIOXIDE: Physical measurements (e.g. its zero dipole moment) show that the three atoms in the molecule are all in one straight line. Now interatomic distances as determined by electron diffraction suggest that carbon dioxide exists as a resonance hybrid (see § 14.11) of the three forms:

$$O \leftarrow C \equiv O \qquad O = C = O \qquad O \equiv C \rightarrow O$$

which helps to account for its great stability.

20.6. Carbonic Acid and Carbonates. When carbon dioxide dissolves in water its solution is slightly acidic owing to the formation of carbonic acid:

$$CO_2 + H_2O \rightleftharpoons H_2CO_3$$

This is a weak acid, too unstable to isolate from solution. Being dibasic it gives two series of salts, carbonates and hydrogen carbonates:

$$H_2CO_3 \rightleftharpoons H^+ + HCO_3^-$$
$$HCO_3^- \rightleftharpoons H^+ + CO_3^{2-}$$

Most carbonates decompose on heating giving carbon dioxide, although the carbonates of sodium, potassium and barium are exceptions. Sodium, potassium and ammonium carbonates are soluble in water, the rest insoluble. Bicarbonates, all of which are soluble in water, decompose readily when heated or when their solutions are boiled, forming carbonates:

$$2NaHCO_3 = Na_2CO_3 + H_2O + CO_2\uparrow$$

Sodium bicarbonate solution should be used to prepare carbonates or the less electropositive metals, because on adding a solution of sodium carbonate, basic carbonates such as $CuCO_3.Cu(OH)_2$ are precipitated.

Carbonates and bicarbonates are detected by the evolution of carbon dioxide when they are treated with dilute acids. They are distinguished by adding magnesium sulphate solution, when a carbonate will give an immediate white precipitate of magnesium carbonate in the cold, whereas a hydrogen carbonate only gives such a precipitate on boiling:

carbonate: $MgSO_4 + Na_2CO_3 = Na_2SO_4 + MgCO_3\downarrow$ } in the cold
bicarbonate: $MgSO_4 + 2NaHCO_3 = Na_2SO_4 + Mg(HCO_3)_2$ } in the cold
$Mg(HCO_3)_2 = MgCO_3\downarrow + H_2O + CO_2\uparrow$ on boiling

To detect a bicarbonate in the presence of a carbonate, magnesium sulphate solution should be added in excess, and the precipitate of magnesium carbonate filtered off; then the filtrate should be boiled, when a second white precipitate will appear owing to the decomposition of the hydrogen carbonate.

20.7. Carbon Monoxide, CO. It is produced in large amounts in various gaseous fuels, details of which are given later in this chapter.

LABORATORY PREPARATION: By the dehydrating action of hot concentrated sulphuric acid on formic acid or a formate, using the apparatus shown in Fig. 20.6.

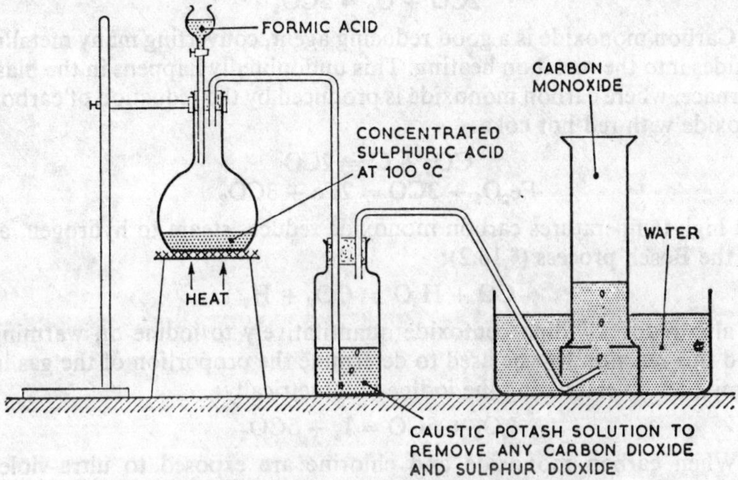

FIG. 20.6 Preparation of Carbon Monoxide

$$H.COOH - H_2O = CO$$

The gas should be bubbled through concentrated sulphuric acid and collected in an evacuated flask if required dry.

Alternatively oxalic acid or its salts may be used, but this method is less good because thorough washing with caustic alkali solution is necessary to remove all the carbon dioxide by-product:

$$\begin{array}{l} COOH \\ | \\ COOH \end{array} - H_2O = CO + CO_2$$

PROPERTIES OF CARBON MONOXIDE: It is a colourless, odourless gas, b.p. −111°C, only slightly soluble in water. It is extremely poisonous, combining irreversibly with the haemoglobin in the blood to form a compound called carboxyhaemoglobin, thereby preventing the red blood corpuscles from acting as oxygen carriers. Breathed for any length of time, even 1% in the air may cause death, which makes it clear how dangerous it can be to run a car engine in a closed garage, or to use a large geyser without an efficient flue, or to neglect leaks of coal gas. Very often the victim collapses without warning, so insidious is its effect.

REACTIONS OF CARBON MONOXIDE: It burns readily with a bright blue flame, giving out much heat and producing a gas which turns lime-water milky, i.e. carbon dioxide:

$$2CO + O_2 = 2CO_2$$

Carbon monoxide is a good reducing agent, converting many metallic oxides into the metal on heating. This undoubtedly happens in the blast furnace, where carbon monoxide is produced by the reduction of carbon dioxide with red-hot coke:

$$CO_2 + C \rightleftharpoons 2CO$$
$$Fe_2O_3 + 3CO = 2Fe + 3CO_2$$

At high temperatures carbon monoxide reduces steam to hydrogen, as in the Bosch process (§ 16.2):

$$CO + H_2O \rightleftharpoons CO_2 + H_2$$

It also reduces iodine pentoxide quantitatively to iodine on warming and this reaction can be used to determine the proportion of the gas in a mixture by estimating the iodine volumetrically:

$$I_2O_5 + 5CO = I_2 + 5CO_2$$

When carbon monoxide and chlorine are exposed to ultra-violet light or passed over a catalyst of activated charcoal at 150°C, they combine to form carbonyl chloride. The product, also known as phosgene, was employed as a poison gas in the First World War; it is now used to manufacture dyestuffs and isocyanates:

$$CO + Cl_2 = COCl_2$$

Carbon monoxide combines with certain metals (e.g. Ni, Fe, Co), when hot, forming volatile carbonyls:

e.g. $$Ni + 4CO \rightleftharpoons Ni(CO)_4$$

This reaction is used in the purification of nickel industrially (see § 27.17).

When carbon monoxide is mixed with sodium hydroxide at 150°C under 5×10^6 N m^{-2} (50 atm) pressure, sodium formate results:

$$CO + NaOH = H.COONa$$

This product decomposes into sodium oxalate and hydrogen at 400°C:

$$2H.COONa = \begin{array}{c} COONa \\ | \\ COONa \end{array} + H_2$$

Carbon monoxide is absorbed readily by solutions of cuprous

chloride in ammonium hydroxide or concentrated hydrochloric acid, forming a compound $CuCl.CO.2H_2O$. Use is made of this to remove carbon monoxide from mixtures of gases.

USES OF CARBON MONOXIDE: Its principal use is as a fuel, but it is also much used to manufacture methyl alcohol (§ 35.3), synthetic petrol (§ 33.11), carbonyl chloride, formates and oxalates, and in the reduction of ores and the refining of nickel. For these purposes pure carbon monoxide is rarely needed, and water gas and producer gas are frequently used.

COMPOSITION OF CARBON MONOXIDE: When carbon monoxide is burned in oxygen it forms carbon dioxide, and it is found that:

2 volumes carbon monoxide + 1 volume oxygen
= 2 volumes carbon dioxide

Using Avogadro's hypothesis we conclude that:

2 molecules carbon monoxide + 1 molecule oxygen
= 2 molecules carbon dioxide

Since oxygen is diatomic and carbon dioxide has the formula CO_2, the molecular formula of carbon monoxide must therefore be CO, which is confirmed by v.d. measurements.

STRUCTURE OF CARBON MONOXIDE: Its structure is believed to be a resonance hybrid (§ 14.10) of $C \overset{\leftarrow}{=} O$, in which the atoms are held together by a double covalent bond and a co-ordinate linkage, and $\underset{\cdot}{C}{=}\overset{\cdot}{O}$, in which the two atoms share two pairs of electrons. This not only accounts for its chemical behaviour, but also explains its very low dipole moment.

20.8. Carbon Disulphide, CS_2. Two methods of manufacture are in current use:

1. By heating sulphur with carbon in an electric furnace (see Fig. 20.7). The sulphur volatilizes and reacts with the white-hot coke or charcoal thus:

   $$C + 2S = CS_2$$

   It should be noted that this is an *electrothermal* process i.e. the electricity is used solely for heating purposes and no electrolysis is involved.
2. By passing sulphur vapour and methane at 650°C over a silica or alumina catalyst, when the following reaction takes place:

   $$CH_4 + 4S = CS_2 + 2H_2S$$

   In countries with an abundant supply of natural gas (§ 33.9), this method is rapidly replacing the other because it is more economical.

It is a colourless liquid, b.p. 46°C, with a dense poisonous vapour. The foul smell of the commercial product is caused by impurities. Because of its volatility and very low ignition temperature, carbon disulphide is exceedingly inflammable; its vapour forms explosive mixtures with air.

It is an excellent solvent dissolving fats, oils, resins, rubber, phosphorus, sulphur, and iodine, and this is its principal use. It is also used to manufacture carbon tetrachloride (see below) and viscose rayon, also known as 'artificial silk'. The latter is made by digesting cellulose

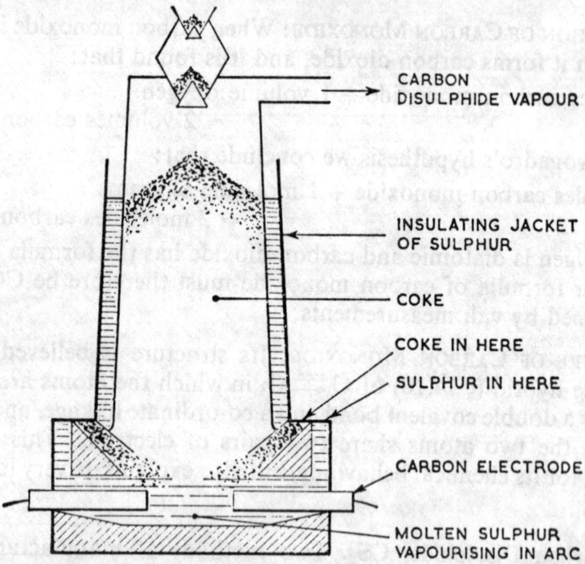

FIG. 20.7 Manufacture of Carbon Disulphide

(§ 42.7) with sodium hydroxide solution and adding carbon disulphide. The product, dissolved in excess sodium hydroxide solution, is forced through jets into a tank of dilute sulphuric acid when the alkali is neutralized and the cellulose is deposited as a continuous fibre. Xanthate flotation agents (see § 12.10) are also made from carbon disulphide.

20.9. Carbon Tetrachloride, CCl_4. This compound is made industrially in two ways:

1. By passing chlorine into boiling carbon disulphide in the presence of iodine or ferric chloride as a catalyst:

$$CS_2 + 3Cl_2 = \underset{\text{b.p. 77°C}}{CCl_4} + \underset{\text{b.p 138°C}}{S_2Cl_2}$$

The two products are separated by fractional distillation and where a ready market for the sulphur monochloride does not exist it is converted into more carbon tetrachloride by heating it with carbon disulphide in the presence of iodine or ferric chloride:

$$CS_2 + 2S_2Cl_2 = CCl_4 + 6S$$

2. Carbon disulphide is also made in countries with a plentiful supply of natural gas (§ 33.9) by the direct chlorination of methane at 400°C using an excess of chlorine:

$$CH_4 + 4Cl_2 = CCl_4 + 4HCl$$

Carbon tetrachloride is a dense, colourless, non-inflammable liquid, immiscible with water. It is unreactive, not being hydrolysed even on boiling with water (contrast with silicon tetrachloride). It is used in fire-extinguishers ('Pyrene') because its dense vapour blankets the burning surface excluding oxygen, and for dry-cleaning clothes because it is an excellent solvent for grease. A new and important use is in the manufacture of the chlorofluoromethanes, as described in § 40.11.

20.10. Cyanogen, C_2N_2. Made by strongly heating mercuric cyanide:

$$Hg(CN)_2 = Hg + C_2N_2\uparrow$$

or by warming potassium cyanide solution with copper sulphate solution, when the cupric cyanide first formed decomposes into cuprous cyanide and cyanogen:

$$CuSO_4 + 2KCN = Cu(CN)_2 + K_2SO_4$$
$$2Cu(CN)_2 = Cu_2(CN)_2\downarrow + C_2N_2\uparrow$$

It is a colourless, poisonous gas with an almond-like smell and is fairly soluble in water.

20.11. Hydrocyanic Acid (Prussic Acid), HCN. A solution of this acid can be prepared by acidifying a solution of sodium or potassium cyanide. The anhydrous acid is made by distilling potassium cyanide with moderately concentrated sulphuric acid, drying the liberated hydrogen cyanide gas with anhydrous calcium chloride and condensing it to a colourless liquid, b.p. 26°C, by using a freezing mixture. It is exceptionally poisonous, even small amounts causing death in a few minutes.

Hydrocyanic acid is a very weak acid ($K = 1.3 \times 10^{-9}$ mol dm^{-3}), being only slightly dissociated into ions in solution:

$$HCN \rightleftharpoons H^+ + CN^-$$

Its salts are called cyanides, the most important of which, sodium cyanide, is considered in § 17.13. The characteristic reaction of cyanides is the formation of complex ions with simple metallic ions. For example, with silver nitrate solution they give a white precipitate of silver cyanide, which is soluble in excess of cyanide:

$$AgNO_3 + KCN = AgCN\downarrow + KNO_3$$
$$AgCN + KCN \rightleftharpoons KAg(CN)_2$$

potassium argento-cyanide

Boiled with ferrous sulphate solution and sodium hydroxide solution, cyanides give sodium ferrocyanide solution, which is recognized by the formation of a dark blue precipitate on adding a ferric solution and hydrochloric acid. This sequence of reactions is used to detect nitrogen in an organic compound (see § 32.7.).

20.12. Cyanates and Thiocyanates. Cyanates are made by heating cyanides to about 700°C with a suitable oxidizing agent, such as manganese dioxide:

e.g. $$KCN + 2MnO_2 = KNCO + Mn_2O_3$$

Sodium and potassium cyanates are colourless solids, soluble in water. Ammonium cyanate is of particular interest because it changes spontaneously into its isomer urea, forming an equilibrium mixture containing about 95% of the latter (see § 41.9):

$$NH_4NCO \rightleftharpoons CO(NH_2)_2$$

This is a good example of tautomerism (see Table 48.1).

Potassium thiocyanate is made by fusing potassium cyanide with sulphur and extracting the product with alcohol, in which it is very soluble. Ammonium thiocyanate is made by heating ammonium hydroxide solution with carbon disulphide under pressure:

$$2NH_3 + CS_2 = NH_4NCS + H_2S$$

Both thiocyanates are colourless solids, extremely soluble in water. With silver nitrate solution they give a white precipitate of silver thiocyanate, insoluble in dilute acids:

e.g. $$AgNO_3 + KNCS = AgNCS\downarrow + KNO_3$$

Thiocyanates are used, therefore, for estimating solutions of silver by Volhard's method.

When added to a solution containing ferric ions, thiocyanates give a blood-red coloration which is used as a sensitive test for ferric iron (see § 30.18). Although ferric thiocyanate is certainly produced in this reaction:

$$3KNCS + FeCl_3 = Fe(NCS)_3 + 3KCl$$

the characteristic coloration is attributed to a complex ion which is formed at the same time.

20.13. Coal Gas. This important gaseous fuel is produced by carbonizing coal, i.e. by heating it in the absence of air. The process, which is shown diagrammatically in Fig. 20.8, is carried out in fireclay retorts heated by producer gas. The products depend upon the quality of the coal and the carbonizing temperature, which usually lies between 500°C and 1300°C, but Table 20.2 shows the principal products formed in a typical case.

Coke and gas carbon remain as residues in the retort, the volatile products distilling over. Coal tar, which is a valuable by-product, is condensed and used as the source of a great range of organic compounds, as described in § 43.2. Ammoniacal liquor is used to make ammonium

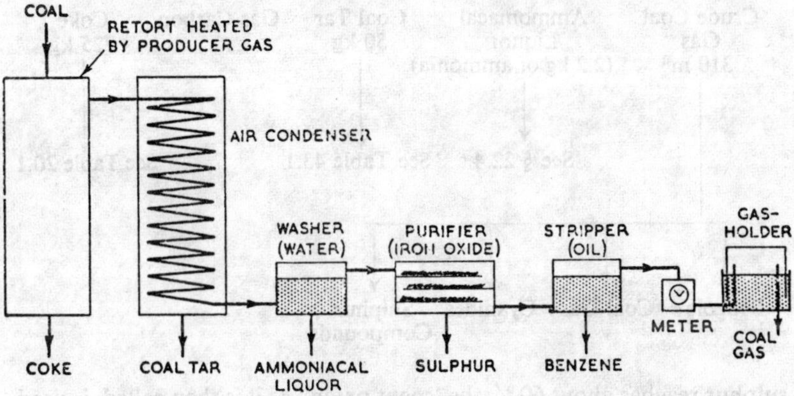

FIG. 20.8 Manufacture of Coal Gas

sulphate, used extensively as a fertilizer. After recovery of the valuable benzole (§ 43.3) by washing with oil, the crude coal gas consists of about 50 % hydrogen, 30 % methane, 10 % carbon monoxide, with small amounts of ethylene, nitrogen, carbon dioxide, hydrogen sulphide, carbon disulphide and hydrogen cyanide. After being washed with water it is purified before use by removing the sulphur compounds from it, as these would produce dangerous amounts of sulphur dioxide during combustion. The carbon disulphide is first converted into hydrogen sulphide by passing the crude coal gas over a nickel catalyst at 400°C, when the following reaction takes place:

$$CS_2 + 4H_2 = CH_4 + 2H_2S$$

Hydrogen sulphide is removed by passing the coal gas over hydrated ferric oxide, which is converted into ferric sulphide. This is exposed to the air periodically, when it is oxidized to ferric oxide and sulphur and can be used again in the purifier. Eventually when the proportion of

TABLE 20.2. TYPICAL PRODUCTS OBTAINED BY CARBONIZING COAL

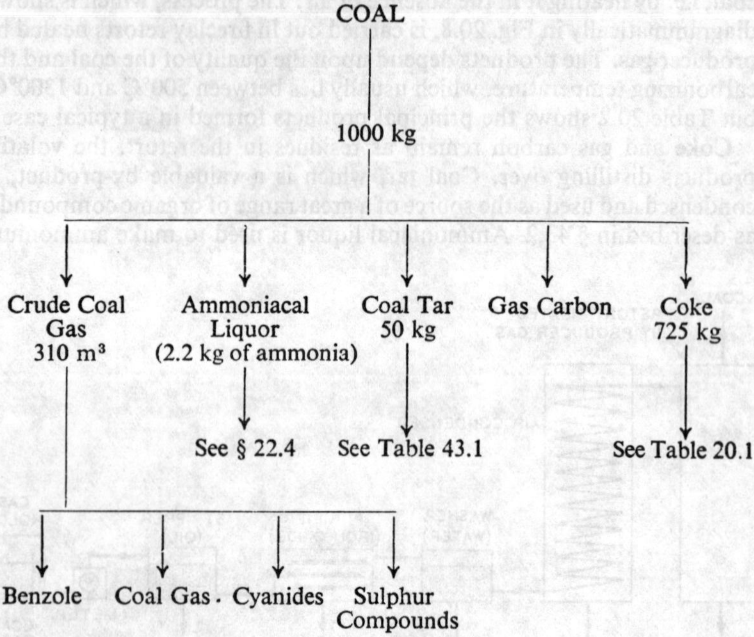

sulphur reaches about 60%, the 'spent oxide', as it is then called, is used for the manufacture of sulphuric acid.

$$Fe_2O_3 + 3H_2S = Fe_2S_3 + 3H_2O$$
$$2Fe_2S_3 + 3O_2 = 2Fe_2O_3 + 6S\downarrow$$

Hydrogen cyanide is removed partly as ammonium thiocyanate and partly as ferrocyanide in the ferric oxide purifiers.

20.14. Producer Gas. This is a mixture of about 35% carbon monoxide and 65% nitrogen made by blowing air over white-hot coke. At the base of the producer, as the plant shown in Fig. 20.9 is called, coke burns to carbon dioxide, which is then reduced by the hot coke to the monoxide:

$$\text{First stage:} \quad C + O_2 = CO_2$$
$$\text{Second stage:} \quad CO_2 + C \rightleftharpoons 2CO$$

The position of equilibrium in the second stage depends upon the temperature, the left-to-right reaction being endothermic; above 1000°C very little carbon dioxide remains in the gas, but below this temperature the proportion rises markedly.

Producer gas has a low heating value because of its high proportion of incombustible nitrogen, which makes it uneconomical to store or distribute. It is a cheap fuel, normally used straight away, whilst still hot, for heating retorts or furnaces.

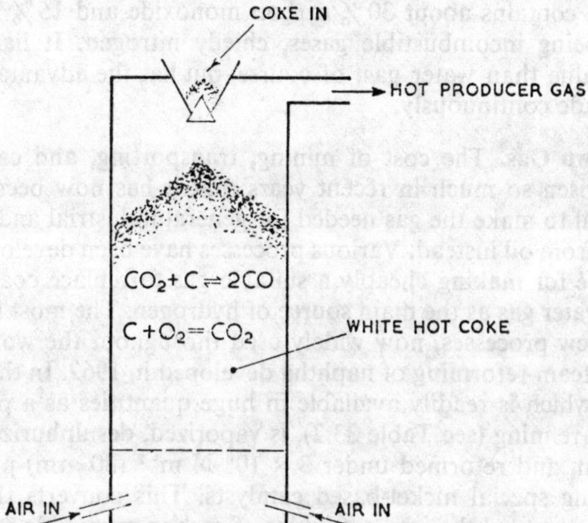

FIG. 20.9 Manufacture of Producer Gas

20.15. Water Gas. This is made by passing steam over white-hot coke. It contains about 45% carbon monoxide, 50% hydrogen, with small amounts of carbon dioxide and nitrogen:

$$C + H_2O = CO + H_2$$

This reaction is endothermic, and the temperature of the coke falls steadily as it proceeds. Provided the temperature is kept above 1000°C the proportion of carbon dioxide produced is very small, depending upon the position of equilibrium of the reversible reaction:

$$CO_2 + C \rightleftharpoons 2CO$$

For this reason water gas and producer gas are usually made intermittently from the same plant by alternating the input every few minutes between steam and air. The resulting producer gas is used to raise more steam. In this way the temperature is kept above 1000°C all the time and the carbon dioxide content is kept low.

Water gas has a high heating value. It is often added to coal gas to boost the supply at periods of peak demand. It is also used to manufacture hydrogen (§ 16.2), and methyl alcohol (§ 35.3), and as a reducing agent.

Another gaseous fuel is made by passing a mixture of steam and air simultaneously over white-hot coke in such proportion that the temperature is maintained above 1000°C. The product, which is called *semi-water gas*, contains about 30% carbon monoxide and 15% hydrogen, the rest being incombustible gases, chiefly nitrogen. It has a lower heating value than water gas, of course, but has the advantage that it can be made continuously.

20.16. Town Gas. The cost of mining, transporting, and carbonizing coal has risen so much in recent years that it has now become more economical to make the gas needed for general industrial and domestic purposes from oil instead. Various processes have been developed in the last decade for making cheaply a suitable gas to replace coal gas as a fuel and water gas as the main source of hydrogen. The most important of these new processes, now widely used throughout the world, is the catalytic steam-reforming of naphtha developed in 1962. In this process naphtha, which is readily available in huge quantities as a product of petroleum refining (see Table 33.2), is vaporized, desulphurized, mixed with steam, and reformed under 3×10^6 N m^{-2} (30 atm) pressure at 800°C using special nickel-based catalysts. This converts the hydrocarbons in the naphtha into a mixture of carbon monoxide and hydrogen, a product known as 'lean gas' because of its relatively low heating value:

e.g. $$C_6H_{14} + 6H_2O = 6CO + 13H_2$$

If a further supply of naphtha, vaporized and freed from sulphur compounds, is mixed with a small proportion of steam and reformed at 500°C and 3×10^6 N m^{-2} (30 atm) pressure over a nickel–alumina catalyst, it is converted to a mixture of gaseous hydrocarbons (mainly methane and ethane) of high heating value known as 'rich gas'. *Town gas*, with the same heating value as coal gas (approximately 1.9×10^7 joule per cubic metre) can be made by mixing these lean and rich gases in suitable proportions. The process is economical because it uses a cheap raw material and can be operated continuously on a large scale with low labour and capital costs.

The lean gas process described above is also extensively used for manufacturing hydrogen. The product is mixed with more steam and passed over an iron catalyst at 450°C, as in the Bosch process (§ 16.2). This converts the carbon monoxide present into additional hydrogen and carbon dioxide, the latter being removed by washing under pressure:

$$CO + H_2O \rightleftharpoons H_2 + CO_2$$

20.17. Silicon, Si. Electronic configuration: 2,8,4.

OCCURRENCE: Elemental silicon does not occur naturally, but in the combined state, principally with oxygen, silicon is the second commonest element in the earth's crust. Silica and silicates abound in many minerals and rocks and in sand and clay.

LABORATORY PREPARATION: Several methods are available:

(1) Silica is heated in a fireclay crucible with powdered magnesium:

$$SiO_2 + 2Mg = Si + 2MgO$$

The product is treated with dilute hydrochloric acid to remove magnesium oxide, unchanged magnesium, and magnesium silicide, and then filtered leaving a brown powder which consists of about 97% silicon, the rest being silica.

(2) Silica is heated with excess of powdered aluminium or zinc; reduction occurs and the silicon produced dissolves in the excess of molten metal, from which it is deposited as crystals on cooling. The excess metal can then be removed with dilute hydrochloric acid, which leaves the silicon undissolved.

(3) Sodium or potassium is heated in an atmosphere of silicon tetrafluoride, when reduction takes place:

$$SiF_4 + 4Na = Si + 4NaF$$

MANUFACTURE: Powdered coke is heated in an electric furnace to a high temperature with excess of sand:

$$SiO_2 + 2C = Si + 2CO\uparrow$$

The process is operated where hydroelectric power is available.

PROPERTIES AND REACTIONS OF SILICON: At one time it was thought that silicon existed in two distinct forms, amorphous and crystalline. X-ray analysis shows that this is not so, the two varieties differing only in particle size and purity, and it is these factors which account for their different physical properties and reactivity.

Crystalline silicon is a hard grey lustrous solid, ρ 2.35 g cm^{-3}, melting at about 1400°C. It conducts electricity with difficulty. It combines with fluorine spontaneously and with the other halogens and oxygen on heating. It is very resistant to acids, being attacked only by a mixture of concentrated nitric and hydrofluoric acids. It dissolves in hot concentrated solutions of caustic alkalis, forming silicates and hydrogen:

$$Si + 2NaOH + H_2O = Na_2SiO_3 + 2H_2\uparrow$$

Silicon combines with many metals on heating, forming silicides, e.g. magnesium silicide, Mg_2Si. Heated strongly with carbon, silicon gives silicon carbide (§ 20.19).

USES OF SILICON: Alloyed with iron, as 'ferrosilicon', it is used as a deoxidizer and to make acid-resistant steels. Very pure silicon, only obtainable in recent years, is of great importance in electronics as a semi-conductor, used to make transistors.

20.18. Silicon Dioxide, Silica, SiO_2. This compound occurs naturally as quartz and sand, and also as flint, opal and agate. Another variety, kieselguhr, consists of deposited skeletons of minute marine organisms called diatoms. Pure silica is colourless, but sand is usually coloured yellow or brown by ferric oxide impurity. Silica exists in three crystalline forms, quartz (stable up to 870°C), tridymite (stable over the range 870°C–1470°C), and cristobalite (stable from 1470°C up to the melting point, 1710°C).

Silicon dioxide is insoluble in water and in all acids except hydrofluoric, which explains why this acid is used to etch glass:

$$SiO_2 + 6HF = \underset{\text{(fluorosilicic acid)}}{H_2SiF_6} + 2H_2O$$

It dissolves in boiling alkalis, forming silicates:

$$SiO_2 + 2NaOH = Na_2SiO_3 + H_2O$$

It also combines with metallic oxides at high temperatures, giving silicates:

e.g. $$SiO_2 + CaO = CaSiO_3$$

This reaction is used in the blast furnace to remove silica as a fusible slag (§ 30.2).

When silica is heated strongly with metallic salts, silicates are formed and the much more volatile oxides are driven off as vapours:

$$SiO_2 + Na_2CO_3 = Na_2SiO_3 + CO_2\uparrow$$
$$SiO_2 + Na_2SO_4 = Na_2SiO_3 + SO_3\uparrow$$
$$3SiO_2 + Ca_3(PO_4)_2 = 3CaSiO_3 + P_2O_5\uparrow$$

The first two examples quoted here are important in glass making and the third in the manufacture of phosphorus (§ 23.2.).

Silica, like diamond, is macro-molecular; atoms of silicon and oxygen are linked together covalently in tetrahedra which are joined in a three-dimensional framework structure so that each crystal of quartz is virtually one huge molecule, consisting of myriads of atoms. This accounts for its non-volatility and hardness. When melted and cooled silica forms a glass-like solid, used for laboratory apparatus, and sometimes called 'quartz glass'. It is resistant to acids, transparent to ultra-violet light, infusible below about 1500°C, and because it has an extremely small coefficient of thermal expansion it can be subjected to sudden temperature changes without fear of cracking.

When a solution of a silicate is acidified a white gelatinous precipitate of hydrated silica, often referred to as silicic acid, slowly appears, although it shows a strong tendency to remain in colloidal solution, particularly if the dissolved ions are removed by dialysis (§ 12.3):

$$Na_2SiO_3 + 2HCl = 2NaCl + SiO_2.H_2O\downarrow$$

If this precipitate is filtered off and heated it gradually loses water, forming a series of hydrates of silica. When the water content is down to a few per cent the solid product is called *silica gel.* It possesses excellent adsorptive properties, presumably because of its porous nature and large surface area. It is used to dry gases, to recover valuable vapours from industrial effluents, and in the refining of petroleum. Heat drives off the adsorbed vapour and reactivates the gel so that it can be used again.

Silica finds many other industrial uses. Sand is used in great quantities to make mortar and cement, and in the manufacture of glass. Large crystals of quartz are used for lenses of optical instruments, being transparent to ultra-violet light, and also for accurately controlling the frequency of radio transmitters. Powdered quartz is used in the manufacture of carborundum, silicon tetrafluoride, and sodium silicate, and to make silica bricks used for lining furnaces. Kieselguhr absorbs liquids readily; saturated with nitroglycerine it is known as dynamite (§ 35.9).

20.19. Silicon Hydrides. When magnesium silicide is treated with dilute hydrochloric acid in the absence of air, a mixture of hydrogen and various silicon hydrides results, e.g. monosilane, SiH_4, disilane Si_2H_6, and trisilane, Si_3H_8:

$$Mg_2Si + 4HCl = SiH_4\uparrow + 2MgCl_2$$

Unlike the paraffins (§ 33.4) these hydrides are very reactive substances. They inflame spontaneously in air and explode when brought into contact with halogens:

$$SiH_4 + 2O_2 = SiO_2 + 2H_2O$$
$$SiH_4 + 4Cl_2 = SiCl_4 + 4HCl$$

They decompose on heating, particularly. those of higher molecular weight, and they react readily with alkalis:

$$SiH_4 + 2NaOH + H_2O = Na_2SiO_3 + 4H_2\uparrow$$

These hydrides are strong reducing agents, decolorizing potassium permanganate solution and reducing ferric salts to ferrous.

20.20. Silicon Carbide (Carborundum), SiC. Made by heating sand and coke in an electric furnace to about 3000°C. The product solidifies when cooled, and is broken up into lumps:

$$SiO_2 + 3C = SiC + 2CO$$

It is an extremely hard substance, second only to diamond in this respect, which accounts for its widespread use as an abrasive. It has a tetrahedrally linked covalent framework structure like diamond.

20.21. Sodium Silicate, Na_2SiO_3. Made by strongly heating powdered quartz with sodium carbonate until the mixture melts:

$$SiO_2 + Na_2CO_3 = Na_2SiO_3 + CO_2\uparrow$$

It is a colourless solid, m.p. 1090°C, which dissolves in hot water to form a syrupy liquid, known as 'water glass'. It is used for sizing paper, for fireproofing wood and textiles, and for preserving eggs by sealing their shells with a layer of impervious silicate. A dilute solution of sodium silicate can be used to make a chemical garden, as described in § 5.18.

20.22. Silicon Tetrafluoride, SiF_4. Made by the action of hydrofluoric acid on silica, usually by heating powdered fluorspar and sand with excess of concentrated sulphuric acid:

$$CaF_2 + H_2SO_4 = CaSO_4 + 2HF$$
$$SiO_2 + 4HF = SiF_4\uparrow + 2H_2O$$

Silicon tetrafluoride is a colourless pungent gas, b.p.—101°C, which fumes in moist air. It is hydrolysed readily, even by cold water, giving hydrogen fluoride, which accounts for its corrosive action on glass when moist:

$$SiF_4 + 3H_2O = SiO_2.H_2O\downarrow + 4HF$$

The hydrogen fluoride converts unchanged tetrafluoride into fluorosilicic acid, which remains in solution:

$$SiF_4 + 2HF = H_2SiF_6$$

This is a strong acid with powerful disinfecting properties. It is used for the fluoridation of municipal water supplies (§ 25.2.), and for making fluorosilicates which are used as insecticides and as hardening agents in concrete, ceramics, and vitreous enamels.

20.23. Silicon Tetrachloride, $SiCl_4$. Made by strongly heating a mixture of silica and carbon in a current of chlorine:

$$SiO_2 + 2C + 2Cl_2 = SiCl_4 + 2CO$$

It is a colourless liquid, b.p. 59°C, which fumes in moist air owing to hydrolysis:

$$SiCl_4 + 3H_2O = SiO_2.H_2O\downarrow + 4HCl\uparrow$$

It is used in the manufacture of silicones (see §20.25).

20.24. Silicates. X-ray analysis shows that the fundamental unit present in silicates is the SiO_4^{4-} ion, in which four oxygen atoms are arranged tetrahedrally with the silicon atom in the middle (see Fig. 20.10). The different silicate structures arise from the different ways in which these SiO_4 tetrahedra are linked together by oxygen atoms into chains, rings, sheets, and three-dimensional frameworks (see Fig. 20.11). The physical properties of the silicates tend to be largely determined by the type of

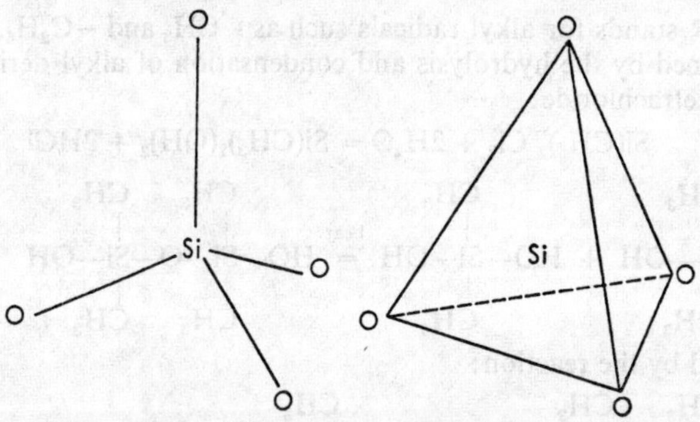

FIG. 20.10 Two Representations of the Tetrahedral SiO_4 Group

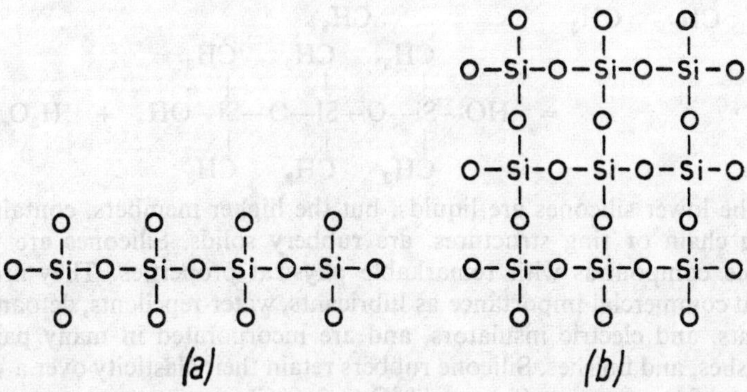

FIG. 20.11 Projections of Two Typical Silicate Structures—(*a*) chains (*b*) sheets

structure they possess. For example, the long chain structures, such as asbestos, are fibrous in character, whilst sheet structures, such as mica, split easily into thin plates or sheets. Intermingled with the negatively-charged silicate groups are various cations, of which Na^+, K^+, and Al^{3+} are the commonest. In some framework structures these can be reversibly exchanged for other cations, which accounts for the use of permutit as a water softener (§ 16.5).

20.25. Silicones. These are compounds of the type:

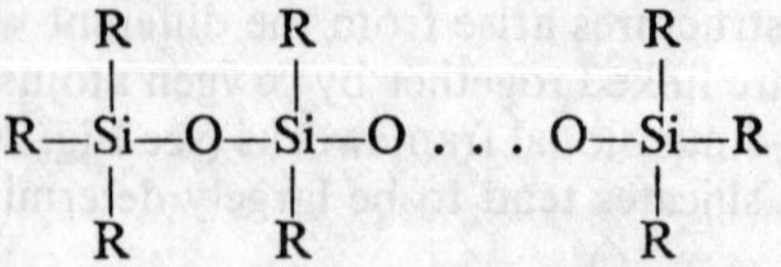

where R stands for alkyl radicals such as $-CH_3$ and $-C_2H_5$. Silicones are formed by the hydrolysis and condensation of alkyl derivatives of silicon tetrachloride:

e.g. $$Si(CH_3)_2Cl_2 + 2H_2O = Si(CH_3)_2(OH)_2 + 2HCl$$

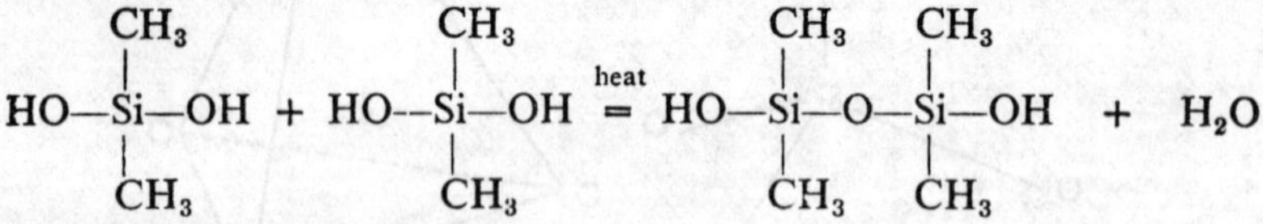

followed by the reaction:

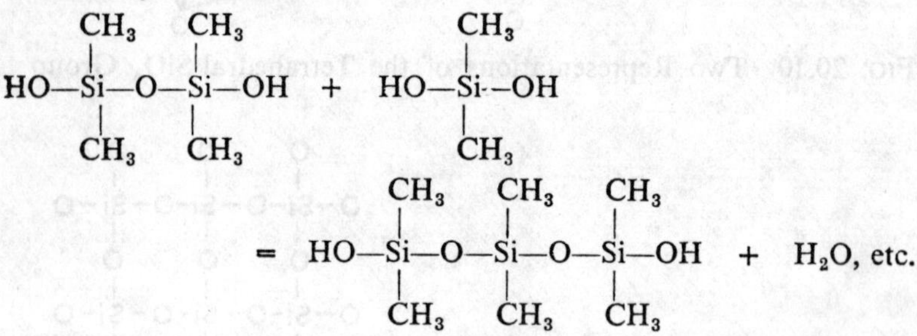

The lower silicones are liquids, but the higher members, containing long chain or ring structures, are rubbery solids. Silicones are very stable compounds with remarkable physical properties. They are of great commercial importance as lubricants, water-repellents, defoaming agents, and electric insulators, and are incorporated in many paints, polishes, and finishes. Silicone rubbers retain their elasticity over a wide range of temperature (from −80°C to 250°C).

20.26. Glass. Whilst glass varies widely in its composition, essentially it consists of a mixture of silicates which have not crystallized out on cooling from the molten state. It is made by melting together silica (i.e. sand) with calcium carbonate or oxide and sodium or potassium salts, usually the sulphate and the carbonate:

e.g. $$SiO_2 + CaCO_3 = CaSiO_3 + CO_2\uparrow$$
$$SiO_2 + K_2SO_4 = K_2SiO_3 + SO_3\uparrow$$

Common glass, such as that used in windows, has the approximate composition: $Na_2SiO_3.CaSiO_3.4SiO_2$. The physical properties of glass depend upon the proportions of the various silicates present. For example, a high proportion of sodium silicate gives a glass which melts at a low temperature, whereas the presence of potassium silicate raises the melting point. Lead silicate, made by incorporating lead oxide into the melt, gives glass with a high refractive index. If the glass contains about 10% of boric oxide, boro-silicates are formed with very low coefficients of thermal expansion, as in the 'Pyrex' glass used in the laboratory and home. Coloured glass is made by adding the oxides of certain transition metals (e.g. copper, nickel, cobalt), to the melt.

The extensive use of glass can largely be attributed to three of its properties—its transparency, its gradual softening at red heat which enables it to be blown and moulded to the desired shape, and its cheapness. The modern 'float process' for making plate glass is referred to in § 21.2 under the uses of tin.

20.27. Clays. These are formed by the weathering of certain silicate rocks, which slowly disintegrate into fine particles of hydrated aluminium silicates (e.g. $Al_2Si_2O_7.2H_2O$), and silica. The product is usually coloured by impurities such as iron oxide, but china clay, which is found in Cornwall, is white. Clay is converted to pottery, porcelain, china-ware, tiles, bricks, and refractories by strong heating or 'firing', which drives off the water of hydration leaving a hard, infusible solid which does not hydrate again on cooling. This product can be glazed to render it non-porous. Clay is also used as a filler for paper, rubber, soap, and paint, to give 'body' to these materials, and in huge amounts for the manufacture of cement (§ 18.25).

20.28. Detection of Silicon. Compounds containing silicon give silicon tetrafluoride vapour when heated with concentrated sulphuric acid and calcium fluoride in a lead crucible. On contact with water the tetrafluoride is hydrolysed to a white deposit of silica:

$$SiF_4 + 2H_2O = SiO_2\downarrow + 4HF$$

Silica and insoluble silicates are converted into sodium silicate by melting with excess of sodium carbonate. On dissolving the product in water and acidifying it, a white precipitate of hydrated silica results, which is insoluble in all acids except hydrofluoric.

20.29. Summary. In certain respects carbon and silicon are very similar. The atoms of both elements contain four electrons in their outermost shells and both are tetravalent in their compounds. Neither forms stable ions bearing four units of charge. Both elements are non-metallic, unreactive solids with high melting points, and both form crystals which are effectively giant molecules. Their dioxides are weakly acidic, reacting with alkalis to give salts. Both carbon and silicon form volatile tetrachlorides, volatile hydrides, and stable alkyl derivatives.

These two elements differ in several important ways. Carbon forms very stable bonds with other carbon atoms (the C–C bond has a high heat of formation), resulting in carbon compounds with long chain and ring structures. The Si–Si bond is much less stable and chains involving more than five silicon atoms linked together are not found at room temperature. Moreover, carbon atoms combine with hydrogen, oxygen, and halogen atoms with about equal ease, giving a great number of stable compounds, whereas silicon–hydrogen bonds are relatively weak, being readily broken by hydrolysis. Unlike carbon, silicon does not form double or triple bonds with itself or with other elements such as oxygen and nitrogen. Silicon has, however, a high affinity for oxygen, readily forming the stable Si–O–Si linkages which are present in many silicate structures.

The two dioxides have completely different structures, which explains the big difference in their physical properties. Carbon dioxide, which consists of small separate covalent molecules bound together by weak van der Waals forces, has the characteristic volatility associated with such a structure. In contrast, silica is macromolecular, each crystal consisting of billions of tetrahedral SiO_4 groups linked together covalently by oxygen atoms into a rigid three-dimensional framework, like that of diamond, giving a compound with characteristic hardness and high melting point.

For carbon the maximum number of electrons in the outermost shell is eight, so that compounds such as methane and carbon tetrachloride are incapable of forming co-ordinate bonds, either as donors or acceptors. Silicon, however, in common with the other elements in its period, can have up to twelve electrons in the outermost shell during chemical combination, so co-ordination can occur, as in H_2SiF_6 and $(NH_4)_2SiF_6$. This also explains why silicon tetrachloride, silicon disulphide, and silane, unlike the corresponding carbon compounds, are readily hydrolysed by even cold water. Presumably the hydrolysis of the tetrachloride proceeds as follows:

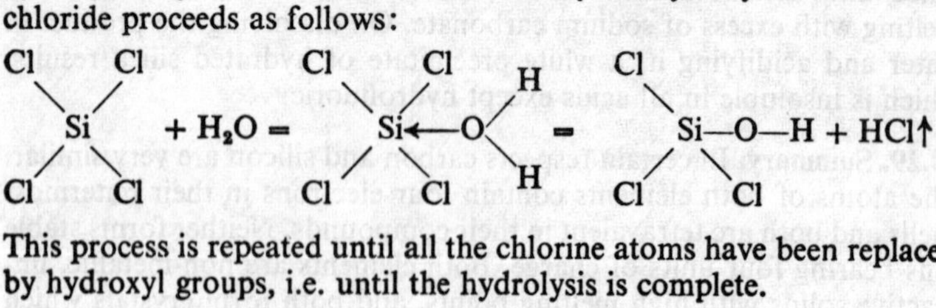

This process is repeated until all the chlorine atoms have been replaced by hydroxyl groups, i.e. until the hydrolysis is complete.

21

Tin and Lead

21.1. Introduction. The remaining elements in Group IVB are germanium, tin, and lead. The first of these is a comparatively rare element, obtainable pure only in recent years. It is of great commercial importance because of its use as a semi-conductor in transistors. The other two are important metals whose chemistry will be considered in detail.

21.2. Tin, Sn. Electronic configuration: 2,8,18,18,4.

OCCURRENCE: The metal itself does not occur naturally. The main source is the mineral cassiterite or tinstone, SnO_2, found as low grade ore and alluvial deposits in certain parts of the world, particularly in Malaysia, Bolivia and Indonesia.

EXTRACTION: The ore is first crushed and concentrated by washing away the less dense material, and then it is roasted in air, when impurities such as sulphur, arsenic, and antimony are removed as their volatile oxides. The product is mixed with powdered coal and heated in a reverberatory furnace to about 1300°C, when reduction of the oxide takes place, the molten tin being tapped off periodically:

$$SnO_2 + 2C = Sn + 2CO\uparrow$$

The metal is refined by heating it gently on an inclined surface, when the molten tin flows away from the less fusible impurities, and by exposing the molten metal to the air. The remaining metallic impurities, being more electropositive than tin, are thereby converted to their oxides and form a scum, leaving tin of 99.9% purity.

PROPERTIES OF TIN: It is a soft malleable metal with a silver-white lustre. It melts at only 232°C and boils at about 2300°C. Three allotropes of tin are known, giving a good example of enantiotropy (see § 6.5):

$$\underset{\rho\ 5.76\ g\ cm^{-3}}{\text{grey tin}} \underset{}{\overset{13.2°C}{\rightleftharpoons}} \underset{\rho\ 7.28\ g\ cm^{-3}}{\text{white tin}} \overset{161°C}{\rightleftharpoons} \underset{\rho\ 6.6\ g\ cm^{-3}}{\text{rhombic tin}}$$

Although white tin is only stable above 13.2°C, the conversion to the grey form takes place so slowly at temperatures just below the transition temperature that in practice the white form persists indefinitely, even at 0°C. Cooling to about −50°C, or rubbing of cooled white tin with crystals of the grey variety, brings about the change rapidly, causing the white tin to swell and crumble to a grey powder.

REACTIONS OF TIN: The metal is unreactive and only combines with oxygen above 1200°C. It does not corrode in the atmosphere, nor does it combine, even when hot, with nitrogen and carbon. When heated it combines with chlorine, giving stannic chloride, with hydrogen chloride, giving stannous chloride, and with sulphur, giving stannous sulphide. Tin reacts with dilute hydrochloric acid only slowly, but it dissolves rapidly in concentrated hydrochloric acid giving hydrogen:

$$Sn + 2HCl = SnCl_2 + H_2\uparrow$$

With dilute sulphuric acid little reaction occurs, but with hot concentrated sulphuric acid, sulphur dioxide is produced:

$$Sn + 2H_2SO_4 = SnSO_4 + SO_2\uparrow + 2H_2O$$

Tin reacts with nitric acid in various ways, depending on the temperature and the concentration of the acid. With concentrated acid, oxides of nitrogen are liberated vigorously, leaving a white residue of hydrated stannic oxide. Tin dissolves in hot concentrated solutions of alkalis, forming stannates and hydrogen:

$$Sn + 2NaOH + H_2O = Na_2SnO_3 + 2H_2\uparrow$$

USES OF TIN: The principal use is for coating steel to prevent its corrosion. The tinplate, as the product is called, is covered with a layer of tin about 2.5 millionths of a metre thick, applied either by dipping the steel into molten tin and rolling, or, more commonly nowadays, by electrolytic deposition. The product is used extensively for the canning of food and drinks. Fig. 29.2 shows how corrosion of tinplate takes place as soon as a scratch exposes the steel to the atmosphere. Unlike zinc the layer of tin offers no sacrificial protection, being less electropositive than iron, and corrosion of the iron proceeds rapidly, causing pitting of the surface. One advantage of tinning over other types of protection, such as galvanizing, is that tin is not poisonous and can be safely brought into contact with food.

A second important use of tin is in alloys, especially solders (e.g. 50% tin, 50% lead) and type metal (10% tin, 75% lead, 15% antimony) where its low melting point is an advantage. Other alloys include those with copper, called bronzes, and pewter (80% tin, 20% lead), and various bearing-metals.

Its low melting point and its resistance to atmospheric corrosion even when molten, account for the use of tin in an important new process for making sheet glass. In this process molten glass is floated on a layer of molten tin in a large thermostatically-controlled bath and allowed to congeal giving plates which are perfectly flat and so smooth that they need no grinding or polishing.

21.3. Compounds of Tin. Tin combines with other atoms in three ways:

(i) covalently by sharing two additional electrons as in the divalent *stannous** compounds; these substances become ionic, forming the Sn^{2+} ion, when dissolved in water, e.g. $SnCl_2$.

(ii) covalently by gaining shared control of four additional electrons, as in the tetravalent *stannic** compounds giving an outermost shell containing eight electrons, e.g. SnO_2, $SnCl_4$.

(iii) by co-ordination of negative ions, forming complex anions, e.g. $[SnCl_6]^{2-}$, $[SnO_3]^{2-}$, $[SnS_3]^{2-}$.

21.4. Tin(II) Oxide (Stannous Oxide), SnO. Made by heating stannous oxalate or hydroxide in an inert atmosphere:

$$SnC_2O_4 = SnO + CO\uparrow + CO_2\uparrow$$

It is a grey solid which oxidizes readily to stannic oxide when heated in air. It is amphoteric, giving stannous salts with acids and stannites with caustic alkalis.

21.5. Tin(II) Sulphide (Stannous Sulphide), SnS. Made by heating tin with sulphur, or by bubbling hydrogen sulphide through a solution of stannous chloride, when it appears as a brown precipitate:

$$SnCl_2 + H_2S = SnS\downarrow + 2HCl$$

It is soluble in hot concentrated hydrochloric acid and in yellow ammonium sulphide, which converts it into ammonium thiostannate $(NH_4)_2SnS_3$.

21.6. Tin(II) Chloride (Stannous Chloride), $SnCl_2$. The anhydrous salt is best made by heating the metal in a stream of dry hydrogen chloride:

$$Sn + 2HCl = SnCl_2 + H_2\uparrow$$

It is a white solid, m.p. 247°C, which dissolves readily in water and many organic solvents. If tin is dissolved in concentrated hydrochloric acid and the solution is concentrated, crystals of the dihydrate, $SnCl_2.2H_2O$, appear on cooling. These dissolve in water, forming a basic chloride by hydrolysis, particularly on dilution:

$$SnCl_2 + H_2O = Sn(OH)Cl\downarrow + HCl$$

Stannous chloride is a powerful reducing agent. Its solution is used in the laboratory to reduce ferric salts to ferrous, cupric to cuprous, chromates to chromic salts, and gold and silver salts to the metals.

* These are often distinguished by referring to them as tin (II) compounds and tin (IV) compounds as explained in Appendix I.

When an excess of stannous chloride solution is added to mercuric chloride solution, a white precipitate of mercurous chloride appears, which slowly turns grey as further reduction to mercury occurs:

$$2HgCl_2 + SnCl_2 = Hg_2Cl_2\downarrow + SnCl_4$$
$$Hg_2Cl_2 + SnCl_2 = 2Hg\downarrow + SnCl_4$$

A solution of stannous chloride gives a white precipitate of stannous hydroxide, soluble in excess, on addition of sodium hydroxide solution:

$$SnCl_2 + 2NaOH = Sn(OH)_2\downarrow + 2NaCl$$

21.7. Tin(IV) Oxide (Stannic Oxide), SnO_2. Occurs naturally as cassiterite. It can be made by heating tin strongly in air, or by dissolving tin in concentrated nitric acid and heating the insoluble product. Stannic oxide is a white solid which is insoluble in water and acids but dissolves on fused alkalis forming stannates:

$$SnO_2 + 2NaOH = Na_2SnO_3 + H_2O$$

When sodium stannate solution is acidified with dilute hydrochloric acid a thick white precipitate is thrown down, which is probably composed of hydrated stannic oxide. Similar products are obtained by adding solutions of sodium hydroxide or sodium carbonate, or even water alone, to stannic chloride:

$$Na_2SnO_3 + 2HCl = 2NaCl + SnO_2\downarrow + H_2O$$
$$SnCl_4 + 2H_2O = SnO_2\downarrow + 4HCl$$

These precipitates are amphoteric, dissolving in concentrated acids and alkalis when freshly prepared, but after standing or heating they become progressively less soluble. It has been suggested that various stannic acids are formed, but the evidence for this is not convincing and it is more likely that the various precipitates consist of hydrates of stannic oxide of differing particle size, the latter depending upon the conditions under which precipitation occurred.

Stannic oxide is used industrially to make white enamels and tiles.

21.8. Tin(IV) Chloride (Stannic Chloride), $SnCl_4$. Made by passing dry chlorine over heated tin and condensing the product:

$$Sn + 2Cl_2 = SnCl_4$$

It is a colourless liquid, b.p. 114°C, which fumes strongly in moist air. The addition of a small amount of water gives white crystals of the pentahydrate, $SnCl_4.5H_2O$, but excess of water causes hydrolysis first to basic chlorides and then to hydrated stannic oxide.

Stannic chloride forms a number of complex salts (e.g. ammonium chlorostannate $(NH_4)_2SnCl_6$), which, like stannic chloride itself, are used as mordants.

21.9. Tin(IV) Sulphide (Stannic Sulphide), SnS_2. Made by passing hydrogen sulphide through a solution of a stannic salt in dilute hydrochloric acid, when it appears as a yellow precipitate. It dissolves in ammonium sulphide solution forming a thiostannate:

$$SnS_2 + (NH_4)_2S = (NH_4)_2SnS_3$$

Crystalline stannic sulphide is used as a gilt paint.

21.10. Detection of Tin. Solutions of *stannous or tin* (*II*) salts give:

(i) a dark brown precipitate of stannous sulphide on passing hydrogen sulphide in the presence of dilute hydrochloric acid; this precipitate dissolves in yellow ammonium sulphide and in hot concentrated hydrochloric acid.

(ii) a white precipitate of mercurous chloride, turning grey on standing, with mercuric chloride solution.

(iii) a white precipitate of stannous hydroxide, insoluble in excess, with ammonium hydroxide solution.

Solutions of *stannic or tin* (*IV*) salts give:

(i) a yellow precipitate of stannic sulphide on passing hydrogen sulphide in the presence of dilute hydrochloric acid; this precipitate dissolves in ammonium sulphide solution and in hot concentrated hydrochloric acid.

(ii) a white precipitate of stannic hydroxide, insoluble in excess, with ammonium hydroxide solution.

21.11. Lead, Pb. Electronic configuration: 2,8,18,32,18,4.

OCCURRENCE: The principal ore of lead is galena, PbS, found in many parts of the world, particularly in Australia, U.S.A. and Mexico. Other lead minerals such as anglesite, $PbSO_4$, and cerussite, $PbCO_3$, are of much less importance. The metal itself does not occur naturally.

EXTRACTION: The first step is the concentration of the crushed ore by flotation methods (see § 12.10). The product is then roasted in a reverberatory furnace at about 600°C in a controlled supply of air, which oxidizes part of the sulphide to sulphate and oxide:

$$PbS + 2O_2 = PbSO_4$$
$$2PbS + 3O_2 = 2PbO + 2SO_2\uparrow$$

The next stage consists of heating more strongly with the air supply cut off completely, when the unchanged lead sulphide reduces the mixture of oxide and sulphate to lead:

$$PbS + PbSO_4 = 2Pb + 2SO_2\uparrow$$
$$PbS + 2PbO = 3Pb + SO_2\uparrow$$

In an alternative process the galena is first oxidized as completely as possible by roasting in air and then reduced by heating with carbon in a small blast furnace or obtained as a bi-product of zinc manufacture (§ 29.2).

The crude lead contains many impurities, especially copper, zinc, iron, arsenic, antimony, sulphur, and a small amount of silver. The recovery of silver, which is important because it is one of the main sources of the metal, is discussed in detail in § 28.14. The other impurities are removed by remelting and aerating, when the more electropositive metals are converted to a scum of oxides on the surface, and the copper forms a relatively infusible alloy which can be separated from the molten lead. Alternatively, the lead can be refined electrolytically by the Betts process. The anode is made of crude lead, the cathode of pure lead, and the electrolyte consists of lead fluorosilicate, $PbSiF_6$, and gelatin, which improves the texture of the lead deposited on the cathode. Impurities remain in solution or fall to the bottom of the cell as a sludge, depending upon their electropositivity, so that only pure lead is deposited:

$$PbSiF_6 \rightleftharpoons Pb^{2+} + SiF_6^{2-}$$

PROPERTIES OF LEAD: It is a soft grey metal, lustrous when freshly cut, but tarnishing on exposure to air. It melts at 327°C, boils at 1750°C, and has a density of 11.3 g cm^{-3}. It is very malleable, but not strong enough to be drawn into wires.

REACTIONS OF LEAD: Lead combines slowly with oxygen on heating, forming lead monoxide as a scum. Continued heating in air gives red lead. Lead also combines directly with sulphur and the halogens when heated. Water containing dissolved oxygen attacks lead very slowly, forming plumbous hydroxide, which is slightly soluble, but water containing dissolved sulphates and bicarbonates soon forms a protective layer of insoluble lead salt which prevents further action. This is important where drinking water is concerned because lead compounds are very poisonous, accumulating in the body over a period of years.

Lead does not dissolve in dilute hydrochloric acid, nor in dilute sulphuric acid, presumably because an insoluble coating protects the metal from further attack. With hot concentrated sulphuric acid it gives sulphur dioxide slowly:

$$Pb + 2H_2SO_4 = PbSO_4 + SO_2\uparrow + 2H_2O$$

Lead is attacked slowly by concentrated hydrochloric acid in the cold and rapidly when heated. It dissolves readily in nitric acid, dilute and concentrated, and in acetic acid, particularly in the presence of air, forming plumbous salts. Lead is not attacked by alkalis.

USES OF LEAD: It is used for roofing and plumbing, and for sheathing cables; its resistance to corrosion, flexibility, low melting point and cheapness are advantages in these applications. Lead is also used in accumulators, in plant for manufacturing sulphuric acid, and for making bullets, shot, and weights. Lead alloys include solders, typemetal, pewter, and various bearing metals. Lead is used for making lead compounds, of which the basic carbonate, the oxides, and lead tetraethyl are amongst the most important.

Another important use is as screening against radiation; lead is chosen because it is by far the cheapest of the very dense metals with high absorbing power. Lead bricks are used to build screening walls around small nuclear reactors, X-ray generators, and other powerful sources of radiation, and lead containers are used for storing and transporting highly radioactive materials such as isotopes.

21.12. Compounds of Lead. Lead forms stable electrovalent plumbous compounds in which it is present as the divalent ion, Pb^{++}, and less stable covalent plumbic compounds in which it is tetravalent. In the presence of high concentrations of suitable negative ions, lead also forms complex anions, e.g. $[PbCl_6]^{2-}$.

A general characteristic of lead compounds is their insolubility in water.

21.13. Lead Monoxide (Litharge), PbO. Litharge is a dense orange solid m.p. 888°C, made industrially by blowing air onto the surface of molten lead at 900°C. At lower temperatures, or on heating the nitrate, carbonate or hydroxide of lead, a powdery yellow form called massicot is produced. Heated in air to about 450°C this is oxidized to red lead, Pb_3O_4.

Lead monoxide is readily reduced to the metal by heating it in hydrogen or carbon monoxide, or by heating it with carbon. It is amphoteric, dissolving in acids to form plumbous salts and in concentrated solutions of alkalis to form plumbites:

$$PbO + 2HNO_3 = Pb(NO_3)_2 + H_2O$$
$$PbO + 2NaOH = Na_2PbO_2 + H_2O$$

When caustic alkalis are added to solutions of plumbous salts, a gelatinous white precipitate of plumbous hydroxide, probably hydrated lead monoxide in fact, appears. Although amphoteric in the sense that it dissolves in acids and in excess of alkali, its basic properties predominate and its suspension in water turns red litmus blue.

Lead monoxide is used in the manufacture of lead glass and various lead compounds, and as a drier in paints and varnishes, when it catalyses the process of atmospheric oxidation involved in hardening.

21.14. Lead Dioxide, PbO_2. Made by adding sodium hypochlorite solution to freshly precipitated lead hydroxide, or by digesting red lead with dilute nitric acid and filtering off the insoluble product:

$$Pb(OH)_2 + NaOCl = PbO_2\downarrow + NaCl + H_2O$$
$$Pb_3O_4 + 4HNO_3 = 2Pb(NO_3)_2 + PbO_2\downarrow + 2H_2O$$

It is a purple-brown solid which is insoluble in water and dilute acids but dissolves in concentrated alkalis giving plumbates:

$$PbO_2 + 2NaOH = Na_2PbO_3 + H_2O$$

Lead dioxide is a powerful oxidising agent. It gives off oxygen when heated alone and when warmed with concentrated sulphuric acid:

$$2PbO_2 = 2PbO + O_2\uparrow$$
$$2PbO_2 + 2H_2SO_4 = 2PbSO_4 + 2H_2O + O_2\uparrow$$

With concentrated hydrochloric acid it gives chlorine on warming:

$$PbO_2 + 4HCl = PbCl_2 + 2H_2O + Cl_2\uparrow$$

When heated with sulphur or red phosphorus it causes their ignition, which explains its use in match-heads. When warmed it combines exothermically with sulphur dioxide and nitrogen dioxide:

e.g. $$PbO_2 + SO_2 = PbSO_4$$

Lead dioxide does not give hydrogen peroxide solution when treated with dilute acids and is not, therefore, a true peroxide. Its principal use is in accumulators (see § 21.22).

21.15. Trilead Tetroxide (Red Lead), Minium, Pb_3O_4. Made by heating powdered lead monoxide in air to about 450°C. Above this temperature it tends to decompose, giving off oxygen rapidly at 550°C:

$$6PbO + O_2 \rightleftharpoons 2Pb_3O_4$$

It is an orange-red solid, practically insoluble in water. Red lead is a compound oxide which behaves chemically as a mixture of composition $PbO_2.2PbO$, as its reactions with dilute nitric acid and concentrated hydrochloric acid demonstrate:

$$Pb_3O_4 + 4HNO_3 = 2Pb(NO_3)_2 + PbO_2\downarrow + 2H_2O$$
$$Pb_3O_4 + 8HCl = 3PbCl_2 + 4H_2O + Cl_2\uparrow$$

X-ray analysis of its crystals indicates, however, that structurally it is lead orthoplumbate, $Pb_2.PbO_4$.

Mixed with linseed oil red lead is used extensively as a paint to prevent the rusting of steel, particularly for objects exposed to the weather such as ships and agricultural machinery. It is also used for making lead glass and special jointing cements.

21.16. Lead(II) Chloride, $PbCl_2$. Made by adding dilute hydrochloric acid to a cold solution of lead salt, when it appears as a white precipitate:

e.g. $$Pb(NO_3)_2 + 2HCl = PbCl_2\downarrow + 2HNO_3$$

It is only slightly soluble in cold water, but appreciably soluble in hot. It dissolves in concentrated hydrochloric acid forming a complex ion (see § 9.9, Group I):

$$PbCl_2 + 2HCl \rightleftharpoons H_2PbCl_4 \text{ (chloroplumbous acid)}$$

21.17. Lead(II) Acetate $(CH_3COO)_2Pb$. Made by dissolving lead monoxide or carbonate in acetic acid, or by adding lead to acetic acid in the presence of air:

$$PbO + 2CH_3COOH = (CH_3COO)_2Pb + H_2O$$

It crystallizes from solution as colourless crystals of the trihydrate $(CH_3COO)Pb.3H_2O$, which dissolve readily in water giving a solution with a very sweet taste. This gave rise to the common name of 'sugar of lead', although it should be noted that it is very poisonous. It is mainly important as one of the few common lead salts soluble in water. It is used, therefore, to prepare other lead salts by double decomposition.

21.18. Lead(II) Nitrate, $Pb(NO_3)_2$. Made by dissolving lead, litharge, of lead carbonate in hot dilute nitric acid:

$$PbO + 2HNO_3 = Pb(NO_3)_2 + H_2O$$

Colourless crystals appear on concentrating the solution and cooling. They are readily soluble in water giving precipitates of other lead compounds by double decomposition.

When heated lead nitrate decomposes, usually with loud popping noises as the liberated gases break open the crystals:

$$2Pb(NO_3)_2 = 2PbO + 4NO_2\uparrow + O_2\uparrow$$

This reaction is usually chosen for preparing nitrogen tetroxide (§22.12) because the lead nitrate crystals contain no water of crystallization.

21.19. Lead(II) Carbonate, $PbCO_3$. Made by adding sodium bicarbonate solution to a solution of a lead salt, when it is obtained as a white precipitate. More important commercially is basic lead carbonate, $Pb(OH)_2.2PbCO_3$, precipitated on adding sodium carbonate solution to a solution of a lead salt. This compound is used as a pigment, when it is known as 'white lead'. It is usually made industrially by some modification of the so-called Dutch process whereby sheets of lead are exposed to the action of moist air, acetic acid vapour, and carbon dioxide simultaneously. Slow conversion to the basic carbonate takes place, the product consisting of particles of the size which gives the maximum covering power when used in paint. On exposure to air containing traces of hydrogen sulphide, white lead paints tend to darken owing to the formation of black lead sulphide, and for this reason such paints are not used in chemistry laboratories. Another disadvantage of lead paints is their toxicity, which makes it inadvisable to use them in the kitchen or nursery. In recent years white lead has been largely replaced as a pigment in paints by titanium dioxide (§ 27.4) on account of its greater opacity and covering power and its cheapness, non-toxicity, and stability.

21.20. Lead(II) Chromate, $PbCrO_4$. Made by adding potassium chromate solution to a solution of a lead salt, when it is precipitated as a bright yellow solid:

$$Pb(NO_3)_2 + K_2CrO_4 = PbCrO_4\downarrow + 2KNO_3$$

It is used as a pigment, called chrome yellow, and so is the basic chromate, $PbCrO_4.Pb(OH)_2$, known as chrome red.

21.21. Lead(II) Sulphide, PbS. This compound occurs naturally as galena, the main ore of lead, forming grey cubic crystals with a metallic lustre. It is prepared by bubbling hydrogen sulphide through a solution of a lead salt, when it is formed as a black precipitate:

$$Pb(NO_3)_2 + H_2S = PbS\downarrow + 2HNO_3$$

Its appearance as a brown-black deposit upon a test paper soaked in lead acetate solution or lead nitrate solution is used as a test for hydrogen sulphide.

21.22. The Lead Accumulator. This consists of two hollow plates, one filled with lead dioxide and the other with spongy lead, standing in sulphuric acid of about 30% concentration. When the two plates are joined by a circuit, electrons flow through it from the lead plate to the lead dioxide, and the following chemical changes take place in the cell to maintain the potential difference between the plates at about two volts:

Positive Plate: $PbO_2 + 4H^+ + SO_4^{2-} + 2e = PbSO_4\downarrow + 2H_2O$
(i.e. $Pb^{4+} + 2e = Pb^{2+}$)

Negative Plate: $Pb + SO_4^{2-} = PbSO_4\downarrow + 2e$
(i.e. $Pb = Pb^{2+} + 2e$)

As a result of these changes both plates become coated with lead sulphate and the concentration and specific gravity of the acid in the cell gradually falls.

If a big enough potential difference is applied to the plates in the reverse direction to the current flow during discharge, the above chemical changes can be reversed. This process of charging is necessary periodically if the accumulator is to continue to provide electrical energy whenever required. The overall effect is to store electrical energy as chemical energy by using the reversible reaction:

$$Pb + PbO_2 + 2H_2SO_4 \underset{\text{charge}}{\overset{\text{discharge}}{\rightleftharpoons}} 2PbSO_4 + 2H_2O$$

21.23. Plumbic Compounds. Most of these are unimportant because they decompose readily on heating and are hydrolysed to lead dioxide by even cold water. An exception is lead tetraethyl, $Pb(C_2H_5)_4$, a colourless liquid which is much used as an anti-knock agent in petrol. It is usually added to the extent of about one part in a thousand, when it acts as a negative catalyst during the combustion of the petrol vapour, moderating its explosive violence. This makes it possible to use a higher compression ratio, with a consequent gain in efficiency (see § 33.9).

21.24. Detection of Lead. Solutions of lead salts give precipitates with many reagents, of which the following are used for detecting lead:

(i) dilute hydrochloric acid gives a white precipitate of lead chloride in the cold, which dissolves on boiling and crystallizes out again on cooling.

(ii) potassium chromate solution gives a yellow precipitate of lead chromate.

(iii) hydrogen sulphide gives a black precipitate of lead sulphide, soluble in dilute nitric acid.

(iv) ammonium hydroxide solution gives a white precipitate of lead hydroxide, insoluble in excess.

(v) potassium iodide solution gives a yellow precipitate of lead iodide, which dissolves in boiling water to give a colourless solution from which scintillating golden-yellow crystals separate on cooling.

On heating strongly on a charcoal block lead compounds mixed with sodium carbonate are reduced to a soft metallic bead which marks paper.

21.25. Summary. Tin and lead are closely related elements, forming many similar compounds. Superimposed upon this similarity are certain well defined trends which apply to the Group as a whole. For example, Group IVB elements become increasingly metallic with rising atomic weight; tin and lead are both metals with amphoteric oxides, although pronounced basic properties are evident in lead monoxide. Both elements can be tetravalent, forming covalent compounds such as hydrides and tetrahalides which are volatile and readily hydrolysed; their thermal stability decreases as the atomic weight rises in keeping with the trend from carbon to lead. Both tin and lead can also be divalent, forming compounds containing the ions Sn^{2+} and Pb^{2+}, but whereas stannous compounds are readily oxidized to the more stable stannic condition, plumbous compounds are much more stable than tetravalent plumbic ones, continuing the trend shown by silicon and germanium. Thus stannous chloride, unlike plumbous chloride, is a reducing agent, and lead dioxide, in contrast to stannic oxide, is an oxidizing agent.

Table 21.1. Comparison of the Elements of Group IVB

Feature \ *Element*		C	Si	Ge	Sn	Pb
Atomic number		6	14	32	50	82
Electronic configuration		2,4	2,8,4	2,8,18,4	2,8,18,18,4	2,8,18,32,18,4
m.p./°C		c. 3600	c. 1400	940	232	327
b.p./°C		c. 4200	c. 3300	c. 2800	c. 2300	1750
Density/g cm^{-3}		graphite 2.3 diamond 3.5	2.35	5.4	grey 5.7 white 7.3	11.3
Latent heat of fusion L_f/kJ mol^{-1}		?	46	31	6.8	4.6
Latent heat of evaporation L_e/kJ mol^{-1}		?	398	340	293	180
First ionization energy ΔU/kJ mol^{-1}		1088	795	753	716	720
Structure of element		framework of tetrahedrally-directed covalent bonds—the diamond structure			grey: diamond white: metallic	cubic close-packing metallic lattice
Chlorides	MCl_2	none	none	$GeCl_2$	$SnCl_2$	$PbCl_2$
		—	—	covalent	covalent/ ionic	ionic
		—	—	reducer	strong reducer	stable
	MCl_4:	CCl_4	$SiCl_4$	$GeCl_4$	$SnCl_4$	$PbCl_4$
	b.p./°C	77	59	86	114	—
	with water	immiscible, not hydrolysed	fumes in moist air and rapidly hydrolysed by cold water			
	stability	very stable	very stable	stable	stable	decomposes at room temperature

TABLE 21.1. COMPARISON OF THE ELEMENTS OF GROUP IVB (*contd.*)

Feature \ Element		C	Si	Ge	Sn	Pb
Oxides	MO:	CO	SiO	GeO	SnO	PbO
	stability	stable	unstable	unstable	oxidises in air	stable
	nature	neutral	amphoteric	amphoteric	amphoteric	amphoteric, even weakly basic
		reducer	reducer	reducer	strong reducer	
	structure	covalent	covalent	covalent	covalent	Pb^{2+} ions
	MO_2:	CO_2	SiO_2	GeO_2	SnO_2	PbO_2
	stability	very stable	very stable	stable	stable	decomposes to PbO at 300°C
	nature	weakly acidic	weakly acidic	amphoteric	amphoteric	amphoteric
	structure	small discrete molecules	framework structure giving non-volatile giant molecules			
Hydrides	MH_4	CH_4	SiH_4	GeH_4	SnH_4	PbH_4
	name	methane	silane	germane	stannane	plumbane
	stability	very stable up to 800°C	unstable above 400°C	unstable above 250°C	unstable above 150°C	unstable even at room temp.
	nature	not reducer	reducer	strong reducer	strong reducer	strong reducer
	structure	tetrahedrally directed covalent bonds				
	Others:	great range of stable hydrocarbon known—ring and chain structures	few hydrides known	—	—	—
Element + hot conc $NaOH^{aq}$		no reaction	forms silicate	forms germanate	forms stannate	forms plumbate
Sulphides	MS:	—	—	—	SnS	PbS
	MS_2	CS_2	SiS_2	GeS_2	SnS_2	

22

Nitrogen

22.1. Introduction. Group VB of the Periodic Classification contains the elements nitrogen, phosphorus, arsenic, antimony, and bismuth. Only nitrogen and its important compounds are studied in this chapter, the other elements being considered in chapter 23.

22.2. Nitrogen, N. Electronic configuration: 2,5.

OCCURRENCE: About 78% by volume of the atmosphere consists of nitrogen. Few compounds of nitrogen occur naturally in quantity, although sodium nitrate is found in certain rainless parts of Chile. All living matter contains nitrogenous organic compounds, usually as proteins; these undergo decomposition into smaller molecules during decay.

MANUFACTURE: By the liquefaction and fractional distillation of air. Details are given in § 24.2. The product usually contains about 1% of impurities, chiefly noble gases and oxygen, which can be removed if the extra cost is justified.

LABORATORY PREPARATION: Impure nitrogen can be made by passing air free from carbon dioxide over red-hot copper, which removes the oxygen by combining with it; the 1% impurity of inert gases is not usually disadvantageous. Purer nitrogen is prepared by warming a concentrated solution of ammonium nitrite, made *in situ* from ammonium chloride and sodium nitrite, until it decomposes:

$$NH_4NO_2 = N_2\uparrow + 2H_2O$$

Since the decomposition is exothermic, care should be taken not to overheat the mixture at first, or the reaction may get out of control. Traces of ammonia and oxides of nitrogen can be removed by washing the product with chromic acid, and it can be dried with concentrated sulphuric acid.

Other less important methods of preparing nitrogen include the thermal decomposition of ammonium dichromate (which is difficult to control once started) and the oxidation of ammonia by passing it over heated copper oxide:

$$(NH_4)_2Cr_2O_7 = Cr_2O_3 + 4H_2O + N_2\uparrow$$
$$3CuO + 2NH_3 = 3Cu + 3H_2O + N_2\uparrow$$

Very pure nitrogen can be obtained by heating pure sodium azide:

$$2NaN_3 = 2Na + 3N_2\uparrow$$

PROPERTIES AND REACTIONS OF NITROGEN: It is a colourless, odourless gas, b.p. −196°C, only slightly soluble in water. It neither burns nor supports combustion. It consists of diatomic molecules, N_2, which are extraordinarily stable, not undergoing any appreciable dissociation below 3000°C. This explains the inertness of nitrogen, which only combines with other elements when heated.

For example, nitrogen combines with oxygen according to the equation:

$$N_2 + O_2 \rightleftharpoons 2NO; \Delta H = +180 \text{ kJ mol}^{-1}.$$

As will be seen from the heat of reaction, the left to right change is highly endothermic (§ 11.1), so by applying Le Chatelier's principle (§ 7.9) we conclude that the yield of nitric oxide is increased by raising the temperature. Experiment shows, in fact, that the proportion of nitric oxide in the equilibrium mixture is negligible at temperatures below 1500°C (if it were not then the oxygen in the air would have combined long ago), reaching about 1% at 2000°C and about 5% at 3000°C. Nitric oxide is formed in this way when flashes of lightning occur in the atmosphere during a thunderstorm and the reaction forms part of the nitrogen cycle described in the next section. The reaction is also the basis of the now obsolete Birkeland–Eyde process for making nitric acid, in which air was momentarily raised to a high temperature by passing it through a huge electric arc and was then cooled rapidly to a temperature at which the nitric oxide formed would combine with excess oxygen before it had time to decompose again into its elements.

Nitrogen combines reversibly with hydrogen when the gases are heated together in the presence of an iron catalyst, as in the Haber process for manufacturing ammonia (§ 22.4). It also combines directly with heated metals forming nitrides:

e.g. $$3Ca + N_2 = Ca_3N_2$$

These nitrides are hydrolysed even by cold water, liberating ammonia:

$$Mg_3N_2 + 6H_2O = 3Mg(OH)_2 + 2NH_3\uparrow$$

In contrast to these electrovalent nitrides which contain the N^{3-} ion, many transition metals (§ 27.1) form nitrides of a different kind in which the nitrogen atoms occupy empty spaces in the crystal lattice. These *interstitial* nitrides, as they are called, resemble the hydrides and carbides of these metals in structure. Their composition is variable, depending upon the amount of nitrogen taken up by the lattice, so it is difficult to represent them by any particular formulae. Advantage is taken of their hardness, stability, and inertness in the industrial process of *nitriding*, in which, for example, an article made of iron is heated strongly in ammonia to give it a hard and corrosion-resistant surface.

When nitrogen is passed over calcium carbide at about 1000°C, calcium cyanamide is formed:

$$CaC_2 + N_2 = CaNCN + C$$

This is an important fertilizer, presumably because it is slowly hydrolysed in the soil to ammonia:

$$CaNCN + 3H_2O = CaCO_3 + 2NH_3$$

USES OF NITROGEN: Huge amounts are used for making ammonia by the Haber process (§ 22.4), and for making calcium cyanamide (§ 18.22) The gas is also used to provide an inert atmosphere during welding and for certain metallurgical processes where oxidation might otherwise occur, and for filling electric lamps.

DETECTION OF NITROGEN: It is recognized by its negative response to the tests for more reactive gases, e.g. it does not support combustion or turn lime water milky, etc. No simple chemical test exists which gives a direct and positive result with nitrogen, but its rapid absorption by red-hot magnesium distinguishes it from the noble gases, especially as the nitride so formed evolves ammonia gas when treated with a few drops of water (§ 18.10).

22.3. The Nitrogen Cycle. This is illustrated diagrammatically in Fig. 22.1. Plants remove nitrogen compounds from the soil during growth, and where the same soil is cropped intensively year after year natural

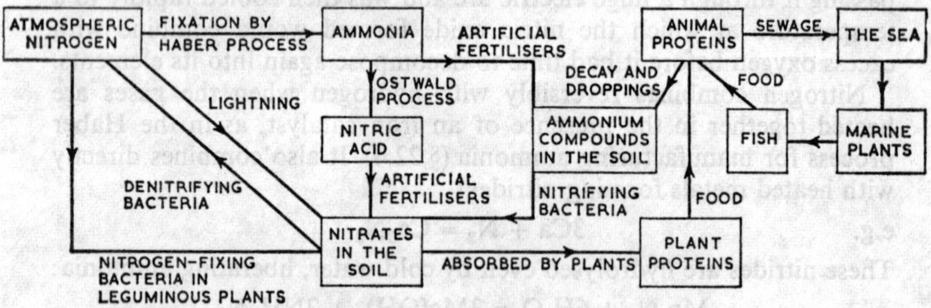

FIG. 22.1 The Nitrogen Cycle

replacement by thunderstorms and by bacterial action is too slow to maintain the supply, which must be supplemented, therefore, by artificial fertilizers. At one time the main sources were the Chilean nitrate deposits and the ammonia produced as a by-product during the carbonization of coal, but nowadays these have been largely superseded by processes using atmospheric nitrogen as the raw material. This *fixation* of nitrogen, as the process of converting it into compounds is called, is capable of providing an inexhaustible supply of nitrogenous fertilizers.

22.4. Ammonia, NH_3.

MANUFACTURE: Several methods are in use, the Haber process being by far the most important:

(1) *The Haber Process:* This involves the synthesis of ammonia directly from its elements:

$$N_2 + 3H_2 \rightleftharpoons 2NH_3; \Delta H = -92 \text{ kJ mol}^{-1}.$$

This reversible reaction is discussed in chapter 7, where it is shown that the conditions giving the maximum yield of ammonia are high pressure and low temperature. In practice the operating temperature is usually about 500°C because at lower temperatures the reaction is too slow, even in the presence of a catalyst. The operating pressure is normally about 2.5×10^7 N m^{-2} (250 atm), although pressures up to 10^8 N m^{-2}

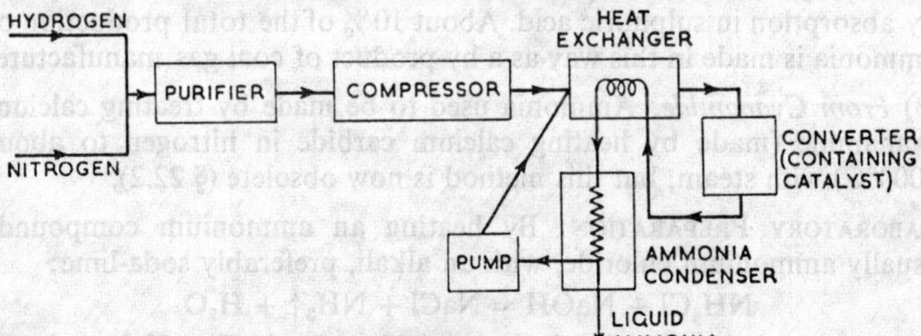

FIG. 22.2 The Haber Process

(1000 atm) have been used where the increased cost is considered justified by the higher yield. The best catalyst is iron mixed with various promoters such as aluminium and potassium oxides to increase its catalytic activity.

The process is shown diagrammatically in Fig. 22.2. The nitrogen is usually obtained by fractional distillation of liquid air (§ 24.2). In most countries the hydrogen is obtained from water gas by the Bosch process (§ 16.2), but where there are abundant supplies of natural gas the latter can be converted into hydrogen by treatment with steam at about 900°C using a nickel catalyst (§ 16.2). In either case thorough purification of the product is necessary to remove sulphur compounds, oxides of carbon, and moisture, as these would poison the catalyst in the converter. A heat exchanger is used so that once the process is started the exothermic reaction maintains the operating temperature.

At 500°C and 2.5×10^7 N m^{-2} (250 atm) pressure the yield of ammonia is about 10%. It is either condensed to a liquid by external cooling or absorbed in water, and the unchanged nitrogen and hydrogen are recirculated over the catalyst as shown in the Figure. From time to time the inert gases which accumulate are expelled from the system.

Because this process involves a continuous flow method and produces a pure product from cheap and readily accessible materials, it has largely superseded competing processes for fixing atmospheric nitrogen. Over eight million tons of ammonia are produced each year by this process.

(2) *From Gas Liquor:* When coal is heated strongly in the absence of air, the nitrogen compounds in it break down and distil over. The cooled distillate condenses to a solution of ammonium compounds known as gas liquor. Lime is added and the mixture is heated to drive off the ammonia, which is usually converted directly into ammonium sulphate by absorption in sulphuric acid. About 10% of the total production of ammonia is made in this way as a by-product of coal gas manufacture.

(3) *From Cyanamide:* Ammonia used to be made by treating calcium cyanamide (made by heating calcium carbide in nitrogen to about 1000°C) with steam, but this method is now obsolete (§ 22.2).

LABORATORY PREPARATION: By heating an ammonium compound, usually ammonium chloride, with an alkali, preferably soda-lime:

$$NH_4Cl + NaOH = NaCl + NH_3\uparrow + H_2O$$

The gas is dried by passing it over quicklime, as in Fig. 22.3, and collected by downward displacement of air. Acidic drying agents (e.g. phosphorus pentoxide and concentrated sulphuric acid) and anhydrous calcium chloride cannot be used because these substances react with ammonia, forming, for example, $CaCl_2.8NH_3$.

PROPERTIES OF AMMONIA: It is a colourless gas, b.p. −33.5°C, with a characteristic pungent smell. It is extremely soluble in water (1 volume of water dissolves about 1200 volumes of the gas at s.t.p. and about 700 volumes at 20°C). A saturated solution at room temperature contains about 35% ammonia by weight; its density is 0.880 g cm^{-3} and for this reason such a solution is often referred to as '880 ammonia'. The ammonia is completely expelled from the solution on boiling (see § 6.6).

The high solubility of the gas can be demonstrated by the fountain experiment. A flask is filled with dry ammonia and inverted over a beaker of water as shown in Fig. 22.4. If a little water is introduced into

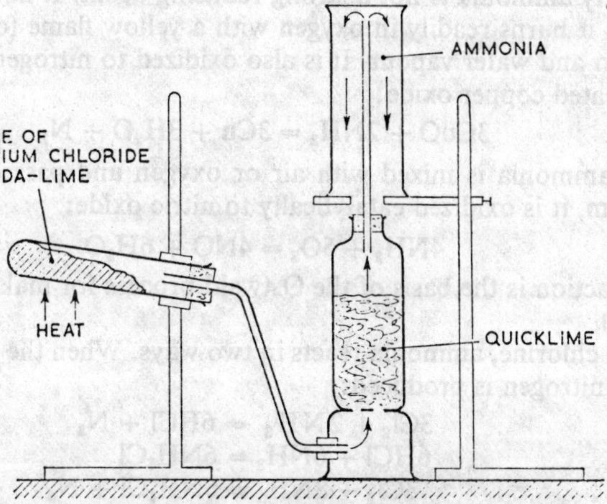

FIG. 22.3 Preparation of Ammonia

the flask by alternate warming and cooling, all the ammonia dissolves in it immediately, leaving a partial vacuum. The remaining water rushes up into the flask, producing a spectacular fountain effect.

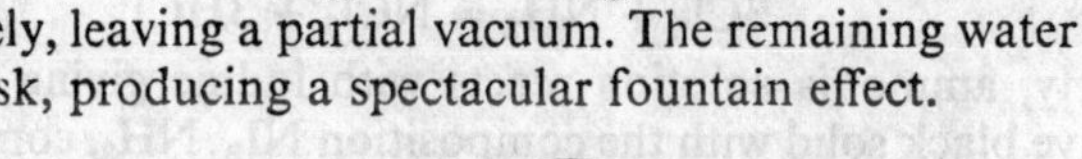

FIG. 22.4 The Fountain Experiment

REACTIONS OF AMMONIA: It is a fairly stable substance which does not decompose into its elements below 500°C, although above this temperature or on prolonged sparking it does so almost completely. Consequently ammonia is not a strong reducing agent. It does not burn in air, but it burns readily in oxygen with a yellow flame forming mainly nitrogen and water vapour. It is also oxidized to nitrogen when passed over heated copper oxide:

$$3CuO + 2NH_3 = 3Cu + 3H_2O + N_2$$

When ammonia is mixed with air or oxygen and passed over heated platinum, it is oxidized catalytically to nitric oxide:

$$4NH_3 + 5O_2 = 4NO + 6H_2O$$

This reaction is the basis of the Ostwald process for making nitric acid § 22.13).

With chlorine, ammonia reacts in two ways. When the ammonia is in excess, nitrogen is produced:

first: $3Cl_2 + 2NH_3 = 6HCl + N_2$
then: $6HCl + 6NH_3 = 6NH_4Cl$
giving: $3Cl_2 + 8NH_3 = 6NH_4Cl + N_2$

When excess of chlorine is present, an explosive, yellow oil called nitrogen trichloride is formed:

$$3Cl_2 + NH_3 = NCl_3 + 3HCl$$

Similarly, ammonia solution reacts with iodine giving a dangerously explosive black solid with the composition $NI_3.NH_3$, commonly known as nitrogen tri-iodide.

Ammonia, being basic, forms salts with acids. Direct absorption in sulphuric acid gives ammonium sulphate, and combination with gaseous hydrogen chloride produces clouds of white fumes composed of suspended particles of ammonium chloride:

$$2NH_3 + H_2SO_4 = (NH_4)_2SO_4$$
$$NH_3 + HCl = NH_4Cl$$

Ammonia combines with electropositive metals when heated. For example, with sodium it gives sodamide and with magnesium it forms the nitride:

$$2Na + 2NH_3 = 2NaNH_2 + H_2\uparrow$$
$$3Mg + 2NH_3 = Mg_3N_2 + 3H_2\uparrow$$

When ammonia is heated with carbon dioxide under high pressure, the ultimate product is urea, an important organic compound described in § 41.9:

$$2NH_3 + CO_2 = (NH_2)_2CO + H_2O$$

Ammonia is absorbed by certain metallic salts, forming solid ammines, such as $CaCl_2.8NH_3$, which give up ammonia again on heating.

USES OF AMMONIA: By far the biggest use is for the manufacture of nitrogenous fertilizers, particularly ammonium salts such as the sulphate, nitrate, and phosphate. Nearly all nitric acid is now made from ammonia by the Ostwald process. The gas is also used to make urea, aniline, nylon, and sodium carbonate (by the Solvay process). The ease with which ammonia is liquefied (only a few times atmospheric pressure is required at room temperature), and its high latent heat of evaporation lead to its widespread use as a refrigerant. It is used in metallurgy to provide an inert or reducing atmosphere during processes such as heat treatment and brazing; the hot ammonia decomposes into a mixture of nitrogen and hydrogen which prevents access of oxygen to the metal. Dilute solutions of ammonia are valuable cleansing agents, used to remove grease. In the laboratory ammonia is used to neutralize acids and as an analytical reagent; liquid ammonia is a useful solvent.

FORMULA OF AMMONIA: Its direct synthesis from nitrogen and hydrogen shows that it contains only these two elements. Its empirical formula can be determined by Hofmann's method, using the apparatus shown in Fig. 22.5. The long glass tube is filled with chlorine and a concentrated solution of ammonia is allowed to drip in from the tap funnel. A vigorous reaction occurs, producing white clouds of ammonium chloride. The addition of ammonia solution is continued until it is present in excess and no further reaction takes place. Then the long tube is inverted over water and the tap opened. After equalizing the pressure and allowing the contents to cool to room temperature, it is found that the water rises to fill two-thirds of the tube, leaving a volume of residual gas (nitrogen) equal to one-third that of the original chlorine. Thus we have the relationship:

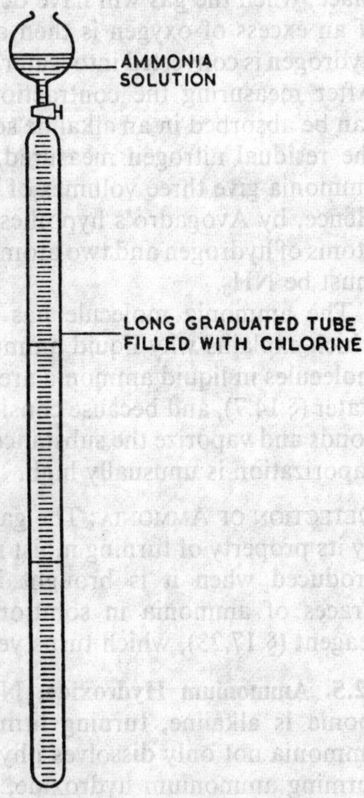

FIG. 22.5 Apparatus for Determining the Formula of Ammonia

ammonia + 3 volumes of chlorine → 1 volume of nitrogen.

Now the chlorine has been used up in reacting with the hydrogen atoms in the ammonia, and we know from § 25.6 that chlorine reacts with an equal volume of hydrogen in forming hydrogen chloride. Hence 3 volumes of chlorine have reacted with 3 volumes of hydrogen for each volume of nitrogen formed. Applying Avogadro's hypothesis, there are therefore 3 molecules of hydrogen to each molecule of nitrogen in ammonia, and, since both gases are diatomic, the empirical formula must be NH_3. Now the vapour density of ammonia is 8.5, so its molecular weight is 17 and its molecular formula is NH_3.

Alternatively a known volume of ammonia can be subjected to prolonged sparking in a eudiometer until no further volume change takes place, when the gas will have decomposed into nitrogen and hydrogen. If an excess of oxygen is then added and the sparking continued, the hydrogen is converted into water, which condenses to negligible volume. After measuring the contraction which occurs, the excess of oxygen can be absorbed in an alkaline solution of pyrogallol and the volume of the residual nitrogen measured. It will be found that two volumes of ammonia give three volumes of hydrogen and one volume of nitrogen. Hence, by Avogadro's hypothesis, two molecules of ammonia give six atoms of hydrogen and two atoms of nitrogen, and its molecular formula must be NH_3.

The ammonia molecule has a fairly large dipole moment (§ 4.9) which explains why liquid ammonia is such a good polar solvent. The molecules in liquid ammonia are associated by hydrogen bonding, as in water (§ 14.7), and because considerable energy is needed to break these bonds and vaporize the substances as single molecules, the latent heat of vaporization is unusually high.

DETECTION OF AMMONIA: The gas is detected by its characteristic smell, by its property of turning moist red litmus blue, and by the white fumes produced when it is brought into contact with hydrogen chloride. Traces of ammonia in solution can be detected by using Nessler's reagent (§ 17.23), which turns yellow or brown.

22.5. Ammonium Hydroxide, NH_4OH. An aqueous solution of ammonia is alkaline, turning litmus blue. The explanation is that the ammonia not only dissolves physically in water, but also reacts with it forming ammonium hydroxide, which undergoes dissociation in solution causing a preponderance of hydroxyl ions over H^+ ions:

$$NH_3 + H_2O \rightleftharpoons NH_4OH \rightleftharpoons NH_4^+ + OH^-$$

Ammonium hydroxide is a weak alkali, only slightly dissociated into ions. It neutralizes acids forming salts, all of which contain the ammonium ion, NH_4^+:

$$NH_4OH + HCl = NH_4Cl + H_2O$$

It precipitates insoluble hydroxides from solutions of many metallic salts, although some such as copper, silver, and zinc, redissolve in excess of the alkali owing to the formation of soluble complex ions with ammonia:

$$3NH_4OH + FeCl_3 = Fe(OH)_3\downarrow + 3NH_4Cl$$

$$\begin{cases} 2NH_4OH + CuSO_4 = Cu(OH)_2\downarrow + (NH_4)_2SO_4 \\ Cu(OH)_2 + 4NH_3 = [Cu(NH_3)_4](OH)_2 \end{cases}$$

22.6. Ammonium Salts. These resemble the salts of the alkali metals in many ways. They are all electrovalent compounds containing the univalent NH_4^+ ion, which is formed by the combination of a molecule of ammonia with a hydrogen ion and has the electronic structure shown in Fig. 22.6.

Ammonium salts are mostly white crystalline solids which dissolve readily in water; they are strong electrolytes, completely dissociated into their ions in solution. Being salts of a weak alkali and a strong acid, the sulphate, chloride, and nitrate are acidic in solution owing to hydrolysis. Some ammonium salts undergo thermal dissociation and most of them decompose when heated strongly:

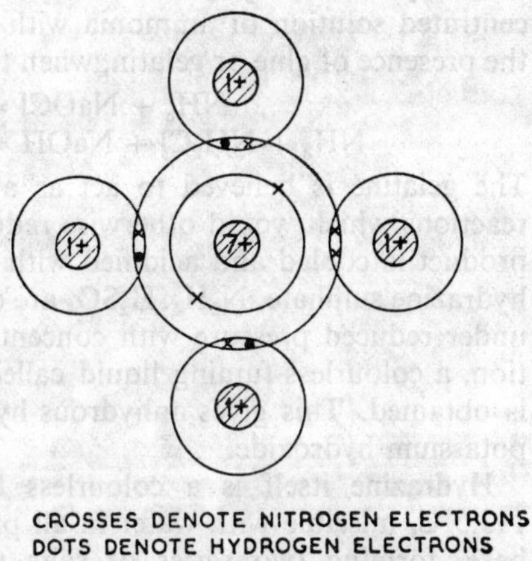

FIG. 22.6 The Ammonium Ion NH_4^+

$$NH_4Cl \rightleftharpoons NH_3 + HCl$$

$$NH_4NO_3 = N_2O\uparrow + 2H_2O$$

All ammonium salts yield ammonia when heated with alkalis:

$$NH_4Cl + NaOH = NaCl + H_2O + NH_3\uparrow$$

This reaction can be used not only to detect ammonium compounds, but also to estimate their solutions or to determine their purity as solids. A known quantity of the ammonium compound, or its solution, is boiled with an excess of standard sodium hydroxide solution until no more ammonia is evolved. The solution is cooled and the residual sodium hydroxide is determined by titration with standard acid in the usual way. Alternatively the ammonia liberated is absorbed in an excess of acid, which is then back-titrated with standard alkali.

USES OF AMMONIUM SALTS: *Ammonium sulphate* is much used as a fertilizer. It is made either directly from ammonia and sulphuric acid, or by passing ammonia and carbon dioxide into a fine suspension of gypsum, filtering off the precipitated calcium carbonate, and evaporating to crystallization:

$$CaSO_4 + 2NH_3 + CO_2 + H_2O = CaCO_3\downarrow + (NH_4)_2SO_4$$

Ammonium nitrate is another important fertilizer. It was also used in warfare as a high explosive mixed with T.N.T. or powdered aluminium. *Ammonium chloride* is used in Leclanché and dry cells (§ 10.6), and as a flux in soldering, and as an analytical reagent in the laboratory.

22.7. Hydrazine, N_2H_4. This compound is prepared by boiling a concentrated solution of ammonia with sodium hypochlorite solution in the presence of glue or gelatin, when the following reactions take place:

$$NH_3 + NaOCl = NH_2Cl + NaOH$$
$$NH_3 + NH_2Cl + NaOH = N_2H_4 + NaCl + H_2O$$

The gelatine is believed to act as a negative catalyst retarding side reactions which would otherwise reduce the yield of hydrazine. If the product is cooled and acidified with dilute sulphuric acid, crystals of hydrazine sulphate, $N_2H_4.H_2SO_4$ are obtained. When these are distilled under reduced pressure with concentrated potassium hydroxide solution, a colourless fuming liquid called hydrazine hydrate, $N_2H_4.H_2O$, is obtained. This gives anhydrous hydrazine when distilled over solid potassium hydroxide.

Hydrazine itself is a colourless fuming liquid, m.p. 1.4°C, b.p. 113.5°C, miscible with water in all proportions. It is a weak, diacidic base, forming two series of salts with acids, e.g. $N_2H_4.HCl$ and $N_2H_4.2HCl$ which contain the ions $N_2H_5^+$ and $N_2H_6^{2+}$. It is a reactive substance which inflames in dry oxygen and combines vigorously with halogens, oxides of sulphur, and the alkali metals. In particular, hydrazine and its salts are very powerful reducing agents, precipitating the metals from solutions of silver, gold, and platinum salts, and reducing cupric hydroxide to cuprous oxide and ferric salts to ferrous. Hydrazine is used as a fuel in rockets and missiles.

22.8. Hydrazoic Acid, HN_3, and the Azides. Sodium azide is formed when nitrous oxide is passed over sodamide at 200°C:

$$2NaNH_2 + N_2O = NaN_3 + NaOH + NH_3$$

If the product is distilled with dilute sulphuric acid, hydrazoic acid is obtained:

$$NaN_3 + H_2SO_4 = NaHSO_4 + HN_3$$

Hydrazoic acid is a colourless liquid, b.p. 37°C. It is poisonous and dangerously explosive. Its sodium and potassium salts are fairly stable, but lead azide and silver azide explode violently when heated or subjected to shock and are widely used as detonators.

22.9. Hydroxylamine, NH_2OH. This is prepared by electrolytic reduction of nitric acid in the presence of concentrated sulphuric acid using an amalgamated lead cathode, when the effective change is:

$$HNO_3 + 6H^+ + 6e = NH_2OH + 2H_2O$$

It forms colourless crystals, m.p. 33°C, which decompose on warming. Because of this it is usually encountered as its hydrochloride, $NH_2OH.HCl$, a crystalline solid which dissolves readily in water. Hydroxylamine and its salts are powerful reducing agents, converting silver, gold, and mercury salts into the metals and cupric salts into the cuprous state. Its most important reaction is condensation with organic compounds containing the carbonyl group (e.g. aldehydes and ketones) to give oximes as described in § 36.2, reaction 4, and § 36.6, reaction 5.

22.10. Nitrous Oxide, N_2O. This is usually prepared by heating ammonium nitrate (or a mixture of ammonium chloride and sodium nitrate) to about 200°C; the reaction is exothermic and may become explosive if too strong heating is used:

$$NH_4NO_3 = N_2O\uparrow + 2H_2O$$

The product is bubbled in turn through ferrous sulphate solution (removes nitric oxide), through sodium hydroxide solution (removes nitrogen dioxide and any chlorine present), and through concentrated sulphuric acid (removes moisture).

Pure nitrous oxide is made by warming a mixture of sodium nitrite and hydroxylamine hydrochloride solutions:

$$NH_2OH.HCl + NaNO_2 = N_2O\uparrow + 2H_2O + NaCl$$

PROPERTIES AND REACTIONS OF NITROUS OXIDE: It is a colourless gas, b.p. −89°C, with a pleasant smell. It is appreciably soluble in cold water, giving a neutral solution. Above 500°C it decomposes rapidly into nitrogen and oxygen, which explains why it supports the combustion of most burning substances and causes a glowing splint to burst into flame:

$$2N_2O = 2N_2 + O_2$$

It may be distinguished from oxygen by its insolubility in an alkaline solution of pyrogallol and by the absence of brown fumes when it is mixed with nitric oxide. When phosphorus burns in nitrous oxide no permanent change in volume occurs, which is a further distinction:

$$4P + 10N_2O = P_4O_{10} + 10N_2$$

(compared with $4P + 5O_2 = P_4O_{10}$).

Nitrous oxide, mixed with oxygen, is frequently used as an anaesthetic, particularly in dental surgery. Smaller amounts of the gas tend to cause hysteria, hence the common name 'laughing gas'.

FORMULA OF NITROUS OXIDE: The decomposition of the gas into nitrogen and oxygen shows it to be composed of only these two elements. If a copper wire is heated electrically in the gas, using the apparatus shown in Fig. 22.7, the volume of nitrogen after cooling to room temperature

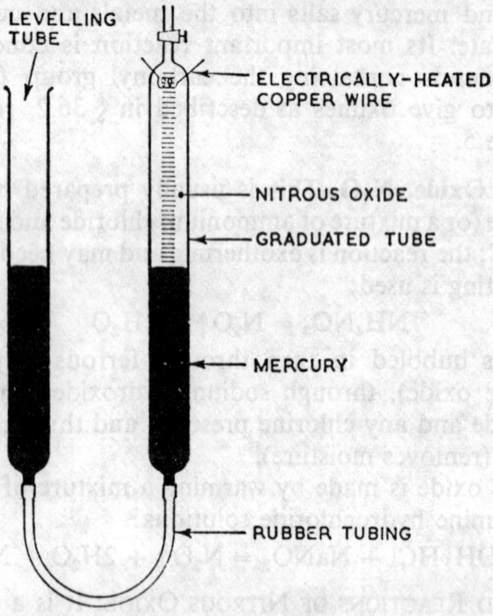

FIG. 22.7 Apparatus for Determining the Formula of Nitrous Oxide

is the same as the original volume of nitrous oxide, i.e. one volume of nitrous oxide, N_xO_y, gives one volume of nitrogen, N_2. Therefore, by Avogadro's hypothesis, x must be 2. The vapour density of nitrous oxide is 22, corresponding to a molecular weight of 44, so y is 1 and the molecular formula is N_2O.

Physical measurements reveal that the molecule is a linear one, probably a resonance hybrid (§ 14.11) of the two structures: $N \leftarrow N{=}O$ and $N{\equiv}N \rightarrow O$.

22.11. Nitrogen Monoxide (Nitric Oxide), NO. This is usually prepared by the action of moderately concentrated nitric acid (one volume of concentrated acid + one volume of water) on copper:

$$3Cu + 8HNO_3 = 3Cu(NO_3)_2 + 4H_2O + 2NO\uparrow$$

The product is not very pure. It can be collected over water, which dissolves the principal impurity, nitrogen dioxide, or the nitric oxide can be absorbed in a cold concentrated solution of ferrous sulphate and released again by heating, and then dried with concentrated sulphuric acid.

Pure nitric oxide is prepared by warming potassium nitrate with dilute sulphuric acid and ferrous sulphate solution, or by shaking mercury with potassium nitrate solution and concentrated sulphuric acid:

$$6FeSO_4 + 2KNO_3 + 4H_2SO_4 = 3Fe_2(SO_4)_3 + K_2SO_4 + 4H_2O + 2NO\uparrow$$
$$2KNO_3 + 6Hg + 4H_2SO_4 = 3Hg_2SO_4 + K_2SO_4 + 4H_2O + 2NO\uparrow$$

PROPERTIES AND REACTIONS OF NITRIC OXIDE: It is a colourless gas, b.p. −150°C, only slightly soluble in water. On exposure to the air it immediately combines with oxygen producing brown fumes of nitrogen dioxide:

$$2NO + O_2 = 2NO_2$$

For this reason its smell is unknown.

Nitric oxide decomposes into its elements rapidly above about 1000°C, but below this temperature it appears stable. It only supports the combustion, therefore, of fiercely burning substances such as phosphorus or magnesium:

$$2NO = N_2 + O_2$$

Nitric oxide dissolves readily in ferrous sulphate solution giving a brown solution of nitroso-ferrous sulphate, which decomposes when heated.

$$FeSO_4 + NO \rightleftharpoons FeSO_4.NO$$

This reaction is used for detecting nitric oxide and for purifying it or removing it from a mixture of gases. It is also the basis of the brown ring test for nitrates (§ 22.14).

Nitric oxide combines with chlorine in the presence of a charcoal catalyst forming nitrosyl chloride:

$$2NO + Cl_2 = 2NOCl$$

It also reacts with heated metals, which reduce it to nitrogen:

$$2Cu + 2NO = 2CuO + N_2$$

FORMULA OF NITRIC OXIDE: Its synthesis from nitrogen and oxygen shows that it contains only these elements. Its formula is determined by a method similar to that used for nitrous oxide. It is found that two volumes of N_xO_y leave one volume of nitrogen. Thus, by Avogadro's hypothesis, $2N_xO_y = N_2$, showing that x is 1. The vapour density being 15, the molecular weight is 30, and y must also be 1, giving the molecular formula NO.

Physical measurements suggest that the structure of nitric oxide is probably a resonance hybrid (§ 14.11) of the two forms N$=$O and N$\leftarrow$O, which accounts for its stability and small dipole moment, and for its reactivity.

22.12. Nitrogen Tetroxide, N_2O_4, and Nitrogen Dioxide, NO_2. Nitrogen dioxide is usually prepared by the action of concentrated nitric acid on copper:

$$Cu + 4HNO_3 = Cu(NO_3)_2 + 2NO_2\uparrow + 2H_2O$$

The gas is collected by upward displacement of air. Other oxides of nitrogen are inevitably produced at the same time and it is difficult to obtain a pure product by this method.

The pure gas is best prepared by heating well-dried and finely-powdered crystals of lead nitrate, which is chosen because it has no water of crystallization:

$$2Pb(NO_3)_2 = 2PbO + 4NO_2\uparrow + O_2\uparrow$$

A cooling bath is used, as in Fig. 22.8, to condense the nitrogen tetroxide, thereby separating it from the oxygen.

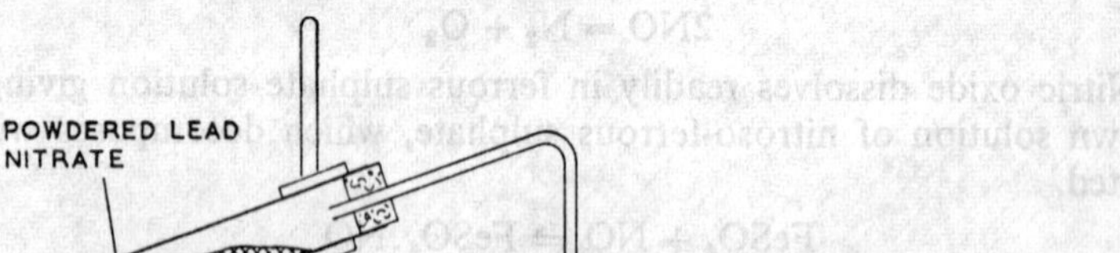

FIG. 22.8 Preparation of Nitrogen Tetroxide

Nitrogen tetroxide and dioxide are in equilibrium with each other:

$$N_2O_4 \rightleftharpoons 2NO_2$$

The mixture is usually called by the name of the form which predominates at that temperature; neither form is a true peroxide (§ 16.6), and it is incorrect to refer to this substance as nitrogen peroxide for that reason. The degree of dissociation is about 20% at the boiling point, but vapour density measurements show that the proportion of dioxide increases rapidly as the temperature is raised, reaching nearly 100% at 140°C. This is also apparent from the change in colour from yellow to dark brown as the temperature rises. At higher temperatures further dissociation occurs, which explains why nitrogen dioxide supports the combustion of strongly burning materials:

$$N_2O_4 \underset{\text{complete at }150°C}{\rightleftharpoons} 2NO_2 \underset{\text{complete at }600°C}{\rightleftharpoons} 2NO + O_2$$

The dark colour, which is caused by the NO_2 molecules, is at its deepest at about 150°C and fades to colourless above 600°C, reappearing on cooling. The arrows stress the reversibility of these changes, which provide good examples of *thermal dissociation* (§ 5.7).

Nitrogen tetroxide is extremely poisonous. It exists as colourless crystals below −10°C and as a straw-coloured liquid boiling at 23°C. It dissolves in water very readily giving an acidic solution:

$$2NO_2 + H_2O = HNO_3 + HNO_2$$

The nitrous acid subsequently decomposes into more nitric acid, so the overall equation is:

$$3NO_2 + H_2O = 2HNO_3 + NO$$

With solutions of alkalis it gives a mixture of nitrate and nitrite:

$$2NO_2 + 2NaOH = NaNO_3 + NaNO_2 + H_2O$$

It is a powerful oxidizing agent, being readily reduced to nitric oxide or nitrogen:

$$4Cu + 2NO_2 = 4CuO + N_2$$
$$H_2S + NO_2 = H_2O + S\downarrow + NO$$
$$CO + NO_2 = CO_2 + NO$$
$$SO_2 + H_2O + NO_2 = H_2SO_4 + NO$$

This last example is the basis of the lead chamber process for making sulphuric acid (§ 24.11).

FORMULA OF NITROGEN DIOXIDE: The reduction of the gas to nitrogen by heated copper or burning charcoal can be used to show that the empirical formula is NO_2. Vapour density measurements can be used to determine the proportion of N_2O_4 molecules present at any given temperature, as in § 5.8. Physical evidence suggests that the dioxide is a resonance hybrid of the two forms 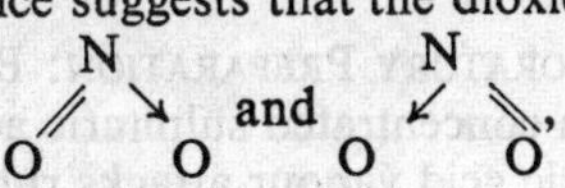, and that the tetroxide probably has the structure 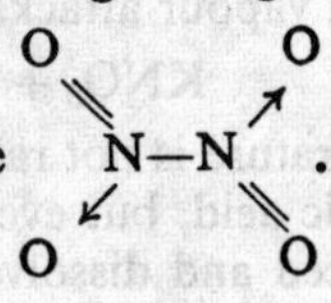.

22.13. Nitric Acid, HNO_3.

MANUFACTURE: The only important method is the Ostwald process, which has rendered other methods (e.g. by heating Chile saltpetre with concentrated sulphuric acid, or by heating nitrogen and oxygen in an electric arc, as in the Birkeland-Eyde process) completely obsolete because of its cheapness and efficiency.

The first stage of the Ostwald process involves the catalytic oxidation of ammonia to nitric oxide:

$$4NH_3 + 5O_2 = 4NO + 6H_2O; \Delta H = -904 \text{ kJ mol}^{-1}$$

Ammonia made by the Haber process (§ 22.4) is mixed with about ten times its volume of air and blown rapidly over the hot catalyst, which consists of a fine-mesh gauze of platinum-rhodium. Under these conditions 96%–98% of the ammonia is converted into nitric oxide. The reaction is exothermic and maintains the temperature at 900°C without external heating provided a heat exchanger is used. The mixture of gases is then cooled and diluted with air, when the nitric oxide combines with oxygen to form nitrogen dioxide, which reacts with water giving nitric acid of about 50% concentration. This second stage takes place in large stainless steel absorption towers designed to ensure thorough mixing of the ascending gases and the descending liquid. The absorption is sometimes carried out at a few times atmospheric pressure, when 60% acid is produced. Equations for the reactions occurring in the second stage are:

$$2NO + O_2 = 2NO_2$$

At first: $2NO_2 + H_2O = HNO_3 + HNO_2$

then: $3HNO_2 = HNO_3 + 2NO\uparrow + H_2O$

giving overall: $3NO_2 + H_2O = 2HNO_3 + NO$

The product can be concentrated to 68% by distillation, when a constant boiling point mixture is formed. More concentrated acid, which is needed for certain uses such as nitration, can be made by distilling this mixture with concentrated sulphuric acid.

LABORATORY PREPARATION: By distilling sodium or potassium nitrate with concentrated sulphuric acid in apparatus with ground glass joints (nitric acid vapour attacks rubber and cork) as in Fig. 22.9:

$$KNO_3 + H_2SO_4 = KHSO_4 + HNO_3\uparrow$$

The temperature is not raised above 150°C to minimize decomposition of the nitric acid, but even so some brown fumes of nitrogen dioxide are produced and dissolve in the condensed acid colouring it yellow. This can be remedied by bubbling warm air through the acid. It is important to realize that this reaction is reversible and only proceeds from left to right because the nitric acid, being much more volatile than the other substances, is driven out of the mixture as vapour when heated. The product is about 98 % HNO_3.

PROPERTIES OF NITRIC ACID: The concentrated acid used in the laboratory is a constant boiling point mixture containing 68 % HNO_3, having a density of 1.41 g cm^{-3} and boiling at 120.5°C under atmospheric pressure. The 98 % acid, often referred to as fuming nitric acid, boils at

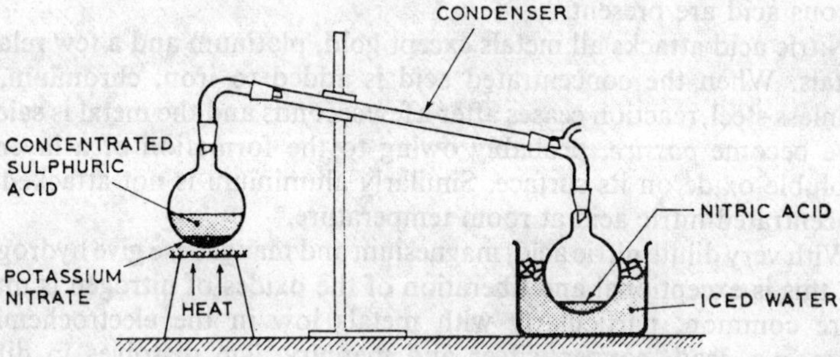

FIG. 22.9 Preparation of Nitric Acid

86°C and has a density of 1.52 g cm^{-3}. When it is cooled to −42°C the pure acid is deposited as colourless crystals. Concentrated solutions are highly corrosive and should be treated with care.

REACTIONS OF NITRIC ACID: It decomposes slowly at room temperature, particularly in sunlight, and rapidly when heated, giving nitrogen dioxide and oxygen:

$$4HNO_3 = 2H_2O + 4NO_2\uparrow + O_2\uparrow$$

Its reactions are best considered under three headings—as an acid, as an oxidizing agent, and as a nitrating agent:

As an acid: It is a strong acid, almost completely dissociated into ions in solution:

$$HNO_3 \rightleftharpoons H^+ + NO_3^-$$

It reacts with metals and with metallic oxides, hydroxides, and carbonates forming salts called nitrates, although the anhydrous acid is reputed to be covalent and relatively inactive in these respects.

The reaction of nitric acid with metals is of particular interest because of the great variety of products obtained (e.g. hydrogen, ammonium nitrate, hydroxylamine, nitrogen, nitrous oxide, nitric oxide, and nitrogen dioxide) in addition to the nitrate or oxide of the metal. In many cases mixtures of two or more of these are produced at the same time.

The main factors determining the products seem to be the nature of the metal, the concentration of the acid, and the temperature. The matter is complicated by the fact that nitric acid, particularly when concentrated, is a powerful oxidizing agent and secondary reactions tend to occur leading to the reduction of the acid. It has also been shown that certain metals, such as copper, are only attacked by nitric acid when traces of nitrous acid are present.

Nitric acid attacks all metals except gold, platinum and a few related metals. When the concentrated acid is added to iron, chromium, or stainless steel, reaction ceases after a few seconds and the metal is said to have become *passive*, probably owing to the formation of a layer of insoluble oxide on its surface. Similarly aluminium is not attacked by concentrated nitric acid at room temperature.

With very dilute nitric acid, magnesium and manganese give hydrogen, but this is exceptional and liberation of the oxides of nitrogen is much more common, particularly with metals low in the electrochemical series, e.g. lead, copper, silver and mercury. Tin dissolves in dilute nitric acid forming stannous nitrate and ammonium nitrate, which decomposes to nitrous oxide when heated, but with concentrated nitric acid it gives a white residue of hydrated stannic oxide and various oxides of nitrogen. Some equations are listed below, but it should be realized that most of them are composite, representing only the overall change and not the actual course of the reaction:

$$Cu + 4HNO_3 = Cu(NO_3)_2 + 2H_2O + 2NO_2\uparrow$$

$$3Cu + 8HNO_3 = 3Cu(NO_3)_2 + 4H_2O + 2NO\uparrow$$

$$4Zn + 10HNO_3 = 4Zn(NO_3)_2 + 3H_2O + NH_4NO_3$$

$$Sn + 4HNO_3 = SnO_2.H_2O\downarrow + H_2O + 4NO_2\uparrow$$

$$Mg + 2HNO_3 = Mg(NO_3)_2 + H_2\uparrow$$

As an oxidizing agent: Nitric acid is a powerful oxidizing agent. The concentrated acid behaves thus with reducing agents:

$$2HNO_3 = H_2O + 2NO_2 + \bar{O}$$

For example, sulphur is oxidized to sulphuric acid, phosphorus to orthophosphoric acid, and iodine to iodic acid when boiled with concentrated nitric acid:

$$S + 6HNO_3 = H_2SO_4 + 2H_2O + 6NO_2\uparrow$$

$$P + 5HNO_3 = H_3PO_4 + H_2O + 5NO_2\uparrow$$

$$I_2 + 10HNO_3 = 2HIO_3 + 4H_2O + 10NO_2\uparrow$$

Concentrated or fuming nitric acid oxidizes many organic compounds with vigour or even violence; some of these reactions (e.g. with alcohol) have been used for propelling rockets.

More dilute nitric acid behaves differently with reducing agents, giving nitric oxide:

$$2HNO_3 = H_2O + 2NO + 3\bar{O}.$$

For example, ferrous salts are oxidized to ferric, stannous to stannic, sulphites to sulphates, and the hydrides of bromine, iodine, and sulphur to the elements:

$$6FeSO_4 + 3H_2SO_4 + 2HNO_3 = 3Fe_2(SO_4)_3 + 4H_2O + 2NO\uparrow$$
$$6HBr + 2HNO_3 = 3Br_2 + 4H_2O + 2NO\uparrow$$
$$3H_2S + 2HNO_3 = 3S\downarrow + 4H_2O + 2NO\uparrow$$

Aqua regia (i.e. 1 volume of concentrated nitric acid + 3 volumes of concentrated hydrochloric acid) attacks and dissolves gold forming chloroauric acid, probably because the nitric acid first oxidizes the hydrochloric acid to chlorine:

$$HNO_3 + 3HCl = NOCl + Cl_2 + 2H_2O$$

As a nitrating agent: Concentrated nitric acid, particularly in the presence of concentrated sulphuric acid, is capable of substituting nitro groups, $-NO_2$, for hydrogen atoms in certain organic molecules:

e.g. $$C_6H_6 + HNO_3 = \underset{\text{(nitrobenzene)}}{C_6H_5NO_2} + H_2O$$

This process, which is called *nitration*, is of great industrial importance. The active nitrating agent is believed to be the nitronium ion, NO_2^+, into which nitric acid dissociates in the presence of the concentrated sulphuric acid (see § 50.6):

$$H_2SO_4 + HNO_3 \rightleftharpoons NO_2^+ + HSO_4^- + H_2O$$

USES OF NITRIC ACID: Much is used to make nitrates, particularly ammonium and calcium nitrates, for use as fertilizers. Many textile dyes and nearly all chemical explosives (e.g. T.N.T., RDX, ammonium nitrate, picric acid, gun-cotton, and dynamite), are made by processes which use concentrated nitric acid in their manufacture. It is also used as a rocket fuel and in the manufacture of phosphoric and oxalic acids. Nitric acid is also used as an oxidizing agent in the production of the important polymers nylon and 'Terylene'. In the laboratory, nitric acid is used as an oxidizing and nitrating agent, particularly in organic chemistry, and for making aqua regia and nitrates.

22.14. Nitrates. These are electrovalent compounds containing the NO_3^- ion. They are all soluble in water. Sodium and potassium nitrates, which are the most important, are described in § 17.11 and § 17.20 respectively. Nitrates decompose in various ways when heated:

(*a*) nitrates of copper, zinc, magnesium, calcium, lead, etc., give the oxide of the metal, nitrogen dioxide and oxygen:

e.g. $$2Cu(NO_3)_2 = 2CuO + 4NO_2\uparrow + O_2\uparrow$$

(*b*) nitrates of the weakly electropositive metals silver and mercury decompose to the metal instead of the oxide:

$$2\,AgNO_3 = 2\,Ag + 2\,NO_2\uparrow + O_2\uparrow$$

(*c*) nitrates of the alkali metals give the nitrite and oxygen (no brown fumes):

$$2KNO_3 = 2KNO_2 + O_2\uparrow$$

(*d*) ammonium nitrate gives nitrous oxide:

$$NH_4NO_3 = N_2O\uparrow + 2H_2O$$

Nitrates are detected by the following reactions:

(1) The brown ring test—freshly prepared ferrous sulphate solution and dilute sulphuric acid are added to a solution of a nitrate, and concentrated sulphuric acid is poured carefully into the tube so that it forms a dense layer at the bottom. At the junction of the two layers a brown ring is visible, provided that the mixture is kept cool. Nitrites give a brown solution before the concentrated acid is added.

(2) Since bromides and iodides tend to give dark rings in the above test resembling those given by nitrates, a nitrate is detected in their presence by boiling with sodium hydroxide solution and aluminium foil, when the nitrate will be reduced to ammonia, which can be detected in the usual ways. The solution should always be boiled with the alkali before adding the aluminium to prove the absence of an ammonium compound. Nitrites also give ammonia by reduction.

(3) When heated with concentrated sulphuric acid, nitrates give brown fumes and a yellow liquid (nitric acid) condenses on the cooler part of the tube.

(4) When heated most nitrates give oxygen and brown fumes of nitrogen dioxide (see above).

22.15. Nitrous Acid, HNO_2. A solution of this acid is formed when a solution of a nitrite is acidified in the cold:

$$NaNO_2 + HCl = NaCl + HNO_2$$

Anhydrous nitrous acid is too unstable to be isolated. In solution it decomposes slowly at room temperature and rapidly when warmed, and for this reason it is invariably prepared *in situ* from a nitrite:

$$3HNO_2 = HNO_3 + 2NO\uparrow + H_2O$$

It is a weak acid, only slightly dissociated into ions in solution, forming salts called nitrites:

$$HNO_2 \rightleftharpoons H^+ + NO_2^-$$

Nitrous acid is very reactive, being both an oxidizing agent and a reducing agent. For example, in acid solution it oxidizes ferrous salts to ferric, stannous salts to stannic, and sulphites to sulphates, and it liberates iodine readily from an acidified solution of potassium iodide:

$$2FeSO_4 + 2HNO_2 + H_2SO_4 = Fe_2(SO_4)_3 + 2H_2O + 2NO\uparrow$$
$$2KI + 2HNO_2 + 2HCl = 2KCl + I_2 + 2H_2O + 2NO\uparrow$$

In each case the essential change is:

$$2HNO_2 = H_2O + 2NO \quad + \bar{O}$$

With powerful oxidizing agents (e.g. potassium permanganate), nitrous acid is oxidized to nitric acid:

$$2KMnO_4 + 5HNO_2 + 3H_2SO_4$$
$$= K_2SO_4 + 2MnSO_4 + 5HNO_3 + 3H_2O$$

This reaction is used to estimate solutions of nitrites volumetrically; the fundamental reaction is:

$$HNO_2 + \bar{O} = HNO_3$$

Nitrous acid reacts with aliphatic primary amines and with amides. replacing an $-NH_2$ group by a $-OH$ group and liberating nitrogen, With aromatic amines such as aniline, it brings about *diazotization*, an extremely important reaction described in detail in § 44.6 and used for making azo dyes. The reaction with urea is used to remove an unwanted excess of nitrous acid from solution:

$$CO(NH_2)_2 + 2HNO_2 = 3H_2O + CO_2\uparrow + 2N_2\uparrow$$

22.16. Nitrites. These are salts of nitrous acid, containing the NO_2^- ion. Only the nitrites of alkali and alkaline earth metals are stable, and of these sodium nitrite is by far the most important. It is made by the thermal decomposition of sodium nitrate, or by absorbing a mixture of nitric oxide and nitrogen dioxide in sodium hydroxide solution:

$$2NaNO_3 = 2NaNO_2 + O_2\uparrow$$
$$NO + NO_2 + 2NaOH = 2NaNO_2 + H_2O$$

It is a pale yellow, hygroscopic solid, m.p. 280°C, which dissolves readily in water with the absorption of heat.

Nitrites are detected by the following reactions:

(1) When a freshly prepared solution of ferrous sulphate is added to a solution of a nitrite acidified with dilute sulphuric acid, the whole solution immediately turns brown (distinction from nitrates).

(2) Nitrites give off brown fumes when treated with dilute acids, especially on warming.

(3) Iodine (detected by blue colour with starch) is liberated immediately when a nitrite is added to acidified potassium iodide solution. This is not, of course, a specific test for nitrites.

22.17. Summary. The atom of nitrogen, like those of the other elements in Group VB, has five electrons in its outermost shell. It forms compounds in several ways. With very electropositive metals it forms a N^{3-} ion, as in the nitrides of sodium and calcium, but more commonly it shares electrons with other atoms by forming three single covalent bonds as in ammonia, nitrogen trichloride and hydroxylamine, or multiple covalent bonds as in elemental nitrogen and its oxides. Under no circumstances does the nitrogen atom give up all its outermost electrons to form a pentavalent cation. In many of its compounds there remains in the outermost shell of the nitrogen atom a pair of electrons not in use in valency bonding. These two electrons, called a *lone pair*, are readily shared with another atom or ion, providing a co-ordinate linkage. This happens in the ammonium ion, NH_4^+, in the oxides and oxy-acids of nitrogen, and in the many complex ions involving ammonia molecules. This can give nitrogen the illusory appearance of penta-covalency.

+5	N_2O_5 and HNO_3 and nitrates, NO_3^-
+4	N_2O_4 and NO_2
+3	N_2O_3 and HNO_2 and nitrates, NO_2^-
+2	NO
+1	N_2O
0	N_2
−1	NH_2OH
−2	N_2H_4
−3	NH_3 and NH_4^+ ion and nitrides, N^{3-}

TABLE 22.1. THE OXIDATION STATES OF NITROGEN

Unlike the other Group VB elements, nitrogen cannot accommodate more than eight electrons in its outermost shell, which explains why nitrogen pentachloride does not exist and also why nitrogen trichloride is hydrolysed less readily than the trichloride of phosphorus and in a different way.

Nitrogen is more electronegative than phosphorus. For this reason ammonia and the ammonium ion are more stable than phosphine and the phosphonium ion, and ammonia is more strongly basic than phosphine, and nitric acid is stronger than orthophosphoric acid.

Another notable difference between nitrogen and the other elements in the Group is its much greater volatility. This arises not only from its lower atomic weight, but also from the fact that its molecules are diatomic in contrast to the tetratomic molecules in the vapours of phosphorus, arsenic, and antimony. The striking difference in volatility between the lightest elements of Group IVB (i.e. carbon and silicon) and those of Group Vb (nitrogen and phosphorus) is due to the tendency of atoms of the former to join together into huge molecules of vastly greater weight than the small discrete molecules of nitrogen and phosphorus.

A special feature of nitrogen is the great stability of its molecule owing to its low energy content. As a result nitrogen combines less readily with oxygen and the halogens than does phosphorus, and the compounds of nitrogen with these elements are endothermic, decomposing when heated.

A detailed comparison of the elements in Group Vb is given in Table 23.5 at the end of the next chapter.

23

Phosphorus, Arsenic, Antimony and Bismuth

23.1. Introduction. Nitrogen having been considered in detail in the previous chapter, this one will be devoted to the remaining elements in Group VB (which in many respects show a closer resemblance to each other) and to a comparative study of the Group as a whole.

23.2. Phosphorus, P. Electronic configuration: 2,8,5.

OCCURRENCE: Elemental phosphorus does not occur naturally, but large deposits of phosphate minerals exist, e.g. phosphorite or 'phosphate rock', $Ca_3(PO_4)_2$, and apatite, $CaF_2.3Ca_3(PO_4)_2$. Phosphorus, usually as phosphates, is an essential constituent of all living matter; bones and teeth consist largely of calcium phosphate.

MANUFACTURE: The main source of the element is calcium phosphate from mineral deposits or bone ash. This is heated to about 1500°C in an electric furnace (Fig. 23.1) with sand and coke. The silica displaces phosphorus pentoxide, which is then reduced by the carbon:

$$2Ca_3(PO_4)_2 + 6SiO_2 = 6CaSiO_3 + P_4O_{10}$$
$$P_4O_{10} + 10C = P_4\uparrow + 10CO\uparrow$$

It is important to realize that this is an *electrothermal* method in which the electricity is used solely to produce a high temperature and not for electrolysis. The molten calcium silicate is run off as a slag. The phosphorus distils over and is condensed to a liquid under water. It is purified by treatment with chromic acid and by filtration through canvas, followed, if necessary, by redistillation in an inert atmosphere.

PROPERTIES OF PHOSPHORUS: It can exist in several allotropic forms of which the white (or yellow) and the red are the best known. The former is a soft, translucent, waxy solid produced whenever phosphorus vapour is condensed or when phosphorus is deposited from its solution in carbon disulphide. It is metastable, changing into the red form at an extremely slow rate at room temperature but rapidly when heated, especially in the presence of a trace of iodine as catalyst. As explained in § 6.5, this allotropy is a good example of *monotropy*, the red allotrope being the stable form at all temperatures. Red phosphorus is made industrially by heating the white allotrope to 260°C in the absence of air; care is needed to avoid overheating, as the reaction is exothermic and may become explosive. The product is freed from traces of white phosphorus by boiling it with sodium hydroxide solution. The two allotropes are compared in Table 23.1.

Evidence also exists of violet phosphorus, formed by heating a solution of red phosphorus in molten lead to 500°C, and black phosphorus,

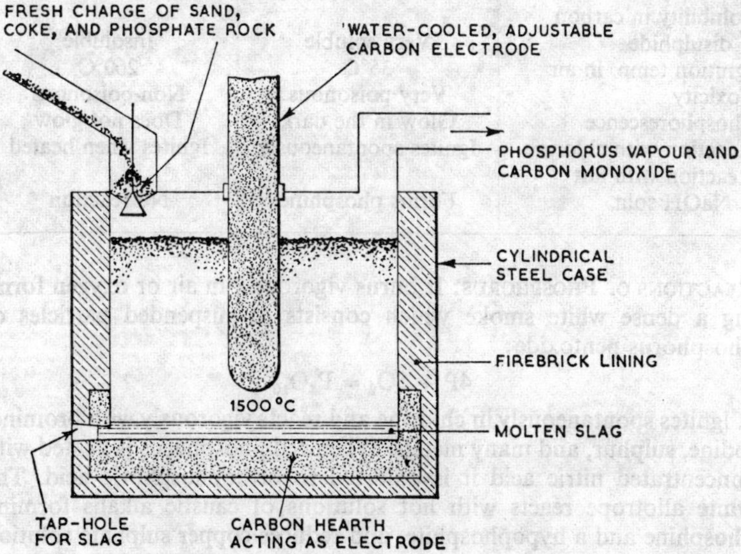

FIG. 23.1 Manufacture of Phosphorus

formed by heating white phosphorus under a pressure thousands of times atmospheric, but these varieties are of little practical importance.

White phosphorus is always stored under water because of its very low ignition temperature. It should never be handled, since it tends to ignite and cause serious burns which are slow to heal. It is also very poisonous, about 0.1 g being sufficient to cause death, and its vapour should not be inhaled. When exposed to the air, white phosphorus emits a greenish glow which is visible in the dark, a phenomenon known as *phosphorescence*. The effect can be used to detect the presence of traces of white phosphorus.

Phosphorus is tetratomic in the vapour (below about 700°C) and in solution in carbon disulphide, the four atoms being linked by single covalent bonds in a tetrahedral arrangement.

TABLE 23.1. COMPARISON OF THE ALLOTROPES OF PHOSPHORUS

Property	*White Phosphorus*	*Red Phosphorus*
Melting point	44°C	600°C (under pressure)
Boiling point	280°C	Sublimes at 400°C
Density	1.83 g cm^{-3}	2.2 g cm^{-3}
Solubility in water	Insoluble	Insoluble
Solubility in carbon disulphide	Very soluble	Insoluble
Ignition temp. in air	35°C	260°C
Toxicity	Very poisonous	Non-poisonous
Phosphorescence	Glow in the dark	Does not glow
Reaction with chlorine	Ignites spontaneously	Ignites when heated
Reaction with hot NaOH soln.	Forms phosphine	No reaction

REACTIONS OF PHOSPHORUS: It burns vigorously in air or oxygen forming a dense white smoke which consists of suspended particles of phosphorus pentoxide:

$$4P + 5O_2 = P_4O_{10}$$

It ignites spontaneously in chlorine and reacts vigorously with bromine, iodine, sulphur, and many metals. When red phosphorus is heated with concentrated nitric acid it is oxidized to orthophosphoric acid. The white allotrope reacts with hot solutions of caustic alkalis forming phosphine and a hypophosphite, and reduces copper sulphate solution to metallic copper.

USES OF PHOSPHORUS: The biggest use is in the manufacture of orthophosphoric acid and its salts (see § 23.10). Smaller amounts are used in making the chlorides of phosphorus and phosphor-bronze alloys. White phosphorus is used in rat poisons and in smoke and incendiary bombs. Its use in match-heads was discontinued because of its toxicity, and friction matches now contain the sesquisulphide P_4S_3 mixed with potassium chlorate, glue, powdered glass, and a dye. Safety matches usually have red phosphorus and antimony sulphide on the box and the oxidizing agent and an abrasive in the head.

23.3. Phosphine, PH_3. When white phosphorus is boiled with sodium hydroxide solution, phosphine and sodium hypophosphite solution are produced:

$$4P + 3NaOH + 3H_2O = 3NaH_2PO_2 + PH_3\uparrow$$

Before the preparation begins the air in the flask must be displaced by coal gas because the phosphine produced is impure containing hydrogen and diphosphine, P_2H_4, which is spontaneously inflammable. Unless the diphosphine is removed by passing the product through a U-tube immersed in freezing mixture, the bubbles of gas burst into flame as they reach the surface of the water in the trough producing rings of white smoke, as in Fig. 23.2.

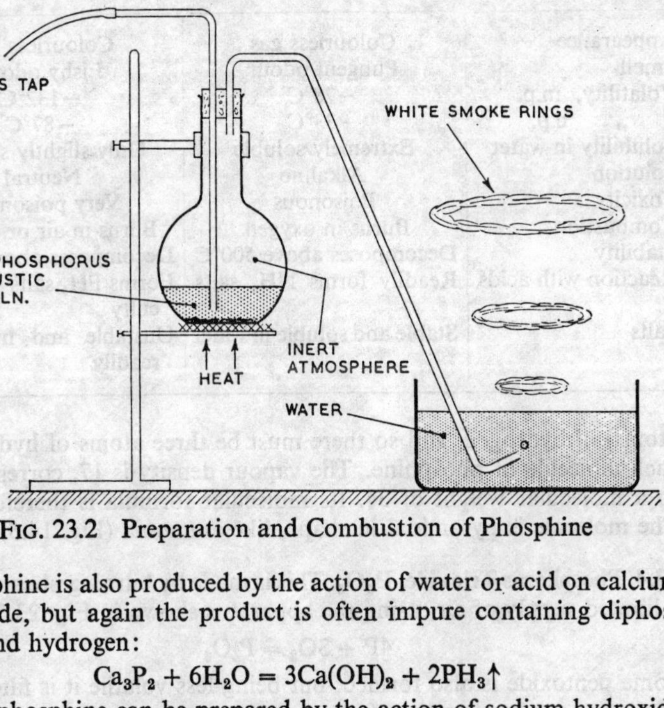

FIG. 23.2 Preparation and Combustion of Phosphine

Phosphine is also produced by the action of water or acid on calcium phosphide, but again the product is often impure containing diphosphine and hydrogen:

$$Ca_3P_2 + 6H_2O = 3Ca(OH)_2 + 2PH_3\uparrow$$

Pure phosphine can be prepared by the action of sodium hydroxide solution on phosphonium iodide, a method similar to that used for preparing ammonia from ammonium salts:

$$PH_4I + NaOH = NaI + H_2O + PH_3\uparrow$$

The phosphonium iodide is made by evaporating to dryness in a retort a solution of white phosphorus and iodine in carbon disulphide, having displaced the air with carbon dioxide; the residue is then hydrolysed with a calculated quantity of cold water yielding phosphonium iodide, which sublimes into the neck of the retort.

PROPERTIES OF PHOSPHINE: It is a colourless gas with a smell like rotting fish. It is compared with ammonia in Table 23.2.

If phosphine is sparked repeatedly it is found that two volumes of it decompose to give three volumes of hydrogen. Hence by Avogadro's hypothesis, two molecules of phosphine give three molecules (i.e. six

TABLE 23.2. COMPARISON OF PHOSPHINE WITH AMMONIA

Property	*Ammonia*	*Phosphine*
Appearance	Colourless gas	Colourless gas
Smell	Pungent odour	Fishy odour
Volatility, m.p.	−78°C	−132°C
„ b.p.	−33°C	−87°C
Solubility in water	Extremely soluble	Only slightly soluble
Solution	Alkaline	Neutral
Toxicity	Poisonous	Very poisonous
Combustion	Burns in oxygen	Burns in air or oxygen
Stability	Decomposes above 500°C	Decomposes above 440°C
Reaction with acids	Readily forms NH_4 salts	Forms PH_4 salts with difficulty
Salts	Stable and soluble in water	Unstable and hydrolysed readily

atoms) of hydrogen, and so there must be three atoms of hydrogen in each molecule of phosphine. The vapour density is 17, corresponding to a molecular weight of 34. Its molecular formula is therefore PH_3. The molecule is pyramidal in shape, like ammonia (Fig. 14.5).

23.4. Phosphorus Trioxide, P_4O_6. This is made by burning phosphorus in a limited supply of air, using the apparatus shown in Fig. 23.3.

$$4P + 3O_2 = P_4O_6$$

Some pentoxide is also formed, but being less volatile it is filtered out of the vapour by the plug of glass wool.

The trioxide is a white solid, m.p. 24°C, b.p. 173°C, with a garlic-like smell. It is easily oxidized, inflaming in chlorine at room temperature and in air above 60°C. It is an acid anhydride, dissolving slowly in cold water to form phosphorous acid:

$$P_4O_6 + 6H_2O = 4H_3PO_3$$

It reacts vigorously, however, with hot water thus:

$$P_4O_6 + 6H_2O = 3H_3PO_4 + PH_3\uparrow$$

Physical measurements show the molecular formula to be P_4O_6 both in the vapour and in solution in organic solvents.

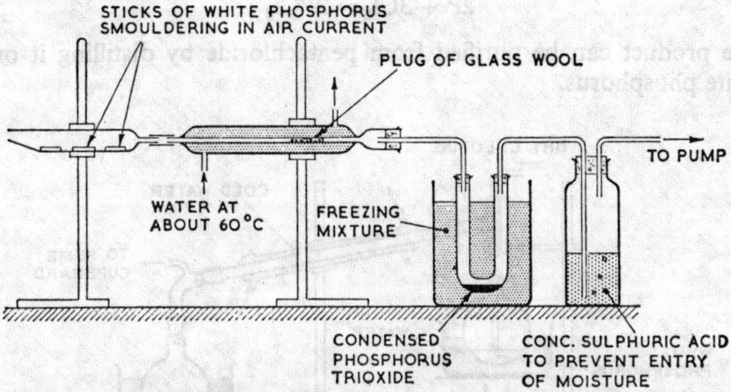

FIG. 23.3 Preparation of Phosphorus Trioxide

23.5. Phosphorus Pentoxide, P_4O_{10}. This oxide is formed by burning phosphorus in a plentiful supply of dry air or oxygen:

$$4P + 5O_2 = P_4O_{10}$$

It is a white solid which sublimes when heated. Being extremely hygroscopic it is a powerful drying agent, used particularly for removing the last traces of moisture from gases. It is so effective because it reacts chemically with water and not merely combines with it loosely and reversibly to form hydrates. Unfortunately it has to be discarded after use as it turns into a sticky mass, whereas another excellent drying agent, magnesium perchlorate (§ 25.8), can be dehydrated by heating and used again. So great is the affinity for water of phosphorus pentoxide that it also acts as a dehydrating agent, removing the elements of water from other compounds:

e.g. $CH_3.CONH_2 - H_2O = CH_3.CN$ (see § 41.10).

It reacts with cold water vigorously, producing much heat and a hissing noise, and forming metaphosphoric acid:

$$P_4O_{10} + 2H_2O = 4HPO_3$$

With hot water, orthophosphoric acid is formed:

$$P_4O_{10} + 6H_2O = 4H_3PO_4$$

Vapour density measurements show the vapour to consist of P_4O_{10} molecules.

23.6. Phosphorus Trichloride, PCl_3. This is prepared, using the apparatus shown in Fig. 23.4 by burning white phosphorus in dry chlorine:

$$2P + 3Cl_2 = 2PCl_3$$

The product can be purified from pentachloride by distilling it over white phosphorus.

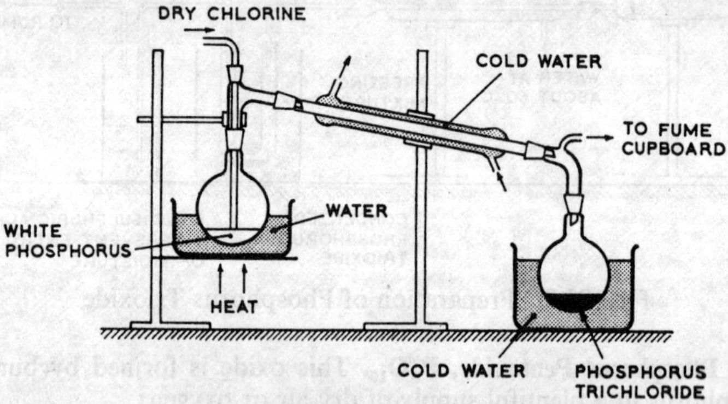

FIG. 23.4 Preparation of Phosphorus Trichloride

It is a colourless liquid, b.p. 76°C, which fumes in moist air. It is instantly hydrolysed by cold water:

$$PCl_3 + 3H_2O = H_3PO_3 + 3HCl\uparrow$$

It is a reactive substance, combining readily with oxygen, sulphur, and chlorine to form compounds in which phosphorus is pentavalent.

23.7. Phosphorus Pentachloride, PCl_5. This is prepared by the action of dry chlorine on phosphorus trichloride, using the apparatus shown in Fig. 23.5:

$$PCl_3 + Cl_2 = PCl_5$$

It is a pale yellow solid which sublimes at about 160°C and dissociates when heated strongly:

$$PCl_5 \rightleftharpoons PCl_3 + Cl_2$$

Vapour density measurements show that this dissociation is complete above 300°C.

It fumes in moist air and reacts violently with water:

first: $PCl_5 + H_2O = POCl_3 + 2HCl\uparrow$

then, with excess of water: $POCl_3 + 3H_2O = H_3PO_4 + 3HCl\uparrow$

It reacts with any compound containing a hydroxyl group, replacing it by a chlorine atom and liberating hydrogen chloride. This is important

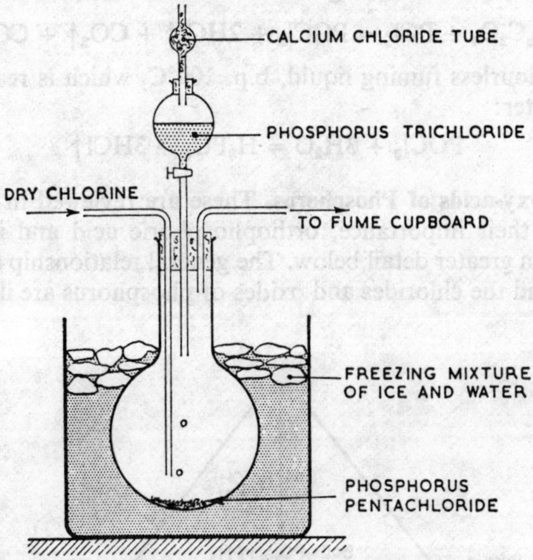

FIG. 23.5 Preparation of Phosphorus Pentachloride

both as a means of chlorinating compounds and as a test for hydroxyl groups. The general reaction is:

$$R.OH + PCl_5 = R.Cl + POCl_3 + HCl\uparrow$$

e.g.

$$\underset{\text{ethyl alcohol}}{C_2H_5OH} + PCl_5 = \underset{\text{ethyl chloride}}{C_2H_5Cl} + POCl_3 + HCl\uparrow$$

$$\underset{\text{acetic acid}}{CH_3.COOH} + PCl_5 = \underset{\text{acetyl chloride}}{CH_3.COCl} + POCl_3 + HCl\uparrow$$

$$H_2SO_4 + PCl_5 = \underset{\text{chlorosulphonic acid}}{Cl.SO_2OH} + POCl_3 + HCl\uparrow$$

It also replaces an oxygen atom in certain compounds by two chlorine atoms, although in these reactions no hydrogen chloride is evolved:

e.g.

$$\underset{\text{acetone}}{CH_3.CO.CH_3} + PCl_5 = \underset{\text{2, 2-dichloropropane}}{CH_3.CCl_2.CH_3} + POCl_3$$

$$SO_2 + PCl_5 = \underset{\text{thionyl chloride}}{SOCl_2} + POCl_3$$

Although the vapour is composed of covalent PCl_5 molecules, X-ray analysis shows that the crystalline solid is built up from the ions $[PCl_4]^+[PCl_6]^-$.

23.8. Phosphorus Oxychloride, Phosphoryl Chloride, $POCl_3$. This is made by the action of phosphorus pentachloride on oxalic acid, chosen because the by-products are gaseous.

$$H_2C_2O_4 + PCl_5 = POCl_3 + 2HCl\uparrow + CO_2\uparrow + CO\uparrow$$

It is a colourless fuming liquid, b.p. 109°C, which is readily hydrolysed by water:

$$POCl_3 + 3H_2O = H_3PO_4 + 3HCl\uparrow$$

23.9. The Oxy-acids of Phosphorus. These are reviewed in Table 23.3. Because of their importance, orthophosphoric acid and its salts are considered in greater detail below. The general relationship between the oxy-acids and the chlorides and oxides of phosphorus are illustrated in Fig. 23.6.

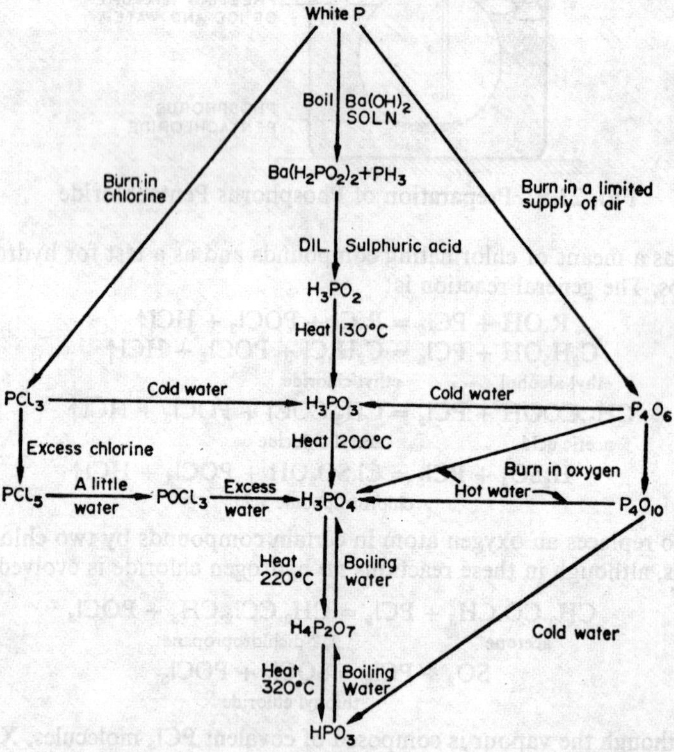

FIG. 23.6 The Relationship between some common Phosphorus Compounds

TABLE 23.3. THE OXY-ACIDS OF PHOSPHORUS

Name	Formula	Structure	Basicity	Salts	Preparation of acid	Physical props.	Chemical properties
Hypophosphorous	H_3PO_2	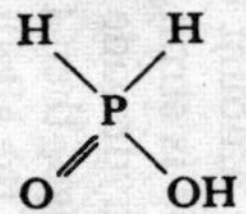	1	Hypophosphites	White P + hot $Ba(OH)_2$ soln., then acidify with dil. H_2SO_4	M.p. 27°C Soluble in water	Powerful reducing agent $3H_3PO_2 \xrightarrow{130°C} 2H_3PO_3 + PH_3\uparrow$
Phosphorous	H_3PO_3	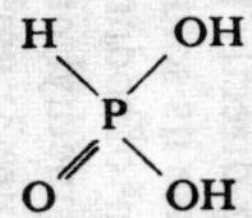	2	Phosphites	Hydrolysis of PCl_3, or cold water on P_4O_6	M.p. 74°C Soluble in water	Strong reducing agent $4H_3PO_3 \xrightarrow{200°C} 3H_3PO_4 + PH_3\uparrow$
Orthophosphoric	H_3PO_4	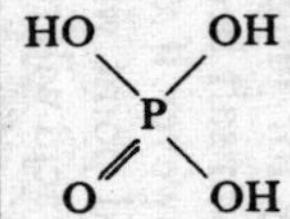	3	Orthophosphates	Conc. HNO_3 on red P (best), or hot water on P_4O_{10}, or dil. H_2SO_4 on $Ca_3(PO_4)_2$	M.p. 42°C Very soluble in water. Deliquescent	$2H_3PO_4 \xrightarrow{220°C} H_4P_2O_7 + H_2O$ $H_3PO_4 \xrightarrow{320°C} HPO_3 + H_2O$
Metaphosphoric	HPO_3	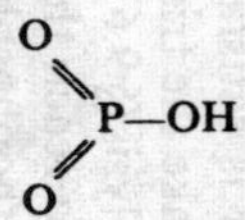	1	Metaphosphates	Heat H_3PO_4 to 320°C, or cold water on P_4O_{10}	Very soluble in water. Deliquescent	$HPO_3 + H_2O \xrightarrow{boil} H_3PO_4$
Pyrophosphoric	$H_4P_2O_7$	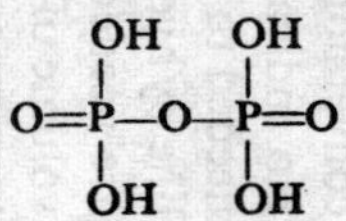	4	Pyrophosphates	Heat H_3PO_4 to 220°C	M.p. 61°C Soluble in water	$H_4P_2O_7 \xrightarrow{320°C} 2HPO_3 + H_2O$ $H_4P_2O_7 + H_2O \xrightarrow{boil} 2H_3PO_4$

23.10. Orthophosphoric Acid, H_3PO_4, and its Salts. The acid is made industrially by treating phosphate rock with dilute sulphuric acid:

$$Ca_3(PO_4)_2 + 3H_2SO_4 = 3CaSO_4\downarrow + 2H_3PO_4$$

The calcium sulphate is filtered off and the solution of phosphoric acid concentrated by evaporation.

Purer acid is made on a large scale by burning white phosphorus in a plentiful supply of air and hydrating the pentoxide:

$$P_4O_{10} + 6H_2O = 4H_3PO_4$$

In the laboratory orthophosphoric acid is prepared by adding moderately concentrated nitric acid to red phosphorus in a flask fitted with a reflux condenser:

$$P + 5HNO_3 = H_3PO_4 + 5NO_2\uparrow + H_2O$$

A trace of iodine catalyses the reaction. When the evolution of brown fumes ceases, the temperature is raised to 180°C to concentrate the solution.

PROPERTIES OF ORTHOPHOSPHORIC ACID: The colourless crystals are very deliquescent and the acid is usually encountered as the syrupy concentrated solution. It is a weak, tribasic acid which dissociates in solution thus:

$$H_3PO_4 \rightleftharpoons H^+ + H_2PO_4^- \rightleftharpoons 2H^+ + HPO_4^{2-} \rightleftharpoons 3H^+ + PO_4^{3-}$$

It gives rise, therefore, to three series of salts:

e.g. $NaH_2PO_4.H_2O$ sodium dihydrogen orthophosphate
$Na_2HPO_4.12H_2O$ disodium hydrogen orthophosphate
$Na_3PO_4.12H_2O$ trisodium or normal sodium orthophosphate

These salts can be prepared from caustic soda solution and orthophosphoric acid by taking advantage of the fact that methyl orange changes colour at the point when all the acid has been converted into the sodium dihydrogen salt and phenolphthalein changes colour when conversion into the disodium salt is complete (see § 9.13 and Fig. 9.4). No indicator is satisfactory for judging directly the amount of alkali needed to make the trisodium salt because its solution is strongly alkaline by hydrolysis (§ 9.6). However, if x cm³ of sodium hydroxide solution are needed to reach the end point with methyl orange, then the same volume of orthophosphoric acid will require $3x$ cm³ of the alkali solution to give a solution of the trisodium salt. Alternatively, if $2x$ cm³ of the sodium hydroxide were added with phenolphthalein as indicator, then $3x$ cm³ of the alkali solution are needed for preparing the trisodium salt. These relationships follow from the equations:

$H_3PO_4 + NaOH = NaH_2PO_4 + H_2O$ Methyl orange end point
$H_3PO_4 + 2NaOH = Na_2HPO_4 + 2H_2O$ Phenolphthalein end point
$H_3PO_4 + 3NaOH = Na_3PO_4 + 3H_2O$

When heated the orthophosphates lose their water of crystallization, and some undergo dehydration thus:

$$2Na_2HPO_4 = H_2O\uparrow + Na_4P_2O_7 \text{ (sodium pyrophosphate)}$$
$$NaH_2PO_4 = H_2O\uparrow + NaPO_3 \text{ (sodium metaphosphate)}$$

The metaphosphate often gives rise to polymers of the type $(NaPO_3)_n$, where *n* may be 100–150. Similarly, sodium ammonium hydrogen orthophosphate, which is present in urine and is called microcosmic salt, decomposes to the metaphosphate and ammonia when heated:

$$NaNH_4HPO_4 = NaPO_3 + NH_3\uparrow + H_2O$$

The ammonium and alkali metal phosphates are soluble in water; the rest are insoluble, but they dissolve in strong acids as explained in § 9.11.

When orthophosphoric acid is heated it progressively loses water as shown in Table 23.3. These changes are reversible, the meta- and pyro-acids giving a solution of orthophosphoric acid when boiled with water:

$$2H_3PO_4 \rightleftharpoons H_4P_2O_7 + H_2O \rightleftharpoons 2HPO_3 + 2H_2O$$

USES OF ORTHOPHOSPHORIC ACID: The main use is in the manufacture of phosphates, particularly triple superphosphate and ammonium phosphate, which are used as fertilizers, and pentasodium triphosphate, $Na_5P_3O_{10}$, which is a constituent of many detergents, and sodium metaphosphate, which is widely used as a water-softener (§ 16.5). It is also used for rustproofing steel articles before painting, and its salts for fireproofing timber and textiles. Orthophosphates are mixed with sodium bicarbonate in baking powder and self-raising flour because when moistened they react releasing carbon dioxide. The acid and its salts are also used in processed foods and drinks.

TABLE 23.4. TESTS TO DISTINGUISH ORTHO-, META- AND PYRO-PHOSPHATES

Reagent	*Orthophosphate*	*Pyrophosphate*	*Metaphosphate*
Neutral silver nitrate soln.	Yellow ppt., Ag_3PO_4	White, crystalline ppt., $Ag_4P_2O_7$	White gelatinous ppt., $AgPO_3$
Albumen and dil. acetic acid	No effect	No effect	Coagulated
Cadmium chloride and acetic acid	No ppt.	White ppt.	No ppt.

23.11. Detection of Phosphates. All phosphates give a bright yellow precipitate of ammonium phosphomolybdate when dissolved in nitric acid and warmed (not boiled) with an excess of ammonium molybdate solution. Arsenates give a similar precipitate only when boiled with these reagents.

Ortho-, meta-, and pyro-phosphates are distinguished by the tests shown in Table 23.4.

23.12. The Phosphorus Cycle. During growth plants take up phosphates from the soil and if the latter is to remain fertile, this loss must be replaced. The natural process of replacement by the weathering of

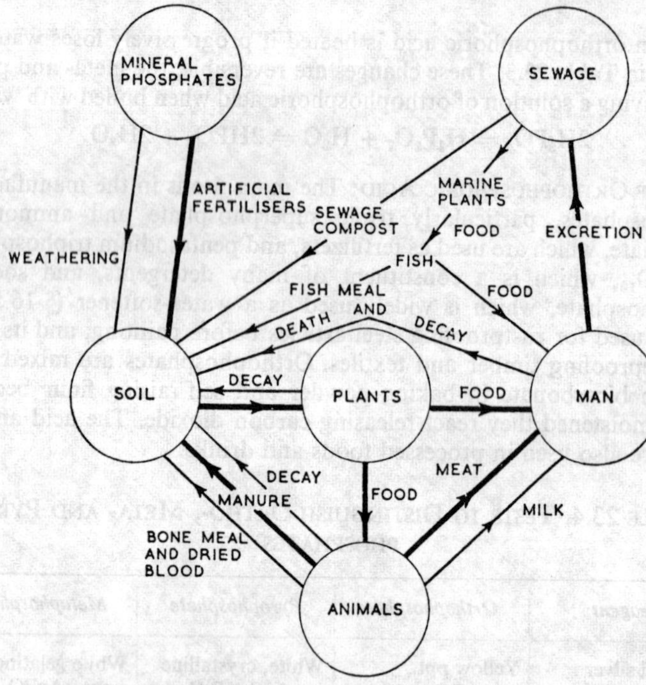

FIG. 23.7 The Phosphorus Cycle

calcium phosphate rock is far too slow when the soil is cultivated intensively, and manure or an artificial fertilizer must be added. Sometimes bone meal or dried blood are available, or basic slag obtained as a by-product in the manufacture of steel, but the main replenishment of phosphates is achieved by using 'superphosphate', which is a mixture of calcium tetra-hydrogen di-orthophosphate and calcium sulphate made by treating mineral phosphate with concentrated sulphuric acid:

$$Ca_3(PO_4)_2 + 2H_2SO_4 = \underbrace{Ca(H_2PO_4)_2 + 2CaSO_4}_{\text{superphosphate}}$$

The diphosphate salt is more soluble than untreated mineral phosphates and is therefore more easily assimilated by plants. Huge quantities of superphosphate are made each year, the process being a major consumer of sulphuric acid.

An improved product is obtained by treating mineral phosphates with orthophosphoric acid instead of sulphuric acid:

$$Ca_3(PO_4)_2 + 4H_3PO_4 = 3Ca(H_2PO_4)_2$$

This fertilizer, called 'triple superphosphate' because its phosphate content is about three times that of ordinary superphosphate fertilizer, is preferred because it lowers freight costs.

The phosphorus cycle is illustrated diagrammatically in Fig. 23.7.

23.13. Arsenic, As. Electronic configuration: 2,8,18,5.

OCCURRENCE: It occurs chiefly as arsenopyrite or mispickel, FeSAs, and as realgar, As_4S_4, and orpiment, As_2S_3, usually as impurities in the sulphide ores of other metals. When these ores are roasted in air to convert them into the oxide, volatile arsenious oxide is produced as a by-product.

MANUFACTURE: Arsenopyrite is heated to 700°C in the absence of air:

$$4FeSAs = 4FeS + As_4\uparrow$$

The arsenic sublimes and is condensed on a cooled surface.

Alternatively, arsenious oxide is reduced by heating it with powdered charcoal:

$$As_4O_6 + 6C = As_4\uparrow + 6CO\uparrow$$

PROPERTIES AND REACTIONS OF ARSENIC: Several allotropes exist. The commonest and the most stable is grey arsenic, which is a brittle metallic variety of density 5.72 g cm^{-3}, which sublimes above 630°C giving a tetratomic vapour. When this vapour is cooled rapidly, a soft waxy yellow allotrope resembling white phosphorus is deposited, having a density of only 1.97 g cm^{-3}. It is soluble in carbon disulphide and reverts rapidly to grey arsenic when gently heated. A third form, called black arsenic, is deposited when arsine is decomposed by heat.

Arsenic burns with a blue flame when heated in air, forming white clouds of arsenious oxide. It inflames in chlorine when powdered and combines with many other elements when heated. Hot dilute nitric acid and hot concentrated sulphuric acid convert it into arsenious acid, but concentrated nitric acid oxidizes it to arsenic acid. It forms hard brittle alloys with many metals.

Metallic arsenic has few applications, most of it being used to make lead shot (it hardens the lead and lowers its melting point).

23.14. Compounds of Arsenic

Arsine, AsH_3. This is prepared by reducing a soluble arsenic compound with nascent hydrogen or by treating an arsenide with a dilute acid:

$$Zn_3As_2 + 3H_2SO_4 = 3ZnSO_4 + 2AsH_3\uparrow$$

It is a colourless, very poisonous gas, b.p. −55°C, which is practically insoluble in water and has no basic properties. It burns with a blue flame forming arsenious oxide:

$$4AsH_3 + 6O_2 = As_4O_6 + 6H_2O$$

When heated above 250°C it decomposes into its elements:

$$4AsH_3 = As_4 + 6H_2$$

Arsine is a strong reducing agent, precipitating silver, for example, from silver nitrate solution (see § 23.15).

Arsenious Oxide, As_4O_6: This substance, which is commonly called 'white arsenic' or even just 'arsenic', is formed when arsenical ores are roasted in air. It sublimes when heated, condensing to a white solid on cooled surfaces. It is reduced to arsenic by heating with carbon and to arsine by nascent hydrogen. Concentrated nitric acid oxidizes it to arsenic acid. Arsenious oxide is extremely poisonous, less than 0.1 g being fatal, although people who take small doses regularly develop some immunity to it. It is used for destroying vermin and weeds.

Arsenious oxide is amphoteric, dissolving in alkalis to give arsenites and in concentrated hydrochloric acid to give arsenic trichloride solution:

$$As_4O_6 + 12NaOH = 4Na_3AsO_3 + 6H_2O$$

$$As_4O_6 + 12HCl = 4AsCl_3 + 6H_2O$$

It is only slightly soluble in water, giving a weakly acidic solution of *arsenious acid*, the salts of which are called *arsenites*:

$$As_4O_6 + 6H_2O = 4H_3AsO_3$$

Arsenites are oxidized quantitatively to arsenates by iodine provided sodium bicarbonate is present in excess to react with the hydriodic acid and prevent the back reaction:

$$Na_3AsO_3 + I_2 + H_2O \rightleftharpoons Na_3AsO_4 + 2HI$$

Arsenic Pentoxide, As_4O_{10}. This is a white, deliquescent solid formed by heating *arsenic acid* to 200°C:

$$4H_3AsO_4 \rightleftharpoons As_4O_{10} + 6H_2O$$

The acid is obtained as colourless crystals by treating arsenious oxide or arsenic itself with concentrated nitric acid and evaporating the resulting solution almost to dryness:

$$As + 5HNO_3 = H_3AsO_4 + H_2O + 5NO_2\uparrow$$

Arsenic acid is a fairly weak acid forming salts called *arsenates*, many of which are isomorphous with the phosphates. Arsenates are used as insecticides in agriculture but their high toxicity is a disadvantage. They are oxidizing agents and are readily reduced to arsenites:

$$As_4O_{10} + 4SO_2 + 4H_2O = As_4O_6 + 4H_2SO_4$$

Arsenic Trichloride, $AsCl_3$. This is a colourless fuming liquid, b.p. 130°C, formed when arsenic burns in chlorine. It is usually prepared by passing hydrogen chloride over arsenious oxide at 200°C:

It dissolves in water readily, being partially hydrolysed in the process:

$$AsCl_3 + 3H_2O \rightleftharpoons H_3AsO_3 + 3HCl$$

23.15. Detection of Arsenic. Arsenates give a yellow precipitate when dissolved in concentrated nitric acid and boiled with excess of ammonium molybdate solution. With neutral silver nitrate solution, arsenates give a red-brown precipitate of silver arsenate and arsenites a yellow precipitate of silver arsenite. These tests distinguish arsenates and arsenites from phosphates. With hydrogen sulphide in the presence of dilute hydrochloric acid, solutions of all arsenic compounds give a yellow precipitate of arsenic sulphide, which is soluble in ammonium sulphide and ammonium carbonate solutions (distinction from tin):

e.g. $$As_4O_6 + 6H_2S = 2As_2S_3 + 6H_2O$$

Several special tests have been devised for detecting traces of arsenic in other materials:

Marsh's Test: The apparatus is shown in Fig. 23.8. Acid is added to arsenic-free zinc in the flask and when all the air has been driven out of the apparatus (this should be checked by igniting a sample in a test-tube first), the hydrogen is burned at the jet. At this stage, heating the tube at X produces no black deposit. Then the substance suspected of containing arsenic is added to the flask; any arsenic present will be converted by the nascent hydrogen into arsine, which can be detected by the appearance of a black deposit of arsenic just beyond X on heating, and by a bluish-white flame at the jet. If a cold surface such as an evaporating dish is put into this flame for a few seconds it becomes coated with a shiny metallic deposit of arsenic which is soluble in sodium hypochlorite solution (distinction from antimony, which is insoluble). In skilled hands this test can be used to detect as little as a microgramme of arsenic and can be adapted for quantitative use.

Gutzeit's Test: Zinc is added to some dilute sulphuric acid in a test-tube and the issuing gas, after passing through a loose wad of cotton wool soaked in lead acetate solution (to remove any hydrogen sulphide

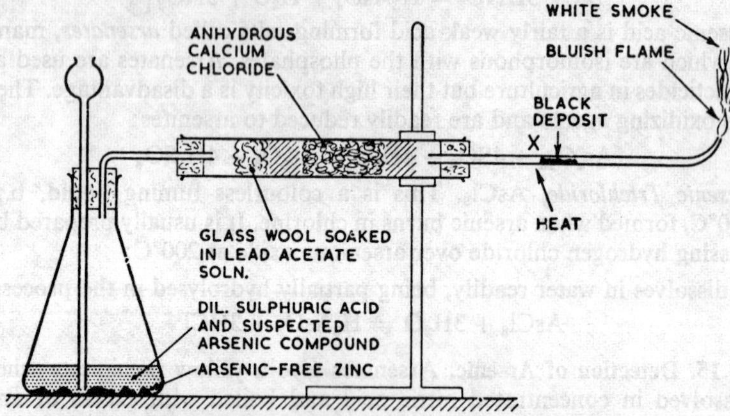

FIG. 23.8 The Marsh Test for Arsenic

formed), is allowed to flow over a test paper soaked in silver nitrate solution as in Fig. 23.9. No change in the test paper should be apparent,

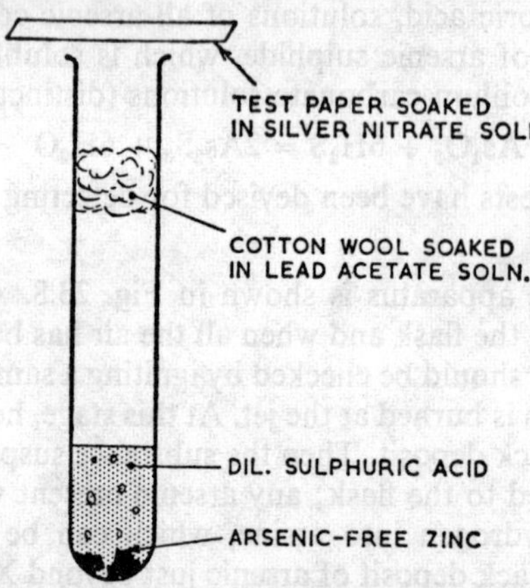

FIG. 23.9 The Gutzeit Test for Arsenic

showing the absence of arsenic in the materials used. If an arsenic compound is then added, arsine is evolved; this rapidly reduces the silver nitrate to silver causing a black deposit on the test paper:

$$6AgNO_3 + AsH_3 + 3H_2O = 6Ag\downarrow + H_3AsO_3 + 6HNO_3$$

Alternatively a test paper soaked in mercuric chloride solution can be used, when arsenic compounds cause a yellow stain and antimony compounds none.

Reinsch's Test: If a solution of an arsenic compound is boiled with hydrochloric acid and a piece of copper foil, a grey deposit forms on the copper. If the latter is washed, dried, and heated strongly, a white sublimate of arsenious oxide appears.

23.16. Antimony, Sb. Electronic configuration: 2,8,18,18,5.

OCCURRENCE: It occurs chiefly as the sulphide, Sb_2S_3, known as stibnite, usually in association with the sulphide ores of copper, silver, and lead.

MANUFACTURE: The sulphide is roasted in air and the resulting tetroxide reduced by heating with charcoal:

$$Sb_2S_3 + 5O_2 = Sb_2O_4 + 3SO_2$$
$$Sb_2O_4 + 4C = 2Sb + 4CO$$

Alternatively, if the ore is rich, antimony is displaced from its sulphide by heating it with iron:

PROPERTIES AND REACTIONS OF ANTIMONY: It is a white, lustrous, brittle metal, m.p. 630°C, with a density of 6.7 g cm^{-3}. It boils at 1440°C, giving a tetratomic vapour. Allotropes, such as yellow antimony, are so unstable that they exist only at very low temperatures.

Antimony burns when heated strongly in air, forming the trioxide, Sb_2O_3. It inflames in chlorine and bromine and reacts vigorously with iodine and sulphur when heated. It is oxidized by concentrated nitric acid to the pentoxide.

USES OF ANTIMONY: Its main use is to make alloys. It hardens lead and makes it more resistant to acids, so the plates of lead accumulators are made of an alloy containing about 15% of antimony. Type metal is a hard alloy of lead, tin, and antimony which melts easily and, because of its antimony content, expands upon solidification giving sharp castings. Other alloys include Britannia metal (90% tin, 8% antimony and 2% copper) and many bearing metals.

23.17. Compounds of Antimony.

Antimony Trioxide, Sb_4O_6: This is formed by burning antimony in air, when it condenses as a white powder on a cooled surface. It melts at 650°C, but if heated strongly it tends to sublime. It is insoluble in water, but being amphoteric it dissolves in concentrated acids forming antimony salts, e.g. $SbCl_3$, and in alkalis giving antimonites, e.g. $NaSbO_2$.

When antimony trioxide is boiled with a solution of potassium hydrogen tartrate and cooled, crystals of potassium antimonyl tartrate $(K.SbO.C_4H_4O_6)_2.H_2O$, known as tartar emetic, are obtained.

Antimony Pentoxide, Sb_4O_{10}: This is a yellow solid made by heating antimony with concentrated nitric acid or by hydrolysing antimony pentachloride, when it is precipitated as a hydrate. It is amphoteric, dissolving in hydrochloric acid and in concentrated alkalis, when it forms antimonates, the salts of the hypothetical antimonic acid. A solution of potassium dihydrogen pyro-antimonate, $K_2H_2Sb_2O_7$, is used for testing for sodium, since the sodium salt formed by double decomposition is only sparingly soluble in water and is precipitated on standing.

Antimony Trichloride, $SbCl_3$: This is a soft, white solid, m.p. 73°C, made by dissolving antimony trisulphide in concentrated hydrochloric acid:

$$Sb_2S_3 + 6HCl = 2SbCl_3 + 3H_2S$$

When added to a small amount of water it dissolves giving a clear solution, but dilution gives a white precipitate of antimony oxychloride (antimonyl chloride) by hydrolysis:

$$SbCl_3 + H_2O \rightleftharpoons SbOCl\downarrow + 2HCl$$

Antimony Pentachloride, $SbCl_5$: This is a fuming, yellow liquid, b.p. 140°C, obtained by treating the molten trichloride with chlorine. It is used as a chlorinating agent in organic chemistry.

Antimony Trisulphide, Sb_2S_3: In the laboratory it is made by passing hydrogen sulphide through a solution of the trichloride, when it appears as an orange precipitate:

$$2SbCl_3 + 3H_2S = Sb_2S_3\downarrow + 6HCl$$

It occurs naturally as the black compound stibnite, which is used not only as the main source of antimony and its compounds, but also in the manufacture of matches and fireworks.

Antimony Pentasulphide, Sb_2S_5: This is made by boiling the trisulphide with sulphur and caustic soda solution and acidifying the product with dilute sulphuric acid, when it is obtained as an orange-red precipitate. It is used to vulcanize rubber.

Stibine, SbH_3: It is prepared by reducing a soluble antimony compound with nascent hydrogen. It is a colourless, poisonous gas, b.p. −18°C, which is slightly soluble in water. It decomposes readily when heated, leaving a black deposit of metallic antimony, which, unlike arsenic, is not soluble in sodium hypochlorite solution.

23.18. Detection of Antimony. Solutions of antimony salts give an orange precipitate when saturated with hydrogen sulphide. This precipitate dissolves in ammonium sulphide solution forming the thio-salt, $(NH_4)_3SbS_3$. Antimony sulphide is distinguished from arsenic sulphide not only by its colour, but also by its solubility in hot concentrated hydrochloric acid and its insolubility in ammonium carbonate solution.

23.19. Bismuth, Bi. Electronic configuration: 2,8,18,32,18,5.

OCCURRENCE: The metal does not occur naturally, but it is found mostly as the sulphide, Bi_2S_3, and the oxide, Bi_2O_3, often with the sulphides of other metals.

MANUFACTURE: The principles are similar to those used for antimony. Either the sulphide is roasted in air and the resulting oxide reduced with carbon, or the sulphide is heated with iron:

$$Bi_2O_3 + 3C = 2Bi + 3CO$$
$$Bi_2S_3 + 3Fe = 2Bi + 3FeS$$

The bismuth can be refined electrolytically if necessary.

PROPERTIES AND REACTIONS OF BISMUTH: It is a white, lustrous, brittle metal, m.p. 271 °C, having a density of 9.80 g cm^{-3}. Its vapour is diatomic.

It burns when heated strongly in air, forming the trioxide, and it combines with the halogens and sulphur when heated.

USES OF BISMUTH: It is used to make low melting point alloys, e.g. Wood's metal, which is used in making automatic sprinklers because it melts at only 70°C, releasing water on to any fire beneath.

23.20. Compounds of Bismuth.

Bismuth Trioxide, Bi_2O_3: This is formed by burning the metal in air or by heating the carbonate or nitrate. It is a yellow solid, m.p. 820°C, with marked basic properties. It dissolves in acids giving bismuth salts, but it is insoluble in dilute alkalis.

Bismuth Chloride, $BiCl_3$: This is prepared by dissolving bismuth oxide in hydrochloric acid and evaporating the solution. It is a white, deliquescent solid, m.p. 232°C, which is hydrolysed by excess of water to bismuth oxychloride (bismuthyl chloride), which appears as a white precipitate:

$$BiCl_3 + H_2O \rightleftharpoons BiOCl\downarrow + 2HCl$$

Bismuth Nitrate, $Bi(NO_3)_3$: When bismuth is dissolved in concentrated nitric acid and the solution is concentrated, crystals of hydrated nitrate

TABLE 23.5. A COMPARISON OF THE ELEMENTS OF GROUP VB

Properties \ Element	Nitrogen	Phosphorus	Arsenic	Antimony	Bismuth
M.p./°C	−210	44 (white)	815 (under press.)	630	271
B.p./°C	−196	280 („)	Sublimes above 630	1440	1550
Density/g cm^{-3}	0.81 (as liquid)	1.83 („)	5.7	6.7	9.8
Colour	None	Yellow, or red	Grey	White	White
Electrical conductor	Non-conductor	Non-conductor	Conducts electricity	Conducts electricity	Conducts electricity
Character	Non-metal	Non-metal	Metalloid	Metalloid	Metal
Vapour at b.p.	N_2	P_4	As_4	Sb_4	Bi_2
Allotropes	Not allotropic	Allotropic	Allotropic	Allotropic	Not allotropic
Solubility in water	Slightly soluble	Insoluble	Insoluble	Insoluble	Insoluble
Combustion in air	Does not burn	Burns readi[illegible]	Burns when heated	B[illegible]	[illegible]

Hydrides	NH_3 basic, very soluble, stable to heat N_2H_4 unstable	PH_3 feebly basic, almost insoluble, decomposes when heated P_2H_4 spont. inflamm.	AsH_3 not basic, insoluble, decomposes readily when heated —	SbH_3 not basic, insoluble, unstable —	BiH_3 not basic, insoluble, very unstable —
Oxides	N_2O neutral NO neutral NO_2 strongly acidic	P_4O_6 acidic P_4O_{10} acidic	As_4O_6 amphoteric As_4O_{10} acidic	Sb_4O_6 amphoteric Sb_4O_{10} amphoteric	Bi_2O_3 weakly basic
Oxy-acids	HNO_2 weak HNO_3 strong	H_3PO_2, H_3PO_3, H_3PO_4 } weak	H_3AsO_3, H_3AsO_4 } weak	H_3SbO_3, H_3SbO_4 } very weak	
Chlorides	NCl_3 only Explosively unstable Hydrolysed rapidly and completely to NH_3 + HOCl	PCl_3 stable PCl_5 dissociates Both hydrolysed rapidly and completely	$AsCl_3$ stable $AsCl_5$ dissociates Hydrolysed reversibly	$SbCl_3$ stable $SbCl_5$ dissociates Hydrolysed reversibly to SbOCl↓	$BiCl_3$ only Stable Hydrolysed reversibly to BiOCl↓
Sulphides	Unstable	Fairly stable	Stable	Stable	Stable
Action of hot conc. nitric acid	None	Forms H_3PO_4	Forms H_3AsO_4	Forms Sb_4O_{10}	Forms Bi $(NO_3)_3$
Ions formed	N^{3-}	P^{3-}	As^{3+}	Sb^{3+}	Bi^{3+}

$Bi(NO_3)_3.5H_2O$, appear. If these are heated they are converted into bismuth oxy-nitrate, sometimes called bismuthyl nitrate, $BiONO_3$. The latter is used medicinally as a disinfectant, and in the manufacture of paint.

Bismuth Hydride, BiH_3: This is a very unstable compound which decomposes slowly even at room temperature.

Sodium Bismuthate, $NaBiO_3$: This is a useful reagent in qualitative analysis; when added to a solution of manganese salt in nitric acid it gives a purple colour of permanganic acid.

23.21. Detection of Bismuth. When hydrogen sulphide is passed through a solution of a bismuth salt in dilute hydrochloric acid, a dark brown precipitate of bismuth sulphide is obtained which is soluble in hot dilute nitric acid but insoluble in ammonium sulphide solution:

$$2BiCl_3 + 3H_2S = Bi_2S_3\downarrow + 6HCl$$

A white precipitate of bismuth hydroxide is obtained when ammonium hydroxide solution is added to a solution of a bismuth salt. If this precipitate is filtered off and a mixture of stannous chloride and sodium hydroxide solutions poured over it, a black stain of metallic bismuth appears.

23.22. Summary. The elements of Group VB are compared and contrasted in Table 23.5. From this Table it is clear that as the atomic number increases, the elements become increasingly metallic in character, varying from the electronegative non-metals nitrogen and phosphorus to the weakly electropositive metal bismuth. Arsenic and antimony are intermediate in character, displaying metallic and non-metallic properties, and are often referred to as *metalloids.*

This gradual increase in metallic character and electropositivity down the Group shows itself in many ways—in the physical properties of the elements, in the increasingly basic nature of the oxides (which vary from strongly acidic through amphoteric to weakly basic), in the increasingly salt-like character of the halides and the changing ease with which they are hydrolysed, in the increasing stability of the sulphides to heat and hydrolysis, and in the marked decrease in the stability of the hydrides, which consequently become increasingly powerful reducing agents.

Several other important trends also become apparent. The elements from phosphorus to bismuth show a decreasing tendency to pentacovalency in their oxides and halides and a corresponding tendency towards trivalency. As explained in § 22.17 nitrogen differs from the other elements in the Group in having a maximum of eight electrons in its outermost shell, so that it is incapable of forming a pentahalide. Again, antimony and bismuth, and to a less extent arsenic, tend to form positively charged ions by the loss of three electrons from the outermost shell of each atom, but nitrogen and phosphorus never do this.

Despite these trends, indeed because of them, the five elements make a clear cut and interesting Group. Their atoms all possess five electrons in their outermost shells and it is upon this fundamental common feature that their many similarities depend.

24

Oxygen and Sulphur

24.1. Introduction. Group VIB of the Periodic Classification contains the elements oxygen, sulphur, selenium, tellurium, and polonium, of which only the first two are sufficiently common and important to be studied here.

24.2. Oxygen, O. Electronic configuration: 2,6.

OCCURRENCE: About 21% by volume of the atmosphere consists of oxygen. The proportion remains constant because the oxygen consumed during respiration and combustion is continually replaced as a result of photosynthesis. The element is by far the commonest in the earth's crust, since nearly all rocks and clays and many minerals contain a high proportion of it. The oceans are composed of about 86% by weight of oxygen. It is an essential constituent of all living matter—over 70% by weight of the human body consists of combined oxygen.

MANUFACTURE: Oxygen is mostly made by the fractional distillation of liquefied air. The process is depicted diagrammatically in Fig. 24.1. Air is compressed to 1.5×10^7 Nm^{-2} (150 atm), cooled, and freed from dust, carbon dioxide, and moisture. It is then passed through a heat exchanger where it is strongly cooled by the outgoing nitrogen gas from the fractionating column. The cooled compressed gas is liquefied in two ways. A large part of it is made to do external work by driving an expansion engine used for supplying power to the compressor. The remainder is passed through an expansion valve where it is cooled by the Joule-Thomson effect (see § 3.14 for an explanation of both methods).

The liquid air enters the lower part of the special fractionating column at a temperature of about −170°C and a pressure of 6×10^5 N m^{-2} (6 atm). Since nitrogen is more volatile than oxygen (the boiling points at atmospheric pressure are −196°C and −183°C respectively), the nitrogen evaporates more rapidly, leaving a liquid containing about 40% of oxygen. This is passed to the upper part of the fractionating column where it is distilled under atmospheric pressure by thermal contact with the nitrogen gas which evaporated in the lower section. Liquid oxygen of over 98% purity (the main impurity is argon) is drawn off and stored in special containers or in cylinders under pressure, whilst the cold nitrogen gas is supplied to the heat exchanger.

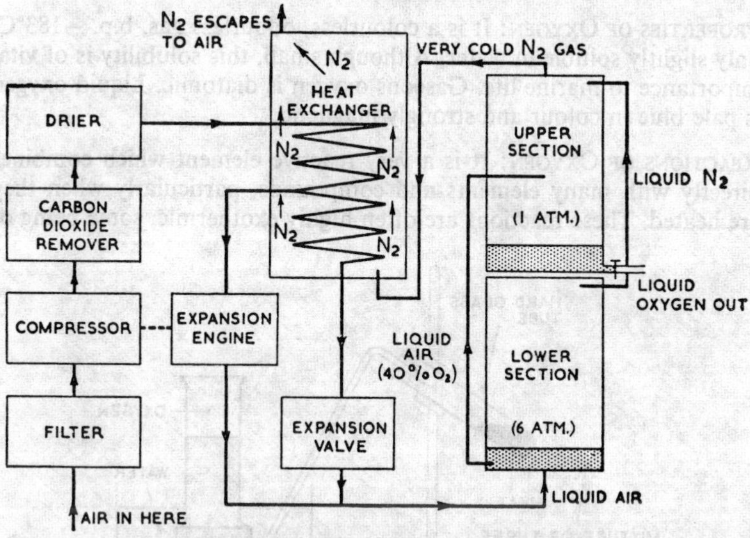

FIG. 24.1 Manufacture of Oxygen

In countries with an abundance of hydro-electric power some oxygen is made by electrolysis of a dilute solution of sodium hydroxide using nickel electrodes. Oxygen is liberated at the anode and hydrogen at the cathode, as explained in § 16.2.

LABORATORY PREPARATION: By the thermal decomposition of a suitable oxide or oxy-salt. Potassium chlorate is often used because it decomposes below 300°C in the presence of manganese dioxide catalyst:

$$2KClO_3 = 2KCl + 3O_2\uparrow$$

The apparatus is shown in Fig. 24.2. The gas can be dried with anhydrous calcium chloride or phosphorus pentoxide after it has been washed with sodium hydroxide solution to remove traces of chlorine which may be present.

Alternatively, the reaction between acidified potassium permanganate solution and hydrogen peroxide solution can be used, no heating being necessary:

$$2KMnO_4 + 5H_2O_2 + 3H_2SO_4 = K_2SO_4 + 2MnSO_4 + 8H_2O + 5O_2\uparrow$$

Very pure oxygen can be prepared by electrolysing a solution of barium hydroxide (chosen because it can be purified by recrystallization) using nickel electrodes or by heating crystals of potassium permanganate:

$$2KMnO_4 = K_2MnO_4 + MnO_2 + O_2\uparrow$$

Properties of Oxygen: It is a colourless, odourless gas, b.p. −183°C, only slightly soluble in water. Although small, this solubility is of vital importance to marine life. Gaseous oxygen is diatomic. Liquid oxygen is pale blue in colour and strongly magnetic.

Reactions of Oxygen: It is a very reactive element which combines directly with many elements and compounds, particularly when they are heated. These reactions are often highly exothermic, some being of

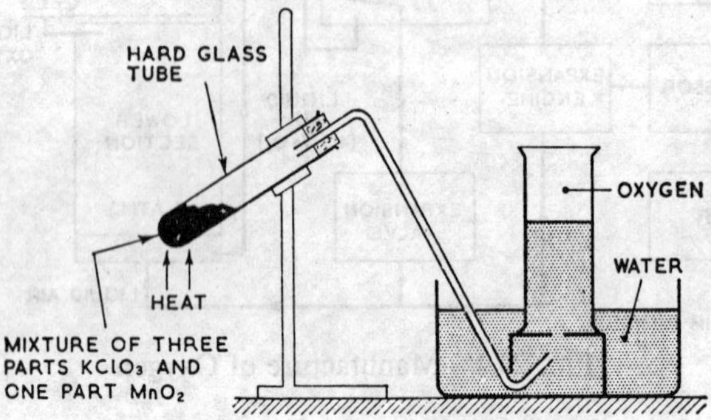

Fig. 24.2 Preparation of Oxygen

great industrial importance as sources of energy (e.g. the burning of coal, coke, oil, and gaseous fuels). In others, such as the rusting of iron or the respiration of plants and animals, the heat evolved is dissipated to the surroundings as fast as it is produced and ignition does not occur.

Combination with oxygen is often accelerated by the use of catalysts. Platinum is particularly active in this way causing hydrogen to unite with oxygen at room temperature and catalysing the oxidation of sulphur dioxide, ammonia, and methyl alcohol:

$$2H_2 + O_2 = 2H_2O$$
$$2SO_2 + O_2 \rightleftharpoons 2SO_3$$
$$4NH_3 + 5O_2 = 4NO + 6H_2O$$
$$2CH_3OH + O_2 = 2H.CHO + 2H_2O$$

Oxygen can be removed from a mixture of gases either by dissolving it in an alkaline solution of pyrogallol or by passing it over red-hot copper, which combines with it forming copper oxide.

TABLE 24.1. SOME INDUSTRIAL USES OF AIR AND OXYGEN

AIR OR OXYGEN

Process	Product
burning carbon	producer gas (§ 20.14)
burning sulphur	sulphur dioxide ⟶ sulphuric acid (§ 24.11)
burning phosphorus	phosphorus pentoxide $\xrightarrow{\text{water}}$ orthophosphoric acid (§ 23.10)
silicon, phosphorus, carbon in pig iron	oxides during steel-making (§ 30.2)
hydrogen or acetylene burning	welding or cutting metals
molten sodium	sodium peroxide (§ 17.6)
lead at 900°C	litharge (§ 21.13) $\xrightarrow{450°C}$ red lead (§ 21.15)
sulphide ores roasted	metallic oxides, e.g. copper, lead, zinc
ammonia + Pt catalyst at 900°C	nitric acid (§ 22.13)
ethylene + Ag catalyst at 300°C	ethylene oxide $\xrightarrow{\text{steam}}$ ethylene glycol (§ 35.8)
methyl alcohol + Cu catalyst at 250°C	formaldehyde (§ 36.3)
ethyl alcohol + Ag catalyst at 300°C	acetaldehyde (§ 36.2)
butane + catalysts	acetaldehyde and acetic acid (§ 37.2)
acetaldehyde + $Mn\overline{Ac}_2$ catalyst	acetic acid (§ 37.2)
benzene, hydrogen chloride + catalyst at 250°C	chlorobenzene (§ 43.6)
toluene + MnO_2 catalyst at 500°C	benzaldehyde (§ 47.3)
cumene (§ 45.2)	phenol (§ 45.2) and acetone (§ 36.6)
naphthalene + V_2O_5 catalyst at 350°C	phthalic anhydride (§ 43.2)

USES OF OXYGEN: Being essential for respiration, it is widely used in breathing apparatus for mountaineers, airmen, cosmonauts, firemen, divers, and mine rescue workers. In hospitals oxygen is supplied to patients who have difficulty in breathing or who are under anaesthetics. Industrially the gas is needed in large amounts for oxy-hydrogen and oxy-acetylene flames used for welding and cutting metals and for melting platinum, silica, and similar materials. Modern processes of steel-making require oxygen on an enormous scale and many steelworks contain special plant for manufacturing the gas from the atmosphere in these huge quantities. Liquid oxygen is used as a fuel in many rockets and missiles. Table 24.1 lists some of the principal industrial uses of air and oxygen.

DETECTION OF OXYGEN: The gas is commonly recognized by its ability to rekindle a glowing splint. It can be distinguished from nitrous oxide, which also shows this property, by the instant appearance of brown fumes when the oxygen is mixed with nitric oxide:

$$2NO + O_2 = 2NO_2$$

24.3. Ozone, O_3. Oxygen shows allotropy, existing in two gaseous forms, ordinary oxygen and ozone.

OCCURRENCE: The concentration of ozone in the atmosphere at ground level is extremely small, never exceeding one part in ten million, but at high altitudes the proportion is much higher. It is formed in the upper atmosphere by the action of intense ultra-violet light on oxygen. Ozone is also produced in traces when electric sparks are passed in air, when white phosphorus smoulders in air, when fluorine reacts with water, and when dilute sulphuric acid is electrolysed using platinum electrodes and a very high current density.

PREPARATION: Ozone is made in the laboratory by passing a silent

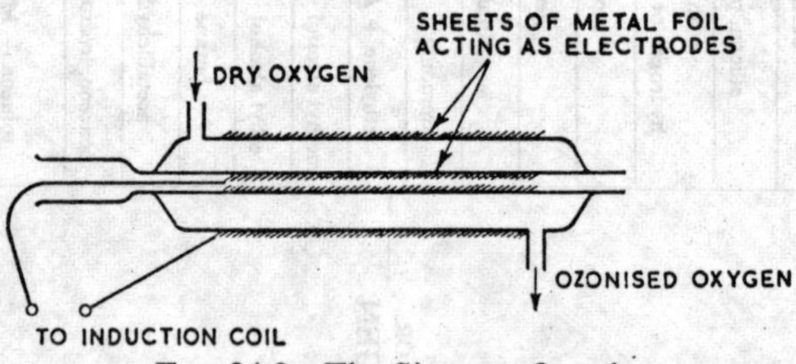

FIG. 24.3 The Siemens Ozonizer

electrical discharge through dry oxygen. The apparatus used is called an *ozonizer*; two common forms are shown in Figs. 24.3 and 24.4. Although different in construction, they both work in essentially the same way. A high potential difference (usually obtained from an induction coil) is applied between the two electrodes, which are separated by the walls of a glass tube to prevent sparking. In the Siemens ozonizer the outer electrode consists of a collar of metallic foil, whereas in the other type an electrolyte is used. The effect of the silent discharge is similar to that

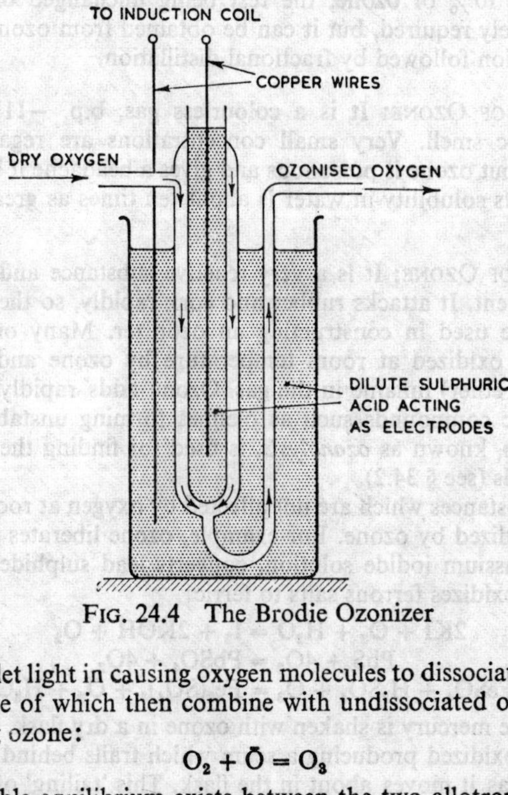

FIG. 24.4 The Brodie Ozonizer

of ultra-violet light in causing oxygen molecules to dissociate into single atoms, some of which then combine with undissociated oxygen molecules giving ozone:

$$O_2 + \bar{O} = O_3$$

A reversible equilibrium exists between the two allotropes, the proportion of ozone depending upon the temperature and being very small at temperatures encountered on earth. For example, there is negligible ozone in the equilibrium mixture at room temperature and only about 1% at 3000°C:

$$3O_2 \rightleftharpoons 2O_3 \qquad \Delta H = +285 \text{ kJ mol}^{-1}$$

Thus when ozonized oxygen is prepared the ozone in excess of the minute equilibrium proportion tends to decompose into oxygen, doing so slowly at room temperature and rapidly when heated. This explains the use of a silent discharge because an ordinary electric spark would heat the ozone and cause it to decompose almost as fast as it was formed. Certain finely divided materials (e.g. platinum and manganese dioxide) catalyse the decomposition of ozone, so care must be taken to exclude such substances.

Ozonizers produce a mixture called ozonized oxygen, which contains about 10% of ozone, the rest being unchanged oxygen. Pure ozone is rarely required, but it can be obtained from ozonized oxygen by liquefaction followed by fractional distillation.

PROPERTIES OF OZONE: It is a colourless gas, b.p. −111°C, with a characteristic smell. Very small concentrations are regarded as invigorating, but ozone is poisonous and gives a headache if breathed for very long. Its solubility in water is about ten times as great as that of oxygen.

REACTIONS OF OZONE: It is a very reactive substance and a powerful oxidizing agent. It attacks rubber and cork rapidly, so these materials must not be used in constructing an ozonizer. Many organic compounds are oxidized at room temperature by ozone and some (e.g. alcohol and ether) inflame in the gas. Ozone adds rapidly to unsaturated organic compounds such as olefines forming unstable *ozonides*. This process, known as *ozonolysis*, is used for finding the position of double bonds (see § 34.2).

Many substances which are not affected by oxygen at room temperature are oxidized by ozone. For example, ozone liberates iodine from neutral potassium iodide solution, converts lead sulphide to the sulphate, and oxidizes ferrous salts to ferric:

$$2KI + O_3 + H_2O = I_2 + 2KOH + O_2$$
$$PbS + 4O_3 = PbSO_4 + 4O_2$$
$$2FeSO_4 + H_2SO_4 + O_3 = Fe_2(SO_4)_3 + O_2 + H_2O$$

When a little mercury is shaken with ozone in a dry flask, some of the mercury is oxidized producing a scum which trails behind the globule of mercury as it moves about in the flask. This 'tailing' of mercury is unique to ozone and is therefore used to distinguish it from other oxidizing agents.

USES OF OZONE: On account of its powerful oxidizing properties ozone is used for purifying the air in tunnels and for sterilizing supplies of drinking water and occasionally for bleaching. In the laboratory it is used for ozonolysis.

FORMULA OF OZONE: The fact that ozone can be made from pure oxygen and that on decomposition it yields only oxygen shows that it must be an allotrope of that element with the formula O_x. To determine x two equal volumes of the same ozonized oxygen are taken and one is exposed to turpentine, which absorbs ozone readily, so that the contraction which occurs reveals the volume of ozone present. The other sample is heated to about 300°C to decompose the ozone. When cooled to its original temperature the volume will be found to be greater than before by an amount equal to half the volume of ozone present.

Thus 1 volume of ozone decomposes into $1\frac{1}{2}$ volumes of oxygen
∴ 2 " " " " " 3 " " "
∴ 2 molecules " " " 3 molecules " "
(by Avogadro's hypothesis)
∴ 2 " " " " " 6 atoms of oxygen
(since oxygen is diatomic)
∴ $x = 3$ and ozone has the molecular formula O_3.

Alternatively the molecular formula can be determined from the vapour density as measured by the effusion method described in § 3.7.

24.4. Oxides. These may be classified as follows:

(1) Acidic Oxides: These react with alkalis forming a salt and water. When they dissolve in water their solutions are acidic. Most oxides of the non-metals are of this type, and so are some of the higher metallic oxides.

Examples: CO_2, SiO_2, NO_2, P_4O_6, P_4O_{10}, SO_2, SO_3, Mn_2O_7, CrO_3.

$$CO_2 + 2NaOH = Na_2CO_3 + H_2O$$
$$CrO_3 + 2NaOH = Na_2CrO_4 + H_2O$$
$$SO_2 + H_2O \rightleftharpoons H_2SO_3$$

(2) Basic Oxides: These are metallic oxides which react with acids forming a salt and water only. When they dissolve in water their solutions are alkaline.

Examples: Na_2O, K_2O, CaO, MgO, BaO, FeO, CuO.

$$Na_2O + H_2O = 2NaOH$$
$$MgO + H_2SO_4 = MgSO_4 + H_2O$$

(3) Neutral Oxides: These have neither acidic nor basic properties and when dissolved in water they give a neutral solution.

Examples: CO, N_2O, NO.

(4) Amphoteric Oxides: These are metallic oxides which show both acidic and basic properties according to their environment. They dissolve in acids giving salts and water and in alkalis too.

Examples: Al_2O_3, ZnO, SnO, PbO, Cr_2O_3, Mn_2O_3, As_4O_6, Sb_4O_6, SnO_2.

$$ZnO + 2HCl = ZnCl_2 + H_2O$$
$$ZnO + 2NaOH = Na_2ZnO_2 + H_2O$$

(5) PEROXIDES: These contain the peroxide ion $(O—O)^{2-}$. They all give hydrogen peroxide solution when treated with dilute acids.

Examples: Na_2O_2, BaO_2.

$$Na_2O_2 + H_2SO_4 = Na_2SO_4 + H_2O_2$$

(6) DIOXIDES: These are covalent metallic oxides which decompose when heated giving the basic oxide and oxygen. They are oxidizing agents but they do not yield hydrogen peroxide solution when acidified, and should not, therefore, be called peroxides (§ 16.6).

Examples: PbO_2, MnO_2.

$$2PbO_2 = 2PbO + O_2\uparrow$$
$$MnO_2 + 4HCl = MnCl_2 + 2H_2O + Cl_2\uparrow$$

(7) COMPOUND OXIDES: These may be regarded as composed of two simpler oxides, the properties of which they usually show.

Examples:

$Pb_3O_4(2PbO.PbO_2)$, $Fe_3O_4(Fe_2O_3.FeO)$, $Mn_3O_4(MnO_2.2MnO)$.

$$Pb_3O_4 + 4HNO_3 = 2Pb(NO_3)_2 + PbO_2\downarrow + 2H_2O$$

24.5. Sulphur, S. Electronic configuration: 2,8,6.

OCCURRENCE: Large deposits of the element are found in U.S.A., Sicily and Japan. Sulphur also occurs as sulphide ores (e.g. iron, copper, lead, zinc), and as sulphates (e.g. calcium, magnesium, barium, and sodium). It is an essential constituent of certain proteins and therefore of living matter. When organisms decay, some of this sulphur escapes as simple compounds such as hydrogen sulphide. Coal and crude petroleum contain small percentages of sulphur compounds and so, sometimes, does natural gas.

EXTRACTION: The American deposits now provide about 70% of the world's supply of elemental sulphur. The sulphur lies about 150 m down beneath layers of quicksand and gravel which make straightforward mining impracticable. The *Frasch* process uses a series of concentric pipes as shown in Fig. 24.5. Superheated water under pressure at 180°C is pumped down the outermost pipe to melt the sulphur underground and to keep it molten during its ascent. Compressed air is pumped down the innermost pipe to change the molten sulphur into a low-density froth and force it to the surface, where it is allowed to solidify in large tanks. The product is about 99.8% pure.

The second most important source of sulphur, now accounting for about 20% of the world's needs, is the large supply of natural gas which has become available since 1957 at Lacq in the French Pyrenees. This gas is exceptional in containing about 15% of hydrogen suphide. The latter is removed from the crude natural gas by washing it with amine

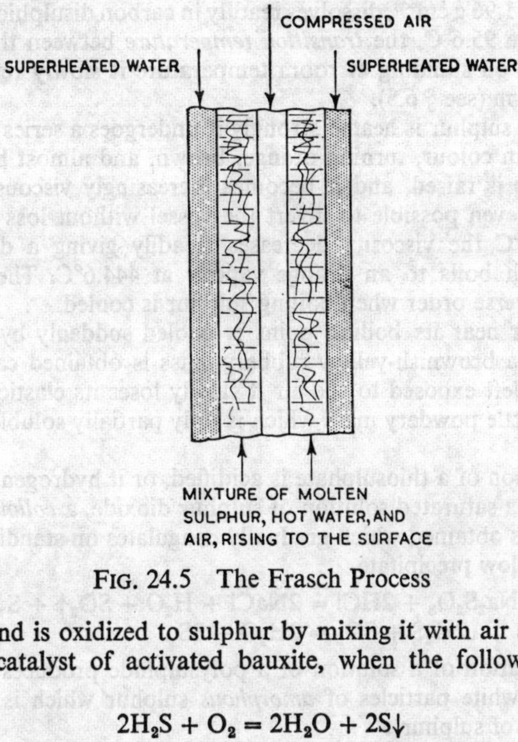

FIG. 24.5 The Frasch Process

solutions and is oxidized to sulphur by mixing it with air and passing it over a catalyst of activated bauxite, when the following change occurs:

$$2H_2S + O_2 = 2H_2O + 2S\downarrow$$

In Sicily the ore, which contains only about 20% of sulphur, is stacked in kilns and ignited. The heat generated by the combustion causes the unburnt sulphur to melt and run into troughs. It is purified by distillation in iron retorts.

The supply obtained from natural deposits is supplemented by sulphur recovered from industrial processes, e.g. the purification of crude coal gas, the refining of petroleum, and the removal of sulphur dioxide from flue gases. Efforts are also being made to use bacterial decomposition of sewage as a source of sulphur.

PROPERTIES OF SULPHUR: Several varieties exist. The stable allotrope at room temperature is *rhombic* sulphur (also known as α-sulphur). It forms yellow octahedral-shaped crystals, with density 2.06 g cm^{-3}, which are insoluble in water but soluble in carbon disulphide and benzene. It is a very poor conductor of heat and electricity. Heated to 115°C it melts to a yellow liquid which, when cooled slowly, solidifies to a mass of needle-shaped or prismatic crystals of *monoclinic* sulphur (also known as β-sulphur). This allotrope, which melts at 120°C and has a density of 1.96 g cm^{-3}, dissolves readily in carbon disulphide. It is only stable above 95.6°C, the *transition temperature* between the two allotropes, and on standing at room temperature it slowly reverts to the rhombic form (see § 6.5).

If molten sulphur is heated strongly it undergoes a series of changes. It darkens in colour, turning orange, brown, and almost black as the temperature is raised, and it becomes increasingly viscous. At about 200°C it is even possible to invert the vessel without loss of sulphur. Above 250°C the viscosity decreases steadily giving a dark mobile liquid which boils to an orange vapour at 444.6°C. These changes occur in reverse order when boiling sulphur is cooled.

If sulphur near its boiling point is cooled suddenly by pouring it into water, a brownish-yellow rubbery mass is obtained called *plastic* sulphur. If left exposed to the air it slowly loses its elasticity and becomes a brittle powdery mass which is only partially soluble in carbon disulphide.

If a solution of a thiosulphate is acidified, or if hydrogen sulphide is passed into a saturated solution of sulphur dioxide, a *colloidal solution* of sulphur is obtained which gradually coagulates on standing giving a white or yellow precipitate.

$$Na_2S_2O_3 + 2HCl = 2NaCl + H_2O + SO_2\uparrow + S\downarrow$$
$$2H_2S + SO_2 = 2H_2O + 3S\downarrow$$

The acidification of a solution of a polysulphide produces a fine suspension of white particles of *amorphous* sulphur which is sometimes called 'milk of sulphur':

$$Na_2S_5 + 2HCl = 2NaCl + H_2S + 4S$$

When sulphur is distilled and the vapour is condensed to a solid on a cold surface the deposit is known as *flowers of sulphur*. It is only partially soluble in carbon disulphide. Molten sulphur is often run into tubular moulds giving the *roll sulphur* familiar in the laboratory.

ALLOTROPY OF SULPHUR:

Crystalline Sulphur: Rhombic and monoclinic sulphur are *enantiotropic* allotropes, i.e. the two forms can be changed into each other at will merely by altering the physical conditions. At atmospheric pressure the rhombic form is stable below 95.6°C and the monoclinic form above, and only at this transition temperature can the two allotropes exist in equilibrium with each other. Although it proceeds in either direction with equal ease, in practice the conversion from one form to the other is often slow, taking several hours, during which time one form is metastable (see § 6.5):

$$\text{rhombic} \rightleftharpoons \text{monoclinic}$$

X-ray analysis reveals that both forms consist of puckered eight-membered rings of atoms as shown in Fig. 24.6. The explanation of the allotropy is believed to lie in the different ways in which these ring molecules are packed together in the crystals, the difference between the two forms being solely one of molecular arrangement.

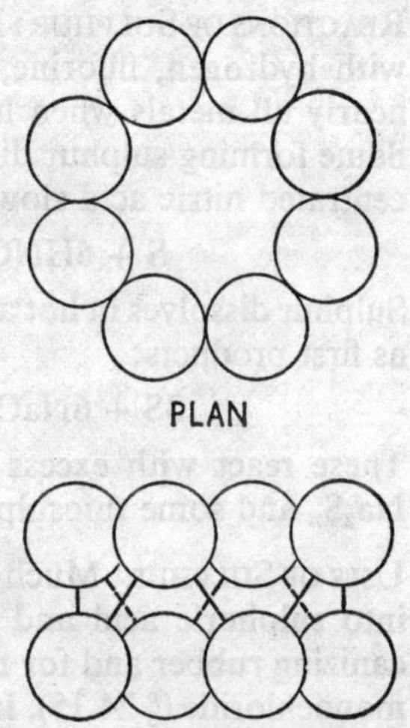

FIG. 24.6 The Sulphur Molecule, S_8

The S–S Bond length is 0.212 nm and the Angle between Bonds is 105°

Liquid Sulphur: The behaviour of molten sulphur when heated can be explained in terms of changes which are believed to occur in the shape and size of the molecules. Just above the melting point the sulphur consists almost entirely of the regular eight-membered ring molecules found in the crystals, but these are now free to move independently giving a mobile liquid. As the temperature is raised increasing thermal agitation causes some of these rings to break open forming short zig-zag chains which tend to link with each other into very long chains containing hundreds or even thousands of atoms. These chains are in constant motion and soon get entangled giving a very viscous liquid. The proportion of molecules in the form of ring molecules diminishes as the temperature rises. At still higher temperatures the long chains tend to break into shorter ones owing to the vigour of their thermal motion, and the viscosity of the liquid decreases steadily. Thus the allotropy of liquid sulphur is based upon a variation in molecular form; one variety consisting of ring molecules (it is often referred to as S_λ), is in equilibrium with the other modification, S_μ, which consists of open chain molecules.

If liquid sulphur is suddenly cooled the element solidifies before all the atoms can rearrange themselves into rings again so that the product, plastic sulphur, consists of a mixture of ring molecules and chain molecules. Its elasticity arises, as does that of rubber, from its capacity to revert into a tangled mass of zig-zag chains after stretching. When left to stand at room temperature the atoms slowly reform the ring molecules and this elasticity disappears.

Vapour Sulphur: Vapour density measurements show that just above the boiling point the vapour consists largely of S_8 molecules, but that at higher temperatures these progressively dissociate until at 750°C sulphur is predominantly diatomic. At much higher temperatures dissociation into single atoms occurs.

REACTIONS OF SULPHUR: It is a reactive element which combines directly with hydrogen, fluorine, chlorine, bromine, carbon, phosphorus, and nearly all metals when heated. It burns in air or oxygen with a blue flame forming sulphur dioxide and a little sulphur trioxide. Hot concentrated nitric acid slowly oxidizes it to sulphuric acid:

$$S + 6HNO_3 = H_2SO_4 + 6NO_2\uparrow + 2H_2O$$

Sulphur dissolves in hot alkalis giving a mixture of sulphide and sulphite as first products:

$$3S + 6NaOH = 2Na_2S + Na_2SO_3 + 3H_2O$$

These react with excess of sulphur giving polysulphides of the type Na_2S_n and some thiosulphate, $Na_2S_2O_3$.

USES OF SULPHUR: Much of it is burnt to sulphur dioxide for conversion into sulphuric acid and sulphites. Other large-scale uses are for vulcanizing rubber and for making carbon disulphide (§ 20.8) and sulphur monochloride (§ 24.15). In powder form or colloidal suspension sulphur is used as a fungicide in agriculture, particularly for dusting grape vines. The element is also used in the manufacture of matches, fireworks, skin ointments, and ultramarine dyes.

24.6. Hydrogen Sulphide, H_2S.

OCCURRENCE: It occurs naturally in volcanic gases and in the water of certain springs. It is formed when organic matter containing sulphur undergoes putrefaction. Traces of the gas can be detected in the atmosphere in industrial areas where coal-burning is prevalent. It also occurs in crude coal gas, from which it is carefully removed (see § 20.13).

PREPARATION: By the action of dilute hydrochloric acid on ferrous sulphide:

$$FeS + 2HCl = FeCl_2 + H_2S\uparrow$$

When the gas is required intermittently, as in analysis, a Kipp generator,

Fig. 24.7, is often used. The only reactions to which this device can be applied, of course, are those which involve a solid and a liquid and which do not require heating. Its use prevents waste and minimizes the escape of gas into the laboratory. The gas is washed with water to remove hydrogen chloride fumes and can be dried with phosphorus pentoxide. This method only gives a pure product if the ferrous sulphide is pure, which is rarely the case as the commercial material is made by melting sulphur with iron and usually contains some free iron which reacts with the acid evolving hydrogen.

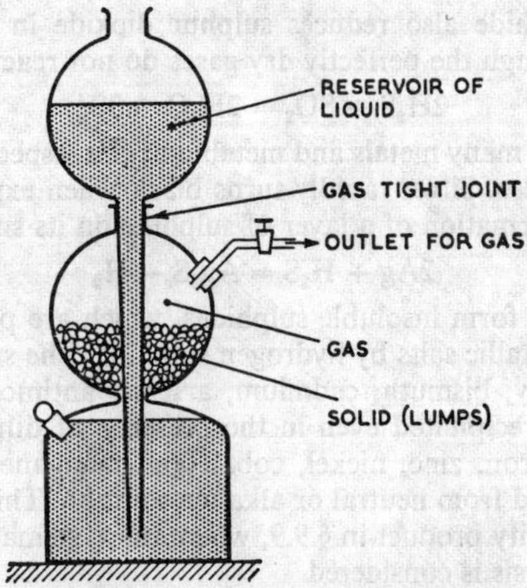

FIG. 24.7 The Kipp Generator

Pure hydrogen sulphide can be obtained by passing sulphur vapour and hydrogen over pumice at 600°C or by the action of hot concentrated hydrochloric acid on antimony sulphide:

$$Sb_2S_3 + 6HCl = 2SbCl_3 + 3H_2S\uparrow$$

PROPERTIES OF HYDROGEN SULPHIDE: It is a colourless, very poisonous gas, b.p. −60°C, with the smell of rotten eggs. It is fairly soluble in water (one volume dissolves three volumes at room temperature).

REACTIONS OF HYDROGEN SULPHIDE: Hydrogen sulphide burns with a blue flame, the products depending upon how much oxygen is available:

with excess of air or oxygen: $2H_2S + 3O_2 = 2H_2O + 2SO_2$
in a limited supply of air: $2H_2S + O_2 = 2H_2O + 2S\downarrow$

Hydrogen sulphide decomposes into its elements when heated strongly or when sparked repeatedly.

It is a powerful reducing agent, as the following examples show:

$$H_2S + Cl_2 = 2HCl + S\downarrow$$
$$3H_2S + K_2Cr_2O_7 + 4H_2SO_4 = K_2SO_4 + Cr_2(SO_4)_3 + 7H_2O + 3S\downarrow$$
$$H_2S + H_2SO_4 = SO_2 + 2H_2O + S\downarrow$$
$$H_2S + 2HNO_3 = 2H_2O + 2NO_2 + S\downarrow$$
$$H_2S + H_2O_2 = 2H_2O + S\downarrow$$
$$H_2S + 2FeCl_3 = 2FeCl_2 + 2HCl + S\downarrow$$

In many cases the fundamental change is:

$$H_2S + \bar{O} = H_2O + S\downarrow$$

Hydrogen sulphide also reduces sulphur dioxide in the presence of moisture, although the perfectly dry gases do not react together:

$$2H_2S + SO_2 = 2H_2O + 3S\downarrow$$

It reacts with many metals and metallic oxides, especially on heating, to form sulphides. Silver rapidly turns black when exposed to the gas owing to the formation of a layer of sulphide on its surface:

$$2Ag + H_2S = Ag_2S + H_2$$

Many metals form insoluble sulphides, which are precipitated from solutions of metallic salts by hydrogen sulphide. The sulphides of lead, copper, mercury, bismuth, cadmium, arsenic, antimony, and tin, for example, are precipitated even in the presence of dilute hydrochloric acid, whereas iron, zinc, nickel, cobalt and manganese sulphides are only precipitated from neutral or alkaline solution. This is explained in terms of solubility product in § 9.9, where the systematic separation of cations in analysis is considered.

DETECTION OF HYDROGEN SULPHIDE: Its smell is easily recognizable, even in very small concentration. The best chemical test is the use of paper soaked in lead acetate solution which blackens in the presence of hydrogen sulphide:

$$(CH_3COO)_2Pb + H_2S = PbS\downarrow + 2CH_3.COOH$$

FORMULA OF HYDROGEN SULPHIDE: If the gas is sparked repeatedly, the volume of the residual gas after cooling to the original temperature is found to be unchanged. Hence one volume of hydrogen sulphide gives one volume of hydrogen when decomposed. Since hydrogen is diatomic, by using Avogadro's hypothesis we conclude that each molecule of hydrogen sulphide must contain two atoms of hydrogen. The vapour density is 17 showing that only one sulphur atom is present, so the molecular formula must be H_2S.

24.7. Sulphides. These are prepared (1) by heating the element with sulphur directly (e.g. FeS), (ii) by reducing a sulphate with hot carbon (e.g. Na_2S), (iii) by precipitation from solutions of metallic salts with hydrogen sulphide (e.g. As_2S_3), or (iv) by saturating a hydroxide solution with hydrogen sulphide (e.g. NH_4HS). The sulphides of the alkali metals are soluble in water, their solutions being strongly alkaline by hydrolysis. They readily take up further sulphur forming polysulphides (e.g. Na_2S_n, where *n* is 2–5). The sulphides of the alkaline earth metals are insoluble in water but slowly undergo hydrolysis. The other metallic sulphides do not dissolve in water but some are instantly hydrolysed by even cold water giving the hydroxide and hydrogen sulphide:

e.g. $$Al_2S_3 + 6H_2O = 2Al(OH)_3\downarrow + 3H_2S\uparrow$$

The sulphides of arsenic, antimony and tin dissolve in sodium or ammonium sulphide solutions forming thiosalts:

e.g. $$As_2S_3 + 3(NH_4)_2S = \underset{\text{ammonium thioarsenite}}{2(NH_4)_3AsS_3}$$

When heated in air metallic sulphides are converted into the oxide of the metal and sulphur dioxide; this oxidation is often the first stage in the smelting of sulphide ores.

Most sulphides are detected by the evolution of hydrogen sulphide when heated with concentrated hydrochloric acid or by the purple colour which their solutions give with sodium nitroprusside solution. Sulphides insoluble in hydrochloric acid (e.g. NiS) must first be fused with sodium carbonate.

24.8. Sulphur Dioxide, SO_2.

OCCURRENCE: The gas is formed in small amounts wherever coal, coke, or even oil are burned, and traces of it are found, therefore, in the air in industrial districts. This pollution is a serious matter because even in very low concentration sulphur dioxide aggravates bronchial troubles and does untold damage to stonework and plant life.

LABORATORY PREPARATION: By heating copper with concentrated sulphuric acid:

$$Cu + 2H_2SO_4 = CuSO_4 + 2H_2O + SO_2\uparrow$$

Side reactions occur and some copper sulphide is formed. The gas is dried by passing it through concentrated sulphuric acid and is collected by upward displacement of air.

Alternatively it can be made by heating a solution of a sulphite with concentrated sulphuric acid:

$$Na_2SO_3 + H_2SO_4 = Na_2SO_4 + H_2O + SO_2\uparrow$$

Siphons of liquefied sulphur dioxide provide a convenient supply of the gas for laboratory use.

MANUFACTURE: Large amounts are made by burning sulphur in air as the first stage in the manufacture of sulphuric acid (§ 24.11):

$$S + O_2 = SO_2$$

The gas is also made by burning the 'spent oxide' obtained from gas works (§ 20.13). Impure sulphur dioxide is produced when sulphide ores are roasted in air:

$$4FeS_2 + 11O_2 = 2Fe_2O_3 + 8SO_2$$
$$2ZnS + 3O_2 = 2ZnO + 2SO_2$$

Some sulphur dioxide is recovered from industrial flue gases. It is trapped chemically in the scrubbing liquid, from which it is released again by heating.

PROPERTIES OF SULPHUR DIOXIDE: It is a colourless, poisonous gas, b.p. −10°C, with a very pungent smell. It is very soluble in water (one volume dissolves about 45 volumes of gas at room temperature), giving an acidic solution from which it is completely expelled by boiling.

REACTIONS OF SULPHUR DIOXIDE: It combines reversibly with oxygen in the presence of a catalyst to give sulphur trioxide:

$$2SO_2 + O_2 \rightleftharpoons 2SO_3$$

This reaction is the basis of the contact process for making sulphuric acid.

Sulphur dioxide is a strong reducing agent in aqueous solution, the essential change being:

$$SO_2 + H_2O + \bar{O} = H_2SO_4$$

Examples are:

$$2FeCl_3 + SO_2 + 2H_2O = 2FeCl_2 + H_2SO_4 + 2HCl$$
$$Cl_2 + SO_2 + 2H_2O = 2HCl + H_2SO_4$$
$$2KMnO_4 + 2H_2O + 5SO_2 = K_2SO_4 + 2MnSO_4 + 2H_2SO_4$$
$$K_2Cr_2O_7 + H_2SO_4 + 3SO_2 = K_2SO_4 + Cr_2(SO_4)_3 + H_2O$$
$$NO_2 + SO_2 + H_2O = H_2SO_4 + NO$$

This last reaction is the basis of the lead chamber process for making sulphuric acid.

Presumably the bleaching effect of sulphur dioxide also depends upon its reducing action, since prolonged exposure to light and air often restores the original colour.

Sulphur dioxide oxidizes hydrogen sulphide to sulphur provided moisture is present. At 1000°C it oxidizes carbon, a reaction used for recovering sulphur from industrial flue gases:

$$SO_2 + C = CO_2 + S\downarrow$$

Being acidic it reacts with alkalis giving sulphites and bisulphites:

$$SO_2 + 2NaOH = Na_2SO_3 + H_2O$$
$$SO_2 + NaOH = NaHSO_3$$

Thus concentrated solutions of alkalis can be used to remove sulphur dioxide from a mixture of non-acidic gases. It also unites exothermically with certain metallic oxides when heated giving sulphates:

$$PbO_2 + SO_2 = PbSO_4$$

Sulphur dioxide combines with chlorine forming sulphuryl chloride and with phosphorus pentachloride forming thionyl chloride (details of both reactions are given in § 24.16):

$$SO_2 + Cl_2 = SO_2Cl_2$$
$$SO_2 + PCl_5 = SOCl_2 + POCl_3$$

USES OF SULPHUR DIOXIDE: The main use is in the manufacture of sulphuric acid. It is also used to make bisulphites, and as a bleaching agent especially for straw, silk, and wool, which are attacked by chlorine. It is used as a food preservative because even small concentrations of it prevent fermentation. Its easy liquefaction and high latent heat of evaporation lead to its use as a refrigerant.

DETECTION OF SULPHUR DIOXIDE: The gas can be recognized by its smell and acidity, by its decolorization of potassium permanganate solution, and by its reduction of an orange solution of potassium dichromate to green chromic sulphate.

24.9. Sulphurous Acid, H_2SO_3, and Sulphites. A solution of sulphur dioxide in water is known as sulphurous acid:

$$SO_2 + H_2O \rightleftharpoons H_2SO_3 \rightleftharpoons H^+ + HSO_3^- \rightleftharpoons 2H^+ + SO_3^{2-}$$

It is a weak, dibasic acid which gives rise to three sodium salts, the normal sulphite Na_2SO_3, the bisulphite $NaHSO_3$, and the metabisulphite $Na_2S_2O_5$. The latter is produced when sodium bisulphite solution is evaporated to dryness in an atmosphere of sulphur dioxide:

$$2NaHSO_3 = Na_2S_2O_5 + H_2O\uparrow$$

Sulphites give sulphur dioxide when heated with dilute acids, and hydrogen sulphide when reduced with zinc and dilute hydrochloric acid. With barium chloride solution they give a white precipitate of barium sulphite which is soluble in hydrochloric acid. When heated, sulphites decompose into sulphides and sulphates:

$$4Na_2SO_3 = Na_2S + 3Na_2SO_4$$

Sodium and calcium bisulphite solutions are much used in making paper from wood because they dissolve the fibrous material, lignin, and facilitate pulping of the cellulose.

24.10. Sulphur Trioxide, SO_3. This is best prepared from sulphur dioxide and oxygen. The reaction forms an essential part of the processes used in manufacturing sulphuric acid and is discussed in detail in the next section. It can be prepared by heating anhydrous ferrous sulphate or by dehydrating concentrated sulphuric acid with phosphorus pentoxide, but neither method is now of much practical value:

$$2FeSO_4 = Fe_2O_3 + SO_2\uparrow + SO_3\uparrow$$
$$2H_2SO_4 + P_4O_{10} = 2SO_3\uparrow + 4HPO_3$$

Sulphur trioxide is a colourless crystalline solid which exists in three polymorphic forms with widely different melting points. It is very deliquescent, combining very vigorously and exothermically with water to form sulphuric acid; it is a good example, therefore, of an *acid anhydride*, i.e. an oxide which reacts with water forming an acid:

$$SO_3 + H_2O \rightleftharpoons H_2SO_4$$

It dissolves readily in concentrated sulphuric acid giving a viscous liquid called oleum, which is also referred to as fuming sulphuric acid:

$$H_2SO_4 + SO_3 \rightleftharpoons H_2S_2O_7$$

24.11. Sulphuric Acid, H_2SO_4.

MANUFACTURE: Two main methods are in use:

1. The Contact Process: This is based upon the reversible combination of sulphur dioxide and oxygen:

$$2SO_2 + O_2 \rightleftharpoons 2SO_3; \Delta H = -192 \text{ kJ mol}^{-1}$$

This reaction has been discussed in the light of Le Chatelier's principle in § 7.10, where the conditions for maximum yield of sulphur trioxide were found to be high pressure and low temperature. In practice the process is operated at atmospheric pressure because the small improve ment in yield resulting from the use of high pressure would not justify the heavy capital cost of the special plant that would be required. In deciding the operating temperature yield is not the only factor, and rate of reaction also has to be considered. Despite the use of a catalyst, only at a temperature of about 500°C is the reaction fast enough to attain equilibrium in the very short time during which, as the name suggests, the reactants are in contact with the catalyst. The process is operated, therefore, at atmospheric pressure and at about 500°C, when the yield is approximately 95%.

The process is shown diagrammatically in Fig. 24.8. The sulphur dioxide is obtained by burning sulphur or by roasting sulphide ores in air. Thorough purification is essential to remove arsenical compounds and other substances which might 'poison' the catalyst, i.e. destroy its catalytic activity. The sulphur dioxide, mixed with an excess of air, is purified and dried and then passed through a series of converters where the catalyst is stored on shelves in a way which exposes the maximum possible surface area to the reacting gases. Vanadium pentoxide is the most suitable catalyst for this reaction as it is cheaper and less susceptible to poisoning than the slightly more efficient platinum. The oxidation is exothermic, and the operating temperature is maintained without external heating by using heat exchangers.

After passing through the converters the gases are cooled and passed into an absorption tower where the sulphur trioxide dissolves in concentrated sulphuric acid. Direct absorption in water is impracticable

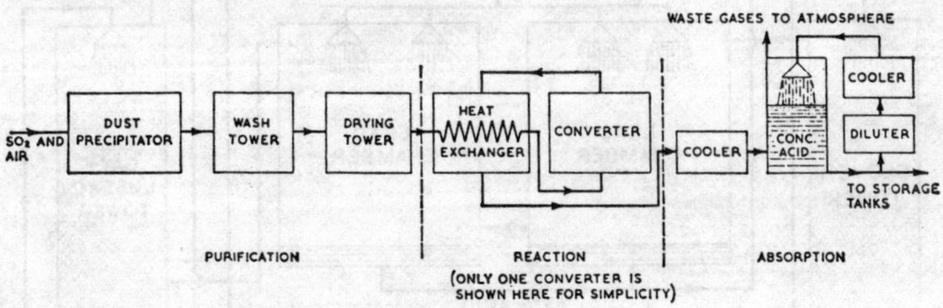

FIG. 24.8 The Contact Process

because the heat evolved causes the solution to boil producing a fog of fine droplets which are slow to condense. The product leaving the absorption tower is normally 100% sulphuric acid. Some of it is diluted with water, cooled, and recirculated through the absorption tower. By controlling the dilution the contact process can be adapted to produce fuming sulphuric acid, which cannot be made by the chamber process. Contact acid has the further advantage that it is purer than that produced by the chamber process and is therefore preferred for uses such as food processing and drug manufacture. At present about two-thirds of the world's production of sulphuric acid is made by the contact process.

2. The Lead Chamber Process: Essentially this consists of oxidizing a mixture of sulphur dioxide and water to sulphuric acid using nitric oxide as an oxygen carrier. The reaction can be expressed in simplest terms by the equation:

$$H_2O + SO_2 + NO_2 = H_2SO_4 + NO$$

The nitric oxide then combines with oxygen to form more nitrogen dioxide, which is used again.

$$2NO + O_2 = 2NO_2$$

The reactions are really much more complicated than this involving various intermediate compounds such as nitrosyl sulphuric acid, $NO.HSO_4$, crystals of which have been isolated from the reaction mixture, but for our purposes the simple theory will suffice.

The process is shown diagrammatically in Fig. 24.9. It may be regarded as consisting of three stages. The first takes place in the Glover tower. This is packed with acid resistant bricks over which trickles a constant stream of 70% sulphuric acid made by mixing the outputs of

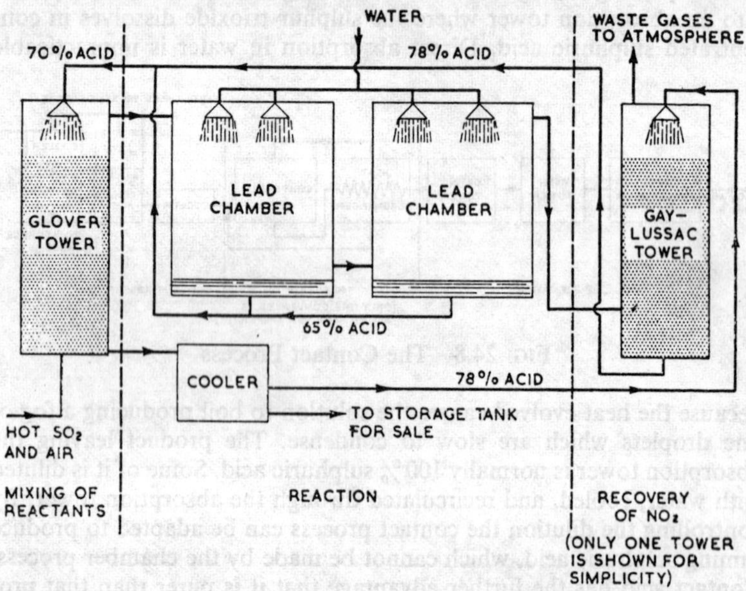

FIG. 24.9 The Lead Chamber Process

the lead chambers (65% acid) and the Gay-Lussac tower (78% acid saturated with oxides of nitrogen). A mixture of hot sulphur dioxide and air from the burners is fed into the base of the Glover tower and comes into intimate contact with the descending acid. As a result the gases from the burners are cooled from about 500°C to about 90°C, and the oxides of nitrogen are extracted from the acid and carried over into the chambers. In addition, the acid undergoes concentration to about 78% by the time it leaves the base of the tower. Some of this acid is returned to the Gay-Lussac tower, but the remainder is available for sale.

The second stage takes place in the lead chambers from which the process derives its name. Whilst water is sprayed from the roof onto the mixture of gases from the Glover tower, they slowly react together under carefully controlled conditions of humidity and temperature, producing 65% sulphuric acid which collects on the floor. Lead is used because it is not attacked by acid of this concentration.

The third stage takes place in the Gay-Lussac towers, which are designed to recover as much as possible of the oxides of nitrogen from the gases leaving the chambers by thoroughly washing them with cold concentrated acid. The original purpose of these towers was to minimize the nuisance caused by the escape of nitrogen dioxide into the atmosphere, but after the invention of the Glover tower this recovery stage became an important part of the process greatly adding to its cheapness and efficiency. A small loss of oxides of nitrogen is inevitable, however, and is made good by introducing additional nitric oxide formed by the catalytic oxidation of ammonia.

The chamber process produces cheap acid of doubtful purity and of about 65%–80% concentration. This is very suitable for making fertilizers, but where more concentrated acid is required it is preferable to use the contact process.

Properties of Sulphuric Acid: When a dilute solution of sulphuric acid is distilled, a constant boiling point mixture (§ 6.11) is obtained containing 98.3% sulphuric acid. This mixture, which boils at 338°C and has a density of 1.84 g cm^{-3}, is the concentrated acid used in the laboratory. If a little sulphur trioxide is dissolved in it, 100% acid is obtained as an oily liquid which freezes to white crystals at 10°C.

Concentrated sulphuric acid is extremely corrosive and should always be treated with care; it causes severe burns if brought into contact with the skin.

Reactions of Sulphuric Acid: It is a strong dibasic acid, reacting with bases to give two series of salts, sulphates and bisulphates:

$$H_2SO_4 \rightleftharpoons H^+ + HSO_4^- \rightleftharpoons 2H^+ + SO_4^{2-}$$

The dilute acid attacks many metals forming sulphates and hydrogen, but it does not react with lead, copper, mercury, and silver:

$$Fe + H_2SO_4 = FeSO_4 + H_2\uparrow$$

The concentrated acid attacks most metals when heated forming a sulphate, sulphur dioxide, and water, but it has no action on gold and platinum and very little on silicon steel, which is used, therefore, for constructing distillation vessels:

$$Cu + 2H_2SO_4 = CuSO_4 + SO_2\uparrow + 2H_2O$$

Concentrated sulphuric acid has a great affinity for water, forming hydrates very readily. Much heat is liberated when the acid is mixed with water and dilution should always be performed cautiously, therefore, by pouring the acid slowly into water with constant stirring. If water is added to the concentrated acid there is a danger that a pocket of steam may be produced which might blow the acid out of the vessel. Because it is very hygroscopic the concentrated acid is used for drying gases, although it must not be used for basic substances such as ammonia, nor for reducing agents such as hydrogen sulphide. So great is its avidity for water that the concentrated acid, particularly when hot, decomposes substances containing hydrogen and oxygen and removes these elements as water. For example, cane sugar is converted into a mass of sugar carbon, and formic and oxalic acids are dehydrated to carbon monoxide:

$$C_{12}H_{22}O_{11} - 11H_2O = 12C$$
$$H.COOH - H_2O = CO\uparrow$$
$$(COOH)_2 - H_2O = CO\uparrow + CO_2\uparrow$$

Similarly, paper, wood, and clothes are rapidly dehydrated and charred by the hot concentrated acid.

Hot concentrated sulphuric acid is an oxidizing agent, behaving thus with reducing agents:

$$H_2SO_4 = SO_2 + H_2O + \bar{O}.$$

Examples of such oxidations are:

$$C + 2H_2SO_4 = CO_2\uparrow + 2SO_2 + 2H_2O$$
$$S + 2H_2SO_4 = 3SO_2\uparrow + 2H_2O$$
$$2HBr + H_2SO_4 = SO_2\uparrow + Br_2\uparrow + 2H_2O$$

Powerful reducing agents such as hydrogen iodide will reduce sulphuric acid to hydrogen sulphide and sulphur.

Hot concentrated sulphuric acid displaces hydrochloric and nitric acids from their salts because these two acids are more volatile than sulphuric acid and distil over when heated:

$$2NaNO_3 + H_2SO_4 = Na_2SO_4 + 2HNO_3\uparrow$$
$$2NaCl + H_2SO_4 = Na_2SO_4 + 2HCl\uparrow$$

Concentrated sulphuric acid takes part in a number of important reactions in organic chemistry, e.g. sulphonation (§ 43.7), nitration (§ 44.2), esterification (§ 38.1), and hydrolysis.

Uses of Sulphuric Acid: The biggest uses are in the manufacture of superphosphate (§ 23.12) and ammonium sulphate. Production of these fertilizers consumes about 40% of all the sulphuric acid made. Other large-scale uses are the manufacture of pigments, especially barium sulphate and titanium dioxide, and the manufacture of viscose rayon (called artificial silk, see § 42.7), detergents, dyestuffs, drugs, explosives, plastics, and sulphates. Sulphuric acid is also used for dissolving unsaturated hydrocarbons during petrol refining, for pickling iron and steel (i.e. removing the oxide layer before galvanizing, tinning, plating or painting), for filling accumulators, and for killing weeds. Indeed there are few industries which do not make some use of sulphuric acid, so many and so diverse are its applications.

STRUCTURE OF SULPHURIC ACID: The chemical evidence (e.g. the dibasicity, and the reaction with phosphorus pentachloride) indicates that the molecular formula is H_2SO_4 and the structural formula $SO_2(OH)_2$, although the molecules are believed to be partially associated by hydrogen bonding in the pure acid. The graphic formula

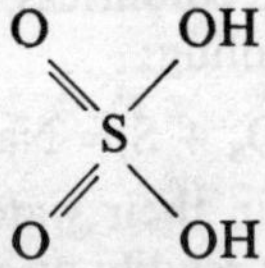

contains two hydroxyl groups, each of which readily dissociates in aqueous solution yielding hydroxonium ions and a sulphate ion, SO_4^{2-}. The latter has been shown by X-ray analysis of sulphates to be tetrahedral in shape in the crystalline state.

24.12. Sulphates. These are prepared by the action of sulphuric acid on metals, bases, and carbonates, or by double decomposition, or by oxidation of sulphides and sulphites. They are colourless unless the cation is coloured and are mostly soluble in water, the chief exceptions being the sulphates of barium, strontium, lead and mercurous mercury. Soluble sulphates usually form hydrated crystals in which one molecule of water is linked to each SO_4^{2-} ion, the others being associated with the metallic ion, e.g. $[Cu(H_2O)_4]^{2+}[SO_4.H_2O]^{2-}$. Many double sulphates exist, of which ferrous ammonium sulphate and the alums (§ 19.12) are the best known. The sulphates of the alkali and alkaline earth metals are very stable to heat, but some sulphates decompose when heated strongly giving sulphur dioxide and trioxide:

$$2FeSO_4 = Fe_2O_3 + SO_2\uparrow + SO_3\uparrow$$

The more important sulphates are considered in the chapters dealing with the metals they contain.

24.13. Persulphuric Acid, $H_2S_2O_8$, and Persulphates. A solution of the acid is prepared by electrolysing cold concentrated sulphuric acid using a platinum electrode and a high current density (see § 16.6). More important are its potassium and ammonium salts which are stable crystalline solids, sparingly soluble in water. They are strong oxidizing agents which liberate iodine from solutions of iodides, oxidize ferrous salts, convert chromic salts into chromates, and manganese and lead salts into their dioxides:

$$2KI + K_2S_2O_8 = 2K_2SO_4 + I_2$$
$$Mn(OH)_2 + K_2S_2O_8 = 2KHSO_4 + MnO_2\downarrow$$

24.14. Thiosulphates. As the name implies, these are sulphates in which an oxygen atom has been replaced by a sulphur atom giving the ion $S_2O_3^{2-}$. Sodium thiosulphate, commonly known as 'hypo', is by far the most important. It is prepared by boiling a solution of sodium sulphite with sulphur for several hours. If the excess of sulphur is filtered off and the filtrate concentrated, colourless crystals of the pentahydrate, $Na_2S_2O_3.5H_2O$, appear on cooling:

$$Na_2SO_3 + S = Na_2S_2O_3$$

These crystals melt at 48°C in their water of crystallization, the melt showing supersaturation very readily on cooling.

Thiosulphates are decomposed by dilute acids giving off sulphur dioxide and slowly depositing sulphur as a yellow precipitate:

$$Na_2S_2O_3 + 2HCl = 2NaCl + H_2O + SO_2\uparrow + S\downarrow$$

With iodine sodium thiosulphate forms sodium tetrathionate quantitatively:

$$2Na_2S_2O_3 + I_2 = 2NaI + Na_2S_4O_6$$

This reaction, which is rapid even at room temperature, is used in volumetric analysis for estimating solutions of iodine.

Sodium thiosulphate solution is used in photography to 'fix' the negative (i.e. to render the film insensitive to light by removing the silver halide which was not changed during the exposure). This it does by forming a soluble complex ion of the type $[Ag(S_2O_3)_2]^{3-}$. It is also used as an 'antichlor' to remove any residual chlorine from the fabric after bleaching.

$$Na_2S_2O_3 + 4Cl_2 + 5H_2O = Na_2SO_4 + 8HCl + H_2SO_4$$

24.15. Halogen Compounds of Sulphur. Only two of these are important:

SULPHUR HEXAFLUORIDE, SF_6: This is formed directly from sulphur and fluorine. It is a colourless, odourless gas, b.p. −64°C, notable for its stability and inertness. Its existence establishes that sulphur can be hexavalent. It is widely used, under pressure, for filling very high voltage cables because it suppresses internal discharges.

SULPHUR MONOCHLORIDE, S_2Cl_2: This is prepared by passing dry chlorine over molten sulphur in a retort. It is an orange-yellow liquid, b.p. 137°C, with a repulsive smell. It fumes in moist air owing to hydrolysis:

$$2S_2Cl_2 + 2H_2O = 4HCl + SO_2 + 3S\downarrow$$

It is used for vulcanizing rubber (§ 49.4), being a good solvent for sulphur.

24.16. Oxy-halogen Compounds of Sulphur. The most important of these are:

THIONYL CHLORIDE, $SOCl_2$: This is prepared by the action of sulphur dioxide on phosphorus pentachloride, the products being separated by distillation:

$$PCl_5 + SO_2 = POCl_3 + SOCl_2$$

It is a colourless liquid, b.p. 78°C, which fumes in moist air owing to hydrolysis:

$$SOCl_2 + H_2O = 2HCl + SO_2$$

Thionyl chloride is a useful chlorinating agent in organic chemistry, often being preferred to phosphorus pentachloride for this purpose because the by-products are gaseous and easily removed:

$$\underset{\text{benzoic acid}}{C_6H_5COOH} + SOCl_2 = \underset{\text{benzoyl chloride}}{C_6H_5COCl} + SO_2\uparrow + HCl\uparrow$$

SULPHURYL CHLORIDE, SO_2Cl_2: This is formed when sulphur dioxide and chlorine react together in sunlight or in the presence of a catalyst of camphor or charcoal:

$$SO_2 + Cl_2 = SO_2Cl_2$$

It is a colourless liquid, b.p. 69°C, with a pungent smell. It is rapidly hydrolysed by water:

$$SO_2Cl_2 + 2H_2O = H_2SO_4 + 2HCl$$

It is used as a chlorinating agent.

CHLOROSULPHONIC ACID, $Cl.SO_2.OH$: This is prepared by the action of phosphorus pentachloride on concentrated sulphuric acid, or by direct combination of hydrogen chloride with sulphur trioxide:

$$PCl_5 + H_2SO_4 = Cl.SO_2.OH + POCl_3 + HCl\uparrow$$
$$HCl + SO_3 = Cl.SO_2.OH$$

Moisture must be excluded or the product will be instantly hydrolysed:

$$Cl.SO_2.OH + H_2O = H_2SO_4 + HCl$$

It is a colourless liquid, b.p. 152°C, which fumes in moist air. It is useful in organic chemistry for sulphonylation, i.e. substituting an $-SO_2Cl$ group in place of an H atom.

24.17. Sulphamic Acid, $NH_2.SO_3H$. This is made by treating urea with fuming sulphuric acid:

$$NH_2.CO.NH_2 + H_2O + 2SO_3 = 2NH_2.SO_3H + CO_2$$

It may also be prepared from sulphur dioxide and hydroxylamine sulphate:

$$NH_2OH.H_2SO_4 + SO_2 = NH_2.SO_3H + H_2SO_4$$

It is a colourless crystalline solid, m.p. 205°C (decomposes) and it dissolves readily in water.

Sulphamic acid reacts with nitrous acid and nitrites and, since the reaction is quantitative, the volume of nitrogen liberated can be used to estimate solutions of nitrites:

$$NH_2.SO_3H + HNO_2 = N_2\uparrow + H_2O + H_2SO_4$$

With concentrated nitric acid it gives nitrous oxide:

$$NH_2.SO_3H + HNO_3 = N_2O\uparrow + H_2O + H_2SO_4$$

Sulphamic acid is a strong monobasic acid, giving salts called sulphamates. It is an excellent primary standard for alkalis, since the crystals are not hygroscopic. Industrially it is used for disposing of excess nitrous acid after diazotization. It is also extensively used in the dairy and brewing industries for descaling and cleaning plant because of the high solubility of sulphamates. Ammonium sulphamate is used as a weed-killer and for flame-proofing fabrics.

24.18. Summary. The marked differences which often occur between the lightest member of a Group and the others are very apparent in Group VIB. Indeed the differences between oxygen and sulphur are much more prominent than the similarities. Apart from the formation of the allotrope ozone, oxygen is invariably diatomic when gaseous, even at low temperatures, and shows little tendency to form O–O linkages. Thus the only chains of oxygen atoms which exist are those found in the peroxides and in ozone, and these break readily on heating. In contrast to this, ring and chain molecules predominate in solid and liquid sulphur and even persist in the vapour, so that sulphur is diatomic only at very high temperatures. This accounts for the big difference in the volatility of oxygen and sulphur. The polysulphides and polythionic acids are further evidence of the pronounced tendency of sulphur to form stable S–S linkages.

Oxygen is more electronegative than sulphur, its smaller atom having a stronger attraction for electrons at the periphery. Thus oxygen is the more reactive element combining more exothermically with electropositive elements such as metals and hydrogen and forming more stable compounds with these elements than does sulphur. On the other hand the compounds that oxygen forms with the electronegative halogens are less stable than those formed by sulphur. The hydrides of the two elements strongly reflect this difference in electronegativity. For example, water is less dissociated than hydrogen sulphide in solution and its molecule is more highly polar giving rise to association by hydrogen bonding (§ 16.4), which is responsible for its relatively high melting point, boiling point, and latent heat.

Another important difference is that oxygen is invariably divalent whereas sulphur displays variable valency (it is divalent in hydrogen sulphide, tetravalent in sulphur dioxide, and hexavalent in sulphur hexafluoride). Both elements contain six electrons in the outermost shells of their atoms, but in contrast to sulphur which can share control of up to six additional electrons, the total number of electrons which can be accommodated in the outermost shell of the oxygen atom is limited to eight. In addition, both elements form divalent anions (e.g. O^{2-} and S^{2-}) by accepting electrons from electropositive atoms.

The various oxidation states of sulphur are shown in Table 24.2.

TABLE 24.2. THE OXIDATION STATES OF SULPHUR

+7	$H_2S_2O_8$ and persulphates, $S_2O_8^{2-}$
+6	SO_3 and H_2SO_4 and sulphates, SO_4^{2-} and SF_6 and SO_2Cl_2
+5	
+4	SO_2 and H_2SO_3 and sulphites, SO_3^{2-} and $SOCl_2$
+3	
+2	$H_2S_2O_3$ and thiosulphates, $S_2O_3^{2-}$ and SCl_2
+1	
0	S_8
−1	
−2	H_2S and sulphides, S^{2-}

25

The Halogens

25.1. Introduction. The elements of Group VIIB, viz. fluorine, chlorine, bromine, iodine, and astatine, are known collectively as the halogens. Astatine is extremely rare, being radioactive with a half-life of only a few hours, but the others are important and will be studied in detail.

25.2. Fluorine, F. Electronic configuration: 2,7.

OCCURRENCE: The element is far too reactive to occur naturally, but it is found as fluorides in certain minerals, e.g. fluorspar or fluorite, CaF_2, and cryolite, Na_3AlF_6, and fluo-apatite, $CaF_2.3Ca_3(PO_4)_2$. Traces of fluorides are found in teeth and bones and in the ash of plants.

PREPARATION: An electrolytic method is employed, but considerable difficulty is encountered because of the very great reactivity of fluorine. The apparatus (Fig. 25.1) must be kept completely free of moisture, oil,

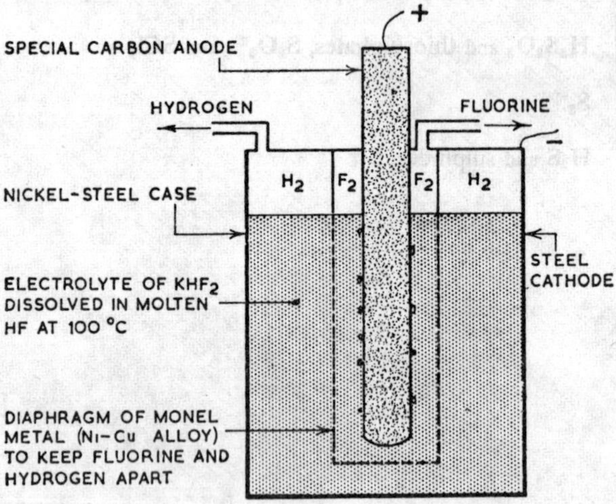

FIG. 25.1 Preparation of Fluorine

and grease, because fluorine attacks these materials vigorously. In the modern process an electrolyte of composition KF.2HF (i.e. a mixture of KHF_2 and HF) is used at about 100°C. The anode is made of special non-graphitic carbon impregnated with copper so that it is not readily attacked by fluorine, and the cathode is of steel. The cell itself is made of nickel-steel, which is fairly resistant to attack, and a diaphragm is fitted to ensure complete separation of the products, since fluorine and hydrogen combine explosively.

The electrolyte may be regarded as a solution of potassium fluoride in anhydrous hydrogen fluoride; it contains, therefore, a high concentration of potassium and fluoride ions and a much lower concentration of hydrogen ions. The changes at the electrodes are:

At the anode: $F^- \rightarrow F + e$ and $2F \rightarrow F_2\uparrow$
At the cathode: $H^+ + e \rightarrow H$ and $2H \rightarrow H_2\uparrow$

Thus the overall effect of the electrolysis is to decompose hydrogen fluoride into its elements, and the cell has to be topped up with more anhydrous hydrogen fluoride from time to time. The fluorine liberated at the anode can be freed from traces of hydrogen fluoride by passing it over anhydrous sodium fluoride:

$$NaF + HF = NaHF_2$$

PROPERTIES OF FLUORINE: It is a greenish-yellow diatomic gas, slightly paler in colour than chlorine, which it resembles in smell. It melts at −220°C and boils at −188°C. Its low critical temperature, −129°C, makes it impossible to store liquid fluorine at ordinary temperatures, but it can be transported in bulk as a liquid by using liquid nitrogen as a coolant.

REACTIONS OF FLUORINE: It is the most reactive element of all, combining directly with every other element except oxygen and some of the noble gases. It is also the most electronegative element (§ 10.7). It explodes with hydrogen, even in the dark at very low temperatures, forming hydrogen fluoride:

$$H_2 + F_2 = 2HF$$

So great is its affinity for hydrogen that fluorine also reacts vigorously with water, hydrogen sulphide, ammonia, and hydrogen chloride and with organic compounds, many of which inflame spontaneously when brought into contact with fluorine:

$$2F_2 + 2H_2O = 4HF + O_2$$
$$F_2 + 2HCl = 2HF + Cl_2$$
$$4F_2 + CH_4 = 4HF + CF_4$$

Fluorine combines readily at room temperature with silicon, phosphorus, and sulphur, forming fluorides in which these elements show their highest covalency, e.g. SiF_4, PF_5 and SF_6. It also reacts with carbon, particularly when heated, giving carbon tetrafluoride, and with boron giving boron trifluoride. The alkali and alkaline earth metals ignite spontaneously in fluorine at room temperature and most other metals, including platinum, react with it when heated, but some metals such as copper and nickel show some resistance to attack, probably because they form impervious fluoride layers on their surfaces which hinder further action.

When fluorine is passed through a very dilute solution of sodium hydroxide some fluorine monoxide is formed as a colourless gas, b.p. −146°C; it is the only oxide of fluorine stable at room temperature:

$$2F_2 + 2NaOH = 2NaF + F_2O\uparrow + H_2O$$

USES OF FLUORINE AND ITS COMPOUNDS: Fluorine is required in great quantities for making uranium hexafluoride, UF_6, which is the only volatile compound of uranium. Its production is the first step in separating uranium into its isotopes by the diffusion process described in § 3.6. Fluorine compounds such as CCl_2F_2 and CCl_3F are being increasingly used as refrigerants because of their inertness, stability, non-toxicity, and thermodynamic efficiency, and also as propellents in aerosols. Boron trifluoride and hydrogen fluoride are used as catalysts in the petroleum industry, and cobalt fluoride and bromine trifluoride as fluorinating agents. Other fluorides are used as disinfectants in the brewing industry and for preventing rot in timber. The fluorocarbon, tetrafluoroethylene, C_2F_4, can be polymerized into a plastic, known as 'P.T.F.E.', which is renowned for its chemical inertness; it is already important as an electrical insulator and as a lubricant (see § 40.11). Sulphur hexafluoride is used as an electrical insulator in gas-filled high voltage cables (§ 24.15). Molten cryolite is used as a solvent for alumina in the manufacture of aluminium (§ 19.8). The mineral was the original source, but nowadays a synthetic product is used.

Fluorosilicic acid, H_2SiF_6 (see § 20.21), is used for fluoridation of drinking water. By adding fluorides artificially in this way so that their concentration is only a few parts per million, it has been found possible to reduce dental decay, particularly in children, presumably by strengthening the flouride-containing enamel which protects the teeth. Toothpastes containing fluorides (e.g. stannous fluoride) are claimed to have the same beneficial effect.

25.3. Hydrogen Fluoride, HF. This is usually made in the laboratory and industrially by heating calcium fluoride with concentrated sulphuric acid in steel vessels (the product attacks glass):

$$CaF_2 + H_2SO_4 = CaSO_4 + 2HF\uparrow$$

It can be prepared in the anhydrous condition by heating dry potassium hydrogen fluoride:

$$KHF_2 = KF + HF\uparrow$$

Anhydrous hydrogen fluoride is a colourless fuming liquid, b.p. 19°C, which does not conduct electricity. Not only is it extremely poisonous, but it causes severe burns when brought into contact with the skin. Vapour density measurements show that the vapour is highly associated just above the boiling point, but that above 90°C it consists of single molecules only. The liquid is also strongly associated, owing to hydrogen bonding between the molecules as in water (see § 14.7 and Fig. 14.9). This explains why its boiling point is much higher than that of the other halogen hydrides and also why its dielectric constant is so large, making it a very good polar solvent. It is mainly used in the manufacture of organic fluorine compounds (see § 40.11) and uranium tetrafluoride, which is an intermediate compound in the extraction of uranium from its ores.

Hydrogen fluoride is very soluble in water, giving a solution of *hydrofluoric acid.* This is a weak, monobasic acid which reacts with metals and bases forming salts called fluorides. For example, sodium fluoride, which is only sparingly soluble in water, is obtained as a white precipitate when hydrogen fluoride is passed into a concentrated solution of sodium hydroxide:

$$NaOH + HF = NaF\downarrow + H_2O$$

When solutions of sodium or potassium fluoride are evaporated in the presence of excess of hydrofluoric acid, crystals of the acid fluorides are obtained, e.g. KHF_2. These salts, of which no counterparts exist amongst the other halogens, arise as a result of hydrogen bonding between the fluorine and hydrogen fluoride and contain the ions K^+ $(F{-}H{-}F)^-$.

Hydrofluoric acid attacks silica and silicates forming silicon tetrafluoride (§ 20.21), which explains its use for etching glass:

$$SiO_2 + 4HF = SiF_4\uparrow + 2H_2O$$

It is usually stored, therefore, in vessels made of lead, polythene, or gutta percha, or in glass vessels which have been coated with a layer of paraffin wax. The solution should be treated with care as it is strongly corrosive and its vapour is very poisonous. When distilled, hydrofluoric acid forms a constant boiling point mixture which contains 36% HF and boils at 120°C.

25.4. Detection of Fluorides. A fluoride is detected by heating it in a lead crucible with concentrated sulphuric acid, when hydrogen fluoride fumes are evolved; if a wet glass rod is held in these fumes it rapidly becomes rough and frosted where it has been attacked by hydrofluoric acid. Solutions of fluorides give a white precipitate of calcium fluoride with calcium chloride solution, but no precipitate with silver nitrate solution, which distinguishes them from other halides.

25.5. Chlorine, Cl. Electronic configuration: 2,8,7.

OCCURRENCE: Chlorine is found only in the combined state, chiefly in the form of the chlorides of sodium, potassium, calcium, and magnesium, of which the first is by far the commonest (see § 17.10). Sea water contains over 3% by weight of dissolved chlorides, and large deposits are found where inland seas have undergone evaporation.

MANUFACTURE: Chlorine is now made almost entirely by electrolysis. It is chiefly obtained as a by-product during the manufacture of sodium hydroxide from brine. Details are given in § 17.7 and § 17.4. Chlorine is easily liquefied since its critical temperature is 144°C (see § 3.14), and it is as a liquid that it is usually transported under pressure in tanks or cylinders.

LABORATORY PREPARATION: The gas is prepared by heating concentrated hydrochloric acid (or a mixture of sodium chloride and concentrated sulphuric acid) with manganese dioxide:

$$4HCl + MnO_2 = MnCl_2 + 2H_2O + Cl_2\uparrow$$

The gas, which is bubbled through water to remove hydrogen chloride and through concentrated sulphuric acid to dry it, is collected by upward displacement of air. Chlorine is also produced by the action of dilute acids on hypochlorites or bleaching powder:

$$NaOCl + 2HCl = NaCl + H_2O + Cl_2\uparrow$$

Pure chlorine is obtained by adding concentrated hydrochloric acid to solid potassium permanganate (no heat is necessary), and then washing and drying the gas:

$$2KMnO_4 + 16HCl = 2KCl + 2MnCl_2 + 8H_2O + 5Cl_2\uparrow$$

PROPERTIES OF CHLORINE: It is a greenish-yellow diatomic gas with a characteristic pungent smell. It boils at −34°C, melts at −102°C, and is fairly soluble in water. It is very poisonous, causing inflammation of the lungs and mucous membranes if inhaled even in small amounts.

REACTIONS OF CHLORINE: It is highly reactive, though less so than fluorine. It reacts directly with all the elements except the inert gases, carbon, oxygen, and nitrogen, but the reactions are often slow unless heat is applied. Dry chlorine reacts with many heated metals giving anhydrous chlorides; where the metal exerts more than one valency the higher chloride is formed:

$$2Fe + 3Cl_2 = 2FeCl_3$$

Chlorine explodes with hydrogen if the two gases are sparked together or exposed to intense ultra-violet light, but in the presence of a charcoal catalyst they combine together smoothly at room temperature forming hydrogen chloride:

$$H_2 + Cl_2 = 2HCl$$

Many substitution reactions occur in which chlorine reacts with saturated hydrocarbons, replacing hydrogen atoms by chlorine atoms (§ 33.2). With unsaturated compounds chlorine readily forms addition compounds (§ 34.2).

When chlorine dissolves in water forming 'chlorine water', some of it reacts producing hypochlorous and hydrochloric acids:

$$Cl_2 + H_2O = HOCl + HCl$$

The hypochlorous acid gradually decomposes on standing, particularly in sunlight, giving hydrochloric acid and oxygen:

$$2HOCl = 2HCl + O_2\uparrow$$

The bleaching action of moist chlorine is attributed to the formation of hypochlorous acid, which oxidizes the organic dyestuff to a colourless substance.

In the presence of a charcoal catalyst chlorine combines with carbon monoxide giving carbonyl chloride or phosgene, and with sulphur dioxide it gives sulphuryl chloride. Both of these reactions are also accelerated by sunlight:

$$CO + Cl_2 = COCl_2$$
$$SO_2 + Cl_2 = SO_2Cl_2$$

Chlorine is a mild oxidizing agent. It oxidizes sulphites to sulphates, ferrous salts to ferric, and it liberates sulphur from a solution of hydrogen sulphide:

$$Cl_2 + H_2O + K_2SO_3 = K_2SO_4 + 2HCl$$
$$2FeCl_2 + Cl_2 = 2FeCl_3$$
$$H_2S + Cl_2 = S\downarrow + 2HCl$$

With ammonia the products depend upon the relative proportions of the reactants—if ammonia is in excess, nitrogen is formed, but with excess of chlorine nitrogen trichloride is produced:

$$8NH_3 + 3Cl_2 = N_2 + 6NH_4Cl$$
$$NH_3 + 3Cl_2 = NCl_3 + 3HCl$$

The first of these reactions is used for determining the formula of ammonia (see § 22.4).

Chlorine dissolves readily in solutions of alkalis forming a chloride and a hypochlorite in the cold and a chloride and a chlorate when heated:

$$Cl_2 + 2NaOH = NaCl + NaOCl + H_2O$$
$$3Cl_2 + 6NaOH = 5NaCl + NaClO_3 + 3H_2O$$

With slaked lime, chlorine gives bleaching powder (see § 18.23).

It displaces less electronegative halogens from solutions of halides, chlorine itself assuming the ionic condition:

$$Cl_2 + 2KBr = 2KCl + Br_2$$

USES OF CHLORINE: Over half the output of chlorine is used in making trichloroethylene (§ 40.10), ethyl chloride (§ 40.5), methylene and ethylene dichlorides (§ 40.7), polyvinyl chloride (the plastic known as PVC), and many other organic compounds valuable as solvents, drugs, antiseptics, insecticides, anaesthetics, and refrigerants. Other important uses are for bleaching linen, cotton, wood-pulp, and paper (either as elemental chlorine or as bleaching powder), for making hydrochloric acid and metallic chlorides, for extracting bromine, for sterilizing drinking water, sewage, and swimming-bath water, and for detinning tin-plate to recover the tin as stannic chloride. Chlorine is also used to make carbon tetrachloride (§ 20.9), silicon tetrachloride (§ 20.22), and sulphur monochloride (§ 24.15); three compounds of growing commercial importance.

25.6. Hydrogen Chloride, HCl.

MANUFACTURE: Considerable quantities are now made by direct synthesis. Chlorine is burnt at a jet in hydrogen or the gases are passed over a catalyst of activated charcoal. Both elements are produced electrolytically as by-products during the manufacture of caustic soda (see § 17.7).

The older method of making hydrogen chloride by heating rock salt with concentrated sulphuric acid to 500°C is still widely used because the sodium sulphate is needed in large amounts for making glass (§ 20.25):

$$2NaCl + H_2SO_4 = Na_2SO_4 + 2HCl\uparrow$$

In both processes the product is usually dissolved in water and sold as hydrochloric acid.

Laboratory Preparation: By heating concentrated sulphuric acid with sodium chloride to about 300°C:

$$NaCl + H_2SO_4 = NaHSO_4 + HCl\uparrow$$

The product is dried with concentrated sulphuric acid and collected by downward delivery.

Properties of Hydrogen Chloride: It is a colourless gas, b.p. −85°C, with a pungent smell. When liquefied it is a covalent compound, being a non-conductor of electricity and having no acidic properties. Hydrogen chloride gas is very soluble in water (1 volume of water dissolves over 500 volumes at s.t.p.). Because of this high solubility and the danger of suck-back when hydrogen chloride dissolves, *hydrochloric acid* (as its solution is called) is prepared in the laboratory by using an inverted funnel device as shown in Fig. 25.2. The concentrated acid used in the laboratory contains about 38% HCl and has a density of 1.20 g cm^{-3}. When solutions of hydrochloric acid are distilled a constant boiling point mixture is formed which boils at 110°C and contains 20.24% HCl. Both hydrogen chloride and its concentrated solutions fume in air because the escaping gas combines with moisture in the air forming minute droplets of acid which rapidly grow in size.

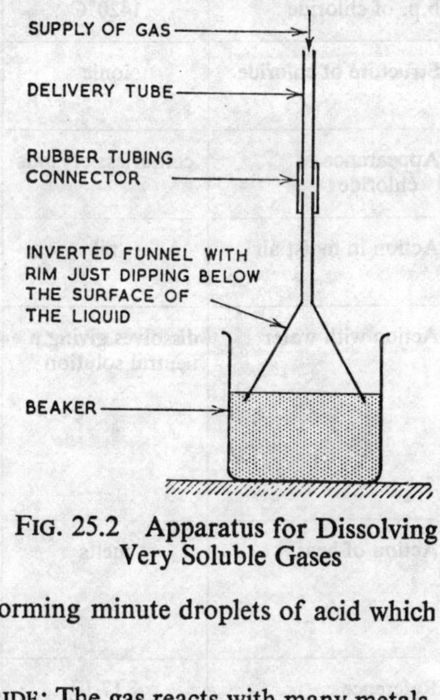

Fig. 25.2 Apparatus for Dissolving Very Soluble Gases

Reactions of Hydrogen Chloride: The gas reacts with many metals, especially on heating, giving chlorides; where the metal exhibits two valencies the lower chloride is formed:

$$Fe + 2HCl = FeCl_2 + H_2$$

It reacts with basic oxides giving a chloride and water and with higher oxides, e.g. MnO_2 or PbO_2, giving chlorine. Hydrogen chloride combines directly with ammonia producing white clouds, which consist of particles of solid ammonium chloride suspended in air:

$$NH_3 + HCl = NH_4Cl$$

TABLE 25.1. THE CHLORIDES OF THE ELEMENTS

Element / *Property*	*Sodium*	*Magnesium*	*Aluminium*
Formula	NaCl	$MgCl_2$ ($6H_2O$)	Al_2Cl_6 ($AlCl_3$)
m.p. of chloride	800°C	715°C	sublimes at 183°C
b.p. of chloride	1420°C	1400°C	—
Structure of chloride	ionic	ionic	covalent
Appearance of chloride	colourless crystals	colourless crystals	colourless crystals
Action in moist air	nil	deliquesces	deliquesces and fumes
Action with water	dissolves giving a neutral solution	dissolves readily giving an acidic solution $MgCl_2 + 2H_2O = Mg(OH)_2 + 2HCl$	vigorously hydrolysed to aluminium hydroxide $Al_2Cl_6 + 6H_2O = 2Al(OH)_3 + 6HCl$
Action of heat	melts	melts, hydrate hydrolyses when heated	sublimes, dissociating into $AlCl_3$ molecules above 400°C
Reference	§ 17.10	§ 18.8	§ 19.10

Hydrochloric acid is a strong, monobasic acid. It reacts with many metals, bases, and carbonates forming chlorides, but mercury, silver, platinum, and gold do not dissolve in it, and copper and lead do so only on heating:

$$Zn + 2HCl = ZnCl_2 + H_2\uparrow$$

IN THE THIRD PERIOD

Silicon	*Phosphorus*		*Sulphur*
$SiCl_4$	PCl_3	PCl_5	S_2Cl_2
—	−90°C	Sublimes at 160°C	—
59°C	76°C	—	137°C
covalent	covalent	solid ionic, vapour covalent	covalent
colourless liquid	colourless liquid	pale yellow solid	orange liquid
fumes	fumes	fumes	fumes
rapidly hydrolysed to hydrated silica $SiCl_4 + 3H_2O =$ $SiO_2.H_2O\downarrow +$ $4HCl\uparrow$	instantly hydrolysed to phosphorous acid $PCl_3 + 3H_2O =$ $H_3PO_3 + 3HCl\uparrow$	violently hydrolysed to orthophosphoric acid $PCl_5 + H_2O =$ $POCl_3 + 2HCl\uparrow$ $POCl_3 + 3H_2O =$ $H_3PO_4 + 3HCl\uparrow$	readily hydrolysed to sulphur $2S_2Cl_2 + 2H_2O =$ $4HCl + SO_2 + 3S\downarrow$
boils	boils	sublimes, dissociating into PCl_3 and Cl_2 above 300°C	boils
§ 20.22	§ 23.6	§ 23.7	§ 24.15

Hydrochloric acid is a mild reducing agent, being oxidized to chlorine by powerful oxidizing agents such as manganese dioxide, permanganates, dichromates, and chlorates.

USES OF HYDROGEN CHLORIDE: The acid is used to prepare chlorides, to 'pickle' metals (i.e. remove from their surfaces any rust before tinning, plating, or galvanizing), and as an important laboratory reagent.

COMPOSITION OF HYDROGEN CHLORIDE: The vessel shown in Fig. 25.3 is filled with equal volumes of hydrogen and chlorine at atmospheric pressure and the mixture is sparked causing the gases to react forming hydrogen chloride. After cooling to the original temperature, one of the taps is opened over mercury. No change occurs, showing that the volume of the product is exactly equal to the total volume of the reactants, i.e. one volume of hydrogen has reacted with one volume of chlorine to form two volumes of hydrogen chloride. This is confirmed by opening the tap over water, which dissolves the hydrogen chloride and rushes in, completely filling the vessel.

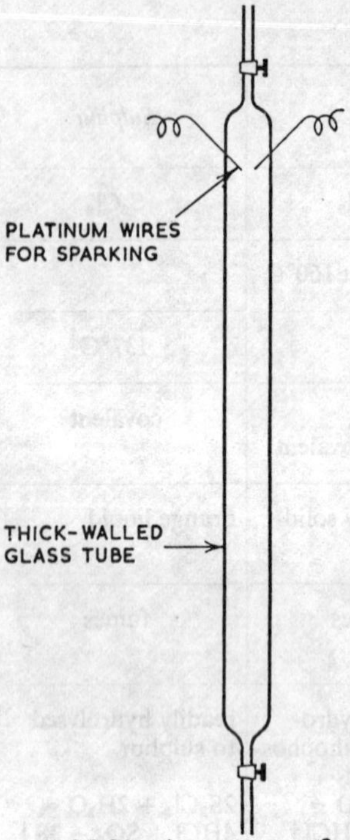

FIG. 25.3 Apparatus for Finding the Formula of Hydrogen Chloride

Hence, by Avogadro's hypothesis, one molecule of hydrogen has combined with one molecule of chlorine to form two molecules of hydrogen chloride. Since the two elements are diatomic, the formula of hydrogen chloride must be HCl, which is confirmed by vapour density measurements:

i.e. $$H_2 + Cl_2 = 2HCl$$

25.7. Detection of Chlorides. Most chlorides (except lead, silver, mercurous and cuprous) are soluble in water. With silver nitrate solution they give a white precipitate of silver chloride, which is insoluble in dilute nitric acid but which dissolves readily in ammonium hydroxide solution. The precipitate coagulates when shaken and gradually turns purple on exposure to light. Chlorides also give chlorine gas (recognized by its colour, smell, and bleaching action) when heated with concentrated sulphuric acid and a little manganese dioxide. When heated with concentrated sulphuric acid and some solid potassium dichromate, chlorides give red chromyl chloride, CrO_2Cl_2, which distils over. When dissolved in caustic soda solution, this gives a yellow precipitate with lead acetate solutions (§ 27.9). The chlorides of the elements in the third period are compared in Table 25.1.

25.8. Oxygen Compounds of Chlorine. The oxides of chlorine are listed in Table 25.2 and the oxy-acids in Table 25.3.

TABLE 25.2. THE OXIDES OF CHLORINE

Name	*Formula*	*Colour*	m.p./°C	b.p./°C
Chlorine monoxide	Cl_2O	Brown gas	−12	3
Chlorine dioxide	ClO_2	Yellow gas	−59	10
Chlorine hexoxide	Cl_2O_6	Red liquid	3	c. 200
Chlorine heptoxide	Cl_2O_7	Colourless liquid	−92	c. 80

TABLE 25.3. THE OXY-ACIDS OF CHLORINE

Name	*Formula*	*Structure*	*Strength*	*Salts*
Hypochlorous acid	HOCl	H—O—Cl	Very weak acid	Hypochlorites
Chlorous acid	$HClO_2$	H—O—Cl→O	Weak acid	Chlorites
Chloric acid	$HClO_3$	O ↗ H—O—Cl ↘ O	Strong acid	Chlorates
Perchloric acid	$HClO_4$	O ↑ H—O—Cl→O ↓ O	Very strong acid	Perchlorates

The oxides of chlorine are all very unstable; they explode when heated or brought into contact with organic matter, sulphur or phosphorus. With the exception of perchloric acid, the oxy-acids of chlorine are too unstable to be isolated from solution. Their more important salts are considered below:

Hypochlorites are made by the action of chlorine on cold solutions of alkalis:

$$Cl_2 + 2NaOH = NaCl + NaOCl + H_2O$$

Sodium hypochlorite is usually made by electrolysing cooled brine in a cell which permits the chlorine evolved at the anode to mix with the hydroxide solution (see § 17.7) produced at the cathode. In the presence of cobalt salts, which act as catalysts, hypochlorites readily decompose releasing oxygen; when heated in solution they are decomposed into the corresponding chlorate and chloride. Because of their instability they are widely used as bleaching agents and disinfectants. They are recognized by their oxidizing properties, by the way they evolve chlorine when acidified, and by the white precipitate they give with silver nitrate solution.

Chlorates are formed when chlorine reacts with hot solutions of alkalis. Sodium chlorate is made by electrolysing a concentrated solution of sodium chloride at 70°C; the anodes and cathodes are placed close together so that the products react with each other. On evaporating the solution sodium chloride is precipitated and filtered off, and the chlorate crystallizes from the filtrate on further evaporation. Sodium chlorate is a colourless, deliquescent solid, very soluble in water. It is used extensively as a weed killer. Potassium chlorate, being much less soluble than the sodium salt at room temperature, crystallizes out when concentrated solutions of sodium chlorate and potassium chloride are mixed:

$$NaClO_3 + KCl = KClO_3\downarrow + NaCl$$

It decomposes above 600°C, giving oxygen:

$$2KClO_3 = 2KCl + 3O_2\uparrow$$

This decomposition is catalysed by a number of metallic oxides, of which manganese dioxide is the most effective; in their presence the chlorate decomposes smoothly above 250°C. Potassium chlorate is used in making safety matches and fireworks. It should always be treated with care as it forms dangerously explosive mixtures with carbon, phosphorus, sulphur, powdered metals, and many organic substances.

Chlorates are detected by their oxidizing properties and by the formation of the explosive gas chlorine dioxide when a small amount of chlorate is cautiously warmed with concentrated sulphuric acid:

$$3KClO_3 + 2H_2SO_4 = KClO_4 + 2KHSO_4 + H_2O + 2ClO_2\uparrow$$

They do not give a precipitate with silver nitrate solution provided they are free of chloride impurity.

Perchlorates are the most stable oxygen compounds of chlorine; they decompose only when heated strongly. Potassium perchlorate is prepared from potassium chlorate by heating it to 400°C:

$$4KClO_3 = 3KClO_4 + KCl$$

Magnesium perchlorate, $Mg(ClO_4)_2$, forms a hydrate with an extremely low water vapour pressure and is consequently a valuable drying agent for use in desiccation. Unlike phosphorus pentoxide it does not become sticky after use and it has the further advantage that it can be rendered anhydrous merely by heating and so can be used again and again.

Perchlorates are detected by the precipitates which their solutions give with concentrated solutions of potassium salts, by their evolution of oxygen on strong heating, and by the absence of a precipitate with silver nitrate solution.

The various oxidation states of chlorine are shown in Table 25.4.

TABLE 25.4. THE OXIDATION STATES OF CHLORINE

+7	Cl_2O_7 and	$HClO_4$	and perchlorates,	ClO_4^-
+6	ClO_3			
+5		$HClO_3$	and chlorates,	ClO_3^-
+4	ClO_2			
+3		$HClO_2$	and chlorites,	ClO_2^-
+2				
+1	Cl_2O and	$HClO$	and hypochlorites,	ClO^-
0	Cl_2			
−1		HCl	and chlorides,	Cl^-

25.9. Bromine, Br. Electronic configuration 2,8,18,7.

OCCURRENCE: None occurs free, but as the bromides of sodium, potassium, and magnesium, it is present in sea-water (which contains about 0.006% by weight of bromine) and in deposits formed by the evaporation of inland seas, e.g. at Stassfurt and in the Dead Sea area.

MANUFACTURE: Much bromine is now obtained from sea-water by treating it with chlorine, which liberates the element from the bromides present:

$$Cl_2 + MgBr_2 = MgCl_2 + Br_2$$

The free bromine is volatilized by blowing air through the liquid, and it is then reduced to bromide by adding sulphur dioxide:

$$SO_2 + Br_2 + 2H_2O = H_2SO_4 + 2HBr$$

The resulting concentrated solution of bromide is treated with chlorine and steam. The liberated bromine is condensed to a liquid, dried with concentrated sulphuric acid, and distilled over potassium bromide to free it from chlorine.

Bromine is extracted in a similar way from the mother liquors obtained from the Stassfurt deposits after crystallization of potassium salts.

LABORATORY PREPARATION: Bromine can be displaced from solutions of bromides by chlorine, but it is usually prepared by heating a mixture of concentrated sulphuric acid, manganese dioxide, and a bromide in a retort. The bromine distils over and condenses in the cooled receiver:

$$2KBr + 3H_2SO_4 + MnO_2 = 2KHSO_4 + MnSO_4 + 2H_2O + Br_2\uparrow$$

It is purified by distilling it with potassium bromide and quicklime.

PROPERTIES OF BROMINE: It is a dense, dark red liquid, m.p. $-7°C$, b.p. 59°C, which gives off a red-brown vapour on standing at room temperature. The liquid causes painful wounds if spilt on the skin and the vapour is poisonous and irritating. Bromine is fairly soluble in water, its brown solution being known as *bromine water*. It is very soluble in chloroform and carbon disulphide.

REACTIONS OF BROMINE: It resembles chlorine chemically, but it is less reactive. Thus it combines directly with most elements, although heat is often necessary to start the reaction. Bromine does not react with hydrogen at room temperature, unless a catalyst of platinum is present, but it combines readily at 200°C giving hydrogen bromide. Like chlorine, bromine takes part in numerous substitution and addition reactions with organic compounds.

When bromine dissolves in water, some of it reacts forming hypobromous acid, which yields oxygen when heated or exposed to sunlight:

$$Br_2 + H_2O = HOBr + HBr$$
$$2HOBr = 2HBr + O_2\uparrow$$

Thus bromine water is a mild oxidizing agent and moist bromine has a gentle bleaching action:

e.g. $$Br_2 + H_2S = 2HBr + S\downarrow$$

Bromine reacts with solutions of alkalis in a similar way to chlorine, forming bromides and either hypobromites or bromates according to the temperature. It also displaces iodine from solutions of iodides:

$$2KI + Br_2 = 2KBr + I_2$$

USES OF BROMINE: The main use of the element, which led to its large-scale production, is for making ethylene dibromide (§ 40.6). This liquid is added to petrol with the anti-knock agent, lead tetraethyl (§ 33.10), so that after combustion lead is not deposited in the cylinders but is expelled in the exhaust gases as volatile lead tetrabromide. Bromine is also used to make bromides, some of which are important in medicine and photography, and organic bromine compounds. In the laboratory, bromine and bromine water are used in testing for unsaturation (§ 34.6).

25.10. Hydrogen Bromide, HBr. This is best prepared in the laboratory by adding bromine to a mixture of red phosphorus and water, using the apparatus shown in Fig. 25.4. Some sand is often added to moderate

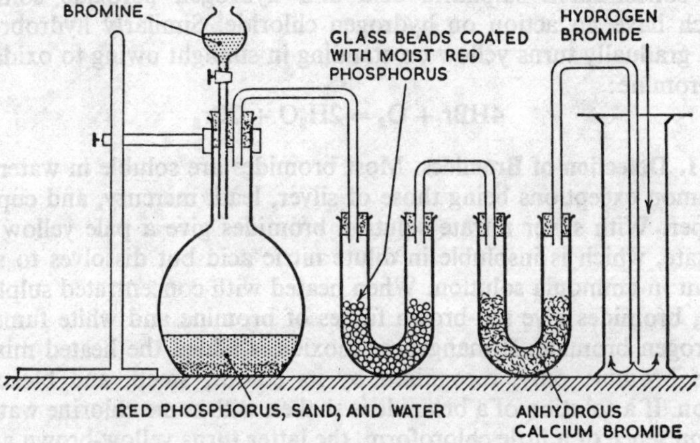

FIG. 25.4 Preparation of Hydrogen Bromide

the reaction. Presumably phosphorus tribromide is formed *in situ* and is immediately hydrolysed:

$$2P + 3Br_2 = 2PBr_3$$
$$PBr_3 + 3H_2O = H_3PO_3 + 3HBr\uparrow$$

The product, which is passed through a U-tube containing moist red phosphorus to remove any bromine vapour, is dried with anhydrous calcium bromide. If a solution of hydrobromic acid is required, an inverted funnel device must be used to dissolve the very soluble gas.

A solution of hydrobromic acid can be obtained by adding sulphuric acid to a bromide, but this method cannot be used to prepare hydrogen bromide because concentrated sulphuric acid tends to oxidize the product to bromine on warming:

$$KBr + H_2SO_4 = KHSO_4 + HBr$$
$$2HBr + H_2SO_4 = SO_2\uparrow + 2H_2O + Br_2\uparrow$$

A solution of hydrobromic acid is also formed if sulphur dioxide is bubbled through bromine under water:

$$SO_2 + 2H_2O + Br_2 = H_2SO_4 + 2HBr$$

PROPERTIES AND REACTIONS OF HYDROGEN BROMIDE: It is a colourless gas, b.p. −69°C, which is very soluble in water (600 volumes dissolve in 1 volume of water at s.t.p.), giving a strongly acidic solution. On heating this solution a constant boiling point mixture is obtained which boils at 126°C and contains 47.7% HBr. Chemically, hydrogen bromide is very similar to hydrogen chloride, but it has stronger reducing properties and is therefore oxidized to bromine by substances such as hot concentrated sulphuric acid and hydrogen peroxide solution, which have no action on hydrogen chloride. Similarly hydrobromic acid gradually turns yellow on standing in sunlight owing to oxidation to bromine:

$$4HBr + O_2 = 2H_2O + 2Br_2$$

25.11. Detection of Bromides. Most bromides are soluble in water, the common exceptions being those of silver, lead, mercury, and cuprous copper. With silver nitrate solution bromides give a pale yellow precipitate, which is insoluble in dilute nitric acid but dissolves to some extent in ammonia solution. When heated with concentrated sulphuric acid, bromides give red-brown fumes of bromine and white fumes of hydrogen bromide. If manganese dioxide is added the heated mixture gives bromine only, recognized by its colour, smell, and bleaching action. If a solution of a bromide is shaken with some chlorine water in the presence of a little chloroform, the latter turns yellow-brown as the liberated bromine dissolves in it.

25.12. Oxygen Compounds of Bromine. The oxides of bromine are very unstable and are prepared only with difficulty. Only two oxy-acids are known, hypobromous acid, HOBr, the salts of which are called hypobromites, and bromic acid, $HBrO_3$, which gives salts called bromates. In their preparations and properties they closely resemble the corresponding compounds of chlorine.

25.13. Iodine, I. Electronic configuration: 2,8,18,18,7.

OCCURRENCE: The element does not occur naturally, but sea-water contains a few parts per million of iodides. Certain seaweeds and sponges extract this iodine during growth and their ash contains up to 2% of iodine as iodides. The chief sources of the element, however, are sodium iodate, $NaIO_3$, which is present to the extent of about 0.2% in Chile saltpetre deposits, and certain oil-well brines, which contain appreciable quantities of dissolved iodide.

MANUFACTURE: Two methods are in use:

(1) *From the Chile deposits:* After crystallization of sodium nitrate, the mother liquor is treated with a calculated amount of sodium bisulphite solution, which reduces and acidifies the iodate:

$$NaIO_3 + 3NaHSO_3 = NaI + 3NaHSO_4$$

$$NaIO_3 + 5NaI + 6NaHSO_4 = 6Na_2SO_4 + 3I_2 + 3H_2O$$

Care must be taken not to add excess of bisulphite or the iodine will be converted into hydriodic acid. The iodine is filtered off, washed, dried, and sublimed.

(2) *From Sea-weed Ash:* The ash is mixed with hot water and the filtered solution evaporated to crystallize the less soluble salts, such as chlorides and sulphates. The iodides remaining in the mother liquor are then heated with concentrated sulphuric acid and manganese dioxide in iron pots, when iodine distils over and condenses. It is purified by sublimation.

LABORATORY PREPARATION: An iodide is heated with concentrated sulphuric acid and manganese dioxide, as in the manufacture described above:

$$2KI + 3H_2SO_4 + MnO_2 = 2KHSO_4 + MnSO_4 + I_2\uparrow + 2H_2O$$

The iodine which condenses is heated with potassium iodide, when pure iodine sublimes.

Very pure iodine can be made by heating pure iodine pentoxide to 300°C:

$$2I_2O_5 = 2I_2 + 5O_2$$

PROPERTIES OF IODINE: It is a lustrous black solid, with a density of 4.94 g cm^{-3}. It vaporizes slightly, even at room temperature, and it sublimes when heated, giving a violet vapour which is diatomic up to about 700°C; above this temperature dissociation into single atoms takes place. If heated rapidly it melts at 114°C and boils at 184°C. Its vapour is irritating and poisonous. Iodine is only sparingly soluble in water, giving a brown solution, but it dissolves readily in many organic solvents forming a brown solution if they contain oxygen (e.g. alcohol and ether) and a violet solution if they do not (e.g. carbon tetrachloride, chloroform, and carbon disulphide).

REACTIONS OF IODINE: It is not as reactive as the other halogens, but it combines directly with many elements, particularly when heated, forming iodides. The reaction with hydrogen is slow, reversible, and incomplete, even in the presence of a platinum catalyst:

$$H_2 + I_2 \rightleftharpoons 2HI$$

Iodine is oxidized to iodic acid by hypochlorites, chlorates, and concentrated nitric acid. It inflames in fluorine giving a pentafluoride, IF_5, and it combines with chlorine giving a monochloride, ICl, and a trichloride, ICl_3. It is also capable of forming salt-like compounds, e.g. a sulphate, $I_2(SO_4)_3$, and a nitrate, $I(NO_3)_3$, in which the iodine is trivalent.

Iodine dissolves readily in potassium iodide solution forming a deep red solution of potassium tri-iodide, KI_3. This reaction is reversible, the tri-iodide breaking up again to yield free iodine in the presence of any substance which reacts with iodine:

$$KI + I_2 \rightleftharpoons KI_3$$

Thus iodine dissolved in potassium iodide solution is a useful reagent in the laboratory, acting as a reservoir of free iodine in solution.

Iodine is a mild oxidizing agent, as the following examples show:

$$As_2O_3 + 2I_2 + 2H_2O \rightleftharpoons As_2O_5 + 4HI$$
$$Na_2SO_3 + I_2 + H_2O = Na_2SO_4 + 2HI$$

Iodine reacts with sodium thiosulphate solution forming sodium tetrathionate:

$$2Na_2S_2O_3 + I_2 = 2NaI + Na_2S_4O_6$$

This important reaction is used for standardizing solutions of iodine (which cannot be made accurately by direct weighing because of the volatility of the element) and also for estimating oxidizing agents, which liberate iodine from an acidified solution of an iodide.

Iodine reacts with alkali solutions in the cold forming an iodide and a hypoiodite, but the latter is unstable and decomposes into iodide and iodate even at room temperature. Although it is displaced from iodides by the other halogens, iodine will displace chlorine and bromine from their oxy-salts:

$$2KClO_3 + I_2 = 2KIO_3 + Cl_2$$

USES OF IODINE: It has powerful germicidal properties and a 2% solution in alcohol, known as 'tincture of iodine', is used as an antiseptic. Iodine, usually as iodides, is needed for making iodoform and many other organic compounds of pharmaceutical value. Iodine is an essential element for the healthy growth and fertility of animals, and iodides are often added to cattle foods or provided as cattle licks in necessary areas. Deficiency of iodine in the diet of humans results in the disease of goitre; to prevent this traces of iodides or iodates are added to table salt.

DETECTION OF IODINE: The element is readily recognized by its violet vapour. In solution iodine is detected, even in low concentration, by the dark blue colour it gives with starch and by its violet solution in chloroform.

25.14. Hydrogen Iodide, HI. This is prepared by adding water from a dropping funnel to a mixture of red phosphorus, sand and iodine in a flask. The apparatus is similar to that shown in Fig. 25.4. The hydrogen iodide is passed through a U-tube containing moist red phosphorus to remove any free iodine, and dried with anhydrous calcium iodide. It is collected by upward displacement of air:

$$2P + 3I_2 = 2PI_3$$
$$PI_3 + 3H_2O = H_3PO_3 + 3HI$$

Its solution can be obtained by dissolving the gas in water using an inverted funnel device, or by passing hydrogen sulphide into a suspension of iodine in water:

$$H_2S + I_2 = 2HI + S\downarrow$$

Where the presence of other ions is no disadvantage, the solution can be made simply by adding dilute hydrochloric or sulphuric acid to an iodide:

$$2KI + H_2SO_4 = K_2SO_4 + 2HI$$

PROPERTIES AND REACTIONS OF HYDROGEN IODIDE: It is very similar to hydrogen chloride and bromide in its physical and chemical properties. Like them it is very soluble in water giving a strongly acidic solution known as *hydriodic acid*. On distillation this yields a constant boiling point mixture which boils at 127°C and contains 57% HI.

Hydrogen iodide is less stable than the other hydrogen halides and decomposes when heated giving nascent hydrogen. This explains its use in organic chemistry to effect powerful reductions, usually in the presence of red phosphorus which reconverts the iodine formed into hydrogen iodide. It also explains why it cannot be prepared by heating an iodide with concentrated sulphuric acid because the latter oxidizes it to iodine, being reduced itself to sulphur dioxide, sulphur, and hydrogen sulphide.

Hydriodic acid is a much stronger reducing agent than hydrochloric or hydrobromic acids. It is oxidized to iodine by almost any oxidizing agent thus:

$$2HI + \bar{O} = I_2 + H_2O$$

This reaction is used to estimate oxidizing agents volumetrically. An excess of acidified potassium iodide is added to a known amount of the oxidizing agent and the liberated iodine is determined by titration against standard sodium thiosulphate solution using starch as indicator. The reaction can also be used as a means of standardizing a solution of sodium thiosulphate and as a general test for oxidizing agents. Even oxygen of the air liberates iodine slowly, causing hydriodic acid solution to turn brown on standing. Other examples of its reducing action are:

$$H_2O_2 + 2HI = I_2 + 2H_2O$$
$$2CuSO_4 + 4HI = 2CuI\downarrow + I_2 + 2H_2SO_4$$
$$HIO_3 + 5HI = 3I_2 + 3H_2O$$
$$2HNO_2 + 2HI = I_2 + 2H_2O + 2NO\uparrow$$
$$Cl_2 + 2HI = I_2 + 2HCl$$
$$2KMnO_4 + 3H_2SO_4 + 10HI = 5I_2 + K_2SO_4 + 2MnSO_4 + 8H_2O$$

25.15. Detection of Iodides. All the common iodides are soluble in water, except those of lead, silver, mercury, and cuprous copper. With silver nitrate solution iodides give a yellow precipitate of silver iodide, which is insoluble in dilute nitric acid and in ammonium hydroxide solution. All iodides give a violet vapour when heated with concentrated sulphuric acid, especially if manganese dioxide is added. If chlorine water is added to a solution of an iodide it displaces iodine, which turns some added chloroform a characteristic violet colour on shaking.

25.16. Oxygen Compounds of Iodine.

Iodine Pentoxide, I_2O_5: This is obtained as a white solid by heating iodic acid to 200°C:

$$2HIO_3 \rightleftharpoons I_2O_5 + H_2O$$

This change is reversible, iodine pentoxide dissolving readily in water to form iodic acid, of which it is therefore the anhydride. It decomposes into iodine and oxygen above 300°C. It is a strong oxidizing agent, converting carbon monoxide quantitatively into the dioxide on warming (see § 20.7):

$$5CO + I_2O_5 = 5CO_2 + I_2$$

Hypoiodous Acid, HIO: This acid and its salts, which are called *hypoiodites*, are less stable than the corresponding compounds of chlorine and bromine. They are prepared in similar ways, but they decompose into iodates and iodides in dilute solution, even at room temperature.

Iodic Acid, HIO_3: This is prepared by heating iodine with concentrated nitric acid to 200°C, when a residue of iodine pentoxide remains. If this is dissolved in a minimum of hot water, crystals of iodic acid are deposited on cooling:

$$I_2 + 10HNO_3 = 2HIO_3 + 4H_2O + 10NO_2\uparrow$$

It can also be made by passing chlorine through a suspension of iodine in water, or by acidifying an iodate:

$$I_2 + 6H_2O + 5Cl_2 = 2HIO_3 + 10HCl$$
$$NaIO_3 + H_2SO_4 = HIO_3 + NaHSO_4$$

It forms colourless crystals, m.p. 110°C, which are very soluble in water. It is a powerful oxidizing agent and a strong acid, giving salts called *iodates*:

$$HIO_3 \rightleftharpoons H^+ + IO_3^-$$

The iodates are more stable than bromates or chlorates. They can be prepared by adding iodine to hot solutions of alkalis, or by heating iodine with a concentrated solution of a chlorate:

$$3I_2 + 6KOH = KIO_3 + 5KI + 3H_2O$$
$$I_2 + 2KClO_3 = 2KIO_3 + Cl_2$$

Iodates act as oxidizing agents. They react with iodides in the presence of an acid to give free iodine:

$$KIO_3 + 5KI + 6H^+ = 3I_2 + 3H_2O + 6K^+$$

This is an important reaction which can be used to estimate acids volumetrically or to standardize a solution of sodium thiosulphate. It is the basis of the method used for extracting iodine from the naturally-occurring sodium iodate in Chile saltpetre.

Periodic Acid, $HIO_4.2H_2O$ is also known. Its salts are called *periodates*.

25.17. Summary. The halogens are a family of elements which show a marked resemblance to each other, making them one of the best defined groups in the periodic classification. They are all non-metals with poisonous, irritating, diatomic vapours. They all combine with hydrogen forming very soluble gases which give acidic solutions and form constant boiling point mixtures when distilled. The halogens are electronegative and very reactive; they combine with most other elements, forming stable electrovalent compounds with the most electro-positive metals and covalent compounds with the rest. In both cases the halogen atoms, all of which have seven electrons in their outermost shells, gain control of an additional electron originating from another atom and thereby acquire the electronic configuration of an inert gas.

Where the halogens differ, as in their physical properties, there tends to be a regular and gradual change with rise in atomic number. For example, melting points, boiling points, and densities rise steadily and there is a progressive deepening in colour (see Table 13.2).

The most important trend of all is the decrease in electronegativity as we go down the Group. Fluorine is the most electronegative element known, chlorine is strongly electronegative, whilst bromine and iodine are markedly less so. Indeed, in some of its compounds such as the salt-like nitrate, phosphate, and trichloride, iodine shows distinct signs of electropositive character. This trend produces many others. It accounts for the fall in the reactivity of the halogens as the atomic number rises and the decrease in their oxidizing power. In particular, it shows itself in the decreasing affinity of the halogens for hydrogen, which is apparent not only from the striking decrease in the vigour with which they combine with the element and the marked fall in the heat of formation of the hydrides, but also from the way the stability of the hydrides decreases and their reducing power increases as the molecular weight rises. The great difference in the reactivity of the various halogens with water and organic compounds is another aspect of the same trend, so is the greater stability of the oxy-compounds of iodine (e.g. the pentoxide and the iodates), compared with the corresponding compounds of the other halogens. It is also because of their greater tendency to exist as negative ions that the lighter halogens displace the heavier ones from their salts.

This variation in electronegativity arises from differences in the size of their atoms and in the magnitude of the charges on their nuclei. The atom of fluorine is very small and its nucleus exerts a strong attractive force upon electrons at the perimeter. As a result fluorine never gives up electrons to other atoms in chemical combination, but always the reverse, causing other atoms to surrender, or at least share, their outermost electrons. The atoms of the other halogens are larger, and despite their higher nuclear charge the electron pull at the circumference gets progressively less.

It has often been found that the lightest element in a Group differs noticeably from the others, and this is true here. Fluorine is invariably univalent in its covalent compounds, whereas the other halogens show variable valency. It forms stable compounds with other elements in which they display their maximum covalency (e.g. PF_5, SF_6, IF_7), and unlike the other halogens it does not form oxy-acids or oxy-salts. Indeed so electronegative is it that fluorine accepts electrons from metal atoms reluctant to part with them, forming electrovalent fluorides such as AlF_3 and HgF_2, whereas the other halides of these metals are covalent compounds. Fluorine differs from the other halogens in combining directly with carbon and nitrogen; nitrogen trifluoride, unlike the other trihalides of nitrogen, is a stable exothermic compound. Fluorinated organic compounds are not hydrolysed by water or alkalis, nor are some fluorine compounds of non-metals such as NF_3, CF_4, PF_5 and SF_6. Hydrogen fluoride differs from the other halogen hydrides in several respects. Owing to the electronegativity of fluorine, hydrogen fluoride is a highly polar molecule which associates by hydrogen bonding (§ 14.7) in the liquid and vapour states. This gives it a much higher melting point and boiling point and an abnormally high dielectric constant. It is a much weaker acid than the others and is capable of forming acid salts of the type KHF_2, presumably also as a result of hydrogen bonding. In contrast to the other halides, silver fluoride is soluble and calcium fluoride insoluble in water.

26

The Noble Gases

26.1. Introduction. The six elements listed in Table 26.1 are very similar physically and chemically and form the well-defined Group O in the Periodic Classification. As the name implies*, their principal characteristic is very low reactivity. Since 1963 various compounds of the noble gases have been prepared at low temperatures, but none of them are stable under normal laboratory conditions.† The noble gases are all monatomic, i.e. their molecules are single atoms. The forces which bind these atoms together in the liquid and solid states are weak ones of the van der Waals type (§ 4.9), which explains their great volatility.

TABLE. 26.1 THE NOBLE GASES

Name	*Symbol*	*Electronic configuration*	m.p./°C	b.p./°C	*Proportion in air (by volume)*
Helium	He	2	−272*	−269	1 : 200 000
Neon	Ne	2,8	−249	−246	1 : 55 000
Argon	Ar	2,8,8	−189	−186	1 : 107
Krypton	Kr	2,8,18,8	−157	−153	1 : 1 000 000
Xenon	Xe	2,8,18,18,8	−112	−108	1 : 11 000 000
Radon	Rn	2,8,18,32,18,8	− 71	− 62	—

* Under 2.5×10^6 N m^{-2} (25 atm) pressure.

26.2. Occurrence. As the Table shows, the noble gases occupy about 1% by volume of the atmosphere, argon being by far the most common. Certain natural spring waters contain small amounts of dissolved helium, neon and argon. Natural gases which issue from the ground in certain parts of the world, particularly in U.S.A., contain up to 2% of helium. Helium is also produced in small quantities during the radioactive decay of uranium and thorium and is expelled from minerals containing these elements by strong heating. Spectroscopy reveals that stars contain immense amounts of helium, the proportion increasing with the age of

* The elements were formerly known as the *inert gases* but they were renamed after the discovery of xenon compounds.

† The formation of clathrates is described in § 26.7.

the star; the thermonuclear fusion of hydrogen atoms into helium is regarded as the principal source of energy in these stars (§ 31.13).

26.3. Discovery. In 1892 Lord Rayleigh found that nitrogen obtained from the air by the removal of oxygen always had a slightly higher density than nitrogen prepared from chemical compounds. The difference was only about $\frac{1}{2}$%, but it was too great to be attributed to experimental error and an explanation of the discrepancy was sought. Ramsay, working in close consultation with Lord Rayleigh, heated magnesium in a sample of atmospheric nitrogen and showed that an inert residue remained which was responsible for the density difference. At first this residual gas was regarded as a single element and called argon, but later work using spectroscopy revealed the presence of small proportions of the other noble gases which were eventually isolated by fractional distillation of liquid air. Meanwhile Ramsay obtained a similar gas by treating the mineral cleveite with sulphuric acid and he identified its spectrum as one which had previously been observed in sunlight and had been ascribed to an unknown element helium.

26.4. Isolation. This depends upon two physical processes, fractional distillation and selective adsorption by charcoal. The first of these separates the elements on the basis of their differing volatility. Where the boiling points are close together complete separation is difficult to achieve by this means alone and use is also made of the second process. This takes advantage of the differing degrees to which gases are adsorbed by activated coconut charcoal cooled with liquid air or liquid hydrogen, adsorption being greatest for elements of highest atomic weight. Neon, argon, krypton and xenon are obtained industrially by liquefying air, distilling it, and submitting the resulting fractions to successive adsorption and desorption at various temperatures, giving complete separation. The Haber process (§ 22.4) also provides argon.

The main source of helium is natural gas, which consists predominantly of hydrocarbons and nitrogen. These are liquefied by cooling under pressure and the residual helium purified by passing it over activated charcoal cooled with liquid air, which adsorbs traces of heavier gases, leaving the helium unaffected.

Radon is obtained by allowing radium or one of its salts to stand for a period of several weeks in a sealed vessel. Each gramme of radium is then in equilibrium with about $\frac{1}{2000}$ cm^3 of radon at s.t.p., the rate of radioactive decay of this amount being just equal to its rate of release.

26.5. Properties and Uses.

HELIUM: Points of particular interest are its low density (second only to hydrogen), its low solubility in water, its great volatility, and the remarkable properties such as high conductivity and very low viscosity which it displays at very low temperatures. Helium is used in balloons and airships, where it is preferred to hvdrogen on account of its non-inflammability. Mixed with oxygen, it is supplied instead of air to divers who work in pressurized suits; helium is less soluble in the blood than nitrogen and diffuses out more quickly through the lungs, which makes it safe to bring the diver to the surface after only short periods of de-compression. A major use of the gas is to produce an inert atmosphere during the welding of metals such as aluminium and titanium; sur-rounding the weld with helium prevents these reactive metals from undergoing atmospheric oxidation whilst hot. Helium is also used as a coolant in some nuclear reactors (§ 31.11), and liquid helium is impor-tant in cryogenics.

NEON: When submitted to an electrical discharge at low pressure it emits a bright orange-red glow. This characteristic accounts for its widespread use in advertising signs. Neon is also used in electronic equipment such as voltage stabilizers and time switches and in cryogenics.

ARGON: Its principal use is for filling metal filament electric lamps, when it is usually mixed with about 7% of nitrogen to prevent arcing. Gas-filled lamps are more efficient than evacuated ones because the gas suppresses the evaporation of the filament which normally occurs during use and thus makes it possible to run the filament at a higher temperature without shortening its life. Argon is chosen because of its stability, inertness, and low thermal conductivity. It is also used in discharge lamps, where its function is to initiate the discharge in the mercury vapour or sodium vapour inside the lamp, and for producing completely inert atmospheres for certain metallurgical processes parti-cularly the arc welding of stainless steel, titanium, and aluminium. Argon is often used as the carrier in gas chromatography (§ 12.15).

KRYPTON AND XENON: Few uses have yet been found for these gases, but they are being employed to some extent in gas-filled filament lamps in place of argon because their lower thermal conductivity confers an even greater efficiency. A discharge lamp containing xenon is used for film projectors. In view of their high absorption of X-rays, it has been suggested that they might be used to facilitate X-ray examination of the lungs.

RADON: This gas is highly radio-active, having a half-life of less than four days. It is used in the treatment of cancer, a capsule of the gas being embedded in the tumour to irradiate it.

26.6. Theoretical Significance. With the discovery of the first noble gases it was realized that there existed a hitherto unsuspected group in the Periodic Classification, later called Group O, the members of which fell between the halogens and the alkali metals when the elements were arranged in order of ascending atomic number. Paradoxically these elements, which form no compounds themselves, have provided the essential clue to an understanding of valency and chemical combination. It was noticed that all the noble gases possessed outermost electron shells containing the maximum number of electrons, and this condition was inevitably associated with their inertness. From this start was developed the electronic theory of valency (§ 14.2), the main postulate of which is that in chemical combination elements tend to acquire the electronic configurations of noble gases by the transfer or sharing of electrons.

Helium provides two further points of theoretical importance. Its discovery on this planet dispelled the belief that an element existed in the sun which was not present on earth, and its exceptional behaviour as a liquid has been of great theoretical significance in low temperature physics.

26.7. Clathrates. When quinol, $C_6H_4(OH)_2$, which is a substance like ice with a very open crystal structure as a result of extensive hydrogen bonding (14.7), is crystallized from solution in the presence of argon, krypton or xenon under high pressure, or of other gases with molecules of comparable size (e.g. hydrogen sulphide or carbon dioxide), then molecules of these gases tend to be trapped in cavities in the quinol crystal lattice (§ 4.2), giving a 'compound' of varying composition depending upon the number of spaces filled. Such a product is called a *clathrate* to distinguish it from a compound in which ordinary valency forces are at work. The enclosed gas is released again when the crystals are melted or dissolved.

26.8. Xenon Tetrafluoride, XeF_4. The noble gases, particularly the lowest members, have very high ionization energies. Because of this only the larger atoms form stable compounds and then only with the most electronegative elements such as fluorine and oxygen. When xenon and fluorine are heated together for several hours under pressure, for example, the two gases combine directly to form compounds such as the difluoride, XeF_2, the tetrafluoride, XeF_4, and the hexafluoride, XeF_6.

The most important of these compounds is *xenon tetrafluoride*, formed by heating a 1 : 5 mixture of xenon and fluorine in a nickel can at 400°C and 9×10^5 N m^{-2} (9 atm) pressure. White crystals of the tetrafluoride are deposited on rapid cooling and can be stored in a vacuum in a nickel vessel. They melt at 114°C and are very susceptible to hydrolysis by moist air, but they can be heated up to 400°C without decomposing.

In xenon tetrafluoride there are four covalent bonds and two lone pairs of electrons, since xenon has the configuration 2, 8, 18, 18, 8. The shape of the XeF_4 molecule is dictated by the tendency of these two lone pairs to arrange themselves in space as far from each other as possible owing to their strong mutual repulsion. One would expect the lone pairs to be at diammetrically opposite ends of an octahedron and the fluorine atoms to lie in one plane with the Xe—F bonds at right angles to each other as shown in Fig. 26.1. X-ray analysis confirms this shape and shows the Xe—F bond length to be 0.195 nm.

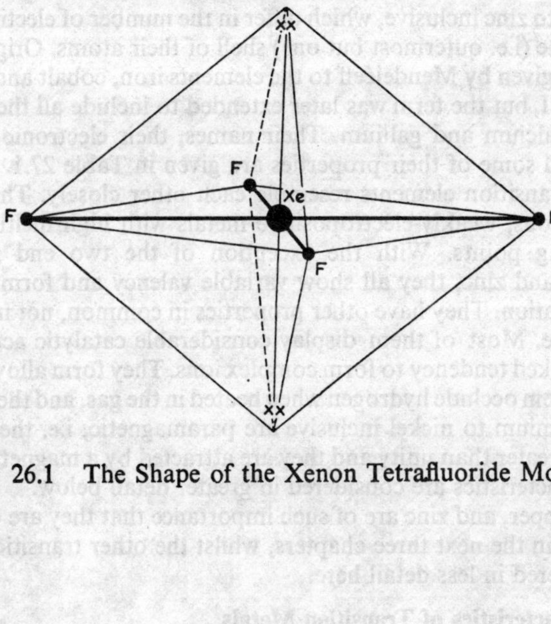

FIG. 26.1 The Shape of the Xenon Tetrafluoride Molecule

27

The Transition Metals

27.1. Introduction. As explained in § 13.8, the first series of transition metals consists of ten successive elements in the fourth period, from scandium to zinc inclusive, which differ in the number of electrons in the penultimate (i.e. outermost but one) shell of their atoms. Originally the name was given by Mendeléeff to the elements iron, cobalt and nickel in Group VIII, but the term was later extended to include all the elements between calcium and gallium. Their names, their electronic configurations, and some of their properties are given in Table 27.1.

These transition elements resemble each other closely. They are all hard, lustrous, weakly-electropositive metals with high melting points and boiling points. With the exception of the two end members, scandium and zinc, they all show variable valency and form coloured ions in solution. They have other properties in common, not mentioned in the table. Most of them display considerable catalytic activity and have a marked tendency to form complex ions. They form alloys readily, many of them occlude hydrogen when heated in the gas, and the elements from chromium to nickel inclusive are paramagnetic, i.e. their permeability is greater than unity and they are attracted by a magnet. Some of these characteristics are considered in greater detail below.

Iron, copper, and zinc are of such importance that they are described separately in the next three chapters, whilst the other transition metals are considered in less detail here.

27.2. Characteristics of Transition Metals

VARIABLE VALENCY: As Table 27.1 shows, this is most marked in the centre of the series. All the transition metals except scandium are capable of divalency, forming doubly-charged positive ions by the loss of two electrons. In addition, some of them can form trivalent cations by losing three electrons, and titanium, vanadium, chromium, and manganese can exert higher valencies by forming covalent compounds. In these cases, one or more electrons from the penultimate shell are used for valency purposes, in addition to the outermost electrons. For

TABLE 27.1. PROPERTIES OF THE TRANSITION METALS OF THE FIRST SERIES

Property / *Element*	Sc	Ti	V	Cr	Mn	Fe	Co	Ni	Cu	Zn
Atomic Number	21	22	23	24	25	26	27	28	29	30
Electronic Configuration	2,8,9,2	2,8,10,2	2,8,11,2	2,8,13,1,	2,8,13,2	2,8,14,2	2,8,15,2	2,8,16,2	2,8,18,1	2,8,18,2
Melting point/°C	1400	1720	1710	1900	1250	1530	1490	1450	1083	419
Boiling point/°C	c. 3800?	c. 3200	c. 3500	c. 2480	c. 2000	2730	c. 2900	2840	2350	907
Density/g cm^{-3}	3.1	4.5	6.1	7.1	7.2	7.9	8.7	8.9	8.9	7.1
Colour of M^{2+} in soln.	—	brown	purple	blue	pink	green	pink	green	blue	none
Valencies (commonest in bold)	3	2,3,4	2,3,4,5	2,3,6	2,3,4,6,7	2,3	2,3	2,3	1,2	2
Oxides	Sc_2O_3	TiO Ti_2O_3 TiO_2	VO V_2O_3 VO_2 V_2O_5	CrO Cr_2O_3 CrO_3	MnO Mn_2O_3 MnO_2 MnO_3 Mn_2O_7	FeO Fe_2O_3	CoO Co_2O_3	NiO Ni_2O_3	Cu_2O CuO	ZnO

example, when iron (2,8,14,2) is in the ferric condition, its atom has lost the two electrons from its outermost shell and also one from the third shell, so that the ferric ion has the configuration 2,8,13. Similarly, manganese (2,8,13,2) can form covalent linkages involving from one to five of the electrons in its penultimate shell, as well as the two in the outermost shell, culminating in heptavalent manganese (2,8,18,4) in the oxide Mn_2O_7 and in the permanganate ion, MnO_4^-.

This variable valency arises from the fact that only small differences exist between the various energy levels of the electrons in the two outer shells of transition metal atoms. Thus varying numbers of these electrons may be involved in forming chemical bonds. This effect is only commonly found in atoms in which the penultimate shell is incompletely filled. When this shell contains eighteen electrons, as it does in zinc and the succeeding elements, variable valency is not so frequently encountered, nor is it found in those elements such as potassium and calcium in which the penultimate shell contains the 'temporary maximum' of eight electrons because in these cases the inner electrons are all retained very firmly by the nucleus.

COLOURED IONS IN SOLUTION: Compounds of the transition elements are often recognized by their characteristic colours. With the exception of zinc, not only do solutions of their divalent cations show the colours listed in Table 27.1, but covalent compounds and anions containing the elements (e.g. chromates and permanganates) are also highly coloured. This contrasts with the absence of colour in compounds of most non-transitional metals.

The explanation again lies in the small differences in energy levels between electron paths in the outer and penultimate shells. When light energy is absorbed by an atom, an electron is 'excited' or temporarily raised to a higher energy level, but it soon falls back to its stable condition of lowest energy, emitting light energy in the process. The wavelength of this light corresponds to the difference in energy levels occupied by the electron; in the case of ions of the transition metals it falls in the waveband of visible light, whereas for most other metals it is shorter and lies in the ultra-violet range.

CATALYTIC ACTIVITY: The transition metals are highly active as catalysts. For example, nickel catalyses a wide range of hydrogenations in organic chemistry, iron is used in the important Haber and Bosch processes, and copper acts as a catalyst in dehydrogenations. Compounds of these elements are active as well as the metals themselves; vanadium pentoxide is used in the contact and other processes, manganese dioxide catalyses the decomposition of hydrogen peroxide solution and also of potassium chlorate, cobalt ions greatly accelerate the decomposition of hypochlorites, and zinc chromite is used in the synthesis of methyl alcohol. Many other examples are given in this book and it is an interesting exercise for the student to make a collection of them.

There appears to be no agreed single explanation of this catalytic activity. For the metals it almost certainly involves *adsorption* of the reacting gases upon their surface, which explains why they are usually finely divided to expose the maximum area, but the activity of compounds of the transition metals seems more likely to arise from the formation of unstable intermediate compounds, perhaps as a result of variable valency.

COMPLEX ION FORMATION: This is one of the most notable characteristics of the transition metals. Their simple ions are often not very stable, but by sharing electrons provided by other ions or molecules through co-ordinate bonds, complex ions of considerable stability can be formed, even though the penultimate shell may still be incompletely filled. Many examples are provided by the elements chromium, iron, cobalt, nickel, copper, and zinc, particularly in their ammines and hydrates and in their anions involving the cyanide and chloride ions.

27.3. Titanium, Ti. Electronic configuration: 2,8,10,2

OCCURRENCE: The only important ores are rutile, TiO_2, and ilmenite, $FeTiO_3$, which are so widely distributed that titanium is one of the commonest metals in the earth's crust.

EXTRACTION: This presents considerable difficulty owing to the great affinity of the molten metal for electronegative elements such as carbon, nitrogen, and oxygen. Two processes are in use:

1. By Chemical Reduction: Titanium tetrachloride (made by heating the dioxide with carbon and chlorine) is heated to 850°C with magnesium or sodium in an atmosphere of argon:

$$TiO_2 + 2C + 2Cl_2 = TiCl_4 + 2CO$$
$$TiCl_4 + 2Mg = Ti + 2MgCl_2$$
$$TiCl_4 + 4Na = Ti + 4NaCl$$

When heated strongly in an electric furnace the by-products vaporize leaving molten titanium, which solidifies on to a rod of titanium used as one pole of the arc.

2. By the Filament Growth Process: Very pure titanium is best made by heating the impure metal with iodine in an evacuated vessel (see Fig. 27.1). The titanium tetraiodide vapour formed decomposes thermally when it comes near the tungsten filament, which is kept at about 1800°C by an electric current. Pure titanium is deposited on the filament, which slowly grows in thickness, and the iodine vapour combines with more of the impure metal. The method depends upon the volatility of the tetraiodide and its low decomposition temperature compared with the boiling point of titanium.

$$TiI_4 \rightleftharpoons Ti + 2I_2$$

PROPERTIES OF TITANIUM: It is a hard, lustrous, grey metal melting at about 1720°C. Its density is 4.5 g cm^{-3}. When pure it is ductile even at

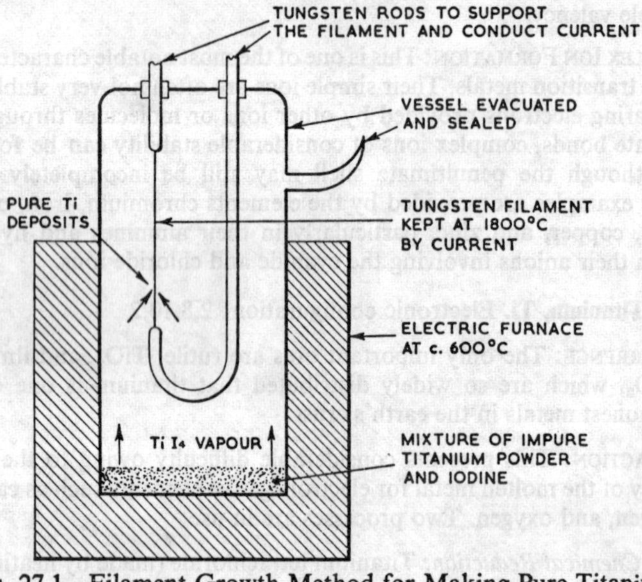

FIG. 27.1 Filament Growth Method for Making Pure Titanium

room temperature. Acids and alkalis have little action on titanium at room temperature but it reacts readily with the halogens forming a tetrahalide when heated and it burns vigorously in air or nitrogen above 1000°C.

USES OF TITANIUM: Its alloy with iron, called ferrotitanium, is added to molten steel to remove dissolved gases. Titanium and its alloys are of growing importance in the construction of aircraft and missiles where its high melting point, great resistance to corrosion, and low density are valuable properties. Its remarkable resistance to attack by sea-water is leading to many marine uses. Another recent use is in surgery for pinning together broken bones or repairing damaged skulls; the light but strong pieces of titanium remain uncorroded in the body throughout the lifetime of the patient.

27.4. Compounds of Titanium. The dioxide, TiO_2, is by far the most important. It is a stable white solid used as a pigment in paints, enamels, and face-powders, and as a filler for paper, soap, rubber, and linoleum. Its widespread use in paints in preference to compounds of lead, barium, and zinc springs from the superior opacity and covering power of titanium dioxide and from its non-toxicity, stability, and cheapness. Titanium carbide is so hard that it is used for making cutting tools and abrasives.

Many titanium salts are readily hydrolysed by water. For example, the tetrachloride is a colourless liquid, b.p. 136°, which fumes in moist air and is used, therefore, in smoke bombs and for sky-writing:

$$TiCl_4 + 2H_2O = TiO_2 + 4HCl$$

It is also an important catalyst in the Ziegler process (see § 34.2.).

Salts of tetravalent titanium are reduced to purple solutions containing the trivalent ion by zinc and hydrochloric acid. Such solutions are powerful reducing agents and can be used to estimate oxidizing agents volumetrically. They must be stored in an inert atmosphere, because they are oxidized by air.

Titanium salts in solution give an intense yellow or orange colour when added to hydrogen peroxide solution (see § 16.6).

27.5. Vanadium, V. Electronic configuration: 2,8,11,2.

OCCURRENCE: The principal minerals are the sulphide, patronite, and the complex uranyl vanadate, carnotite, which is also an important source of uranium.

EXTRACTION: The metal is usually obtained as an alloy with iron by heating vanadium–iron slags with powdered aluminium as in the thermite reaction (§ 19.7). Pure vanadium can be made by the filament growth method.

PROPERTIES AND REACTIONS OF VANADIUM: It is a very hard, grey metal, m.p. 1710°C, ρ 6.1 g cm^{-3}. When heated strongly in oxygen it burns to the pentoxide:

$$4V + 5O_2 = 2V_2O_5$$

Vanadium is unaffected at room temperature by air, water, alkalis, or non-oxidizing acids, but it dissolves in concentrated nitric acid and aqua regia. At high temperatures it combines directly with nitrogen, carbon, and silicon.

USES OF VANADIUM: It is mostly used for making steel alloys which are renowned for their great hardness and strength even when the proportion of vanadium is as low as 0.1%. Vanadium is also used for making its pentoxide which is an important catalyst, e.g. in the contact process for making sulphuric acid (§ 24.11), and in the oxidation of naphthalene to phthalic anhydride (§ 43.2).

27.6. Compounds of Vanadium. The metal exerts valencies of 2,3,4, and 5 in its principal compounds. In its lower valency states vanadium forms highly coloured ions (e.g. greens, purples, and blues) and is a strong reducing agent. It is tetravalent in the vanadyl or vanadium(IV) salts and hypovanadates, and pentavalent in the vanadates and in the pentoxide, which is an orange-red solid, m.p. about 660°C, made by strongly heating ammonium vanadate:

$$2NH_4VO_3 = V_2O_5 + 2NH_3\uparrow + H_2O$$

The pentoxide is only slightly soluble in water, but it dissolves readily in alkalis giving solutions of vanadates.

27.7. Chromium, Cr. Electronic configuration: 2,8,13,1.

OCCURRENCE: The metal does not occur naturally and the only important ore is chromite, a double oxide of chromium and iron having the composition, $FeO.Cr_2O_3$.

EXTRACTION: Two methods are in use:

1. If pure chromium is needed, chromite is first fused with an alkali in the presence of air, as described in § 17.25. The resulting dichromate, which crystallizes out from a concentrated solution, is reduced to pure chromic oxide by heating with carbon:

$$Na_2Cr_2O_7 + 2C = Cr_2O_3 + Na_2CO_3 + CO\uparrow$$

This oxide is then reduced by the highly exothermic Goldschmidt or thermite process (§ 19.8) by igniting a mixture of it with aluminium powder:

$$Cr_2O_3 + 2Al = Al_2O_3 + 2Cr$$

2. If pure chromium is not required, but only a ferro-chromium alloy for steelmaking, chromite can be reduced directly by heating it with carbon in an electric furnace:

$$FeCr_2O_4 + 4C = Fe + 2Cr + 4CO\uparrow$$

PROPERTIES AND REACTIONS OF CHROMIUM: It is a hard, white, lustrous metal with a remarkable resistance to atmospheric corrosion which is probably due to an impervious coating of oxide on its surface. The metal dissolves in dilute hydrochloric and sulphuric acids, but not in nitric acid and aqua regia. After immersion in these oxidizing acids, however, chromium is rendered *passive* and is not then attacked by dilute hydrochloric and sulphuric acids even on boiling. It reacts with halogens, sulphur, nitrogen, and carbon when heated.

USES OF CHROMIUM: The metal is used to make special steels which are renowned for their hardness, strength, and corrosion resistance. Most stainless steels contain between 12% and 18% of chromium. An alloy of chromium, nickel, and iron is used to make the heating element in furnaces and electric fires. Chromium plating is extensively used for cars, bicycles, etc., because of its hardness, high lustre, and resistance to corrosion.

27.8. Compounds of Chromium. The element is divalent, trivalent, and hexavalent. Chromium is divalent in the *chromous* or *chromium(II)* salts, which are powerful reducing agents readily oxidized to chromic compounds even by the air. The most important is *chromous chloride*, $CrCl_2$, which is prepared as a blue solution by reducing chromic chloride solution with zinc and dilute hydrochloric acid in the absence of air.

Trivalent chromium is more common, occurring in the following compounds:

Chromic oxide, chromium(III) oxide, Cr_2O_3, is a green, insoluble solid usually made by heating a dichromate with carbon or by heating ammonium dichromate alone:

$$(NH_4)_2Cr_2O_7 = Cr_2O_3 + N_2\uparrow + 4H_2O$$

The hydrated oxide is amphoteric, dissolving in acids to form *chromic* salts and in concentrated solutions of alkalis forming *chromites*:

$$Cr_2O_3 + 3H_2SO_4 = Cr_2(SO_4)_3 + 3H_2O$$
$$Cr_2O_3 + 2NaOH = 2NaCrO_2 + H_2O$$

Chromic chloride, chromium(III) chloride, $CrCl_3$, is made in the anhydrous form as purple crystals by passing dry chlorine over the heated metal or over a hot mixture of chromic oxide and carbon:

$$2Cr + 3Cl_2 = 2CrCl_3$$
$$Cr_2O_3 + 3C + 3Cl_2 = 2CrCl_3 + 3CO\uparrow$$

Several crystalline hexahydrates can be prepared by adding hot concentrated hydrochloric acid to chromic oxide and evaporating the solution. They are of interest because although they all possess the formula $CrCl_3.6H_2O$, they differ in colour and in structure, as their reactions with silver nitrate solution show:

$[Cr(H_2O)_6]^{3+}3Cl^-$	violet crystals	all the chlorine precipitated at once as silver chloride.
$[Cr(H_2O)_5Cl]^{2+}2Cl^-.H_2O$	bright green crystals	two-thirds of the chlorine precipitated at once as silver chloride.
$[Cr(H_2O)_4Cl_2]^+Cl^-.2H_2O$	dark green crystals	only one-third of the chlorine precipitated at once as silver chloride.

Several hydrates of *chromic sulphate*, $Cr_2(SO_4)_3$, are known, of which the violet ($18H_2O$) and the green ($3H_2O$) are the most important. By mixing equimolecular solutions of the violet salt and potassium sulphate, it is possible to obtain chrome alum, $K_2SO_4.Cr_2(SO_4)_3.24H_2O$ (see § 19.13) as violet octahedral crystals. This compound is important in tanning leather and as a mordant in dyeing.

The chromic ion, Cr^{3+}, readily forms very stable complexes with ammonia. The *ammines*, as they are called, bear a triple positive charge because the co-ordinated molecules of ammonia are themselves uncharged, e.g. $[Cr.(NH_3)_6]^{3+}$.

Chromium is hexavalent in the *trioxide*, *chromium(VI) oxide*, CrO_3, which is formed by adding excess of concentrated sulphuric acid to a concentrated solution of a dichromate:

$$K_2Cr_2O_7 + H_2SO_4 = K_2SO_4 + 2CrO_3 + H_2O$$

The dark red crystals are deliquescent and very soluble in water, giving a solution of *chromic acid*, H_2CrO_4. Chromium trioxide is a very powerful oxidizing agent. It ignites many organic materials (e.g. alcohol and paper) and it reacts with alkalis giving chromates and dichromates.

Chromates and Dichromates: Most chromates are yellow (silver chromate is an exception), whereas dichromates are orange-red. The chromates of the alkali metals are soluble in water but those of lead, barium, and silver are insoluble and are often used to confirm the presence of these metals in analysis. Most dichromates dissolve readily in water, and some, such as those of sodium and calcium, are deliquescent. Chromate and dichromate ions are related by the equilibrium:

$$2CrO_4^{2-} + 2H^+ \rightleftharpoons Cr_2O_7^{2-} + H_2O$$

Thus when a solution of a chromate is acidified this equilibrium is driven to the right by the high concentration of hydrogen ions and conversion into dichromate occurs. Both chromates and dichromates are powerful oxidizing agents. They are used extensively industrially for tanning leather and for chromium plating, and also as pigments and as starting materials for making chromic compounds. Potassium dichromate is described in § 17.25.

27.9. Detection of Chromium. Solutions of chromic salts give a greenish-blue precipitate of chromic hydroxide when treated with ammonium hydroxide solution:

$$Cr_2(SO_4)_3 + 6NH_4OH = 2Cr(OH)_3\downarrow + 3(NH_4)_2SO_4$$

This precipitate dissolves in hot caustic soda solution and is converted to a yellow chromate by adding sodium peroxide. Chromic salts give a green coloration in the borax bead test (§ 19.4).

Chromates and dichromates are usually recognized by their colour and by their yellow precipitates with lead acetate solution and barium chloride solution. They can be detected in dilute solution by adding dilute sulphuric acid and hydrogen peroxide solution in the presence of a little ether, when a deep blue colour appears in the ether layer.

TABLE 27.2. OXIDATION STATES OF VANADIUM, CHROMIUM, AND MANGANESE

+5	V_2O_5	vanadates	
+4	VO_2	hypovanadates	vanadyl or vanadium(IV) salts
+3	V_2O_3		vanadic or vanadium(III) salts
+2	VO		vanadous or vanadium(II) salts
+1			
0	V		
+6	CrO_3	chromates and dichromates	
+5			
+4			
+3	Cr_2O_3	chromites	chromic or chromium(III) salts
+2	CrO		chromous or chromium(II) salts
+1			
0	Cr		
+7	Mn_2O_7	permanganates	
+6	MnO_3	manganates	
+5			
+4	MnO_2	manganites	
+3	Mn_2O_3		manganic or manganese(III) salts
+2	MnO		manganous or manganese(II) salts
+1			
0	Mn		

27.10. Manganese, Mn. Electronic configuration: 2,8,13,2.

OCCURRENCE: The most important ore is pyrolusite, MnO_2.

EXTRACTION: Pyrolusite is first heated strongly to convert it into mangano-manganic oxide, which is then mixed with aluminium powder and ignited, when it is reduced to the metal by the Goldschmidt or thermite reaction (§ 19.8):

$$3MnO_2 = Mn_3O_4 + O_2\uparrow$$
$$3Mn_3O_4 + 8Al = 9Mn + 4Al_2O_3$$

When manganese is required for steelmaking, however, it is usually obtained as ferro-manganese by heating a mixture of manganese oxide and iron oxide with carbon.

PROPERTIES AND REACTIONS OF MANGANESE: It is a hard, brittle metal, similar to iron in appearance. It reacts with hot water and dissolves readily in dilute acids, liberating hydrogen and forming manganous salts:

e.g. $$Mn + 2HCl = MnCl_2 + H_2\uparrow$$

It combines directly with the halogens, sulphur, nitrogen, phosphorus, carbon, and silicon when heated with these elements.

USES OF MANGANESE: By far the biggest use is as a deoxidant for iron and steel. The manganese, usually in the form of ferro-manganese alloys, is added to the molten steel to increase its tensile strength and, by removing dissolved oxygen, prevent the occurrence of blow-holes in the solidified metal. Manganese is also used to make alloys with many metals; steels containing about 10% of manganese show great hardness and are used for making armour plate, steel-helmets, and safes. Manganese is an important constituent of soil; minute traces of it are essential for healthy plant growth.

27.11. Compounds of Manganese. The element exerts various valencies, of which two, four, and seven are the most important. It forms divalent ions, Mn^{2+}, in *manganous* or *manganese(II) salts*, e.g. $MnCl_2$ and $MnSO_4$, and trivalent ions, Mn^{3+}, in the less stable *manganic* or *manganese(III) salts*, e.g. $Mn_2(SO_4)_3$. The corresponding oxides, manganous oxide MnO, and manganese sesquioxide, Mn_2O_3, are basic oxides. Other oxides are the compound oxide mangano-manganic oxide Mn_3O_4, which is formed by heating any oxide of manganese to 100°C in air, the unstable manganese heptoxide Mn_2O_7, which is produced by the action of concentrated sulphuric acid on potassium permanganate, and manganese dioxide MnO_2, which, in view of its importance, is described in detail in the next section. Manganese is hexavalent in the *manganates*, which contain the MnO_4^{2-} ion, and heptavalent in the *permanganates*, which contain the MnO_4^- ion. The most important of these is potassium permanganate, which is described in § 17.24.

27.12. Manganese Dioxide, Manganese(IV) Oxide, MnO_2. This occurs naturally as pyrolusite. It is a black solid, insoluble in water, with amphoteric properties. Thus when hydrated it dissolves in strong acids forming salts, e.g. $Mn_2(SO_4)_3$, and in alkalis giving manganites, e.g. $CaMnO_3$. More important are its strong oxidizing properties. It converts hot concentrated hydrochloric acid to chlorine and it liberates oxygen when heated with concentrated sulphuric acid:

$$MnO_2 + 4HCl = MnCl_2 + Cl_2\uparrow + 2H_2O$$
$$2MnO_2 + 2H_2SO_4 = 2MnSO_4 + O_2\uparrow + 2H_2O$$

Its use as a depolarizer in Leclanché cells and dry batteries (see § 10.6) is based upon its oxidizing power. When heated strongly it decomposes:

$$3MnO_2 = Mn_3O_4 + O_2\uparrow$$

Manganese dioxide is an important catalyst for various reactions, particularly the decomposition of hydrogen peroxide, the thermal decomposition of potassium chlorate, and the atmospheric oxidation of linseed oil. This last reaction takes place during the drying and hardening of paints and varnishes and accounts for the extensive use of manganese dioxide as a drier. Another use is for decolorizing glass which often tends to have a yellow colour owing to the presence of traces of iron salts. The dioxide confers a purple shade to counteract this and make the glass appear colourless. The dioxide is also used to make permanganates.

27.13. Detection of Manganese. When treated with ammonium sulphide, solutions of manganous salts give a buff precipitate of manganese sulphide, soluble in acetic acid (distinction from zinc.):

$$MnCl_2 + (NH_4)_2S = MnS\downarrow + 2NH_4Cl$$

Sodium hydroxide solution gives a white precipitate of manganese hydroxide:

$$MnCl_2 + 2NaOH = Mn(OH)_2\downarrow + 2NaCl$$

Unlike zinc hydroxide, this precipitate is insoluble in excess of alkali and turns brown on exposure to air.

A characteristic purple colour (due to the permanganate ion) is obtained whenever sodium bismuthate is added to a solution of a manganese salt in dilute nitric acid.

Permanganates are recognized by the purple colour of their solutions and by their oxidizing action on hydrogen peroxide, oxalates (at 70°C), and concentrated hydrochloric acid. Manganese compounds give a purple colour in the borax bead test with an oxidizing flame (§ 19.4).

27.14. Cobalt, Co. Electronic configuration: 2,8,15,2.

OCCURRENCE: The most important ores are smaltite $CoAs_2$, and cobaltite or cobalt glance CoAsS, which are usually found associated with ores of nickel and copper.

EXTRACTION: The cobalt ores are roasted in air giving a mixture of oxides and arsenates. These are dissolved in hydrochloric acid and treated with hydrogen sulphide to precipitate copper, lead, and bismuth. Calcium carbonate is added to precipitate arsenic and iron. The addition of a calculated quantity of bleaching powder precipitates hydrated cobalt oxide, which is reduced to the metal by heating it in hydrogen.

PROPERTIES AND REACTIONS OF COBALT: It is a very hard, ferromagnetic metal resembling iron in appearance. At room temperature it is not attacked by air, water, or alkalis and only slowly by dilute hydrochloric or sulphuric acids, but dilute nitric acid dissolves it readily. Concentrated nitric acid renders it passive, as it does iron. Cobalt combines directly with most electronegative elements when heated.

USES OF COBALT: It is principally used for making alloys which retain their extreme hardness at high temperatures and which are therefore suitable for high-speed cutting tools. Steel containing about 35% of cobalt is used for making permanent magnets. Cobalt appears to be an essential trace element in the diet of animals, particularly sheep, although the amount required is extremely small. Cobalt atoms are present in vitamin B_{12}.

27.15. Compounds of Cobalt. The element shows divalency and trivalency. It is divalent in the stable *cobaltous* or *cobalt(II) salts* which contain the Co^{2+} ion. They are mostly soluble in water, giving pink solutions. The hydrated salts are usually pink or red, e.g. $CoCl_2.6H_2O$ and $CoSO_4.7H_2O$, but they turn blue when their water of crystallization is driven off by heating. Cobaltous salts are used as paint, varnish, and ink driers, as catalysts in petroleum refining, and for colouring ceramics and glass a deep blue.

Simple *cobaltic* or *cobalt(III) salts* containing the Co^{3+} ion are unstable, but many stable complexes are formed of which the hexa-amminocobalt(III) ion, e.g. $[Co(NH_3)_6]Cl_3$, and hexanitrocobaltate(III) ion, e.g. $Na_3[Co(NO_2)_6]$, are the best known. The latter is used as a test for potassium salts (see § 17.26), since the potassium compound is only sparingly soluble in cold water.

Cobalt provides a good example of the readiness of transition metal ions to exchange their ligands in complex ions according to their environment. The hydrated cobalt(II) ion has six water molecules linked by co-ordinate bonds in an octahedral arrangement around the central divalent ion giving a complex ion with a double positive charge overall, $[Co(H_2O)_6]^{2+}$. It is this ion which confers the characteristic pink colour on the hydrated crystals of the chloride, nitrate, and sulphate, and on aqueous solutions of cobalt(II) salts. If such a solution is treated with a high concentration of Cl^- ions (by adding hydrochloric acid, for example, or potassium chloride solution), a tetrahedrally co-ordinated tetrachlorocobaltate(II) anion is formed with a blue colour and the formula $[CoCl_4]^{2-}$. The movement of the blue coloration towards the anode during electrolysis demonstrates conclusively that an anion is formed in this case. The same change can be brought about by concentrating a solution of cobalt(II) chloride, the solution changing from pink to blue, or by heating cobalt chloride paper to drive off the water ligands as steam.

On the other hand, if an excess of ammonium hydroxide solution is added to an aqueous solution of a cobalt(II) salt and air is bubbled through the solution, oxidation occurs and successive ammonia molecules are substituted for the water ligands giving the six co-ordinated complex hexa-amminocobalt(III) cation, $[Co(NH_3)_6]^{3+}$ with a characteristic yellow colour. This complex ion, formerly known as a cobalt-ammine, is very stable because in it the central cobalt atom has the same electronic structure as krypton (2, 8, 18, 8) with all its orbitals fully occupied and all the electrons paired.

These changes can be summarized thus:

$$\underset{\text{yellow}}{[Co(NH_3)_6]^{3+}} \xleftarrow[O_2]{NH_3} \underset{\text{pink}}{[Co(H_2O_6]^{2+}} \underset{H_2O}{\overset{Cl^-}{\rightleftharpoons}} \underset{\text{blue}}{[CoCl_4]^{2-}}$$

Similar changes in ligand and colour are described for chromium in § 27.8 and for copper in § 28.10.

27.16. Detection of Cobalt. When treated with ammonium sulphide, solutions of cobalt salts give a black precipitate of cobalt sulphide. If the precipitate is dissolved in a little aqua regia and the solution is diluted, the addition of α-nitroso-β-naphthol solution gives a brown precipitate (distinction from nickel, which gives no precipitate in acid solution). Cobalt compounds give a blue colour in the borax bead test.

27.17. Nickel, Ni. Electronic configuration: 2,8,16,2.

OCCURRENCE: Ores containing nickel combined with sulphur, arsenic, and antimony, are fairly common, but the main source of nickel is pentlandite, a nickel iron sulphide which is found in Ontario, Canada and which contains about 2% of nickel and a similar amount of copper. Nickel is also found as a silicate called garnierite.

EXTRACTION: The ore is crushed, concentrated by flotation (§ 12.10), and roasted in air, which removes much of the sulphur as its dioxide. It is then heated in a Bessemer converter (§ 30.2) with sand so that the iron is oxidized and removed as a silicate slag. The residue, a mixture of nickel and copper sulphides called 'matte', is converted into the oxides by roasting in air and then reduced to the metals by heating in water gas to about 350°C. Carbon monoxide is passed over the product at 60°C, when volatile nickel carbonyl (b.p. 43°C) is formed and carried over as vapour, leaving the other metals behind (this residue is the main source of platinum and palladium). When heated to 180°C the nickel carbonyl decomposes, depositing the pure metal on nickel pellets and providing carbon monoxide for use again:

$$Ni + 4CO \rightleftharpoons Ni(CO)_4$$

PROPERTIES AND REACTIONS OF NICKEL: It is a hard, lustrous metal which is very malleable and ductile and has ferromagnetic properties. It is unattacked by air or water at room temperature. Like cobalt, it dissolves only slowly in dilute hydrochloric and sulphuric acids and readily in dilute nitric acid, but concentrated nitric acid renders it passive. It is not attacked by molten alkalis or by their concentrated solutions.

USES OF NICKEL: It is used for making crucibles, spatulas, and electrodes because of its resistance to alkalis. In a special finely-divided form made by heating nickel oxide in hydrogen it is an important catalyst for hydrogenation reactions such as the hardening of oils for making margarine (§ 38.4). Nickel forms many important alloys including stainless steels (8%—25% Ni), invar (35% Ni) which is useful because of its very small coefficient of thermal expansion, and special nickel steels (1%—5% Ni) which are used in gas turbines because of their strength at high temperatures and their resistance to corrosion. With copper it forms monel metal (70% Ni) and coinage metals, and with brass it gives nickel silver (10%—20% Ni) used extensively for making plated tableware. Nichrome (60% Ni) is an alloy with chromium which is used for making the heating elements for electric fires and furnaces. Nickel plating is an important protection against corrosion and usually precedes chromium plating because it helps the latter to adhere firmly to the article.

27.18. Compounds of Nickel. The only stable salts are the *nickelous* or *nickel(II)* ones containing the Ni^{2+} ion, which is usually hydrated with six molecules of water to form a green hexahydrate, e.g. $Ni(NO_3)_2.6H_2O$ and $NiCl_2.6H_2O$. Like cobalt, nickel readily forms ammines, i.e. complex ions with ammonia, of the type $[Ni(NH_3)_6]Cl_2$ and $[Ni(NH_3)_6]SO_4$. Nickel compounds are used for electroplating, for colouring ceramics, and in alkaline storage batteries.

27.19. Detection of Nickel. When treated with ammonium sulphide, solutions of nickel salts give a black precipitate of nickel sulphide. If this is dissolved in aqua regia and the solution made alkaline with ammonium hydroxide, then a pink precipitate is obtained on boiling with dimethylglyoxime solution (distinction from cobalt, which gives no precipitate). Nickel compounds give a brownish colour in the borax bead test in an oxidizing flame.

27.20. Uranium, U. Electronic configuration 2,8,18,32,21,9,2.

OCCURRENCE: The commonest ore is the oxide U_3O_8 which occurs widely in ores of low concentration as uranite or pitchblende. The ore carnotite, a complex uranyl vanadate, is used as a source of both uranium and vanadium (§ 27.5).

EXTRACTION: In the first stage the low grade ores are concentrated by crushing and leached with dilute sulphuric acid to give a solution of uranyl sulphate. The uranium is recovered from this by ion-exchange (§ 9.17) or solvent extraction, or it is precipitated as ammonium diuranate $(NH_4)_2U_2O_7$, reduced to the dioxide by roasting in hydrogen, and converted to the tetrafluoride by gaseous hydrogen fluoride.

Metallic uranium is made by the reduction of the oxide or tetrafluoride, usually by heating with a more electropositive metal such as magnesium or calcium:

$$UO_2 + 2Ca = U + 2CaO$$
$$UF_4 + 2Mg = U + 2MgF_2$$

Such reductions are highly exothermic and have to be performed in special closed containers lined with calcium fluoride or magnesia to prevent the molten uranium from reacting with the container or with oxygen or nitrogen from the atmosphere.

Very pure uranium can be obtained by the filament growth method, as described in § 27.3:

$$UI_4 \rightleftharpoons U + 2I_2$$

PROPERTIES OF URANIUM: It is a lustrous white metal, m.p. 1132°C, ρ 19.1 g cm^{-3}. It exists in three allotropic forms which differ markedly in softness and strength. Uranium is a very reactive metal which corrodes readily and unites with many elements when heated. Its most important compounds are the dioxide UO_2 (m.p. 2800°C) readily formed by heating the metal in air, the carbide UC (m.p. 2350°C) formed by heating the dioxide to 1600°C with graphite, and the volatile hexafluoride UF_6 made by heating the tetrafluoride in fluorine. A sample of natural uranium consists of three isotopes uranium-238 (99.28%), uranium-235 (0.715%), and uranium-234 (0.005%), all three of which are radioactive (§ 31.2).

USES OF URANIUM: As a source of nuclear energy in reactors and bombs (§ 31.10 and § 31.11).

27.21. Summary. As we have seen in chapter 13, the properties of an element are chiefly determined by its electronic configuration and particularly by the number of electrons in the outermost shell of its atom. The differences in chemical character caused by differences in the number of electrons in the penultimate shell are much less noticeable. The transition metals illustrate this point very well. With the exception of chromium and copper they all have two electrons in the outermost shells of their atoms, and it is this common feature which is mainly responsible for their general similarity. Their electronegativities and their atomic sizes show little variation through the series, because as the atomic number rises the stronger inward pull of the nucleus is

TABLE 27.3. THE ELECTRONIC CONFIGURATIONS OF THE TRANSITION METALS

(Ar stands for $1s^2\ 2s^2\ 2p^6\ 3s^2\ 3p^6$ in each case)

Symbol		Configuration 3d					4s	Oxidation Numbers
Sc	Ar	↑					↑↓	3
Ti	Ar	↑	↑				↑↓	2, 3, 4
V	Ar	↑	↑	↑			↑↓	2, 3, 4, 5
Cr	Ar	↑	↑	↑	↑	↑	↑	2, 3, 6
Mn	Ar	↑	↑	↑	↑	↑	↑↓	2, 3, 4, 6, 7
Fe	Ar	↑↓	↑	↑	↑	↑	↑↓	2, 3
Co	Ar	↑↓	↑↓	↑	↑	↑	↑↓	2, 3
Ni	Ar	↑↓	↑↓	↑↓	↑	↑	↑↓	2, 3
Cu	Ar	↑↓	↑↓	↑↓	↑↓	↑↓	↑	1, 2
Zn	Ar	↑↓	↑↓	↑↓	↑↓	↑↓	↑↓	2

just counterbalanced by the screening effect and mutual repulsion of the additional electrons.

It is sometimes convenient to indicate the electronic configurations of atoms by means of a series of small squares or 'boxes', each of which represents a separate orbital. An arrow is used to represent each electron present and these are shown with heads pointing in opposite directions where two electrons with opposite spin occupy the same orbital.

This method of representation is employed for the metals of the first transition series, scandium-zinc, in Table 27.3. To save space and repetition, only the 3*d* and 4*s* orbitals are shown, because all these elements have identical configurations for the innermost 18 electrons and these can be simply represented by the symbol Ar for argon standing for $1s^2$ $2s^2$ $2p^6$ $3s^2$ $3p^6$ and sometimes known as the argon core. From this Table it will be clear that the ten elements in the first transition series differ in the number of electrons in the 3*d* block of orbitals. For this reason they are often known as the *d-block elements*, although this term is sometimes restricted to elements forming ions in which the 3*d* subshell is not completely filled, which excludes scandium and zinc.

It will be noticed that the 3*d* and 4*s* levels are so close in energy that a more stable arrangement is sometimes achieved by having only one electron in the 4*s* subshell and an additional electron in the 3*d* level, as happens in chromium and copper. Another interesting feature is the relationship between electronic configuration and oxidation number (§ 15.8). In the first five transition elements (scandium–manganese) the maximum oxidation number is equal to the total number of electrons in the 3*d* and 4*s* levels. This is explained by the fact that to involve any more electrons would mean taking some from the 2, 8, 8 argon core and this would be unlikely to happen because of the very high ionization energy required. For the remaining five elements, iron–zinc, the 4*s* electrons are the least tightly bound, so oxidation numbers of + 2 for iron, cobalt, nickel, and zinc, and +1 for copper, are common.

28

Copper, Silver, and Gold

28.1. Introduction. The three metals in Group IB, copper, silver, and gold, are commonly known as the coinage metals, although other metals are being increasingly used in their place for this purpose. They share many of the characteristics of the transition metals, but their chemical and commercial importance justifies treatment in a separate chapter.

28.2. Copper, Cu. Electronic configuration: 2,8,18,1.

OCCURRENCE: Metallic copper does occur naturally in a few places, but the commonest minerals are chalcopyrite, $CuFeS_2$, which is the main source of the metal, copper glance or chalcocite, Cu_2S, bornite, $Cu_2S.CuS.FeS$, and malachite, $CuCO_3.Cu(OH)_2$.

MANUFACTURE: Most copper ores contain only a few per cent of copper and the first stage is to crush them and concentrate them by flotation methods (see § 12.10). The concentrates are roasted in air, which converts at least part of the sulphides into oxides, sulphur dioxide being produced as a by-product:

$$2CuFeS_2 + 4O_2 = Cu_2S + 2FeO + 3SO_2\uparrow$$

The product is then mixed with limestone and silica and melted in a reverberatory furnace. Most of the iron present reacts with the silica, forming a fusible slag which is run off, leaving a molten mixture of cuprous and ferrous sulphides called *matte*, which is tapped periodically:

$$FeO + SiO_2 = FeSiO_3$$

Copper is obtained by melting the matte in a converter similar to the Bessemer type shown in Fig. 30.3 with some flux and subjecting it to a prolonged blast of air. The ferrous sulphide undergoes oxidation first and is removed as a slag, then part of the cuprous sulphide is converted into cuprous oxide, which reacts with unchanged sulphide to give the molten metal:

$$2Cu_2S + 3O_2 = 2Cu_2O + 2SO_2\uparrow$$
$$2Cu_2O + Cu_2S = 6Cu + SO_2\uparrow$$

The product is called *blister copper* because it tends to release bubbles of dissolved gas whilst solidifying. It still contains 2%–3% of impurities, mainly iron and sulphur.

REFINING OF COPPER: For some purposes furnace refining is adequate, but where copper of high purity is required for electrical use or where the crude copper contains appreciable amounts of precious metals, the electrolytic method is used as well.

Thermal Method: The blister copper is remelted in a reverberatory furnace, *ca.* 2.5×10^5 kg at a time, and exposed to air, which oxidizes the sulphur to sulphur dioxide and the other impurities to a slag. This is skimmed off the surface, leaving molten copper saturated with oxygen. The latter is removed by stirring the molten metal for several hours with poles made from green wood, which releases reducing gases in the melt. This gives copper of about 99.5% purity.

Electrolytic Method: A cell is constructed using slabs of impure copper as anodes and thin sheets of pure copper as cathodes, the electrolyte being copper sulphate solution acidified with sulphuric acid. On passing the current, copper goes into solution from the anodes and pure copper is deposited upon the cathodes. Those metals which are more electropositive than copper (e.g. nickel, zinc, cobalt, and iron), go into solution as ions and accumulate there, provided their concentration is low, the copper ions being preferentially discharged (§ 10.4). Metals less electropositive than copper (e.g. gold and silver), fall from the anode as it dissolves and collect as a sludge at the base of the cell, to be removed at intervals for further refining.

$$CuSO_4 \rightleftharpoons Cu^{2+} + SO_4^{2-}$$

At the cathode: $Cu^{2+} + 2e \rightarrow Cu\downarrow$

At the anode: $Cu \rightarrow Cu^{2+} + 2e$

PROPERTIES OF COPPER: It is a fairly soft, lustrous metal with a characteristic red colour. It melts at 1083°C and has a density of 8.94 g cm^{-3}. Copper is renowned for its ductility and malleability and for its high conductivity of heat and electricity.

REACTIONS OF COPPER: On standing in air at room temperature it is only slowly tarnished, but when heated strongly in air it is converted into its oxides rapidly. It reacts readily with the halogens and sulphur. On exposure to the atmosphere for long periods, copper forms a green coating or 'patina', which is probably a basic copper sulphate or chloride. Being low in the electrochemical series, copper is not capable of displacing hydrogen from acids. It is unattacked by dilute hydrochloric and sulphuric acids in the absence of air, but it dissolves readily in nitric acid of all concentrations:

dilute acid: $3Cu + 8HNO_3 = 3Cu(NO_3)_2 + 4H_2O + 2NO\uparrow$
concentrated acid: $Cu + 4HNO_3 = Cu(NO_3)_2 + 2H_2O + 2NO_2\uparrow$

With hot concentrated sulphuric acid, copper gives sulphur dioxide and copper sulphate, but side reactions also occur producing the sulphide as well; the equation is usually written as:

$$Cu + 2H_2SO_4 = CuSO_4 + 2H_2O + SO_2\uparrow$$

Copper also dissolves in the presence of air in ammonium hydroxide and potassium cyanide solutions, with the formation of complex ions (see § 9.16 and § 28.7).

USES OF COPPER: The multifarious uses of copper spring directly from its workability, its high thermal and electrical conductivity, its resistance to corrosion, its attractive appearance, and, above all, its readiness to form alloys with other metals. Thus it is used to make electrical conductors of all kinds from the very thin wires used in electronics to the busbars used in power transmission. Sheets of the metal are used for roofing large buildings, and it is widely used for plumbing and for making boilers and condensers in locomotives and ships. When alloyed with other metals copper provides a great range of materials. For instance, at least three-fifths of every piece of brass is copper, the rest being mainly zinc. Bronzes, essential as bearing metals in machinery, are made from copper and tin, sometimes with other metals added or even small amounts of phosphorus (phosphor-bronze). Copper also alloys readily with nickel (one such alloy is used for British 'silver' coinage), and with aluminium, beryllium, and gold (9 carat gold is nearly two-thirds copper).

28.3. Compounds of Copper. Copper forms two series of compounds, *cuprous,* in which it is univalent and *cupric,* in which it is divalent.* Cuprous compounds are formed either electrovalently by the loss of the outermost electron, or much more commonly, covalently, by a sharing of that electron and one from another atom. Cupric compounds are formed either by a double covalency involving the outermost electron and one electron from the penultimate shell, or, less commonly, by the loss of two electrons from each atom giving the ion Cu^{2+}. Both cuprous and cupric ions show a strong tendency to form complexes by co-ordination of suitable ions or molecules, the complex ions so formed being much more stable than the simple copper ions from which they are derived (see § 9.16).

* These are often distinguished by referring to them as copper(I) compounds and copper(II) compounds respectively (see Appendix I).

At high temperatures cuprous salts are more stable than cupric, and if heated strongly cupric oxide, sulphide, fluoride, chloride, and bromide all change into the corresponding cuprous compounds.

The two ions are related to each other in solution in the following way:

$$2Cu^{+} \rightleftharpoons Cu^{2+} + Cu$$

For example, soluble cuprous salts such as the sulphate decompose rapidly in solution into the cupric salt and copper, which is deposited, causing the equilibrium to move completely to the right. Insoluble cuprous salts show no such tendency because in their case the concentration of cuprous ions in solution is extremely low.

Cupric salts containing weakly electronegative anions (e.g. I^-, CN^-, CNS^-), are unstable; when they are formed by double decomposition they spontaneously change into the corresponding cuprous salts.

28.4. Copper(I) Oxide (Cuprous Oxide), Cu_2O. This is best prepared by the reducing action of glucose on boiling *Fehling's solution*, when it appears as a red precipitate which can be filtered off, washed, and dried. Fehling's solution, which is deep blue in colour, is made by adding copper sulphate solution to a solution of sodium potassium tartrate in excess of sodium hydroxide solution. This reaction is used in organic chemistry as a test for reducing agents such as aldehydes and sugars.

Cuprous oxide is insoluble in water, but it dissolves readily in ammonia solution and in acids forming cuprous salts. Some of the latter are unstable in solution and decompose into the corresponding cupric salt and copper:

$$Cu_2O + H_2SO_4 = CuSO_4 + Cu\downarrow + H_2O$$

With hydrochloric acid, however, cuprous chloride dissolves in excess of the acid, forming cuprochlorous acid thus:

$$Cu_2O + 2HCl = 2CuCl\downarrow + H_2O$$
$$CuCl + HCl = HCuCl_2$$

Cuprous oxide is used for making red glass and in the manufacture of metal rectifiers.

28.5. Copper(I) Chloride (Cuprous Chloride), CuCl. This is prepared by heating cupric oxide and concentrated hydrochloric acid with metallic copper:

$$CuO + Cu + 2HCl = 2CuCl\downarrow + H_2O$$

When the product is poured into a large excess of cold water, cuprous chloride is thrown down as a white precipitate which can be filtered off, washed with water containing a little sulphur dioxide (to prevent oxidation), and dried in a vacuum desiccator. Alternatively, sulphur dioxide can be used to reduce a mixture of cupric sulphate and sodium chloride solutions:

$$2CuSO_4 + 2NaCl + SO_2 + 2H_2O = 2CuCl\downarrow + 2NaHSO_4 + H_2SO_4$$

Although it is almost insoluble in water, it dissolves readily in those reagents which form complex ions with cuprous copper, e.g. in concentrated hydrochloric acid forming $HCuCl_2$ and in ammonia solution forming $Cu(NH_3)_2Cl$. These solutions are used to absorb carbon monoxide, with which they form the compound $CuCl.CO.2H_2O$. Ammoniacal solutions of cuprous chloride also give a red precipitate of cuprous acetylide with acetylene (§ 34.4, reaction 4).

Cuprous chloride is predominantly a covalent substance. It is a very poor conductor when fused, its vapour density at 1700°C corresponds to the double molecules Cu_2Cl_2, and its crystal structure is similar to zinc blende and diamond, suggesting covalent linkages between its atoms. It is an important reagent in the Sandmeyer reaction (§ 44.6). Oxidizing agents convert it into cupric chloride:

$$2CuCl + 2HCl + \bar{O} = 2CuCl_2 + H_2O$$

28.6. Copper(I) Iodide (Cuprous Iodide), CuI. This is precipitated as a buff solid when potassium iodide solution is added to copper sulphate solution:

$$2CuSO_4 + 4KI = 2CuI\downarrow + I_2 + 2K_2SO_4$$

This reaction is used for estimating copper salts volumetrically by titrating the liberated iodine with standard sodium thiosulphate solution (§ 24.14).

28.7. Copper(I) Cyanide (Cuprous Cyanide), CuCN. This is obtained when potassium cyanide solution is added dropwise to copper sulphate solution:

$$2CuSO_4 + 4KCN = 2CuCN\downarrow + C_2N_2\uparrow + 2K_2SO_4$$

A greenish-yellow precipitate of cupric cyanide first appears, but this decomposes rapidly into a white precipitate of cuprous cyanide and the poisonous gas cyanogen (§ 20.10).

If excess of potassium cyanide solution is added, cuprous cyanide redissolves forming the complex cuprocyanide or tetracyanocuprate(I) ion:

$$CuCN + 3KCN = K_3Cu(CN)_4$$

This ion is important in copper plating and in qualitative analysis because the high stability of the cuprocyanide ion ensures a very low concentration of simple cuprous ions in equilibrium with it:

$$K_3Cu(CN)_4 \rightleftharpoons 3K^+ + [Cu(CN)_4]^{3-}$$
$$[Cu(CN)_4]^{3-} \rightleftharpoons Cu^+ + 4CN^-$$

Thus if hydrogen sulphide is passed into a solution containing both copper and cadmium salts in the presence of excess of potassium cyanide solution, no precipitate of cuprous sulphide is obtained because the concentration of copper ions is too low to exceed its solubility product (§ 9.9), but cadmium sulphide will be precipitated because the corresponding cadmi-cyanide ion is much less stable and provides an appreciable concentration of simple cadmium ions:

$$K_2Cd(CN)_4 \rightleftharpoons 2K^+ + [Cd(CN)_4]^{2-}$$
$$[Cd(CN)_4]^{2-} \rightleftharpoons Cd^{2+} + 4CN^-$$

In this way it is possible to detect cadmium in the presence of copper, which is difficult under normal conditions because the yellow precipitate of cadmium sulphide is masked by the black precipitate of copper sulphide.

28.8. Copper(II) Oxide (Cupric Oxide), CuO. This is a hygroscopic black solid made by heating cupric hydroxide, carbonate or nitrate, or by heating the metal to about 800°C in air or oxygen:

$$2Cu(NO_3)_2 = 2CuO + 4NO_2\uparrow + O_2\uparrow$$

It decomposes above 1000°C into cuprous oxide and oxygen:

$$4CuO \rightleftharpoons 2Cu_2O + O_2$$

It is a mild oxidizing agent. For example, on heating it oxidizes hydrogen to water and carbon and carbon monoxide to carbon dioxide:

$$CuO + H_2 = Cu + H_2O$$
$$2CuO + C = 2Cu + CO_2$$

Use is made of these reactions in organic analysis (§ 32.7).

Cupric oxide is insoluble in water, but being a basic oxide it dissolves readily in acids giving cupric salts and in ammonia solution forming a complex ion.

28.9. Copper(II) Hydroxide (Cupric Hydroxide), $Cu(OH)_2$. This is formed as a gelatinous pale blue precipitate when sodium hydroxide solution is added to a solution of a cupric salt:

$$CuSO_4 + 2NaOH = Cu(OH)_2\downarrow + Na_2SO_4$$

When suspended in water and boiled, it changes to black hydrated cupric oxide. It dissolves in ammonium hydroxide solution, giving a deep blue solution called *Schweitzer's reagent*, which contains the complex ion $[Cu(NH_3)_4]^{2+}$ (see § 14.5 and Fig. 14.8). This reaction is used as a test for copper in analysis, and commercially as a way of making rayon or artificial silk because this solution dissolves cellulose. The solution of cellulose is squirted through fine jets into a bath of acid when the cellulose is precipitated as long threads of artificial fibre which are washed, dried, and made into cloth (see § 49.7).

28.10. Copper(II) Chloride (Cupric Chloride), $CuCl_2$. The brown anhydrous salt is made by the action of dry chlorine on heated copper. If cupric oxide is dissolved in concentrated hydrochloric acid and the solution is concentrated, green crystals of the deliquescent dihydrate, $CuCl_2.2H_2O$, are obtained.

When cupric chloride is dissolved in concentrated hydrochloric acid it gives a brownish yellow solution because of a preponderance of $[CuCl_4]^{2-}$ ions under these conditions. On adding water the colour of the solution changes to green because of the presence of a second complex ion $[Cu(H_2O)_4]^{2+}$, which is blue in colour. On further dilution the solution turns pale blue because the large excess of water causes the hydrate ion to predominate and show its characteristic colour. The changes can be reversed by adding concentrated hydrochloric acid.

In conc. HCl	In dilute HCl		In water
$Cu^{2+} + 4Cl^-$	$Cu^{2+} + 4Cl^-$	$Cu^{2+} + 4H_2O$	$Cu^{2+} + 4H_2O$
$\rightleftharpoons$	$\rightleftharpoons$	$\rightleftharpoons$	$\rightleftharpoons$
$[CuCl_4]^{2-}$	$[CuCl_4]^{2-}$	$[Cu(H_2O)_4]^{2+}$	$[Cu(H_2O)_4]^{2+}$
yellow	yellow and blue, i.e. green		blue

28.11. Copper(II) Sulphate (Cupric Sulphate), $CuSO_4$. This is the most important copper salt. It is made commercially by spraying hot dilute sulphuric acid on scrap copper in the presence of a current of air, when the metal slowly dissolves:

$$2Cu + 2H_2SO_4 + O_2 = 2CuSO_4 + 2H_2O$$

In the laboratory it is prepared by dissolving copper oxide in dilute sulphuric acid and concentrating the solution, when deep blue crystals of the pentahydrate, known as blue vitriol, are obtained. These behave in the following way when heated:

$$CuSO_4.5H_2O \rightleftharpoons CuSO_4.3H_2O \overset{100°C}{\rightleftharpoons} CuSO_4.H_2O \overset{250°C}{\rightleftharpoons} CuSO_4 \xrightarrow{750°C} CuO + SO_3$$

These changes are reversible, the colourless anhydrous salt turning blue on contact with water to form crystals of $[Cu(H_2O)_4]^{2+} SO_4^{2-}.H_2O$.

Copper sulphate solution gives precipitates of insoluble copper salts with many reagents:

$$CuSO_4 + 4KI = 2CuI\downarrow + I_2 + 2K_2SO_4$$
$$2CuSO_4 + 4KCN = 2CuCN\downarrow + C_2N_2 + 2K_2SO_4$$
$$CuSO_4 + NaHCO_3 = CuCO_3\downarrow + NaHSO_4$$
$$CuSO_4 + 2NaOH = Cu(OH)_2\downarrow + Na_2SO_4$$

It is used in large amounts as a fungicide and timber preservative. Its solution with slaked lime, known as *Bordeaux mixture*, is used as a spray for preventing blight in potatoes and vines. Copper sulphate is also used for electroplating and as the starting point in the preparation of many compounds of copper.

28.12. Copper(II) Nitrate (Cupric Nitrate), $Cu(NO_3)_2$. This is made by the action of nitric acid on the metal, or on its oxide, hydroxide, or carbonate. It forms deep blue crystals of the trihydrate, $Cu(NO_3)_2.3H_2O$, which are deliquescent and decompose when heated strongly:

$$2Cu(NO_3)_2 = 2CuO + 4NO_2\uparrow + O_2\uparrow$$

28.13. Detection of Copper. When saturated with hydrogen sulphide, solutions of copper salts give a black precipitate which is insoluble in dilute hydrochloric and sulphuric acids:

$$CuSO_4 + H_2S = CuS\downarrow + H_2SO_4$$

Solutions of cupric salts give a brown gelatinous precipitate of cupric ferrocyanide with potassium ferrocyanide solution:

$$2CuSO_4 + K_4Fe(CN)_6 = Cu_2Fe(CN)_6\downarrow + 2K_2SO_4$$

Copper salts can also be recognized by their green flame coloration, by their greenish blue borax bead (in the oxidizing flame), and by the characteristic deep blue colour formed with excess of ammonium hydroxide solution.

28.14. Silver, Ag. Electronic configuration: 2,8,18,18,1.

OCCURRENCE: Some silver occurs native, but it is chiefly found as the sulphide called argentite or silver glance, Ag_2S, either alone or in association with other sulphides (e.g. those of lead, copper, and zinc), particularly in Mexico and South America. Deposits of horn silver, AgCl, are also known.

MANUFACTURE: Silver is obtained from its low grade ores by leaching them with a dilute solution of sodium cyanide, when the silver compounds dissolve forming the soluble complex argentocyanide ion:

$$Ag_2S + 4NaCN \rightleftharpoons 2NaAg(CN)_2 + Na_2S$$

Air is blown into the mixture to oxidize the sodium sulphide to sulphate and thereby make the reversible reaction proceed completely to the right. The silver is precipitated from the solution by adding some electropositive metal, usually zinc or aluminium (§ 10.4).

Considerable quantities of silver are recovered as a by-product during the smelting of argentiferous lead and copper ores despite the fact that the proportion of silver in these ores rarely exceeds 0.1%. Copper is usually refined electrolytically as described in § 28.2, the silver being recovered as a by-product from the anode mud, but two distinct processes are in use for recovering silver from the ores of lead:

The Pattinson Process: The molten mixture of lead and silver is cooled slowly. At first pure lead crystallizes and is removed, the melt becoming progressively richer in silver until it contains about 2.6%, when a eutectic mixture (§ 6.15) of this composition solidifies at 303°C. This is then subjected to *cupellation*, i.e. it is melted in a bone ash crucible and exposed to a sustained blast of air. The lead is converted into a scum of litharge which can be skimmed off the surface of the melt, leaving a residue of silver and gold which can be refined electrolytically.

The Parkes Process: This depends upon the partition law discussed in § 6.14. Advantage is taken of the immiscibility of molten lead and molten zinc and the fact that silver is very much more soluble in the latter. Thus when a little molten zinc is added to the molten argentiferous lead and the mixture is stirred, most of the silver and a little lead passes into the upper zinc layer. When cooled slightly this layer solidifies and is removed; the zinc is distilled off and used again and the silver separated from the remaining lead by cupellation.

Silver is refined electrolytically by a method similar to that used for copper. The impure silver is made the anode and silver nitrate solution the electrolyte. Copper and lead impurities pass into solution and accumulate there, only pure silver being deposited on the cathode. Gold, which is usually present in small amount, collects beneath the anode.

PROPERTIES AND REACTIONS OF SILVER: It is a lustrous, white, very ductile metal melting at 960°C. Its density is 10.5 g cm^{-3}. It is the best conductor of heat and electricity known, and it readily forms alloys with other metals.

Silver is not attacked by air or moisture at room temperature, but when molten it can dissolve considerable amounts of oxygen. As the metal solidifies the gas is expelled again causing the metal to 'spit' violently. The metal tarnishes rapidly when exposed to sulphur compounds, forming a layer of black silver sulphide. Thus silver articles are affected by traces of hydrogen sulphide in the atmosphere in industrial districts, and silver spoons by sulphur compounds present in eggs.

Being low in the electrochemical series (§ 10.4), silver is not attacked by hydrochloric acid, nor by dilute sulphuric acid. Oxidizing agents such as nitric acid or hot concentrated sulphuric acid dissolve the metal readily, so do solutions of alkali cyanides in the presence of air, but molten alkalis have little effect on it:

$$Ag + 2HNO_3 = AgNO_3 + H_2O + NO_2\uparrow$$
$$2Ag + 2H_2SO_4 = Ag_2SO_4 + 2H_2O + SO_2\uparrow$$

USES OF SILVER: In the past the main uses of silver have been for coinage (usually alloyed with copper), for making jewellery and mirrors, and for electroplating (see below), all uses arising from its high lustre and its resistance to corrosion. In recent years, however, the metal has become increasingly important in other ways, as a catalyst in the manufacture of acetaldehyde (§ 36.2) and ethylene oxide (§ 34.2) for example, and as a constituent of special low temperature brazing alloys with zinc, copper, and sometimes cadmium. Alkalis and organic acids are often contained in silver-lined vats owing to their resistance to attack, and the metal is also used for electrical contacts in switches and in silver–zinc batteries. Much silver is converted into its compounds, which are of considerable commercial importance particularly in photography.

28.15. Silver Plating. The article to be electroplated, which is usually made of German silver (50% copper, 30% zinc, 20% nickel), is made the cathode of a cell in which the electrolyte is a solution of potassium argentocyanide and the anodes are pure silver. Argentocyanide is used because this complex ion is in equilibrium with an extremely small concentration of simple silver ions, which are discharged at the cathode during the electrolysis and form a firm, coherent, lustrous deposit:

$$KAg(CN)_2 \rightleftharpoons K^+ + [Ag(CN)_2]^- \rightleftharpoons K^+ + Ag^+ + 2CN^-$$

The supply of silver ions in the electrolyte is maintained by solution of silver from the anodes. If silver nitrate is used as the electrolyte, the high concentration of silver ions in its solution results in the rapid displacement of silver by the more electropositive metals present in the article to be plated and leads to the production of a soft, spongy deposit.

28.16. Compounds of Silver. In all its important compounds silver is univalent. These argentous compounds are formed either by the loss of the single electron from the outermost shell of the silver atom, when they are electrovalent and contain the Ag^+ ion, or by the sharing of a pair of electrons with another atom in a single covalent linkage. They are mostly colourless and insoluble in water. Complex ions are also formed (§ 9.16).

28.17. Silver Oxide, Ag_2O. This is obtained as a brown precipitate when sodium hydroxide solution is added to a solution of silver nitrate:

$$2AgNO_3 + 2NaOH = Ag_2O\downarrow + 2NaNO_3 + H_2O$$

It decomposes rapidly at 200°C, which prevents its preparation from the nitrate by heating:

$$2Ag_2O = 4Ag + O_2\uparrow$$

It is a basic oxide, dissolving in acids to give silver salts, and turning litmus blue. Because of this and because most silver halides are insoluble, moist silver oxide is much used in organic chemistry for replacing halogen atoms by oxygen or hydroxyl groups.

Silver oxide dissolves in ammonium hydroxide solution forming the complex ion $[Ag(NH_3)_2]^+$. This solution is called *Tollens's reagent* or *ammoniacal silver oxide*. It is prepared by adding ammonium hydroxide solution drop by drop to silver nitrate solution until the precipitate of silver oxide has almost entirely redissolved. It is an important reagent in organic chemistry for detecting reducing agents such as aldehydes and glucose, which cause it to deposit silver as a lustrous mirror on the inside of the test tube.

28.18. Silver Chloride, AgCl. This is obtained as a white precipitate by adding hydrochloric acid or any soluble chloride to silver nitrate solution:

$$AgNO_3 + HCl = AgCl\downarrow + HNO_3$$

The precipitate coagulates into large lumps when boiled or shaken vigorously and gradually turns purple on exposure to light. Although insoluble in water and nitric acid, it dissolves readily in potassium cyanide solution forming the argentocyanide ion $[Ag(CN)_2]^-$, in ammonium hydroxide solution forming the complex cation $[Ag(NH_3)_2]^+$, and in sodium thiosulphate solution forming the complex anion $[Ag(S_2O_3)_2]^{3-}$ (see § 24.14). Silver chloride is reduced to the metal by heating it in hydrogen or by the action of zinc and hydrochloric acid:

$$2AgCl + H_2 = 2Ag\downarrow + 2HCl$$

The silver halides are of great importance in photography owing to their sensitivity to light, which brings about their reduction to metallic silver. *Silver iodide* is used in artificially inducing rainfall because its crystals when scattered from a high-flying aircraft provide suitable nuclei for the crystallization of ice and the formation of raindrops.

28.19. Silver Nitrate, $AgNO_3$. This is made by the action of nitric acid on metallic silver, colourless crystals appearing when the solution is concentrated and cooled. It melts at 208°C and decomposes when heated strongly thus:

$$2AgNO_3 = 2Ag + 2NO_2\uparrow + O_2\uparrow$$

Silver nitrate is important as the commonest silver salt which is soluble in water. Its solution is used volumetrically for estimating solutions of chlorides, bromides, iodides, cyanides, and thiocyanates, and in qualitative analysis for detecting those anions which give precipitates with silver ions, e.g. halides, cyanides, sulphides, oxalates, chromates, phosphates, etc. It is also used to prepare insoluble silver salts by double decomposition, and to reveal the presence of arsine in the Gutzeit test (§ 23.15). Its use to prepare ammoniacal silver oxide solution for detecting reducing agents has already been discussed.

28.20. Detection of Silver. When treated with hydrochloric acid, solutions of silver salts give a white precipitate which is insoluble in nitric acid but which dissolves readily in solutions of ammonium hydroxide, sodium thiosulphate, or potassium cyanide. With potassium chromate solution, soluble silver salts give a crimson precipitate of silver chromate which is insoluble in acetic acid but which dissolves readily in dilute nitric acid:

$$2AgNO_3 + K_2CrO_4 = Ag_2CrO_4\downarrow + 2KNO_3$$

28.21. Gold, Au. Electronic configuration: 2,8,18,32,18,1.

OCCURRENCE: The metal occurs naturally, usually as very fine particles embedded in quartz.

EXTRACTION: The first stage consists of crushing the gold-bearing minerals and concentrating the gold in them by purely mechanical means, taking advantage of its high density. The concentrate is then leached with a very dilute solution of sodium or potassium cyanide, which dissolves the gold by forming a soluble complex ion, from which it is precipitated by adding zinc. Any residual zinc is removed from the precipitate by treatment with dilute sulphuric acid, and the product is then melted into plates and refined electrolytically. The impure gold is made the anode, chloroauric acid (§ 28.22) the electrolyte, and a sheet of pure gold the cathode. Only gold is deposited upon the cathode, silver being precipitated as its chloride and platinum remaining in solution in the electrolyte.

Some gold is obtained as a by-product during the refining of copper and silver.

PROPERTIES AND REACTIONS OF GOLD: It is a soft, lustrous, yellow metal of m.p. 1063°C and density 19.3 g cm^{-3}. It is extremely ductile and malleable and is a very good conductor of heat and electricity. Being very low in the electrochemical series (§ 10.4), it is not affected by air or water, nor by any single mineral acid. It does dissolve, however, in aqua regia (1 volume concentrated nitric acid + 3 volumes concentrated hydrochloric acid), forming a solution of chloroauric acid. It also dissolves in alkali cyanide solutions forming aurocyanides, and in mercury forming a liquid amalgam. Gold combines readily with moist chlorine forming gold chloride solution:

$$2Au + 3Cl_2 = 2AuCl_3$$

USES OF GOLD: Most of the world's gold is stored in the various national banks, the metal being accepted as an international currency. It is used for jewellery and coinage, but since pure gold is too soft it is usually alloyed with other metals such as copper and silver, the gold content being expressed in carats where twenty-four carats corresponds to pure gold. It is also used for electroplating cheaper metals, for filling teeth, and for making compounds of gold, which are used in photography and in making ruby glass.

28.22. Compounds of Gold. The element forms two series of salts, aurous or gold(I), and auric or gold(III), in which it is univalent and trivalent respectively.

When gold is dissolved in aqua regia and the solution is concentrated, yellow crystals of hydrated chloroauric acid, $HAuCl_4.4H_2O$, separate on cooling. Heated to 200°C in chlorine, these give red crystals of auric chloride, $AuCl_3$, which decompose when heated to 175°C in air, forming aurous chloride and chlorine. Gold readily forms complex anions with high concentrations of CN^- and Cl^- ions, e.g. $[Au(CN)_2]^-$ and $[AuCl_4]^-$.

When a solution of chloroauric acid is treated with a strong reducing agent (e.g. formaldehyde, tartaric acid, or stannous chloride), gold is obtained as a colloidal sol which is blue, red, or purple in colour, depending upon the particle size.

28.23. Summary. Copper, silver, and gold are very similar in many respects. All three elements are feebly electropositive metals; they are so resistant to atmospheric corrosion that they occur native, they do not displace hydrogen from acids, and they are displaced from solutions of their salts by more electropositive metals such as zinc, iron, and aluminium. The metals themselves all crystallize in cubic close-packed structure (§ 4.7). They are highly lustrous and capable of taking a high polish, they melt at about 1000°C, they are extremely malleable and ductile and are excellent conductors of heat and electricity. They are all attacked by chlorine with the formation of a metallic chloride.

All three elements form colourless univalent compounds. They also exert a higher valency to form coloured compounds in which the penultimate shell is incomplete. Some of their compounds are predominantly covalent in character because their high nuclear charges and the weakness of the screening provided by their penultimate shells causes them to have small atomic volumes and to exert strong attractive forces upon electrons at the peripheries of their atoms. For the same reason they encourage the formation of complex ions by co-ordination of other ions (e.g. Cl^- or CN^-) or groups of atoms (e.g. H_2O or NH_3) which have electron-donating tendencies.

Despite these marked resemblances and the similar electronic configurations which give rise to them (there is a single electron in the outermost shell of each atom), these elements differ in certain ways. The oxides of copper are much more stable than those of silver and gold, for example, and silver oxide is alone in being strongly alkaline. Copper and silver dissolve readily in nitric acid, but gold does not, and silver nitrate decomposes at a much higher temperature than copper nitrate. Again, copper and silver chlorides are much more stable thermally than the chloride of gold.

The most striking difference is in the greater stability of silver in the univalent state compared with gold. For example, silver(I) compounds such as argentous nitrate and argentous sulphate are stable substances showing no tendency to decompose to the metal. On the other hand, compared with copper and gold, silver is much less stable in its higher valency state, the fluoride being the only common argentic or silver(II) compound.

29

Zinc, Cadmium, and Mercury

29.1. Introduction. Group IIB of the Periodic Classification comprises the elements zinc, cadmium, and mercury, which are of particular interest because they are the last members of their respective series of transition metals.

29.2. Zinc, Zn. Electronic configuration: 2,8,18,2.

OCCURRENCE: The metal does not occur native, but zinc compounds are widely distributed, the commonest ores being zinc blende, ZnS, and calamine, $ZnCO_3$. Other minerals are zincite, ZnO, franklinite, $ZnO.Fe_2O_3$, and siliceous zinc ore, $Zn_2SiO_4.H_2O$.

MANUFACTURE: Three methods are in use:

(1) *Vertical Retort Process*: The first stage consists of concentrating the ores (see § 12.10) and converting them into zinc oxide. This is achieved by heating calamine strongly or by roasting zinc blende in air in a reverberatory furnace, the sulphur dioxide being a valuable by-product which is used for making sulphuric acid (§ 24.11), giving about two kg of sulphuric acid for every kg of zinc:

$$ZnCO_3 = ZnO + CO_2\uparrow$$
$$2ZnS + 3O_2 = 2ZnO + 2SO_2\uparrow$$

The second stage consists of reducing the zinc oxide by heating it to 1350°C with carbon (see § 11.13):

$$ZnO + C = Zn + CO$$

This is done by mixing the oxide with anthracite or coke and making briquettes which are fed into a vertical retort heated by producer gas as in Fig. 29.1. The zinc distils over and condenses to a liquid in the receiver. After solidification the crude zinc, called spelter, is purified by remelting, the product still containing about 1% of lead. Zinc of at least 99.8% purity can be obtained from this by redistillation.

(2) *Blast Furnace Method:* The concentrated ores are first roasted to a porous sinter and then after preheating and mixing with hot coke are

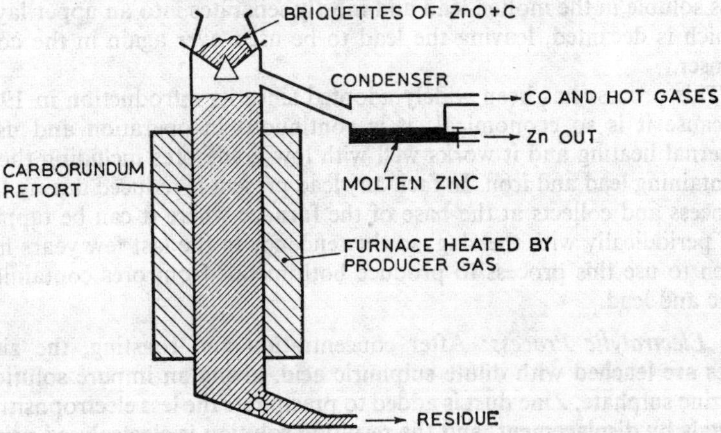

FIG. 29.1 Vertical Retort Process for Zinc

fed into the top of a blast furnace similar to that used in the manufacture of iron, but smaller (see Fig. 29.2). The burning coke maintains the temperature of the furnace above 1000°C and provides a supply of carbon monoxide which helps with the reduction:

$$ZnO + CO \rightleftharpoons Zn + CO_2$$

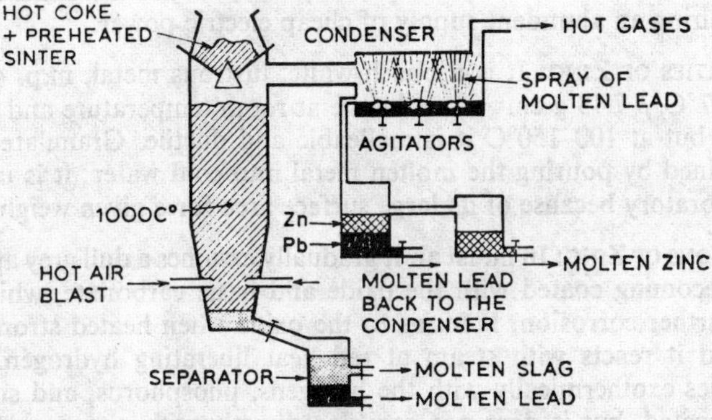

FIG. 29.2 Blast Furnace Process

The zinc vaporizes and escapes from the furnace with the hot gases. It is cooled very rapidly to 600°C by passing it through sprays of molten lead; this condenses and dissolves the zinc and prevents it from being re-oxidized by the back reaction. The liquid from the condenser is run off and cooled to about 450°C at which temperature the zinc is much less soluble in the molten lead and mostly separates into an upper layer which is decanted, leaving the lead to be used over again in the condenser.

This process has been widely adopted since its introduction in 1950 because it is so economical; it is continuous in operation and uses internal heating and it works well with low grade ores including those containing lead and iron. In fact, any lead present is reduced during the process and collects at the base of the furnace where it can be tapped off periodically with the slag, so the tendency in the last few years has been to use this process to produce both metals from ores containing zinc and lead.

(3) *Electrolytic Process:* After concentration and roasting, the zinc ores are leached with dilute sulphuric acid, giving an impure solution of zinc sulphate. Zinc dust is added to precipitate the less electropositive metals by displacement, and the resulting solution is electrolysed using an aluminium cathode and a high current density. Under these conditions only zinc ions are discharged at the cathode and the pure metal is deposited. Despite their lower electropositivity, hydrogen ions are not discharged because of the overpotential (§ 10.5) of the cathode.

The electrolytic method offers several advantages over its rivals. It supplies zinc of high purity which does not need refining, it can be applied to low grade zinc ores, and it permits the recovery of any precious metals which are present as impurities. It has the disadvantage of requiring an abundant supply of cheap electric power.

PROPERTIES OF ZINC: It is a bluish-white, lustrous metal, m.p. 419°C, b.p. 907°C, ρ 7.13 g cm^{-3}. It is brittle at room temperature and above 200°C, but at 100–150°C it is malleable and ductile. Granulated zinc is obtained by pouring the molten metal into cold water; it is used in the laboratory because of its large surface area for a given weight.

REACTIONS OF ZINC: In moist air it gradually assumes a dull grey appearance becoming coated with the oxide and basic carbonate, which inhibit further corrosion. It burns to the oxide when heated strongly in air, and it reacts with steam at red heat liberating hydrogen. Zinc combines exothermically with the halogens, phosphorus, and sulphur when heated, but it does not combine directly with nitrogen. At red heat it decomposes ammonia forming the nitride, Zn_3N_2.

Provided that the zinc is not extremely pure, it dissolves in dilute hydrochloric and sulphuric acids with the evolution of hydrogen:

$$Zn + 2HCl = ZnCl_2 + H_2\uparrow$$
$$Zn + H_2SO_4 = ZnSO_4 + H_2\uparrow$$

Impurities such as copper or arsenic discharge the hydrogen ions in the acid by setting up small galvanic couples on the surface of the metal, which explains why the addition of a few drops of copper sulphate solution has such a striking effect upon the rate at which acid attacks a sample of very pure zinc.

Zinc reacts with nitric acid giving zinc nitrate solution and a variety of products depending upon the temperature and the concentration of the acid (§ 22.13). With dilute acid, ammonium nitrate is formed, but with concentrated acid the main products are the oxides of nitrogen. With hot concentrated sulphuric acid, zinc gives sulphur dioxide:

$$Zn + 2H_2SO_4 = ZnSO_4 + 2H_2O + SO_2\uparrow$$

It dissolves in strong alkalis giving hydrogen and a solution of a zincate:

$$Zn + 2NaOH = Na_2ZnO_2 + H_2\uparrow$$

USES OF ZINC: The metal is used for roofing because of its resistance to corrosion, and in die castings. It is also used for making the casing of dry batteries, and for desilvering lead by the Parkes process (§ 28.14), and for precipitating gold and silver after cyanide leaching. Zinc forms a large number of alloys, particularly brasses, where the zinc content is often as high as 40%. The metal is used to make zinc compounds, many of which are of commercial importance. Zinc dust is a useful reducing

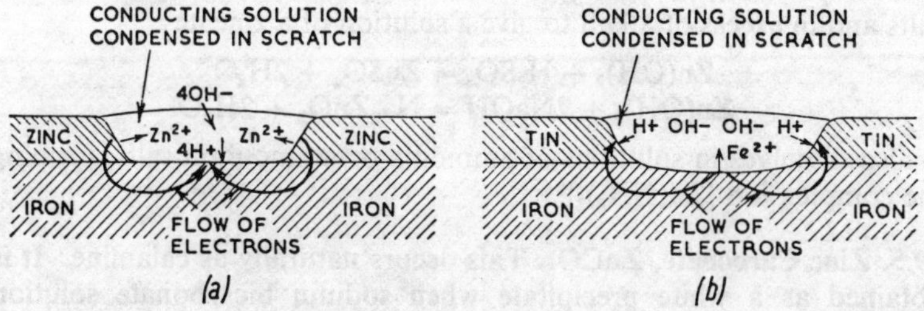

FIG. 29.3 Corrosion of Galvanized Iron (*a*) and Tinplate (*b*) Compared

agent. Plates of zinc are frequently used in the lithographic method of printing.

Large amounts of zinc are used for *galvanizing* iron and steel to prevent rusting. In the hot-dip process the iron is cleaned by pickling it in acid, washed, coated with flux (usually ammonium chloride), and then dipped into molten zinc. Alternatively the article may be sprayed with powdered zinc or coated electrolytically. Even if the layer of zinc is removed at one place by scratching, rusting will not readily occur because the zinc, being more electropositive than the iron (§ 10.4) sacrificially protects the latter. As shown in Fig. 29.3*a*, a galvanic cell is set up around the scratch and the zinc corrodes preferentially, going into solution as zinc ions and leaving the iron intact. A coating of zinc hydroxide and basic zinc carbonate soon forms, inhibiting further corrosion. This protective action is in direct contrast to tinning, (§ 21.2), where a scratch on the tinplate results in accelerated rusting of the iron because the latter is more electropositive than the tin (see Fig. 29.3*b*). Galvanized vessels cannot be used for storing food, however, because of the danger of zinc poisoning.

29.3. Zinc Oxide, ZnO. This is made commercially by burning zinc in air; it is prepared in the laboratory by heating the carbonate, hydroxide, or nitrate. It is a white solid which turns yellow when hot. It is insoluble in water, but being amphoteric it dissolves readily in acids and alkalis forming salts. Zinc oxide is used as a pigment in paints and as a filler for rubber (§ 49.4). The very pure product is used in cosmetic powders and creams and in medicinal ointments.

29.4. Zinc Hydroxide, $Zn(OH)_2$. This is obtained as a gelatinous white precipitate when a solution of a zinc salt is treated with an alkali:

$$ZnSO_4 + 2NaOH = Zn(OH)_2\downarrow + Na_2SO_4$$

It is amphoteric, dissolving in acids to give the corresponding zinc salts and in excess of alkali to give a solution of a zincate:

$$Zn(OH)_2 + H_2SO_4 = ZnSO_4 + 2H_2O$$
$$Zn(OH)_2 + 2NaOH = Na_2ZnO_2 + 2H_2O$$

It also dissolves in solutions of ammonia or ammonium salts, forming the complex ion $[Zn(NH_3)_4]^{2+}$.

29.5. Zinc Carbonate, $ZnCO_3$. This occurs naturally as calamine. It is obtained as a white precipitate when sodium bicarbonate solution is added to a solution of a zinc salt. Addition of sodium carbonate solution gives a precipitate of *basic zinc carbonate* of composition $2ZnCO_3.3Zn(OH)_2$. Zinc carbonate is present in calamine lotion, used for treating inflammation of the skin.

29.6. Zinc Sulphate, $ZnSO_4$. This is obtained as colourless crystals of the heptahydrate, $ZnSO_4.7H_2O$, known as white vitriol, by concentrating a solution of zinc or its oxide or carbonate in dilute sulphuric acid and cooling below 39°C. It is isomorphous with the corresponding sulphates of magnesium, iron, manganese, nickel, cobalt, and cadmium. When heated it changes thus:

$$ZnSO_4.7H_2O \rightarrow ZnSO_4.6H_2O \xrightarrow{100°C} ZnSO_4.H_2O$$
$$\xrightarrow{250°C} ZnSO_4 \xrightarrow{750°C} ZnO + SO_3$$

Zinc sulphate is chiefly used to manufacture lithopone (§ 29.8) for paints.

29.7. Zinc Chloride, $ZnCl_2$. The monohydrate is made by dissolving zinc or its oxide, hydroxide, or carbonate in excess of hydrochloric acid and evaporating the solution. The anhydrous salt is made by the action of dry chlorine or dry hydrogen chloride on heated zinc. It is a very deliquescent, white solid which dissolves in water exothermically giving a solution which is acidic by hydrolysis. It is also soluble in alcohol, ether, and acetone. The low conductivity of molten zinc chloride and its solubility in organic solvents suggests that it consists of two forms in equilibrium, one ionic and the other covalent.

Anhydrous zinc chloride is used in organic chemistry as a dehydrating agent (§ 40.2). Mixed with moist zinc oxide it gives a hard mass of zinc oxychloride, Zn (OH)Cl, suitable for filling teeth. In solution zinc chloride is used for preserving timber and as a flux in soldering.

29.8. Zinc Sulphide, ZnS. This compound is dimorphous, occurring naturally as zinc blende and as wurtzite. It is precipitated when solutions of zinc salts are treated with ammonium sulphide solution. Although white when freshly precipitated, it tends to turn yellow on exposure to light.

Zinc sulphide is a constituent of *lithopone*. This important pigment, which is extensively used in white paints, is made by mixing solutions of zinc sulphate and barium sulphide, filtering off the precipitates, and heating them to about 700°C in the absence of air:

$$ZnSO_4 + BaS = BaSO_4\downarrow + ZnS\downarrow$$

Lithopone is preferred to basic lead paints because it is cheaper, less poisonous, and does not darken when exposed to hydrogen sulphide (§ 21.19), but it is inferior in covering power and durability, particularly for outdoor use. In recent years, however, titanium dioxide has largely replaced both, because of its superior qualities as a pigment (see § 27.4).

29.9. Detection of Zinc. Solutions of zinc compounds give a white precipitate when treated with ammonium sulphide solution. This precipitate is soluble in dilute hydrochloric acid but insoluble in acetic acid. Zinc hydroxide, precipitated by adding sodium hydroxide solution a drop at a time to a solution of a zinc salt, dissolves readily in excess of alkali, which distinguishes it from the hydroxide of manganese.

When heated on a charcoal block zinc compounds give an incrustation which is white when cold and yellow when hot. When moistened with cobalt nitrate solution and reheated, the residue turns green.

29.10. Cadmium, Cd. Electronic configuration: 2,8,18,18,2.

OCCURRENCE: Zinc ores invariably contain small amounts of cadmium compounds as impurities.

MANUFACTURE: The metal is obtained as a by-product during the extraction of zinc. Being more volatile than zinc, the cadmium distils over in the early stages of the distillation process and condenses with the zinc dust in the receiver or is recovered as flue dust from the furnace gases. It is purified by fractional distillation or by electrolysis.

PROPERTIES AND REACTIONS OF CADMIUM: It is a lustrous white metal. m.p. 321°C, b.p. 767°C. Although malleable and ductile, it is too soft for constructional use. When exposed to moist air it turns dull grey, forming a coating of oxide which prevents further corrosion, but it burns when heated strongly in air.

USES OF CADMIUM: The chief use is for plating iron and steel to prevent rusting, even a thin film providing excellent protection; it cannot be used for food containers, however, because cadmium compounds are poisonous. The metal absorbs neutrons very readily and rods of it or its alloys are used in slowing down the activity of atomic reactors (§ 31.11):

$$^{113}_{48}Cd + ^{1}_{0}n = ^{114}_{48}Cd + \gamma$$

Cadmium is increasingly use for making nickel–cadmium batteries which have a long life and are very compact.

29.11. Compounds of Cadmium. These closely resemble the compounds of zinc; except for the oxide and sulphide they are colourless and mostly soluble in water. When solutions of cadmium salts are treated with zinc dust, cadmium is precipitated because the metal is less electropositive than zinc (§ 10.4).

The oxide, CdO, and hydroxide, $Cd(OH)_2$, dissolve in acids giving cadmium salts, but not in dilute alkalis. Like the corresponding compounds of zinc, they dissolve in ammonia solution forming the complex ion $[Cd(NH_3)_4]^{2+}$. The sulphide, CdS, is a bright yellow solid used as a pigment. The sulphate, $3CdSO_4.8H_2O$, is used in Weston standard cells (§ 10.6). The cyanide, $Cd(CN)_2$, is precipitated from solutions of cadmium salts by adding potassium cyanide solution. Like zinc cyanide it dissolves in excess of the reagent, giving a complex ion $[Cd(CN)_4]^{2-}$, which is used for electroplating.

29.12. Detection of Cadmium. When saturated with hydrogen sulphide, solutions of cadmium salts give a yellow precipitate which is insoluble in dilute hydrochloric acid and in ammonium sulphide solution but which dissolves in warm dilute nitric acid and in hot dilute sulphuric acid. As explained in § 28.7, cadmium is detected in the presence of copper by precipitating its sulphide from cyanide solution by hydrogen sulphide.

29.13. Mercury, Hg. Electronic configuration: 2,8,18,32,18,2.

OCCURRENCE: Although small amounts of the element have been found native, the only important source is the mineral cinnabar, HgS.

MANUFACTURE: The ore, which usually contains only a few per cent of mercury, is roasted in retorts with air or iron:

$$HgS + O_2 = Hg\uparrow + SO_2\uparrow$$
$$HgS + Fe = Hg\uparrow + FeS$$

The mercury distils over and is condensed in water-cooled earthenware tubes. It is purified by filtration through chamois leather, by washing with very dilute nitric acid, and by distillation under reduced pressure.

PROPERTIES OF MERCURY: At room temperature it is a lustrous, silvery-white liquid. It freezes at −39°C and boils at 357°C. Its density is 13.59 g cm^{-3}. Its vapour is monatomic and very poisonous.

REACTIONS OF MERCURY: It is hardly affected by the atmosphere at room temperature, but when heated in air it is gradually converted into its oxide, HgO. It combines readily with sulphur and the halogens and 'tails' when exposed to ozone (§ 24.3). Mercury is not attacked by hydrochloric acid, nor by dilute sulphuric acid, but it dissolves in nitric acid giving nitrates and oxides of nitrogen (see § 29.21). Hot concentrated sulphuric acid attacks it, evolving sulphur dioxide, and it dissolves readily in aqua regia forming the chloride, $HgCl_2$.

Mercury forms alloys known as *amalgams* with many metals, especially with silver, gold, and the alkali metals. For example, small pieces of sodium dissolve in mercury forming a liquid amalgam which reacts steadily with water giving nascent hydrogen and is therefore a useful reducing agent.

USES OF MERCURY: The metal is used in mercury vapour lamps and rectifiers, and in scientific equipment such as thermometers, barometers, and high vacuum pumps. It is also used in the manufacture of mercury compounds, many of which are of considerable importance, and to make amalgams, particularly for filling teeth, and in the manufacture of sodium hydroxide (§ 17.7).

29.14. Compounds of Mercury. Two series of compounds exist. In *mercurous* compounds the metal appears to be univalent, whereas in fact it is divalent; X-ray analysis and other physical evidence shows that each molecule contains two mercury atoms linked covalently to each other, so that the structure of the mercurous ion is $^{+}Hg—Hg^{+}$ or Hg_2^{2+} and mercurous chloride, for example, is $Cl^{-}\ ^{+}Hg—Hg^{+}\ ^{-}Cl$, written Hg_2Cl_2 not HgCl. In *mercuric* compounds the metal is also divalent; the mercuric ion has the structure Hg^{2+}, but the formula of mercuric chloride, for example, is written Cl—Hg—Cl or $HgCl_2$, since like most mercuric compounds it is largely covalent.

29.15. Mercurous Chloride (Calomel), Hg_2Cl_2. This is obtained by adding dilute hydrochloric acid or a soluble chloride to mercurous nitrate solution:

$$Hg_2(NO_3)_2 + 2HCl = Hg_2Cl_2\downarrow + 2HNO_3$$

Alternatively it can be made by heating a mixture of mercuric chloride and mercury, when the mercurous chloride sublimes:

$$HgCl_2 + Hg = Hg_2Cl_2$$

It is a white solid, insoluble in water and dilute acids. It is easily reduced to mercury when heated with carbon or sodium carbonate or when treated with an excess of stannous chloride solution:

$$Hg_2Cl_2 + SnCl_2 = 2Hg\downarrow + SnCl_4$$

With sodium hydroxide solution, mercurous chloride gives a black precipitate consisting of mercuric oxide and mercury, and with ammonium hydroxide solution it gives a black mixture of metallic mercury and mercuric amido chloride, NH_2HgCl.

29.16. Mercuric Oxide, HgO. This is slowly formed as a red powder by heating mercury in air to about 350°C; stronger heating causes it to dissociate again into its elements (descriptions of the classical experiments of Priestley and Lavoisier in 1774 will be found in most elementary textbooks of chemistry):

$$2Hg + O_2 \rightleftharpoons 2HgO$$

If sodium hydroxide solution is added to mercuric chloride solution, the oxide is obtained as a yellow precipitate:

$$HgCl_2 + 2NaOH = HgO\downarrow + H_2O + 2NaCl$$

As X-ray analysis shows, the red and yellow forms differ only in their particle size.

Mercuric oxide is insoluble in water, but being weakly basic it dissolves in acids giving mercuric salts.

29.17. Mercuric Chloride, $HgCl_2$. This salt, which is commonly known as 'corrosive sublimate', is made either by dissolving mercuric oxide in hydrochloric acid or by heating a mixture of mercuric sulphate and sodium chloride in a retort, when it sublimes:

$$2NaCl + HgSO_4 = HgCl_2\uparrow + Na_2SO_4$$

it is also formed by passing dry chlorine over heated mercury. It can be purified by recrystallization, since it is much more soluble in hot water than in cold.

It is a very poisonous, white solid, soluble in alcohol and ether. Its aqueous solution is acidic by hydrolysis. It is a weak electrolyte, only slightly dissociated into ions in solution, as shown by its low conductivity (§ 8.5). For this reason it does not give hydrogen chloride when boiled with concentrated sulphuric acid and it is only incompletely precipitated as silver chloride by adding silver nitrate solution.

Mercuric chloride is noticeably more soluble in hydrochloric acid and solutions of chlorides than in water because it readily forms complex ions such as $[HgCl_4]^{2-}$. It is easily reduced to mercurous chloride, or even to metallic mercury, by reducing agents such as stannous chloride or formaldehyde.

$$2HgCl_2 + SnCl_2 = Hg_2Cl_2\downarrow + SnCl_4$$

When boiled with ammonium hydroxide solution, mercuric chloride solution gives a white precipitate of mercuric amido chloride, NH_2HgCl, which is known as 'infusible white precipitate' because it decomposes below its melting point. However, if an excess of ammonium chloride is also present, a white precipitate of mercuric diammino chloride, $Hg(NH_3)_2Cl_2$, is obtained which is known as 'fusible white precipitate' because it melts without decomposition.

Mercuric chloride is used as an antiseptic and disinfectant, and for preserving timber and hides, and for making fungicides.

29.18. Mercuric Iodide, HgI_2. This is obtained as a yellow then scarlet precipitate when potassium iodide solution is added dropwise to mercuric chloride solution:

$$HgCl_2 + 2KI = HgI_2\downarrow + 2KCl$$

The precipitate dissolves in excess of potassium iodide solution forming potassium mercuri-iodide, K_2HgI_4 (see § 17.23).

Mercuric iodide is dimorphic, being stable in the yellow form above 126°C and in the scarlet form below. Although only sparingly soluble in water, it is appreciably soluble in organic solvents. So slight is its dissociation into ions in solution that it gives no precipitate of silver iodide with silver nitrate solution. It is used in ointments for treating skin infections.

29.19. Mercuric Sulphide, HgS. This salt exists in two forms. One occurs naturally as cinnabar and can be made by heating mercury with potassium pentasulphide solution:

$$Hg + K_2S_5 = HgS\downarrow + K_2S_4$$

It is scarlet in colour and is used as an artist's pigment (vermilion). The other form, which is black, is prepared directly from its elements or by saturating a solution of a mercuric salt with hydrogen sulphide:

$$HgCl_2 + H_2S = HgS\downarrow + 2HCl$$

29.20. Mercuric Sulphate, $HgSO_4$. This is obtained as colourless crystals of the monohydrate by dissolving mercury or mercuric oxide in hot concentrated sulphuric acid and evaporating the solution:

$$Hg + 2H_2SO_4 = HgSO_4 + SO_2\uparrow + 2H_2O$$

It is hydrolysed in solution and readily forms a basic salt. It is used industrially as a catalyst for converting acetylene into acetaldehyde.

29.21. Nitrates of Mercury.

Mercurous nitrate is made by treating an excess of metallic mercury with dilute nitric acid. It crystallizes as the colourless dihydrate, $Hg_2(NO_3)_2.2H_2O$, which effloresces on standing in air. It is very soluble in water giving a solution which is acidic by hydrolysis and from which a basic salt is precipitated on dilution.

Mercuric nitrate is made by dissolving mercury in hot concentrated nitric acid. It crystallizes from the evaporated solution as colourless deliquescent crystals of monohydrate. Its solution is acidic by hydrolysis.

29.22. Detection of Mercury. When saturated with hydrogen sulphide, solutions of mercury salts give a black precipitate which is insoluble in dilute nitric acid or in ammonium sulphide solution. With dilute hydrochloric acid solutions of mercurous salts give a white precipitate which turns black when treated with ammonium hydroxide solution.

29.23. Summary. All three elements are fairly volatile, divalent metals. They occur naturally as their sulphides, which are insoluble in water. They form many similar compounds, their oxides are easily reduced to the metals, and they readily form basic salts, e.g. carbonates. Their electronic configurations are very similar, each atom possessing eighteen electrons in its penultimate shell and two in its outermost shell.

Two other characteristics common to all three elements are their tendencies towards covalency and complex ion formation. In these elements the nucleus exerts a strong attraction upon the electrons at the periphery because of the high nuclear charge and the small size of the atom. The penultimate shells, although they contain ten additional electrons, possess much less screening power than the penultimate shells in the alkaline earth metals, and consequently many of the compounds of the Group IIB metals are predominantly covalent and are hydrolysed to give acidic solutions. For the same reason they readily accept co-ordinated groups such as ammonia, forming complex ions.

Unlike the majority of the transition metals, they do not show variable valency or paramagnetism, and solutions of their ions are mostly colourless. This is easily understood when it is remembered that these are the very properties which depend upon 'incomplete' penultimate electron shells, whereas in the atoms of these elements the penultimate shell contains the highest possible number of electrons.

Despite all these similarities, the Group is characterized by several notable differences between its members. They differ considerably in electropositivity, zinc being strongly electropositive, cadmium less so, and mercury being so weakly electropositive as to be almost noble in character. This is apparent from the widely differing heats of formation of the oxides, from the inability of mercury to liberate hydrogen from acids, and from the thermal instability of mercuric oxide. Moreover, zinc oxide is amphoteric, whereas the other oxides are basic.

Mercury is of especial interest because of its unique features. It is the only metal which is liquid at room temperature, which accounts for many of its uses. Mercurous compounds contain the very unusual covalent linkage between the atoms of the metal, giving the illusion of univalency. No hydroxide of mercury is known, and when mercury compounds react with ammonia they tend to form substitution products rather than complex ions like those of zinc and cadmium.

30

Iron and Steel

30.1. Introduction. Although a transition metal of the first series and a member of Group VIII, iron demands a special chapter to itself because of its very great economic and industrial importance.

30.2. Iron, Fe. Electronic configuration: 2,8,14,2.

OCCURRENCE: Apart from in meteorites, iron is not found as the element, but its compounds are abundant, the commonest being pyrites, FeS_2, which is unsuitable for smelting because of its high sulphur content, haematite, Fe_2O_3, magnetite, Fe_3O_4, limonite or brown iron ore, $Fe_2O_3 . H_2O$, and siderite or spathic iron ore, $FeCO_3$. Iron is also present in silicates, clays and soils, and in all living matter, being essential for the production of haemoglobin in blood and chlorophyll in plants.

MANUFACTURE OF IRON: The first stage is the preparation of the iron ores for reduction in a blast furnace. Carbonate ores are calcined in kilns in the presence of excess of air, when they are converted into ferric oxide:

$$4FeCO_3 + O_2 = 2Fe_2O_3 + 4CO_2$$

The oxide ores such as magnetite and haematite are crushed and sintered, where necessary, to produce lumps about the size of one's fist. They are also often pre-heated with hot gas from the blast furnace in order to drive off moisture and volatile impurities and render the lumps porous.

The coke used must be very hard and must have a low sulphur content. It is made by carbonizing specially selected coal in multiple retorts; the volatile by-products such as gas, tar, and ammonia are all recovered and sold. About one kg of coke is required for each kg of iron produced.

The iron ores contain between 20% and 65% of iron, the chief impurities being silica and alumina. Both have such high melting points that they would not melt in the blast furnace but would form an ash which would have to be removed mechanically. To overcome this difficulty the ore is mixed with limestone. In the blast furnace this is decomposed into calcium oxide, which combines with the silica and alumina, removing them as a fusible slag:

$$CaO + SiO_2 = CaSiO_3$$

The last raw material to be considered is air, about six kg of which is needed for every kg of iron made. The blast of air which gives the

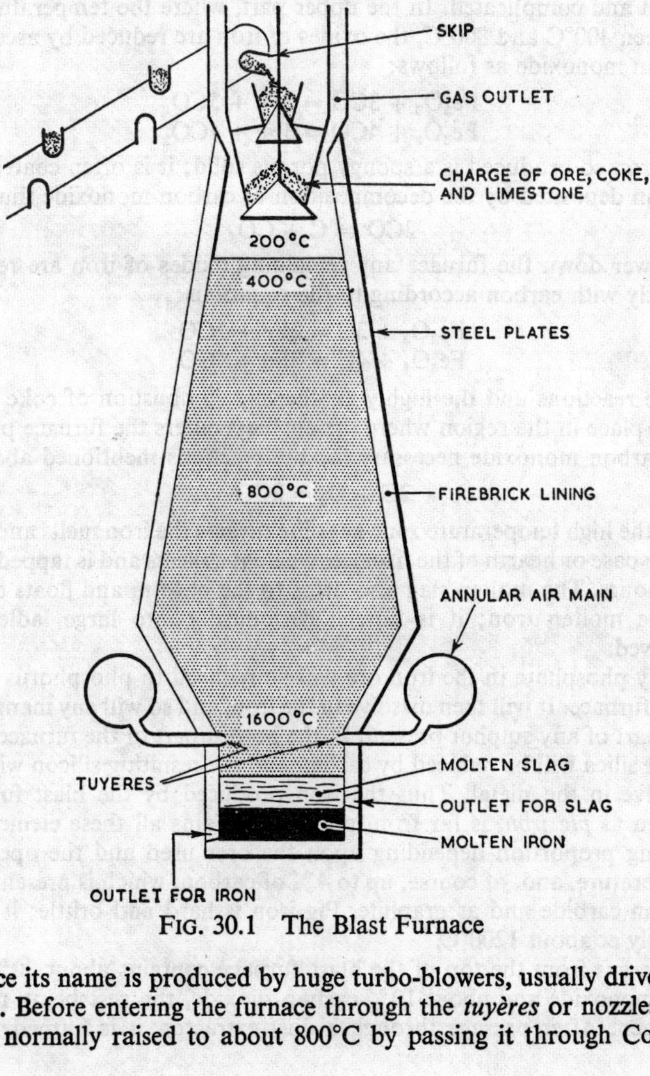

FIG. 30.1 The Blast Furnace

furnace its name is produced by huge turbo-blowers, usually driven by steam. Before entering the furnace through the *tuyères* or nozzles this air is normally raised to about 800°C by passing it through Cowper stoves previously heated to redness by burning blast furnace gas in them. This pre-heating results in a considerable saving of the expensive coke.

The blast furnace itself (Fig. 30.1) consists of a tapered cylindrical tower about 30 m in height. It is made of steel and lined with refractory bricks. The furnace is fed mechanically in such a way that no gas escapes during charging. The chemical reactions taking place inside are varied and complicated. In the upper part, where the temperature lies between 400°C and 800°C, the oxides of iron are reduced by ascending carbon monoxide as follows:

$$Fe_2O_3 + 3CO = 2Fe + 3CO_2$$
$$Fe_3O_4 + 4CO = 3Fe + 4CO_2$$

The iron so produced is a spongy porous solid; it is often coated with carbon deposited by the decomposition of carbon monoxide thus:

$$2CO \rightleftharpoons C + CO_2$$

Lower down the furnace any remaining oxides of iron are reduced directly with carbon according to the equations:

$$Fe_2O_3 + 3C = 2Fe + 3CO$$
$$Fe_3O_4 + 4C = 3Fe + 4CO$$

These reactions and the highly exothermic combustion of coke which takes place in the region where the air blast enters the furnace provide the carbon monoxide necessary for the reactions mentioned above:

$$2C + O_2 = 2CO$$

In the high temperature zone near the tuyères the iron melts and flows to the base or hearth of the furnace, where it collects and is tapped every 4–6 hours. The molten slag also drips to the bottom and floats on top of the molten iron; it is tapped periodically into large ladles and removed.

Any phosphate in the iron ore will be reduced to phosphorus in the blast furnace. It will then dissolve in the iron, and so will any manganese and part of any sulphur present. In the hottest part of the furnace some of the silica will be reduced by carbon and the resulting silicon will also dissolve in the metal. Thus the iron produced by the blast furnace, known as *pig iron*, is far from pure. It contains all these elements in varying proportion depending upon the ores used and the operating temperature, and, of course, up to 4% of carbon, which is present both as iron carbide and as graphite. Pig iron is hard and brittle; it melts sharply at about 1200°C.

The gas from the top of the blast furnace contains about 30% carbon monoxide and about 10% carbon dioxide, the rest being mostly nitrogen. After passage through a dust extractor, it is burned, being used for pre-heating the air blast and the iron ore, for raising the steam for the turbo-blower, and for firing the coke ovens, as shown in Fig. 30.2. A typical blast furnace produces over a million kilogramme of iron every twenty-four hours. It is normally run continuously for years on end until the lining needs replacement.

TYPES OF IRON:

Cast iron is pig iron which has been remelted, mixed with steel scrap, and then cooled in moulds so that it assumes a definite shape. It has the

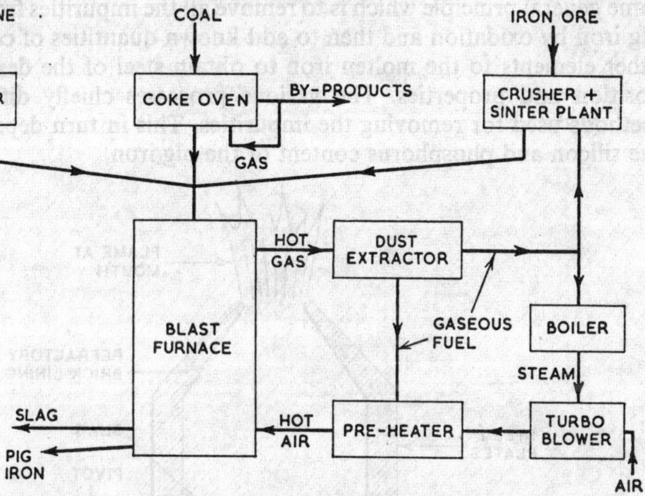

FIG. 30.2 The Manufacture of Iron

same physical properties as pig iron. It is used for making articles such as drain pipes, lamp posts, and fire grates, where cheapness is more important than strength.

Wrought iron is almost pure iron, made by heating pig iron in a reverberatory furnace lined with haematite and stirring in a current of air until all the impurities have been oxidized. It is soft but very tough and can be easily welded and forged. It is used to make chains and ornamental gates.

Steel is an alloy of iron with carbon and other elements, especially manganese, silicon, and phosphorus. The carbon content usually lies between 0.1% and 1.5%, and the other elements are often present in only small proportions. Alloy steels also exist; they may contain nickel, chromium, vanadium, tungsten, cobalt, molybdenum, etc., sometimes in appreciable amounts. They have special properties such as great hardness or resistance to corrosion which lead to specific uses. Most stainless steels contain 12%–18% of chromium and some nickel; they are used increasingly for cutlery, sink units, car bumpers, machinery, and chemical and nuclear plant. The properties of a steel depend not only on its chemical composition, but also on its heat treatment. Quenching (heating to redness and then cooling suddenly) and tempering (heating to 250°C–300°C and cooling slowly) have a big effect upon its hardness and toughness.

MANUFACTURE OF STEEL: Several processes are in use, but they share the same general principle which is to remove all the impurities from molten pig iron by oxidation and then to add known quantities of carbon and other elements to the molten iron to obtain steel of the desired composition and properties. The various processes chiefly differ in the methods used for removing the impurities. This in turn depends upon the silicon and phosphorus content of the pig iron.

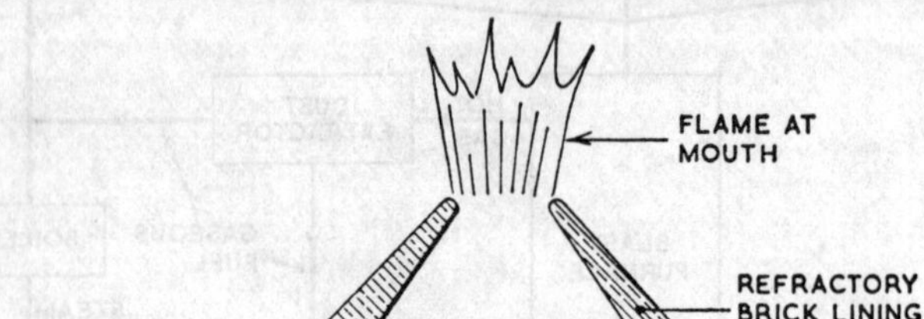

FIG. 30.3 The Bessemer Converter

The Bessemer Process: This takes place in a converter, which looks like an enormous concrete mixer, as shown in Fig. 30.3. It is made of steel lined with refractory material, and can be tilted about its centre. Its base is perforated with small holes through which air is blown during operation. Two types exist; an *acid* converter, which is lined with silica bricks, is used for dealing with pig iron with a very low phosphorus content, and a *basic* converter, which has a lining made from calcined dolomite, is used for making steel from pig iron containing between 1.6% and 2% of phosphorus.

In the acid process molten pig iron is introduced into the converter whilst in a horizontal position. It is then turned to the vertical and air is blown in for about twenty minutes. The silicon and manganese present are converted into their oxides and form a slag which floats on the surface of the molten metal. These oxidations are highly exothermic and the heat evolved maintains the contents of the converter in a molten condition despite the rise in the melting point as the impurities are removed. The carbon is oxidized to carbon monoxide which burns at the mouth of the converter. When the process is complete an iron alloy containing carbon and manganese is added to de-oxidize the steel and leave the required amounts of these elements in excess. The slag is then skimmed off and the molten steel poured into a ladle ready for moulding into ingots. In the basic process quicklime is added to convert the silica and phosphorus pentoxide formed by the air blast into a silicate and phosphate slag, which is poured off before adding the ferromanganese. This *basic slag*, as it is called, is a useful fertilizer owing to its high phosphorus content (§ 23.12).

In recent years the basic Bessemer process has been modified in two important ways. In some plants the air blast has been replaced by a mixture of oxygen and superheated steam. This gives a steel with improved physical properties because it is almost free of dissolved nitrogen. The other modification involves replacing the air blast by a jet of oxygen directed forcefully on to the surface of the molten pig iron. The oxygen penetrates into the melt and rapidly oxidizes the impurities, giving a relatively cheap and nitrogen-free product. It has the further advantage of being applicable to pig iron of only moderate phosphorus content. These new Kaldo and L.D. converters are fitted with oxygen lances which require a huge supply of oxygen provided by special plants constructed on the site (§ 24.2).

The Siemens-Martin or Open Hearth Process: A mixture of pig iron, steel scrap, and haematite ore is melted in a shallow hearth furnace as shown in Fig. 30.4. Such furnaces have a capacity of about 2×10^5 kg and are lined with either silica or calcined dolomite depending upon the phosphorus content of the charge. Producer gas and air are preheated to 1000°C–1300°C (by drawing them through firebricks over which the very hot exit gases have recently passed) and burned in the furnace. By reversing the direction of flow at frequent intervals the melt is raised to a temperature of about 1600°C. An excess of air is provided so that the silicon, phosphorus, and manganese in the melt are oxidized and converted into slag (limestone is added to phosphatic charges to facilitate this). The amount of carbon remaining in the iron is lowered to the desired level by adding haematite, which oxidizes it to the monoxide. The whole process from charging to tapping takes about ten hours, unless oxygen is used as described below. After tapping, ferromanganese is added to de-oxidize the steel as in the Bessemer process.

The open hearth process accounts for over 80% of the steel made in Great Britain and over half the world production. Its advantages over

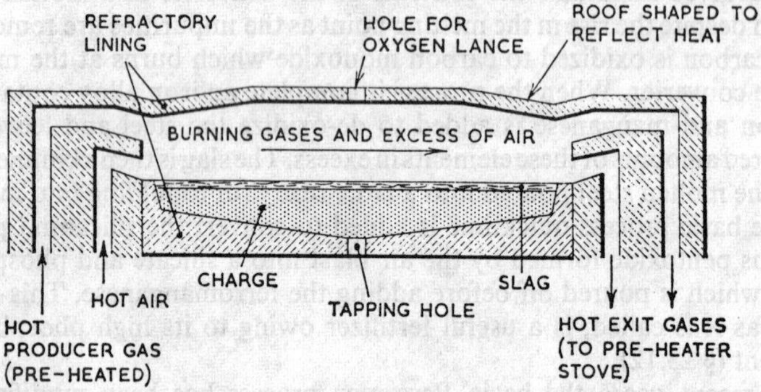

FIG. 30.4 The Open Hearth Furnace

the Bessemer process are that it is not restricted to pig iron of particular composition, that it uses large amounts of steel scrap, and that its slowness allows better control and more complete separation of the iron from the slag. It has been modified recently by using oxygen to assist in the refining, either for removing most of the easily oxidized silicon from the pig iron before introduction into the open hearth furnace, or by injecting it into the melt towards the end of the process for removing the residual carbon.

The Electric Furnace Process: This is essentially a recovery rather than a steelmaking process because almost all its raw material is steel scrap. The direct arc furnaces consist of huge crucibles lined with basic refractory material and fitted with carbon electrodes. The heat of the arc melts the scrap, which is then treated with oxygen to oxidize the impurities. Alternatively, high-frequency induction heating may be used. Electric furnaces are particularly suitable for making high quality alloy steels since the operating temperature can reach 1800°C and the composition of the steel can be accurately controlled.

The Spray Process: In this recently developed process, molten pig iron from a blast furnace is poured through high velocity jets of oxygen. The metal is broken up into a fine spray thereby exposing a large surface area to the oxygen and bringing about very rapid oxidation of the impurities. As a result the droplets of steel, dusted with powdered lime to form slag, are hot enough to melt a high proportion of scrap iron in the receiver. The process offers several big advantages and although still in the pioneering stage, it is likely to be widely adopted in future years. Oxidation is so quick that it can be operated continuously instead of as a batch process; this and the simplicity of the plant leads to low capital costs. The process lends itself to automatic control and to a minimum of refractory wear, thereby reducing labour costs.

Properties of Iron: Pure iron is a white, lustrous, fairly soft metal with a melting point of 1530°C and a density of 7.86 g cm^{-3}. It is malleable and ductile, although the presence of even small amounts of certain impurities (e.g. sulphur) make it brittle. Up to 768°C iron is ferromagnetic (i.e. it becomes strongly magnetized when placed in a magnetic field), but above this temperature it loses its magnetic properties.

Iron exists in two allotropes, γ-iron, which has the face-centred cubic structure and is stable between 906°C and 1400°C, and α-iron or δ-iron, which is body-centred cubic in structure and is stable below 906°C and over the range 1400°C–1530°C.

Reactions of Iron: On exposure to moist air, iron slowly rusts, forming a surface layer of hydrated oxide, $2Fe_2O_3 . 3H_2O$, which is unlike those formed by nickel, chromium and aluminium in being porous and not preventing further corrosion. *Rusting* is a complicated process, being primarily electrolytic in nature. Moisture and oxygen are essential, and carbon dioxide and metallic salts appear to accelerate the process. Various methods of preventing rusting are referred to in this book, e.g. painting (§ 21.15), galvanizing (§ 29.2), tinplating (§ 21.2), cadmium plating (§ 29.10), chromium plating (§ 27.7), phosphating (see below), and providing sacrificial protection (§ 18.4).

When heated above 150°C in air, iron becomes coated with the oxide, Fe_3O_4. At higher temperatures it burns in oxygen. The reversible reaction of red-hot iron with steam is used for manufacturing hydrogen, as explained in § 16.2:

$$3Fe + 4H_2O \rightleftharpoons Fe_3O_4 + 4H_2$$

Iron combines readily with chlorine (§ 30.15), sulphur (§ 30.9), and phosphorus when heated, but not with nitrogen. It unites with carbon forming a carbide, Fe_3C, which is present in steel and contributes largely to its hardness.

Iron dissolves in dilute hydrochloric and sulphuric acids liberating hydrogen and forming ferrous salts:

$$Fe + H_2SO_4 = FeSO_4 + H_2\uparrow$$

With dilute nitric acid it forms ferrous nitrate and either ammonium nitrate or the oxides of nitrogen, but concentrated nitric acid has only a momentary action on iron, rendering it *passive* (see § 22.13). Treatment with chromic acid or orthophosphoric acid has a similar effect and use is made of this to prevent corrosion (§ 23.10).

If finely divided iron is heated to 200°C with carbon monoxide under 2×10^7 N m^{-2} (200 atm) pressure, it gives a pale yellow liquid, iron pentacarbonyl, which decomposes into very pure iron when heated to 400°C:

$$Fe + 5CO \rightleftharpoons Fe(CO)_5$$

30.3. Compounds of Iron. Two series exist, *ferrous* compounds in which the iron is divalent and *ferric* compounds in which it is trivalent, the two being of similar stability.* Thus ferrous compounds are reducing agents, being easily converted into the corresponding ferric form, even by the air, and ferric compounds act as mild oxidizing agents, being readily reduced to the ferrous condition.

30.4. Iron(II) Oxide (Ferrous Oxide), FeO. This is a black powder obtained by heating ferrous oxalate in an inert atmosphere (it inflames in air):

$$(COO)_2Fe.3H_2O = FeO + CO\uparrow + CO_2\uparrow + 3H_2O$$

It is a basic oxide, dissolving in acids to form ferrous salts.

30.5. Iron(II) Hydroxide (Ferrous Hydroxide), $Fe(OH)_2$. This is obtained as a white gelatinous precipitate (which oxidizes rapidly on standing, turning green then brown), when an alkali is added to a solution of a ferrous salt:

$$FeSO_4 + 2NaOH = Fe(OH)_2\downarrow + Na_2SO_4$$

It is only incompletely precipitated by ammonium hydroxide in the presence of ammonium chloride because its solubility product is usually only just exceeded under these conditions (§ 9.9). To avoid this difficulty the solution is boiled with a little concentrated nitric acid at the beginning of Group III of the Qualitative Analysis Scheme, so that any iron present is precipitated as the less soluble ferric hydroxide.

Ferrous hydroxide dissolves in acids giving ferrous salts, including a carbonate, bicarbonate, and sulphide.

30.6. Iron(II) Sulphate (Ferrous Sulphate), $FeSO_4$. This is obtained as

* These are often distinguished by referring to them as iron(II) compounds and iron(III) compounds respectively, as explained in Appendix I.

pale green crystals of the heptahydrate, $FeSO_4.7H_2O$, commonly known as green vitriol, by dissolving iron in dilute sulphuric acid and then concentrating and cooling the solution. Use of air-free acid minimizes formation of the ferric salt. The crystals slowly effloresce in air and turn yellow owing to partial oxidation to basic ferric sulphate. On heating the following changes occur:

$$FeSO_4.7H_2O \rightleftharpoons FeSO_4.H_2O + 6H_2O$$
$$FeSO_4.H_2O \rightleftharpoons FeSO_4 + H_2O$$
$$2FeSO_4 = Fe_2O_3 + SO_2\uparrow + SO_3\uparrow$$

This decomposition is of historical interest in that the earliest sulphuric acid was made in this way.

Ferrous sulphate forms double salts; one of these, ferrous ammonium sulphate, $FeSO_4.(NH_4)_2SO_4.6H_2O$, known as Mohr's salt, is used in volumetric analysis for standardizing oxidizing agents. It is not a complex salt (§ 9.16), behaving in solution merely as a mixture of its component sulphates, but it is less readily oxidized by the air than ferrous sulphate.

A solution of ferrous sulphate absorbs nitric oxide readily forming a brown compound, $FeSO_4.NO$, which decomposes when heated releasing nitric oxide again (§ 22.11).

Ferrous sulphate is used in the manufacture of ink, in tanning and dyeing, and for killing moss.

30.7. Iron(II) Chloride (Ferrous Chloride), $FeCl_2$. The anhydrous salt is obtained by passing dry hydrogen chloride over red-hot iron filings:

$$Fe + 2HCl = FeCl_2 + H_2$$

It is a white solid which is deliquescent and very soluble in water. When iron is dissolved in hydrochloric acid in the absence of air and the solution is concentrated, blue-green crystals of the tetrahydrate, $FeCl_2.4H_2O$, are obtained.

30.8. Iron(II) Carbonate (Ferrous Carbonate), $FeCO_3$. This occurs naturally as siderite or spathic iron ore. It is obtained as a white precipitate by adding sodium carbonate solution to a solution of a ferrous salt in the absence of air. Although insoluble in water, it dissolves in the presence of carbon dioxide forming the bicarbonate:

$$FeCO_3 + CO_2 + H_2O \rightleftharpoons Fe(HCO_3)_2$$

On exposure to the air this solution gives up carbon dioxide again and the resulting carbonate is hydrolysed and oxidized, depositing red-brown hydrated ferric oxide. It is probable that changes such as these occur during the rusting of iron.

30.9. Iron(II) Sulphide (Ferrous Sulphide), FeS. This compound is made commercially by melting together iron and sulphur; when so prepared it is grey-black in colour and usually contains appreciable amounts of uncombined iron. Pure ferrous sulphide is precipitated when a sulphide solution is added to a solution of a ferrous salt:

$$Na_2S + FeSO_4 = FeS\downarrow + Na_2SO_4$$

It is soluble in dilute acids liberating hydrogen sulphide, which is its main use in the laboratory:

$$FeS + H_2SO_4 = FeSO_4 + H_2S\uparrow$$

30.10. Ferrous Disulphide, FeS_2. This occurs naturally as pyrites and marcasite. It is like brass in appearance, being yellow and lustrous. When heated in air it is converted into ferric oxide and sulphur dioxide, the latter being extensively used for making sulphuric acid:

$$4FeS_2 + 11O_2 = 2Fe_2O_3 + 8SO_2\uparrow$$

30.11. Iron(III) Oxide (Ferric Oxide), Fe_2O_3. This occurs as haematite and as the hydrate, limonite; it is very widely distributed being responsible for the red-brown colour of the soil in many areas. It is obtained by strongly heating ferrous sulphate or ferric hydroxide. It is used as a red pigment and as an abrasive polishing powder (Jeweller's rouge).

30.12. Iron(III) Hydroxide (Ferric Hydroxide), $Fe(OH)_3$. This is obtained as a red-brown gelatinous precipitate by adding an alkali to a solution of a ferric salt:

$$FeCl_3 + 3NaOH = Fe(OH)_3 + 3NaCl$$

It is a very weak base, dissolving in strong acids to form ferric salts which are extensively hydrolysed in solution.

30.13. Tri-iron Tetroxide (Ferroso-ferric Oxide), Fe_3O_4. This oxide, which occurs naturally as magnetite or lodestone, is highly magnetic. It is made by heating iron in air or steam:

$$3Fe + 4H_2O \rightleftharpoons Fe_3O_4 + 4H_2$$

It is inert, and concentrated nitric acid has no action on it, but it is easily reduced when heated with carbon or carbon monoxide.

30.14. Iron(III) Sulphate (Ferric Sulphate), $Fe_2(SO_4)_3$. This is made by heating ferrous sulphate with concentrated sulphuric acid:

$$2FeSO_4 + 2H_2SO_4 = Fe_2(SO_4)_3 + SO_2\uparrow + 2H_2O$$

When heated strongly it decomposes thus:

$$Fe_2(SO_4)_3 = Fe_2O_3 + 3SO_3\uparrow$$

A solution of ferric sulphate is acidic owing to hydrolysis. It readily forms sparingly soluble double salts called alums (§ 19.13), of which ammonium iron alum, $(NH_4)_2SO_4 . Fe_2(SO_4)_3 . 24H_2O$, and potash iron alum, $K_2SO_4 . Fe_2(SO_4)_3 . 24H_2O$, are the best known. They are extensively used as mordants in dyeing.

30.15. Iron(III) Chloride (Ferric Chloride), $FeCl_3$. The anhydrous salt is made by passing dry chlorine over heated iron in the apparatus shown in Fig. 30.5:

$$2Fe + 3Cl_2 = 2FeCl_3$$

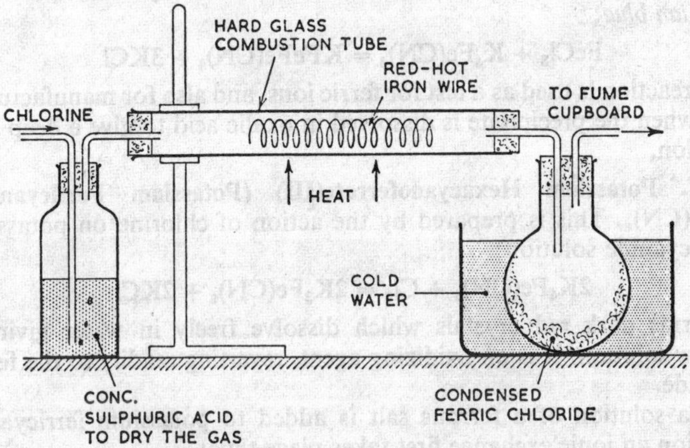

FIG. 30.5 Preparation of Ferric Chloride

The chloride, which sublimes at 315°C, is condensed in the cooled receiver. Density measurements show that the vapour consists of Fe_2Cl_6 molecules at 400°C, but that these dissociate into $FeCl_3$ molecules at higher temperatures. The hexahydrate, $FeCl_3 . 6H_2O$, crystallizes when iron is dissolved in hydrochloric acid in the presence of chlorine and the solution is concentrated and cooled. Ferric chloride is highly deliquescent and very soluble in water, forming a brown solution which is acidic as a result of hydrolysis and which often gives a colloidal solution of ferric hydroxide on standing:

$$FeCl_3 + 3H_2O \rightleftharpoons Fe(OH)_3 + 3HCl$$

30.16. Potassium Heracyanoferrate(II) (Potassium Ferrocyanide), $K_4Fe(CN)_6$. This is obtained from spent oxide (§ 20.13), which contains Prussian blue formed by the absorption of hydrogen cyanide from the crude coal gas. Heating with slaked lime gives calcium ferrocyanide, which is leached out and treated with potassium carbonate, giving yellow crystals of the trihydrate, $K_4Fe(CN)_6 . 3H_2O$.

The complex ion $[Fe(CN)_6]^{4-}$ is so stable that its solution does not give the reactions of simple ferrous or cyanide ions (see § 9.16 and § 14.5). For the same reason it is not poisonous. With solutions of copper salts it gives a brown precipitate of cupric ferrocyanide. If potassium ferrocyanide is boiled with dilute nitric acid, crystals of sodium nitroprusside, $Na_2[Fe(CN)_5NO].2H_2O$, are obtained on neutralizing the solution with sodium carbonate.

With solutions of a ferric salt, potassium ferrocyanide solution gives a dark blue precipitate of potassium ferric ferrocyanide, known as *Prussian blue*:

$$FeCl_3 + K_4Fe(CN)_6 = KFeFe(CN)_6 + 3KCl$$

This reaction is used as a test for ferric ions, and also for manufacturing ink, when the precipitate is dissolved in oxalic acid to give a deep blue solution.

30.17. Potassium Hexacyanoferrate(III) (Potassium Ferricyanide), $K_3Fe(CN)_6$. This is prepared by the action of chlorine on potassium ferrocyanide solution:

$$2K_4Fe(CN)_6 + Cl_2 = 2K_3Fe(CN)_6 + 2KCl$$

It forms dark red crystals which dissolve freely in water giving a yellow solution. It is an oxidizing agent, reverting readily to the ferrocyanide.

If a solution of a ferrous salt is added to potassium ferricyanide solution an ionic exchange first takes place thus:

$$Fe^{2+} + [Fe(CN)_6]^{3-} = Fe^{3+} + [Fe(CN)_6]^{4-}.$$

As a result a dark blue precipitate of potassium ferric ferrocyanide, known as *Turnbull's blue*, is obtained. This precipitate, which is used to detect ferrous salts, is now known to be identical with the Prussian blue referred to in the previous section.

30.18. Detection of Iron. Compounds of iron are detected and distinguished as ferrous or ferric by the tests shown in Table 30.1. In remembering the reactions with cyanide complexes it is helpful to realize that the dark blue precipitate is obtained only when the different valency states of iron are mixed, i.e. ferr*ous* with ferr*i*cyanide or ferr*ic* with ferr*o*cyanide solutions. The reaction with potassium thiocyanate solution is a particularly sensitive test for ferric ions (see § 20.12) and use is often made of it to determine when a ferric salt has been completely reduced to the ferrous condition.

30.19. Summary. The world production of iron far exceeds that of all the other metals added together. There are several reasons for this. Its ores are abundant, rich and readily accessible. They are easily reduced

TABLE 30.1. TESTS TO DISTINGUISH FERROUS FROM FERRIC SALTS

Reagent	*Ferrous Soln. Iron(II)*	*Ferric Soln. Iron(III)*
Sodium hydroxide soln.	White ppt., turning green and then brown	Red-brown gelatinous ppt.
Potassium ferrocyanide soln.	White or pale blue ppt.	Dark blue ppt.
Potassium ferricyanide soln.	Dark blue ppt.	Greenish-brown soln. (no ppt.)
Potassium thiocyanate soln.	No pronounced colour	Blood-red colour
Hydrogen sulphide	Black ppt.	Whitish-yellow ppt. of sulphur

to the metal by processes which can be operated continuously on a large scale with consequent low labour costs. The versatility of the product when alloyed with other elements in steel gives it a great range of uses. Its biggest drawback is its susceptibility to corrosion. It is an interesting thought that the principal use of several other metals, notably tin, zinc, cadmium, and chromium, is to prevent this corrosion.

In its chemistry iron is in every respect a typical transition metal. It gives coloured ions in solution, it readily forms stable complexes, it is an active catalyst, it shows variable valency and it is strongly magnetic. Many of its compounds show a resemblance to those of manganese, cobalt, and nickel, although iron differs from them in forming ferrous and ferric salts of about equal stability, whereas in these neighbouring elements the divalent salts are much more stable.

31

Nuclear Chemistry

31.1. Introduction. The view was put forward in chapter 14 that all chemical reactions could be explained in terms of rearrangements of electrons brought about by their transfer or sharing between the atoms concerned, and that throughout all these reactions the atomic nuclei remain unchanged. In this chapter we shall consider certain special changes which directly involve the nucleus of the atom (e.g. radioactivity, atomic transmutations, and nuclear fission and fusion) and the products which are formed as a result of these changes.

31.2. Natural Radioactivity. In 1896 Becquerel, whilst investigating the property of fluorescence, discovered that a crystal of uranium salt spontaneously emitted radiation which could penetrate matter opaque to light and affect a photographic plate. Experiment showed that metallic uranium itself and all its compounds possessed this remarkable property of radioactivity. Soon afterwards it was found that thorium also was *radioactive*. In 1898 Pierre and Mme Curie discovered that a mineral of uranium called pitchblende was considerably more radioactive than could be explained from its uranium content. They concluded that another radioactive substance must be present as well and set out to isolate it. Eventually, as a result of prodigious effort under very trying circumstances they obtained specimens of two new and intensely radioactive elements, polonium and radium. The latter is several million times as radioactive as uranium, mass for mass, but 10^3 kg of pitchblende contains less than 200 milligramme of radium. In 1899 another radioactive element, called actinium, was detected in pitchblende.

Investigation of these substances showed that they were invariably associated with other radioactive substances which were formed from them during radioactive changes and which could be separated from them by chemical means. On isolation of these products it was found that for each of them the intensity of the radiation emitted decreased with time, but at widely different rates. About fifty naturally occurring radioactive substances are now known, although some of them are only weakly active (like the isotope potassium-40) and others are comparatively rare. Most of them belong to three distinct series known as the uranium–radium series, the thorium series, and the actinium series, which involve isotopes of the twelve elements from thallium to uranium. All isotopes with an atomic number of 84 or above are radioactive.

This phenomenon of radioactivity, although at first regarded as a mere curiosity, soon became a subject of tremendous importance and interest, for few studies have provided more convincing proof of the existence of atoms or thrown more light on their structure.

31.3. Characteristics of Radioactivity. Radioactive substances continually and spontaneously emit penetrating radiation at a rate which, unlike any chemical change, is completely independent of temperature and pressure. This radiation affects a photographic plate, causes gases through which it passes to ionize, and makes certain substances (e.g. crystalline zinc sulphide) fluoresce. It also has powerful physiological effects, some of them cumulative with time. For example, living cells suffer irreparable damage when exposed to intense radiation, and milder doses can cause anaemia or leukaemia in mammals or can induce cancerous growths. Even highly attenuated radiation may have serious genetic consequences. For these reasons great care should always be taken when working with radioactive substances to avoid exposure to dangerous amounts of radiation.

Radioactivity is always associated with the escape of a large amount of energy considering the mass of material taken—the energy released is, weight for weight, about a million times as great as that liberated during any chemical reaction. That it is a property of the atom is shown by the fact that if an element is radioactive, so are all its compounds to a degree which is proportional to the content of the element.

31.4. Types of Radiation. Naturally radioactive substances have been shown to emit three distinct kinds of radiation:

ALPHA RAYS (α-rays): Experiment shows these to be fast-moving streams of positively-charged particles, each having a mass four times that of the hydrogen atom and bearing two units of positive charge. Thus each α-particle is, in fact, the nucleus of a helium atom. This has been demonstrated experimentally by collecting α-particles and identifying them as helium spectroscopically. Careful measurements have shown that one gramme of radium, together with the decay products naturally associated with it, emits 1.4×10^{11} α-particles per second, providing 0.16 cm^3 of helium gas in a year.

α-rays have very little penetrating power and a sheet of paper or a layer of air about 7–8 cm thick is sufficient to absorb them completely.

They do, however, have a very powerful ionizing effect upon any gas through which they pass and it is this property which is most used for detecting them and measuring their intensity. Textbooks of physics should be consulted for details of the various electrical counting devices based upon this principle. It must suffice here to mention the important cloud chamber method developed by Wilson for detecting the actual paths followed by individual α-particles. In this apparatus the α-particle is allowed to pass through a gas which has just previously been supersaturated with water vapour. The ions formed in the track of the α-particle act like dust in serving as centres for the condensation of the water vapour, so that the path of the particle is revealed in much the same way that a vapour trail makes visible the course of a high-flying aircraft too small to be seen with the naked eye. The tracks persist long enough to be photographed so that a permanent record can be obtained of the movement of each α-particle through the gas.

Beta Rays (β-rays): Investigation shows that these are very fast-moving streams of electrons. They are about a hundred times as penetrating as α-rays, but they are much less effective in ionizing gases.

Gamma Rays (γ-rays): These are a form of electromagnetic radiation and consequently they travel with the speed of light. They resemble X-rays, but have an even shorter wavelength. They possess very great

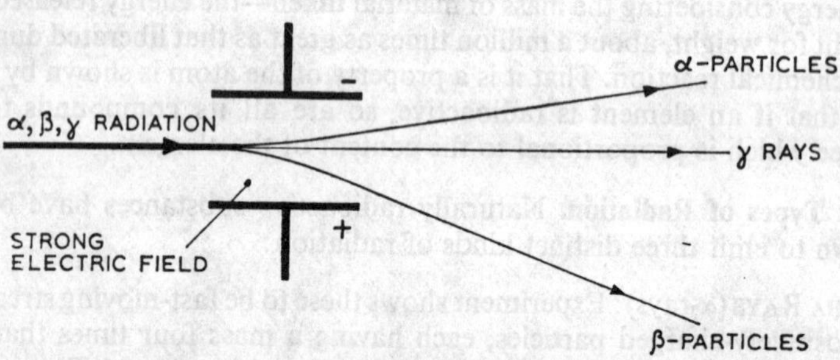

Fig 31.1 Separation of the Three Types of Atomic Radiation

powers of penetration. For example, to completely absorb the γ-radiation emitted by a sample of radium requires a layer of lead about 5 cm thick or correspondingly greater thicknesses of other substances, since absorption is roughly proportional to the density of the material.

The properties of these three forms of radiation are compared and contrasted in Table 31.1, and their separation illustrated in Fig. 31.1.

TABLE 31.1. TYPES OF RADIATION

	α-Rays	*β-Rays*	*γ-Rays*
Nature	Helium nuclei, He^{2+}	Electrons	Electromagnetic radiation
Velocity	About $\frac{1}{20}$th speed of light	Varies widely—up to speed of light	Speed of light
Mass	4 units	$\frac{1}{1836}$th of a unit	None
Charge	2 units (positive)	1 unit (negative)	None
Magnetic field	Deflected one way	Strongly deflected the opposite way	Not deflected
Electric field	Deflected one way	Strongly deflected the opposite way	Not deflected
Fluorescent screen	Scintillates	Scintillates	Scintillates
Photographic plate	Affected	Affected	Affected
Effects on gases	Rapid ionization	Ionization	Ionization
Relative penetration	1	*ca.* 100	5 000–10 000

31.5. The Scattering of α-Particles. If thin sheets of metal foil are bombarded with α-particles, results of great importance are obtained which have been of tremendous help in arriving at our present theories of atomic structure. It is found, provided the foil is thin enough, that most of the α-particles pass straight through it, but that some are deflected through small angles and a very small proportion are 'reflected' back from the foil (see Fig. 31.2). It was these few remarkable deflections through large angles* which led Rutherford to conclude in 1909 that

* One can recapture something of the excitement of this experiment from Rutherford's remark, made many years later in a lecture, to the effect that it was as if you had fired a 15-in shell at a piece of tissue paper and it bounced back and hit you.

atoms must largely consist of empty space with their mass concentrated into a very small positively-charged central particle which he

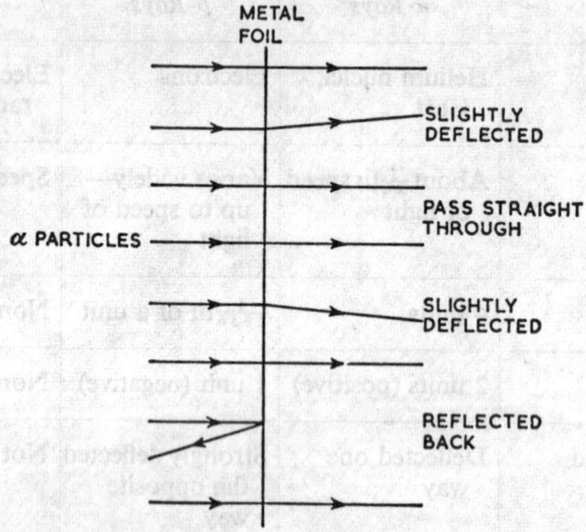

FIG. 31.2 The Scattering of α-Particles

called the *nucleus*. From his measurements on the deflected α-particles he calculated that in any case the radius of the nucleus was less than 10^{-14} m, i.e. less than $\frac{1}{10\,000}$th of the radius of the atom as a whole.

31.6. Explanation of Radioactivity. In 1903 Rutherford and Soddy put forward the theory of nuclear disintegration to explain the phenomenon of radioactivity. They suggested that in a radioactive element the atomic nucleus was unstable and tended to disintegrate spontaneously (i.e. without the necessity of excitation of any kind) into the nucleus of a different element. Thus when an atom emits an α-particle it changes into another with atomic number two less than before and with atomic weight four units less; emission of a β-particle raises the atomic number by one but hardly affects the atomic weight. The new element formed in this way is itself often unstable and after an interval which may vary from a few microseconds to millions of years it undergoes further disintegration with the release of more energy. A series of changes may occur, therefore, until finally a stable nucleus is produced. By reasoning along these lines it was predicted that the uranium and thorium series of naturally radioactive substances would both end in the production of stable isotopes of lead of weight 206 and 208 respectively. The experimental confirmation of these predictions was one of the triumphs of the disintegration theory.

The theory also solved the problem of the source of the enormous amount of energy released during radioactivity as this was now seen to come from the nucleus of the unstable atom. Exactly when any given atom disintegrates releasing its excess of energy is purely a matter of chance and defies prediction, but clearly for any one element the number of nuclei breaking up per minute, say, is likely to be a fixed proportion of the number of unstable atoms present. This explains the experimental observation that the intensity of radiation from radioactive substances decreases exponentially with time, since as the nuclei disintegrate fewer and fewer unstable atoms remain. The rate at which radiation decays varies very widely from one radioactive substance to another and can therefore be used to characterize a substance. It is usually expressed in terms of its *half-life*, which is the time taken for the intensity of radiation to fall to half of its original value, or, what amounts to the same thing, the time taken for half of the total number of atoms in the sample to disintegrate. It may be regarded as an 'expectation of life' for any one atom of that substance, indicating how long on statistical grounds it can be expected to survive before undergoing disintegration. For a very unstable radioelement the half-life may be measured in fractions of a second (e.g. for polonium-212 it is 3.0×10^{-7} second), but half-lives of hours, days, or years are more common and some important isotopes have immensely long half-lives (e.g. for uranium-238 it is 4.51×10^{9} years and for thorium-232 it is 1.39×10^{10} years). For such substances as these the rate of emission shows no detectable change over any measurable period and the half-life has to be determined by indirect means.

Lastly, in attributing radioactivity to sub-atomic changes, the theory explains why it is entirely unaffected by physical conditions such as temperature and pressure which have such a marked effect upon the rates of ordinary chemical changes.

31.7. Atomic Transmutations. In 1919 Rutherford observed that when nitrogen gas was exposed to a strong source of α-rays, a supply of fast-moving protons was obtained. He concluded that these protons had been expelled from the nuclei of the nitrogen atoms by the bombarding action of the α-particles, an interpretation which was later confirmed by cloud chamber photographs. These showed that the other product of the collision was an isotope of oxygen weighing 17 units, so that the change can be represented by the equation:

$$^{14}_{7}N + ^{4}_{2}He = ^{17}_{8}O + ^{1}_{1}H$$

where the superscript numbers indicate mass numbers* and the sub-

* In the U.S.A. it is usual to write the mass numbers after the symbol, thus N^{14}, but this book follows the practice recommended by the International Union of Pure and Applied Chemistry in 1953.

script numbers refer to atomic number. This was the first time that an artificially induced atomic *transmutation*, as the change of one element into another is called, had been observed, although spontaneous transmutations were continually occurring, of course, in radioactive substances. Artificial transmutations such as this were, however, very difficult to achieve with particles of such low energy and only about one in every quarter of a million α-particles brought about a nuclear change.

In 1930 it was found that bombardment of certain light elements such as beryllium or boron with α-particles did not give protons but a new type of radiation instead. Chadwick showed that this consisted of a stream of uncharged particles called *neutrons* with a mass approximately that of a proton, so that the transmutation taking place was as follows:

$$^{9}_{4}Be + ^{4}_{2}He = ^{12}_{6}C + ^{1}_{0}n$$

The neutrons, which are ejected with great speed and are consequently called *fast neutrons*, are extraordinarily penetrating and will pass through up to 50 cm of lead. They are not easily detected because being uncharged they have little or no ionizing action upon gases through which they pass.

Many hundreds of different atomic transmutations can now be carried out by bombarding various elements with fast-moving atomic particles such as neutrons, protons, deuterons (nuclei of deuterium atoms), and α-particles. In general, neutrons, especially slow-moving ones, are the most useful projectiles because they are so readily available in the atomic pile (see § 31.11) and because, being uncharged, they are not repelled electrostatically by the nuclei of atoms with high atomic numbers. More advanced books should be consulted for details of these changes and it must suffice here to stress three important points. Firstly, in every case there is a change in either the mass or the charge (and often in both) of the nucleus of the atom undergoing bombardment. Secondly, many of the products formed are isotopes which do not exist naturally; some 700, most of them radioactive, have now been made artificially in this way. Here is the near-fulfilment of the alchemist's dream. Admittedly it is not economical to make gold in this way, but many of these radioactive isotopes have proved of greater value than gold in view of their industrial importance and uses, some of which are described in § 31.12. Lastly, by suitable transmutations we can make atoms with atomic numbers above 92, thereby synthesizing new elements not previously known to man. In the last twenty-five years with the help of powerful cyclotrons eleven such elements have been made, neptunium, plutonium, americium, curium, berkelium, californium, einsteinium, fermium, mendelevium, nobelium, and lastly, in 1961, lawrencium, although as yet some of them exist only in exceptionally small quantities. These *transuranic elements*, as they are called, are members of the actinon series (§ 13.8) and are therefore very similar chemically. They are all radioactive with half-lives which are short compared with the age of the earth, so that even if they did occur naturally at one time, they have long since disintegrated into lighter, stabler substances.

31.8. The Group Displacement Law. If an atom of a radioactive element emits an α-particle, its mass decreases by four units and its nuclear charge decreases by two units, since the escape of the α-particle removes two protons and two neutrons. If a β-particle is lost there is negligible change in mass, but the positive charge on the nucleus increases by one owing to the loss of a negatively-charged electron from the nucleus. Thus the loss of an α-particle causes an element to change into another with an atomic number two less than before and occupying a space two places before in the Periodic Table, whereas the loss of a β-particle causes an element to change into another with an atomic number one higher than before. This Group Displacement Law, as it is called, can be illustrated by considering the natural decay of uranium-235, the first three stages of which are:—

$$\underset{\text{uranium-235}}{{}^{235}_{92}U} = \underset{\text{thorium-231}}{{}^{231}_{90}Th} + \underset{\alpha\text{-particle}}{{}^{4}_{2}He} \quad \text{(half-life: } 7.1 \times 10^8 \text{ years)}$$

$$\underset{\text{thorium-231}}{{}^{231}_{90}Th} = \underset{\text{protactinium-231}}{{}^{231}_{91}Pa} + \underset{\beta\text{-particle}}{{}^{0}e^{-1}} \quad \text{(half-life: 25.6 hours)}$$

$$\underset{\text{protactinium-231}}{{}^{231}_{91}Pa} = \underset{\text{actinium-227}}{{}^{227}_{89}Ac} + \underset{\alpha\text{-particle}}{{}^{4}_{2}He} \quad \text{(half-life: } 3.4 \times 10^4 \text{ years)}$$

These changes can be summarized in the following way:—

$${}^{235}_{92}U \xrightarrow[(7.1 \times 10^8 \text{ yrs})]{-\alpha} {}^{231}_{90}Th \xrightarrow[(25.6 \text{ hrs})]{-\beta} {}^{231}_{91}Pa \xrightarrow[(3.4 \times 10^4 \text{ yrs})]{-\alpha} {}^{227}_{89}Ac$$

Similarly the 238 isotope of uranium undergoes radioactive decay with the loss of an α-particle and the production of thorium-234; this is an extremely slow change, the half-life being about 4.5×10^9 years:

$$\underset{\text{uranium-238}}{{}^{238}_{92}U} = \underset{\text{thorium-234}}{{}^{234}_{90}Th} + \underset{\alpha\text{-particle}}{{}^{4}_{2}He}$$

It should be noted that for nuclear changes such as these there must be a balance with respect to mass number and atomic number on both sides of the equation. The mass number, which is written as a superscript, is the relative weight of the atom to the nearest whole number and is equal to the sum of the neutrons and protons in the nucleus.

31.9. The Neutron–Proton Ratio. For the lighter elements with atomic number below twenty, the numbers of neutrons and protons present in the nucleus of the stable isotopes is either identical or very similar and the neutron–proton ratio is near unity. In elements of higher atomic number, however, the most stable nuclei contain more neutrons than protons and the neutron–proton ratio rises steadily with atomic number as shown in Fig. 31.3. From this Figure it is clear that the ratio of

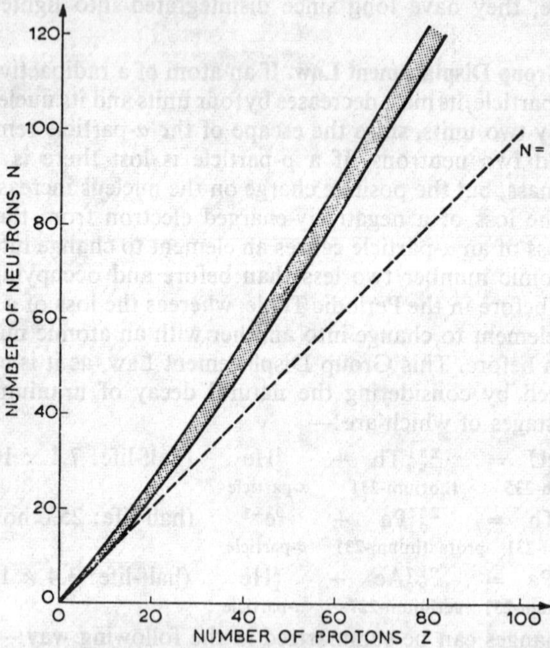

FIG. 31.3 The Neutron–Proton Ratio in Stable Nuclei

neutrons to protons bears a definite relationship to the stability of the nucleus and for stable atoms varies between unity for the elements of low atomic number and a value of about 1.5 for those with atomic number around eighty. Atoms with a neutron–proton ratio lying outside the band of stability shown in the Figure tend to be unstable and to undergo radioactive decay.

It is interesting to note that nearly all stable nuclei contain either an even number of protons or an even number of neutrons, or, as is true in the majority of cases, even numbers of both. In fact, only four stable nuclei are known in which the numbers of protons and neutrons are both odd, ${}^{2}_{1}H$, ${}^{6}_{3}Li$, ${}^{10}_{5}B$, and ${}^{14}_{7}N$.

31.10. Nuclear Fission. When certain heavy atoms (e.g. uranium-235 or plutonium-239) are bombarded with neutrons of suitable energy, their nuclei break up into two smaller nuclei with the release of two or more new fast-moving neutrons and a large amount of energy. This process is known as *nuclear fission* and the fragments formed, which are intensely radioactive, as *fission products*. If things are so arranged that at least one of the neutrons expelled during each fission collides with another fissionable atom, then a chain reaction can be set up which will maintain itself indefinitely. This is the principle of the *atomic pile* or *reactor*, and also, when the chain reaction builds up explosively quickly, of the so-called *atomic bomb*. The same amount of energy may be released in both instances, but in the reactor the rate of nuclear fission is carefully controlled so that the heat evolved can be dissipated as it is produced.

To maintain a chain reaction in some fissile material it is essential to have so much of it present in one lump that only a small proportion of the neutrons escape from the surface without causing further fission. This minimum quantity of fissile material is called the *critical mass*; for pure uranium-235 it is of the order of 30 kilogramme. Thus an atomic bomb can be exploded by suddenly bringing together two portions of the material each weighing just over half the critical mass.

If every neutron released during nuclear fission is to bring about the fission of other atoms so that the chain reaction builds itself up at maximum speed, as it must in an atomic bomb, then the fissile material must be free of all neutron-absorbing impurities. Now ordinary uranium as extracted from its ores consists of three isotopes, uranium-238 99.28%, uranium-235 0.715%, and uranium-234 0.005%. Thus ordinary uranium cannot be used for making a bomb and it is first necessary to separate the fissile uranium-235 from the neutron-absorbing isotopes. This presents a very difficult problem, since all the isotopes are identical chemically and even their physical differences are only slight. The problem has been solved by converting uranium into its volatile hexafluoride UF_6 and separating the fluorides of the isotopes by means of gaseous diffusion (see § 3.6).

31.11. The Atomic Pile. In a nuclear reactor a rapid build up of the chain reaction is not an essential or even desirable feature. It is possible to obtain a self-sustaining chain reaction by using ordinary uranium which has been considerably enriched with additional uranium-235 provided some material is present in the reactor to slow down the fast neutrons released during fission so that not too many of them are captured by the atoms of 238 isotope, which is a strong absorber of fast neutrons, before they can collide with the fissile uranium-235 which only captures slow neutrons readily. The best *moderators*, as such materials are called, are graphite and heavy water, D_2O, since they are both low absorbers of neutrons. Graphite is cheaper, but when deuterium oxide is used the reactor can be made small enough to be used for propelling ships or submarines.

Most of the neutrons not used in causing further fission are absorbed by the uranium-238 converting it into uranium-239, which then changes into neptunium and plutonium by radioactive decay thus:

$$^{238}_{92}U + ^{1}_{0}n \rightarrow ^{239}_{92}U \xrightarrow{-\beta} ^{239}_{93}Np \xrightarrow{-\beta} ^{239}_{94}Pu$$

In this way plutonium, which is a valuable fissile material for use in bombs and reactors, can be made from uranium in large quantities. After a suitable period the uranium fuel is removed from the reactor and the plutonium recovered by chemical methods, but the intense radioactivity of the plutonium and the various fission products mixed with it makes the whole process hazardous and expensive.

The fissionable material in the reactor is usually contained in metal cans to facilitate loading and removal and to prevent the escape of the highly radioactive fission products. The metal chosen must be strong and easy to machine and weld and it must be a good conductor of heat. It should have as high a melting point as possible and, above all, it must be a low absorber of neutrons. An alloy of magnesium has been widely used, and zirconium alloys and stainless steel are also suitable. At one time beryllium promised to be the best metal for the purpose, but brittleness after cooling eventually led to its rejection in 1962 (§ 18.2).

A coolant is essential in a pile to remove the considerable heat energy produced as a result of nuclear fission; air, water, carbon dioxide, helium, and liquid sodium have all been employed for this purpose. Where the pile is being used for producing power a heat exchanger is fitted so that the heat can be used for generating steam to drive turbines. An important part of a nuclear reactor is the system for preventing the fission process from getting out of control. The usual method is to insert rods of boron steel or cadmium into the pile, since these are strong absorbers of neutrons and will rapidly bring the chain reactions to a halt. Lastly, the inside of a pile is an intense source of neutrons and γ-rays, so it is necessary to protect the operators by surrounding it with a thick shield of steel and concrete. Unfortunately this makes it a heavy and very bulky object, ill suited for propelling cars or aircraft.

Apart from its uses for producing plutonium or for generating electric power, a pile can also be used for effecting artificial transmutations. Elements inserted temporarily into the pile are subjected to intense irradiation which converts them into radioactive isotopes, many of which, as described in the next section, are of great industrial and medical importance.

The typical processes taking place in an atomic pile containing enriched uranium are shown diagrammatically in Fig. 31.4.

If the fissile material in the core of a reactor is surrounded with a layer of uranium-238, then many of the neutrons escaping from the core are captured by the uranium-238 producing uranium-239 which soon decays through neptunium into plutonium-239, itself a valuable fissile

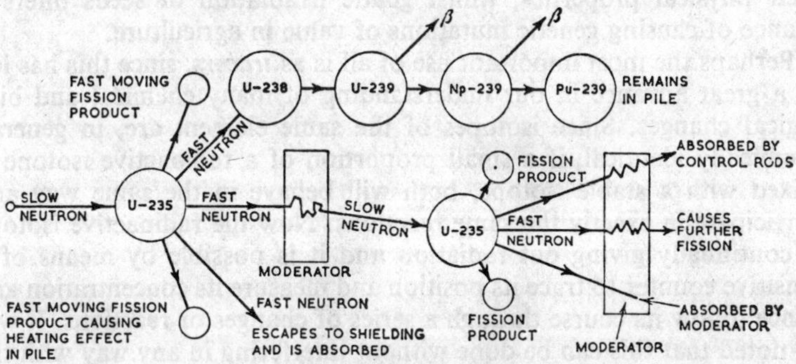

FIG. 31.4 Diagrammatic Representation of Fission Processes in an Atomic Pile

material. Similarly, if an outer layer of thorium-232 is used for surrounding the core of a reactor, it is converted into thorium-233 which then decays through protactinium into the highly fissile isotope uranium-233. In these ways it is possible to produce more fissile material than is consumed. Such a system is referred to as a *breeder reactor* and is of great interest because in effect it makes use of the more abundant isotopes such as uranium-238 and thorium-232 as sources of nuclear energy.

31.12. Uses of Radioisotopes. Those isotopes which are intense sources of γ-rays have found extensive use in medicine for the radiation treatment of cancer. In particular cobalt-60 (half-life 5.25 years) has largely replaced the more expensive radium for this purpose. Small pellets of other radioisotopes, e.g. gold-198 (half-life 2.73 days), have been implanted in tumours where their radiation has served to destroy the cancerous cells around them. Iodine-131 (8 days) is especially useful for treating cancer of the thyroid gland because of its natural tendency to accumulate there after ingestion.

Cobalt-60 is also widely used for industrial radiography since it makes possible rapid inspection of welds and castings for cracks or flaws. Other isotopes which are less powerful sources are used for very accurately measuring the thickness of materials such as paper, plastic, or metal foil by measuring the attenuation of their radiation on passing through the material. Such sources are also used for ionizing the air and thereby preventing the accumulation of static electricity during industrial processes where there is a risk of fire, and for detecting the position of leaks in buried pipelines, and for sterilizing food, drugs, and bandages. Irradiation of certain plastics has been found to improve their physical properties, whilst gentle irradiation of seeds offers a chance of causing genetic mutations of value in agriculture.

Perhaps the most important use of all is as *tracers*, since this has led to a great advance in our understanding of many chemical and biological changes. Since isotopes of the same element are, in general, chemically identical, if a small proportion of a radioactive isotope is mixed with a stable isotope, both will behave in the same way and participate in exactly the same reactions. Now the radioactive isotope is continually giving out radiation and it is possible by means of a sensitive counter to trace its position and measure its concentration and hence follow its course through a series of changes or reactions. It will be noted that this can be done without interfering in any way with the normal processes taking place in the system. Employed in this way radioisotopes can be used to study the mechanisms and kinetics of chemical reactions, to investigate complex changes like photosynthesis, to control industrial processes particularly in metallurgy, to measure the wear caused by friction, and to facilitate a study of the circulation of the blood and of the metabolism of plants and animals. For example, because a growing tumour preferentially absorbs iodine, injected radioactive iodine compounds can be used to locate the exact position of a brain tumour in the skull by means of its radiation. When administered to the body the radioisotopes are usually incorporated in ordinary compounds, e.g. ${}^{24}_{11}Na$ (15 hours) in sodium chloride solution, ${}^{32}_{15}P$ (14 days) in phosphates, ${}^{131}_{53}I$ (8 days) in sodium iodide, ${}^{14}_{6}C$ (5600 years) as a sugar or amino-acid, ${}^{3}_{1}H$ (12.3 years) as tritium oxide, and so on.

The ${}^{32}_{15}P$ isotope of phosphorus has proved particularly useful in studying the metabolism of plants. By using it as a tracer in phosphates, it has been possible to discover at what rate the growing plant absorbs phosphate fertilizers from the soil under various conditions and how the absorbed phosphates are distributed throughout the stems and leaves. Phosphorus-32 has also been used in the treatment of certain types of cancer because it tends to be concentrated by the body at the centres of malignant growth.

The use of carbon-14 for dating organic material is referred to in § 20.4. Isotopes with even longer half-lives have been used by archaeologists and geologists for estimating the age of rocks and even the earth itself.

31.13. Nuclear Fusion. From the relative values of the mass defects (§ 2.19) it is clear that the synthesis of helium from atoms of hydrogen would result in the release of an enormous amount of energy. It is almost certain that transmutations of this kind are the main source of energy in the stars, many of which consist principally of hydrogen and helium. Direct fusion of four hydrogen atoms cannot yet be achieved on earth, but by exploding a fission bomb it has been found possible in the so-called *hydrogen bomb* to produce high enough temperatures (e.g. 10 000 000°C) locally to enable tritium to fuse with deuterium thus:

$$^{2}_{1}D + ^{3}_{1}H = ^{4}_{2}He + ^{1}_{0}n$$

The tritium, a radioactive isotope of hydrogen, is made *in situ* by neutron bombardment of lithium deuteride:

$$^{7}_{3}Li + ^{1}_{0}n = 2.^{3}_{1}H + ^{2}_{1}D$$

Devices of this kind are not subject to any critical size and can therefore be used to produce explosions of terrifying power; one tested in 1961 was equivalent to the simultaneous explosion of 5×10^{10} kg of T.N.T.

Just as the energy provided by nuclear fission can be released in a controlled manner in an atomic pile and used for peaceful purposes, so it may eventually become possible to obtain from thermo-nuclear changes an abundance of energy for the benefit of all mankind.

32

Introduction to Organic Chemistry

32.1. Organic Chemistry. At first this term was applied only to substances of plant or animal origin—the name was taken from their association with living *organisms*. Substances such as sugars, fats, and alcohols, were imagined to differ from the others (called *inorganic*) in that they were endowed with a mysterious *vital force* which made it impossible for man to prepare them from purely inorganic materials. In 1828 Wohler prepared urea, which was indisputably organic, from ammonium cyanate (which could be obtained from inorganic sources) and showed that there was no essential difference between organic and inorganic substances. In the next thirty years other similar preparations followed and the vital force theory was eventually abandoned. Nevertheless the term organic chemistry was retained, gradually acquiring its modern meaning as *the chemistry of the compounds of carbon.* Apart from a few compounds such as the oxides, the disulphide, and the tetrachloride, which it is conventional to regard as inorganic, carbon compounds are grouped together in this separate branch of chemistry because of the important features they have in common and because many of them are of great medical and biochemical interest.

32.2. Characteristics of Organic Compounds. Despite the great number of organic compounds (there are so many that they far outnumber all the compounds of all the other elements added together), it is a striking feature of organic chemistry that only a very few elements are involved. Carbon is always present, by definition, and so usually is hydrogen. Compounds composed of these two elements alone are known as *hydrocarbons*; there are many thousands of them, including such important compounds as methane CH_4, ethylene C_2H_4, acetylene C_2H_2, benzene C_6H_6, and toluene C_7H_8. Many organic compounds contain oxygen as well, and sometimes nitrogen or halogen atoms are present. Compounds containing phosphorus, sulphur, and metals are occasionally encountered, but other elements are rare in organic chemistry and most organic compounds are composed of carbon, hydrogen and oxygen only. That these few elements can form such an immense number of different compounds arises principally from the ability of carbon atoms to combine with each other to form stable chain and ring molecules of great size and complexity. No other element possesses this property to anything like the same extent. In all these compounds carbon is invariably tetravalent, forming covalent bonds with neighbouring atoms so that the outermost shell of its atom is completely filled. It is this fact which makes its compounds stable despite their high molecular weights, which reach 60 000–80 000 in some proteins and polymers and even higher values in certain carbohydrates. The variety of organic compounds is much increased because carbon combines with hydrogen, oxygen, nitrogen, and the halogens with almost equal facility.

The vast majority of organic compounds are entirely covalent (§ 14.4). It follows that in general they are volatile (most of them boil below 300°C), non-conducting, and immiscible with water. Many of them dissolve readily, however, in non-polar solvents such as benzene or ether. Being covalent their reactions tend to be much slower than the ionic reactions which are so common in inorganic chemistry and this difference has several important practical consequences. For example, their preparations often involve heating, thorough mixing, and the use of a catalyst, and even so they may take several hours to complete. Side reactions often occur at the same time, reducing the yield and making the purification a lengthy and difficult process. In such circumstances great care has to be exercised to choose just those conditions which promote the desired reaction and ensure that it predominates over its rivals. Moreover some organic reactions are reversible, and in these the yield obtained is also greatly influenced by conditions.

One characteristic of covalency is that the bonds are directed in space. For example, the four single bonds associated with the carbon atom are known to be arranged tetrahedrally at an angle of 109° 28′ to each other, so that they point towards the corners of a regular tetrahedron when the carbon atom is placed at its centre. This important conclusion was first reached by van't Hoff and Le Bel in 1874, and was confirmed many years later by X-ray analysis (§ 4.5), and electron diffraction. As a result of this directional effect, some carbon compounds give rise to *stereoisomerism*, a phenomenon of such importance that it is considered separately in chapter 48.

Most organic compounds are inflammable and burn exothermically in air. Indeed most fuels are organic (e.g. coal, wood, oil, petrol, and natural gas), and their combustion provides our main source of chemical energy. If the proportion of carbon in an organic compound is high, it burns with a smoky, luminous flame and leaves a sooty residue. Another characteristic of organic compounds is their thermal instability; many organic compounds decompose into simpler molecules when heated above 500°C. This tendency is sometimes of considerable commercial importance as in the cracking of petroleum (§ 33.11) and the carbonization of coal (§ 20.13).

32.3. Organic Formulae. In view of the great number of organic compounds, it is convenient to classify them into two main groups according to their structure. All compounds whose molecules are composed of *chains* of carbon atoms are termed *aliphatic*, whereas those whose molecules contain six-membered *rings* of carbon atoms are known as *aromatic* compounds. Certain well-defined characteristics of the latter are referred to in § 43.8.

The simplest formula of a compound is called its *empirical formula*. It is derived directly from the percentage composition as determined by analysis and it indicates only the proportions in which the elements are present. For example, acetic acid has the empirical formula CH_2O because analyis shows that it contains 12 g of carbon to every 2 g of hydrogen and 16 g of oxygen, i.e. carbon, hydrogen, and oxygen atoms in the ratio 1 : 2 : 1 respectively.

More useful is the *molecular formula* because this conveys the additional information of how many atoms are present in the molecule. It is obtained from the empirical formula by multiplying by a whole number, usually one, two, or three. Thus in our example the molecular formula of acetic acid is $C_2H_4O_2$ (i.e. twice the empirical formula), and the molecule contains eight atoms, two of carbon, four of hydrogen, and two of oxygen.

In inorganic chemistry, a molecular formula is usually sufficient to distinguish a particular compound from any other, but in organic chemistry this is not so and several or even many different substances may exist with the same molecular formula. For instance, the formula $C_2H_4O_2$ may equally well refer to acetic acid or to methyl formate, a completely different substance with the same molecular formula. This phenomenon is known as *isomerism*, and compounds having the same molecular formulae are called *isomers* or *isomerides*. The various types of isomerism are listed in chapter 48.

Because of the prevalence of isomerism in organic chemistry, a molecular formula is often not sufficient to identify a given substance, and some kind of *structural formula* which reflects the arrangement of the atoms within the molecule has to be employed. Two such formulae are in common use. The *graphic formula* shows which atoms are linked together by using conventional symbols for single, double, and triple bonds. Except in cases of stereoisomerism, it serves to distinguish one substance from another at a glance. Thus the graphic formula of acetic

acid is usually written as

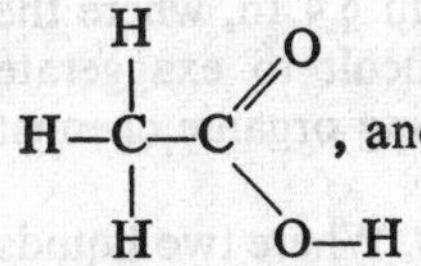

, and that of methyl formate as

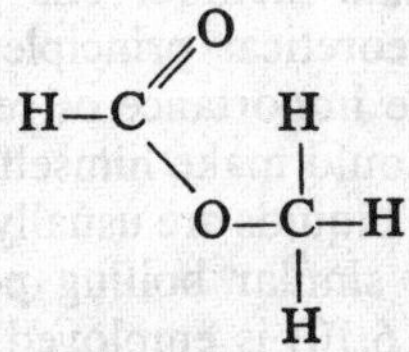

It is important to realize what the graphic formula does not do. It does not, for example, show the actual arrangement of the atoms in space and the true angles between the bonds. Neither of these molecules is planar, as the graphic formula would suggest, but for convenience three-dimensional molecules are projected on to the plane of the paper in this way.

The graphic formula is cumbersome and except where it is desired to show the mechanism of a reaction, the *condensed structural formula* (also known as the *constitutional formula*) is generally preferred because it takes up much less space and time. The way the atoms are linked together is shown by writing their symbols in conventional groups separated by dots, e.g. $CH_3.COOH$ and $H.COOCH_3$ for acetic acid and methyl formate respectively. Another advantage of a condensed formula is that it emphasizes the groupings of the atoms and makes it easier to recognize which functional groups are present in the molecule.

32.4. Investigation of an Organic Compound. When a new organic compound is discovered it is systematically investigated in the following way:

1. Purification—it is essential to obtain a pure sample of the substance before it can be analysed.

2. Purity Check—a criterion of purity is applied to confirm that the purification, which may be long and tedious, is complete.

3. Qualitative Analysis—this determines which elements are present.

4. Quantitative Analysis—this provides the percentage composition, from which the empirical formula can be calculated.

5. Determination of Molecular Weight—this decides the molecular formula by settling what multiple it is of the empirical formula.

6. Determination of Structure—this stage may prove the most difficult of all for large or complicated molecules. It enables the structural formula to be decided.

7. Synthesis—i.e. preparation of the substance from smaller, simpler molecules of known structure.

Each of these steps will now be considered in turn.

32.5. Purification. If the compound is a solid, recrystallization is the usual method. The process is described in detail in § 4.16, where the theoretical principles are also discussed. It is difficult to exaggerate the importance of recrystallization and every aspiring organic chemist should make himself thoroughly proficient in it.

Liquids are usually purified by distillation (§ 3.18). Where two liquids of similar boiling point have to be separated, fractional distillation (§ 6.10) is employed. If the substance decomposes when heated to its normal boiling point, distillation under reduced pressure (see Fig. 3.16) is used. Steam-distillation (§ 6.13) is valuable when the required substance is immiscible with water. Chromotography, which is described in §§ 12.11 to 12.15, is also sometimes a useful method. Most organic compounds have to be dried, even when one of the above-mentioned methods is used. If chemical drying is applied, care must be taken to select a drying agent which does not react with the substance (e.g. anhydrous calcium chloride must not be used for drying alcohols, phenols, or amines).

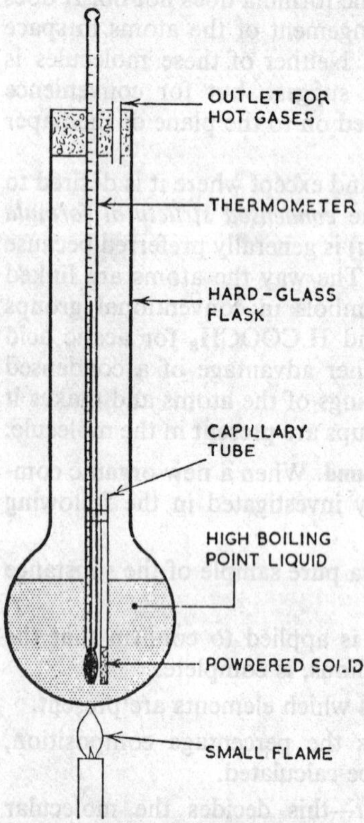

FIG. 32.1 Melting Point Apparatus

32.6. Purity Check. It is a matter of experience that a pure compound has a sharp melting point, whereas impure substances melt gradually over a range of several degrees. Thus sharpness of melting point is a useful criterion of purity. The apparatus used is shown in Fig. 32.1. The compound is finely powdered and a little of it is introduced into a thin-walled capillary tube which has previously been sealed at one end. This tube is attached to a thermometer as shown and dipped into a bath of suitable liquid (e.g. medicinal paraffin), which is then heated gently. If the substance is pure it melts completely within a range of about one kelvin as the temperature slowly rises.

A pure liquid has a sharp and constant boiling point at any particular pressure, and boils away completely leaving no residue. These features are used, therefore, as criteria of purity. There is a slight danger that a constant boiling point mixture (§ 6.11), which also has a sharp and constant boiling point at constant pressure, might be confused with a pure substance. This difficulty is best overcome by redistilling at a different pressure; a mixture will then give a distillate of different composition which when distilled again at the original pressure will have a changed boiling point.

32.7. Qualitative Analysis. The elements in an organic compound are detected in the following ways:

CARBON AND HYDROGEN: The presence of these two elements is usually assumed, but it can be demonstrated by heating a little of the compound with an excess of powdered cupric oxide as in Fig. 32.2. The carbon present is oxidized to carbon dioxide, which turns lime-water milky on

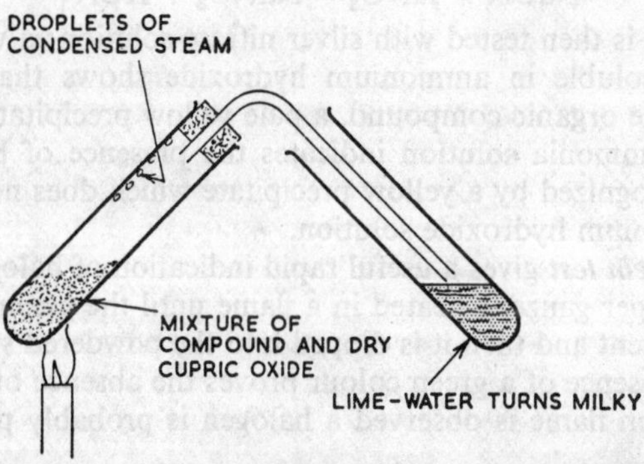

FIG. 32.2 Detection of Carbon and Hydrogen

shaking, and the hydrogen is oxidized to steam which condenses on the cooler part of the tube. Cupric oxide is hygroscopic and must be thoroughly dried beforehand by heating it in a crucible.

NITROGEN: This is detected by the *Lassaigne test*. A little of the substance is heated with a small lump of metallic sodium in an ignition tube, gently at first and then to red heat. The tube is then plunged into a mortar half full of distilled water and ground to a powder. The contents are filtered and the filtrate is tested for the presence of sodium cyanide in the following way. First it is boiled with a freshly prepared solution of ferrous sulphate to form sodium ferrocyanide solution:

$$6NaCN + FeSO_4 = Na_4Fe(CN)_6 + Na_2SO_4$$

A few drops of ferric chloride solution are then added and the solution is acidified with concentrated hydrochloric acid when the dark blue colour of Prussian blue (§ 30.16) is apparent if nitrogen is present in the compound:

$$Na_4Fe(CN)_6 + FeCl_3 = NaFe[Fe(CN)_6] + 3NaCl$$

Sometimes only a greenish solution is obtained, but when it is filtered small particles of Prussian blue are usually visible on the filter paper, confirming the presence of nitrogen.

HALOGENS: The Lassaigne test is also used for detecting these; chloride, bromide, and iodide are identified in the filtrate from the sodium fusion by acidifying with dilute nitric acid and adding silver nitrate solution in the usual way. However, if nitrogen has been found to be present, the filtrate from the sodium fusion is first acidified with dilute nitric acid and boiled to half bulk in a fume cupboard to expel all the cyanide as volatile hydrogen cyanide:

$$NaCN + HNO_3 = NaNO_3 + HCN\uparrow$$

The product is then tested with silver nitrate solution; a white precipitate freely soluble in ammonium hydroxide shows that chlorine is present in the organic compound, a pale yellow precipitate moderately soluble in ammonia solution indicates the presence of bromine, and iodine is recognized by a yellow precipitate which does not dissolve at all in ammonium hydroxide solution.

The *Beilstein test* gives a useful rapid indication of halogens. A piece of clean copper gauze is heated in a flame until the green colour is no longer apparent and then it is dipped into the powdered substance and reheated. Absence of a green colour proves the absence of halogens; if a bright green flame is observed a halogen is probably present in the compound.

SULPHUR: Any sulphur present in the organic compound will be converted into sodium sulphide solution by the Lassaigne test and can be recognized by adding a freshly prepared solution of sodium nitroprusside to the filtrate, when an intense purple colour appears.

METALS: Any metals present in an organic compound can be detected by heating the compound strongly in air and performing the usual tests upon the solid residue.

OXYGEN: No satisfactory method exists for detecting oxygen in an organic compound and its presence is usually inferred from the reactions of the compound and from the percentage composition of the other elements.

32.8. Quantitative Analysis. A textbook of practical organic chemistry should be consulted for full experimental details. The aim here is to describe the methods in outline so that their principles are understood.

CARBON AND HYDROGEN: The simplified apparatus is shown in Fig. 32.3. A known weight of the organic compound is heated in a stream of pure dry oxygen until it is completely oxidized. The hydrogen present is converted into water which is absorbed in weighed tubes containing anhydrous calcium chloride. The carbon is oxidized to its dioxide (dried cupric oxide is usually added to the combustion tube to ensure

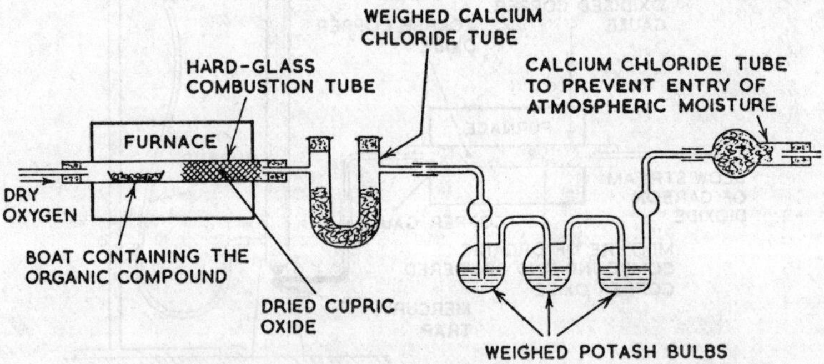

FIG. 32.3 Apparatus for Estimation of Carbon and Hydrogen

this) and the gas is absorbed in weighed bulbs containing caustic potash solution or soda lime. If the organic compound contains nitrogen, halogens, or sulphur, the method has to be modified to prevent absorption of by-products. The calculation of the percentage composition follows from the results because $\frac{2}{18}$ths of the mass of water absorbed is hydrogen and $\frac{12}{44}$ths of the mass of carbon dioxide absorbed is carbon, and the mass of the sample taken is known.

NITROGEN: Two methods are in use:

Dumas Method: A known mass of the substance is heated with an excess of powdered cupric oxide in a combustion tube from which all air is excluded by carbon dioxide. The apparatus is shown in Fig. 32.4. Carbon, hydrogen, and sulphur present in the organic compound are converted into their oxides and absorbed in alkali, which also takes up any free halogen. The nitrogen present comes over as the element, any oxides of nitrogen being immediately decomposed by the heated copper gauze. The nitrogen is collected in a nitrometer and its volume measured under known conditions of temperature and pressure so that its mass can be calculated.

Kjeldahl's Method: A known mass of the substance is boiled with concentrated sulphuric acid and anhydrous sodium sulphate until the solution is colourless, which converts the nitrogen in the organic com-

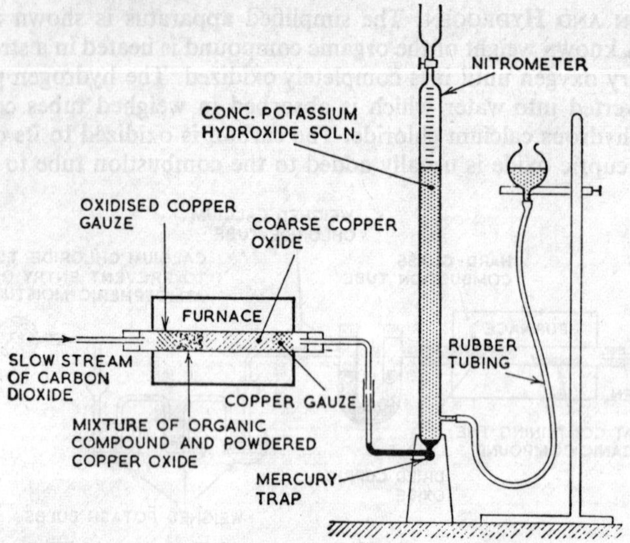

FIG. 32.4 Apparatus for Estimation of Nitrogen by the Dumas Method

pound into ammonium sulphate. This is decomposed by boiling it with an excess of sodium hydroxide solution. The ammonia liberated is absorbed in a known volume of standard acid and estimated by back titration. This method is inaccurate when applied to compounds containing nitrogen atoms linked directly to oxygen or nitrogen, but it works well for amines, proteins, and similar substances.

HALOGENS: Carius' method is used, in which a known mass of the substance is heated for several hours with fuming nitric acid and silver nitrate solution in a sealed tube. Under these conditions the chlorine, bromine, or iodine present is converted into a precipitate of silver halide, which is filtered off, washed, dried, and weighed.

SULPHUR: Again Carius' method is used. A known mass of the organic compound is heated strongly in a sealed tube with fuming nitric acid, which oxidizes the sulphur present to sulphuric acid. After cooling, an excess of barium chloride solution and some hydrochloric acid are added and the precipitate of barium sulphate is filtered off, washed, dried, and weighed.

OXYGEN: It is difficult to estimate the percentage of oxygen in an organic compound experimentally, and it is usual to determine it by subtracting the percentages of all the other elements from one hundred.

32.9. Determination of Empirical Formula. When analysis is complete and the percentage composition of the substance is known, it is possible to calculate its empirical formula. The method is illustrated by the following example:

Example: What is the empirical formula of an organic compound whose percentage composition is C 52.2%, H 13.1%?

Since the percentages do not add up to 100, we can assume that the remainder is oxygen, i.e. O 34.7%. To find the ratio of the atoms present we must divide the percentage of each element by its respective atomic weight and then divide the lowest number obtained into the rest thus:

Element	*% by mass*	*at. wt.*	$\frac{\%\ by\ mass}{at.\ wt.}$	*ratio of atoms*
C	52.2	12	4.35	2.00
H	13.1	1	13.1	6.03
O	34.7	16	2.17	1.00

The substance contains, therefore, carbon, hydrogen, and oxygen in the ratio 2 : 6 : 1 (within the limits of experimental error) and its empirical formula is C_2H_6O.

32.10. Determination of Molecular Formula. In most cases this is done by finding the molecular weight in order to decide what multiple the molecular formula is of the empirical formula. Since this multiple is necessarily a whole number, even an approximate value of the molecular weight will usually suffice. If the substance is volatile, and most organic compounds are, its molecular weight is best determined by measuring its vapour density by one of the methods described in chapter 5. Some organic compounds decompose, however, when heated (e.g. urea and cane sugar), and for such substances a colligative method is used such as depression of freezing point or elevation of boiling point. These methods are described in detail in §§ 5.11–5.14. The measurement of osmotic pressure (§ 5.17) is a method of particular value for compounds of high molecular weight, such as proteins, for which other methods are not suitable.

The molecular formulae of gaseous hydrocarbons are generally determined by *eudiometry*. A known volume of the gas is mixed with a known volume of oxygen, which must be present in excess, and sparked repeatedly in a eudiometer (Fig. 32.5) until no further change in volume occurs. After cooling again to room temperature the volume of gas is noted. On shaking with alkali solution carbon dioxide is absorbed, and the contraction which takes place is measured. The residue is assumed to be excess oxygen. The following example shows how the molecular formula is calculated from typical experimental results:

Example: 10 cm^3 of a gaseous hydrocarbon were mixed with 60 cm^3 of oxygen. After sparking and cooling, 45 cm^3 of gas remained, and when shaken with

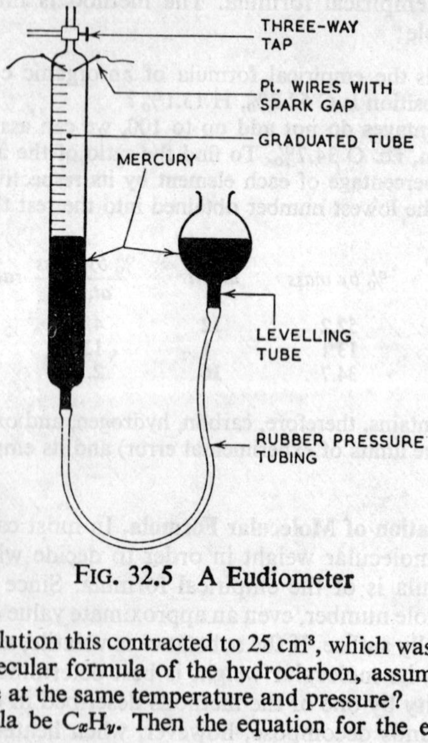

FIG. 32.5 A Eudiometer

caustic potash solution this contracted to 25 cm^3, which was entirely oxygen. What is the molecular formula of the hydrocarbon, assuming all measurements were made at the same temperature and pressure?

Let the formula be C_xH_y. Then the equation for the explosion can be written:

$$C_xH_y + \left(x + \frac{y}{4}\right)O_2 = xCO_2 + \frac{y}{2}H_2O$$

$$1 \text{ molecule} + \left(x + \frac{y}{4}\right) \text{ molecules give } x \text{ molecules} + \frac{y}{2} \text{ molecules}$$

$$\therefore 1 \text{ volume} + \left(x + \frac{y}{4}\right) \text{ volumes give } x \text{ volumes} + \frac{y}{2} \text{ volumes} \quad \text{(i)}$$

by the converse of Avogadro's hypothesis (§ 1.13).

But from the data given, 10 cm^3 of hydrocarbon combine with 60–25, i.e. 35 cm^3 of oxygen to give 20 cm^3 of carbon dioxide.

∴ 1 cm^3 of hydrocarbon + 3.5 cm^3 of oxygen give 2 cm^3 of carbon dioxide (ii)

* At room temperature the steam will condense to water and occupy negligible volume.

Equating coefficients in (i) and (ii):

$$x + \frac{y}{4} = 3.5 \quad \text{and} \quad x = 2$$

$$\therefore \frac{y}{4} = 1.5 \quad \text{and} \quad y = 6$$

giving the molecular formula C_2H_6 (ethane).

Special methods are available for determining the molecular weights of organic acids. These substances are often not amenable to the methods previously mentioned because they tend to decompose when heated, to dissociate into ions in aqueous solution, and to associate in organic solvents. An acid is first converted into its silver salt, which is insoluble and readily obtainable in a pure condition. Such salts leave only metallic silver on strong heating. A weighed amount of this salt is then heated in a crucible to constant weight, and the mass of the residue noted. The equivalent weight of the acid is the number of parts by mass of it corresponding to that mass of silver salt which yields 107.9 parts by mass of silver. If the basicity of the acid is known, then the molecular weight is given by the relationship: basicity × equivalent weight = molecular weight.

A similar method is used for finding the molecular weight of organic bases such as amines. The chloroplatinate (formula $B_2H_2PtCl_6$, where B is a monoacidic base), is prepared and a known mass of it is ignited in air. From the mass of the residual metallic platinum, the equivalent weight of the base can be calculated. If the acidity is known the molecular weight can be deduced.

32.11. Determination of Structural Formula. Once its molecular formula is known, the structure of a simple organic compound can be inferred from its reactions, having regard to valency considerations. For example, if it reacts with phosphorus pentachloride giving hydrogen chloride, then it contains a hydroxyl group, –OH, and if it undergoes a series of addition reactions, it is unsaturated and its molecule contains double or triple bonds. In this way the arrangement of the atoms within the molecule can be decided. When the molecule is large and consists of many atoms, the observation of its chemical reactions, whilst helping to reveal the grouping of those atoms, is rarely decisive enough to settle the structure, and in such cases physical methods can provide very valuable aid. For solids the most important of these is undoubtedly X-ray crystallography, described briefly in § 4.5, but for gaseous substances electron diffraction is particularly useful. Other methods which are frequently used include the study of infra-red and Raman spectra and the measurement of magnetic susceptibility and dipole moment, (§ 4.9), but details of these are beyond the scope of this book.

The process of determining structure is made considerably easier by the existence of a number of *functional groups*, the characteristic properties of which are well known, and which are therefore readily recognized. These functional groups are of great importance in organic chemistry because the chemical properties of a substance are essentially those of the groups it contains. When two or more functional groups occur in one molecule, the properties of one are often modified or influenced by the presence of the other, but despite this it is possible to predict the likely chemical behaviour of a substance from a knowledge of its structural formula and even to 'design' molecules with desired properties by building them from appropriate groups. It is for these reasons that so much emphasis is placed upon these functional groups in the following chapters.

32.12. Synthesis. This is the ultimate aim in investigating an organic compound because by successful synthesis the structural formula is confirmed beyond doubt. Many organic compounds which were originally derived solely from plant or animal sources (e.g. alcohols, dyes, vitamins and drugs), have since been synthesized and thereby made more cheaply and in abundant quantities. Complete synthesis of a compound from its elements is very rarely practicable, however, and the usual practice is to devise a means of making it from readily available materials such as those in natural gas or the products provided by the refining or cracking of petroleum or the carbonization of coal.

33

Saturated Hydrocarbons and Petroleum

33.1. Introduction. An organic compound is said to be *saturated* when all the covalent linkages in its molecule are single bonds; when double or triple bonds are present it is termed *unsaturated.* The most important saturated hydrocarbons are the *paraffins* or *alkanes*, a series with the general molecular formula C_nH_{2n+2}, e.g. methane CH_4, ethane C_2H_6, and so on.

33.2. Methane, CH_4.

OCCURRENCE: Methane is given off by swamps and stagnant ponds, when it is known as 'marsh gas', and it is found in badly ventilated coal mines as the dreaded 'fire damp'. Natural gas (§ 33.9), which issues from the ground in great quantities in certain parts of the world, consists mostly of methane.

LABORATORY PREPARATION: Several methods are available:

1. Pure methane is prepared by reducing methyl iodide with nascent hydrogen obtained by the action of a zinc–copper couple on aqueous alcohol:

$$CH_3I + 2\overline{H} = CH_4\uparrow + HI$$

The couple can be made by immersing granulated zinc in copper sulphate solution until it is covered with deposited copper. Fig. 33.1 shows the apparatus used; the purpose of the U-tube is to remove any methyl iodide vapour which may be carried over with the methane. The gas is normally collected over water, but if necessary it can be dried with phosphorus pentoxide and collected by downward displacement of air.

2. Less pure methane, containing some hydrogen and ethylene, is prepared by heating a finely-powdered mixture of anhydrous sodium acetate and soda-lime (quicklime which has been slaked with sodium hydroxide solution):

$$CH_3.COONa + NaOH = CH_4\uparrow + Na_2CO_3$$

The soda-lime behaves as sodium hydroxide in this reaction, but is

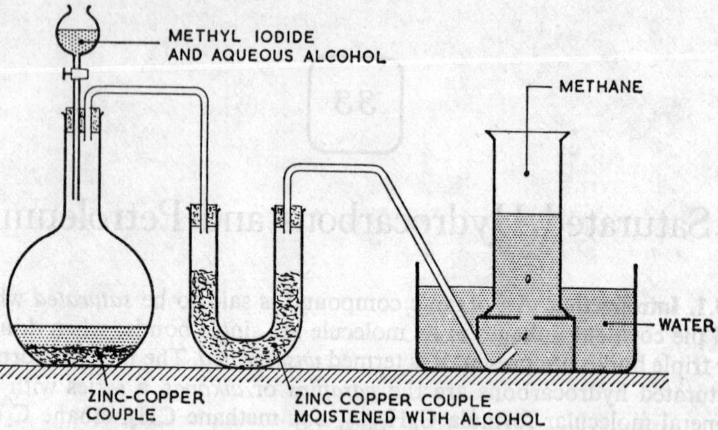

FIG. 33.1 Preparation of Methane from Methyl Iodide

preferred because it is not deliquescent and does not attack the glass. The apparatus is shown in Fig. 33.2.

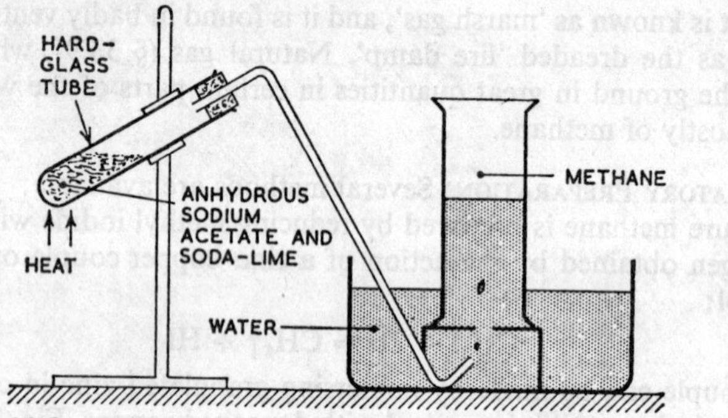

FIG. 33.2 Preparation of Methane from Sodium Acetate

3. Impure methane is also obtained when aluminium carbide is treated with cold water:

$$Al_4C_3 + 12H_2O = 3CH_4\uparrow + 4Al(OH)_3$$

MANUFACTURE: The main source is natural gas (§ 33.9), from which it is obtained by liquefaction. It is also made by carbonizing coal (about 40% of coal gas is methane), and by 'cracking' petroleum (see § 33.11), and by the bacterial decomposition of sewage.

PROPERTIES AND REACTIONS OF METHANE: It is a colourless, odourless gas, b.p. −162°C, only slightly soluble in water.

1. It burns in air or oxygen with a non-luminous flame:

$$CH_4 + 2O_2 = CO_2 + 2H_2O$$

Mixtures of methane with air or oxygen can explode violently when ignited; this has been the cause of many accidents in mines.

2. Methane reacts explosively with chlorine when the gases are exposed to sunlight:

$$CH_4 + 2Cl_2 = C + 4HCl$$

3. In diffused light methane reacts quietly with chlorine or bromine (but not with iodine) to give a series of products by the successive replacement of hydrogen atoms by halogen atoms:

$$CH_4 + Cl_2 = HCl + CH_3Cl \text{ (methyl chloride or chloromethane)}$$
$$CH_3Cl + Cl_2 = HCl + CH_2Cl_2 \text{ (methylene dichloride)}$$
$$CH_2Cl_2 + Cl_2 = HCl + CHCl_3 \text{ (chloroform or trichloromethane)}$$
$$CHCl_3 + Cl_2 = HCl + CCl_4 \text{ (carbon tetrachloride)}$$

Although used to some extent for manufacturing these compounds, especially the tetrachloride, this is a poor way of preparing them in the laboratory because it gives a mixture of products from which the isolation of any one is difficult and tedious.

These four reactions are good examples of *substitution*, which may be defined as the replacement of hydrogen by other atoms or groups. Substitution is a characteristic reaction of saturated compounds since in them all the available valency linkages are already in use and the formation of any new compound necessarily involves replacing atoms already in the molecule. To emphasize the idea of substitution these compounds are sometimes named chloromethane, dichloromethane, and so on, but the names given above are widely used.

Apart from these reactions, methane is very inert. It does not react at ordinary temperatures and pressures with fuming sulphuric acid, concentrated nitric acid, alkalis, bromine water, potassium permanganate solution, or phosphorus pentoxide.

USES OF METHANE: Its main use is as a fuel, either by itself or mixed with other gases, as in coal gas. It can be pumped from its sources through pipelines, or it can be liquefied and transported in bulk. It is also used for making hydrogen, carbon black, carbon tetrachloride, acetylene, carbon disulphide, methyl chloride, and hydrocyanic acid, as described in § 33.9.

STRUCTURE OF METHANE: If the tetravalency of carbon and the univalency of hydrogen is accepted, only one structure is possible, which is represented graphically thus:

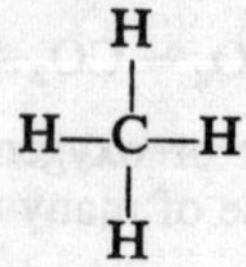

As explained in chapter 32, this formula does not imply a planar molecule. The carbon bonds are in fact directed tetrahedrally as in Fig. 33.3, which accounts for the existence of only one form of methylene dichloride, all four hydrogen atoms being equivalent.

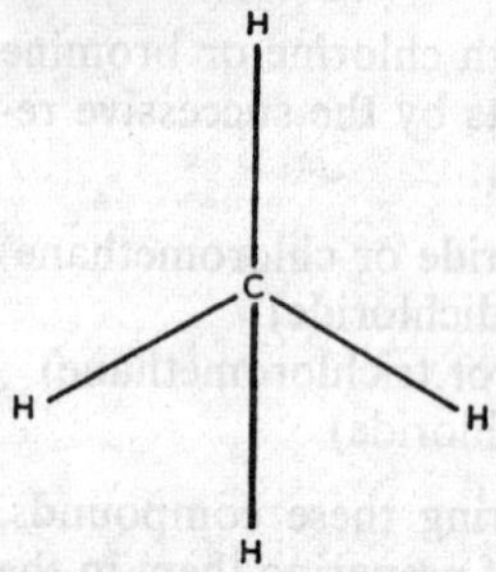

FIG. 33.3 The Methane Molecule

33.3. Ethane, C_2H_6.

OCCURRENCE: Ethane, like methane, is present in natural gas (§ 33.9). It is also obtained as a by-product during the 'cracking' of petroleum (§ 33.11).

LABORATORY PREPARATION: Several methods are available:

1. By reducing ethyl iodide with nascent hydrogen (from a zinc–copper couple and aqueous alcohol):

$$C_2H_5I + 2H = C_2H_6\uparrow + HI$$

2. By heating a mixture of sodium propionate and soda lime:

$$C_2H_5.COONa + NaOH = C_2H_6\uparrow + Na_2CO_3$$

3. By Wurtz's synthesis, from metallic sodium and a dry ethereal solution of methyl iodide:

$$2CH_3I + 2Na = C_2H_6\uparrow + 2NaI$$

This is a useful general method of preparing a paraffin containing an even number of carbon atoms in its molecule; the sodium removes the halogen atoms causing the residual alkyl radicals to combine together thus:

$$CH_3.\boxed{I + Na} + \boxed{Na + I.}CH_3 = CH_3 — CH_3 + 2NaI$$

4. By Kolbe's method; a concentrated solution of sodium acetate is electrolysed using platinum electrodes. Hydrogen is evolved at the cathode and a mixture of ethane and carbon dioxide at the anode. The carbon dioxide is removed by bubbling the product through caustic potash solution.

$$CH_3.COONa \rightleftharpoons CH_3COO^- + Na^+$$
$$H_2O \rightleftharpoons H^+ + OH^-$$

At the cathode: $2H^+ + 2e \rightarrow 2H \rightarrow H_2\uparrow$

At the anode: $2CH_3.COO^- - 2e \rightarrow C_2H_6\uparrow + 2CO_2$

5. By passing a mixture of ethylene or acetylene with hydrogen over finely-divided nickel at about 200°C:

$$C_2H_4 + H_2 = C_2H_6$$

PROPERTIES AND REACTIONS OF ETHANE: It is a colourless, odourless gas, b.p. −89°C, almost insoluble in water.

Chemically ethane is very similar to methane, being generally inert. Like methane, it undergoes substitution by chlorine and bromine and it burns in air or oxygen:

$$2C_2H_6 + 7O_2 = 4CO_2 + 6H_2O$$

USES OF ETHANE: As a fuel.

STRUCTURE OF ETHANE: Having regard to valency considerations, there is only one possible structure, which is represented graphically thus:

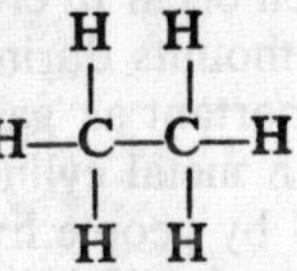

33.4. The Paraffin Series. Some details of the first ten members of this series are given in Table 33.1. All have the same general molecular formula, C_nH_{2n+2}, each member of the series differing in formula from its neighbours by CH_2. They are all colourless, odourless, and insoluble in water. As the table shows, there is a gradual change in their

TABLE 33.1. THE FIRST TEN MEMBERS OF THE PARAFFIN SERIES

Name	*Molecular Formula*	*Structural Formula*	m.p./°C	b.p./°C	*(as liquid)* ρ/g cm^{-3}
Methane	CH_4	CH_4	−183	−162	0.42
Ethane	C_2H_6	$CH_3.CH_3$	−183	−89	0.55
Propane	C_3H_8	$CH_3.CH_2.CH_3$	−188	−42	0.58
n-Butane	C_4H_{10}	$CH_3.(CH_2)_2.CH_3$	−138	−1	0.58
n-Pentane	C_5H_{12}	$CH_3.(CH_2)_3.CH_3$	−130	36	0.63
n-Hexane	C_6H_{14}	$CH_3.(CH_2)_4.CH_3$	−95	69	0.66
n-Heptane	C_7H_{16}	$CH_3.(CH_2)_5.CH_3$	−91	98	0.68
n-Octane	C_8H_{18}	$CH_3.(CH_2)_6.CH_3$	−57	126	0.70
n-Nonane	C_9H_{20}	$CH_3.(CH_2)_7.CH_3$	−54	151	0.72
n-Decane	$C_{10}H_{22}$	$CH_3.(CH_2)_8.CH_3$	−30	174	0.73

other physical properties with rising molecular weight. Thus the first four members are gases at room temperature, the next thirteen are liquids, and the higher paraffins are waxy solids. They are all very similar chemically, being generally unreactive; their characteristic reactions are combustion, substitution by halogens, and thermal decomposition. Certain general methods of preparation, such as the reduction of the appropriate alkyl halide or the action of hot soda lime on the appropriate sodium salt, can be applied to them all. Such a series is called a *homologous series* and the individual members are known as *homologues*.

Homologous series are common in organic chemistry and many other examples will be studied, e.g. alcohols, aldehydes, and fatty acids. They make the learning of organic chemistry very easy because as soon as the preparation and properties of one or two members are known, those of their homologues can be inferred. Admittedly the first member of a homologous series occasionally differs from the others, but in general it will be sufficient in the following chapters to study only the lowest members of a series and to regard them as typical of the rest.

Propane and butane, which occur in crude petroleum and in natural gas, are obtained in large amounts during the refining and 'cracking' of petroleum. They are important as gaseous fuels, because they are easily liquefied and stored in metal cylinders under pressure, and are widely used industrially and by people living in camps, caravans, and boats. In countries where it is plentiful, butane is used for manufacturing acetaldehyde (§ 36.2), acetic acid (§ 37.2), and butadiene (§ 34.5).

The liquid paraffins are important constituents of petrol, kerosene, and oils of all kinds, whilst the solid paraffins are used in making candles and greases (see Table 33.2).

33.5. Isomerism in the Paraffins. Whereas there is only one possible structure for each of the first three paraffins, the four carbon atoms and ten hydrogen atoms in the butane molecule can be linked in two different ways, as represented by the following formulae:

(*a*) 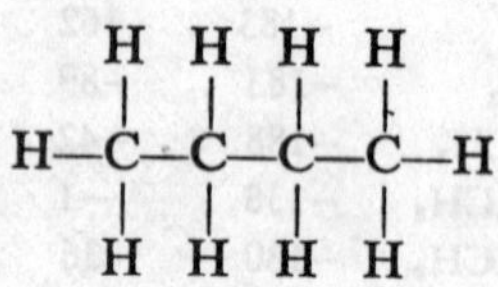 and (*b*) 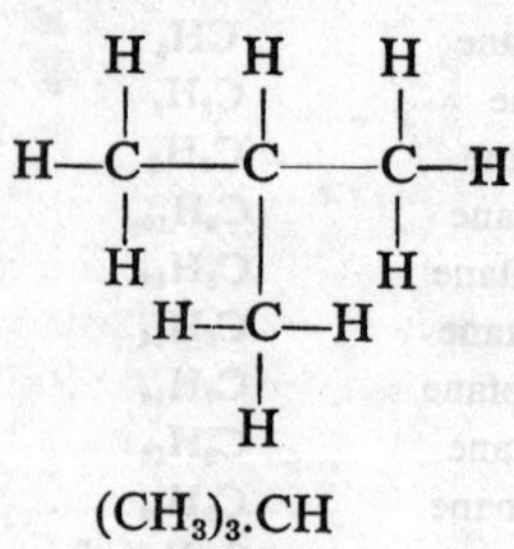

$CH_3.CH_2.CH_2.CH_3$
normal butane

$(CH_3)_3.CH$
iso-butane

The student should satisfy himself that no other arrangement of these atoms is possible involving the accepted valencies of carbon and hydrogen.

This is a good example of *structural isomerism*, i.e. the existence of two or more substances having the same molecular formula but different structural formulae. Structural isomers are not necessarily compounds of the same chemical type, however. For example, ethyl alcohol and dimethyl ether both have the molecular formula C_2H_6O, having structural formulae C_2H_5OH and $CH_3.O.CH_3$ respectively, but they are very dissimilar compounds belonging to two different homologous series. The structural isomerism shown by butane is sometimes called *chain* or *nuclear isomerism* since the isomers differ in the chains of carbon atoms they contain. Form (*a*) is said to be a *straight chain* compound because it is possible, starting from one end of the molecule, to pass through every carbon atom in turn without retracing one's steps, whereas form (*b*) is described as having a *branched chain* of carbon atoms.

Since the two isomers of C_4H_{10} are both forms of butane, they are distinguished in name by using a prefix. Form (*a*), the straight chain isomer, is called normal butane (abbreviated to n-butane) and form (*b*) is known as iso-butane. Both isomers occur in natural gas and in crude petroleum. They differ slightly in their physical properties (e.g. n-butane boils at −1°C and iso-butane at −12°C), but chemically they are very similar.

Similar nomenclature is conventionally applied to the higher paraffins, which explains the use of the prefix n- in Table 33.1, where the data given refer to the straight chain isomer in every case. The term 'straight

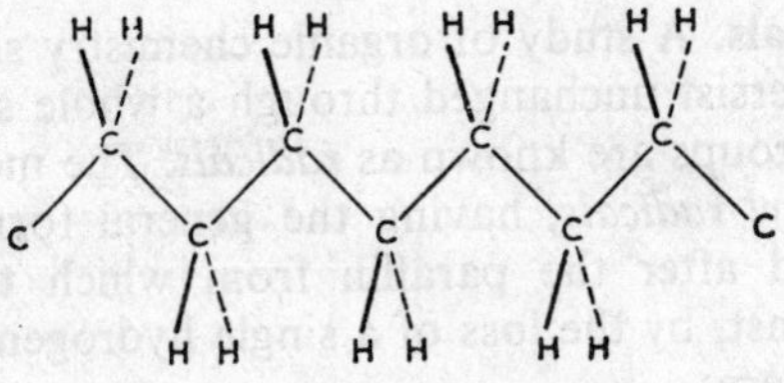

FIG. 33.4 Spatial Arrangement of Atoms in a Paraffin

chain' is rather misleading when applied to those paraffins which are solid at room temperature because, as X-ray analysis reveals, owing to the tetrahedrally directed bonds of the carbon atoms, their carbon chains are in fact zig-zag in shape as shown in Fig. 33.4.

Structural isomerism becomes more common in the higher paraffins. Thus there are three isomers of pentane, C_5H_{12}, with the following graphic formulae:

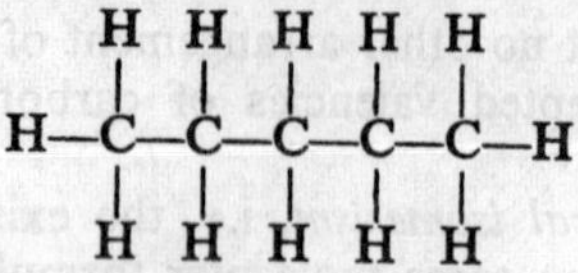

n-pentane (b.p. 36°C)

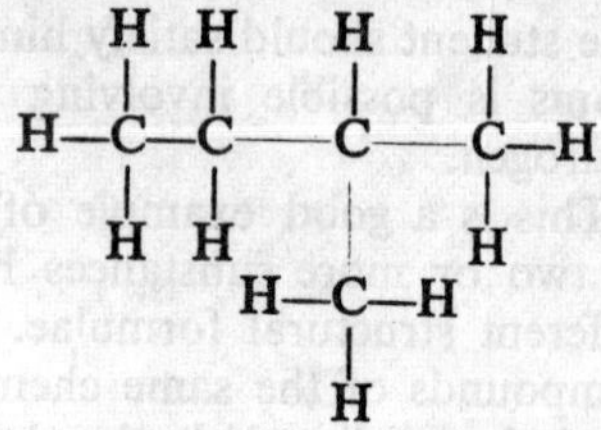

iso-pentane (b.p. 28°C)

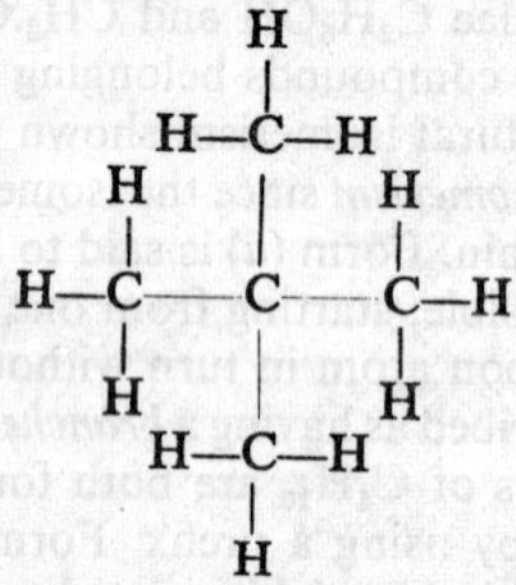

neo-pentane (b.p. 10°C)

Five isomers of hexane exist, and above that the number of isomers increases rapidly to 75 of decane, $C_{10}H_{22}$, and 366 319 of eicosane, $C_{20}H_{42}$. Of course, only a few of these forms have been isolated, but theoretically they are all capable of existence and they illustrate very well the great variety of different compounds which can be formed from only a few atoms of carbon and hydrogen and the consequent inadequacy of the molecular formula in organic chemistry.

33.6. Alkyl Radicals. A study of organic chemistry shows that certain groups tend to persist unchanged through a whole series of chemical reactions. Such groups are known as *radicals*. The most important are the univalent *alkyl radicals*, having the general formula C_nH_{2n+1}—, which are named after the paraffin from which they are derived, theoretically at least, by the loss of a single hydrogen atom. The commonest examples are:

methyl CH_3—, ethyl C_2H_5—, propyl C_3H_7—, butyl C_4H_9—

When it is desired to refer to alkyl radicals in general and not to any particular one of them, it is convenient to use the symbol R—. Thus the general methods of preparing paraffins may be represented by the following equations:

$$RI + 2\overline{H} = R.H + HI$$
$$R.COONa + NaOH = R.H + Na_2CO_3$$
$$2RI + 2Na = R.R + 2NaI$$

33.7. Petroleum. This occurs as a dark, evil-smelling, viscous liquid in huge subterranean deposits in certain parts of the world, particularly in U.S.A., Russia, Venezuela, Kuwait, Saudi Arabia, Iraq, Iran, the Sahara, and the East Indies. The origin of these deposits is uncertain, but they are believed to have been formed by bacterial decomposition of animal and plant remains under high pressure. The composition of crude petroleum varies with its country of origin, but in general it consists of a mixture of paraffins (containing anything from one to about forty carbon atoms per molecule), olefines, cycloparaffins or naphthenes, and aromatic hydrocarbons, together with up to 1% of sulphur compounds and traces of oxygen compounds and nitrogen compounds. The complexity of this mixture is such that a given sample may contain well over a hundred different compounds, most of them liquids at room temperature, with some gases and solids dissolved in them. The petroleum, which is often mixed with water, sand, and brine, is extracted from its deposits by drilling deep holes into the earth. At first it usually issues from the ground under its own pressure, but pumping may become necessary later.

33.8. Refining of Petroleum. After the removal of suspended solids and dissolved gases, the crude petroleum is separated by fractional distillation into a number of portions, each corresponding to a particular range of boiling points. This is done by heating the crude mixture to about 400°C and passing the resulting vapours up a tall column where they condense into various *fractions*, as they are called, according to their volatility. The principal fractions obtained are shown in Table 33.2 with some of their uses.

The naphtha fraction is separated by further distillation into very volatile liquids such as petroleum ether and ligroin on the one hand and a mixture of liquids boiling over the range 50°C–230°C on the other. The latter is suitable for use in internal combustion engines and is known as *petrol* or *gasoline*. It is refined chemically to remove sulphur compounds, which give it an obnoxious smell and which would form corrosive acids during combustion, by treating it with 98% sulphuric acid or a concentrated solution of sodium hydroxide. Either adsorbent alumina or concentrated sulphuric acid is used to remove those olefines which would tend to polymerize into gums on standing in air.

33.9. Natural Gas. This is a mixture of gases which is found in vast quantities underground in certain parts of the world. It is frequently associated with known deposits of petroleum, as in parts of Russia and the U.S.A., North Africa, and the Middle East, but enormous reserves of natural gas have also been discovered beneath the North Sea and

TABLE 33.2. PRINCIPAL PRODUCTS OBTAINED BY THE FRACTIONAL DISTILLATION OF CRUDE PETROLEUM

	Fraction	Composition	b.p. Range	Uses
CRUDE PETROLEUM	Gaseous Hydrocarbons	C_1—C_4 paraffins	Below 20°C	As fuels. For syntheses.
	Naphtha: Petrol ethers and Ligroin	C_5—C_7 paraffins	20–60°C	As solvents.
	Naphtha: Petrol or Gasoline	C_6–C_{12} paraffins, Naphthenes, Aromatic hydrocarbons	50–230°C	As fuel in internal combustion engines. For making town gas and hydrogen.
	Kerosene or Paraffin Oil	C_{11}–C_{16} paraffins, Naphthenes, Aromatic hydrocarbons	200–300°C	As fuel for jet engines and tractors. For cracking into petrol, etc. As fuel in paraffin oil stoves.
	Gas Oil	C_{13}–C_{20} paraffins and other hydrocarbons of similar molecular weight	280–360°C	As fuel in diesel engines. As fuel for domestic and industrial heating. For cracking into petrol, etc.
	Lubricating Oil	Paraffins with more than 20 carbon atoms and other high m.w. hydrocarbons	350–430°C	As lubricants. Paraffin wax for candles. Vaseline and cosmetic creams and emulsions.
	Fuel Oil Residue	Hydrocarbons of high molecular weight	Above 400°C	Asphalt for roads. Greases.

in Holland. The composition of natural gas varies with its origin, but it predominantly consists of methane together with amounts of ethane, propane, and the butanes, and it may also contain small proportions of hydrogen, nitrogen, carbon dioxide, hydrogen sulphide, and helium. The natural gas at Lacq in France, for example, is exceptional in containing about 15% of hydrogen sulphide, making it a major source of the world's supply of sulphur (see § 24.5).

Natural gas is an excellent fuel of high heating value (3.7×10^7 J m^{-3} approx). With its ready availability, its low cost, its cleanliness, and its ease of transportation and control, it is already a serious rival to coal and oil as a primary fuel, and in many countries it is likely in the next decade to supersede coal gas (§ 20.13) for industrial and domestic use. Because it requires much more air than coal gas does for its complete combustion and because its burning velocity is much lower than coal gas or town gas (§ 20.16), both of which contain about 50% of fast-burning hydrogen, natural gas can only be efficiently burned in appliances which have been specially converted for its use.

Valuable though it is as a gaseous fuel, it is not this feature of natural gas which is likely, in the end, to be most important, but rather its use as a source of various other chemicals which can be manufactured cheaply from it. The production of elemental sulphur has already been referred to above and in § 24.5. Other examples of industrial importance are:

1. Acetylene (§ 34.4) can be made by heating natural gas momentarily to about 1500°C, when the following reaction occurs:

$$2CH_4 \rightleftharpoons C_2H_2 + 3H_2$$

The high temperature needed is usually provided by burning some natural gas or by means of an electric arc. This process is more economical than the carbide method and is rapidly replacing it wherever there is an abundant supply of natural gas.

2. Hydrogen is made on a large scale in some countries by passing natural gas mixed with steam and a limited supply of oxygen over a nickel-based catalyst at 900°C, when the following reactions take place:

$$CH_4 + H_2O = CO + 3H_2$$
$$2CH_4 + O_2 = 2CO + 4H_2$$

As explained in § 16.2, another stage is then used to convert the carbon monoxide to further hydrogen and carbon dioxide.

3. Carbon Disulphide (§ 20.8) is increasingly made by passing sulphur vapour and methane (from natural gas) at 600°C over a silica or alumina catalyst:

$$CH_4 + 4S = CS_2 + 2H_2S$$

4. Methyl Chloride (§ 40.5) and *Carbon Tetrachloride* (§ 20.9) are sometimes made by direct chlorination of methane at 400°C:

$$CH_4 + Cl_2 = CH_3Cl + HCl$$
$$CH_4 + 4Cl_2 = CCl_4 + 4HCl$$

TABLE 33.3. THE USES OF NATURAL GAS

Natural gas treatment	Product
Burned in a plentiful supply of air	Heat (*ca.* 3.7×10^7 J m^{-3})
Burned in a limited supply of air	Carbon Black (§ 20.3)
Heated momentarily to 1500°C	Acetylene (§ 34.4)
Steam-reforming at 900°C over Ni catalyst; then more steam at 450°C over iron catalyst	Hydrogen (§ 16.2)
Steam-reforming at 900°C over Ni catalyst; then 300°C and 3×10^7 N m^{-2} (300 atm) over ZnO catalyst	Methyl Alcohol (§ 35.3)
H_2S (from Lacq gas) + air over bauxite catalyst	Sulphur (§ 24.5)
With sulphur at 600°C over silica catalyst	Carbon Disulphide (§ 20.8)
With limited supply of chlorine	Methyl Chloride (§ 40.5)
With large excess of chlorine	Carbon Tetrachloride (§ 20.9)
With ammonia + air at 1300°C over catalyst	Hydrocyanic Acid (§ 20.11)
Butane (from natural gas) at 600°C over alumina catalyst	Butadiene (§ 34.5)

In the first reaction the methane is in large excess, in the second an excess of chlorine is used.

5. *Butadiene* (§ 34.5) is made on a huge scale by the dehydrogenation of *n*-butane (obtained from natural gas) by passing it over chromic oxide or alumina catalyst at 600°C:

$$CH_3.CH_2.CH_2.CH_3 = CH_2{:}CH.CH{:}CH_2 + 2H_2$$

6. *Hydrocyanic Acid* (§ 20.11) is now manufactured by the catalytic reaction of methane (from natural gas) with ammonia and air at 1300°C when the following reaction takes place:

$$2CH_4 + 2NH_3 + 3O_2 = 2HCN + 6H_2O$$

The product is added to caustic soda to give a cheap supply of sodium cyanide (§ 17.13).

7. *Carbon Black* (§ 20.3) is made by burning natural gas in a limited supply of air, when it is deposited as a fine soot.

8. *Methyl Alcohol* is made by the steam-reforming of natural gas as described in reaction 2 above; the resulting mixture of carbon monoxide and hydrogen is catalytically converted to methyl alcohol (see § 35.3).

33.10. Knocking. If a mixture of petrol vapour and air is compressed in a cylinder and sparked, very rapid combustion occurs, driving the piston downwards. When the compression ratio is high there is a tendency for the combustion to become almost explosive in character as it proceeds, so that the piston receives a sharp and violent blow rather than a strong push. This phenomenon, which is known as *knocking*, causes loss of efficiency and increased wear.

Since the efficiency of petrol engines increases with the compression ratio, the prevention of knocking has become an important matter. Experiment has shown that the addition of certain chemicals to the petrol greatly reduces the tendency to knocking: such substances are known as *anti-knock agents*, the commonest being lead tetraethyl, $Pb(C_2H_5)_4$. Between 1 and 3 cubic centimetre of this volatile liquid added to each gallon of petrol acts as a negative catalyst moderating the rate of combustion. Its manufacture is described in § 40.5. Early trials showed that lead oxide was deposited on the walls of the cylinder after combustion, so small amounts of another chemical, ethylene dibromide, $C_2H_4Br_2$ (§ 40.6), are added to petrol in order to convert the lead compounds into volatile lead bromide, which is swept out of the cylinder in the exhaust gases. It has also been found that engines knock less when the petrol contains a high proportion of branched chain hydrocarbons. Certain unsaturated compounds and most aromatic hydrocarbons also have a beneficial effect.

Two substances, n-heptane, a straight chain paraffin which is particularly prone to knocking, and iso-octane (actually 2.2.4. trimethylpentane), a branched chain paraffin which knocks only under extreme conditions, have been chosen as standards by which the knocking tendency of any given petrol can be assessed. The *octane number* of a petrol is the percentage of iso-octane which must be present in a mixture of n-heptane and iso-octane, to give it the same knocking tendency as the petrol under test, so that the higher the octane rating the better.

33.11. Cracking Processes. The demand for petrol increased tremendously during the first half of this century and to meet it the oil companies greatly increased their output of crude petroleum and refined products. However, the demand for the less volatile fractions such as kerosene and gas oil, which cannot be used in petrol engines, did not increase to the same extent and the refiners were left with large surpluses of these fractions. This led them to devise various *cracking* processes for converting these less volatile fractions into material suitable for use in petrol engines. Such processes involve splitting the large paraffin molecules into molecules of lower molecular weight and greater volatility by subjecting them to high temperatures and pressures, usually in the presence of a catalyst. Besides increasing the yield of petrol from petroleum, cracking has two other important effects. It provides a mixture rich in branched chain hydrocarbons, with a consequent high octane rating, and it yields large amounts of very useful by-products such as ethylene, propylene, the butylenes, and the lower paraffins. These are used for making plastics, synthetic rubber, detergents, and such important chemicals as ethyl alcohol, phenol and acetone, or they are converted into additional high-octane petrol by catalytic polymerization.

Processes have also been devised for cracking the ordinary petrol fraction obtained by distilling petroleum, in order to improve its octane number. These processes, known as *catalytic reforming*, consist of passing the vapours at high temperatures and pressures over catalysts such as aluminium oxide, platinum, anhydrous aluminium chloride, and boron trifluoride, when straight chain paraffin molecules are changed into branched chain and aromatic hydrocarbons. These processes have become so successful that in an adapted form they have now become an important source of aromatic compounds, particularly in U.S.A. For example, large amounts of benzene, toluene, and xylenes are obtained in this way to supplement the supplies obtained from the carbonization of coal (see § 43.3).

In recent years the demand for kerosene and gas oil has risen much more rapidly than the demand for petrol owing to the widespread use of these less volatile fractions in jet aircraft and for industrial and domestic heating. If this trend continues we may see in the near future a surplus of petrol and a shortage of the less volatile fractions—the very reverse of the situation in the nineteen thirties. It is important to realize that the refining of crude petroleum by fractional distillation is only a physical process for separating the constituents and that unlike the various cracking, reforming, and polymerization processes it does not alter their relative proportions to meet the changing demand.

33.12. Synthetic Petrol. In one meaning of the term, the products of the cracking and reforming processes referred to in the previous section are synthetic petrols in that the molecules in them do not occur as such in crude petroleum but have been made by chemical changes from naturally occurring materials. The phrase synthetic petrol, however, is normally reserved for petrol made from materials such as coal, coke, and hydrogen which do not occur in crude petroleum. Two processes have been operated on a large scale. Neither of them is economic when there is an abundant supply of crude petroleum, but both were operated successfully in Germany during the Second World War.

Bergius Process: Powdered coal is converted into a mixture of hydrocarbons by heating it with hydrogen to 500°C under a pressure of about 2×10^7 N m^{-2} (200 atm) in the presence of a catalyst of iron or tin. The oily product is separated by distillation into a petrol fraction boiling at less than 200°C and a heavy oil residue which is mixed with fresh coal and treated again. About 0.5 kg of petrol is obtained from each kg of coal.

Fischer-Tropsch Process: This process depends upon the discovery that water gas (a mixture of carbon monoxide and hydrogen made by passing steam over coke at 1000°C) can be hydrogenated to a mixture of hydrocarbons by adding hydrogen and passing it over a finely-divided nickel catalyst at 200°C. About half of the product can be used as petrol, the less volatile fraction being suitable as a fuel for diesel engines.

33.13. Alicyclic Compounds. A homologous series of saturated hydrocarbons exists in which the carbon atoms in the molecule are linked together to form a ring. Such compounds, which have the general molecular formula C_nH_{2n}, are known as aliphatic cyclic or *alicyclic* compounds. The first four members of the series are:

C_3H_6

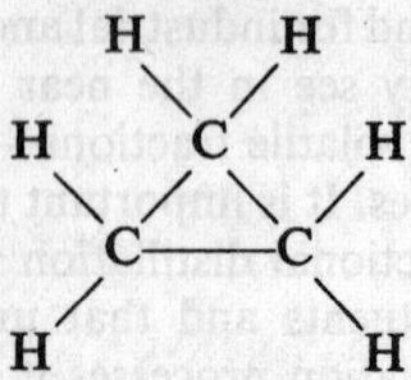

cyclopropane
b.p. −34°C

C_4H_8

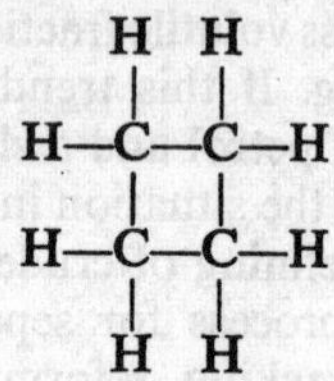

cyclobutane
b.p. 13°C

C_5H_{10}

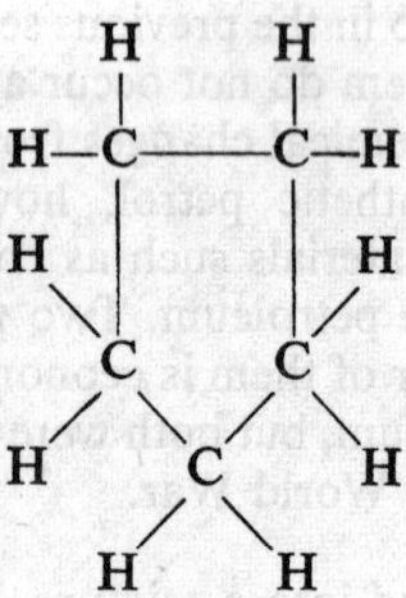

cyclopentane
b.p. 49°C
ρ 0.75 g cm^{-3}

C_6H_{12}

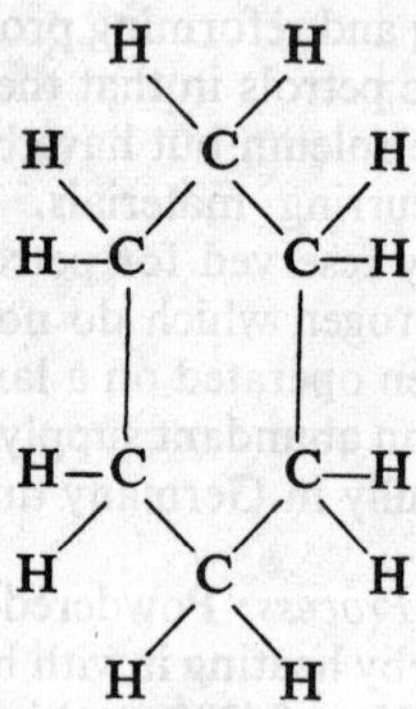

cyclohexane
b.p. 81°C
ρ 0.78 g cm^{-3}

Cyclopropane, which is widely used as a general anaesthetic because of its freedom from the side-effects associated with chloroform and ether, is a reactive compound; it combines with bromine, hydrogen bromide, hydrogen iodide, concentrated sulphuric acid, and hydrogen (in the presence of a nickel catalyst) to give open-chain propyl compounds. The higher members are much less reactive and closely resemble the paraffins in properties. Cyclopentane and cyclohexane occur in crude petroleum and are sometimes known as *naphthenes*. Cyclohexane is an important intermediate in the manufacture of nylon.

The reactivity of cyclopropane is understandable because in its planar ring the angles between the C—C bonds are only 60° compared with the usual tetrahedral value of 109° 28′ found in the paraffins (see Fig. 33.4). In cyclohexane and the higher members, physical evidence such as electron diffraction measurements has established that the carbon atoms in the rings do not all lie in the same plane but that the molecule adopts various 'conformations' or shapes which allow the C—C covalent bonds to be tetrahedrally directed. For example, cyclohexane is believed to exist in the two inter-convertible conformations shown in Fig. 33.5, known as the boat (I) and chair (II) forms respectively, the latter predominating because it is the more stable arrangement.

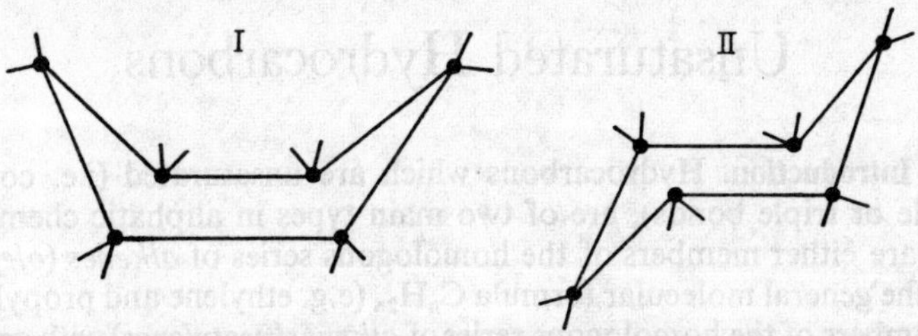

As indicated in Table 43.3, cyclohexane does not show the addition reactions of unsaturated compounds (§ 34.6) or the typically aromatic properties of the benzene ring (§ 43.9). As explained in § 48.6, when two or more hydrogen atoms in a molecule of cyclohexane are substituted by other atoms or groups, the derivatives formed display cyclic isomerism.

34

Unsaturated Hydrocarbons

34.1. Introduction. Hydrocarbons which are unsaturated (i.e. contain double or triple bonds), are of two main types in aliphatic chemistry. They are either members of the homologous series of *alkenes* (*olefines*) with the general molecular formula C_nH_{2n} (e.g. ethylene and propylene), or members of the homologous series of *alkynes* (*acetylenes*) with general molecular formula C_nH_{2n-2}, of which acetylene itself is the simplest member.

34.2. Ethylene, Ethene, $H_2C{=}CH_2$.

LABORATORY PREPARATION: Ethyl alcohol is mixed with excess of con-

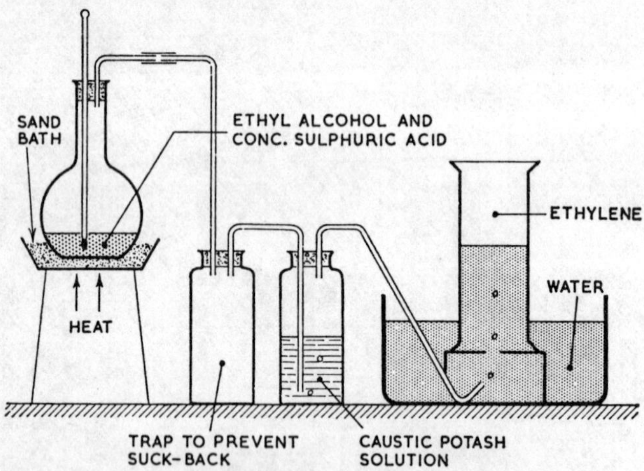

FIG. 34.1 Preparation of Ethylene

centrated sulphuric acid. Ethyl hydrogen sulphate, $C_2H_5OSO_2.OH$, is formed in the cold, but on heating to about 170°C it decomposes into ethylene and sulphuric acid:

$$C_2H_5OH + H_2SO_4 \rightleftharpoons C_2H_5.HSO_4 + H_2O$$
$$C_2H_5.HSO_4 = C_2H_4\uparrow + H_2SO_4$$

This preparation may be regarded as a dehydration in that the overall change is the removal of the elements of water from the alcohol:

$$C_2H_5OH - H_2O = C_2H_4$$

In practice charring and slight oxidation by the hot concentrated acid produces small amounts of carbon dioxide and sulphur dioxide, which are removed by washing the gas with potassium hydroxide solution. The apparatus is shown in Fig. 34.1.

MANUFACTURE: As explained in the previous chapter, great quantities of ethylene are produced during the cracking of petroleum. In some countries it is still made by passing ethyl alcohol vapour over a catalyst of alumina at about 350°C.

PROPERTIES OF ETHYLENE: It is a colourless gas, b.p. −104°C, with a sweetish smell. It is only sparingly soluble in water.

REACTIONS OF ETHYLENE:

1. Combustion: It burns readily in air with a smoky, luminous flame and forms explosive mixtures with air and oxygen. When burned in oxygen it is completely oxidized thus:

$$C_2H_4 + 3O_2 = 2CO_2 + 2H_2O$$

2. Addition: Ethylene is highly reactive, its most important reaction being 'addition' across the double bond. Addition, which is typical of all unsaturated compounds, is a type of reaction in which two substances combine to give a single new substance without forming any other product. For example, if ethylene undergoes addition with a substance XY, the reaction is essentially*:

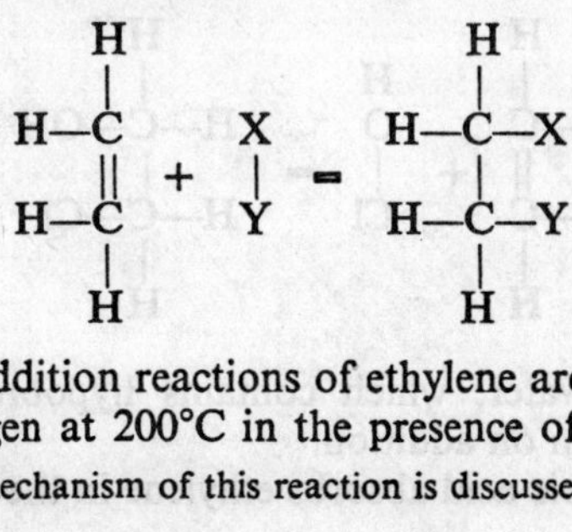

Examples of the addition reactions of ethylene are:

(*a*) With hydrogen at 200°C in the presence of a catalyst of finely-

* The mechanism of this reaction is discussed in § 50.6.

divided nickel, ethane is produced:

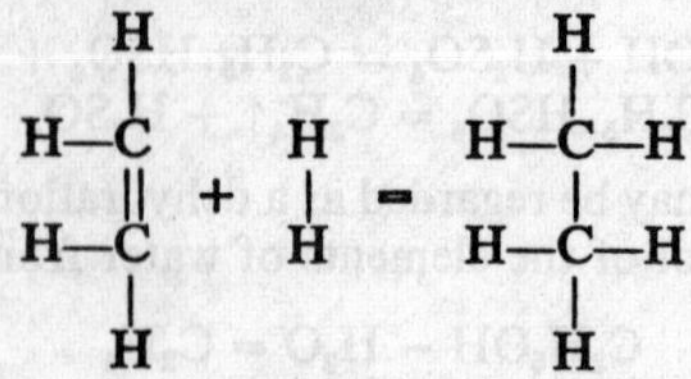

(*b*) Chlorine and bromine add readily at room temperature giving ethylene dichloride and ethylene dibromide respectively:

e.g.

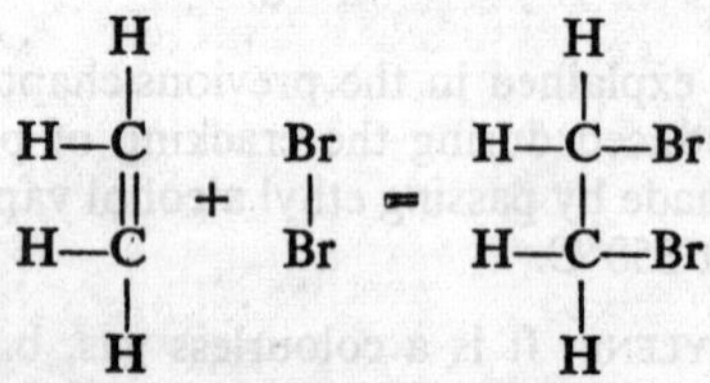

(*c*) Hydrogen iodide and hydrogen bromide add rapidly at room temperature forming ethyl iodide and ethyl bromide respectively:

e.g.

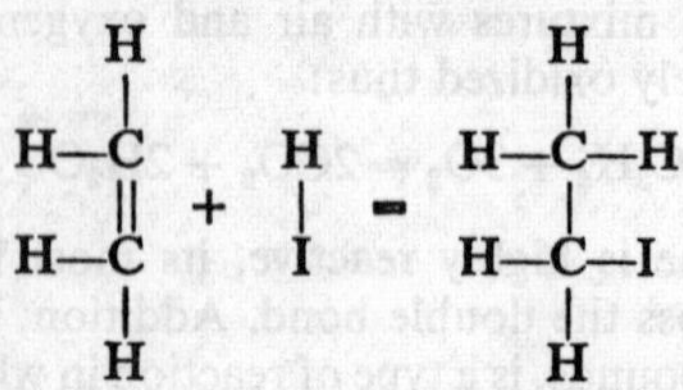

(*d*) Chlorine water, which is effectively hypochlorous acid, HOCl, adds to ethylene giving ethylene chlorohydrin:

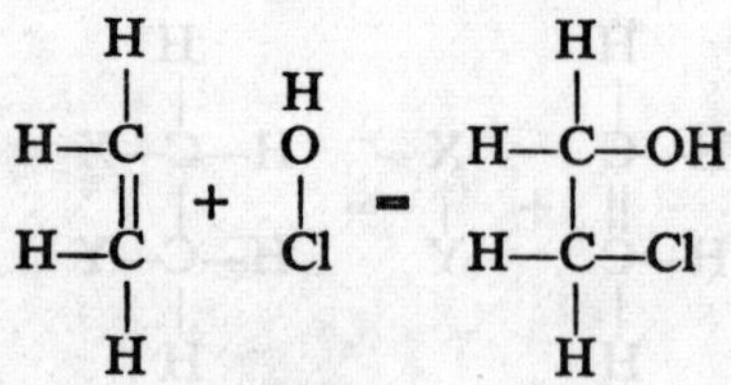

Similarly bromine water, which contains hypobromous acid, gives ethylene bromohydrin on addition.

(*e*) Fuming sulphuric acid absorbs ethylene in the cold forming ethyl hydrogen sulphate:

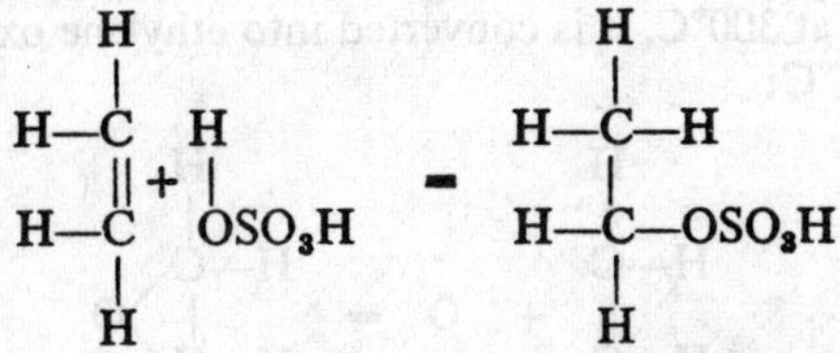

If the product is boiled with water it is hydrolysed to ethyl alcohol and sulphuric acid.

(*f*) On bubbling ozone into a solution of ethylene in carbon tetrachloride, an ozonide is formed which is unstable and decomposes on treatment with water:

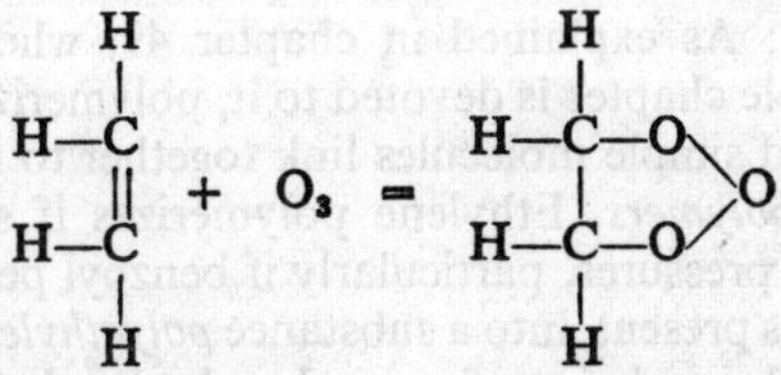

Addition of ozone in this way is called *ozonolysis*. The process is used to detect the position of a double bond in a chain of carbon atoms by analysing the fragments remaining after hydrolysis of the ozonide.

e.g.

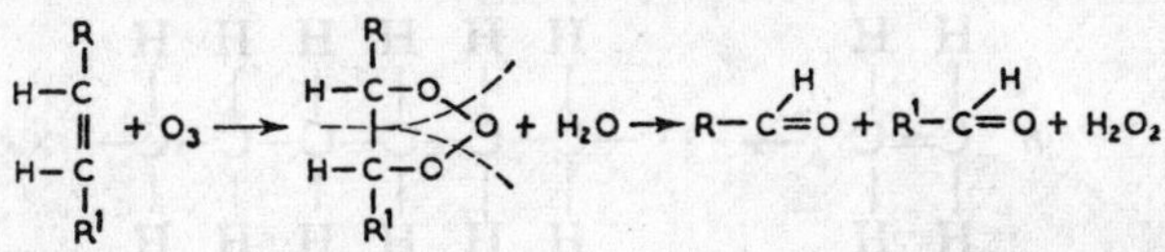

(*g*) Ethylene reacts with a dilute solution of potassium permanganate producing ethylene glycol:

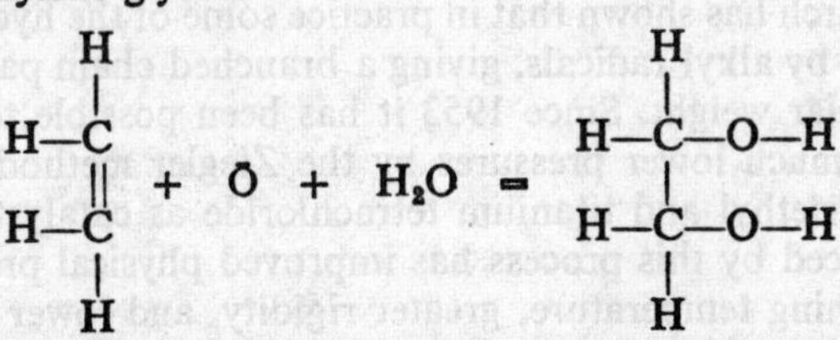

Potassium permanganate solution made alkaline with sodium carbonate solution is first reduced to a green solution of potassium manganate, and then, on standing, to a brown precipitate of manganese dioxide, whereas acidified permanganate solution turns colourless on shaking with ethylene.

(*h*) When ethylene is mixed with air or oxygen and passed over a silver catalyst at 300°C, it is converted into ethylene oxide, a colourless liquid, b.p. 11°C:

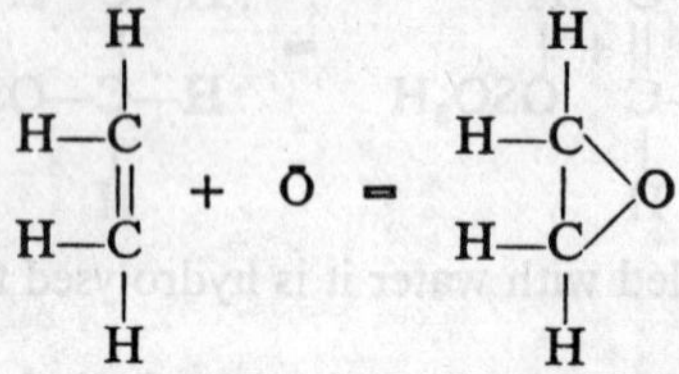

Ethylene oxide is an intermediate in the manufacture of ethylene glycol (§ 35.8), and is also used to make liquid (non-ionic) detergents.

3. *Polymerization*: As explained in chapter 49, where in view of its importance a whole chapter is devoted to it, polymerization is a process whereby small and simple molecules link together to form much larger units known as *polymers*. Ethylene polymerizes if subjected to high temperatures and pressures, particularly if benzoyl peroxide (§ 47.5) or a trace of oxygen is present, into a substance *polyethylene*, or '*polythene*', which consists of long chains of many hundreds of ethylene molecules joined together. This polymerization, which is usually carried out industrially at about 2×10^8 N m^{-2} (2000 atm) and 200°C, may be represented very simply in the following way:

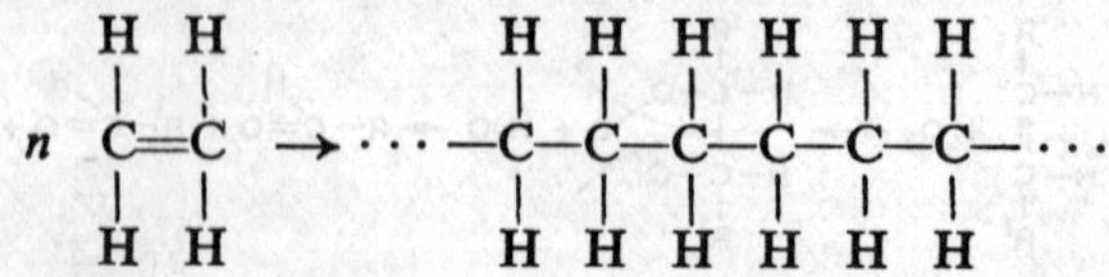

i.e. $n\ C_2H_4 \longrightarrow (C_2H_4)_n$

Recent research has shown that in practice some of the hydrogen atoms are replaced by alkyl radicals, giving a branched chain paraffin of very high molecular weight. Since 1953 it has been possible to polymerize ethylene at much lower pressures by the Ziegler method, which uses aluminium triethyl and titanium tetrachloride as catalysts. The polythene produced by this process has improved physical properties (e.g. higher softening temperature, greater rigidity, and lower permeability to gases) owing to its partially crystalline character (see § 49.2).

USES OF ETHYLENE: Very large amounts are used to manufacture polythene, referred to above. This plastic is tough, light, easily moulded and coloured, very resistant to acids and alkalis, and it has excellent electrical insulating properties. In the form of thin transparent sheets it is very useful for packaging. Another major use of ethylene is in the manufacture of ethylbenzene, which is converted into styrene and then either polymerized to the important plastic polystyrene (§ 43.3) or mixed with butadiene and polymerized to synthetic rubber (see § 34.5). Apart from the production of polymers, ethylene is of immense importance as a source of many other organic compounds, as shown in Table 34.1. For example, most industrial supplies of ethyl alcohol are now obtained from ethylene, as described in § 35.2. An idea of the scale of these processes can be gained from the fact that each year several hundred million kg each of ethyl chloride and ethylene dichloride are made from ethylene in the U.S.A. alone and the annual production of ethylene oxide in that country is well over 5×10^8 kg. A small-scale but interesting use of ethylene is for ripening fruit, even low concentrations of the gas having a marked accelerating effect.

STRUCTURE OF ETHYLENE: Eudiometry establishes the molecular formula as C_2H_4. Assuming that carbon is tetravalent and hydrogen univalent, the only acceptable structure in the light of the above reactions is one containing a double covalent bond thus:

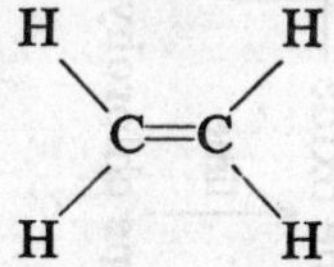

This double bond, sometimes called the 'ethylenic linkage', is believed to consist of four shared electrons, two from each carbon atom, as shown in Fig. 14.7. Electron diffraction reveals that the carbon atoms in ethylene are closer together than in, say, ethane, where they are linked by a single covalent bond, because the double bond causes distortion of the electron orbits (see § 34.7).

34.3. Propylene, Propene, $CH_3.CH{=}CH_2$.

LABORATORY PREPARATION: By the dehydration of isopropyl alcohol with concentrated sulphuric acid, or by heating n-propyl iodide with an alcoholic solution of potassium hydroxide:

$$CH_3.CH_2.CH_2I + KOH = CH_3.CH{:}CH_2 + KI + H_2O$$

MANUFACTURE: Huge quantities are produced cheaply during the cracking of petroleum.

PROPERTIES AND REACTIONS OF PROPYLENE: It is a colourless gas boiling at $-48°C$. It closely resembles ethylene in its physical and chemical properties, its principal reaction being addition across the double bond.

TABLE 34.1. PRINCIPAL PRODUCTS MANUFACTURED FROM ETHYLENE

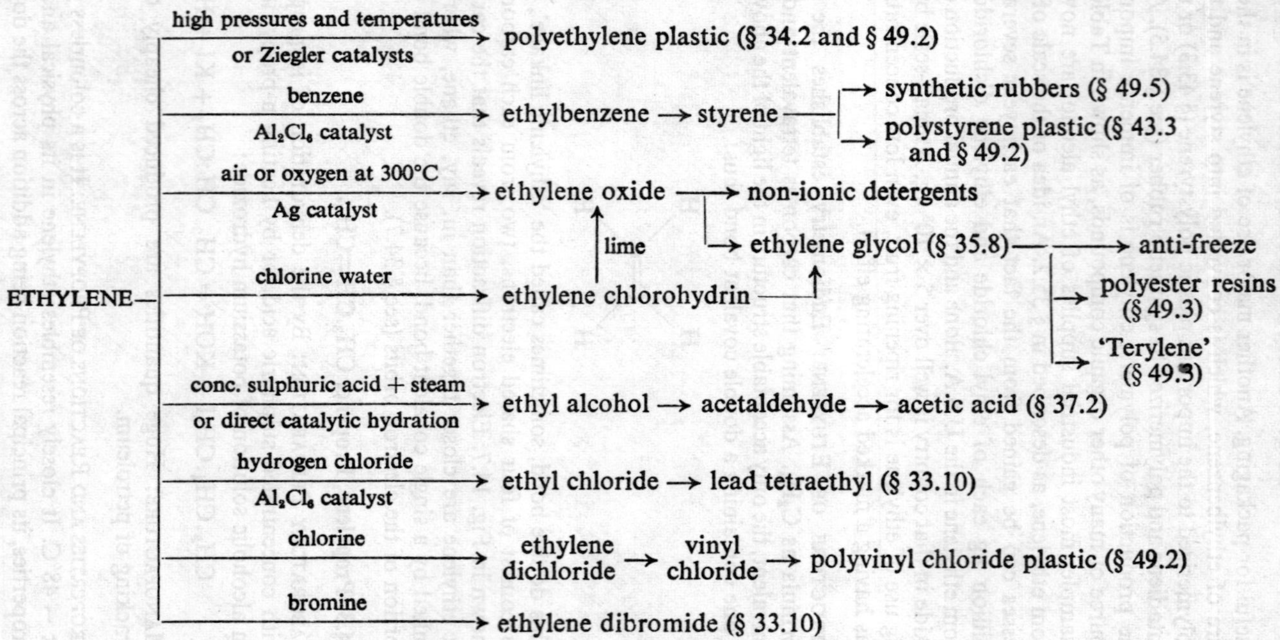

Table 34.2. Principal Products Manufactured from Propylene

PROPYLENE	Ziegler catalysts →	polypropylene plastic and synthetic fibre (§ 49.2)
PROPYLENE	chlorine water →	propylene chlorohydrin → propylene oxide → propylene glycol → polyester resins (§ 49.3)
PROPYLENE	conc. sulphuric acid + steam →	iso-propyl alcohol → acetone (§ 36.6)
PROPYLENE	benzene, Al_2Cl_6 catalyst →	cumene → acetone and phenol (§ 45.2)
PROPYLENE	chlorine at 500°C →	allyl chloride → cyclopropane (anaesthetic); allyl chloride → epichlorohydrin → epoxy resins; allyl chloride → glycerol
PROPYLENE	catalytic oxidation →	acrolein → glycerol (§ 35.9) → alkyd resins
PROPYLENE	water gas, $2\times10^7 N\ m^{-2}$ (200 atm), Co catalyst 150°C →	normal and iso-butyl alcohols → solvent
PROPYLENE	200°C, H_3PO_4 catalyst →	propylene tetramer → dodecyl benzene → detergents

When hydrogen iodide adds to propylene, two products,

$$CH_3.CHI.CH_3 \quad \text{and} \quad CH_3.CH_2.CH_2I$$

are theoretically possible depending upon which carbon atom links to the iodine atom. In practice isopropyl iodide (2-iodopropane) is formed almost exclusively:

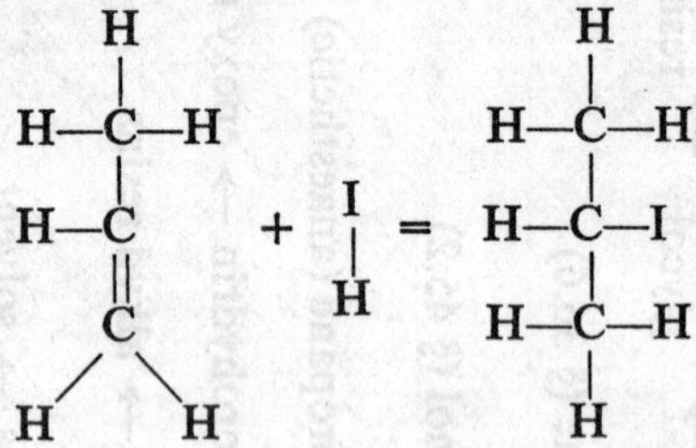

Similarly when hydrogen bromide or concentrated sulphuric acid add to propylene they give the isopropyl derivative in each case. This behaviour has been summarized in a generalization known as *Markownikoff's Rule* which states that *when a hydrogen compound HX adds across the double bond of an unsymmetrical olefine, the X atom tends to attach itself to the carbon atom linked to the smaller number of hydrogen atoms.* This rule is explained in terms of mechanism in § 50.6, section D.

USES OF PROPYLENE: These are summarized in Table 34.2. The plastic polypropylene promises to rival polyethylene in its usefulness and importance. As well as the advantage of a higher melting point, it is very suitable for use as a fibre in clothing, ropes, and fishing nets.

34.4. Acetylene, Ethyne, H—C≡C—H.

LABORATORY PREPARATION: Usually by the action of cold water on calcium carbide (§ 18.22):

$$CaC_2 + 2H_2O = Ca(OH)_2 + C_2H_2\uparrow$$

The product often contains small amounts of phosphine, hydrogen sulphide and arsine, formed from impurities in the commercial carbide, but it can be purified by bubbling it through a solution of copper sulphate before collection, as in Fig. 34.2.

Pure acetylene is prepared by boiling ethylene dibromide with an *alcoholic* solution of potassium hydroxide:

$$C_2H_4Br_2 + 2KOH = C_2H_2\uparrow + 2KBr + 2H_2O.$$

MANUFACTURE: Two methods are in current use:

1. From Calcium Carbide: By the action of water on calcium carbide,

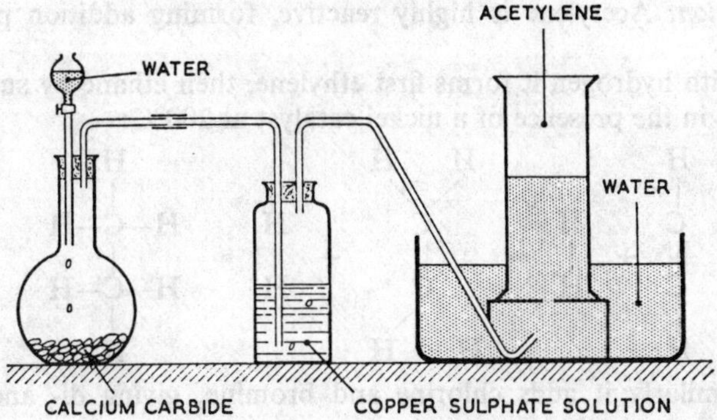

FIG. 34.2 Preparation of Acetylene

which is made by heating quicklime with coke to 2000°C in an electric furnace:

$$CaO + 3C = CaC_2 + CO; \Delta H = -452 \text{ kJ mol}^{-1}$$

2. From Natural Gas or Petroleum Gases: If natural gas (§ 33.9) or a mixture of hydrocarbon gases from the refining of petroleum are raised to a temperature of about 1500°C for about a thousandth of a second, the main reaction is:

$$2CH_4 \rightleftharpoons C_2H_2 + 3H_2; \Delta H = -377 \text{ kJ mol}^{-1}$$

The yield of acetylene varies with the conditions but usually amounts to about 20%, with unchanged methane, hydrogen, and carbon black as the principal byproducts. The mixture must be cooled very rapidly by spraying with water to freeze the equilibrium and to minimize decomposition of the acetylene, which is absorbed under pressure in a suitable solvent, usually dimethyl formamide.

PROPERTIES OF ACETYLENE: It is a colourless, almost odourless gas b.p. – 84°C, sparingly soluble in water. Although liable to explode on compression to a liquid, its solution in acetone under pressure is stable and it is in this form that it is usually stored in cylinders.

REACTIONS OF ACETYLENE:

1. Combustion: It burns in air with a luminous but smoky flame, owing to the high proportion of carbon it contains, and was once used for lighting vehicles. Although mixtures of acetylene with air or oxygen explode violently on ignition, when the gas issues from a jet into an atmosphere of oxygen it burns quickly but very exothermically, producing a temperature of about 3000°C.

2. *Addition:* Acetylene is highly reactive, forming addition products readily:

(*a*) With hydrogen it forms first ethylene, then ethane by successive addition in the presence of a nickel catalyst at 200°C:

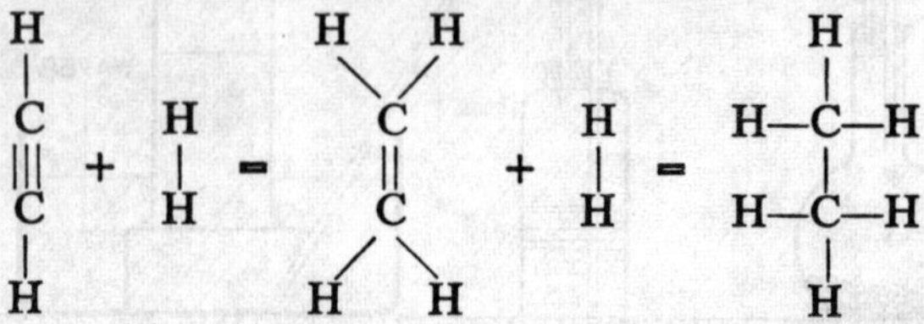

(*b*) Similarly it adds chlorine and bromine, giving di- and tetrahalides:

e.g.

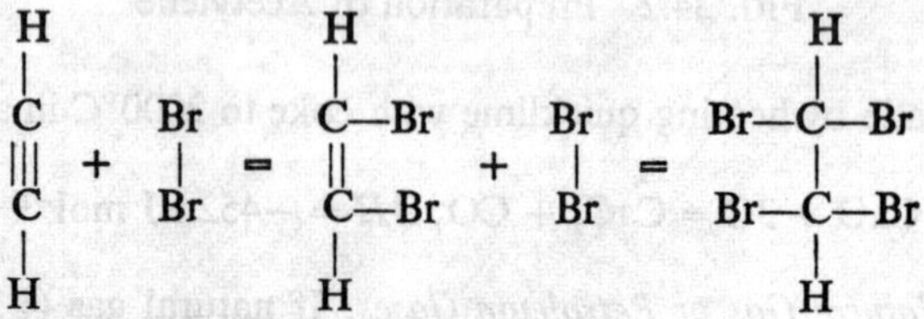

In the absence of a catalyst, however, chlorine tends not to add to acetvlene but to react explosively with it, thus:

$$C_2H_2 + Cl_2 = 2HCl + 2C$$

(*c*) Hydrogen iodide, bromide, and chloride add to acetylene in two stages:

e.g.

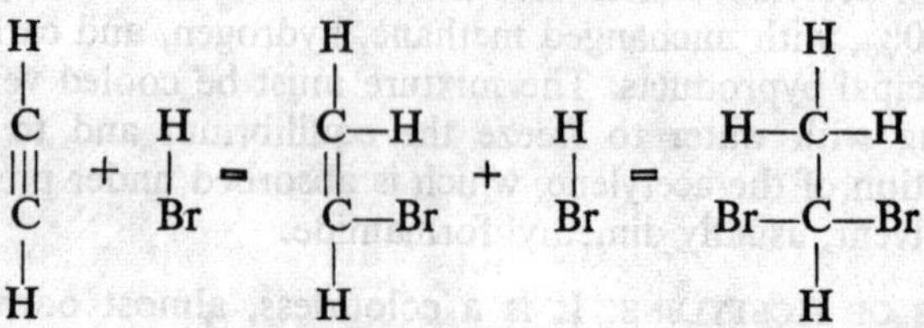

vinyl bromide ethylidene dibromide

(*d*) If acetylene is bubbled into dilute sulphuric acid at 60°C in the presence of mercuric sulphate, which acts as a catalyst, addition of water takes place forming acetaldehyde:

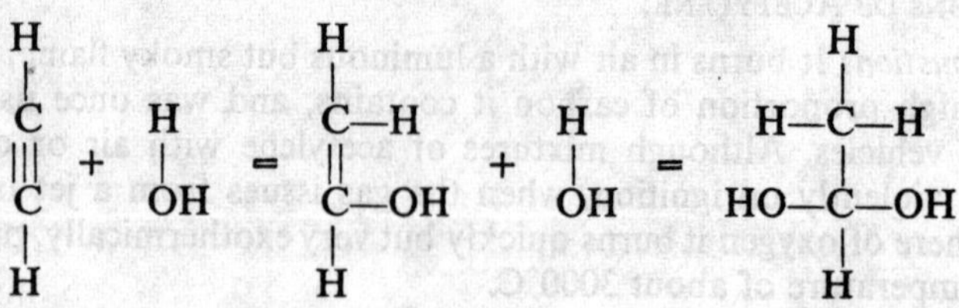

The first stages are probably successive additions, and then, because a compound containing two hydroxyl groups attached to one carbon atom is unstable, a further change occurs:

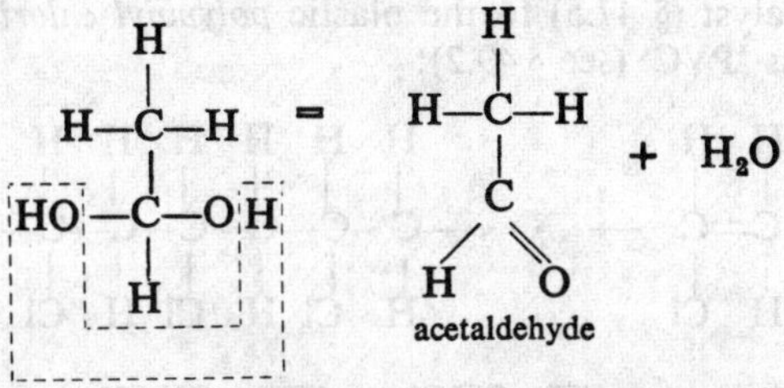

This reaction is of considerable industrial importance.

(*e*) Acetylene adds to ozone, giving an unstable ozonide.

(*f*) Like ethylene, acetylene decolorizes an alkaline solution of potassium permanganate, eventually giving a brown precipitate of manganese dioxide. The acetylene is converted to oxalic acid:

$$C_2H_2 + 4\bar{O} = \begin{array}{c} COOH \\ | \\ COOH \end{array}$$

(*g*) Under pressure and in the presence of a catalyst of barium cyanide hydrogen cyanide adds to acetylene giving acrylonitrile or viny cyanide:

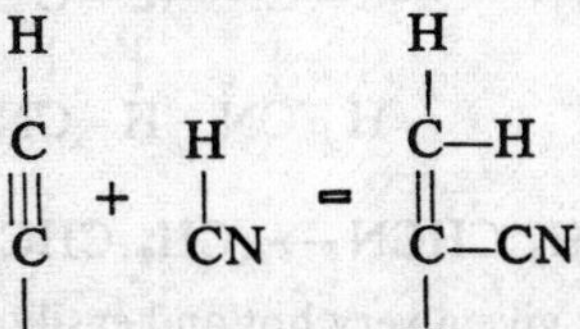

3. Polymerization: On passing through a hot tube, acetylene polymerizes to some extent to benzene:

$$3C_2H_2 = C_6H_6$$

4. Substitution: When acetylene is bubbled into ammoniacal solutions of cuprous chloride and silver nitrate, precipitates of cuprous and silver acetylides appear. The compounds are explosive when dry. These reactions are important because they readily distinguish acetylene from ethylene, which gives no precipitates with these reagents:

$$C_2H_2 + 2CuCl = \underset{\text{crimson ppt.}}{C_2Cu_2\downarrow} + 2HCl$$

$$C_2H_2 + 2AgNO_3 = \underset{\text{white ppt.}}{C_2Ag_2\downarrow} + 2HNO_3$$

Uses of Acetylene: Its most important use is to manufacture acetaldehyde by reaction 2(*d*) above (see § 36.2). It is also used to make vinyl chloride, $CH_2 = CHCl$, which polymerizes in the presence of benzoyl peroxide catalyst (§ 47.5) to the plastic *polyvinyl chloride*, commonly referred to as 'PVC' (see § 49.2):

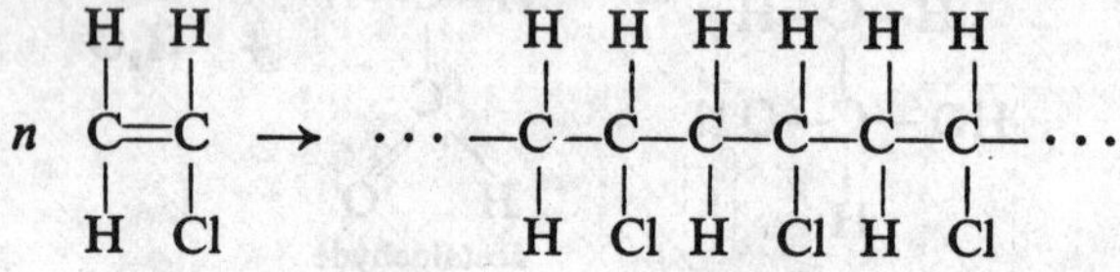

i.e. $$n\ CH_2{:}CHCl \rightarrow (CH_2.CHCl)_n$$

Vinyl chloride is also made to some extent from ethylene, as shown in Table 34.1.

Another major use of acetylene is to make acrylonitrile as in reaction 2(*g*). On polymerization in the presence of a peroxide catalyst this gives *polyacrylonitrile*, which is a valuable synthetic fibre; yarns spun from it are used in knitwear under the trade names 'Courtelle', 'Acrilan', and 'Orlon'. The polymerization can be represented simply as:

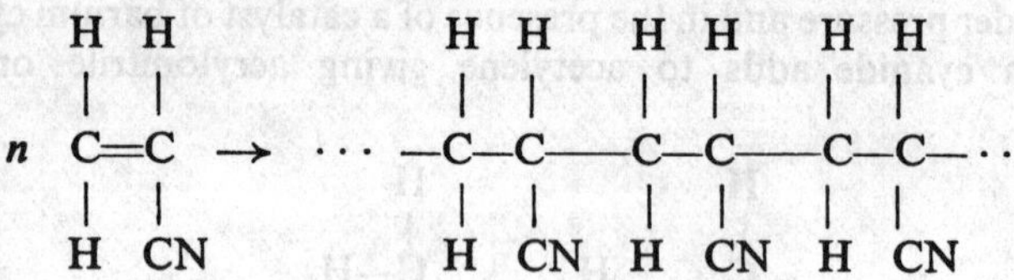

i.e. $$n\ CH_2{:}CHCN \rightarrow (CH_2.CHCN)_n$$

Oxy-acetylene burners give a very hot and easily controllable flame and are widely employed for cutting and welding metals. Acetylene is also used for making synthetic rubbers such as *neoprene* (§ 49.5) and the important solvents trichloroethylene (§ 40.10) and tetrachloroethylene, $CCl_2 : CCl_2$ (often known as perchloroethylene), which are extensively used in industry and in dry-cleaning for dissolving grease, oils, and fats.

Structure of Acetylene: Eudiometry shows the molecular formula to be C_2H_2. From valency considerations and having regard to its many addition reactions, which clearly indicate unsaturation, the molecule is believed to contain a triple covalent bond composed of six shared electrons, three from each carbon atom, as in Fig. 14.7. This structure conforms with physical evidence which shows the carbon atoms to be even closer together than in ethylene (see § 34.7), and it is also in keeping with the highly endothermic nature and reactivity of acetylene.

34.5. Butadiene, $CH_2{=}CH{-}CH{=}CH_2$. In addition to the higher members of the homologous series of olefines and acetylenes, other aliphatic hydrocarbons exist containing double or triple bonds. One of the most important of these is the di-olefine, butadiene, b.p. – 4°C, which is made in large amounts by dehydrogenating n-butane using a catalyst of alumina or chromic oxide:

$$CH_3.CH_2.CH_2.CH_3 \overset{600^\circ C}{=\!=} CH_2{:}CH.CH{:}CH_2 + 2H_2$$

Butadiene polymerizes readily in the presence of sodium to give a tough rubber-like substance. Better synthetic rubbers, which are used extensively for making the treads of tyres because of their high resistance to abrasion, can be obtained by mixing butadiene with another compound such as styrene (§ 43.3) and polymerizing the two together as described in § 49.5. Butadiene is also used in some countries in the manufacture of nylon.

34.6. Tests for Unsaturation. Unsaturated compounds are best recognized by their addition reactions. The most convenient are the reactions with bromine or bromine water and with potassium permanganate solution, since these give readily observable colour changes, but the decolorization of permanganate solution is also brought about by reducing agents and so is not by itself a distinctive test for unsaturation.

34.7. Bond Lengths. Mention has already been made in this chapter of the different distances between carbon atoms in various molecules. Electron diffraction and spectroscopy, which have made it possible to determine bond lengths with accuracy, show that the C—C bond (as in ethane) is 0.154 nm in length,* whereas the C≡C (as in ethylene) is 0.134 nm long and the C=C bond (as in acetylene) is only 0.120 nm in length. These bond lengths are of particular interest in view of the fact that the distance between adjacent carbon atoms in benzene is 0.139 nm (see § 43.3). They also clearly illustrate how the concept of atomic size (referred to in §§ 4.11 and 13.9) depends upon the type of bonding involved, the values usually quoted being those for an atom exerting only single covalent bonds.

Also of interest is the molecular shape of alkenes and alkynes. All the atoms in ethylene are *coplanar*, i.e. they all lie in one plane, a fact which explains the occurrence of geometrical isomerism (§ 48.5) in its disubstituted derivatives because of the resistance of the double bond to free rotation about its axis. The molecule of acetylene has been shown by physical methods to be linear in shape, with all four atoms lying in one straight line.

* Bond lengths are conventionally expressed in nanometre; 1 nm is 10^{-9} m.

35

Aliphatic Alcohols

35.1. Introduction. Aliphatic alcohols are compounds containing *hydroxyl groups* (—OH) linked to alkyl radicals. Although they cannot in fact be prepared in that way, they may be regarded as being derived from paraffins by replacing hydrogen atoms by hydroxyl groups. *Monohydric* alcohols contain one hydroxyl group in each molecule; those containing more are described as *polyhydric* alcohols.

The monohydric alcohols form a homologous series with the general formula C_nH_{2n+1} OH. They are named either by substituting the ending -ol for -e in the name of the corresponding paraffin (e.g. methanol, ethanol), or by adding the word alcohol to the alkyl radical it contains (e.g. methyl alcohol, ethyl alcohol). Ethyl alcohol, the second member of the series, is by far the most important (indeed it is often referred to simply as 'alcohol'), so it will be studied first.

35.2. Ethyl Alcohol, Ethanol, C_2H_5OH.

LABORATORY PREPARATION: It is so readily available commercially that its preparation in the laboratory is rarely performed. Various methods can be used:

1. By hydrolysing ethyl iodide or bromide with alkali:

$$C_2H_5I + NaOH = C_2H_5OH + NaI$$

2. By hydrolysing ethyl esters with hot alkali:

$$CH_3.COOC_2H_5 + NaOH = C_2H_5OH + CH_3COONa$$

3. By reducing acetaldehyde with nascent hydrogen from sodium amalgam and water or sodium borohydride (§ 19.7):

$$CH_3.CHO + 2\bar{H} = C_2H_5OH$$

4. By treating ethylamine with nitrous acid:

$$C_2H_5NH_2 + HNO_2 = C_2H_5OH + N_2\uparrow + H_2O$$

5. By reducing acetic acid with lithium aluminium hydride in ethereal solution (see § 19.14):

$$CH_3COOH + 4\bar{H} = C_2H_5OH + H_2O$$

MANUFACTURE: Ethyl alcohol is made industrially in two main ways:

1. From Ethylene: This gas, which is available in large quantities as a by-product of the cracking of petroleum, is absorbed in 95% sulphuric acid at 80°C under 3×10^6 N m^{-2} (30 atm) pressure, forming ethyl hydrogen sulphate:

$$C_2H_4 + H_2SO_4 \rightleftharpoons C_2H_5HSO_4$$

This is then hydrolysed by boiling it with water:

$$C_2H_5HSO_4 + H_2O \rightleftharpoons C_2H_5OH + H_2SO_4$$

The resulting alcohol is distilled off, leaving the sulphuric acid to be concentrated and used again. Much of the alcohol required for industrial purposes is now made in this way.

In recent years a process has been devised for hydrating ethylene directly by mixing it with steam at 300°C and about 10^7 N m^{-2} (100 atm) pressure and passing the mixture over a catalyst of phosphoric acid:

$$C_2H_4 + H_2O = C_2H_5OH$$

2. By Fermentation: The name *fermentation* is given to chemical reactions brought about by micro-organisms such as yeast, bacteria, or moulds. These lowly forms of life produce certain complex chemical substances called *enzymes*, which are highly effective catalysts. Not only are they very specific in their activity, but they are also very sensitive to conditions of temperature and acidity, which must only vary within narrow limits.

The fundamental reaction underlying the production of ethyl alcohol by fermentation is the catalytic conversion of the sugar glucose into alcohol by the enzyme *zymase* which is present in yeast:

$$C_6H_{12}O_6 = 2C_2H_5OH + 2CO_2\uparrow$$

Research has shown that this change is in fact a very complicated one involving a number of stages, but this equation represents the overall reaction which occurs. The yeast used is a specially selected strain known as brewer's yeast, which has been cultivated under carefully controlled conditions. The sugar solution is maintained at about 27°C for two days, by which time the alcohol content will have reached about 8%. As the carbon dioxide is liberated it produces a frothing effect in the solution; being a valuable by-product the gas is collected, dried, and condensed to a solid (§ 20.5). After fermentation the solution is filtered and the alcohol in it is concentrated by fractional distillation.

In practice glucose is not often used as the starting material for making alcohol, a cheaper substance being preferred. The choice varies from one country to another, depending upon what is most readily available.

In Great Britain and the United States of America, molasses is the main source. This is the syrupy liquid remaining after the crystallization of cane sugar; it contains about 50% of sugars, mainly sucrose. It is diluted until the concentration of sugar is about 15%, acidified slightly, and seeded with yeast. The latter contains an enzyme *invertase* which converts the sucrose into a mixture of two isomeric sugars glucose and fructose, both of which can be fermented to alcohol by the action of zymase:

$$C_{12}H_{22}O_{11} + H_2O = C_6H_{12}O_6 + C_6H_{12}O_6$$

In other countries potatoes, cereals, and rice are used as the chief sources of alcohol. These materials are crushed and treated with superheated steam to extract the starch from them. The products are warmed to 50°C for an hour with malt (made from partially germinated barley), when the enzyme *diastase* present in the malt converts the starch into maltose thus:

$$2(C_6H_{10}O_5)_n + nH_2O = nC_{12}H_{22}O_{11}$$

On adding yeast, which contains the enzyme *maltase*, the maltose is converted into glucose which can then be fermented to alcohol. In countries where wood is plentiful, sawdust is heated with mineral acids under pressure, which converts the cellulose into glucose:

$$(C_6H_{10}O_5)_n + nH_2O = nC_6H_{12}O_6$$

After neutralizing the excess of acid, the product is fermented by yeast as before.

Fermentation methods have been used since ancient times for making alcoholic beverages of all kinds, the sources of alcohol being barley, fruit (especially grapes, apples, and pears), honey, molasses, etc. When the concentration of alcohol in the ferment reaches about 15% the yeast is killed and the process ceases. Consequently beverages such as whisky and gin, which contain up to 50% alcohol, are made by distillation of the fermented solution.

Purification: When dilute solutions of alcohol are concentrated by fractional distillation, the product is a mixture containing 95.57% by weight of ethyl alcohol and 4.43% by weight of water. This mixture called *rectified spirit*, has a constant boiling point of 78.2°C and comes over unchanged when redistilled. Anhydrous calcium chloride cannot be used to remove the residual water because it reacts with alcohol forming compounds of the type $CaCl_2{\cdot}nC_2H_5OH$, but on distilling over quicklime (which reacts with the water but not with the alcohol), a distillate is obtained containing over 99.5% alcohol, known as *absolute* alcohol. The last traces of water can be removed from this by refluxing it with metallic calcium and redistilling.

Alternatively absolute alcohol can be obtained from rectified spirit by an azeotropic method (§ 6.11). Some benzene is added to the rectified spirit and the mixture is distilled. At first a constant boiling point mixture of benzene and water and alcohol distils over at 64.8°C. When all the water has been removed another azeotropic mixture containing benzene and alcohol comes over at 68.2°C until all the benzene has volatilized, when the residue of absolute alcohol distils over at 78.3°C. This method is used industrially.

Absolute alcohol is very hygroscopic and must be kept from atmospheric moisture if it is to remain anhydrous. The much less expensive rectified spirit is suitable for most purposes and is generally used.

PROPERTIES OF ETHYL ALCOHOL: It is a colourless neutral liquid which is miscible with water in all proportions. Its boiling point (78.3°C) is much higher than that of paraffins of similar molecular weight, which suggests that alcohol is associated by hydrogen bonding between the hydroxyl groups as in water (§ 16.4).

REACTIONS OF ETHYL ALCOHOL: These are summarized in Table 35.1.

1. Combustion: It burns with a pale-blue flame giving carbon dioxide and water vapour:

$$C_2H_5OH + 3O_2 = 2CO_2 + 3H_2O$$

2. Reaction with Metals: It reacts quietly with metallic sodium or potassium, forming a solution of an ethoxide and liberating hydrogen:

$$2C_2H_5OH + 2Na = 2C_2H_5ONa + H_2\uparrow$$

The sodium ethoxide, also known as sodium ethylate, is obtained as a white solid on evaporating the solution. It is rapidly hydrolysed by cold water giving an alkaline solution:

$$C_2H_5ONa + H_2O \rightleftharpoons C_2H_5OH + NaOH$$

A mixture of sodium and alcohol makes a useful reducing agent in organic chemistry, particularly for the reduction of substances such as cyanides which tend to be hydrolysed by reducing agents containing water or acids.

3. Halogenation: Phosphorus pentachloride reacts in the cold with anhydrous ethyl alcohol forming ethyl chloride, phosphorus oxychloride, and fumes of hydrogen chloride:

$$C_2H_5OH + PCl_5 = C_2H_5Cl + POCl_3 + HCl\uparrow$$

A mixture of red phosphorus with either bromine or iodine converts alcohol into ethyl bromide or iodide respectively, as described in § 40.2.

TABLE 35.1. REACTIONS OF ETHYL ALCOHOL

Reactant	Reagent	Product
ETHYL ALCOHOL	sodium	sodium ethoxide, C_2H_5ONa and hydrogen, H_2
	phosphorus pentachloride	ethyl chloride, C_2H_5Cl
	chlorine	chloral, $CCl_3.CHO$
	mild oxidizing agents	acetaldehyde, $CH_3.CHO$
	strong oxidizing agents	acetic acid, $CH_3.COOH$
	acetic acid	ethyl acetate, $CH_3.COOC_2H_5$
	potassium iodide soln. and a soln. of sodium hypochlorite	iodoform, CHI_3
	acetyl chloride	ethyl acetate, $CH_3.COOC_2H_5$
	benzoyl chloride	ethyl benzoate, $C_6H_5.COOC_2H_5$

If chlorine is bubbled into ethyl alcohol, the final product is chloral, $CCl_3.CHO$ (see § 36.4).

4. Oxidation: Ethyl alcohol is readily oxidized to acetaldehyde by warming it with potassium dichromate solution and dilute sulphuric acid:

$$C_2H_5OH + \bar{O} = CH_3.CHO + H_2O$$

In the presence of excess of the oxidizing agent the oxidation proceeds further to acetic acid:

$$CH_3.CHO + \bar{O} = CH_3.COOH$$

Wine may go sour on prolonged exposure to the air owing to bacterial oxidation of alcohol to acetic acid; the process is used to make vinegar.

Ethyl alcohol can be converted into acetaldehyde catalytically by passing its vapour over finely-divided copper at 300°C, an example of *dehydrogenation* (i.e. removal of hydrogen):

$$C_2H_5OH = CH_3.CHO + H_2$$

Ethyl alcohol is also oxidized by concentrated or fuming nitric acid on warming. The reaction is sometimes dangerously violent and the mixture has been used as a propellent for rockets.

5. Esterification: Ethyl alcohol reacts reversibly with acids forming ethyl esters. This process, known as esterification, is discussed in detail in § 38.1. The general reaction is:

$$\text{alcohol} + \text{acid} \rightleftharpoons \text{ester} + \text{water}$$

e.g. $$\underset{}{C_2H_5OH} + \underset{\text{acetic acid}}{CH_3COOH} \rightleftharpoons \underset{\text{ethyl acetate}}{CH_3.COOC_2H_5} + H_2O$$

When ethyl alcohol reacts with concentrated sulphuric acid the ester formed is ethyl hydrogen sulphate, $C_2H_5HSO_4$, which has the structure $C_2H_5OSO_2.OH$:

$$C_2H_5OH + H_2SO_4 \rightleftharpoons C_2H_5OSO_2.OH + H_2O$$

If excess of alcohol is present, this reacts forming diethyl ether:

$$C_2H_5OSO_2.OH + C_2H_5OH = C_2H_5.O.C_2H_5 + H_2SO_4$$

On the other hand, if the sulphuric acid is in excess and the temperature is raised above 170°C, the ester decomposes yielding ethylene:

$$C_2H_5OSO_2.OH = C_2H_4\uparrow + H_2SO_4$$

6. *Iodoform Reaction:* If potassium iodide solution and a little sodium hypochlorite solution are gently warmed with ethyl alcohol, pale yellow crystals of iodoform appear on standing. The same result is achieved, though rather more slowly, if alcohol is warmed with iodine and caustic soda solution. The reaction is discussed in § 40.9.

7. *Acetylation:* Acetyl chloride converts ethyl alcohol into ethyl acetate as described in § 37.6:

$$C_2H_5OH + CH_3COCl = CH_3.COOC_2H_5 + HCl\uparrow$$

8. *Benzoylation:* In the presence of sodium hydroxide solution benzoyl chloride reacts with ethyl alcohol giving ethyl benzoate, as described in § 47.5:

$$C_2H_5OH + C_6H_5COCl = C_6H_5.COOC_2H_5 + HCl$$

USES OF ETHYL ALCOHOL: It is an important solvent, widely used industrially to dissolve resins, varnishes, soaps, perfumes, dyes, and drugs. It is used in the manufacture of many other organic compounds particularly acetaldehyde, ethyl esters and halides, diethyl ether, chloral and chloroform. It is also used as a fuel, either alone or mixed with petrol, in racing cars, working models, and in rockets. Owing to its low freezing point it is sometimes used as an anti-freeze and for filling thermometers. Ethyl alcohol is, of course, present in many beverages, e.g. beers, wines, and spirits.

STRUCTURE OF ETHYL ALCOHOL: Analysis and vapour density measurements indicate a molecular formula of C_2H_6O, which is confirmed by combustion measurements. There are two ways of arranging these nine atoms so that they have the accepted valencies, viz.:

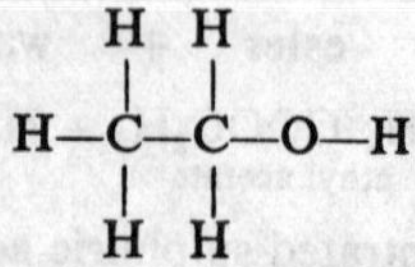

I

and

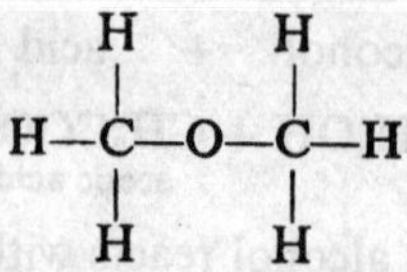

II

The reaction of ethyl alcohol with sodium shows that only one atom of hydrogen in each molecule can be replaced by the metal, suggesting that there is one hydrogen atom which is specially placed compared with the others. Similarly the reaction with phosphorus pentachloride involves the replacement of the oxygen atom and one hydrogen atom in each molecule by a single chlorine atom. Both of these reactions strongly support structure I rather than II. Moreover the preparation of ethyl alcohol by hydrolysing ethyl halides and the analogous structure of methyl alcohol confirm I to be the structure of ethyl alcohol. Structure II is that of dimethyl ether, an isomer with very different properties.

35.3. Methyl Alcohol, Methanol, CH_3OH.

LABORATORY PREPARATION: It is rarely prepared in the laboratory because, like ethyl alcohol, it is cheaply available commercially. It could be prepared by hydrolysing methyl iodide or a methyl ester with a hot solution of an alkali, but not by the action of nitrous acid on methylamine (see § 41.4, reaction 6), nor by the reduction of formaldehyde with nascent hydrogen.

MANUFACTURE: Two methods are in use:

1. From Water Gas: A mixture of water gas (§ 20.15) and hydrogen at 300°C and 3×10^7 N m^{-2} (300 atm) pressure is passed over a catalyst of zinc oxide containing a chromic oxide promoter, when a good yield of the alcohol results:

$$CO + 2H_2 = CH_3OH$$

Most methyl alcohol is now made by this method or by a variant of it which uses carbon monoxide and hydrogen produced by the action of steam on methane (§ 33.9). An alternative process using a copper catalyst and operating at 250°C and only 5×10^6 N m^{-2} (50 atm) pressure has recently been developed.

2. From Wood: When wood is heated strongly in the absence of air it gives a brown watery distillate called pyroligneous acid containing a few per cent of methyl alcohol, which can be separated by fractional distillation over lime. This method is almost obsolete, but it explains the old name of 'wood spirit' for methyl alcohol. When made in this way the alcohol invariably contains some acetone.

PROPERTIES OF METHYL ALCOHOL: It is a colourless hygroscopic liquid, b.p. 64°C. It is miscible with water in all proportions, but unlike ethyl alcohol it does not form a constant boiling point mixture. It burns with a pale-blue, non-luminous flame. It is poisonous, causing blindness, insanity, and death.

REACTIONS OF METHYL ALCOHOL: These resemble the reactions of ethyl alcohol in many respects:

1. With metallic sodium it gives sodium methoxide and hydrogen:

$$2CH_3OH + 2Na = 2CH_3ONa + H_2\uparrow$$

2. With phosphorus pentachloride it forms methyl chloride and gives off fumes of hydrogen chloride:

$$CH_3OH + PCl_5 = CH_3Cl + POCl_3 + HCl\uparrow$$

3. It reacts reversibly with acids forming methyl esters:

e.g. $$\underset{}{CH_3OH} + \underset{\text{acetic acid}}{CH_3.COOH} \rightleftharpoons \underset{\text{methyl acetate}}{CH_3.COOCH_3} + H_2O$$

4. When methyl alcohol vapour is mixed with oxygen and passed over a platinum catalyst, it is oxidized to formaldehyde:

$$CH_3OH + \bar{O} = H.CHO + H_2O$$

If a catalyst of platinum black is used, the methyl alcohol vapour is oxidized to formic acid:

$$CH_3OH + 2\bar{O} = H.COOH + H_2O$$

The exothermic nature of these reactions can be demonstrated by heating a spiral of platinum wire and suspending it over some methyl alcohol in a beaker, when the wire continues to glow as the reaction proceeds.

USES OF METHYL ALCOHOL: Large amounts are converted catalytically into formaldehyde. In some countries methyl alcohol is used as an anti-freeze. It is also used to make methyl compounds, e.g. methyl methacrylate, which polymerizes readily in the presence of a catalyst of benzoyl peroxide (§ 47.5) to the glass-like plastic 'Perspex', and is also important as the raw material from which most artificial teeth and dentures are made. *Methylated spirit*, which is a valuable solvent for paints and varnishes, is made by adding up to 10% of methyl alcohol to rectified spirit to denature it, i.e. make it unfit for drinking. An unpleasant-tasting substance such as pyridine and a dye are often added as well. The mixture can then be sold for industrial use without charging the heavy excise duty which is usually levied on potable ethyl alcohol, since there is no way of removing the poisonous methyl alcohol except by distillation, which requires a special licence.

STRUCTURE OF METHYL ALCOHOL: Analysis and vapour density measurements show that its molecular formula is CH_4O. If the elements are given their usual valencies there is only one possible way of arranging these six atoms, viz.:

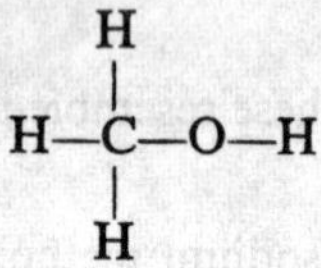 or CH_3OH

This structure is in accord with all the reactions of methyl alcohol and is consequently accepted.

35.4. Propyl Alcohols, C_3H_7OH. Whereas only one methyl and one ethyl alcohol exist, two isomeric propyl alcohols are known having the structures:

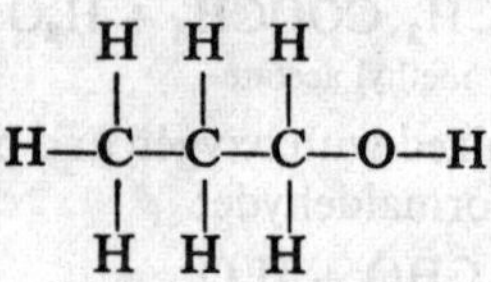 and 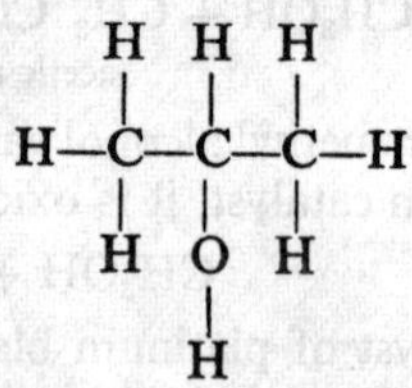

n-propyl alcohol (n-propanol)
propan-1-ol*

iso-propyl alcohol
propan-2-ol*

$CH_3.CH_2.CH_2OH$
or $C_2H_5.CH_2OH$

$CH_3.CH(OH).CH_3$
or $(CH_3)_2.CHOH$

Both isomers are alcohols because both contain hydroxyl groups joined to alkyl radicals, but there is an important difference in their structures which causes them to behave differently when oxidized. In normal propyl alcohol the carbon atom bearing the hydroxyl group has two hydrogen atoms linked to it, as in methyl and ethyl alcohol. Such alcohols are called *primary alcohols*; when oxidized they yield aldehydes and then acids. This is in contrast to alcohols such as iso-propyl alcohol in which the carbon atom bearing the hydroxyl group is linked directly to only one hydrogen atom. Such alcohols are called *secondary alcohols*; when oxidized they yield ketones which are not easily oxidized further. In *tertiary alcohols*, a third type, there is no hydrogen atom linked

* These names distinguish the isomers by indicating to which carbon atom in the chain the hydroxyl group is attached, numbering from either end.

directly to the carbon atom to which the hydroxyl group is attached. When such alcohols are treated with powerful oxidizing agents they give mixtures of acids each of which contains fewer carbon atoms in its molecule than the original alcohol. These ideas are summarized by the following relationships:

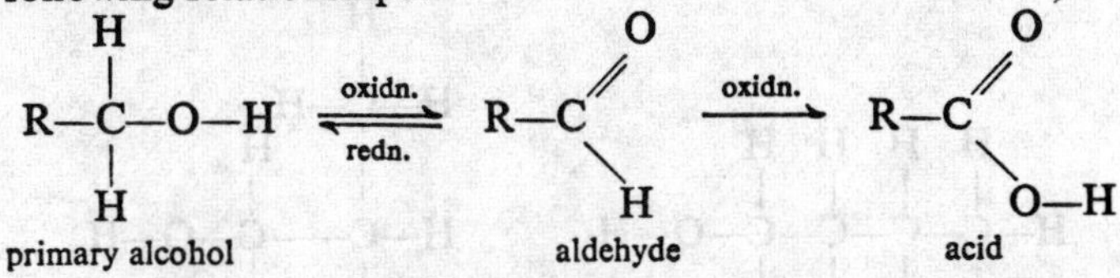

i.e. $$R.CH_2OH \underset{\text{redn.}}{\overset{\text{oxidn.}}{\rightleftharpoons}} R.CHO \xrightarrow{\text{oxidn.}} R.COOH$$

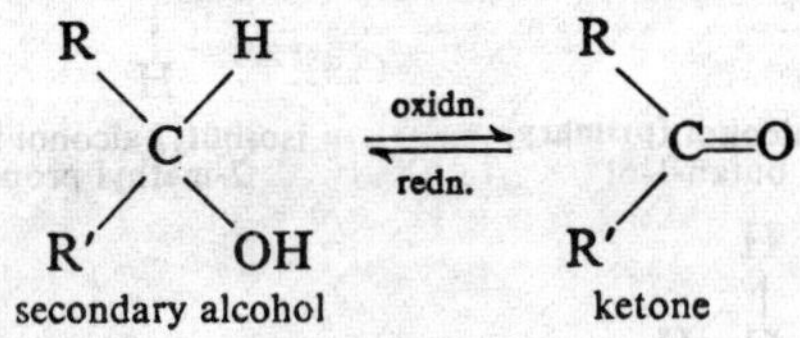

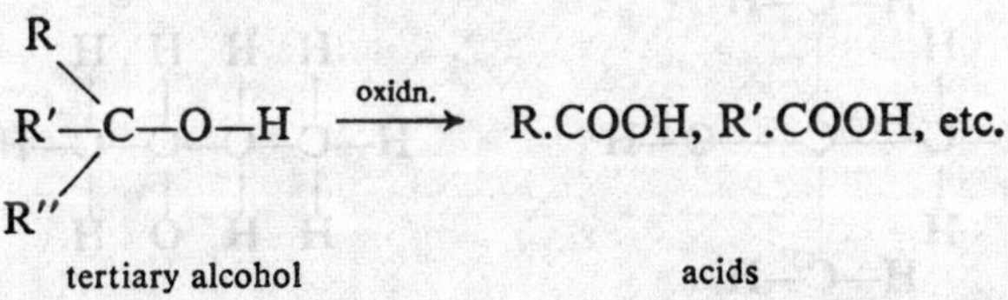

Normal Propyl Alcohol, Propan-1-ol, is made by similar methods to those used for making ethyl alcohol. It has similar properties, yielding propionaldehyde and propionic acid on oxidation and normal propyl esters on treatment with acids.

Iso-Propyl Alcohol, Propan-2-ol, is prepared in the laboratory by reducing acetone with sodium amalgam and water or sodium borohydride (§ 19.7):

$$CH_3.CO.CH_3 + 2\bar{H} = CH_3.CH(OH).CH_3$$

It is made industrially by dissolving propylene, a by-product of the cracking of petroleum, in concentrated sulphuric acid and boiling with water:

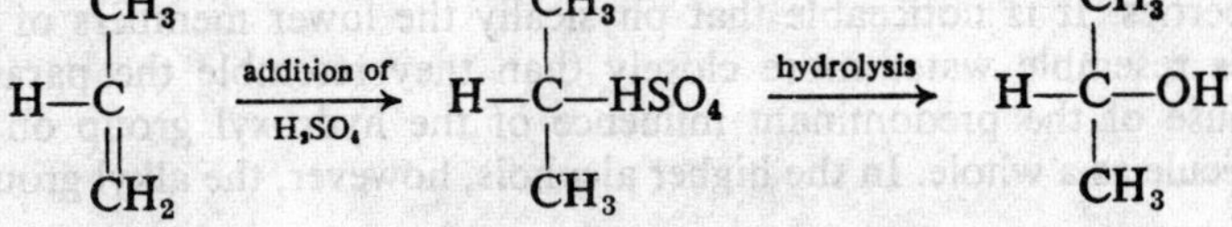

Unlike its normal isomer, iso-propyl alcohol gives the iodoform reaction (see § 40.9). Its main industrial uses are as a solvent and for making acetone by catalytic oxidation.

35.5. Butyl Alcohols, C_4H_9OH. Four isomeric butyl alcohols exist:

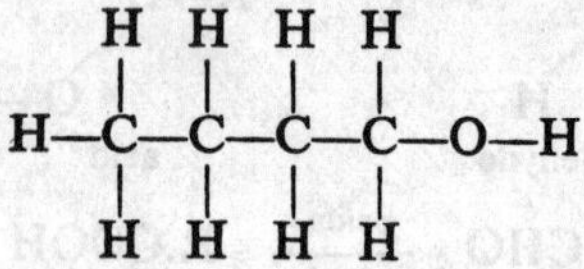

n-butyl alcohol (primary)
butan-1-ol

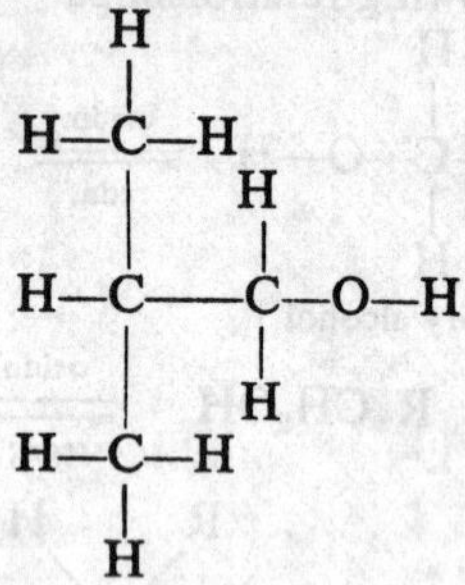

iso-butyl alcohol (primary)
2-methyl propan-1-ol

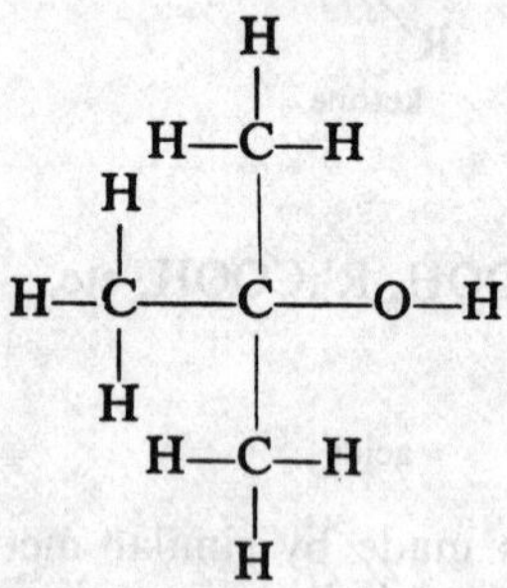

tertiary butyl alcohol
2-methyl propan-2-ol

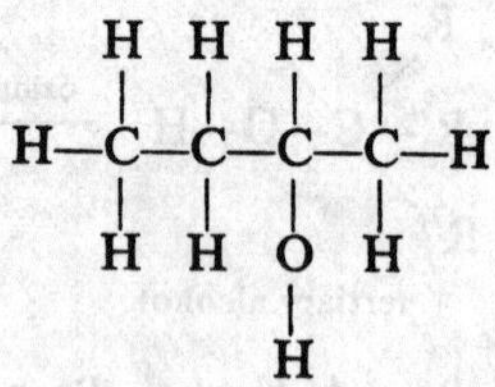

secondary butyl alcohol
butan-2-ol

Normal and iso-butyl alcohols are important solvents for vegetable oils, fats, waxes, and resins, and consequently they are extensively used in paints and varnishes. They are made from propylene (see Table 34.2).

Butan-2-ol is of special interest because it is the first compound we have met containing an *asymmetric carbon atom* (one joined to four different atoms or groups). It therefore displays *optical isomerism* (see § 48.2).

35.6. The Higher Alcohols. These show increasing isomerism as the molecular weight rises. They form a well-defined homologous series with similar methods of preparation and similar physical and chemical properties. It is noticeable that physically the lower members of this series resemble water more closely than they resemble the paraffins because of the predominant influence of the hydroxyl group on the molecule as a whole. In the higher alcohols, however, the alkyl group is larger and the physical properties become progressively more like those of the higher paraffins. Cetyl alcohol, $C_{16}H_{33}OH$, for example, is a waxy solid which is insoluble in water and burns with a smoky, luminous flame.

Some of the higher alcohols are important in the manufacture of detergents. Their esters, which occur in certain natural oils, are reduced catalytically to the alcohols, which are then converted into the sodium derivatives of their hydrogen sulphates. For example, lauryl alcohol, $C_{12}H_{25}OH$, which occurs as its glyceryl ester in coconut oil, is converted into sodium lauryl sulphate, $C_{12}H_{25}SO_4Na$, which is much used for making shampoos. Detergents such as these are often preferred to soaps because their calcium and magnesium salts are soluble in water and so their cleansing action is not diminished as it is in soaps by precipitation reactions with hard water.

35.7. Detection of Alcohols. In general alcohols are recognized by tests for the hydroxyl group they contain, as this is their characteristic functional group. For example, they all evolve hydrogen when treated with metallic sodium, leaving a sodium derivative, and they give fumes of hydrogen chloride when treated with phosphorus pentachloride or acetyl chloride in the cold. Unlike acids their solutions in water are neutral.

Methyl alcohol is distinguished from the other lower alcohols by two tests. It does not give a yellow precipitate of iodoform when warmed with potassium iodide solution and sodium hypochlorite solution (ethyl and isopropyl alcohol do), and on oxidation with potassium dichromate solution and dilute sulphuric acid it gives formic acid, carbon dioxide, and water, whereas ethyl alcohol gives acetaldehyde and acetic acid when oxidized, and iso-propyl alcohol gives acetone.

35.8. Ethylene Glycol, Ethane-1,2-diol, $(CH_2OH)_2$. This is the most important *dihydric* alcohol (i.e. alcohol containing two hydroxyl groups in its molecule).

Laboratory Preparation: It is prepared by boiling ethylene dibromide (made directly from ethylene and bromine) under reflux for several hours with potassium carbonate solution and separating the glycol by fractional distillation:

$$\begin{matrix} CH_2Br \\ | \\ CH_2Br \end{matrix} + K_2CO_3 + H_2O = \begin{matrix} CH_2OH \\ | \\ CH_2OH \end{matrix} + 2KBr + CO_2\uparrow$$

Glycol is also formed when ethylene is bubbled into a dilute alkaline solution of potassium permanganate:

$$\begin{array}{c} CH_2 \\ \| \\ CH_2 \end{array} + \bar{O} + H_2O = \begin{array}{c} CH_2OH \\ | \\ CH_2OH \end{array}$$

MANUFACTURE: It is made by hydrolysing ethylene oxide (made by passing ethylene and air over a silver catalyst at 300°C) with steam and dilute hydrochloric acid, or by boiling ethylene chlorohydrin with milk of lime:

$$\begin{array}{c} CH_2 \\ | \quad \diagdown \\ | \quad\quad O \\ | \quad \diagup \\ CH_2 \end{array} + H_2O = \begin{array}{c} CH_2OH \\ | \\ CH_2OH \end{array}$$

$$2\begin{array}{c} CH_2OH \\ | \\ CH_2Cl \end{array} + Ca(OH)_2 = 2\begin{array}{c} CH_2OH \\ | \\ CH_2OH \end{array} + CaCl_2$$

PROPERTIES AND REACTIONS OF ETHYLENE GLYCOL: It is a colourless, viscous liquid with a sweet taste. It freezes at − 13°C and boils at 197°C. It is miscible with water in all proportions, giving mixtures with low freezing points.

In its reactions it behaves as a typical primary alcohol. There is often a tendency for one of the —CH_2OH groups to react more readily than the other, so that mono- and di-derivatives of glycol are formed in separate stages. For example, with metallic sodium, glycol forms $CH_2OH.CH_2ONa$ at room temperature and $CH_2ONa.CH_2ONa$ on heating, and with acetic acid it gives mono-acetate and di-acetate esters. Phosphorus pentachloride converts it directly into ethylene dichloride, $CH_2Cl.CH_2Cl$, with the evolution of hydrogen chloride. Oxidation produces a variety of products depending upon the agent used; nitric acid converts both of the primary alcohol groups into carboxyl groups forming oxalic acid.

$$\begin{array}{c} CH_2OH \\ | \\ CH_2OH \end{array} + 4\,\bar{O} = \begin{array}{c} COOH \\ | \\ COOH \end{array} + 2H_2O$$

USES OF ETHYLENE GLYCOL: It is extensively used as an anti-freeze in the radiators of vehicles and aircraft and as the source of a number of ethers and esters which are valuable solvents. Its major use, however, is in the manufacture of *polyester resins* as described in § 49.3.

When the polyester resin made by condensing ethylene glycol with dimethyl terephthalate (obtained from *p*-xylene) is polymerized, the synthetic fibre 'Terylene' (§ 49.3) is obtained:

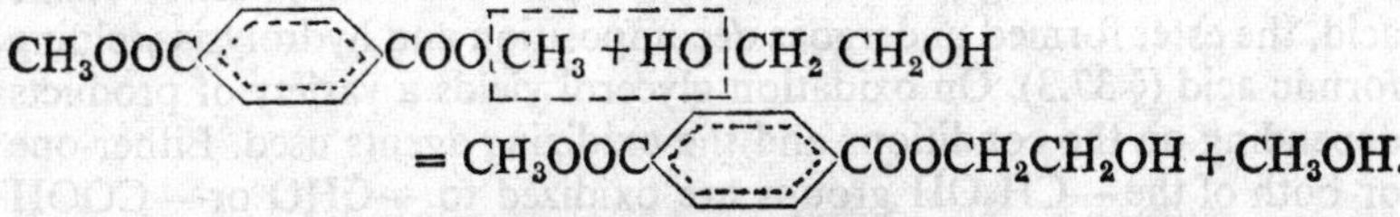

35.9. Glycerol, Propane-1,2,3-triol, $CH_2OH.CH[OH].CH_2OH$. This is the only important trihydric alcohol. It is often known by its common name, glycerine. Its esters with certain of the higher aliphatic acids (e.g. palmitic, stearic, and oleic acids) occur naturally in most animal and vegetable oils and fats. It is chiefly obtained, therefore, as a by-product during soap-making, when these esters are hydrolysed by boiling them with alkalis as described in § 38.5. To meet the large demand this source is supplemented in some countries, particularly in the U.S.A., by glycerol which has been synthesized from propylene (see Table 34.2).

PROPERTIES AND REACTIONS OF GLYCEROL: It is a colourless, viscous liquid which is hygroscopic and has a sweet taste. Its melting point is 17°C, but it supercools readily. It boils at 290°C, often decomposing slightly. It is miscible with water and alcohol in all proportions.

The molecule of glycerol contains two primary alcohol groups and one secondary alcohol group and consequently gives the reactions of both types. With sodium first one and then, on heating, both primary alcohol groups are attacked with the liberation of hydrogen. Phosphorus pentachloride replaces all three hydroxyl groups by chlorine atoms, producing fumes of hydrogen chloride:

$$\begin{array}{l} CH_2OH \\ | \\ CHOH \\ | \\ CH_2OH \end{array} + 3PCl_5 = \begin{array}{l} CH_2Cl \\ | \\ CHCl \\ | \\ CH_2Cl \end{array} + 3POCl_3 + 3HCl\uparrow$$

Organic acids react with one, two, or all three of the alcohol groups in each molecule of glycerol, giving esters called *glycerides*. When glycerol is treated with a cold mixture of concentrated nitric and sulphuric acids,* it is converted into the ester glycerol trinitrate, better known as nitroglycerine, which explodes with great violence if detonated:

$$\begin{array}{l} CH_2OH \\ | \\ CHOH \\ | \\ CH_2OH \end{array} + 3HNO_3 = \begin{array}{l} CH_2.O.NO_2 \\ | \\ CH.O.NO_2 \\ | \\ CH_2.O.NO_2 \end{array} + 3H_2O.$$

* Students should not attempt to carry out this reaction, as it is far too dangerous.

Absorbed in kieselguhr or sawdust it is known as dynamite and mixed with cellulose nitrate (gun-cotton) as gelignite and cordite.

When anhydrous glycerol is heated to 120°C with crystals of oxalic acid, the ester formed undergoes decomposition and hydrolysis yielding formic acid (§ 37.3). On oxidation glycerol yields a variety of products depending on the conditions and the oxidizing agents used. Either one or both of the $-CH_2OH$ groups are oxidized to $-CHO$ or $-COOH$ groups, and the $\rangle$CHOH group may be oxidized to $\rangle$CO as well.

Uses of Glycerol: Its chief use is in the manufacture of *alkyd resins* (§ 49.3) which are important constituents of many modern varnishes and paints because they give a hard, glossy, protective finish. It is also used as a moistening agent in tobacco and cosmetics, as a softening agent in 'Cellophane' (see § 42.7), and for making explosives, some of which are extensively employed in mining, quarrying, and civil engineering.

36

Aliphatic Aldehydes and Ketones

36.1. Introduction. The functional group common to all aldehydes (and known therefore as the *aldehyde group*) is —CHO or $-\overset{H}{\overset{/}{C}}{=}O$.

Aliphatic aldehydes form a homologous series with the general formula R.CHO, e.g. HCHO formaldehyde, CH_3.CHO acetaldehyde, C_2H_5.CHO propionaldehyde, etc. All ketones contain the functional group $\rangle C{=}O$ joined to two hydrocarbon radicals. Aliphatic ketones form a homologous series with the general formula R.CO.R′, the first member being dimethyl ketone, commonly known as acetone. When R and R′ are the same the ketone is said to be simple and when they are different it is called a mixed ketone.

Aliphatic aldehydes and ketones have the same general molecular formula, $C_nH_{2n}O$. Both classes of compound contain the *carbonyl group*, $\rangle C{=}O$, in their molecules, which gives them many properties in common, but in aldehydes there is also a reactive hydrogen atom which, being readily oxidized, confers reducing properties upon the molecule. Both aldehydes and ketones are made by oxidizing alcohols, aldehydes from primary alcohols and ketones from secondary alcohols, as described in the last chapter.

In considering the aldehydes, acetaldehyde will be studied first because it is the most important member of the series and is in many ways typical of the others.

36.2. Acetaldehyde, Ethanal, CH_3.CHO.

Laboratory Preparation: It is prepared by oxidizing ethyl alcohol with a mixture of sodium dichromate and moderately concentrated sulphuric acid using the apparatus shown in Fig. 36.1.

$$Na_2Cr_2O_7 + 4H_2SO_4 + 3C_2H_5OH = 3CH_3.CHO + Na_2SO_4 + Cr_2(SO_4)_3 + 7H_2O$$

Sodium dichromate is used because it is more soluble than the potassium salt in the presence of alcohol. The reaction is exothermic and heating is not necessary once it has started. By arranging that the oxidizing agent is not in excess and by distilling off the volatile acetaldehyde as fast as it is formed, further oxidation to acetic acid is largely prevented. The distillate is impure, containing water, alcohol, and acetic acid. It can be purified by fractional distillation or by conversion into crystals of acetaldehyde-ammonia, which can be dried and distilled with moderately concentrated sulphuric acid.

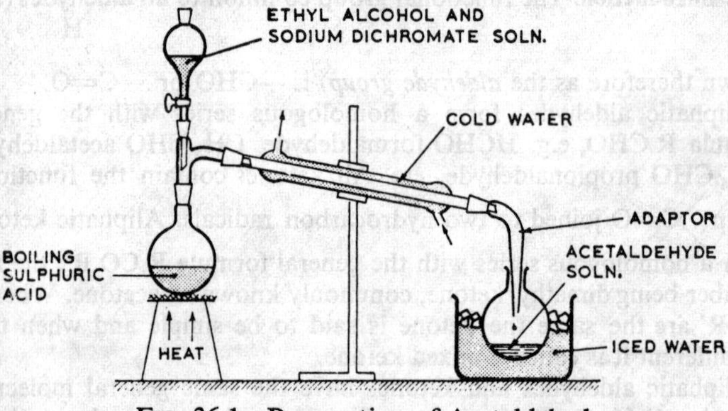

FIG. 36.1 Preparation of Acetaldehyde

Some acetaldehyde is obtained when a mixture of calcium formate and anhydrous calcium acetate is heated strongly, but the yield is poor and the product impure:

$$(CH_3.COO)_2Ca + (H.COO)_2Ca = 2CH_3.CHO + 2CaCO_3$$

MANUFACTURE: Several methods are used:

1. When acetylene, obtained by the action of water on calcium carbide, is bubbled into dilute sulphuric acid at 80°C in the presence of mercuric sulphate and ferric sulphate catalysts, it is converted into acetaldehyde:

$$C_2H_2 + H_2O = CH_3.CHO$$

2. Ethyl alcohol is oxidized by passing its vapour, mixed with air, over a silver catalyst at 300°C:

$$C_2H_5OH + \bar{O} = CH_3.CHO + H_2O$$

3. When ethyl alcohol vapour is passed over a finely-divided copper catalyst at 300°C, dehydrogenation occurs:

$$C_2H_5OH = CH_3.CHO + H_2$$

The name aldehyde is derived from *al*cohol *dehyd*rogenated.

4. Butane, obtained from natural gas or crude petroleum, is oxidized to acetaldehyde by heating it with air to 400°C under carefully controlled conditions.

PROPERTIES OF ACETALDEHYDE: It is a colourless volatile liquid, b.p. 21°C, with a pungent smell. It is miscible with water in all proportions and is frequently encountered in the laboratory as its aqueous solution.

REACTIONS OF ACETALDEHYDE: These are listed in Table 36.1.

1. Oxidation: It is readily oxidized to acetic acid:

$$CH_3.CHO + \bar{O} = CH_3.COOH$$

Dilute nitric acid can be used, or acidified solutions of potassium dichromate or potassium permanganate, or the aldehyde can be oxidized catalytically by passing its vapour mixed with air over finely-divided platinum or copper. Because it is oxidized so easily, acetaldehyde has marked reducing properties, reacting with even mild oxidizing agents. For example, it gives a silver mirror with ammoniacal silver oxide (§ 28.17):

$$CH_3.CHO + Ag_2O = CH_3.COOH + 2Ag\downarrow$$

It also reduces Fehling's solution (§ 28.4) to a red precipitate of cuprous oxide on boiling, and it restores the colour to Schiff's reagent (a solution of a magenta dye, rosaniline hydrochloride, which has been decolorized with sulphur dioxide).

2. Reduction: It is reduced to ethyl alcohol by treating it with sodium amalgam and water or by passing its vapour mixed with hydrogen over a finely-divided nickel catalyst at 150°C:

$$CH_3.CHO + 2\bar{H} = CH_3.CH_2OH$$

Amalgamated zinc and concentrated hydrochloric acid, which is a very strong reducing agent, reduces acetaldehyde on warming to the paraffin ethane by the Clemmensen reaction:

$$CH_3.CHO + 4\bar{H} = CH_3.CH_3 + H_2O$$

3. Addition: Acetaldehyde undergoes a number of addition reactions. These may be represented by the following general equation, where

TABLE 36.1. REACTIONS OF ACETALDEHYDE

ACETALDEHYDE + reagent	Product	Type of reaction
chromic acid, or potassium permanganate soln.	acetic acid, CH_3COOH	Oxidations
Fehling's soln.	red ppt. of cuprous oxide, Cu_2O	Oxidations
Tollens's reagent	silver mirror, Ag	Oxidations
Schiff's reagent	purple colour	Oxidations
sodium amalgam and water, or Ni catalyst and hydrogen at 150°C	ethyl alcohol, C_2H_5OH	Reductions
amalgamated zinc and conc. HCl	ethane, C_2H_6	Reductions
ammonia	acetaldehyde ammonia $CH_3.CH(OH)HN_2$	Additions
hydrogen cyanide	acetaldehyde cyanohydrin $CH_3.CH(OH)CN$	Additions
conc. sodium bisulphite soln.	acetaldehyde bisulphite $CH_3.CH(OH)SO_3Na$	Additions
phenylhydrazine	acetaldehyde phenylhydrazone $CH_3.CH{:}N.NH.C_6H_5$	Condensations
hydroxylamine	acetaldoxime, $CH_3.CH{:}NOH$	Condensations
phosphorus pentachloride	ethylidene dichloride, $CH_3.CHCl_2$	Halogenations
chlorine	chloral, $CCl_3.CHO$	Halogenations
KI soln. and a soln. of NaOCl	iodoform, CHI_3	Halogenations
dilute caustic soda soln.	aldol, $CH_3.CH(OH).CH_2CHO$	Polymerizations
conc. caustic soda soln.	brown resin	Polymerizations
conc. sulphuric acid	paraldehyde $(C_2H_4O)_3$	Polymerizations
hydrogen chloride below 0°C	metaldehyde $(C_2H_4O)_4$	Polymerizations

H—X is the substance with which it forms an addition compound.*

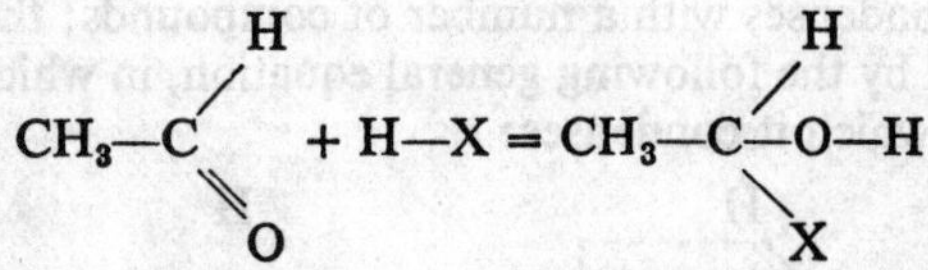

For example, if gaseous ammonia is passed through a dry ethereal solution of acetaldehyde, colourless crystals of acetaldehyde-ammonia are precipitated:

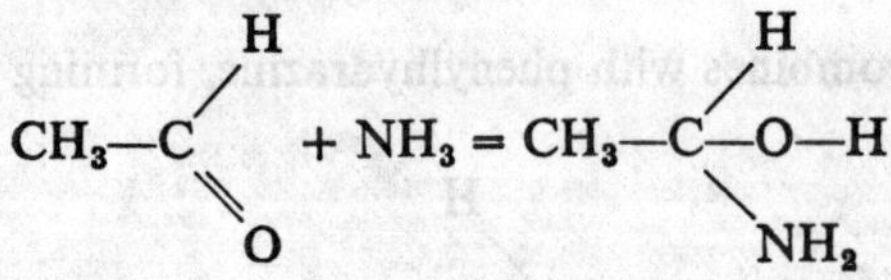

(H—X is $H—NH_2$)

Acetaldehyde combines additively with hydrogen cyanide forming a colourless liquid, acetaldehyde-cyanohydrin:

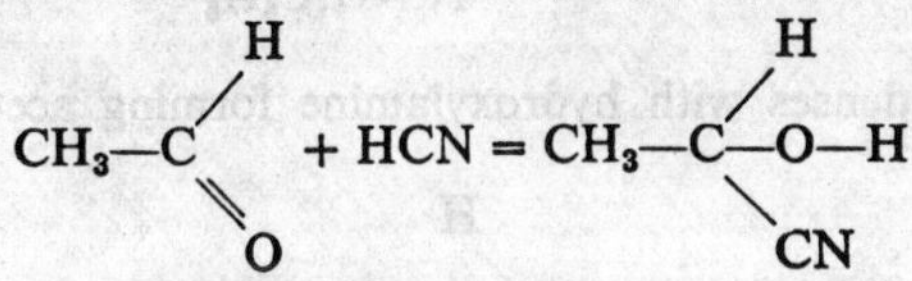

(H—X is H—CN)

Similarly, with a cold saturated solution of sodium bisulphite, acetaldehyde adds forming acetaldehyde-bisulphite, although the colourless crystals are not easily obtained because they are so soluble in water:

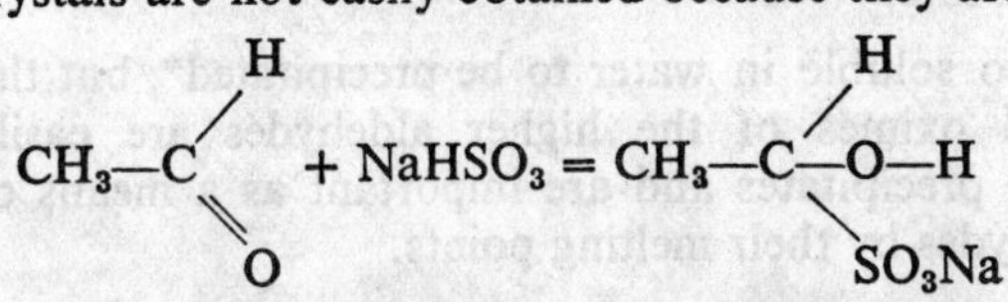

(H—X is $H—SO_3Na$)

The reduction of acetaldehyde to ethyl alcohol, referred to above, can be regarded as another example of addition thus:

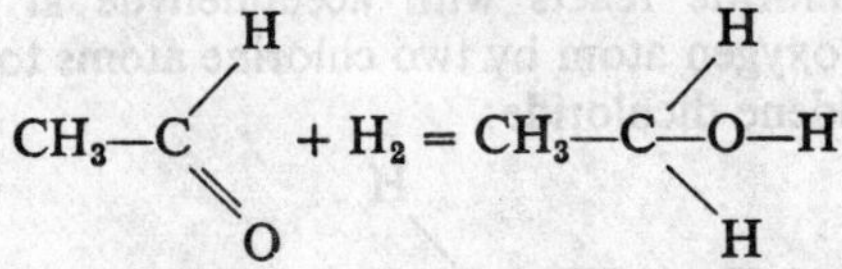

(H—X is H—H)

4. Condensation: This is strictly† defined as a reaction in which two or more molecules combine to form a new compound with the elimination

* The mechanism of this reaction is discussed in §50.7.

† Unfortunately this term is not always interpreted so precisely and it is customary to refer to certain reactions as condensations which are really polymerizations. The 'aldol condensation' is an example of this.

of some simple molecule such as water, ammonia, or hydrogen chloride. Acetaldehyde condenses with a number of compounds; these reactions are summarized by the following general equation, in which H_2X is the substance with which it condenses:

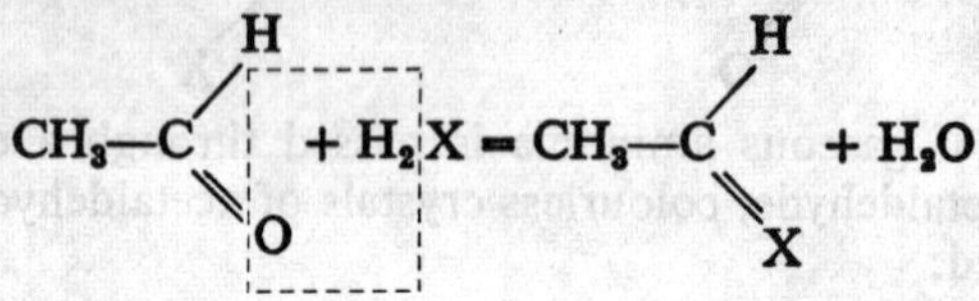

For example, acetaldehyde combines with phenylhydrazine, forming a phenylhydrazone:

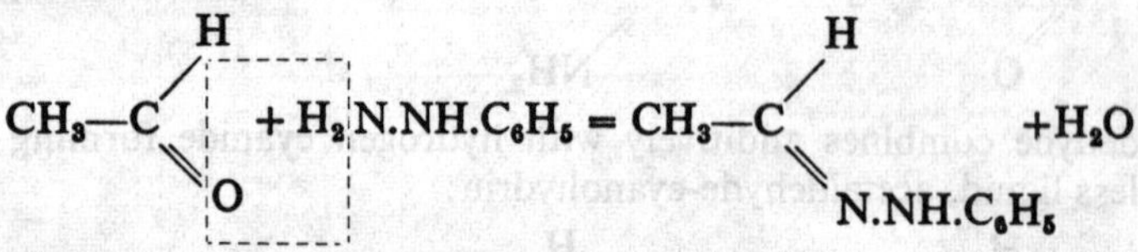

Similarly acetaldehyde condenses with hydroxylamine forming acetaldoxime:

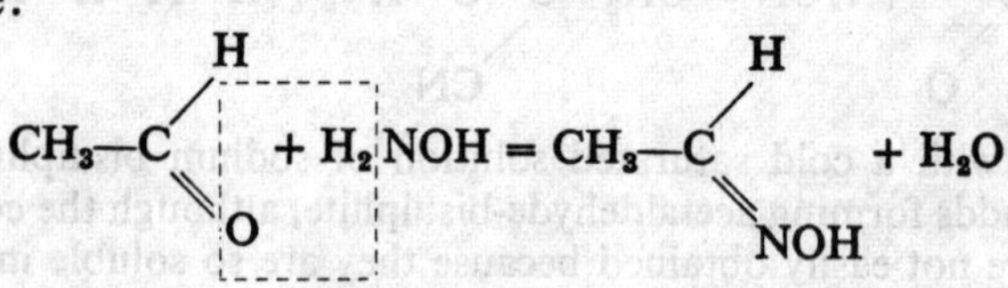

These products are too soluble in water to be precipitated*, but the phenylhydrazones and oximes of the higher aldehydes are easily obtained as crystalline precipitates and are important as a means of identifying those aldehydes by their melting points.

5. Chlorination: Acetaldehyde can be chlorinated in two ways, depending upon the reagent used:

(*a*) Phosphorus pentachloride reacts with acetaldehyde at room temperature replacing the oxygen atom by two chlorine atoms to form the colourless liquid ethylidene dichloride:

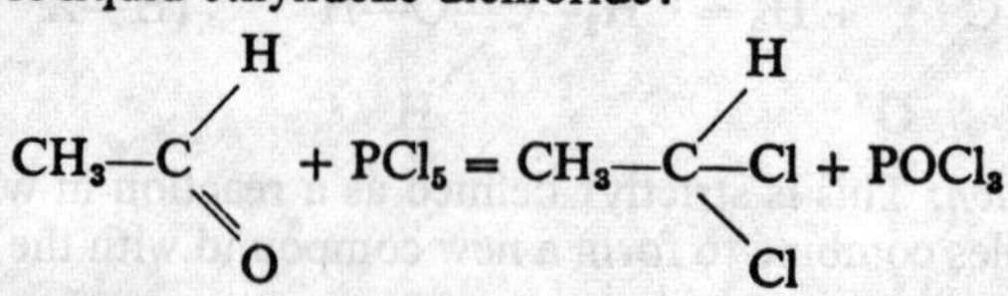

No hydrogen chloride is liberated in this reaction, provided the aldehyde is anhydrous.

* For this reason 2,4-dinitrophenylhydrazine is often preferred.

(*b*) When chlorine is bubbled into acetaldehyde, substitution of chlorine atoms for hydrogen atoms takes place and trichloro-acetaldehyde, commonly known as chloral, is formed:

$$CH_3.CHO + 3Cl_2 = CCl_3.CHO + 3HCl$$

6. *Iodoform Reaction:* Acetaldehyde gives a yellow precipitate of iodoform when warmed with potassium iodide solution and a little sodium hypochlorite solution, or with iodine and caustic soda solution (see § 40.9)

7. *Polymerization:* When acetaldehyde is left standing in contact with a very dilute solution of sodium hydroxide it slowly polymerizes into molecules of aldol, a colourless liquid so named because it contains both aldehyde and alcohol groups:

$2CH_3.CHO =$ 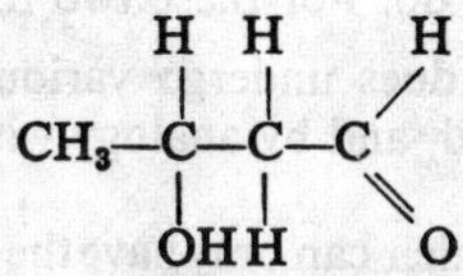

This polymerization proceeds much further if acetaldehyde is heated with a concentrated solution of sodium hydroxide, when a viscous, brown resin is obtained which smells like rotting apples and has a very high molecular weight.

When acetaldehyde is treated with a few drops of concentrated sulphuric acid a vigorous and very exothermic reaction takes place and a colourless liquid called paraldehyde is obtained. This is a polymer of acetaldehyde with the formula $(C_2H_4O)_3$. Acetaldehyde can also be made to polymerize by bubbling hydrogen chloride through it below 0°C, when a white solid called metaldehyde with the formula $(C_2H_4O)_4$ is obtained. Both paraldehyde and metaldehyde can be reconverted to acetaldehyde by distilling them with dilute sulphuric acid.

USES OF ACETALDEHYDE: It is made in large quantities for conversion into acetic acid and acetic anhydride. Normal butyl alcohol, chloral, and butadiene (used to make synthetic rubber), are also made from acetaldehyde. Mixed with bran metaldehyde is the active constituent of most proprietary slug-killers; it is also used as a solid fuel. Paraldehyde is a powerful sedative.

STRUCTURE OF ACETALDEHYDE: The molecular formula is C_2H_4O. There are only three possible ways of arranging these atoms so that they exert their accepted valencies:

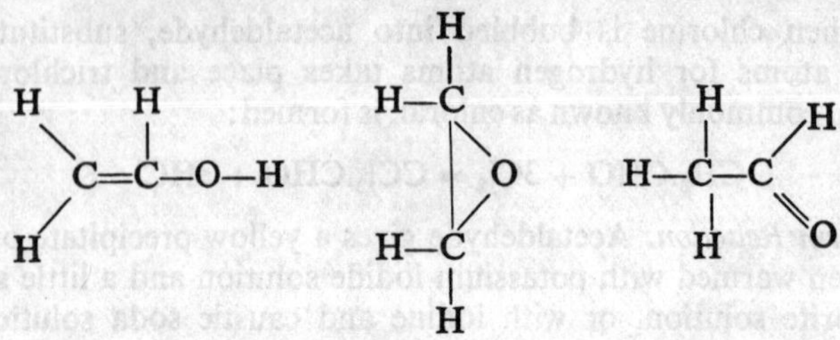

I II III

There can be no hydroxyl group in its molecule, as no fumes of hydrogen chloride are liberated when it reacts with phosphorus pentachloride. Moreover, acetaldehyde does not show addition reactions with hydrogen bromide or bromine water, as substances containing an ethylenic >C=C< bond do. For these two reasons formula I is unacceptable.

Acetaldehyde does undergo various addition reactions, showing it to be unsaturated, and by analogy with formaldehyde, which it closely resembles and which can only have the structure H—C(H)=O, we conclude that formula III represents the structure of acetaldehyde, and not formula II. This conclusion is confirmed by its preparation from ethyl alcohol and its oxidation to acetic acid, both of which are known to contain a methyl group in their molecules, and by the fact that chlorine can substitute only three of the hydrogen atoms in each molecule, suggesting that the fourth is differently placed from the others.

36.3. Formaldehyde, Methanal, H.CHO.

LABORATORY PREPARATION: If necessary it can be made by oxidizing methyl alcohol with sodium dichromate and dilute sulphuric acid, but it is rarely prepared in the laboratory because it is so readily available commercially.

MANUFACTURE: It is made by passing a mixture of methyl alcohol vapour and air or oxygen over a silver catalyst at 250°C:

$$CH_3OH + \dot{O} = H.CHO + H_2O$$

The product is condensed giving a mixture of water, formaldehyde, and unchanged methyl alcohol. A 40% solution, known as *formalin*, is obtained from this by distillation.

PROPERTIES OF FORMALDEHYDE: It is a colourless gas, b.p. −21°C, with a pungent, unpleasant smell. It is very soluble in water and is usually stored and used in aqueous solution for convenience.

Reactions of Formaldehyde: In many of its reactions it resembles acetaldehyde. For example, it is a powerful reducing agent, being readily oxidized to formic acid. Thus it gives a silver mirror with an ammoniacal solution of silver oxide, restores the colour to Schiff's reagent, and reduces Fehling's solution to red cuprous oxide. It also reduces mercuric chloride solution, giving a white precipitate of mercurous chloride, and reduces auric chloride solution to colloidal gold.

Like acetaldehyde it forms addition compounds with hydrogen cyanide and sodium bisulphite. It condenses with phenylhydrazine and hydroxylamine giving a phenylhydrazone and an oxime respectively. Formaldehyde polymerizes very readily giving a variety of polymers. When an aqueous solution of formaldehyde is evaporated to dryness it leaves a white residue of paraformaldehyde, $(CH_2O)_n$, and when formaldehyde is distilled from a 60% solution containing a few per cent of sulphuric acid it polymerizes to trioxymethylene $(CH_2O)_3$. If formaldehyde solution is allowed to stand in contact with lime-water for several days polymerization takes place to a mixture of sugars called formose.

Formaldehyde differs from acetaldehyde in certain important ways. It does not react with phosphorus pentachloride, nor does it give the iodoform reaction. It does not form an addition compound with ammonia, but it gives instead a white crystalline compound, hexamethylene tetramine, known as hexamine for brevity:

$$6CH_2O + 4NH_3 = (CH_2)_6N_4 + 6H_2O$$

Unlike other aliphatic aldehydes, when treated with a concentrated solution of caustic alkali formaldehyde undergoes *Cannizzaro's reaction*, being converted into methyl alcohol and sodium or potassium formate:

$$2CH_2O + NaOH = CH_3OH + H.COONa.$$

This behaviour is typical of aromatic aldehydes (see § 47.3).

Uses of Formaldehyde: Its chief use is to make various plastics and synthetic resins (§ 49.3). For example, formaldehyde condenses with urea giving resins which are valuable as adhesives, and with phenol it gives a thermo-setting plastic known as 'Bakelite', which is widely used for making articles such as telephone receivers, toys, and electrical fittings. Formaldehyde is a powerful germicide and is used for fumigation, for embalming, and for preserving anatomical specimens. Hexamine is used as an internal antiseptic. When treated with fuming nitric acid it is converted into the important high explosive cyclonite or RDX, which was used extensively in the Second World War.

Structure of Formaldehyde: Its molecular formula is CH_2O. Assuming the carbon, oxygen and hydrogen exert valencies of four, two, and one, respectively, only one structure is possible, viz.: 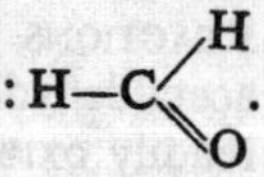 .

Since this conforms with the reactions of formaldehyde, it is confidently accepted.

36.4. Chloral, Trichloro-acetaldehyde, $CCl_3.CHO$. This is made by bubbling chlorine into ethyl alcohol or acetaldehyde solution, excess of calcium carbonate being added to prevent the acid produced from causing its polymerization:

$$CH_3.CHO + 3Cl_2 = CCl_3.CHO + 3HCl$$

If rectified spirit is used the reaction takes several days to complete and the product is an alcoholate which yields chloral when distilled with concentrated sulphuric acid.

PROPERTIES AND REACTIONS OF CHLORAL: It is a colourless oily liquid, b.p. 98°C, which combines readily with water forming colourless crystals of chloral hydrate, m.p. 57°C. These crystals yield chloral again when distilled with concentrated sulphuric acid. They have the structure

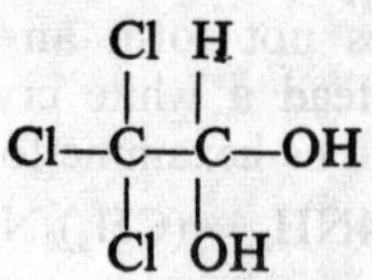

which is of interest because it is rare to find a stable compound with two –OH groups linked to the same carbon atom. Presumably the strongly electronegative CCl_3— radical makes the molecule highly polar and gives stability to this unusual structure.

Chemically chloral resembles acetaldehyde. It reduces an ammoniacal solution of silver oxide to silver, and Fehling's solution to cuprous oxide, and it restores the purple colour to Schiff's reagent. In these reactions it is oxidized to trichloro-acetic acid:

$$CCl_3.CHO + \bar{O} = CCl_3.COOH$$

It forms addition compounds with ammonia, hydrogen cyanide, and sodium bisulphite, and it condenses with phenylhydrazine and hydroxylamine. In the presence of acids it polymerizes to a white solid, metachloral.

Chloral differs from other aldehydes in its reaction with alkalis. When boiled with potassium hydroxide solution chloral is hydrolysed to chloroform and potassium formate:

$$CCl_3.CHO + KOH = CHCl_3 + H.COOK$$

USES OF CHLORAL: The hydrate is used medicinally for inducing sleep. The biggest use of chloral is in the manufacture of the insecticide D.D.T. (*di*chloro-*di*phenyl-*tri*chloroethane) which is made by condensing a molecule of chloral with two of chlorobenzene in the presence of concentrated sulphuric acid:

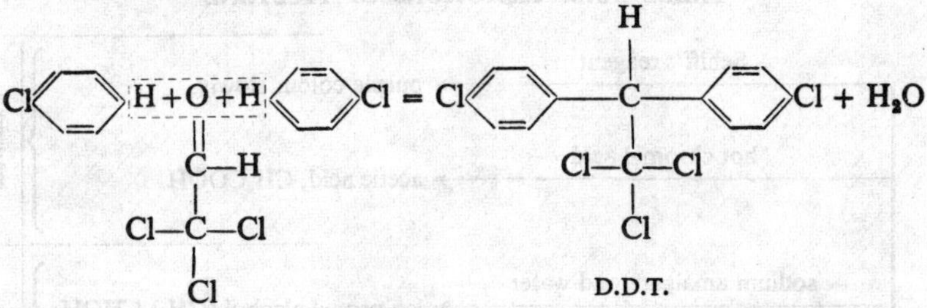

D.D.T.

36.5. Detection of Aldehydes. Aldehydes are generally detected by their reducing properties, particularly by their reduction of Fehling's solution and an ammoniacal solution of silver oxide, and by their addition and condensation reactions with sodium bisulphite, phenylhydrazine, and hydroxylamine.

A solution of formaldehyde is distinguished from solutions of other aldehydes by its very characteristic smell, by its ability to reduce mercuric chloride solution to a white precipitate of mercurous chloride, and by the absence of a resin with sodium hydroxide solution. Acetaldehyde gives the iodoform reaction, which distinguishes it from all other aldehydes. It also forms a brown resin when heated with concentrated caustic alkali solution. Chloral is recognized by the formation of a crystalline hydrate with a little water and by its positive response to the isocyanide or carbylamine test (see § 40.8).

36.6. Acetone, Dimethyl Ketone, Propanone, $CH_3.CO.CH_3$.

LABORATORY PREPARATION: Because it is readily available commercially, its preparation in the laboratory is rarely necessary, but it can be made in poor yield by strongly heating anhydrous calcium acetate, when its vapour distils over and can be condensed:

$$(CH_3COO)_2Ca = CH_3.CO.CH_3 + CaCO_3$$

MANUFACTURE: 1. It is chiefly made by dehydrogenation of iso-propyl alcohol by passing it over a copper catalyst at 300°C:

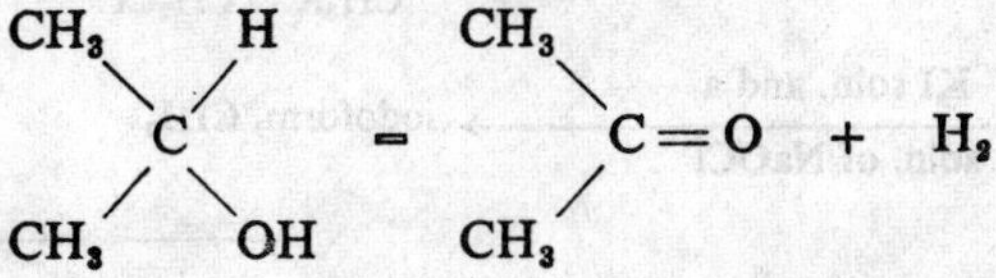

As described in § 35.4 the iso-propyl alcohol is made from propylene which is obtained by cracking petroleum.

2. In recent years acetone has been made in increasing amounts as a

TABLE 36.2. REACTIONS OF ACETONE

ACETONE + reagent	Product	Type
Schiff's reagent	purple colour slowly	*Oxidations*
hot chromic acid	acetic acid, CH_3COOH	*Oxidations*
sodium amalgam and water	iso-propyl alcohol $(CH_3)_2CHOH$	*Reductions*
amalgamated zinc and conc. HCl	propane, C_3H_8	*Reductions*
hydrogen cyanide	acetone cyanohydrin $(CH_3)_2.C(OH)CN$	*Additions*
conc. sodium bisulphite soln.	acetone bisulphite $(CH_3)_2.C.(OH)SO_3Na$	*Additions*
phenylhydrazine	acetone phenylhydrazone $(CH_3)_2C{:}N.NH.C_6H_5$	*Condensations*
hydroxylamine	acetoxime $(CH_3)_2.C{:}NOH$	*Condensations*
hot conc. sulphuric acid	trimethylbenzene $C_6H_3(CH_3)_3$	*Condensations*
phosphorus pentachloride	2:2-dichloropropane $(CH_3)_2.CCl_2$	*Halogenations*
chlorine	chloroacetones, e.g. $CH_3.CO.CH_2Cl$	*Halogenations*
KI soln. and a soln. of NaOCl	iodoform, CHI_3	*Halogenations*

by-product of the manufacture of phenol from cumene. This important process is described in § 45.2.

3. Some acetone is made by passing acetic acid vapour over a catalyst of manganous oxide at about 300°C:

$$2CH_3.COOH = CH_3.CO.CH_3 + CO_2 + H_2O$$

4. Its manufacture by the fermentation of starch and by the destructive distillation of wood are both obsolete.

Properties of Acetone: It is a colourless liquid, b.p. 56°C, with a pleasant smell. It is miscible with water in all proportions.

Reactions of Acetone: These are summarized in Table 36.2.

1. Combustion: It burns in air forming carbon dioxide and water:

$$CH_3.CO.CH_3 + 4O_2 = 3CO_2 + 3H_2O$$

2. Oxidation: It is not affected by mild oxidizing agents, but when boiled with chromic acid or concentrated nitric acid it is slowly oxidized to acetic acid:

$$CH_3.CO.CH_3 + 4\bar{O} = CH_3.COOH + CO_2\uparrow + H_2O$$

3. Reduction: Treatment with a mild reducing agent such as sodium amalgam and water converts acetone into iso-propyl alcohol:

$$CH_3.CO.CH_3 + 2\bar{H} = CH_3.CH(OH).CH_3$$

If amalgamated zinc and concentrated hydrochloric acid are used, then acetone is reduced to the paraffin propane by Clemmensen's reaction:

$$CH_3.CO.CH_3 + 4\bar{H} = CH_3.CH_2.CH_3 + H_2O$$

4. Addition: Acetone forms various addition compounds, the general change being:

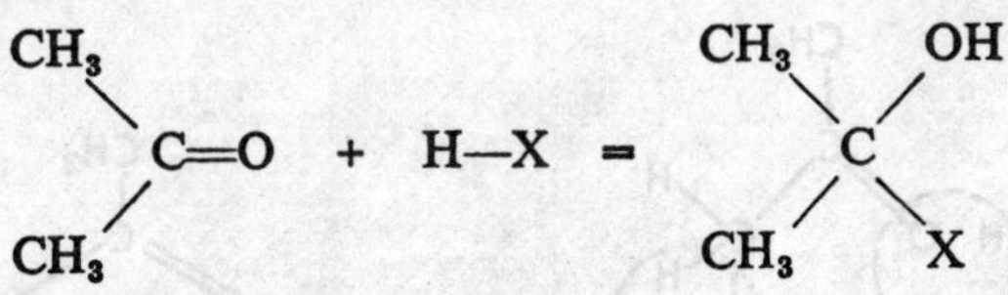

(where H—X is the compound which adds to acetone).

Thus hydrogen cyanide gives a cyanohydrin:

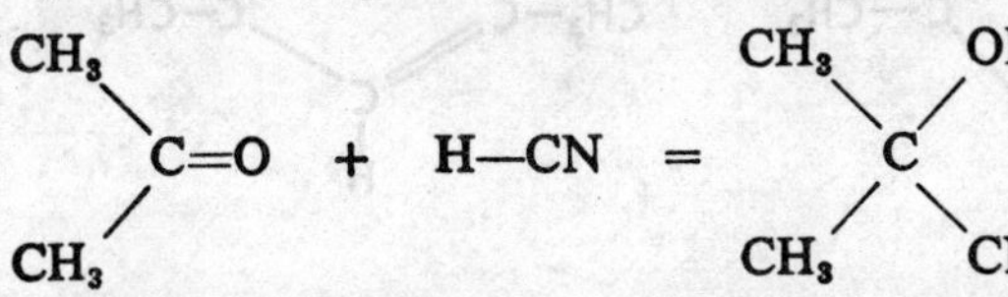

(H—X is H—CN)

Z

Similarly a saturated solution of sodium bisulphite gives a white crystalline precipitate of acetone-bisulphite:

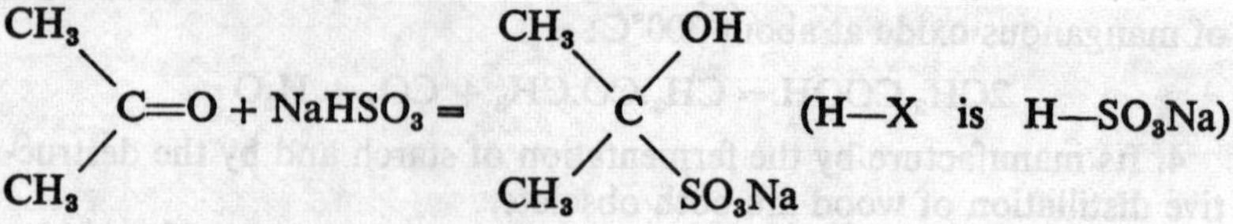

This addition compound can be reconverted into acetone by heating it with sodium carbonate solution and can be used, therefore, for purifying acetone.

Acetone adds to ammonia also, but the product is too unstable to exist at room temperature.

5. Condensation: Like acetaldehyde, acetone forms condensation products readily. Phenylhydrazine gives a phenylhydrazone which is too soluble in water to be precipitated, and hydroxylamine condenses to acetoxime which appears as a white precipitate on cooling:

$$(CH_3)_2C\boxed{O + H_2}N.NH.C_6H_5 = (CH_3)_2.C{=}N.NH.C_6H_5 + H_2O$$

$$(CH_3)_2C\boxed{O + H_2}NOH = (CH_3)_2.C{=}NOH + H_2O$$

Acetone condenses with itself when distilled with concentrated sulphuric acid, three molecules of acetone uniting to form each molecule of trimethylbenzene:

$$3(CH_3)_2CO = C_6H_3(CH_3)_3 + 3H_2O$$

The product, which is also known as mesitylene, has a ring structure thus:

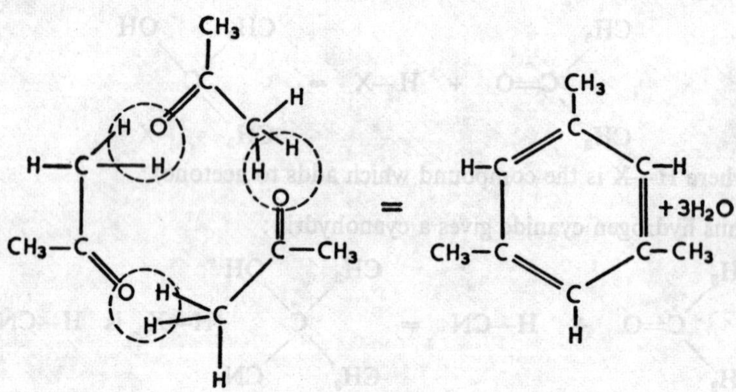

6. Chlorination: Acetone can be chlorinated in two ways. It reacts with phosphorus pentachloride at room temperature forming 2, 2-dichloropropane in which the oxygen atom has been replaced by two chlorine atoms:

$$CH_3.CO.CH_3 + PCl_5 = CH_3.CCl_2.CH_3 + POCl_3$$

No hydrogen chloride is liberated provided the acetone is anhydrous. On the other hand, if chlorine is bubbled into acetone, particularly on warming, the hydrogen atoms are successively replaced by chlorine atoms giving chloroacetones:

e.g. $$CH_3.CO.CH_3 + Cl_2 = CH_3.CO.CH_2Cl + HCl$$

7. Iodoform Reaction: Acetone gives a yellow precipitate of iodoform (§ 40.9) when warmed with potassium iodide solution and sodium hypochlorite solution, or with iodine and caustic soda solution. By a similar reaction it is converted into chloroform (§ 40.8).

8. Polymerization: Acetone does not polymerize readily.

USES OF ACETONE: Its main use is as a solvent. It is used to dissolve acetylene, gun-cotton, nitro-glycerine, cellulose acetate rayon, celluloid, and lacquers and varnishes of all kinds. It is also the starting point for the manufacture of chloroform, iodoform, and the glass-like plastic 'Perspex' (see § 49.2).

STRUCTURE OF ACETONE: The molecular formula is C_3H_6O. Since acetone does not give fumes of hydrogen chloride with phosphorus pentachloride we conclude that it does not contain an —OH group. The replacement of its oxygen atom by two chlorine atoms in this reaction and its numerous addition and condensation reactions strongly suggest the presence in the acetone molecule of a carbonyl group, >C=O, as in aldehydes. Only two such structures are possible:

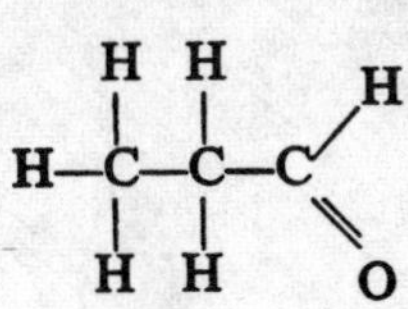

(I)

and

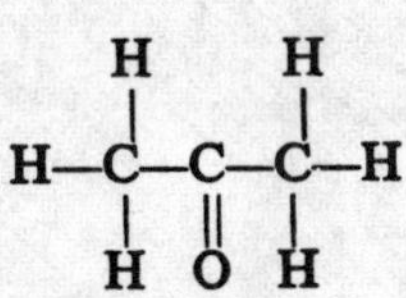

(II)

Structure I is that of an aldehyde, R.CHO, and would certainly show strong reducing properties, which acetone does not. Moreover, on reduction structure I would give a primary alcohol whereas a secondary alcohol is obtained from acetone. We conclude that the structure of acetone is best represented by formula II.

36.7. Detection of Acetone. It is recognized by the white precipitate of the acetone-bisulphite addition compound, by the yellow precipitate of iodoform, by its slow restoration of the purple colour to Schiff's reagent, and by its inability to reduce Fehling's solution to cuprous oxide or an ammoniacal solution of silver oxide to a silver mirror.

36.8. Comparison of Aldehydes and Ketones. It will be seen from this chapter that aldehydes and ketones show many similar reactions, such as additions and condensations, which involve the common carbonyl group, $>C{=}O$. The principal difference which distinguishes them from each other is that aldehydes are strong reducing agents capable of reducing even mild oxidizing agents such as Fehling's solution or ammoniacal silver oxide. Under the influence of the carbonyl group to which it is directly linked the hydrogen atom in the aldehyde group readily takes up an oxygen atom and is converted into a hydroxyl group.

In ketones there is no hydrogen atom which can be rendered active in this way. Other differences, which are less easy to demonstrate experimentally, are the greater readiness of aldehydes to polymerize and the differing types of alcohol formed on reduction.

37

Aliphatic Acids and Acid Derivatives

37.1. Introduction. All organic acids contain the *carboxyl group*,—COOH (the name is derived from *car*bonyl and hydr*oxyl*), as their functional group. Of those which contain one carboxyl group in each molecule the commonest aliphatic ones are the so-called *fatty acids*, which form a homologous series of general molecular formula $C_nH_{2n+1}.COOH$. The lowest members are formic acid H.COOH, and acetic acid CH_3COOH. We shall study acetic acid first because it is much the most important of the fatty acids and in many respects typical of the series.

37.2. Acetic Acid, Ethanoic Acid, $CH_3.COOH$.

LABORATORY PREPARATION: By oxidizing ethyl alcohol with sodium dichromate and moderately concentrated sulphuric acid, using the apparatus shown in Fig. 37.1:

$$CH_3.CH_2OH + 2\bar{O} = CH_3COOH + H_2O$$

By dropping the alcohol into a large excess of the oxidizing agent and by distilling under reflux so that the reactants are brought into thorough and prolonged contact, complete oxidation to acetic acid is ensured. This is a good example of the way in which conditions can be adjusted to obtain the maximum yield of the desired product in an organic reaction, since exactly the same reagents are used to prepare acetaldehyde (§ 36.2). When the reaction is complete the condenser is changed to the normal position (as in Fig. 36.1) and the distillate collected. It consists largely of acetic acid with water, acetaldehyde, ether, and sulphurous acid as impurities. Pure acetic acid is best obtained by forming crystals of cupric acetate (by adding excess of copper carbonate, filtering, and crystallizing), and then, after heating them gently to drive off their water of crystallization, distilling these crystals with concentrated sulphuric acid.

Acetic acid can also be prepared by boiling methyl cyanide with an acid or by boiling an acetate ester under reflux with an alkali and then acidifying the product:

$$CH_3CN + HCl + 2H_2O = CH_3.COOH + NH_4Cl$$
$$CH_3.COOC_2H_5 + NaOH = C_2H_5OH + CH_3.COONa$$
$$CH_3.COONa + H_2SO_4 = CH_3.COOH + NaHSO_4$$

Although these methods are of little practical value for preparing acetic acid, they are useful general methods for preparing the higher fatty acids which are not so readily available commercially.

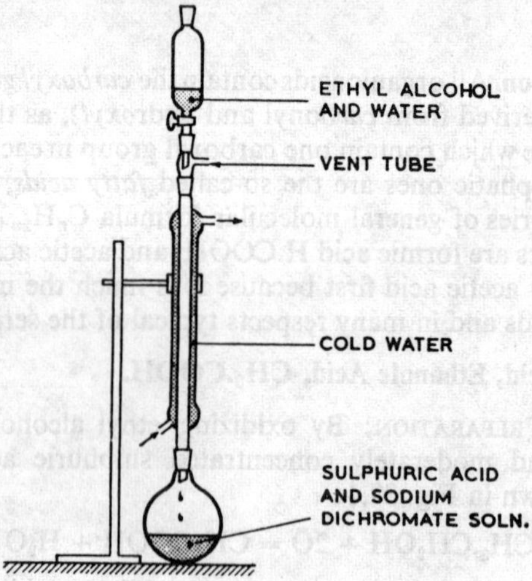

FIG. 37.1 Preparation of Acetic Acid

MANUFACTURE: 1. By oxidation of acetaldehyde; air is blown into acetic acid containing about 5% of dissolved acetaldehyde and a little manganous acetate catalyst at 60°C, and pure acetic acid is obtained by distilling the product:

$$CH_3.CHO + \bar{O} = CH_3.COOH$$

In the absence of the catalyst peracetic acid tends to accumulate and decompose explosively.

2. In recent years butane has been oxidized directly to acetic acid by mixing it with air and passing it into acetic acid kept at 160°C under 2×10^6 N m^{-2} (20 atm) pressure, in the presence of manganese acetate catalyst. The acetic acid formed is recovered from the unreacted gases and purified by fractional distillation. The butane is obtained cheaply by refining and cracking petroleum.

3. *Vinegar*, which is a 4–5% solution of acetic acid, is made by bacterial oxidation of alcohol. Poor quality wines or alcohol produced by fermenting barley are the usual raw materials.

PROPERTIES OF ACETIC ACID: It is a colourless liquid with a very pungent smell. It dissolves freely in water, the two liquids being miscible in all proportions. The anhydrous acid, which melts at 17°C, is known as *glacial* acetic acid because on cold days it freezes into ice-like crystals. It is hygroscopic and corrosive. Its boiling point (118°C) is much higher than the boiling points of hydrocarbons of similar molecular weight, which suggests that it is associated by hydrogen bonding (§ 14.7). This is confirmed by vapour density measurements, which show that double molecules of acetic acid persist even in the vapour, and by colligative measurements which reveal the presence of dimers of molecular weight 120 in a solution of acetic acid in benzene. The association of carboxylic acids in non-polar solvents can also be demonstrated by applying the partition law (§ 6.14).

REACTIONS OF ACETIC ACID: These are listed in Table 37.1.

1. As an Acid: It is a weak acid ($K = 1.8 \times 10^{-5}$ mol dm^{-3}), being only slightly dissociated into ions in aqueous solution:

$$CH_3.COOH \rightleftharpoons CH_3.COO^- + H^+$$

Acetic acid is monobasic, i.e. only one hydrogen atom in each molecule can be replaced by a metal. It turns blue litmus red, liberates carbon dioxide from carbonates, attacks metals evolving hydrogen, and neutralizes bases forming salts called *acetates*:

$$2CH_3.COOH + Na_2CO_3 = 2CH_3.COONa + H_2O + CO_2\uparrow$$
$$CH_3.COOH + NaOH = CH_3.COONa + H_2O$$

2. Esterification: Acetic acid combines reversibly with alcohols forming acetate esters.

e.g. $$CH_3.COOH + C_2H_5OH \rightleftharpoons \underset{\text{ethyl acetate}}{CH_3.COOC_2H_5} + H_2O$$

Esterification is discussed in detail in § 38.1.

3. Halogenation: The anhydrous acid reacts vigorously with phosphorus pentachloride in the cold giving acetyl chloride, phosphorus oxychloride, and fumes of hydrogen chloride:

$$CH_3.COOH + PCl_5 = CH_3COCl + POCl_3 + HCl\uparrow$$

TABLE 37.1. REACTIONS OF ACETIC ACID

Reagent (with ACETIC ACID)	Product
sodium hydroxide or carbonate soln.	sodium acetate, $CH_3.COONa.3H_2O$
lead	lead acetate $(CH_3COO)_2Pb.3H_2O$
ethyl alcohol	ethyl acetate, $CH_3COOC_2H_5$
phosphorus pentachloride	acetyl chloride, CH_3COCl
chlorine (red P or I_2 catalyst)	chloroacetic acids $\begin{cases} CH_2Cl.COOH \\ CHCl_2.COOH \\ CCl_3.COOH \end{cases}$
lithium aluminium hydride	ethyl alcohol, C_2H_5OH
ferric chloride soln.	red soln. (complex ferric acetate)

Acetic acid is chlorinated in a similar way by phosphorus trichloride and by thionyl chloride.

When chlorine is bubbled into boiling acetic acid, particularly in the presence of sunlight or a catalyst of iodine or red phosphorus, substitution takes place giving monochloroacetic acid:

$$CH_3.COOH + Cl_2 = CH_2Cl.COOH + HCl$$

If chlorination is prolonged, di- and tri-chloroacetic acids are produced. Bromine reacts similarly, though less readily, with acetic acid.

4. *Reduction:* Acetic acid is resistant to reduction, being unaffected by all ordinary reducing agents. However, it is reduced to ethyl alcohol by using lithium aluminium hydride (made by adding lithium hydride to a dry ethereal solution of aluminium chloride):

$$4CH_3.COOH + 3LiAlH_4 + 2H_2O = 4C_2H_5OH + 3LiAlO_2 + 4H_2\uparrow$$

Dry ethereal solutions of the acid and hydride are mixed together and after a few minutes water is added to decompose the excess of hydride.

5. *Stability:* Acetic acid is a very stable compound, not easily decomposed by heat. It is unaffected by even the strongest oxidizing agents (e.g. chromic acid or potassium permanganate solution) or by powerful dehydrating agents (e.g. concentrated sulphuric acid).

USES OF ACETIC ACID: The biggest use is in the manufacture of cellulose acetate, which is made by treating cotton waste with a mixture of glacial acetic acid, acetic anhydride, and concentrated sulphuric acid. The product is used in packaging and in varnishes and as a fibre (acetate rayon or artificial silk) for making clothes and filter tips for cigarettes. Acetic acid is also used as a solvent and for making acetic anhydride, acetone, white lead (§ 21.19), various acetates, and dyes. It is used for coagulating rubber latex and, as vinegar, for preserving and flavouring food.

STRUCTURE OF ACETIC ACID: The molecular formula is $C_2H_4O_2$. The reaction of glacial acetic acid with phosphorus pentachloride evolving hydrogen chloride and with sodium liberating hydrogen, clearly indicates the presence of a hydroxyl group, and the fact that only one hydrogen atom is replaced from each molecule by these reagents shows that each contains only one hydroxyl group. Direct chlorination replaces up to three hydrogen atoms in every molecule. This, and the preparation of acetic acid from methyl cyanide, its electrolysis to give ethane, and the decarboxylation of sodium acetate to methane, all strongly suggest the presence of a methyl radical in the molecule. Assuming that the elements exert their usual valencies, the only structure which conforms to these requirements is the one represented by the formula:

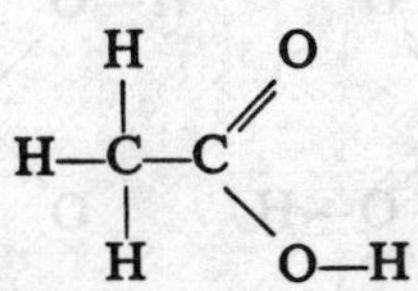

This structure is adopted, therefore, but its acceptance raises certain problems. For example, acetic acid shows none of the properties expected of a compound containing the $>C{=}O$ group in its molecule, i.e. it does not undergo addition with hydrogen cyanide or sodium bisulphite, nor does it condense with phenylhydrazine or hydroxylamine, and it is not reduced by nascent hydrogen from sodium amalgam and water.

This is usually explained in two ways. The hydroxyl group attached to the same carbon atom may be regarded as modifying the characteristic reactions of the carbonyl group. The reverse is certainly the case, the carbonyl group by attracting electrons from the oxygen of the hydroxyl group facilitates its dissociation into ions, which explains why carboxylic compounds are acidic whereas alcohols are neutral. The same effect will be met in even greater degree in the chloroacetic acids (see § 37.9) and is explained in detail in § 50.8.

The second explanation is that the carbonyl group as such does not exist in carboxylic acids and their salts, but that a resonance structure occurs (see § 14.11) which is intermediate between the two forms

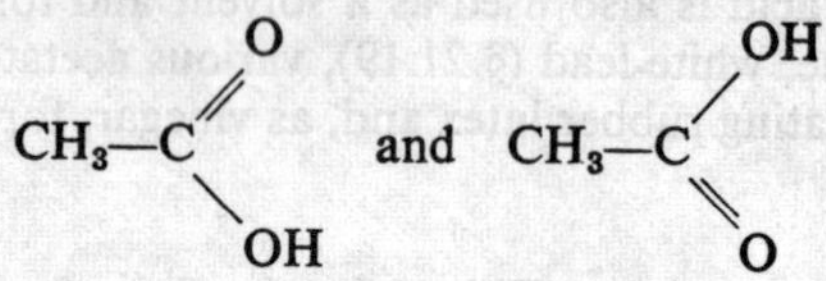

Evidence for this hybrid structure is provided by physical measurements such as X-ray analysis which show the two oxygen atoms in acetates to be equidistant from the carbon atom and at a distance in between those associated with C=O and C—O. This phenomenon of resonance is fairly common in organic chemistry and a number of examples will be encountered in this book, the most important being benzene (§ 43.3). No simple method has yet been devised, however, for representing resonance structures in terms of conventional symbols.

There is a further complication that acetic acid, like its homologues, is associated by hydrogen bonding into double molecules which probably have the cyclic structure:

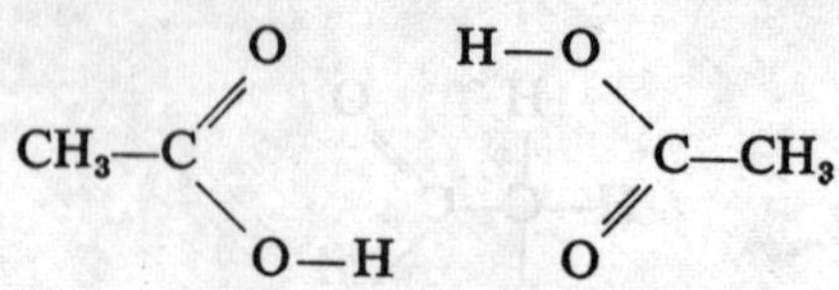

37.3. Formic Acid, Methanoic Acid, H.COOH.

LABORATORY PREPARATION: By heating crystals of oxalic acid with anhydrous glycerol to about 120°C, when a mixture of formic acid and water distils over and condenses in the receiver. The glyceryl mono-oxalate formed initially decomposes into glyceryl mono-formate on heating and is then hydrolysed to formic acid and glycerol, which can be used over again. Thus the overall change can be represented by the simple equation:

$$(COOH)_2 = H.COOH + CO_2\uparrow$$

It is impossible to separate formic acid from water by distillation alone as the two form a constant b.p. mixture, nor is it possible to remove the water by heating the mixture with concentrated sulphuric acid, as dehydration to carbon monoxide then occurs. The impure formic acid is converted into crystals of lead formate by adding an excess of lead carbonate, filtering, and crystallizing:

$$2H.COOH + PbCO_3 = (H.COO)_2Pb + H_2O + CO_2\uparrow$$

These crystals, which are chosen because they contain no water of crystallization, are then heated in a stream of hydrogen sulphide, when double decomposition occurs and the anhydrous acid distils over:

$$(H.COO)_2Pb + H_2S = 2H.COOH + PbS\downarrow$$

The product can be freed from traces of hydrogen sulphide by redistilling it over lead formate.

Formic acid can also be prepared by oxidizing methyl alcohol or formaldehyde with sodium dichromate and dilute sulphuric acid, although the same difficulty over its purification occurs.

MANUFACTURE: Sodium hydroxide is heated to 150°C with carbon monoxide under a pressure of about 5×10^6 N m^{-2} (50 atm), producing sodium formate:

$$NaOH + CO = H.COONa$$

A solution of formic acid is obtained from this by distilling with dilute sulphuric acid:

$$2H.COONa + H_2SO_4 = 2H.COOH + Na_2SO_4$$

PROPERTIES OF FORMIC ACID: It is a colourless liquid, m.p. 8°C, with a pungent smell. It is hygroscopic and mixes with water in all proportions. Concentrated solutions of the acid are corrosive. Like acetic acid it is associated by hydrogen bonding, which accounts for its abnormally high boiling point (101°C) for such a small molecular weight.

REACTIONS OF FORMIC ACID:

1. As an Acid: It is a weak, monobasic acid ($K = 1.8 \times 10^{-4}$ mol dm^{-3}), which combines with alkalis to form salts (formates) and with alcohols to give formate esters:

$$H.COOH + NaOH = H.COONa + H_2O$$
$$C_2H_5OH + H.COOH \rightleftharpoons H.COOC_2H_5 + H_2O$$

2. Reducing Action: Formic acid differs from acetic in being a powerful reducing agent. It reduces an ammoniacal solution of silver oxide to a silver mirror, and mercuric chloride solution to a white precipitate of mercurous chloride or even to a grey precipitate of mercury. It also reduces potassium permanganate solution, causing its decolorization:

$$H.COOH + \bar{O} = H_2O + CO_2$$

3. Dehydration: A further difference is that formic acid is dehydrated to carbon monoxide when heated to 100°C with concentrated sulphuric acid:

$$H.COOH - H_2O = CO\uparrow$$

4. Halogenation: When formic acid is treated with phosphorus pentachloride, carbon monoxide and hydrogen chloride result since formyl chloride is too unstable to exist at room temperature:

$$H.COOH + PCl_5 = CO\uparrow + POCl_3 + 2HCl\uparrow$$

Similarly, direct chlorination gives carbon dioxide and hydrogen chloride instead of chloroformic acid:

$$H.COOH + Cl_2 = CO_2 + 2HCl$$

5. Stability: Formic acid is less stable than acetic acid and decomposes above 150°C:

$$H.COOH = CO_2 + H_2$$

STRUCTURE OF FORMIC ACID: The molecular formula is CH_2O_2. There are only two possible ways of linking these atoms together in accord with the accepted valencies, viz.:

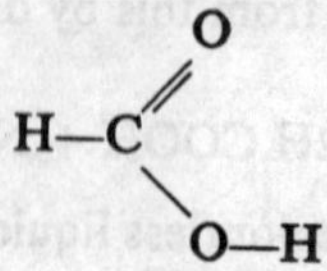

and

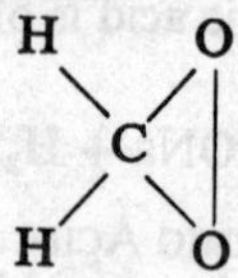

(I) (II)

Since anhydrous formic acid evolves hydrogen when treated with metallic sodium and gives fumes of hydrogen chloride when it reacts with phosphorus pentachloride, its molecule must contain a hydroxyl group. Moreover, its monobasicity suggests that one hydrogen atom in the molecule is differently placed from the other, since only one hydrogen atom in each molecule can be replaced by a metal. For these reasons, structure I is accepted. The fact that this molecule contains the aldehyde group, —CHO, may explain why formic acid, unlike all the other fatty acids, has strong reducing properties. The absence of the usual reactions of the carbonyl group (e.g. addition, condensation, and reduction), can be explained in terms of resonance and association, as with acetic acid.

USES OF FORMIC ACID: It is used in electroplating, tanning, and textile processing. It is sometimes used as a disinfectant and for coagulating rubber latex.

37.4. Detection of Formic and Acetic Acids. These compounds are recognized by their acidic properties and by the deep red colour which is obtained when a neutral solution of a formate or an acetate is added to ferric chloride solution. Formic acid can be distinguished from acetic by its action on potassium permanganate solution, or on an ammoniacal solution of silver oxide, or on mercuric chloride solution, none of which are reduced by acetic acid. The acids can also be distinguished by their characteristic smells and by heating them with concentrated sulphuric acid, when formic acid gives a gas which burns with a bright blue flame.

37.5. The Higher Fatty Acids. After formic and acetic acids, the next member of the series is *propionic acid*, $CH_3.CH_2.COOH$. This is similar to acetic acid in its methods of preparation and in its properties and reactions. When it is chlorinated, only those hydrogen atoms linked to the carbon atom adjacent to the carboxyl group are substituted thus:

$$CH_3.CH_2.COOH + Cl_2 = \underset{\alpha\text{-chloropropionic acid}}{CH_3.CHCl.COOH} + HCl$$

Butyric acid, the next highest member, exists in two isomers, the normal acid, $CH_3.CH_2.CH_2.COOH$, and the iso form $(CH_3)_2.CH.COOH$. This acid is present in stale perspiration and in rancid butter, being largely responsible for their unpleasant smells.

The fatty acids provide another good example of a homologous series. As the molecular weight increases, there is a gradual rise in boiling point and fall in density. The increasingly large paraffin chain with its hydrophobic character predominates over the hydrophilic carboxyl group as the series is ascended, causing the miscibility with water to decrease noticeably. For example, members above C_{10} are completely insoluble solids at room temperature, with little smell.

Many of the higher fatty acids occur naturally as their glyceryl esters in oils and fats (see § 38.4), from which they are obtained by hydrolysis. The most important are *palmitic acid* $C_{15}H_{31}.COOH$, and *stearic acid* $C_{17}H_{35}.COOH$. These are colourless, waxy, odourless solids used for making candles. Their striking resemblance to fats led to the whole homologous series being known as the *fatty acids*.

37.6. Acetyl Chloride, $CH_3.COCl$. This is the most important of the aliphatic *acid chlorides*, a class of substances with the general formula $C_nH_{2n+1}.COCl$, and with the structure

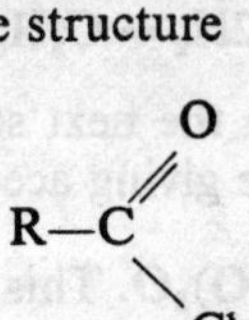

LABORATORY PREPARATION: It is made by treating glacial acetic acid with phosphorus pentachloride, phosphorus trichloride, or thionyl chloride:

$$CH_3.COOH + PCl_5 = CH_3.COCl + POCl_3 + HCl\uparrow$$
$$3CH_3.COOH + PCl_3 = 3CH_3.COCl + H_3PO_3$$
$$CH_3.COOH + SOCl_2 = CH_3.COCl + SO_2\uparrow + HCl\uparrow$$

PROPERTIES AND REACTIONS OF ACETYL CHLORIDE: It is a colourless liquid, b.p. 52°C, with a pungent smell.

1. Hydrolysis: It reacts vigorously with cold water giving off hydrogen chloride, which explains why acetyl chloride fumes strongly when exposed to moist air:

$$CH_3.COCl + H_2O = CH_3COOH + HCl\uparrow$$

With alkalis the hydrolysis is explosive.

2. Acetylation: This, the most important reaction of acetyl chloride, consists of replacing hydrogen atoms in hydroxyl (—OH) or amino (—NH_2) groups by univalent acetyl (CH_3CO—) radicals. For example, it reacts vigorously with ethyl alcohol at room temperature giving ethyl acetate and hydrogen chloride:

$$C_2H_5OH + CH_3.COCl = CH_3.COOC_2H_5 + HCl\uparrow$$

Similarly, with phenol it gives phenyl acetate and hydrogen chloride:

$$C_6H_5OH + CH_3.COCl = CH_3.COOC_6H_5 + HCl\uparrow$$

Acetyl chloride acetylates amines readily, converting aniline into acetanilide:

$$C_6H_5NH_2 + CH_3.COCl = C_6H_5NH.COCH_3 + HCl\uparrow$$

With a concentrated solution of ammonia it gives acetamide:

$$2NH_3 + CH_3.COCl = CH_3.CONH_2 + NH_4Cl$$

Acetylation is important as a means of converting volatile substances containing amino or hydroxyl groups into crystalline solids which are easily purified and can be identified by their melting points. It is also used to 'protect' amino groups during nitration (see § 44.5, reaction 9).

3. Reduction: In the presence of a palladium catalyst acetyl chloride is reduced by bubbling hydrogen into its boiling solution in xylene:

$$CH_3.COCl + 2H_2 = C_2H_5OH + HCl$$

4. With Salts: As described in the next section, acetyl chloride reacts with anhydrous sodium acetate giving acetic anhydride.

37.7. Acetic Anhydride, $(CH_3CO)_2O$. This is the most important of the aliphatic *acid anhydrides*, a class of organic compound with the general formula $(C_nH_{2n+1}.CO)_2O$, having the structure

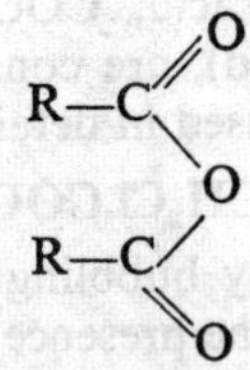

Laboratory Preparation: By adding acetyl chloride to anhydrous sodium acetate and distilling:

$$CH_3.COONa + CH_3COCl = (CH_3CO)_2O + NaCl$$

Manufacture: It is made from acetylene, acetaldehyde, or glacial acetic acid by various catalytic processes.

Properties and Reactions of Acetic Anhydride: It is a colourless liquid, b.p. 139°C, with a pungent smell like that of acetic acid.

1. Hydrolysis: It reacts slowly with cold water and rapidly on warming, giving acetic acid:

$$(CH_3.CO)_2O + H_2O = 2CH_3.COOH$$

2. Acetylation: It acetylates alcohols, phenols, and amines, replacing hydrogen atoms by acetyl radicals, although it does so less vigorously than acetyl chloride:

$$C_2H_5OH + (CH_3CO)_2O = CH_3.COOC_2H_5 + CH_3.COOH$$
$$C_6H_5NH_2 + (CH_3CO)_2O = C_6H_5NH.COCH_3 + CH_3.COOH$$

Uses of Acetic Anhydride: It is used industrially, mixed with glacial acetic acid, as an acetylating agent in the manufacture of cellulose acetate (acetate rayon), aspirin, acetanilide, and other acetyl compounds. It is preferred to acetyl chloride for tnis purpose because it is cheaper and less readily hydrolysed by moisture.

37.8. Substituted Acids. Many different acids can be formed by substituting various atoms or groups in place of hydrogen atoms in fatty acids, as the following examples show:

$CH_2Cl.COOH$	monochloroacetic acid
$CHCl_2.COOH$	dichloroacetic acid
$CCl_3.COOH$	trichloroacetic acid
$CH_2(NH_2).COOH$	aminoacetic acid or glycine
$CH_2(CN).COOH$	cyanoacetic acid
$CH_2(OH).COOH$	hydroxyacetic acid or glycollic acid.

Of these only monochloroacetic acid will be studied in this chapter. Glycine is dealt with in § 41.14. Two other important substituted acids, α-hydroxypropionic acid, $CH_3CH(OH).COOH$ (better known as lactic acid) and dihydroxysuccinic acid, $COOH.CH(OH).CH(OH).COOH$ (commonly called tartaric acid), are considered in chapter 48, where their stereoisomerism is discussed in detail.

37.9. Monochloroacetic Acid, $CH_2Cl.COOH$.

LABORATORY PREPARATION: By bubbling chlorine into boiling glacial acetic acid in sunlight and in the presence of a catalyst of iodine or red phosphorus:

$$CH_3.COOH + Cl_2 = CH_2Cl.COOH + HCl$$

The process should be stopped when the calculated increase in weight has been achieved to prevent the formation of the di- and trichloro acids. The product is isolated from the reaction mixture by fractional distillation.

PROPERTIES AND REACTIONS OF CHLOROACETIC ACID: It is a colourless crystalline solid, m.p. 61°C, b.p. 189°C, which dissolves readily in water. The crystals should not be handled as they blister the skin.

Like acetic acid it is monobasic forming salts with bases and carbonates and reacting with alcohols to form esters. However it is a stronger acid than acetic, the substituted chlorine atom attracting electrons and causing the carboxyl group to dissociate to a greater extent. The di- and trichloro acids are stronger acids still, the latter being comparable to the mineral acids in strength (see § 50.8).

The chlorine atom in chloroacetic acid is reactive and is easily replaced by groups such as —OH, —CN, and —NH_2. For example, chloroacetic acid yields glycollic acid when boiled with water and gives glycine when treated with a concentrated solution of ammonia.

Chloroacetic acid is used for synthesizing the dye indigo and for making selective weedkillers.

37.10. Dicarboxylic Acids. These are acids containing two carboxyl groups in each molecule. The commonest ones form a homologous series with the general formula $COOH.(CH_2)_n.COOH$, the first three members being:

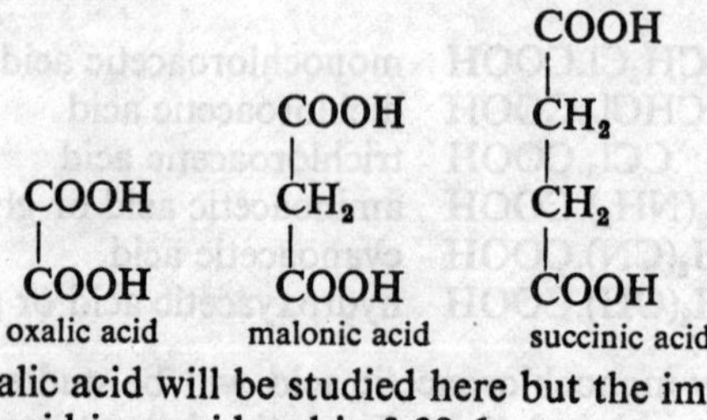

oxalic acid   malonic acid   succinic acid

Of these only oxalic acid will be studied here but the important diethyl ester of malonic acid is considered in § 38.6.

37.11. Oxalic Acid $(COOH)_2$. The acid and its salts occur naturally in the leaves of rhubarb and beet.

LABORATORY PREPARATION: By heating cane sugar (sucrose) with excess of concentrated nitric acid until the reaction ceases and then concentrating the solution, when crystals of the acid appear on cooling:

$$C_{12}H_{22}O_{11} + 18\bar{O} = 6(COOH)_2 + 5H_2O$$

Many other carbohydrates and also ethylene glycol are oxidized to oxalic acid in the same way and may be used instead of cane sugar. The preparation must be performed in a fume cupboard because large quantities of poisonous nitrogen dioxide are evolved.

MANUFACTURE: Sodium formate (made from carbon monoxide and sodium hydroxide as described in § 37.3) is heated to 400°C, when it decomposes into sodium oxalate and hydrogen:

$$2H.COONa = (COONa)_2 + H_2\uparrow$$

The sodium oxalate is dissolved in water and boiled with slaked lime. The resulting calcium oxalate is filtered off and treated with dilute sulphuric acid, giving a solution from which oxalic acid can be crystallized after removal of the insoluble by-product:

$$\left.\begin{array}{l} COO \\ | \\ COO \end{array}\right\rangle Ca + H_2SO_4 = \begin{array}{l} COOH \\ | \\ COOH \end{array} + CaSO_4\downarrow$$

PROPERTIES OF OXALIC ACID: It is fairly soluble in water, crystallizing from solution as the colourless dihydrate $(COOH)_2.2H_2O$. These crystals melt at 101°C and give off their water of crystallization. Oxalic acid and its salts are poisonous.

REACTIONS OF OXALIC ACID:

1. As an Acid: It is dibasic, forming both acid and normal salts with bases:

$$\begin{array}{l} COOH \\ | \\ COOH \end{array} + NaOH = H_2O + \begin{array}{l} COONa \\ | \\ COOH \end{array} \quad \text{sodium hydrogen oxalate}$$

$$\begin{array}{l} COOH \\ | \\ COOH \end{array} + 2NaOH = 2H_2O + \begin{array}{l} COONa \\ | \\ COONa \end{array} \quad \text{normal sodium oxalate}$$

It also forms crystals of tetroxalate,

$$\text{e.g.} \quad \begin{array}{l} COOH \\ | \\ COOK \end{array} \cdot \begin{array}{l} COOH \\ | \\ COOH \end{array} \cdot 2H_2O$$

known as 'salts of lemon'.

Oxalic acid is one of the strongest organic acids owing to the marked influence which one carboxyl group exerts upon the other, inducing its ready dissociation (see § 50.8).

2. *Esterification:* It reacts with alcohols forming esters:

$$\text{e.g.}\quad \begin{array}{c}COOH\\|\\COOH\end{array} + 2C_2H_5OH \rightleftharpoons 2H_2O + \begin{array}{l}COOC_2H_5\\|\\COOC_2H_5\end{array}\quad \text{diethyl oxalate}$$

3. *Stability:* When heated above 180°C the anhydrous acid melts and decomposes into formic acid, oxides of carbon, and steam. This thermal instability is characteristic of compounds containing two carboxyl groups linked either together or to the same carbon atom, and malonic acid shows the same effect.

4. *Oxidation:* Oxalic acid and its salts are mild reducing agents. They are oxidized to carbon dioxide and water by an acidified solution of potassium permanganate, slowly at room temperature but rapidly at 60°C:

$$\begin{array}{c}COOH\\|\\COOH\end{array} + \bar{O} = 2CO_2\uparrow + H_2O$$

$$2KMnO_4 + 5(COOH)_2 + 3H_2SO_4 = K_2SO_4 + 2MnSO_4 + 10CO_2 + 8H_2O$$

Use is made of this reaction to estimate solutions of oxalic acid and oxalates volumetrically.

5. *Dehydration:* When oxalic acid (or one of its salts or esters) is heated to 100°C with concentrated sulphuric acid, the elements of water are removed and carbon monoxide and carbon dioxide are produced:

$$\begin{array}{c}COOH\\|\\COOH\end{array} - H_2O = CO\uparrow + CO_2\uparrow$$

USES OF OXALIC ACID: It is used in the manufacture of inks, metal polishes, and oxalates, and for textile printing.

DETECTION OF OXALIC ACID AND OXALATES: When a solution of calcium chloride is added to a solution of oxalic acid or an oxalate, a white precipitate of calcium oxalate is obtained which is insoluble in acetic acid but soluble in dilute hydrochloric acid. The acid also decolorizes warm acidified potassium permanganate solution and evolves carbon monoxide (which burns with a blue flame) when heated with concentrated sulphuric acid. Unlike formic acid, it does not reduce an ammoniacal solution of silver oxide.

38

Esters, Fats, and Oils

38.1. Introduction. When an alcohol reacts with an acid, water and a substance called an *ester* are produced. It is usual to restrict the term ester to the products of organic acids, classifying those obtained from inorganic acids in other ways (e.g. as alkyl halides). *Esterification*, as the process of making esters is called, is a reversible reaction, and esters are readily hydrolysed to mixtures of alcohols and acids:

$$\text{ALCOHOL} + \text{ACID} \rightleftharpoons \text{ESTER} + \text{WATER}$$

The law of mass action is applied to this equilibrium in § 7.8.

Esterification is normally a very slow reaction, even when heat is applied, but it is catalysed by a high concentration of hydrogen ions. It is usual, therefore, to carry out esterification in the presence of concentrated sulphuric acid. Not only does this acid act as a catalyst, but it also absorbs the water formed, thereby preventing the backward reaction and displacing the position of equilibrium to the right so that a higher yield of ester is obtained. Occasionally concentrated sulphuric acid cannot be used because it reacts with the organic acid, and in such cases hydrogen chloride is bubbled into the reaction mixture instead. This is known as the Fischer-Speier method of esterification.

Although esterification bears a superficial resemblance to neutralization, the two reactions are really very different in type. As explained in § 8.6, neutralization is essentially a reaction between ions and so proceeds almost instantaneously to completion, whereas esters are formed by the slow and incomplete combination of covalent molecules.

Esters of monocarboxylic acids have the structural formula R.COOR′ or

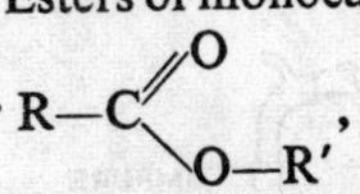

, where R and R′ may be the same or different. They can be pictured as being formed by replacing the active hydrogen atom of an acid by an alkyl or aryl group such as $—C_2H_5$ or $—C_6H_5$.* When

* In fact research has shown that in esterification the water molecules are formed from the hydroxyl groups of the acid and the hydrogen atoms of the alcohol.

looked at in this way the formulae of esters should give no difficulty because, as with salts of organic acids, the acid part is written first. This point is made clear by the following examples:

Formulae			*Names*
$CH_3.COOC_2H_5$	$CH_3{-}C\begin{smallmatrix}{\nearrow}O \\ {\searrow}OC_2H_5\end{smallmatrix}$	acetate $CH_3.COO$ ethyl C_2H_5	ethyl acetate
$C_2H_5.COOCH_3$	$C_2H_5{-}C\begin{smallmatrix}{\nearrow}O \\ {\searrow}OCH_3\end{smallmatrix}$	propionate $C_2H_5.COO$ methyl CH_3	methyl propionate

Esters are of considerable importance in nature. Not only are they largely responsible for the fragrance of flowers and the flavours of fruits, but they also form the main constituent of many naturally occurring fats and oils. Many perfumes and artificial flavourings are made by synthesizing esters.

Since all esters are very similar chemically, it is only necessary to study one of them in detail. Ethyl acetate is chosen for this purpose because it is one of the simplest and best known.

38.2. Ethyl Acetate, $CH_3.COOC_2H_5$.

LABORATORY PREPARATION: 1. The usual method is to heat ethyl alcohol with glacial acetic acid to 140°C in the presence of concentrated sulphuric acid:

$$C_2H_5OH + CH_3.COOH \rightleftharpoons CH_3.COOC_2H_5 + H_2O$$

The apparatus is shown in Fig. 38.1. The mixture is added from the tap funnel at about the same rate as the liquid distils over into the receiver, The product is impure containing unchanged alcohol, acetic acid, water.

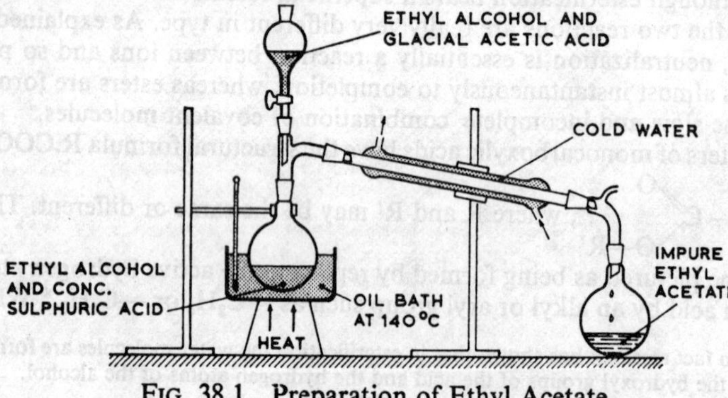

FIG. 38.1 Preparation of Ethyl Acetate

ether, and sulphurous acid. The acidic impurities are removed by shaking with sodium carbonate solution in a separating funnel (it is important to release the pressure from time to time) until no further effervescence occurs. The upper layer is then shaken with a concentrated solution of calcium chloride to remove the alcohol, separated, and dried by leaving it to stand over lumps of anhydrous calcium chloride until the liquid is clear. It is then filtered and carefully distilled on a water bath, the fraction boiling between 74°C and 79°C being almost free of ether.

2, Ethyl acetate can also be prepared by acetylating ethyl alcohol with acetyl chloride or acetic anhydride:

e.g. $$C_2H_5OH + CH_3COCl = CH_3.COOC_2H_5 + HCl$$

3. A good yield of ethyl acetate is obtained when ethyl iodide is heated with silver acetate, but the method is expensive:

$$C_2H_5I + CH_3.COOAg = CH_3.COOC_2H_5 + AgI\downarrow$$

PROPERTIES OF ETHYL ACETATE: It is a colourless liquid, b.p. 77°C ρ 0.90 g cm^{-3}, with a pleasant fruity smell. It is only slightly soluble in water but it dissolves readily in organic solvents.

REACTIONS OF ETHYL ACETATE:

1. Hydrolysis: With water alone the reaction is extremely slow, even on boiling:

$$CH_3.COOC_2H_5 + H_2O \rightleftharpoons C_2H_5OH + CH_3.COOH$$

It is catalysed by hydrogen ions from mineral acids and by hydroxyl ions from alkalis, the rate of reaction being proportional to the concentration of these ions. Hydrolysis is best effected by heating the ester under reflux with a 10% solution of an alkali for about thirty minutes:

$$CH_3.COOC_2H_5 + NaOH = CH_3.COONa + C_2H_5OH$$

The reaction is then known as *saponification* by analogy to the process of soap-making described in § 38.5.

2. With Ammonia: When ethyl acetate is shaken with a concentrated solution of ammonia, acetamide is formed:

$$CH_3.COOC_2H_5 + NH_3 = CH_3.CONH_2 + C_2H_5OH$$

3. Reduction: Ethyl acetate is reduced to ethyl alcohol by nascent hydrogen from, for example, lithium aluminium hydride (§ 19.14):

$$CH_3.COOC_2H_5 + 4\bar{H} = 2C_2H_5OH$$

USES OF ETHYL ACETATE: It is chiefly used as a solvent for cellulose nitrate in quick-drying lacquers and in nail varnish.

38.3. Identification of Esters. An ester is sometimes recognized by its characteristic smell, but the general procedure is to hydrolyse it and identify the fragments produced. If an ester is boiled under reflux with an alkali for about half an hour, it is converted into an alcohol and a salt. If the condenser is then connected in the normal position and the flask heated, the alcohol, being volatile, distils over into the receiver and can be recognized by the usual tests, leaving the non-volatile salt in solution in the flask. The residue is acidified with dilute sulphuric acid and reheated, when the organic acid distils over if it is aliphatic or appears as a white precipitate in the flask if it is aromatic. In either case it can be identified by applying the usual tests. For example, ethyl acetate treated in this way would yield ethyl alcohol in the first distillate and acetic acid in the second, from which its identity could be deduced.

38.4. Fats and Oils. All fats and oils of vegetable or animal origin* are esters of the trihydric alcohol glycerol (§ 35.9). Those which are solids at room temperature are called fats, whereas the liquid ones are known as oils. They are chiefly composed of the glyceryl esters (or *glycerides*) of the higher fatty acids such as palmitic acid $CH_3.(CH_2)_{14}.COOH$, and stearic acid $CH_3.(CH_2)_{16}.COOH$, and of certain unsaturated acids such as oleic acid $CH_3.(CH_2)_7.CH{=}CH.(CH_2)_7.COOH$, and linoleic acid $CH_3.(CH_2)_4.CH{=}CH.CH_2.CH{=}CH.(CH_2)_7.COOH$.

The glycerides in fats and oils are generally mixed esters in which each molecule of glycerol has combined with two or even three different acids, thus:

$$\begin{array}{lllll} R.CO\,\boxed{OH \quad H}\,OCH_2 & & R.COOCH_2 & & \\ & & \quad\;\; | & & \\ R'.CO\,\boxed{OH \quad H}\,OCH & = & R'.COOCH & + & 3H_2O \\ & & \quad\;\; | & & \\ R''.CO\,\boxed{OH \quad H}\,OCH_2 & & R''.COOCH_2 & & \end{array}$$

where R, R′, and R″ are often different (e.g. $C_{17}H_{35}$—for stearates, $C_{15}H_{31}$—for palmitates, $C_{17}H_{33}$—for oleates, etc.).

Any particular fat or oil consists of a complex mixture of these glycerides, its properties largely depending upon the proportions of each constituent. For example, the presence of a high proportion of esters of unsaturated acids gives the substance a low melting point, and

* A clear distinction should be drawn between these oils and mineral oils such as paraffin oil, which are complex mixtures of hydrocarbons obtained from crude petroleum by fractional distillation.

in oils this type of ester predominates. An oil can be changed into a fat therefore, by catalytic hydrogenation, and since fats are more acceptable for cooking and eating and also for soap-making, this process, which is known technically as *hardening*, is carried out on a large scale. Oils derived from olives, soya-beans, peanuts, and whales, for example, are heated to about 200°C in the presence of a catalyst of finely divided nickel and hydrogen is bubbled in under about 3×10^5 N m^{-2} pressure. The unsaturated radicals in the oils take up the hydrogen by addition across their double bonds, the process being continued until a product with the desired physical properties is obtained. Hydrogenations of this kind are known as Sabatier and Senderens reductions after their discoverers. *Margarine* is made by mixing these hardened oils with various fats and with colouring matter, vitamins, and skimmed milk to simulate butter. Butter itself consists largely of the glyceryl esters of palmitic and oleic acid, together with a few per cent of the glycerides of butyric and other similar fatty acids.

Animal fats and oils are extracted by heating the tissue with water or steam, when the fat rises to the surface and is skimmed off. Tallow or dripping is made in this way from beef or mutton, and lard from pigmeat. Most vegetable oils are obtained by compressing the fruits or seeds of the plant so that the oil is squeezed out, although extraction with a solvent is sometimes used. The principal sources are olives, coconuts, soya-beans, cottonseed, linseed, peanuts, palms, and castor beans.

The chief use of fats is as food; they account for up to half of the energy content of the human diet. A second major use is for making soaps of all kinds as described in the next section. Fats and oils are also the main sources of glycerol, and of drying oils used in paints and varnishes, and of the higher fatty acids used to make candles and greases.

38.5. Soaps. These are sodium and potassium salts of stearic, palmitic, and oleic acids. They are made by boiling fats and oils for many hours in large vats with solutions of sodium hydroxide or potassium hydroxide until they are completely converted into glycerol and the alkali salts. If we represent fats by glyceryl tristearate for simplicity, the equation for the hydrolysis is:

$$\begin{array}{lccc}
C_{17}H_{35}.COOCH_2 & & & CH_2OH \\
\quad\quad\quad\quad | & & & | \\
C_{17}H_{35}.COOCH & + \; 3NaOH & = \; 3C_{17}H_{35}.COONa \; + & CHOH \\
\quad\quad\quad\quad | & & & | \\
C_{17}H_{35}.COOCH_2 & & & CH_2OH \\
\text{a fat} & \text{an alkali} & \text{a soap} & \text{glycerol}
\end{array}$$

The soap is separated from the mixture by a process known as 'salting out'. When a concentrated solution of salt is added it causes the soap to collect in a layer at the surface making it possible to run off the mixture of glycerol, water, and brine from the bottom of the vat. The salt can be recovered and used again and pure glycerol can be obtained by distillation under reduced pressure.

In recent years various continuous processes have been developed for soap-making in which fats and oils are rapidly hydrolysed by water at high temperatures and pressures. The resulting acids are then converted into soaps by treatment with sodium carbonate or sodium hydroxide solutions.

The physical properties of soap depend upon the alkali used, the fats or oils chosen, and the proportion of glycerol which remains in the product. Hard soaps such as those used for laundering are chiefly composed of sodium stearate and sodium palmitate, whereas the softer soaps used for toilet purposes are mostly potassium salts of unsaturated acids. After separation the soap is washed, dried, coloured, scented, and pressed into tablets. It is effective as a cleansing agent because it lowers the surface tension of water and causes thorough wetting of greasy surfaces. This action arises from the particular chemical nature of soaps; the long hydrocarbon chains make them miscible with oil and grease, whilst their hydrophilic carboxylate ions confer solubility in water. Despite competition from synthetic detergents, which have the advantage of retaining their efficiency in hard water (see § 16.5), soap-making is still a major chemical industry.

38.6 Diethyl Malonate, Malonic Ester, $CH_2(COOC_2H_5)_2$.

LABORATORY PREPARATION: If the sodium salt of chloroacetic acid (§ 37.9) is heated with potassium cyanide, sodium cyanoacetate is formed:

$$CH_2Cl.COONa + KCN = CH_2(CN).COONa + KCl$$

When heated with ethyl alcohol and concentrated sulphuric acid, this compound is hydrolysed to malonic acid and esterified to the diethyl ester in one operation:

$$\begin{array}{ccccc} COONa & & COOH & & COOC_2H_5 \\ | & & | & & | \\ CH_2 & \longrightarrow & CH_2 & \longrightarrow & CH_2 \\ | & & | & & | \\ CN & & COOH & & COOC_2H_5 \end{array}$$

PROPERTIES OF DIETHYL MALONATE: It is a colourless liquid, b.p. 198°C, with a pleasant smell. Although only slightly soluble in water, it dissolves readily in alcohol and ether.

REACTIONS OF DIETHYL MALONATE: On mixing an alcoholic solution of diethyl malonate with sodium ethoxide, a sodium derivative is obtained. When this is boiled with an alkyl halide (see § 40.1), the alkyl group R replaces the sodium to give an alkyl derivative of malonic ester:

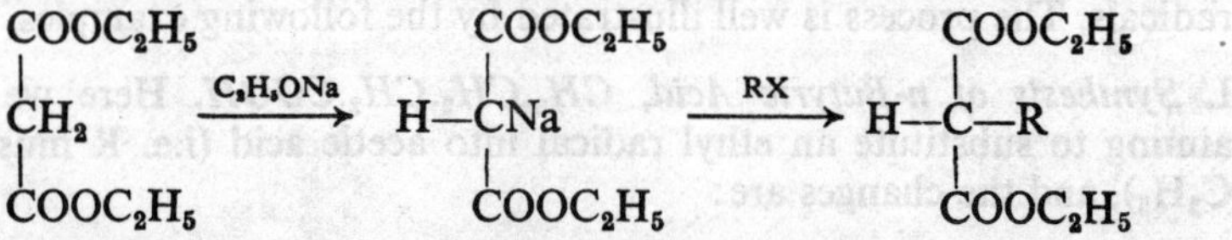

If more sodium ethoxide is added and the product is boiled again with an alkyl halide, the other hydrogen atom can be replaced by a second alkyl radical R′ thus:

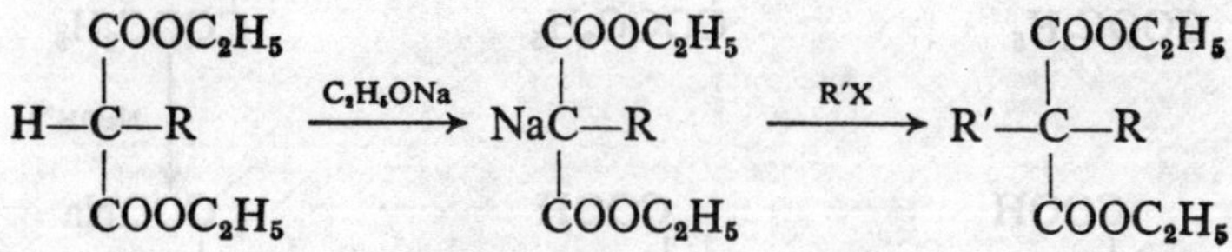

When they are boiled with alkali these alkyl-substituted esters are readily hydrolysed, giving the sodium salts of the corresponding acids. Acidification then gives the free acids themselves which, because they contain two carboxyl groups linked to the same carbon atom, readily decompose when heated evolving carbon dioxide:

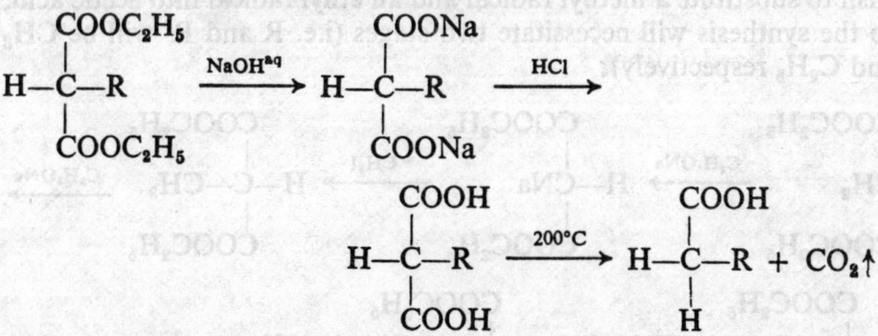

and

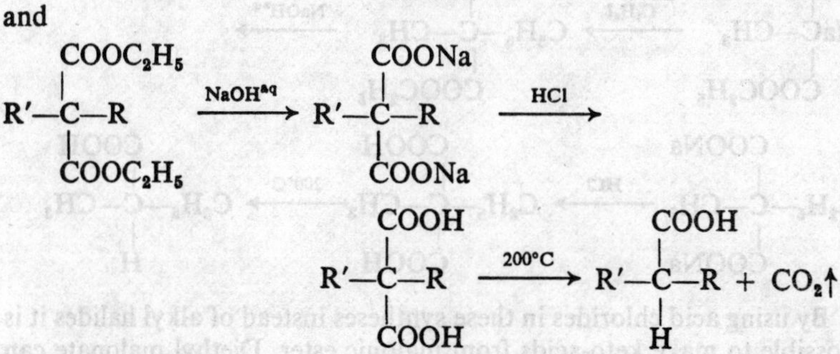

By using malonic ester in this way we can synthesize in good yield a whole range of substituted acetic acids of the type $CH_2R.COOH$ and

CHRR′.COOH, where R and R′ are either the same or different alkyl radicals. The process is well illustrated by the following examples.

I. *Synthesis of n-Butyric Acid, $CH_3.CH_2.CH_2.COOH$.* Here we are aiming to substitute an ethyl radical into acetic acid (i.e. R must be C_2H_5), and the changes are:

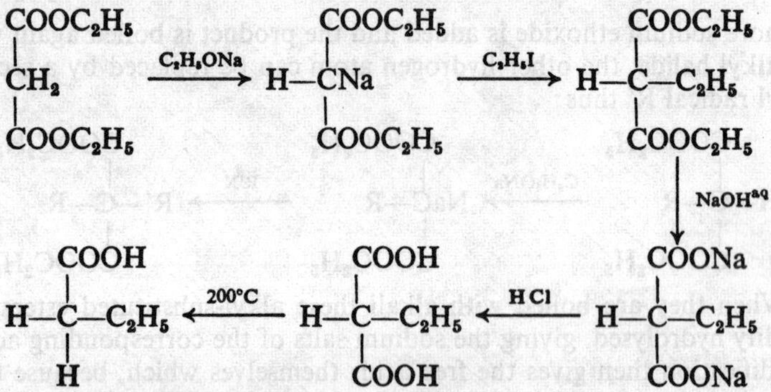

II. *Synthesis of Valeric Acid, $C_2H_5.CH(CH_3).COOH$.* In this case we wish to substitute a methyl radical and an ethyl radical into acetic acid, so the synthesis will necessitate two stages (i.e. R and R′ will be CH_3 and C_2H_5 respectively):

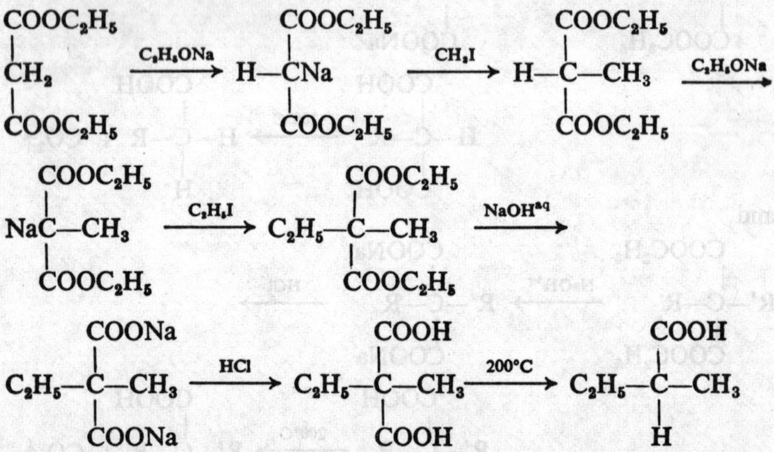

By using acid chlorides in these syntheses instead of alkyl halides it is possible to make keto-acids from malonic ester. Diethyl malonate can also be made to condense with urea (§ 41.9) to give barbituric acid, from which a number of important sedatives and anaesthetics are prepared.

39

Ethers

39.1. Introduction. Aliphatic ethers are compounds consisting of two alkyl radicals linked directly by an oxygen atom. They form a homologous series with the general molecular formula $C_nH_{2n+2}O$, and are therefore isomeric with the alcohols. They have the structure R—O—R′, where R and R′ are the same in a simple ether and different in a mixed ether. The simplest members are:

$CH_3.O.CH_3$	dimethyl ether
$CH_3.O.C_2H_5$	methyl ethyl ether
$C_2H_5.O.C_2H_5$	diethyl ether.

Diethyl ether is by far the commonest and most important of these, so much so that it is frequently referred to simply as 'ether'. Since it is in every respect typical of the other members of the series, it alone will be studied.

39.2. Diethyl Ether, $C_2H_5.O.C_2H_5$.

LABORATORY PREPARATION: 1. Concentrated sulphuric acid is heated to 140°C with an excess of ethyl alcohol. The essential change is a partial dehydration of the alcohol thus:

$$2C_2H_5OH = (C_2H_5)_2O + H_2O$$

It is probable that the sulphuric acid first reacts with the alcohol to give ethyl hydrogen sulphate and that this then combines with further alcohol to form ether, as follows:

$$C_2H_5OH + H_2SO_4 = C_2H_5HSO_4 + H_2O$$
$$C_2H_5HSO_4 + C_2H_5OH = C_2H_5.O.C_2H_5 + H_2SO_4$$

This explains how a given amount of acid can be used to convert a much larger quantity of alcohol into ether because theoretically the acid can be used over and over again. Indeed the method is often described as the 'continuous etherification process' for this reason. In practice, however, side reactions occur and the acid is gradually diluted by the water produced.

Concentrated sulphuric acid is added a little at a time to the alcohol, with shaking and cooling, and the mixture is heated to about 140°C in the apparatus shown in Fig. 39.1. Additional alcohol is run in from the tap funnel at about the same rate as the ether distils over. Ether is so volatile that a very efficient condenser and a well-cooled receiver are necessary to condense it. Even so some of its vapour will tend to escape into the laboratory, and rubber tubing is used to lead it to the floor where, being dense, it is likely to remain out of reach of flames. The liquid which collects in the receiver contains alcohol, water, and dissolved sulphur dioxide as impurities. It is purified by shaking it with

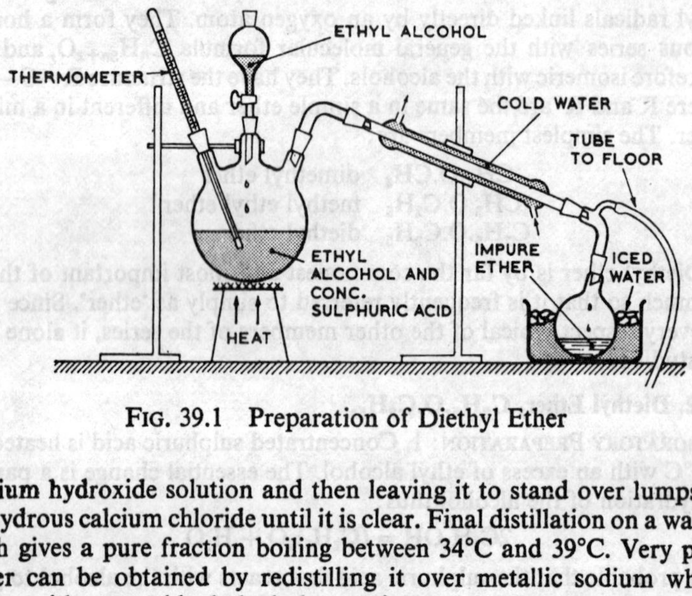

FIG. 39.1 Preparation of Diethyl Ether

sodium hydroxide solution and then leaving it to stand over lumps of anhydrous calcium chloride until it is clear. Final distillation on a water-bath gives a pure fraction boiling between 34°C and 39°C. Very pure ether can be obtained by redistilling it over metallic sodium which reacts with any residual alcohol or moisture.

2. Ether may also be prepared by *Williamson's synthesis*, by heating under reflux a mixture of ethyl iodide and sodium ethoxide in alcoholic solution:

$$C_2H_5I + C_2H_5ONa = C_2H_5.O.C_2H_5 + NaI$$

This method is a general one, used particularly for preparing higher ethers.

3. Ether is also formed when ethyl iodide is warmed with dry silver oxide:

$$2C_2H_5I + Ag_2O = C_2H_5.O.C_2H_5 + 2AgI$$

MANUFACTURE OF DIETHYL ETHER: Two methods are in use:

1. Ethyl alcohol is heated to 140°C with concentrated sulphuric acid, as in the laboratory preparation.
2. Ethyl alcohol vapour, at 250°C and under pressure, is passed over a catalyst of alumina (§ 19.9), when direct dehydration takes place:

$$2C_2H_5OH = (C_2H_5)_2O + H_2O$$

PROPERTIES OF DIETHYL ETHER: It is a colourless liquid, b.p. 35°C, ρ 0.72 g cm^{-3}, with a characteristic sweet smell. Ether is miscible with alcohol in all proportions. When mixed with water it is not completely immiscible, some ether dissolving in the water and a little water dissolving in the ether, but this limited miscibility can be reduced by adding salt. When ether vapour is inhaled it causes unconsciousness.

REACTIONS OF DIETHYL ETHER: Ether is highly inflammable and very volatile and its vapour forms explosive mixtures with air:

$$C_2H_5.O.C_2H_5 + 6O_2 = 4CO_2 + 5H_2O$$

Pockets of the dense vapour have been known to roll surprising distances along a bench, so care should be taken not to allow any flames in the vicinity during its use.

Apart from its combustion, ether is fairly inert. It is unaffected by alkalis or dilute acids, or by dehydrating or reducing agents. It does not react with metallic sodium, nor with phosphorus pentachloride in the cold, although when heated under reflux with it some reaction does occur.

Exposed to air, especially in sunlight, ether slowly absorbs oxygen forming a peroxide which may explode when heated. This oxidation can be prevented by storing ether in dark, well-sealed bottles in the presence of a little alcohol. When refluxed with a concentrated solution of hydriodic acid, ether is converted into ethyl iodide:

$$C_2H_5.O.C_2H_5 + 2HI = 2C_2H_5I + H_2O$$

USES OF DIETHYL ETHER: It is an excellent solvent, widely used for dissolving organic compounds or for extracting substances from aqueous solution. Many syntheses are carried out in ethereal solution. It is also used as an anaesthetic.

STRUCTURE OF DIETHYL ETHER: Its molecular formula is $C_4H_{10}O$. The failure to react with sodium or to evolve hydrogen chloride when treated with phosphorus pentachloride proves the absence of a hydroxyl group, whilst its synthesis from ethyl iodide strongly suggests the existence of two ethyl radicals in the molecule. Consequently the following structure is accepted:

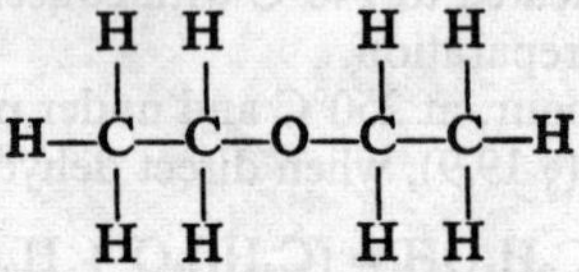

39.3. Isomerism. The ethers afford a good example of a particular type of isomerism called *metamerism*, in which the isomers not only have the same molecular formula but are members of the same homologous series and differ only in the nature of the radicals which are linked to the functional atom or group (in this case the oxygen atom). Thus the following ethers are metamers:

$CH_3.CH_2.O.CH_2.CH_3$	diethyl ether
$CH_3.O.CH_2.CH_2.CH_3$	methyl n-propyl ether
$CH_3.O.CH(CH_3)_2$	methyl iso-propyl ether.

40

Halogen Derivatives of the Paraffins

40.1. Introduction. Although not prepared in this way in the laboratory, these compounds may be regarded as being derived theoretically from the paraffins by the successive replacement of hydrogen atoms by halogen atoms. The monohalogen derivatives, known as the *alkyl halides*, have the general formula $C_nH_{2n+1}X$, where X stands for a halogen atom; the commonest members are listed in Table 40.1. Ethylene dibromide.

TABLE 40.1. THE SIMPLEST ALKYL HALIDES

Names	*Formula*	m.p./°C	b.p./°C	(*as liquid*) ρ/g cm^{-3}
methyl chloride or chloromethane	CH_3Cl	−93	−24	0.95
,, bromide or bromomethane	CH_3Br	−93	5	1.73
,, iodide or iodomethane	CH_3I	−66	42	2.28
ethyl chloride or chloroethane	C_2H_5Cl	−142	12.5	0.92
,, bromide or bromoethane	C_2H_5Br	−125	38	1.45
,, iodide or iodoethane	C_2H_5I	−105	72	1.92

$C_2H_4Br_2$, is the principal dihalogen derivative, and chloroform $CHCl_3$ and iodoform CHI_3 are the best known trihalogen compounds, although trichloroethylene C_2HCl_3 is of considerable industrial importance. Of the polyhalogen derivatives carbon tetrachloride CCl_4 is the commonest, but since it is conventionally regarded as inorganic it is studied in chapter 20. Fluorine derivatives are dealt with separately in § 40.11.

40.2. Preparation of Alkyl Halides

GENERAL METHODS: Two main methods are available:

1. By the action of a halogen hydride on an alcohol, thus:

$$ROH + HX \rightleftharpoons RX + H_2O$$

This reaction, which is essentially an esterification, is slow and reversible. The rate is increased by heating and by using a catalyst. The yield of alkyl halide is improved by adding a substance which removes the water formed and prevents the backward reaction.

2. By the action of a phosphorus halide on an alcohol, thus:

$$3ROH + PX_3 = 3RX + H_3PO_3$$

In practice the phosphorus halide is often made *in situ* by adding the halogen to a mixture of red phosphorus and alcohol, as this gives a better yield.

Other methods, such as direct substitution of halogens in paraffins and addition of halogen hydrides to olefines, are of little practical value and are rarely used. Which general method is chosen depends upon the particular alkyl halide required. This is well illustrated by considering in detail the laboratory preparations of the three main ethyl halides:

PREPARATION OF ETHYL CHLORIDE: This is best prepared by the Groves method. Dry hydrogen chloride is bubbled into absolute ethyl alcohol in the presence of anhydrous zinc chloride, which acts not only as a catalyst but also as a dehydrating agent:

$$C_2H_5OH + HCl \rightleftharpoons C_2H_5Cl + H_2O$$

When the alcohol has been saturated with hydrogen chloride, it is heated under reflux on a water-bath. The ethyl chloride vapour which issues from the top of the condenser is purified by passing it through a series of wash bottles containing water to remove hydrogen chloride, sodium carbonate solution to remove alcohol, and concentrated sulphuric acid to remove moisture. The pure ethyl chloride is then condensed in a U-tube surrounded by freezing mixture.

Ethyl chloride can also be prepared by treating ethyl alcohol with phosphorus trichloride or pentachloride, or with thionyl chloride, but side reactions occur and the yield is poor.

PREPARATION OF ETHYL BROMIDE: This is best prepared by heating potassium bromide and concentrated sulphuric acid with ethyl alcohol, using the apparatus shown in Fig. 40.1. The sulphuric acid reacts with the bromide liberating hydrogen bromide *in situ*; it also catalyses the main reaction and improves the yield by acting as a dehydrating agent:

$$KBr + H_2SO_4 = KHSO_4 + HBr$$
$$C_2H_5OH + HBr \rightleftharpoons C_2H_5Br + H_2O$$

The acid is slowly added to the alcohol with continual shaking. Then powdered potassium bromide is added and the flask is immediately connected to the condenser. When the flask is heated ethyl bromide distils over and collects in the receiver beneath the layer of water. The latter absorbs the escaping hydrogen bromide and also prevents the volatile ethyl bromide from evaporating. The product is impure containing water, alcohol, ether, bromine, and also dissolved sulphur dioxide and hydrogen bromide. After separation from the water it is shaken with successive quantities of sodium carbonate solution until no further effervescence occurs, and then with water to remove residual carbonate. It is then separated again and allowed to stand over lumps of anhydrous calcium chloride until it is clear. In the final distillation on a water-bath

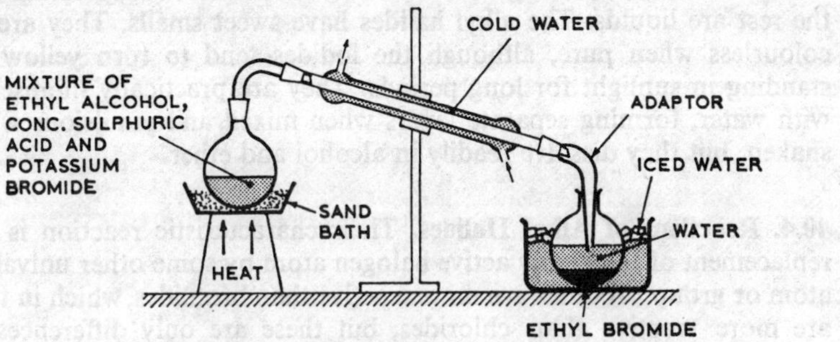

FIG. 40.1 Preparation of Ethyl Bromide

the fraction boiling between 35°C and 40°C is collected; it contains a little ether but it is otherwise pure.

Ethyl bromide can also be prepared by heating red phosphorus and bromine with ethyl alcohol, the details being similar to those for ethyl iodide described below.

Preparation of Ethyl Iodide: This cannot be prepared by heating an iodide with concentrated sulphuric acid and ethyl alcohol because the hydrogen iodide produced, being a strong reducing agent, would be oxidized to iodine by the sulphuric acid. It is usually prepared, therefore, by treating ethyl alcohol with red phosphorus and iodine:

$$2P + 3I_2 = 2PI_3$$
$$3C_2H_5OH + PI_3 = 3C_2H_5I + H_3PO_3$$

Alcohol and red phosphorus are mixed in a flask fitted with a reflux condenser and iodine crystals are added a little at a time, the flask being cooled if necessary. When all the iodine has been added the flask is heated on a boiling water-bath for about an hour. The condenser is then changed from the reflux to the normal position and the ethyl iodide distilled over. It is impure, containing alcohol, water, hydrogen iodide, and a small amount of iodine, which gives it a yellow-brown colour. The ethyl iodide is separated from the water and shaken in a separating funnel with sodium carbonate solution and water in turn. Finally it is dried with anhydrous calcium chloride and distilled on a water-bath, the fraction boiling between 68°C and 73°C being collected.

40.3. Properties of Alkyl Halides. Physical properties of the simplest are given in Table 40.1, from which it is clear that methyl chloride, methyl bromide, and ethyl chloride are gases at room temperature and the rest are liquids. The alkyl halides have sweet smells. They are all colourless when pure, although the iodides tend to turn yellow on standing in sunlight for long periods. They are practically immiscible with water, forming separate layers when mixed and emulsions when shaken, but they dissolve readily in alcohol and ether.

40.4. Reactions of Alkyl Halides. Their characteristic reaction is the replacement of the highly active halogen atom by some other univalent atom or group.* Iodides are more reactive than bromides, which in turn are more reactive than chlorides, but these are only differences of degree and alkyl halides are so similar chemically that it will suffice in the following reactions to write equations for only one of them (e.g. ethyl bromide) as representative of the others. Their principal reactions, which are summarized in Table 40.2, are:

1. Hydrolysis: Alkyl halides are hydrolysed only slowly by boiling water but rapidly by hot aqueous solutions of alkalis giving alcohols:

e.g. $$C_2H_5Br + NaOH = \underset{\text{ethyl alcohol}}{C_2H_5OH} + NaBr$$

2. Reduction: Nascent hydrogen (from a zinc–copper couple and water, for instance), reduces alkyl halides to paraffins:

e.g. $$C_2H_5Br + 2\overline{H} = \underset{\text{ethane}}{C_2H_6} + HBr$$

3. With Potassium Cyanide: When an alkyl halide is boiled with an alcoholic solution of potassium cyanide, an alkyl cyanide is formed:

e.g. $$C_2H_5Br + KCN = \underset{\substack{\text{ethyl cyanide or}\\ \text{propionitrile}}}{C_2H_5CN} + KBr$$

* The mechanism of this reaction is discussed in § 50.7.

TABLE 40.2. REACTIONS OF ETHYL BROMIDE

ETHYL BROMIDE →

boiling water, aqueous alkalis or moist silver oxide	ethyl alcohol, C_2H_5OH
nascent hydrogen	ethane, C_2H_6
boiling alcoholic KCN	ethyl cyanide, C_2H_5CN
conc. soln. of ammonia	mixture of ethylamines
sodium in dry ether	butane, C_4H_{10}
sodium ethoxide	diethyl ether, $C_2H_5.O.C_2H_5$
dry silver oxide	diethyl ether, $C_2H_5.O.C_2H_5$
boiling alcoholic alkalis	diethyl ether, $C_2H_5.O.C_2H_5$
silver acetate	ethyl acetate, $CH_3.COOC_2H_5$
silver nitrite	ethyl nitrite, C_2H_5ONO + nitroethane, $C_2H_5NO_2$
silver cyanide	ethyl isocyanide, C_2H_5NC
magnesium and dry ether	ethyl magnesium bromide C_2H_5MgBr
benzene and anhydrous Al_2Cl_6 (Friedel-Craft reaction)	ethylbenzene, $C_6H_5.C_2H_5$

4. With Ammonia: A concentrated aqueous solution of ammonia reacts with an alkyl halide giving the salt of a primary amine:

e.g. $$C_2H_5Br + NH_3 = \underset{\text{ethylamine hydrobromide}}{C_2H_5NH_3Br}$$

The reaction does not stop there and the salts of secondary and tertiary amines and even quaternary ammonium compounds are also formed:

e.g.
$$C_2H_5NH_3Br + C_2H_5Br = (C_2H_5)_2NH_2Br + HBr$$
$$(C_2H_5)_2NH_2Br + C_2H_5Br = (C_2H_5)_3NHBr + HBr$$
$$(C_2H_5)_3NHBr + C_2H_5Br = (C_2H_5)_4NBr + HBr$$

The difficulty of separating these products makes this an unattractive method for preparing amines.

5. *With Sodium:* When a solution of an alkyl halide in dry ether is warmed with metallic sodium it is converted into a paraffin by the Wurtz synthesis:

e.g. $$2C_2H_5Br + 2Na = \underset{\text{butane}}{C_2H_5.C_2H_5} + 2NaBr$$

6. *With Sodium Ethoxide:* Alkyl halides react with an alcoholic solution of sodium ethoxide on warming giving ethers:

e.g. $$C_2H_5Br + C_2H_5ONa = \underset{\text{diethyl ether}}{C_2H_5.O.C_2H_5} + NaBr$$

This is an example of Williamson's synthesis.

7. *With Alcoholic Alkalis:* Potassium hydroxide dissolves in ethyl alcohol giving a mixture in which potassium ethoxide predominates, so that when an alkyl halide is boiled with this reagent, the main product is an ether, as in reaction 6:

e.g. $$C_2H_5Br + C_2H_5OK = C_2H_5.O.C_2H_5 + KBr$$

At the same time a trace of ethylene is also formed by a side reaction. This is one of the few instances when the behaviour of ethyl bromide is not typical of that of the higher alkyl halides, many of which react with alcoholic potassium hydroxide to give a good yield of olefine and very little ether.

8. *With Silver Oxide:* When an alkyl halide is boiled with a suspension of silver oxide in water, rapid hydrolysis to the alcohol occurs:

e.g. $$2C_2H_5Br + Ag_2O + H_2O = \underset{\text{ethyl alcohol}}{2C_2H_5OH} + 2AgBr\downarrow$$

However alkyl halides react very differently with dry silver oxide, forming ethers thus:

e.g. $$2C_2H_5Br + Ag_2O = \underset{\text{diethyl ether}}{C_2H_5.O.C_2H_5} + 2AgBr$$

9. *With Silver Salts:* In general, when alcoholic solutions of silver salts are warmed with an alkyl halide, an ester is formed and a silver halide is precipitated. Thus *silver acetate* gives a good yield of an acetate ester:

e.g. $$C_2H_5Br + CH_3.COOAg = \underset{\text{ethyl acetate}}{CH_3.COOC_2H_5} + AgBr\downarrow$$

With *silver nitrite* a mixture of nitrite and nitroparaffin is produced:

e.g. $$C_2H_5Br + AgNO_2 = AgBr\downarrow + \underset{\text{nitroethane}}{C_2H_5{-}N{\lower{0.5ex}{\overset{\nearrow O}{\searrow O}}}}$$

$$C_2H_5Br + AgNO_2 = AgBr\downarrow + \underset{\text{ethyl nitrite}}{C_2H_5{-}O{-}N{=}O}$$

The two isomers can be separated by fractional distillation. *Silver cyanide* is exceptional, reacting with alkyl halides to form isocyanides:

e.g. $$C_2H_5Br + AgNC = C_2H_5NC + AgBr\downarrow$$

This happens because silver cyanide is a covalent compound with a chain structure in which the silver atoms are linked directly to both carbon and nitrogen atoms.

10. With Magnesium: Alkyl halides in dry ethereal solution combine with magnesium giving alkyl magnesium halides known as Grignard reagents:

e.g. $$C_2H_5Br + Mg = \underset{\text{ethyl magnesium bromide}}{C_2H_5.Mg.Br}$$

These compounds are of great value in syntheses because they are readily converted into a wide range of organic compounds. For example, when the ethereal solution of a Grignard reagent is treated with water, rapid hydrolysis occurs giving a good yield of the corresponding alkane (paraffin):

$$C_2H_5MgBr + H_2O = \underset{\text{ethane}}{C_2H_6\uparrow} + Mg(OH)Br$$

If a dry ethereal solution of a Grignard reagent is poured on to lumps of solid carbon dioxide and the product hydrolysed by adding water or acid, a caboxylic acid is obtained:

$$C_2H_5MgBr + CO_2 = C_2H_5.C{\overset{\displaystyle O}{\diagup\!\!\diagup}}\atop{\diagdown OMgBr}$$

$$C_2H_5.C(=O)\text{—}OMgBr + H_2O = C_2H_5.COOH + Mg(OH)Br$$

11. Friedel-Crafts Reaction: This important reaction of alkyl halides is described in § 46.2.

40.5. Uses of Alkyl Halides. The value of the alkyl halides in the laboratory arises from their ease of preparation and their reactivity and from the diversity of products that can be made from them. Methyl chloride is used industrially as a refrigerant and as a methylating agent, e.g. in the manufacture of silicones (§ 20.25) and dyes. Methyl bromide is used as a fire extinguisher in aircraft and, owing to its toxicity, as a powerful fumigant for pests in stored grain or fruit. Ethyl chloride is used on a large scale for making lead tetraethyl, the petrol additive (§ 21.23), from a sodium–lead alloy:

$$4C_2H_5Cl + Na_4Pb = (C_2H_5)_4Pb + 4NaCl$$

It is also used as a local anaesthetic in minor operations because of the cooling effect when it evaporates on the skin, and for making the important plastic ethyl cellulose.

40.6. Ethylene Dibromide, 1,2-Dibromoethane, $CH_2Br.CH_2Br$. This is the most important of the dihalogen derivatives, of which it is typical.

LABORATORY PREPARATION: By bubbling ethylene into bromine. Addition takes place readily in the cold and the liquid bromine, which is covered with a layer of water to minimize its evaporation, is gradually decolorized:

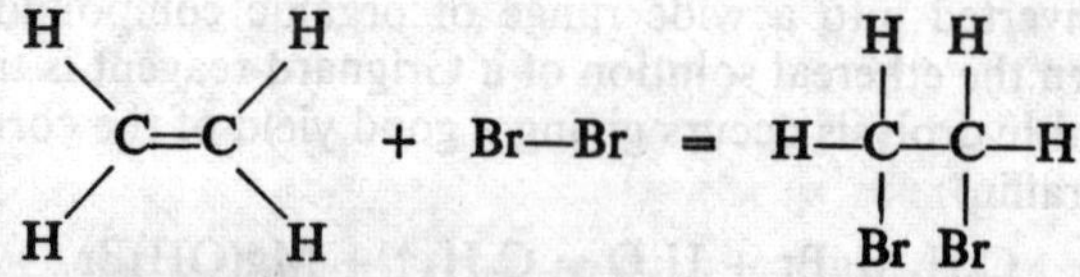

The dense dibromide can be separated from the water layer and purified by shaking it in turn with sodium carbonate solution and water. It can then be dried with anhydrous calcium chloride and distilled.

PROPERTIES AND REACTIONS OF ETHYLENE DIBROMIDE: It is a sweet-smelling colourless liquid, b.p. 132°C, ρ 2.19 g cm^{-3}. It is immiscible with water.

In its reactions ethylene dibromide resembles the alkyl halides, the two bromine atoms being readily replaceable by other atoms or groups. Thus it is hydrolysed to ethylene glycol by boiling it with aqueous alkalis:

$$\begin{matrix} CH_2Br \\ | \\ CH_2Br \end{matrix} + 2NaOH^{aq} = \begin{matrix} CH_2OH \\ | \\ CH_2OH \end{matrix} + 2NaBr$$

Similarly, each bromine atom may be replaced by a cyanide or amino group by boiling ethylene dibromide with alcoholic potassium cyanide or concentrated ammonia solution respectively, and nascent hydrogen reduces it to ethane. An alcoholic solution of potassium hydroxide, however, removes two molecules of hydrogen bromide from each molecule of the dibromide on boiling, leaving acetylene:

$$\begin{matrix} CH_2Br \\ | \\ CH_2Br \end{matrix} + \underset{\text{(alc.)}}{2KOH} = \begin{matrix} CH \\ ||| \\ CH \end{matrix} + 2KBr + 2H_2O$$

USES OF ETHYLENE DIBROMIDE: It is made on a large scale for adding to petrol, as explained in § 33.10.

40.7. Other Dihalogen Derivatives. *Methylene dichloride*, or *dichloromethane* CH_2Cl_2, is a valuable solvent used extensively as a paint remover. It is a colourless non-inflammable liquid, b.p. 40°C, made from chlorine and methyl chloride.

Ethylene dichloride, *1,2-dichloroethane* $CH_2Cl.CH_2Cl$, is a colourless liquid resembling ethylene dibromide. It is made by adding ethylene to chlorine. It is used as a solvent and for manufacturing polyvinyl chloride plastic (PVC). Ethylene dichloride should be clearly distinguished from its isomer *ethylidene dichloride*, *1,1-dichloroethane* $CH_3.CHCl_2$, in which both chlorine atoms are linked to the same carbon atom, and which is usually formed by the action of phosphorus pentachloride on acetaldehyde:

$$CH_3.CHO + PCl_5 = CH_3.CHCl_2 + POCl_3$$

These dihalides are generally less reactive than the alkyl halides.

40.8. Chloroform, Trichloromethane, $CHCl_3$.

LABORATORY PREPARATION: (1) By the action of bleaching powder upon acetone. The chemical nature of bleaching powder is obscure but in aqueous suspension it behaves as if it were a mixture of hypochlorous acid and calcium hydroxide. The first stage of the reaction is probably the chlorination of one of the methyl groups in each molecule thus:

$$CH_3.CO.CH_3 + 3HOCl = CH_3.CO.CCl_3 + 3H_2O$$

The resulting trichloroacetone is hydrolysed by the alkali present, giving chloroform and calcium acetate solution:

$$2CH_3.CO.CCl_3 + Ca(OH)_2 = 2CHCl_3 + (CH_3COO)_2Ca$$

A quantity of bleaching powder is ground with water in a mortar until a creamy paste is obtained. This is transferred to a large flask fitted with a reflux condenser and the acetone added slowly with shaking. The reaction usually starts spontaneously and is so exothermic that the flask may require cooling during the addition. When all the acetone has been added the condenser is changed to the normal position and the flask is heated on a boiling water-bath until no further chloroform distils over into the receiver, where it collects as a dense liquid beneath a layer of water. The product is impure, containing acetone, water, and dissolved chlorine and hydrogen chloride. These impurities are removed by shaking it in turn with caustic soda solution and with water and then drying it with anhydrous calcium chloride. The final distillation gives a fraction of pure chloroform boiling between 60°C and 62°C.

(2) Alternatively, ethyl alcohol can be used in place of acetone. In this case the alcohol is first oxidized to acetaldehyde, which then

undergoes chlorination to chloral and hydrolysis to chloroform and calcium formate, and follows:

$$C_2H_5OH + \bar{O} = CH_3.CHO + H_2O$$
$$CH_3.CHO + 3HOCl = CCl_3.CHO + 3H_2O$$
$$2CCl_3.CHO + Ca(OH)_2 = 2CHCl_3 + (H.COO)_2Ca$$

(3) The first two methods give a lower yield than expected owing to side reactions, and pure chloroform is best prepared by warming caustic soda solution with chloral:

$$CCl_3.CHO + NaOH = CHCl_3 + H.COONa$$

PROPERTIES OF CHLOROFORM: It is a colourless liquid with a sweet smell. It boils at 61°C and has a density of 1.49 g cm^{-3}. Its vapour causes unconsciousness when inhaled. Chloroform is immiscible with water but readily soluble in alcohol and ether.

REACTIONS OF CHLOROFORM: It is much less reactive than the alkyl halides, conforming to the general trend whereby the reactivity of the compounds methyl chloride, methylene dichloride, chloroform and carbon tetrachloride diminishes as the number of chlorine atoms linked to the carbon atom is increased. Its principal reactions are:

1. The Isocyanide Reaction: If chloroform is warmed with a primary amine and an alcoholic solution of potassium hydroxide, an isocyanide is formed, recognised by its characteristic and very unpleasant smell. For example, aniline gives phenyl isocyanide:

$$C_6H_5NH_2 + CHCl_3 + 3KOH = C_6H_5NC + 3KCl + 3H_2O$$

This is known variously as the *Hofmann* or *carbylamine reaction.* It provides an important test for primary amines because only they give isocyanides when treated in this way. It should be noted, however, that it is not a specific test for chloroform, since chloral and carbon tetrachloride also respond to it. After performing the isocyanide test the product should always be hydrolysed by adding concentrated hydrochloric acid in excess, since the smell tends to persist for long periods if the contents of the test tube are poured down the drain.

2. Hydrolysis: When chloroform is boiled with aqueous alkali it undergoes hydrolysis forming a compound with three hydroxyl groups attached to one carbon atom:

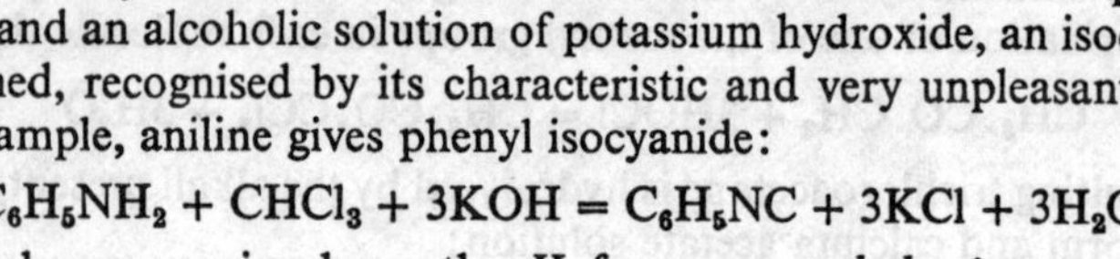

As might be expected, this is unstable and decomposes into formic acid with the elimination of a molecule of water:

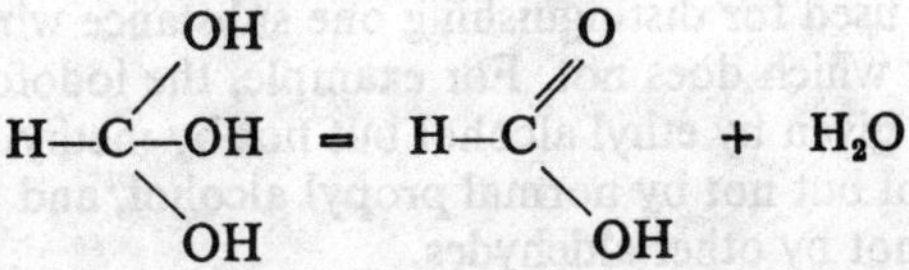

The acid is then neutralized by excess of alkali, so that the overall change is:

$$CHCl_3 + 4KOH = H.COOK + 3KCl + 2H_2O$$

3. Oxidation: On standing in air, particularly in sunlight, chloroform is gradually oxidized to carbonyl chloride, better known as the poisonous gas phosgene:

$$2CHCl_3 + O_2 = 2COCl_2 + 2HCl$$

This can be prevented by storing chloroform in dark-glass bottles with well-fitting stoppers. Sometimes about 1% of alcohol is added to convert any phosgene into harmless ethyl carbonate.

USES OF CHLOROFORM: It is still used as a general anaesthetic, although other compounds with less serious after-effects are now preferred. It is a valuable solvent, used particularly for the extraction of penicillin. A recent use is for manufacturing the important fluorocarbon, tetrafluoroethylene (see § 40.11).

40.9. Iodoform, Triiodomethane, CHI_3.

LABORATORY PREPARATION: By the action of a hypoiodite solution on ethyl alcohol or acetone. The hypoiodite, being unstable, is made *in situ* from potassium iodide solution and freshly prepared sodium hypochlorite solution, or alternatively by dissolving iodine in sodium carbonate solution or sodium hydroxide solution:

$$NaOCl + KI = KIO + NaCl$$
$$CH_3.CO.CH_3 + 3KIO = CHI_3 + CH_3.COOK + 2KOH$$

The iodoform is precipitated slowly when the solutions are mixed. It is filtered off, washed thoroughly with water and recrystallized from methylated spirit.

This preparation is of particular interest because the organic reactant must have one of two specific structures, viz.:

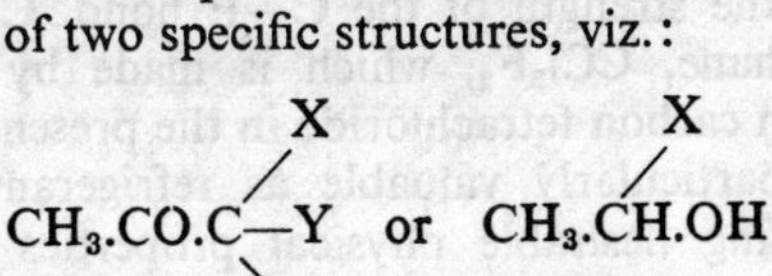

Because iodoform is so easily recognized, its preparation on a test tube scale is often used for distinguishing one substance which does form it from another which does not. For example, the iodoform reaction, as it is called, is given by ethyl alcohol but not by methyl alcohol, by iso-propyl alcohol but not by normal propyl alcohol, and by acetaldehyde solution but not by other aldehydes.

PROPERTIES AND REACTIONS OF IODOFORM: It is a pale yellow, crystalline solid, m.p. 119°C, ρ 4.1 g cm^{-3}, with a characteristic smell. It is insoluble in water, but it dissolves freely in warm alcohol or in ether. In its reactions it closely resembles chloroform. Thus it is hydrolysed to a formate when boiled with alkalis. Iodoform decomposes when heated strongly giving off iodine as a purple vapour.

USES OF IODOFORM: Its main use is as an antiseptic dressing for wounds and ulcers.

40.10. Trichloroethylene, CHCl:CCl$_2$. This compound is made industrially by adding chlorine to acetylene and warming the resulting acetylene tetrachloride with milk of lime:

$$HC{\equiv}CH + 2Cl_2 = CHCl_2.CHCl_2$$

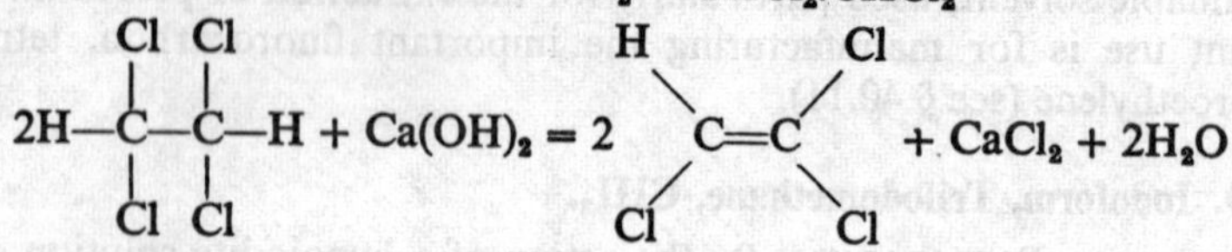

Trichloroethylene is a colourless liquid, b.p. 87°C, ρ 1·48 g cm^{-3}. It is an excellent industrial solvent, being stable, non-corrosive, and non-inflammable. It is much used for dry-cleaning clothes and for extracting oils from seeds, and for degreasing metals. Pure trichloroethylene is used as an anaesthetic during childbirth, when it is known as 'Trilene'.

40.11. Fluorine Derivatives. These are considered separately because of their distinctive properties. The monofluorides are of little importance because most of them decompose readily when heated eliminating hydrogen fluoride, but the poly-fluoro derivatives, particularly those also containing chlorine atoms, are characterized by great stability and inertness owing to the strength of the C—F bond. Compounds like dichlorodifluoromethane, CCl_2F_2, which is made by the action of hydrogen fluoride on carbon tetrachloride in the presence of antimony pentaflouride, are particularly valuable as refrigerants* because in addition to possessing desirable physical properties (e.g. low b.p.

* These have the trade names 'Freon' and 'Arcton'.

and high specific latent heat), they are not toxic, corrosive, or inflammable. They are also much used as propellants in aerosol sprays. Tetrafluoroethylene, $F_2C=CF_2$, which is made from chloroform and hydrogen fluoride, polymerizes under pressure in the presence of beyzoyl peroxide (§ 47.5) to the plastic $(C_2F_4)_n$, known as P.T.F.E. and sold under the trade names 'Fluon' and 'Teflon' (§ 49.2). The compound $CF_3CClBrH$ is of increasing importance as an anaesthetic under the trade name 'Fluothane'; besides being safe and effective, it is non-inflammable and free of unpleasant after effects.

41

Aliphatic Nitrogen Compounds

41.1. Introduction. This chapter deals with amines, amides, nitriles, isonitriles, amino-acids, polypeptides, and proteins.

Aliphatic *amines* may be regarded as derivatives of ammonia formed by replacing one or more hydrogen atoms by alkyl radicals. They are classified as primary (R—NH_2), secondary $\left(\begin{smallmatrix}R\\R\end{smallmatrix}\!>\!NH\right)$, or tertiary $\left(\begin{smallmatrix}R\\R\\R\end{smallmatrix}\!>\!N\right)$, according to the number of alkyl radicals contained in each molecule.

The *primary amines* form a homologous series with general formula $C_nH_{2n+1}.NH_2$, the lowest members being methylamine CH_3NH_2 and ethylamine $C_2H_5NH_2$. Their functional group is $—N\!<\!\begin{smallmatrix}H\\H\end{smallmatrix}$, known as the *amino group*, and it is this which determines their reactions, the alkyl radicals being relatively inert.

The *secondary* and *tertiary amines* are less important. They are described as simple when the alkyl radicals are the same, as in dimethylamine $(CH_3)_2NH$ and trimethylamine $(CH_3)_3N$, and mixed when they are different as in methylethylamine $CH_3.NH.C_2H_5$. *Quaternary ammonium compounds* are also known in which four alkyl radicals are linked to one nitrogen atom thus

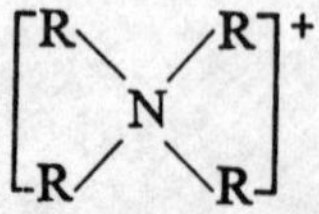

Amides, or *acid amides* as they are often called, are derived theoretically by replacing the hydroxyl part of a carboxyl group by an amino group. Their functional group is the *amido group*,—$CONH_2$. The most important amides are acetamide $CH_3.CONH_2$, and urea NH_2CONH_2.

Nitriles, or *alkyl cyanides*, form a homologous series with the general formula $C_nH_{2n+1}.CN$. The isomeric *isonitriles*, known also as *carbylamines* or *alkyl isocyanides*, are of the type $C_nH_{2n+1}.NC$. The methyl members of both series are studied in this chapter.

Amino-acids are so called because they contain amino groups and carboxyl groups within the same molecule. They are represented here by their simplest members, glycine $NH_2.CH_2.COOH$ and alanine.

Polypeptides and *proteins* are nitrogen compounds of immense biochemical and biological importance since they are essential to the structure and functioning of all plants and animals.

41.2. Preparation of Primary Amines. Several methods are available:

1. The best is *Hofmann's degradation reaction*, so called because an amide is converted into an amine with one less carbon atom. The reaction therefore affords a means of descending one step of a homologous series (see § 41.13). To prepare methylamine, acetamide is first treated with bromine when substitution occurs thus:

$$CH_3.CONH_2 + Br_2 = CH_3.CONHBr + HBr.$$

If dilute potassium hydroxide solution is added until the colour is yellow and the product is heated to 70°C with concentrated alkali, the bromo derivative gives up hydrogen bromide and undergoes an internal rearrangement to methyl isocyanate:

$$CH_3.CONHBr + KOH = CH_3.NCO + KBr + H_2O$$

The isocyanate is then hydrolysed by the hot alkali to methylamine:

$$CH_3.NCO + 2KOH = CH_3.NH_2\uparrow + K_2CO_3$$

The overall equation for the reaction is therefore:

$$CH_3.CONH_2 + Br_2 + 4KOH = CH_3NH_2\uparrow + 2KBr + K_2CO_3 + 2H_2O$$

2. Methylamine can also be prepared by the action of nascent hydrogen on hydrogen cyanide:

$$HCN + 4\bar{H} = CH_3NH_2$$

The best reducing agents are sodium and absolute alcohol or lithium aluminium hydride (§ 19.13), since there is then no chance of hydrolysing the cyanide at the same time.

3. Methylamine is obtained as its hydroiodide salt when methyl iodide is heated to 100°C in a sealed tube with a large excess of an alcoholic solution of ammonia:

$$CH_3I + NH_3 = CH_3NH_3I$$

This method is not a good one because further reaction usually occurs giving secondary and tertiary amines as well.

4. Methylamine can be made by reducing nitromethane with zinc and hydrochloric acid or by passing its vapour mixed with hydrogen over finely-divided nickel at 150°C:

$$CH_3NO_2 + 6\bar{H} = CH_3NH_2 + 2H_2O$$

This method is unimportant because nitroparaffins are not easily prepared. This should be contrasted with the situation in aromatic chemistry, where the reduction of a nitro compound is the normal way of preparing a primary amine (§ 44.5).

Methylamine prepared in any of these ways, all of which are general methods, can be dried with quicklime or potassium hydroxide (it reacts with calcium chloride and with sulphuric acid) and collected by downward delivery. Alternatively it can be dissolved in water, from which it is expelled on boiling, or it can be absorbed in hydrochloric acid as its hydrochloride, from which it is liberated by excess of alkali.

41.3. Properties of Primary Amines. Methylamine (aminomethane), b.p. −7°C, and ethylamine (aminoethane), b.p. 17°C, are colourless substances smelling like ammonia. They are very soluble in water and in organic solvents. Unlike ammonia they burn readily in air with a blue flame.

41.4. Reactions of Primary Amines. Since they are very similar chemically, only methylamine will be considered:

1. As a Base: It is weakly basic giving an alkaline solution which turns litmus blue and which precipitates insoluble metallic hydroxides. This basic character arises from the lone pair of electrons on the nitrogen atom which enables it to accept a proton and form an ion resembling the ammonium ion:

$$CH_3NH_2 + H_2O \rightleftharpoons CH_3NH_3OH \rightleftharpoons CH_3NH_3^+ + OH^-$$

(Compare with $NH_3 + H_2O \rightleftharpoons NH_4OH \rightleftharpoons NH_4^+ + OH^-$)

Methylamine readily forms salts with strong acids, e.g. methylamine hydrochloride, CH_3NH_3Cl. These salts are colourless crystalline compounds with electrovalent structures, which may be emphasized by writing their formulae as $[CH_3NH_3]^+Cl^-$ and $[CH_3NH_3]^+ HSO_4^-$. They dissolve readily in water and decompose when heated with caustic alkalis, liberating the amine:

$$CH_3NH_3Cl + NaOH = CH_3NH_2 + NaCl + H_2O$$

Amines combine with certain complex acids forming sparingly soluble salts, e.g. methylamine combines with chloroplatinic acid giving $(CH_3NH_3)_2PtCl_6$. These salts leave a residue of pure metal when ignited and can be used to determine the molecular weight of the amine (see § 32.10).

2. Isocyanide Reaction: When methylamine (as its hydrochloride) is warmed with chloroform and alcoholic potassium hydroxide solution, the characteristic foul smell of an isocyanide soon becomes apparent (see reaction 1 of § 40.8):

$$CH_3NH_2 + CHCl_3 + 3KOH = CH_3NC + 3KCl + 3H_2O$$

3. Acetylation: Acetyl chloride and acetic anhydride acetylate methylamine readily in the cold:

e.g. $$CH_3NH_2 + CH_3COCl = \underset{\text{methyl acetamide}}{CH_3NH.COCH_3} + HCl\uparrow$$

4. Benzoylation: Benzoyl chloride combines with methylamine in the presence of caustic soda solution thus:

$$CH_3NH_2 + C_6H_5.COCl = CH_3NH.COC_6H_5 + HCl$$

5. With Alkyl Halides: When heated together in a sealed tube an alcoholic solution of methylamine reacts with methyl iodide, for example, giving the following products:

$$CH_3NH_2 + CH_3I = \underset{\text{dimethylamine hydroiodide}}{(CH_3)_2NH_2I}$$

$$(CH_3)_2NH_2I + CH_3I = HI + \underset{\text{trimethylamine hydroiodide}}{(CH_3)_3NHI}$$

$$(CH_3)_3NHI + CH_3I = HI + \underset{\text{tetramethyl ammonium iodide}}{(CH_3)_4NI}$$

The process of introducing alkyl radicals in this way is known as *alkylation.*

6. With Nitrous Acid: When methylamine hydrochloride is treated with nitrous acid (i.e. an acidified solution of sodium nitrite) dimethyl ether and nitrogen are produced:

$$2CH_3NH_2 + 2HNO_2 = CH_3.O.CH_3 + 2N_2\uparrow + 3H_2O$$

The higher primary amines behave differently, giving alcohols and nitrogen:

e.g. $$C_2H_5NH_2 + HNO_2 = C_2H_5OH + N_2\uparrow + H_2O$$

41.5. Secondary and Tertiary Amines. These are prepared by heating an alcoholic solution of an alkyl halide with a primary amine in a sealed tube. A mixture of amine salts is obtained which can be separated by adding alkali and fractionally distilling, or by purely chemical means.

Secondary and tertiary amines closely resemble primary amines in their physical properties, being colourless, highly inflammable, and soluble in water and having strong fishy smells. Like primary amines they are basic, giving alkaline solutions and forming salts with acids. Secondary amines can be acetylated, benzoylated, and alkylated, but they differ from primary amines in two important respects. They do not give the isocyanide reaction and they react differently with nitrous acid, forming yellow oils known as nitrosoamines, thus:

e.g. $$(C_2H_5)_2NH + HNO_2 = \underset{\text{diethyl nitrosoamine}}{(C_2H_5)_2N.NO} + H_2O$$

Tertiary amines are much less reactive. They do not give the isocyanide reaction, nor do they react with nitrous acid. They can be neither acetylated nor benzoylated, since they contain no active hydrogen atom which can be replaced. Almost their sole reaction, other than salt formation, is alkylation to quaternary ammonium salts as described in the next section.

41.6. Quaternary Ammonium Compounds. These are prepared by mixing an ethereal solution of a tertiary amine with excess of the appropriate alkyl halide:

e.g. $$(C_2H_5)_3N + C_2H_5I = \underset{\text{tetraethyl ammonium iodide}}{(C_2H_5)_4NI}$$

They are colourless crystalline compounds, readily soluble in water, with no smell. They can be distinguished from ammonium salts, which they closely resemble structurally, because they do not evolve ammonia when heated with alkalis. Heated with an aqueous suspension of silver oxide they are converted into quaternary ammonium hydroxides, which are alkalis comparable to caustic soda and caustic potash in strength:

$$2(CH_3)_4NI + Ag_2O + H_2O = 2(CH_3)_4NOH + 2AgI\downarrow$$
$$(CH_3)_4NOH \rightleftharpoons (CH_3)_4N^+ + OH^-$$

41.7. Detection and Uses of Amines. Amines may be recognized by their ability to turn litmus solution blue and to form salts with acids, and by their combustion in air. Primary amines are distinguished by the isocyanide reaction, which only they give. When treated with nitrous acid (sodium nitrite solution and dilute hydrochloric acid) secondary amines give a yellow oil. They can also be acetylated to crystalline solids, whereas tertiary amines show no reaction with either reagent.

The lower amines are used for manufacturing detergents, dyes, insecticides, drugs, weed-killers, and chemicals for vulcanizing rubber.

41.8. Acetamide, $CH_3.CONH_2$.

LABORATORY PREPARATION: 1. By heating ammonium acetate under reflux with excess of glacial acetic acid, when dehydration of the ammonium salt occurs:

$$CH_3.COONH_4 = CH_3.CONH_2 + H_2O$$

Without the acid the yield is poor because the ammonium salt dissociates thus:

$$CH_3.COONH_4 \rightleftharpoons CH_3.COOH + NH_3$$

In practice the ammonium acetate is often prepared *in situ* from ammonium carbonate and excess of glacial acetic acid.

An air condenser is used, as in Fig. 41.1a, because acetamide would

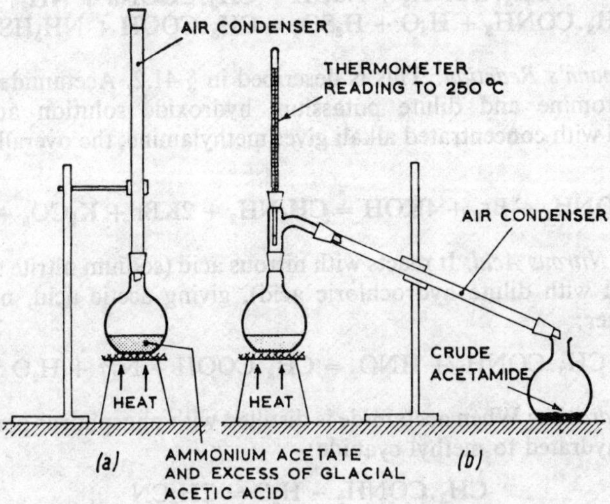

FIG. 41.1 Preparation of Acetamide

condense to a solid in a water condenser and block it. After about an hour the heating is stopped and the mixture is transferred to the apparatus shown in Fig. 41.1b. On heating the flask water and acetic acid distil over first, followed by impure acetamide between 190°C and 210°C. The product can be recrystallized from acetone.

2. Acetamide is also formed by the action of a concentrated solution

of ammonia on acetyl chloride, acetic anhydride, or ethyl acetate:

$$2NH_3 + CH_3COCl = CH_3.CONH_2 + NH_4Cl$$
$$2NH_3 + (CH_3CO)_2O = CH_3.CONH_2 + CH_3.COONH_4$$
$$NH_3 + CH_3.COOC_2H_5 = CH_3.CONH_2 + C_2H_5OH$$

Properties of Acetamide: It is a colourless crystalline solid melting at 82°C and boiling at 222°C. It dissolves in water giving a neutral solution. Pure acetamide is odourless, but commercial samples usually have a characteristic mousy smell caused by traces of methylacetamide impurity.

Reactions of Acetamide:

1. *Hydrolysis:* When boiled with caustic soda solution it gives sodium acetate and ammonia, whereas it is hydrolysed to acetic acid when heated with dilute acids:

$$CH_3.CONH_2 + NaOH = CH_3.COONa + NH_3\uparrow$$
$$CH_3.CONH_2 + H_2O + H_2SO_4 = CH_3.COOH + NH_4HSO_4$$

2. *Hofmann's Reaction:* This is described in § 41.2. Acetamide treated with bromine and dilute potassium hydroxide solution and then warmed with concentrated alkali gives methylamine, the overall change being:

$$CH_3.CONH_2 + Br_2 + 4KOH = CH_3NH_2 + 2KBr + K_2CO_3 + 2H_2O$$

3. *With Nitrous Acid:* It reacts with nitrous acid (sodium nitrite solution acidified with dilute hydrochloric acid), giving acetic acid, nitrogen, and water:

$$CH_3.CONH_2 + HNO_2 = CH_3.COOH + N_2\uparrow + H_2O$$

4. *Dehydration:* When acetamide is distilled with phosphorus pentoxide, it is dehydrated to methyl cyanide:

$$CH_3.CONH_2 - H_2O = CH_3CN$$

Structure of Acetamide: The molecular formula is C_2H_5ON. Its reaction with nitrous acid yielding acetic acid, which is known to have the structure $CH_3{-}C\begin{smallmatrix}{\diagup\!\!\!\diagup O}\\ {\diagdown OH}\end{smallmatrix}$ suggests the formula $CH_3{-}C\begin{smallmatrix}{\diagup\!\!\!\diagup O}\\ {\diagdown NH_2}\end{smallmatrix}$, since the usual action of nitrous acid is to replace $—NH_2$ by $—OH$. The preparation of acetamide by acetylation of ammonia and the following relationship to ammonium salts and cyanides are further evidence in favour of this structure:

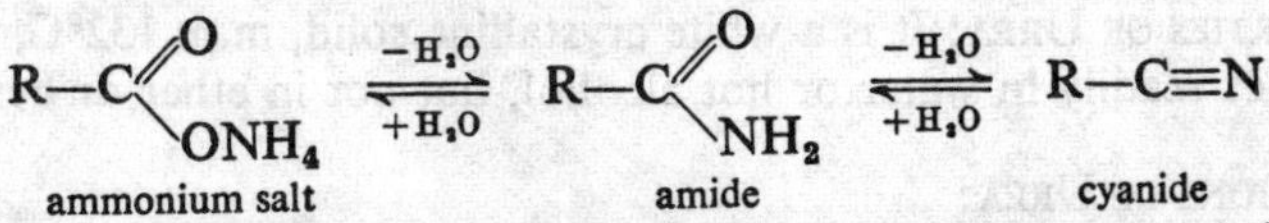

The melting point and boiling point of acetamide are much higher, however, than might be expected from its molecular weight, indicating association by hydrogen bonding. Moreover the carbonyl and amino groups do not behave in their characteristic ways. For example, acetamide shows none of the addition reactions expected of the $>C=O$ group, nor does it give the isocyanide reaction as all primary amines do. This implies that acetamide is another case of resonance, which is confirmed by spectroscopic evidence.

DETECTION OF ACETAMIDE: It is principally recognized by the liberation of ammonia when it is boiled with alkalis and by the evolution of nitrogen with nitrous acid. Acetamide is distinguished from ammonium acetate by not giving off ammonia when treated with caustic soda solution in the cold and by giving no red colour with neutral ferric chloride solution.

41.9. Urea, $NH_2.CO.NH_2$. This is the diamide of carbonic acid, H_2CO_3, which is why it is sometimes called *carbamide* (originally carbonamide). It is formed by some animals as a waste product during metabolism and is eliminated from their bodies in urine (human adults excrete about 30 gramme of urea per day in this way). It has a resonance structure (see § 14.11).

LABORATORY PREPARATION: By evaporating to dryness a solution containing ammonium sulphate and potassium cyanate in equimolecular proportions. Ammonium cyanate is first formed by double decomposition and then undergoes an isomeric change into urea (see § 20.12):

$$(NH_4)_2SO_4 + 2KNCO \rightleftharpoons 2NH_4NCO + K_2SO_4$$
$$NH_4NCO \rightleftharpoons NH_2.CO.NH_2$$

Although theoretically reversible, in practice these changes proceed continuously to the right because urea crystallizes out as the solution is evaporated and is removed from the equilibrium. The solid residue consists, therefore, of urea and potassium sulphate only. The urea is extracted from this mixture with hot absolute alcohol (in which potassium sulphate is insoluble) and is deposited from the alcoholic solution on cooling.

MANUFACTURE: By mixing ammonia and carbon dioxide and heating the product to 150°C under about 1.5×10^7 N m^{-2} (150 atm) pressure:

$$2NH_3 + CO_2 = NH_2.CO.NH_2 + H_2O$$

Properties of Urea: It is a white crystalline solid, m.p. 132°C, which dissolves readily in water or hot alcohol, but not in ether or benzene.

Reactions of Urea:

1. Hydrolysis: It is hydrolysed when boiled with acids or alkalis:

$$NH_2.CO.NH_2 + 2NaOH = 2NH_3\uparrow + Na_2CO_3$$

When treated with *urease*, an enzyme found in soya-beans and jack beans, urea is converted into ammonium carbonate:

$$NH_2.CO.NH_2 + 2H_2O = (NH_4)_2CO_3$$

Since this reaction is quantitative it can be used for estimating urea in solution.

2. Action of Heat: When heated urea melts to a colourless liquid which decomposes on further heating into ammonia and *biuret:*

$$2NH_2.CO.NH_2 = NH_3\uparrow + NH_2.CO.NH.CO.NH_2$$

If the residue is cooled and dissolved in dilute sodium hydroxide solution it gives a violet-pink colour with one drop of very dilute (1%) copper sulphate solution in what is known as the *biuret reaction.*

3. With Nitrous Acid: Urea is converted into nitrogen, carbon dioxide, and water when treated with acidified sodium nitrite solution:

$$NH_2.CO.NH_2 + 2HNO_2 = 2N_2\uparrow + CO_2\uparrow + 3H_2O$$

4. With Strong Acids: Although urea gives a solution which is neutral to litmus, it forms urea nitrate $NH_2.CO.NH_2.HNO_3$ with nitric acid and urea oxalate $2NH_2.CO.NH_2.(COOH)_2.2H_2O$ with oxalic acid. These salts are only sparingly soluble in water and provided concentrated solutions are used they appear as white crystalline precipitates.

5. With Hypohalites: If urea is treated with excess of sodium hypochlorite solution (or sodium hypobromite solution) it is decomposed thus:

$$NH_2.CO.NH_2 + 3NaOCl = CO_2\uparrow + N_2\uparrow + 3NaCl + 2H_2O$$

This reaction has been used for estimating urea (in urine, for example), but the yield of nitrogen is, in fact, slightly less than would be expected from the equation and the method is only an approximate one.

Uses of Urea: It is used extensively as a fertilizer, presumably because it undergoes slow hydrolysis by urease in the soil, and as a constituent of cattle foods. The biggest use of urea is to make urea–formaldehyde resins, which are needed for plastic mouldings, and for imparting crease resistance to textiles, and which are valuable adhesives used in plywood and laminated plastics. Various important sleep-inducing drugs (e.g. phenobarbitone) are made from urea.

DETECTION OF UREA: It is recognized by the biuret reaction (which is given by any substance containing two adjacent peptide linkages, —CO.NH—), and by its reactions with alkalis, nitrous acid, and hypohalites. A specific test for urea is to grind some soya-beans with water and add the aqueous extract to a solution of urea containing a few drops of phenol red, when the course of the hydrolysis can be followed as the colour of the indicator slowly changes from yellow to red.

41.10. Acetonitrile, Methyl Cyanide, Cyanomethane, CH_3CN.

LABORATORY PREPARATION: 1. When acetamide is heated with phosphorus pentoxide it is dehydrated to acetonitrile, which distils over and condenses:

$$2CH_3.CONH_2 + P_4O_{10} = 2CH_3CN + 4HPO_3$$

2. Acetonitrile is also prepared by boiling methyl iodide with a solution of potassium cyanide in methyl alcohol:

$$CH_3I + KCN = CH_3CN + KI$$

PROPERTIES OF ACETONITRILE: It is a colourless, poisonous liquid, b.p. 82°C, ρ 0.78 g cm^{-3} with a pleasant smell. It disolves in water and in alcohol.

REACTIONS OF ACETONITRILE:

1. *Hydrolysis:* Nitriles are hydrolysed very slowly by boiling water and much more rapidly by acids or alkalis. Indeed nitriles are named from the carboxylic acid into which they are converted by hydrolysis. With hot concentrated hydrochloric or sulphuric acids acetonitrile gives acetic acid:

$$CH_3CN + HCl + 2H_2O = CH_3.COOH + NH_4Cl$$

When it is boiled with caustic soda solution, sodium acetate is formed and ammonia is evolved:

$$CH_3CN + NaOH + H_2O = CH_3.COONa + NH_3\uparrow$$

2. *Reduction:* When metallic sodium is added to a warm solution of acetonitrile in absolute alcohol, rapid reduction to ethylamine occurs:

$$CH_3CN + 4\overline{H} = CH_3.CH_2.NH_2$$

This reaction shows that the two carbon atoms in acetonitrile must be linked directly together. Thus its structural formula is $CH_3—C{\equiv}N$, the nascent hydrogen adding across the triple bond between the carbon and nitrogen atoms.

41.11. Methyl Isocyanide, CH_3NC.

LABORATORY PREPARATION: 1. By heating a primary amine with chloroform and alcoholic potash, as in the 'isocyanide reaction' (§ 40.8). In this case the amine hydrochloride is used because methylamine is a gas:

$$CH_3NH_3Cl + CHCl_3 + 4KOH = CH_3NC + 4KCl + 4H_2O$$

2. By boiling methyl iodide with silver cyanide:

$$CH_3I + AgNC = CH_3NC + AgI\downarrow$$

Some methyl cyanide is formed as well since the atom of silver is linked directly to both carbon and nitrogen, but the two isomers can be separated by fractional distillation.

PROPERTIES OF METHYL ISOCYANIDE: It is a colourless, very poisonous liquid, b.p. 60°C, only slightly soluble in water. All isocyanides have a very unpleasant sickly smell which can be detected even in low concentration.

REACTIONS OF METHYL ISOCYANIDE:

1. Hydrolysis: It is hydrolysed readily when boiled with acids, giving a methylamine salt and formic acid:

e.g. $$CH_3NC + HCl + 2H_2O = CH_3NH_3Cl + H{\cdot}COOH$$

Unlike cyanides, isocyanides are not hydrolysed by alkalis.

2. Reduction: Nascent hydrogen (from sodium and alcohol) reduces it to dimethylamine, a secondary amine:

$$CH_3NC + 4\bar{H} = CH_3.NH.CH_3$$

This reaction establishes that the nitrogen atom is linked directly to the methyl radical, giving the isocyanide the structure $CH_3—N\stackrel{\rightarrow}{=}C$.

41.12. Ascent and Descent of a Homologous Series. The reactions described in this chapter complete the sequences needed to go up or down a homologous series. The following chains of reactions show how ethyl alcohol could be prepared from its methyl and propyl homologues:

methyl alcohol ⟶ ethyl alcohol:

CH_3OH methyl alcohol

$\downarrow$ red $P + I_2$

CH_3I methyl iodide

$\downarrow$ heat | alc. KCN

CH_3CN methyl cyanide

$\downarrow$ Na/EtOH

$CH_3CH_2NH_2$ ethylamine

$\downarrow$ HNO_2

CH_3CH_2OH ethyl alcohol

n-propyl alcohol $\longrightarrow$ ethyl alcohol:

$C_2H_5CH_2OH$ propyl alcohol

$\downarrow$ heat | chromic acid

C_2H_5COOH propionic acid

$\downarrow$ $(NH_4)_2CO_3$

$C_2H_5COONH_4$ ammonium propionate

$\downarrow$ heat | $-H_2O$

$C_2H_5CONH_2$ propionamide

$\downarrow$ Br_2 + dil. KOH + conc. KOH

$C_2H_5NH_2$ ethylamine

$\downarrow$ HNO_2

C_2H_5OH ethyl alcohol

41.13. Amino-acids. These compounds each contain at least one amino group and one carboxyl group within the same molecule, with the result that they show the reactions of both amines and carboxylic acids. This is well illustrated by the simplest member glycine, $NH_2.CH_2.COOH$, the preparation and properties of which are considered in the next section.

Altogether more than eighty different amino-acids are known to occur naturally, although many of them are uncommon. Twenty-six of them are of special biochemical importance because they are the components from which are made all the complicated protein molecules which are essential to plant and animal life. Ten of the best known are listed in Table 41.1. All these are α-amino-acids with the general formula

$$\begin{array}{c} NH_2 \\ | \\ R\text{—}C\text{—}COOH \\ | \\ H \end{array}$$

TABLE 41.1. SOME IMPORTANT AMINO-ACIDS

Name	*Radical R*
Alanine	CH_3—
Valine	$(CH_3)_2CH$—
Leucine	$(CH_3)_2.CH.CH_2$—
Serine	$HO.CH_2$—
Aspartic acid	$HOOC.CH_2$—
Glutamic acid	$HOOC.CH_2.CH_2$—
Lysine	$H_2N.(CH_2)_3.CH_2$—
Cysteine	$HS.CH_2$—
Phenylalanine	$C_6H_5.CH_2$—
Tyrosine	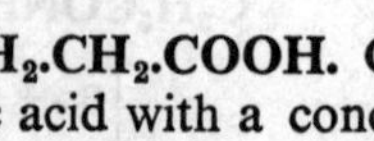

in which the amino and carboxyl groups are linked to the same carbon atom. They closely resemble glycine in physical and chemical properties, but since they contain an asymmetric carbon atom they are all optically active.

41.14. Glycine, Amino-acetic Acid, $NH_2.CH_2.COOH$. Glycine is best prepared by treating monochloroacetic acid with a concentrated solution of ammonia:

$$Cl.CH_2.COOH + 3NH_3 = NH_2.CH_2.COONH_4 + NH_4Cl$$

The copper derivative of glycine is formed by adding excess of copper carbonate, filtering, and crystallizing. The crystals are redissolved in water and treated with hydrogen sulphide to decompose the copper salt into glycine and a precipitate of copper sulphide. The latter is filtered off and the filtrate concentrated until colourless crystals of glycine separate on cooling.

$$2NH_2.CH_2.COONH_4 + CuCO_3$$
$$= (NH_2.CH_2.COO)_2Cu + 2NH_3\uparrow + H_2O + CO_2\uparrow$$

$$NH_2.CH_2.COO)_2Cu + H_2S$$
$$= CuS\downarrow + 2NH_2.CH_2.COOH$$

Glycine gives all the individual reactions of primary amines. For example, it forms salts with acids, it gives the isocyanide reaction, it can be acetylated and benzoylated, and it reacts with nitrous acid evolving nitrogen. It also behaves as a typical carboxylic acid, forming salts with alkalis and carbonates and forming esters with alcohols. Thus both functional groups in glycine are capable of acting independently in their characteristic ways.

The high melting point of glycine (235°C) and its solubility in water and insolubility in organic solvents are marks of a highly polar compound and suggest ionic character. It is believed that two forms exist in solution in equilibrium:

$$\underset{\text{I}}{NH_2.CH_2.COOH} \rightleftharpoons \underset{\text{II}}{{}^{+}NH_3.CH_2.COO^{-}}$$

Form II, which is believed to predominate, is produced by the migration by a proton from the carboxyl group to the amino group, forming what is known as a *zwitterion*. Thus glycine provides a good example of the phenomenon of *tautomerism* or dynamic equilibrium of two isomeric forms.

41.15. Alanine, α-Aminopropionic Acid, $NH_2.CH.CH_3COOH$.

LABORATORY PREPARATION: 1. By treating α-bromopropionic acid with excess of a concentrated solution of ammonia:

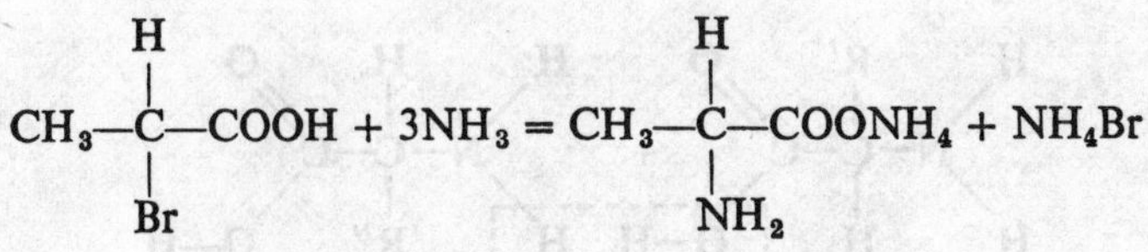

2. By treating acetaldehyde with a mixture of potassium cyanide and ammonium chloride solutions and hydrolysing the resulting α-aminocyanohydrin by boiling it with dilute sulphuric acid:

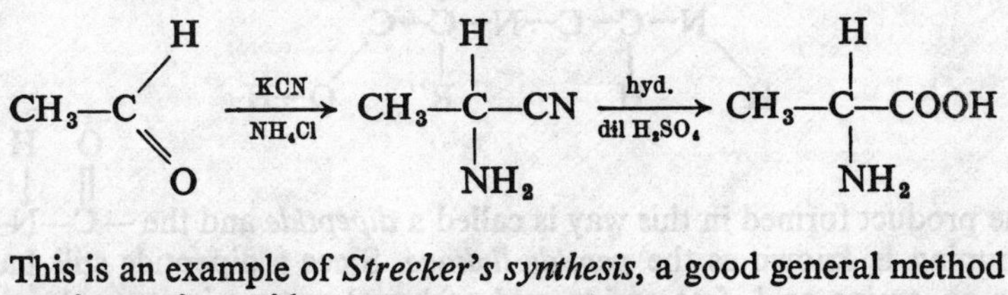

This is an example of *Strecker's synthesis*, a good general method of preparing amino-acids.

PROPERTIES AND REACTIONS OF ALANINE: It is a colourless crystalline solid, soluble in water but insoluble in organic solvents. Like glycine, it forms a zwitterion thus $^{+}NH_3.CH.CH_3.COO^{-}$.

The alanine molecule contains a primary amino group, $—NH_2$, and a carboxyl group, —COOH, and like glycine it undergoes all the characteristic reactions of both functional groups. With nitrous acid it gives lactic acid:

$$CH_3—\underset{NH_2}{\underset{|}{\overset{H}{\overset{|}{C}}}}—COOH + HNO_2 = CH_3—\underset{OH}{\underset{|}{\overset{H}{\overset{|}{C}}}}—COOH + N_2\uparrow + H_2O$$

Unlike glycine, alanine displays *optical isomerism* and can exist as two optical isomers which rotate the plane of polarized light. As explained in § 48.2 this is because alanine contains in its molecule an *asymmetric carbon atom*, i.e. one joined to four different atoms or groups. When these groups are arranged tetrahedrally around the central carbon atom, two alternative spatial configurations are possible, giving rise to two *enantiomorphs*.

41.16. The Peptide Linkage and Polypeptides. Two molecules of amino-acid can condense together with the elimination of a molecule of water from the amino group of one and the carboxyl group of the other thus:—

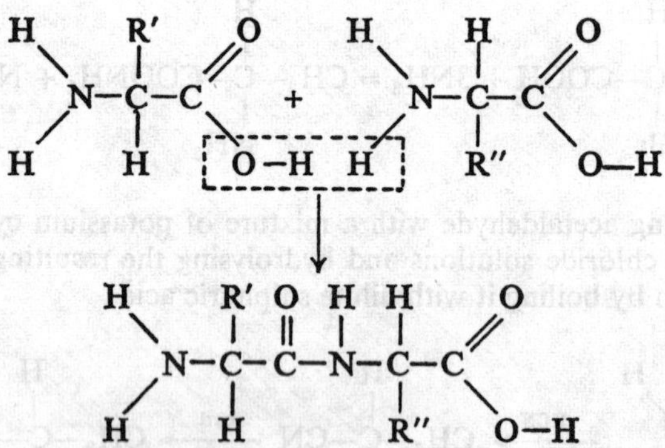

The product formed in this way is called a *dipeptide* and the $—\overset{O}{\overset{\|}{C}}—\overset{H}{\overset{|}{N}}—$ grouping is known as the *peptide linkage*. Since a dipeptide still has, like an amino-acid, free amino and carboxyl groups in its molecule, it is capable of forming further peptide linkages with other molecules of amino-acids until eventually a long chain molecule called a *polypeptide* is formed. Polypeptides may be built up from molecules of the same amino-acid or from different amino-acids; since many different sequences are possible, a great number of polypeptides can exist.

Polypeptides are white solids which decompose below their melting points. They are soluble in water and, like the amino-acids, are amphoteric, behaving as both acids and bases according to the pH of their environment. They all contain asymmetric carbon atoms and peptide linkages and consequently are optically active (§ 48.2) and give the biuret reaction (§ 41.9).

41.17. Proteins. These are larger molecules than polypeptides, although the borderline is ill-defined and arbitrary. Proteins, which may have

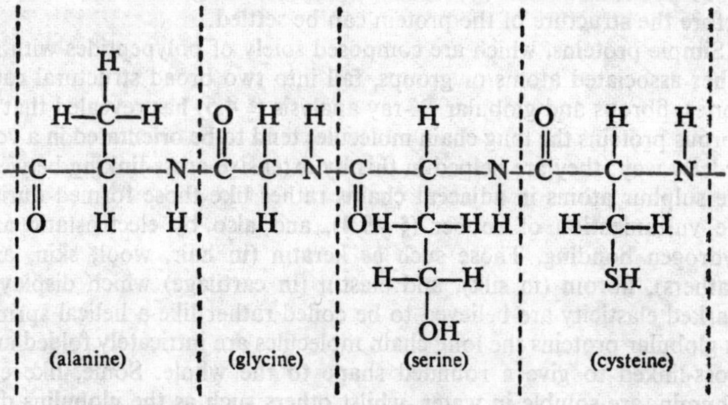

FIG. 41.2 A Fragment of a Protein Molecule

Although the molecule is depicted in this Figure as straight and in one plane, X-ray analysis (§ 4.5) reveals that in fact the valency bonds of each carbon atom are directed tetrahedrally and those of nitrogen trigonally (§ 14.8), so that the long chain molecule can adopt a particular stereochemical configuration by means of a series of folds or twists.

molecular weights of anything from a few thousand to several million, are composed of hundreds or even thousands of amino-acid molecules joined together by peptide linkages as in Fig. 41.2. When a protein is heated with mineral acids or alkalis it is hydrolysed i.e. a number of the peptide linkages are broken and a series of polypeptides or amino-acids are produced. It is not unusual for a given protein to yield between ten and twenty different amino-acids when fragmented in this way, revealing something of the complexity of its composition. Hydrolysis can also be effected by various enzymes, which are themselves proteins, and this is one of the fundamental changes which occurs during the process of digestion.

This hydrolysis of a protein into polypeptides and amino-acids affords valuable information about its structure provided that the resulting mixture can be analysed into its constituents. Since the amino-acids are so similar in reactions and properties, their separation and identification by chemical means presents great difficulties, and chromatography (§§ 12.11–12.13) or paper electrophoresis (§ 12.16) are normally used. Even when the resulting amino-acids have been identified and estimated quantitatively, their sequence in the molecular chain has to be determined and their stereochemical configuration worked out before the structure of the protein can be settled.

Simple proteins, which are composed solely of polypeptides with no other associated atoms or groups, fall into two broad structural categories, fibrous and globular. X-ray analysis (§ 4.5) has revealed that in fibrous proteins the long chain molecules tend to be orientated in a very orderly way; they are helped in this by extensive cross-linking between the sulphur atoms in adjacent chains rather like those formed during the vulcanization of rubber (§ 49.4), and also by electrostatic and hydrogen bonding. Those such as keratin (in hair, wool, skin, and feathers), fibroin (in silk), and elastin (in cartilage) which display a marked elasticity are believed to be coiled rather like a helical spring. In globular proteins the long chain molecules are intricately folded and cross-linked to give a rounded shape to the whole. Some, like egg albumin, are soluble in water, whilst others such as the globulins dissolve in salt solution or in dilute solutions of acids. Many of them, for example enzymes, hormones, blood proteins, and viruses, are of great physiological importance.

Although the proteins have very varied and complicated structures, all of them have approximately the same overall composition by weight of 48%–52% of carbon, 6%–7% of hydrogen, 24%–26% of oxygen, 16%–18% of nitrogen, and 0.2%–0.5% of sulphur. Some proteins contain other elements in the molecules as well; for example, haemoglobin in blood contains iron, and chlorophyll in plants contains magnesium. The soluble proteins are colloidal in character (§ 12.4) owing to their very high molecular weights. They are readily coagulated by gentle heating or by treatment with acids or alkalis. Like the amino-acids and polypeptides from which they are made, all proteins are amphoteric and optically active (§ 48.2). They give the biuret reaction i.e. a violet-pink colour appears when one drop of very dilute (1%) copper sulphate solution is added to a solution of the protein in sodium hydroxide solution. This reaction is not a specific test for proteins, however, as any compound with two adjacent peptide links will respond to it (§ 41.9).

41.18. Nucleic Acids. The nucleic acids are proteins of immense biological importance because they influence the process by which other proteins are formed and are believed to incorporate hereditary information in their structures—the so-called genetic code. Nucleic acids have very large molecules and the molecular weights of some of them exceed a hundred million. They consist of chains of different *nucleotides* linked together in a particular sequence. Each nucleotide comprises an acidic triphosphate group joined by a sugar ring to a cyclic base. The two sugars concerned are ribose in *ribonucleic acid*, commonly known as RNA, and deoxyribose in *deoxyribonucleic acid* or DNA. The bases commonly found in the nucleic acids are adenine, guanine, cytosine, thymine, and uracil. The various nucleotides condense together to form a chain molecule of nucleic acid as in Fig. 41.3.

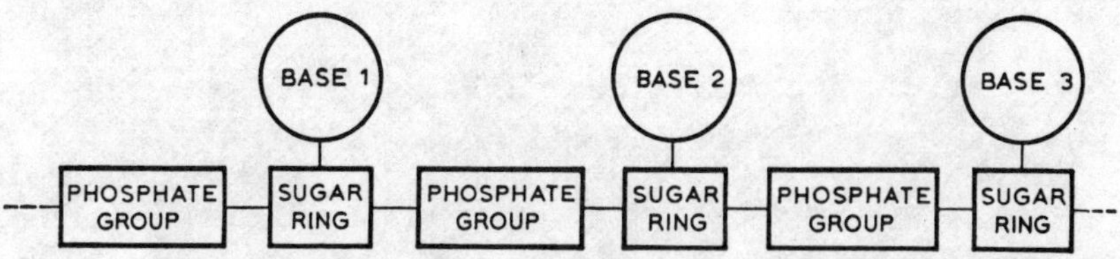

FIG. 41.3 Fragment of a Nucleic Acid Molecule

Recent work has demonstrated that DNA molecules consist of two chains of nucleotides twisted round each other in the form of a double helix or coil and held together by hydrogen bonding between pairs of adjacent bases inside the helix. Such a structure appears as a long thread under the electron microscope. Since only certain bases can pair with each other in this way, for example adenine pairs only with thymine and guanine with cytosine, the two chains must be complementary to each other, each reflecting the sequence of bases in the other. Thus if when cell division occurs the two identical chains in the DNA molecule uncoil and separate from each other, each one of them can act as a template in the new cells for the construction of a new strand and two exact replicas of the original DNA molecule will result. The sequence of bases in the DNA molecule then acts as a guide to the order in which the various amino-acids are linked together to form protein molecules in the new cell.

42

Carbohydrates

42.1. Introduction. Carbohydrates are naturally-occurring compounds containing carbon, hydrogen, and oxygen, the last two elements being present in the same proportions as in water. Thus all carbohydrates have the general molecular formula $C_xH_{2y}O_y$. Not all compounds with this formula are carbohydrates, however, e.g. formaldehyde and acetic acid are not.

Carbohydrates are of two main types. Either they are crystalline, soluble in water, and sweet in taste, when they are known collectively as *sugars* (e.g. glucose, fructose, sucrose), or they are non-crystalline, insoluble, tasteless substances of very high molecular weight (e.g. starch and cellulose). Sugars containing six or less carbon atoms in each molecule are known as *monosaccharides*, and those with twelve carbon atoms as *disaccharides*. Those carbohydrates with larger molecules are known as *polysaccharides*, reflecting the fact that their molecules are composed of numbers of sugar molecules linked together by oxygen atoms into chains.

42.2. Glucose, Dextrose, $C_6H_{12}O_6$.

OCCURRENCE: It is found in ripe grapes (hence the name grape-sugar) and many other fruits, and in honey, and also in the sap of plants and in the blood of animals.

LABORATORY PREPARATION: By warming cane sugar with a dilute acid:

$$\underset{\text{sucrose}}{C_{12}H_{22}O_{11}} + H_2O = \underset{\text{glucose}}{C_6H_{12}O_6} + \underset{\text{fructose}}{C_6H_{12}O_6}$$

If an excess of alcohol is added the fructose dissolves but the glucose is gradually deposited as crystals, particularly if the mixture is 'seeded' with a few crystals of glucose to prevent supersaturation.

MANUFACTURE: By heating starch with dilute hydrochloric acid under pressure:

$$(C_6H_{10}O_5)_n + nH_2O = nC_6H_{12}O_6$$

Properties of Glucose: It crystallizes from warm solutions as colourless anhydrous crystals, m.p. 146°C, and from cold solutions as a monohydrate, m.p. 86°C. Both forms are very soluble in water but only slightly soluble in alcohol. Its solution has a sweet taste.

Reactions of Glucose:

1. Dehydration: When heated strongly it decomposes and chars, leaving a residue of carbon. If glucose is warmed with concentrated sulphuric acid a vigorous reaction takes place and a black mass of sugar carbon is formed:

$$C_6H_{12}O_6 - 6H_2O = 6C$$

2. Oxidation: Glucose is a strong reducing agent. It reduces Fehling's solution to a red precipitate of cuprous oxide on boiling and Tollens's reagent (an ammoniacal solution of silver oxide) to a silver mirror. It is oxidized to gluconic acid and saccharic acid by bromine water and nitric acid respectively.

3. Acetylation: When glucose is heated with an excess of acetic anhydride in the presence of anhydrous zinc chloride as catalyst, penta-acetyl glucose is formed.

4. With Phenylhydrazine: At room temperature it forms a phenylhydrazone which remains in solution, but when warmed to 100°C with an excess of phenylhydrazine glucose is converted into phenyl glucosazone, which appears as a yellow precipitate.

5. With Hydrogen Cyanide: Glucose forms an addition compound, glucose cyanohydrin.

6. With Hydroxylamine: Glucose condenses forming an oxime.

7. With Alkalis: When boiled with caustic soda solution, glucose forms a yellow-brown liquid with a smell of caramel.

8. Fermentation: A solution of glucose is readily fermented by the enzyme zymase, present in yeast, giving ethyl alcohol and carbon dioxide (see § 35.2):

$$C_6H_{12}O_6 = 2C_2H_5OH + 2CO_2\uparrow$$

9. Reduction: Glucose is reduced by sodium amalgam and water to a hexahydric alcohol called sorbitol, and by concentrated hydriodic acid and red phosphorus at 100°C to iodo-hexane and even n-hexane.

Structure of Glucose: The molecular formula is $C_6H_{12}O_6$. Its reduction shows that the carbon atoms are linked in an unbranched chain and its acetylation that there are five hydroxyl groups in each molecule.

Its addition and condensation reactions indicate that a $>C=O$ group is present and its reducing properties suggest that this is part of an aldehyde group. These considerations lead to the following structure for glucose:

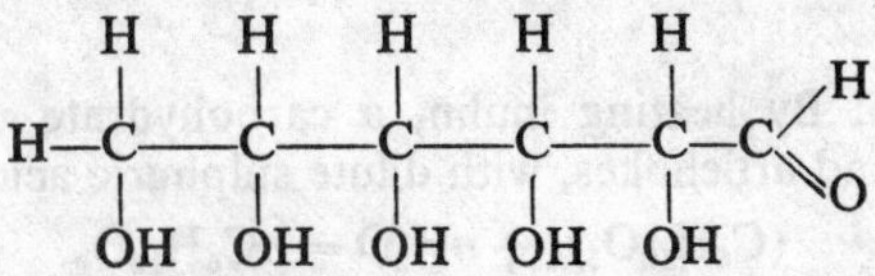

Such a structure contains four asymmetric carbon atoms (see § 48.2) and would be expected to form sixteen optically active isomers. All sixteen have in fact been made, although only four exist naturally.

X-ray analysis of glucose reveals, however, that it has a six-membered ring structure in the crystals. This accounts for the existence of two geometric isomers, α and β glucose, which are both dextro-rotatory but in differing degrees, having different numbers of OH groups on one side of the plane of the ring. It seems highly probable that in a solution of glucose both cyclic isomers are in dynamic equilibrium with each other and with a small proportion of the open-chain form, as in Fig. 42.1. This explains the phenomenon of *mutarotation* in which the

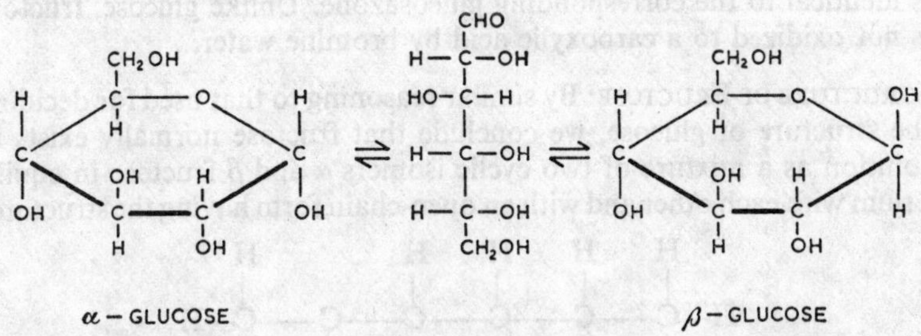

FIG. 42.1 The Equilibrium Forms of Glucose in Solution

specific rotations of freshly prepared solutions of α and β glucose change to a constant value of + 52.5° on standing as the above equilibrium is established.

42.3. Fructose, Laevulose, $C_6H_{12}O_6$.

OCCURRENCE: It is found, usually mixed with glucose, in fruit juices and in honey.

LABORATORY PREPARATION: By warming cane sugar with dilute sulphuric acid and adding slaked lime to the strongly cooled product. A fructosate of composition $C_6H_{12}O_6.CaO$ is precipitated, leaving the more soluble glucosate in solution. The precipitate is filtered off, washed, and suspended in water through which carbon dioxide is bubbled. The fructosate decomposes to a precipitate of calcium carbonate, which can be removed, and to a solution of fructose which can be concentrated and crystallized.

MANUFACTURE: By heating inulin, a carbohydrate which occurs in dahlia tubers and artichokes, with dilute sulphuric acid:

$$(C_6H_{10}O_5)_n + nH_2O = nC_6H_{12}O_6$$

PROPERTIES OF FRUCTOSE: It forms colourless crystals, m.p. 95°C, which dissolve readily in water giving a sweet solution which is laevo-rotatory.

REACTIONS OF FRUCTOSE: In many ways it resembles its isomer glucose. For example, it reduces Fehling's solution and Tollens's reagent and it forms a penta-acetyl derivative when acetylated. It also forms a cyano-hydrin with hydrogen cyanide and an oxime with hydroxylamine. Like glucose, fructose is reduced to sorbitol by sodium amalgam and water, dehydrated to carbon by concentrated sulphuric acid, and fermented to alcohol by the enzyme zymase in yeast. When fructose is heated with an excess of phenylhydrazine it forms phenyl fructosazone, which is identical to the corresponding glucosazone. Unlike glucose, fructose is not oxidized to a carboxylic acid by bromine water.

STRUCTURE OF FRUCTOSE: By similar reasoning to that used for deciding the structure of glucose, we conclude that fructose normally exists in solution as a mixture of two cyclic isomers α and β fructose in equilibrium with each other and with an open-chain form having the structure:

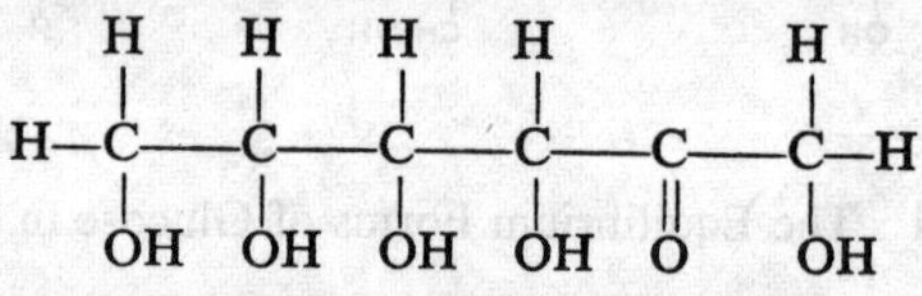

It is also believed to exist in solution in the form of five-membered rings known as γ fructose. The reducing properties of fructose are attributed to the hydroxylketo group:

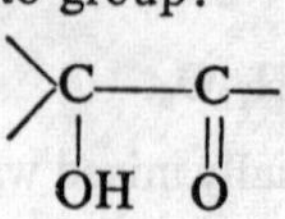

42.4. Sucrose, Cane Sugar, $C_{12}H_{22}O_{11}$.

OCCURRENCE: It occurs naturally in many plants and fruits, particularly in sugar canes and sugar beet which contain 15%–20% of sucrose and are its principal sources.

MANUFACTURE: The canes or beet are shredded and the sugar extracted with water. The solution is purified by treatment with slaked lime and carbon dioxide, the resulting precipitate of calcium carbonate adsorbing and carrying down most suspended impurities. The solution is then filtered and evaporated under reduced pressure until crystallization of sucrose occurs. When sugar canes are used the mother liquor, a dark syrupy liquid known as molasses, is used to make treacle or is fermented to alcohol (§ 35.2.). After crystallization the raw sugar is dissolved in water again and refined by treatment with more lime and carbon dioxide. It is decolorized with animal charcoal before the final crystallization.

PROPERTIES OF SUCROSE: It is a colourless crystalline solid, m.p. about 160°C, with a very sweet taste. It dissolves freely in water but is almost insoluble in alcohol.

REACTIONS OF SUCROSE:

1. Dehydration: When heated above its melting point it loses water and forms a brown substance, caramel, used for flavouring and colouring foods. Stronger heating causes charring. When it is warmed with concentrated sulphuric acid a vigorous reaction takes place leaving a black mass of sugar carbon:

$$C_{12}H_{22}O_{11} - 11H_2O = 12C$$

2. Oxidation: Sucrose is oxidized to oxalic acid when warmed with concentrated nitric acid (see § 37.11), but it is not a reducing sugar and it has no effect upon Fehling's solution or Tollens's reagent.

3. Acetylation: When boiled with acetic anhydride it yields an octa-acetyl derivative.

4. Hydrolysis: When sucrose is boiled with dilute hydrochloric or sulphuric acid, it is hydrolysed to a mixture of glucose and fructose:

$$C_{12}H_{22}O_{11} + H_2O = C_6H_{12}O_6 + C_6H_{12}O_6$$

This reaction is of special interest for several reasons. Sucrose is dextrorotatory (§ 48.2), and so is glucose, but fructose is even more strongly laevorotatory. Consequently as the hydrolysis proceeds the specific rotation decreases and eventually changes from dextro to laevo. For this reason the reaction is often referred to as the *inversion* of cane sugar and the resulting mixture of glucose and fructose is known as *invert sugar.* Moreover the inversion of cane sugar is catalysed by acids, the rate of reaction being directly proportional to the concentration of H^+ ions in the solution. This fact has been used as a means of measuring the pH of a solution. The inversion of cane sugar can also be effected by the enzyme invertase present in yeast.

5. *With Alkalis:* A concentrated solution of sucrose gives a precipitate of a sucrosate (e.g. $C_{12}H_{22}O_{11}.2SrO$) with calcium or strontium hydroxide. The sucrosate decomposes into sucrose again when carbon dioxide is bubbled into its suspension. Use is made of these reactions to obtain the highest possible yield of sucrose from molasses.

STRUCTURE OF SUCROSE: Its acetylation shows that each molecule contains eight hydroxyl groups. The hydrolysis of sucrose implies that its molecule contains a molecule of glucose linked to a molecule of fructose. The absence of reducing properties and its inability to form an oxime, cyanohydrin, or osazone, all point to the absence of a $>C{=}O$ group in sucrose, which suggests that the glucose and fructose molecules are linked through their respective aldehyde and keto groups. This is confirmed by X-ray analysis of the crystals.

42.5. Detection of Sugars. All carbohydrates are dehydrated to a black mass of carbon when heated with concentrated sulphuric acid.

Sucrose can easily be distinguished from glucose and fructose because it does not reduce Fehling's solution or Tollens's reagent, nor does it form an osazone or oxime. A positive test is to boil it with dilute acid, neutralize, and add Fehling's solution, when a red precipitate of cuprous oxide is obtained.

Glucose and fructose can be distinguished from each other in three ways. Firstly glucose is dextrorotatory and fructose laevorotatory. Secondly fructose forms the osazone much more quickly than glucose does under identical conditions. Thirdly the rapid furfural test can be applied in which the sugars are treated with an alcoholic solution of α-naphthol and concentrated hydrochloric acid. When the mixture is heated to boiling, fructose gives an immediate violet colour, whereas the colour only appears after one or two minutes boiling with glucose.

Other reducing sugars such as maltose and lactose can be distinguished from each other by determining the melting points and characteristic shapes of their osazone crystals.

42.6. Starch, $(C_6H_{10}O_5)_n$. This occurs naturally in most plants, but the principal sources are maize (corn), wheat, barley, potatoes, and rice. The plant cells are broken by crushing and the starch is extracted with water giving a pasty suspension from which the water can be removed by use of a centrifuge or by evaporation.

Starch is a white powder with no taste or smell. Its melting point is not known as it decomposes when heated. As normally prepared it does not dissolve in cold water. So called 'starch solution' is usually prepared by adding an aqueous suspension of starch to some boiling water and continuing to boil for a further minute. If the product is filtered and the fitrate treated with alcohol, a form of starch is precipitated which will dissolve in cold water. Because the molecules of starch are very large having a molecular weight of the order of a million, solutions of starch are inevitably colloidal.

Starch does not reduce Fehling's solution or Tollens's reagent, nor does it form an osazone. It reacts with iodine solution, however, giving a deep blue colour which disappears when heated and reappears on cooling. This is a sensitive test for starch or iodine. When starch is boiled with dilute acids, it is hydrolysed to glucose:

$$(C_6H_{10}O_5)_n + nH_2O = nC_6H_{12}O_6$$

Treatment with the enzyme diastase (also known as amylase) which is present in malt and also in saliva, converts starch into dextrin and maltose. These are the first stages of important processes occurring in brewing and in digestion.

Apart from its principal use as a food, starch is used industrially as a source of glucose and alcohol and as a stiffening agent and adhesive. The structure of starch is extraordinarily complex. It is believed to consist of huge branched molecules each containing thousands of glucose molecules linked together by oxygen atoms.

42.7. Cellulose, $(C_6H_{10}O_5)_n$. This material is abundant in the stems and roots of plants. Cotton, hemp, flax, sisal, jute, paper, and straw all consist mainly of cellulose; the principal industrial source is wood, about 50% of which is cellulose.

Celluose is a white solid, insoluble in water, alcohol, or ether. It is the raw material from which *rayon*, the most important of all man-made fibres is obtained. In the commonest process wood pulp is treated with sodium hydroxide solution and carbon disulphide to give sodium cellulose xanthate, which dissolves in caustic alkali giving viscose solution. When this is extruded through a fine nozzle into a bath of dilute sulphuric acid the cellulose comes out of solution and forms a continuous filament known as viscose rayon or artificial silk. World output of this fibre equals that of all the other synthetic fibres added together; it finds use not only in clothing (in linings, for example, where it provides a cheap substitute for cotton) but also for making tyre-cord, tufted carpets, and hose-pipes. If the viscose solution is extruded into the bath of acid through a fine slit, the cellulose forms a thin transparent sheet known as 'Cellophane' which, after treatment with glycerol to increase its flexibility, is widely used as a material for wrappings. Other forms of rayon are made by dissolving cellulose in Schweitzer's reagent (see § 28.9) or by treating it with glacial acetic acid and acetic anhydride (see § 37.2).

When cellulose is treated with 20 % solutions of alkalis it is altered in appearance; the process of 'mercerizing' cotton to make it lustrous and more easily dyed depends upon this effect. A mixture of concentrated nitric and sulphuric acids converts cellulose into various nitrate esters of which the best known is gun-cotton or cellulose trinitrate, used to make cordite and gelignite explosives.

43

Benzene and its Derivatives

43.1. Introduction to Aromatic Chemistry. As explained in § 32.3, organic chemistry is sub-divided into aliphatic, which is concerned with open-chain compounds, and aromatic, which is the chemistry of compounds containing six-membered ring structures such as benzene and toluene. This sub-division arises naturally from the distinctive properties and reactions which the ring structure confers upon compounds possessing it. After studying the chemistry of benzene and some of its derivatives we shall be in a position to consider these general characteristics of aromatic compounds in § 43.8.

TABLE 43.1. THE MAIN PRODUCTS FROM COAL TAR

The percentage composition is only given approximately because it varies with the type of coal and the carbonising temperature.

COAL TAR

- up to 170°C → Light Oil (5% by wt.)
  - → benzene, C_6H_6 (§43.3)
  - → toluene, $C_6H_5.CH_3$. (§ 46.2)
  - → xylenes, $C_6H_4.(CH_3)_2$
  - → pyridine, C_5H_5N
- 170—230°C → Middle Oil (10% by wt.)
  - → phenol, $C_6H_5.OH$ (§ 45.2)
  - → cresols, $CH_3.C_6H_4.OH$
  - → naphthalene, $C_{10}H_8$
- 230—270°C → Heavy Oil (10% by wt.) → creosote
- 270—400°C → Anthracene Oil (20% by wt.) → anthracene, $C_{14}H_{10}$
- residue → Pitch (55% by wt.)

43.2. Coal Tar. When coal is carbonized (§ 20.13), one of the principal products is a black viscous liquid called coal tar. It is a mixture of over two hundred compounds, many of them aromatic. After separation from the ammoniacal liquid which condenses with it, the coal tar is distilled into five main fractions as shown in Table 43.1. The light and middle oils are then further separated into various constituents as indicated in the table. Thus coal tar is a principal source not only of some of the most important aromatic compounds but also of a vast range of dyes, drugs, perfumes, explosives, antiseptics, plastics, resins, insecticides, weedkillers, synthetic fibres, solvents, etc., which are manufactured from them. To quote just one example, when naphthalene (from which the naphthols and hence many dyestuffs are made) is mixed with air and passed over a catalyst of vanadium pentoxide at 350°C, it is oxidized to phthalic anhydride, $C_8H_4O_3$. This compound is used for making alkyd resins which form the basis of many paints and varnishes and are largely responsible for their durability, ease of flow, and high gloss. Phthalic anhydride is also extensively used to make plasticizers (chemicals added to a plastic to give it the right consistency and the desired physical properties), and for the manufacture of dyestuffs.

43.3. Benzene, C_6H_6.

MANUFACTURE: Two methods are in use:

1. From Benzole and Coal Tar: Liquid benzole, recovered from crude coal gas and coke oven gas by washing with oil (see § 20.13) is the main source, supplemented by the light oil fraction of coal tar. These are shaken in turn with dilute sodium hydroxide solution (to remove acidic compounds such as phenol), water, cold 25% sulphuric acid (to remove basic compounds such as pyridine, aniline, and ammonia), and water again. The product is then dried and redistilled, giving a mixture of benzene and toluene from which benzene can be obtained by fractional distillation. Although satisfactory for many industrial uses, this product always contains some thiophene C_4H_4S, which cannot be separated by fractional distillation because its boiling point (84°C) is so similar to that of benzene. This impurity can be removed by repeated shaking with cold concentrated sulphuric acid. Very pure benzene, suitable for use in freezing-point depression experiments, separates as colourless crystals when this purified benzene is cooled in an ice-bath.

In Great Britain liquid benzole and coal tar provide all the benzene required, but in the U.S.A. only about half of the demand can be met from this source.

2. From Petroleum: If petrol vapour rich in aliphatic hydrocarbons is passed over a platinum catalyst at about 500°C and 1.5×10^6 N m^{-2}

(15 atm) pressure, many of the paraffins present are converted into ring molecules:

e.g.
$$\underset{\text{n-hexane}}{C_6H_{14}} = \underset{\text{benzene}}{C_6H_6} + 4H_2$$

These catalytic reforming processes were originally devised to improve the octane rating (§ 33.10) of the petrol by increasing its aromatic content. The U.S.A. now obtains about half of its benzene and more than three-quarters of its toluene in this way; the products are separated by fractional distillation and need little further purification.

LABORATORY PREPARATION: Owing to its ready availability commercially, benzene is rarely prepared in the laboratory, and the following methods are of theoretical interest only:

1. By heating sodium benzoate strongly with soda lime, when benzene vapour distils over and condenses:

$$C_6H_5COONa + NaOH = C_6H_6\uparrow + Na_2CO_3$$

This *decarboxylation* resembles the preparation of methane from sodium acetate (§ 33.2).

2. By passing phenol vapour over heated zinc dust:

$$C_6H_5OH + Zn = C_6H_6 + ZnO$$

3. Acetylene, heated to 400°C under pressure, polymerizes to benzene and other aromatic hydrocarbons:

$$3C_2H_2 = C_6H_6$$

PROPERTIES OF BENZENE: It is a colourless poisonous liquid, b.p. 80°C, m.p. 5.5°C, ρ 0.88 g cm^{-3}, with a typical 'aromatic' smell. It is immiscible with water forming two distinct layers, but it mixes with alcohol and ether in all proportions. It burns with a smoky, luminous flame.

REACTIONS OF BENZENE: These are summarized in Table 43.2.

1. Substitution by Halogens: When chlorine is bubbled into benzene at room temperature in diffused light in the presence of a particular kind of catalyst known as a halogen carrier (e.g. iron filings or iodine), monochlorobenzene is formed:

$$C_6H_6 + Cl_2 = C_6H_5Cl + HCl$$

Prolonged chlorination under these conditions causes further substitution, forming dichlorobenzene $C_6H_4Cl_2$, trichlorobenzene $C_6H_3Cl_3$, etc. Bromine behaves similarly, forming bromobenzenes, but iodine is not readily substituted into benzene in this way.

2. Substitution by Sulphuric Acid: No reaction occurs with concentrated sulphuric acid at room temperature, but benzene sulphonic acid

TABLE 43.2. REACTIONS OF BENZENE

Starting material	Reagent / conditions	Product	Type
BENZENE	chlorine (halogen carrier present)	chlorobenzene, C_6H_5Cl	SUBSTITUTIONS
	bromine (halogen carrier present)	bromobenzene, C_6H_5Br	
	hot conc. sulphuric acid or cold fuming sulphuric acid	benzene sulphonic acid, $C_6H_5SO_3H$	
	conc. nitric and sulphuric acids	nitrobenzene, $C_6H_5NO_2$	
	methyl chloride (Al_2Cl_6 catalyst) (Friedel-Craft reaction)	toluene, $C_6H_5.CH_3$	
	hydrogen (Ni catalyst at 200°C)	cyclohexane, C_6H_{12}	ADDITIONS
	chlorine in intense u.v. light	benzene hexachloride, $C_6H_6Cl_6$	
	bromine in intense u.v. light	benzene hexabromide, $C_6H_6Br_6$	
	ozone	benzene triozonide, $C_6H_6O_9$	

is formed when the two compounds are refluxed together for about twenty hours:

$$C_6H_6 + H_2SO_4 = C_6H_5.SO_3H + H_2O$$

If fuming sulphuric acid is used, *sulphonation* as this process of substituting an —SO_3H group in place of an H atom is called, takes place rapidly at room temperature (§ 43.7).

3. Substitution by Nitric Acid: Benzene hardly reacts with concentrated nitric acid alone, even on heating, but substitution of a nitro group, —NO_2, in place of an H atom (known as *nitration*) occurs rapidly when benzene is mixed with concentrated nitric acid and concentrated sulphuric acid:

$$C_6H_6 + HNO_3 = C_6H_5NO_2 + H_2O$$

This reaction is discussed in detail in § 44.2. Prolonged treatment with the hot nitrating mixture gives dinitrobenzene $C_6H_4(NO_2)_2$. Trinitrobenzene $C_6H_3(NO_2)_3$, is eventually formed if benzene is heated with a mixture of fuming nitric acid and concentrated sulphuric acid.

4. The Friedel-Craft Reaction: When benzene is treated with an olefine or an alkyl chloride in the presence of anhydrous aluminium chloride catalyst, a higher homologue is formed (see § 46.2):

e.g.
$$C_6H_6 + C_2H_5Cl = \underset{\text{ethylbenzene}}{C_6H_5.C_2H_5} + HCl$$

$$C_6H_6 + C_2H_4 = C_6H_5.C_2H_5$$

5. Addition of Hydrogen: If benzene vapour is mixed with hydrogen and passed over a catalyst of finely-divided nickel at 200°C, cyclohexane is formed:

$$C_6H_6 + 3H_2 = C_6H_{12}$$

Cyclohexane is a colourless liquid, b.p. 81°C, which is used as a solvent and for making nylon.

6. Addition of Halogens: When chlorine is bubbled into boiling benzene or benzene exposed to strong ultra-violet light, benzene hexachloride is formed:

$$C_6H_6 + 3Cl_2 = C_6H_6Cl_6$$

This compound, which is a white solid, exists in a number of isomeric forms. One of these, known as the γ form, is a powerful insecticide ('Gammexane' or 'BHC').

Bromine adds under similar conditions to give benzene hexabromide, but iodine does not react in this way.

7. Addition of Ozone: When ozonized oxygen is bubbled into benzene an unstable triozonide is formed:

$$C_6H_6 + 3O_3 = C_6H_6(O_3)_3$$

USES OF BENZENE: Its biggest use is to make ethylbenzene by reaction (4) above. The product is then dehydrogenated by passing it at 600°C over a catalyst composed of metallic oxides:

$$C_6H_5.CH_2.CH_3 = \underset{\text{styrene}}{C_6H_5.CH{=}CH_2} + H_2$$

Styrene is required in huge quantities for two purposes; either it is mixed with butadiene (§ 34.5) and co-polymerized into synthetic rubber (§ 49.5) or it is converted with the help of a peroxide catalyst into the important thermoplastic *polystyrene* described in § 49.2. This polymerization may be represented in the following way:

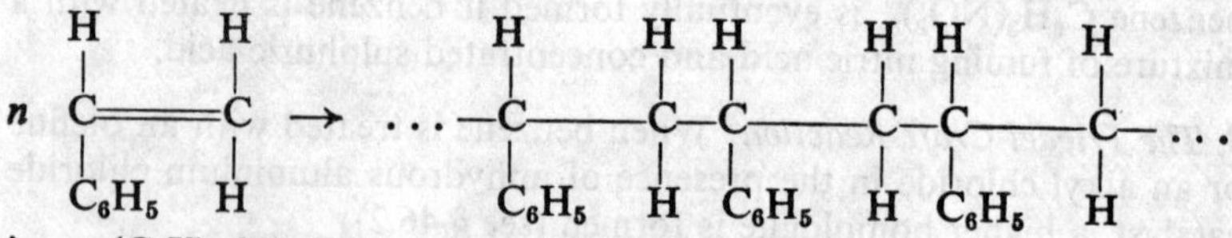

i.e. $n\ (C_6H_5.CH{:}CH_2) \rightarrow (C_6H_5.CH.CH_2)_n$

Large amounts of benzene are used in the manufacture of phenol, aniline, nylon, D.D.T., benzene hexachloride, various dyestuffs, and the range of alkyl aryl sulphonate detergents. Benzene is also a good solvent for oils, fats, resins, iodine, sulphur, and rubber.

STRUCTURE OF BENZENE: The molecular formula is C_6H_6. It has been established beyond any reasonable doubt that the six carbon atoms are arranged in a *ring* with one hydrogen atom linked to each. The chemical evidence for this rests firstly upon the demonstration that all six hydrogen atoms occupy equivalent positions in the molecule, and secondly upon the fact that three, and only three, different isomers can be obtained when any two of the hydrogen atoms are substituted by chlorine atoms. The physical evidence for the ring structure is overwhelming; X-ray analysis (§ 4.5) and electron diffraction of benzene and its derivatives shows the six carbon atoms to be at the corners of a regular hexagon.

The main controversy over the structure of benzene centres on the way in which the carbon atoms are linked together. The difficulty is to devise a structure incorporating tetravalent carbon which takes account of the very unusual chemical properties of benzene. The student is referred to more advanced books for an account of the many and diverse formulae which have been suggested. Although some of them are highly ingenious, none has found universal acceptance. Perhaps the most important, and the only one considered here, is that put forward by Kekulé in 1866 in which the carbon atoms are linked by alternate single and double bonds (a). This formula was rejected because experiment proved that only one derivative could be prepared when two

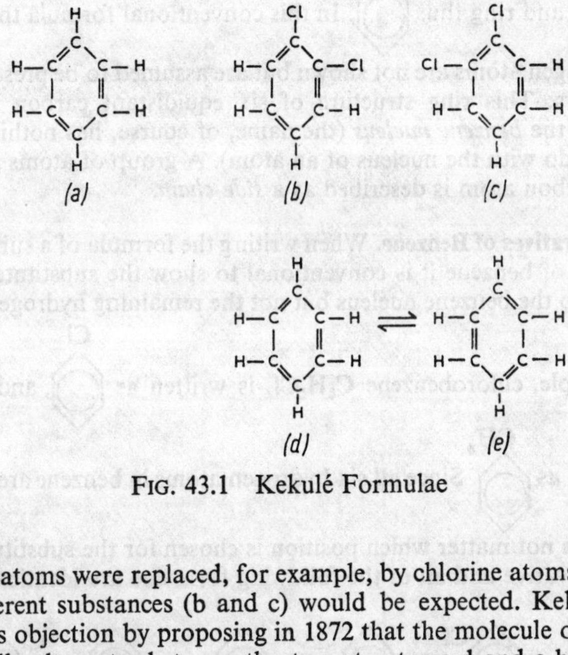

FIG. 43.1 Kekulé Formulae

adjacent atoms were replaced, for example, by chlorine atoms, whereas two different substances (b and c) would be expected. Kekulé overcame this objection by proposing in 1872 that the molecule of benzene continually alternates between the two structures d and e by a rapid oscillation of the double bonds, so that when two adjacent hydrogen atoms are substituted only one product is formed containing two alternative forms in dynamic equilibrium.

Kekulé's structures for benzene account for its addition reactions, but they fail to explain why benzene does not show the usual reactions of substances containing the ethylenic bond, $>C=C<$, e.g. decolorization of bromine water or potassium permanganate solution, or addition of halogen hydrides. Further important evidence against the Kekulé formulae is provided when crystals of benzene and its derivatives are analysed by X-rays; this shows that all adjacent carbon atoms are

0.139 nm apart, a distance intermediate between the 0.154 nm found in C—C linkages and the 0.134 nm measured for double bonds (see § 34.7).

Nowadays it is believed that benzene has a resonance structure which is a hybrid of the two Kekulé forms (see § 14.11). This accounts for its stability and for the absence of certain addition reactions normally associated with double bonds. Since there is no simple way of expressing this resonance structure we shall portray the benzene ring by a hexagon and ring thus . In this conventional formula the carbon and hydrogen atoms are not shown but are assumed to be present at the six corners. This ring structure of six equidistant carbon atoms is known as the *benzene nucleus* (the name, of course, has nothing whatsoever to do with the nucleus of an atom). A group of atoms linked to it by a carbon atom is described as a *side-chain.*

43.4. Derivatives of Benzene. When writing the formula of a substitution derivative of benzene it is conventional to show the substituted atoms attached to the benzene nucleus but not the remaining hydrogen atoms. For example, chlorobenzene C_6H_5Cl, is written as 

, and toluene $C_6H_5.CH_3$, as 

. Since all six hydrogen atoms in benzene are equivalent it does not matter which position is chosen for the substituent in a mono-derivative and all of the following formulae are identical:

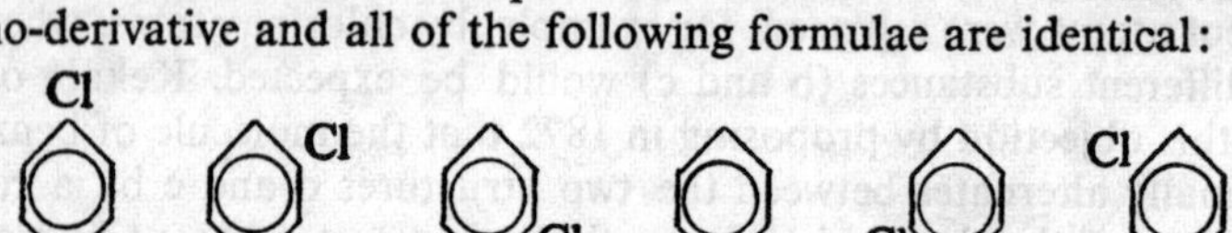

A different situation arises when di-derivatives are considered; a type of isomerism then occurs in which the isomers differ in the relative positions of the substituent atoms. Thus there are three isomeric dichlorobenzenes represented by the formulae:

ortho-dichlorobenzene

meta-dichlorobenzene

para-dichlorobenzene

Di-derivatives with two adjacent substituent atoms or groups are named *ortho-* (*o-*) to distinguish them from *meta-* (*m-*), in which the substituents are in the next-but-one position in the ring, and *para-* (*p-*) in which the two atoms or groups are diametrically opposite each other. These *position isomers*, as they are called, are very similar chemically, although they have different physical properties.

Three isomeric trichlorobenzenes are known. To indicate the position of the chlorine atoms in the ring the carbon atoms are numbered in a clockwise direction (with unity so chosen as to give the lowest possible numbers) thus:

1,2,3-trichlorobenzene

1,2,4-trichlorobenzene

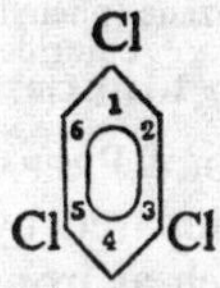

1,3,5-trichlorobenzene

All three isomers have the molecular formula $C_6H_3Cl_3$, but by convention the carbon and hydrogen atoms are omitted from these ring formulae.

C_6H_5— is called the *phenyl radical.* Radicals like it which offer direct linkage to a carbon atom in the benzene nucleus are known as *aryl radicals* to distinguish them from the alkyl radicals of the aliphatic series. Benzene and its homologues are known as *arenes.*

43.5. Orientating Influences. If we substitute a second atom or group of atoms into a given mono-substitution derivative of benzene, three isomeric compounds may be formed, denoted by ortho, meta, and para. Which of these products is obtained in any particular case is decided by the atom or group already substituted in the nucleus, and not by the nature of the group being introduced. Several rules exist for predicting the orientation (i.e. the relative positions of the substituted groups) of which *Vörlander's Rule* is the best:

If the group already substituted in the nucleus is unsaturated (*e.i. contains double or triple bonds*) *then the product will be predominantly the meta- isomer with only traces of ortho- and para-. On the other hand, if the existing substituent group is saturated, it will give a mixture of ortho- and para-isomers with very little meta-derivative.*

Thus the groups 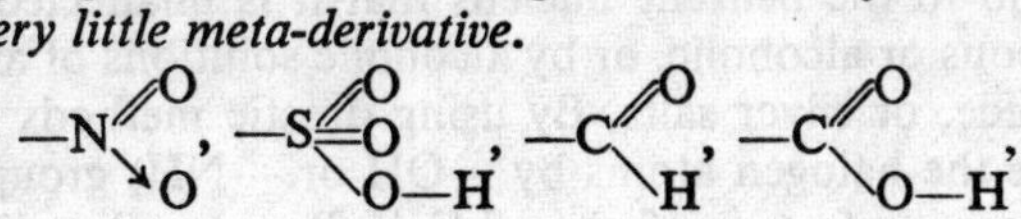 and —C≡N are meta-directing, whereas the halogens, —OH, 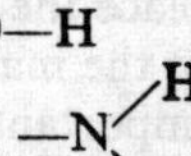,

$$-\mathrm{C}\begin{matrix}\mathrm{H}\\\mathrm{H}\\\mathrm{Cl}\end{matrix}$$

, and the alkyl groups are ortho- and para-directing. In general ortho- and para-directing groups facilitate further substitution and meta-directing groups retard it. For example, toluene, phenol, and aniline readily form substitution products, but nitrobenzene is only further nitrated with difficulty.

43.6. The Aryl Halides.

C_6H_5Cl	C_6H_5Br	C_6H_5I
chlorobenzene (phenyl chloride)	bromobenzene (phenyl bromide)	iodobenzene (phenyl iodide)
b.p. 132°C, ρ 1.11 g cm^{-3}	b.p. 156°C, ρ 1.50 g cm^{-3}	b.p. 188°C, ρ 1.84 g cm^{-3}

LABORATORY PREPARATION: 1. Chlorobenzene and bromobenzene are usually prepared by direct halogenation in the presence of a halogen carrier such as iron filings or pyridine. The reaction is exothermic and cooling may be necessary to prevent formation of di-derivatives. The experiment should be performed in ordinary diffused light to avoid addition reactions. When the theoretical increase in mass has occurred the product is poured into an excess of cold water and separated. It is then shaken in turn with sodium hydroxide solution and with water, dried with anhydrous calcium chloride, and fractionally distilled.

2. Iodobenzene is best prepared by treating a solution of benzene diazonium chloride with potassium iodide solution, as direct iodination of benzene is difficult to effect:

$$C_6H_5N_2Cl + KI = C_6H_5I + KCl + N_2\uparrow$$

Chlorobenzene and bromobenzene can also be prepared from the diazonium salt by adding its solution to cuprous chloride and concentrated hydrochloric acid or cuprous bromide and concentrated hydrobromic acid respectively, as described in § 44.6.

PROPERTIES AND REACTIONS: These are colourless, pleasant-smelling liquids, immiscible with water but soluble in organic solvents. The main feature of their chemical properties is their inertness. In marked contrast to the alkyl halides in which the halogen atom is readily replaced by other atoms or groups, in these compounds it is so firmly bound to the benzene nucleus that it is unaffected by boiling alkalis, aqueous or alcoholic, or by alcoholic solutions of ammonia, potassium cyanide, or silver salts. By using drastic methods it is possible to replace the halogen atoms by —OH or —NH_2 groups, and this is done in the manufacture of phenol (§ 45.2) and aniline (§ 44.5), but the only important reactions which can be carried out under laboratory conditions are:

1. Fittig's Reaction: When metallic sodium is added to a dry ethereal solution of chlorobenzene and methyl iodide, toluene is slowly produced:

$$C_6H_5Cl + CH_3I + 2Na = C_6H_5.CH_3 + NaCl + NaI$$

This is a general reaction, analogous to the Wurtz synthesis of paraffins.

2. Nuclear Substitution: These compounds can be nitrated, sulphonated, and further halogenated. In all these substitution reactions a mixture of ortho- and para-derivatives is obtained in accordance with the orientating influence of the halogen atom already present:

e.g.

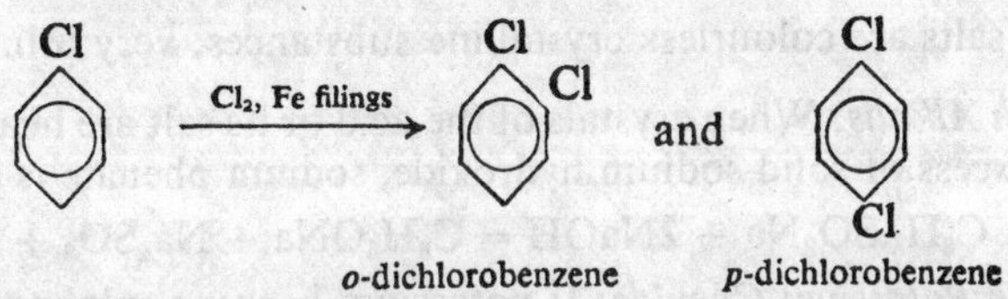

3. Grignard Reagents: When bromobenzene is added to a suspension of magnesium in dry ether, phenyl magnesium bromide is formed:

$$C_6H_5Br + Mg = C_6H_5.Mg.Br$$

This reagent can be used to synthesize many other organic compounds:

USES: Large amounts of chlorobenzene are used as an intermediate in the manufacture of phenol and aniline. It is also used to make D.D.T. (§ 36.4), and *p*-dichlorobenzene, which is a powerful moth-repellent. The more expensive bromo- and iodo-benzenes are little used industrially.

43.7. Benzene Sulphonic Acid, $C_6H_5SO_3H$.

LABORATORY PREPARATION: By boiling benzene under reflux with excess of concentrated sulphuric acid for about twenty hours:

$$C_6H_6 + H_2SO_4 = C_6H_5.SO_3H + H_2O$$

It is much better to use fuming sulphuric acid because sulphonation then proceeds rapidly in the cold and is complete in about fifteen minutes. On pouring the product into a saturated solution of sodium chloride, sodium benzene sulphonate is precipitated by the common ion effect (§ 9.7); the crystals can be filtered off, washed, and dried. The acid itself is very difficult to crystallize, but crystals can be obtained by acidifying a solution of the sodium salt with the calculated quantity of dilute sulphuric acid and evaporating the product in a vacuum desiccator.

PROPERTIES AND REACTIONS OF BENZENE SULPHONIC ACID: It forms colourless, very deliquescent crystals of formula $2(C_6H_5.SO_3H).3H_2O$.

It melts below 50°C; when heated more strongly it loses its water of crystallization and then decomposes. The acid is very soluble in water and alcohol. Its principal reactions are:

1. As an Acid: It is a strong monobasic acid, dissociating thus in solution:

$$C_6H_5.SO_3H \rightleftharpoons C_6H_5SO_3^- + H^+$$

It readily forms salts with solutions of alkalis and carbonates:

e.g. $$C_6H_5.SO_3H + NaOH = \underset{\text{sodium benzene sulphonate}}{C_6H_5.SO_3Na} + H_2O$$

These salts are colourless crystalline substances, very soluble in water.

2. With Alkalis: When crystals of the acid or its salt are heated to 300°C with excess of solid sodium hydroxide, sodium phenate is produced:

$$C_6H_5.SO_3Na + 2NaOH = C_6H_5ONa + Na_2SO_3 + H_2O$$

3. With Potassium Cyanide: If potassium benzene sulphonate is heated to 250°C with solid potassium cyanide, benzonitrile (phenyl cyanide) is formed:

$$C_6H_5.SO_3K + KCN = C_6H_5CN + K_2SO_3$$

4. With Phosphorus Pentachloride: The acid reacts with phosphorus pentachloride in the cold giving benzene sulphonyl chloride and fumes of hydrogen chloride:

$$C_6H_5.SO_3H + PCl_5 = C_6H_5.SO_2Cl + POCl_3 + HCl\uparrow$$

This confirms that the acid molecule contains an —OH group and has the structure $C_6H_5.SO_2.OH$.

5. Nuclear Substitution: Meta-derivatives are slowly formed when benzene sulphonic acid is halogenated, nitrated, or further sulphonated.

USES: The acid is used as an intermediate in the manufacture of phenol. Dyes are often sulphonated to make them water-soluble, and the sodium salts of certain alkylated benzene sulphonic acids are important as detergents.

43.8. Acetophenone, Methyl Phenyl Ketone, $CH_3.CO.C_6H_5$.

LABORATORY PREPARATION: By adding acetyl chloride to a mixture of anhydrous aluminium chloride and dry benzene:

$$CH_3.COCl + C_6H_6 = CH_3.CO.C_6H_5 + HCl$$

This is another example of the Friedel–Craft reaction. The product is poured into water, separated, dried, and purified from excess benzene by distillation.

PROPERTIES OF ACETOPHENONE: It is a colourless compound, m.p. 20°C, b.p. 202°C, ρ 1.03 g cm^{-3}. It dissolves readily in organic solvents but is only sparingly soluble in water.

REACTIONS OF ACETOPHENONE:

1. Oxidation: Mild oxidizing agents have no effect, but when it is boiled with chromic acid it is oxidized to benzoic acid:

$$C_6H_5.CO.CH_3 + 4\bar{O} = C_6H_5.COOH + CO_2 + H_2O$$

2. Reduction: A mild reducing agent such as sodium and alcohol converts it into a secondary alcohol:

$$C_6H_5.CO.CH_3 + 2\bar{H} = C_6H_5.CH(OH).CH_3$$

If acetophenone is boiled with amalgamated zinc and concentrated hydrochloric acid, it is reduced to ethylbenzene by Clemmensen's reaction:

$$C_6H_5.CO.CH_3 + 4\bar{H} = C_6H_5.CH_2.CH_3 + H_2O$$

3. Addition: Acetophenone undergoes addition reactions much less readily than acetone because the adjacent phenyl radical reduces the activity of the carbonyl group. It does not form a bisulphite addition compound and gives only a poor yield of the cyanohydrin:

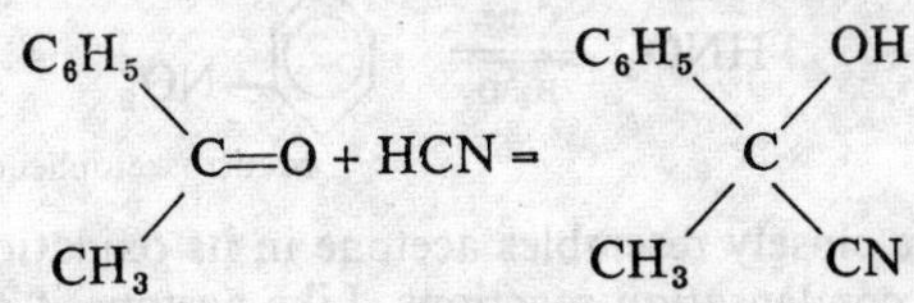

4. Condensation: Acetophenone readily forms condensation products with phenylhydrazine and hydroxylamine; the melting points of the crystalline precipitates can be used for identifying the original ketone:

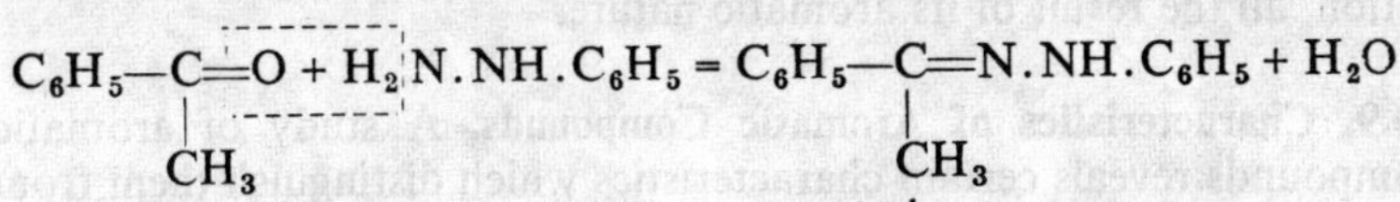

acetophenone phenylhydrazone m.p. 105°C

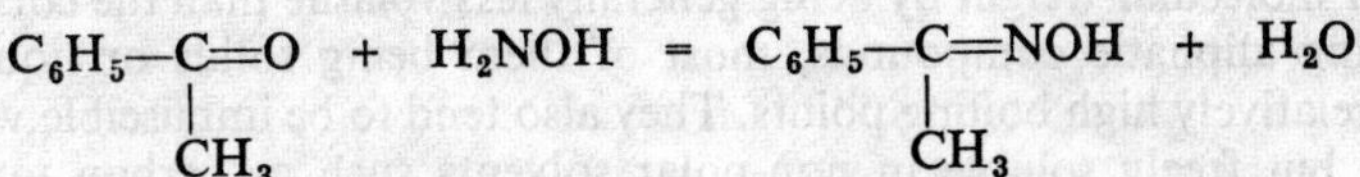

acetophenone oxime m.p. 59°C

When acetophenone is boiled with concentrated sulphuric acid it undergoes condensation with itself to form 1,3,5-triphenylbenzene; this reaction corresponds to the condensation of acetone to trimethylbenzene:

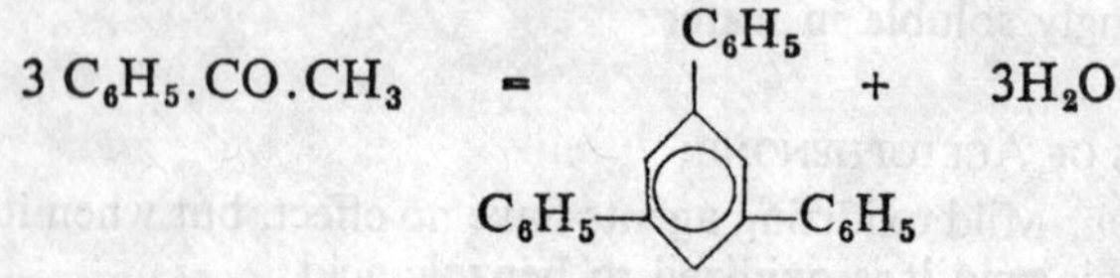

5. Chlorination: If chlorine is bubbled into warm acetophenone, the hydrogen atoms of the methyl group are successively replaced by chlorine atoms to give chloroacetophenones:

e.g. $C_6H_5.CO.CH_3 + Cl_2 = C_6H_5.CO.CH_2Cl + HCl$

6. Iodoform Reaction: A yellow precipitate of iodoform is slowly formed when acetophenone is shaken with potassium iodide solution and sodium hypochlorite solution.

7. Nuclear Substitutions: When treated with the appropriate reagents acetophenone can be nitrated, sulphonated, and halogenated by nuclear substitution. In each case the $CH_3.CO$— group is meta-directing (see § 43.5):

e.g.

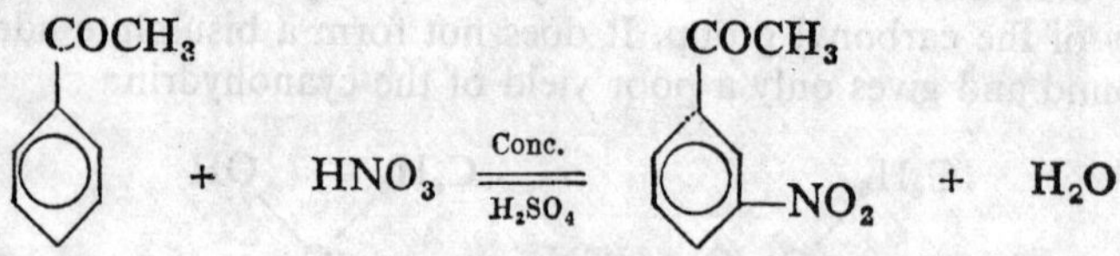

m-nitroacetophenone

Acetophenone closely resembles acetone in its oxidation and reduction, and in its condensation reactions. Like acetone it can be chlorinated and gives the iodoform reaction. It differs in its relative reluctance to take part in addition reactions, in the lower volatility and solubility in water of its derivatives, and in its ability to undergo nuclear substitution, all the result of its aromatic nature.

43.9. Characteristics of Aromatic Compounds. A study of aromatic compounds reveals certain characteristics which distinguish them from aliphatic compounds of the same type. For example, they reflect their higher molecular weight by being generally less volatile than the corresponding aliphatic compounds, most of them being solids or liquids with relatively high boiling points. They also tend to be immiscible with water but freely soluble in non-polar solvents such as carbon tetrachloride or benzene. Owing to their high carbon content they burn with a smoky, luminous flame.

The most striking feature of their chemical properties is the stability

Table 43.3 COMPARISONS OF FOUR C_6 HYDROCARBONS

	Benzene	*Hexane*	*Cyclohexane*	*Cyclohexene*
Molecular formula	C_6H_6	C_6H_{14}	C_6H_{11}	C_6H_{10}
Structural formula (H atoms not shown)	resonance hybrid of 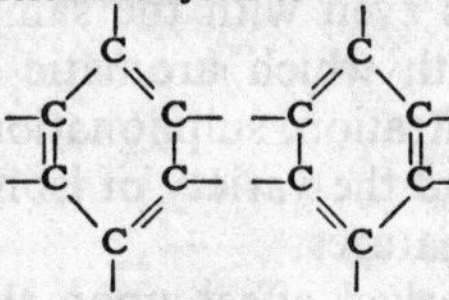	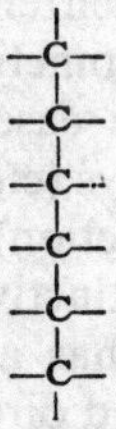	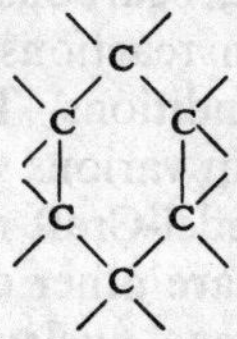	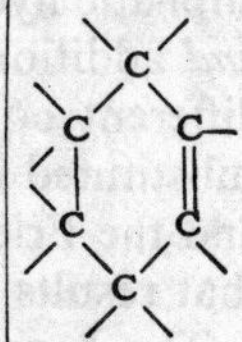
m.w.	78	86	84	82
b.p./°C	80°	69°	81°	83°
Density/g cm^{-3}	0.88	0.66	0.78	0.81
How it burns	smoky luminous flame	blue flame	smoky luminous flame	smoky luminous flame
With $KMnO_4$aq	no decolorization occurs			decolorization occurs rapidly
With HBr gas	no reaction occurs			rapid addition
With Br_2 in CCl_4	no decolorization occurs			rapid decolorization to 1,2-dibromo compd.
Catalytic oxidation with air at 150°C	no reaction occurs		gives cyclohexanol and cyclohexanone	good yield of adipic acid
Catalytic hydrogenation with H_2 at 200°C	cyclohexane formed slowly	no reaction	no reaction	cyclohexane formed rapidly
Boiled with halogen in strong u.v. light	addition occurs giving hexahalide	no addition reaction occurs		rapid addition giving dihalogenocyclohexane
With conc H_2SO_4	no reaction in cold; slow sulphonation when heated	no sulphonation	slow sulphonation when heated	hydrocarbon slowly absorbed
With conc HNO_3 + conc H_2SO_4	substitution giving nitrobenzenes	no reaction	no reaction	oxidation occurs to adipic acid
Friedel-Craft reaction	yes	no	no	no

of the benzene nucleus and its tendency to persist unbroken through a whole series of reactions involving the side-chains. For example, oxidizing agents convert homologues of benzene into benzoic acid but have no effect upon the nucleus at ordinary temperatures. Unlike aliphatic hydrocarbons, benzene and toluene undergo both substitution *and* addition reactions, sometimes even with the same reagent under different conditions. The ease with which aromatic compounds are substituted in various ways (e.g. nitration, sulphonation, halogenation, and the Friedel-Craft reaction), and the variety of isomeric derivatives that results are other distinctive features.

The benzene nucleus has a marked effect upon the properties of functional groups attached directly to it. For example, the aryl halides are far less reactive than the alkyl halides, and aromatic amines such as aniline can be diazotized whereas aliphatic amines do not form diazonium compounds with nitrous acid. The phenyl radical, being partially unsaturated and electron-attracting, enhances the acidity of groups with which it is linked. Thus hydroxy-compounds such as phenols have acidic properties not shown by aliphatic alcohols, benzoic acid is a stronger acid than acetic, and aromatic amines are weaker bases than their aliphatic counterparts (see § 50.8).

44

Aromatic Nitrogen Compounds

44.1. Introduction. Substances containing the $-NO_2$ group, known as *nitro compounds*, are of considerable importance in aromatic chemistry not only as explosives but also as the main sources of amines, of which the most important is aniline. This is a typical *aryl amine* having the amino group linked directly to a carbon atom of the benzene ring. It gives rise to the highly reactive *diazonium salts*, from which the azo dyes can be made. These three classes of aromatic compound are considered in detail in this chapter.

44.2. Nitrobenzene, $C_6H_5NO_2$.

LABORATORY PREPARATION: Benzene is nitrated with a mixture of concentrated nitric acid and concentrated sulphuric acid:

$$C_6H_6 + HNO_3 = C_6H_5NO_2 + H_2O$$

The sulphuric acid is essential for two reasons. Firstly, by providing a high concentration of hydrogen ions it forces the following equilibria over to the right and so produces a good supply of the nitronium ion, NO_2^+, which is believed to be the active nitrating agent*:

$$H^+ + NO_3^- \rightleftharpoons HNO_3 \rightleftharpoons OH^- + NO_2^+$$

Secondly, it prevents dilution of the nitric acid by taking up the water as it is formed. Nitration, unlike esterification, is not a reversible reaction and there is no question here of inhibiting a back reaction.

In practice the two concentrated acids are first mixed together cautiously with constant shaking and cooling. Then the benzene is added a little at a time with thorough shaking between each addition. A considerable amount of heat is evolved and the flask has to be cooled in a water-bath to prevent the temperature exceeding 50°C at this stage, or dinitrobenzene will be formed as well. When all the benzene has been added the flask is fitted with a reflux water condenser and heated for half an hour in a water-bath maintained at 60°C. When the flask has cooled the contents are poured into a large excess of cold water. The

* The mechanism of this reaction is discussed in § 50.6.

nitrobenzene, which settles to the bottom as a dense oily layer, should be stirred vigorously and the aqueous layer decanted. The product is then shaken in turn in a separating funnel with portions of water and sodium carbonate solution until no further effervescence occurs. It is dried with anhydrous calcium chloride, filtered, and distilled in a flask fitted with an air condenser, the fraction which boils between 207°C and 211°C being collected.

PROPERTIES AND REACTIONS OF NITROBENZENE: It is a pale yellow, poisonous liquid, m.p. 5.5°C, b.p. 210°C, ρ 1.20 g cm^{-3}, with a pleasant almond smell. Although immiscible with water, it dissolves in alcohol, ether, and benzene. Its main reactions are:

1. Reduction: When nitrobenzene is treated with tin (or iron) and concentrated hydrochloric acid, it is reduced to aniline:

$$C_6H_5NO_2 + 6\overline{H} = C_6H_5NH_2 + 2H_2O$$

This important reaction is described in detail in § 44.5.

When reduction is carried out under neutral or alkaline conditions, various products are obtained depending upon the reducing agent used. For example, when nitrobenzene is refluxed with a solution of caustic soda in methyl alcohol, azoxybenzene is produced, and this can be further reduced to azobenzene by heating it with iron filings, whereas zinc and alcoholic sodium hydroxide solution reduces nitrobenzene to hydrazobenzene:

$$\begin{array}{ccc} C_6H_5{-}N & C_6H_5{-}N & C_6H_5{-}NH \\ \| & \| & | \\ C_6H_5{-}NO & C_6H_5{-}N & C_6H_5{-}NH \\ \text{azoxybenzene} & \text{azobenzene} & \text{hydrazobenzene} \end{array}$$

Reduction with an aluminium–mercury couple gives phenylhydroxylamine as the main product:

$$C_6H_5NO_2 + 4\overline{H} = C_6H_5.NHOH + H_2O$$

2. Nuclear Substitution: The benzene nucleus can be halogenated, nitrated, and sulphonated in the usual ways. In each case the meta-derivative is formed.

USES OF NITROBENZENE: It is used on a large scale as an intermediate in the manufacture of aniline.

44.3. m-Dinitrobenzene, $C_6H_4(NO_2)_2$.

LABORATORY PREPARATION: Nitrobenzene is added a little at a time with continual shaking to a previously prepared mixture of concentrated sulphuric acid and fuming nitric acid in a flask fitted with a reflux air condenser. The flask is then heated on a boiling water-bath in a fume cupboard for about an hour to complete the nitration, with thorough shaking from time to time. When cool the contents are poured into a large excess of cold water; the dinitrobenzene separates as a solid which can be filtered off, washed thoroughly with water, and recrystallized from alcohol.

PROPERTIES AND REACTIONS: It is a pale yellow solid, m.p. 90°C. Although insoluble in water, it dissolves readily in hot alcohol.

Its reactions are similar to those of nitrobenzene. For example, the nitro groups are reduced to amino groups by nascent hydrogen from tin and concentrated hydrochloric acid, and it undergoes further substitution with difficulty.

44.4. The Nitrotoluenes. If toluene is treated with a mixture of concentrated nitric acid and concentrated sulphuric acid, it is nitrated readily to ortho- and para-nitrotoluene.

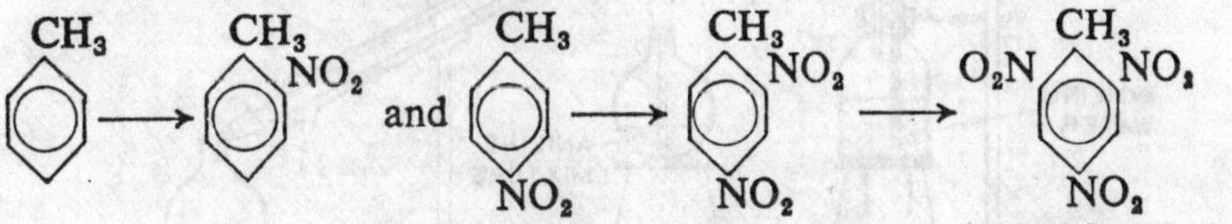

o-nitrotoluene *p*-nitrotoluene 2,4-dinitrotoluene 2,4,6-*trinitrotoluene* (T.N.T.)

Prolonged nitration with fuming nitric acid and concentrated sulphuric acid gives dinitrotoluene and trinitrotoluene. The latter is a pale yellow solid, m.p. 81°C, better known by its initial letters of T.N.T. It is a powerful explosive used extensively in two World Wars either alone or mixed with ammonium nitrate. The nitrotoluenes are very similar chemically to the nitrobenzenes.

44.5. Aniline, Aminobenzene, $C_6H_5NH_2$.

LABORATORY PREPARATION: By the reduction of nitrobenzene under acidic conditions:

$$C_6H_5NO_2 + 6\bar{H} = C_6H_5NH_2 + 2H_2O$$

The best source of nascent hydrogen is tin and concentrated hydrochloric acid; the tin reacts to give stannous chloride, itself a powerful reducing agent:

$$Sn + 2HCl = SnCl_2 + 2H$$
$$SnCl_2 + 2HCl = SnCl_4 + H2$$

Concentrated hydrochloric acid is added a little at a time to nitrobenzene and granulated tin in a flask fitted with a reflux water condenser. The flask is shaken intermittently and cooled if necessary. When all the acid has been added the condenser is removed and the open flask is heated in a boiling water-bath for about half an hour. At this stage the aniline is present as a non-volatile double salt with stannic chloride of composition $(C_6H_5NH_3)_2SnCl_6$, from which it is released by adding an excess of concentrated sodium hydroxide solution to the cooled mixture. The contents of the flask are then steam distilled using the apparatus shown in Fig. 44.1 until the milky emulsion of aniline no longer appears in the distillate. The principle of steam distillation is described in § 6.13; its purpose here is to separate the aniline from a large excess of liquid and solid impurities which would char and bump badly if heated

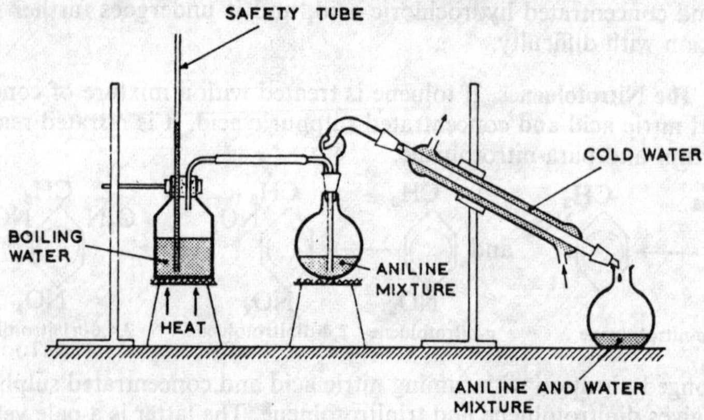

FIG. 44.1 Steam Distillation of Aniline

directly. Sodium chloride is added to the distillate to 'salt out' the aniline, which is then extracted with ether. In this process, which depends upon the partition law described in § 6.14, the aqueous distillate is shaken with successive portions of diethyl ether in a separating funnel and the upper ethereal layers transferred to a dry conical flask. This stage of the preparation must, of course, be performed well away from any flames owing to the danger of explosion. The ethereal solution is dried with pellets of potassium hydroxide (aniline reacts with calcium chloride and acidic drying agents), and when the liquid is clear it is filtered into a distilling flask fitted with a water condenser and the special receiver shown in Fig. 39.1. Hot water-baths are brought to the apparatus in turn to drive off all the ether. Finally the aniline is distilled by direct heating over a gauze using an air condenser, and the fraction boiling over the range 180–185°C is collected.

MANUFACTURE: Two methods are in use:

1. Nitrobenzene is reduced with concentrated hydrochloric acid and iron filings (instead of the more expensive tin used in the laboratory).

When the reduction is complete, the mixture is made alkaline with slaked lime and steam distilled. It is not economical to carry out an ether extraction, but aniline is recovered by distillation under reduced pressure.

2. Chlorobenzene is heated with a large excess of aqueous ammonia to 200°C under a pressure of about 6×10^6 N m^{-2} (60 atm) in the presence of cuprous oxide. Under these drastic conditions the chlorine atoms are replaced by amino groups:

$$C_6H_5Cl + 2NH_3 = C_6H_5NH_2 + NH_4Cl$$

PROPERTIES OF ANILINE: It is a colourless, poisonous liquid, b.p. 184°C, ρ 1.02 g cm^{-3}, with a typical aromatic smell. On exposure to light and air it slowly turns yellow, brown, or even black, owing to oxidation. Aniline is only slightly soluble in water, but it dissolves readily in organic solvents.

REACTIONS OF ANILINE: These are summarized in Table 44.1.

1. As a Base: Like all amines, aniline has basic properties, behaving in solution thus:

$$C_6H_5NH_2 + H_2O \rightleftharpoons C_6H_5NH_3OH \rightleftharpoons C_6H_5NH_3^+ + OH^-$$

However, the influence of the electron-attracting phenyl radical is to make aniline a weaker base than aliphatic amines, and although it readily forms salts with acids, its aqueous solution does not affect litmus. The most important salt of aniline is the hydrochloride, $C_6H_5NH_3Cl$, which is a colourless, crystalline solid, m.p. 198°C, which dissolves freely in water, giving an acidic solution by hydrolysis. These salts are decomposed to aniline by adding excess of sodium hydroxide solution in the cold:

$$C_6H_5NH_3Cl + NaOH = C_6H_5NH_2 + NaCl + H_2O.$$

They also form sparingly soluble double salts with the chlorides of gold, platinum, and tin, e.g. $(C_6H_5NH_3)_2PtCl_6$.

2. The Isocyanide Reaction: Aniline, like aliphatic primary amines, gives an isocyanide when warmed with chloroform and an alcoholic solution of potassium hydroxide:

$$C_6H_5NH_2 + CHCl_3 + 3KOH = C_6H_5NC + 3KCl + 3H_2O$$

3. Acetylation: Aniline is readily acetylated by acetyl chloride or acetic anhydride, giving acetanilide (see § 44.7):

e.g. $$C_6H_5NH_2 + CH_3COCl = C_6H_5.NH.COCH_3 + HCl$$

TABLE 44.1: REACTIONS OF ANILINE

Starting material	Reagent / conditions	Product	Behaviour
ANILINE	acids	salts, e.g. $C_6H_5NH_3Cl$	LIKE ALIPHATIC AMINES
ANILINE	chloroform and alcoholic potash	phenyl isocyanide, C_6H_5NC	LIKE ALIPHATIC AMINES
ANILINE	acetyl chloride or acetic anhydride	acetanilide, $C_6H_5NH.COCH_3$	LIKE ALIPHATIC AMINES
ANILINE	benzoyl chloride	benzanilide, $C_6H_5NH.COC_6H_5$	LIKE ALIPHATIC AMINES
ANILINE	heat with methyl halides	methylaniline, $C_6H_5NH.CH_3$; dimethylaniline, $C_6H_5N.(CH_3)_2$	LIKE ALIPHATIC AMINES
ANILINE	nitrous acid and hydrochloric acid at 5–10°C	benzene diazonium chloride, $C_6H_5N_2Cl$	UNLIKE ALIPHATIC AMINES
ANILINE	chlorine water	2,4,6-trichloroaniline, $NH_2.C_6H_2Cl_3$	UNLIKE ALIPHATIC AMINES
ANILINE	bromine water	2,4,6-tribromoaniline, $NH_2.C_6H_2Br_3$	UNLIKE ALIPHATIC AMINES
ANILINE	conc. sulphuric acid at 200°C	sulphanilic acid, $NH_2.C_6H_4.SO_3H$	UNLIKE ALIPHATIC AMINES
ANILINE	oxidizing agents	various coloured products	UNLIKE ALIPHATIC AMINES

4. *Benzoylation:* When aniline is shaken with benzoyl chloride and excess of sodium hydroxide solution, benzanilide separates as a white solid, m.p. 163°C (see § 44.8 for details):

$$C_6H_5NH_2 + C_6H_5COCl = C_6H_5NH.COC_6H_5 + HCl$$

5. *Alkylation:* If aniline is heated in a sealed tube with methyl iodide, methylaniline (a secondary amine) and dimethylaniline (a tertiary amine) are formed as their hydroiodides:

$$C_6H_5NH_2 + CH_3I = [C_6H_5.NH_2.CH_3]^+I^-$$
$$[C_6H_5NH_2.CH_3]^+I^- + CH_3I = [C_6H_5.NH(CH_3)_2]^+I^- + HI$$

6. *With Nitrous Acid:* Aniline reacts with nitrous acid in a different way from aliphatic primary amines. When aniline is treated with sodium nitrite solution and excess of dilute hydrochloric acid at 5–10°C, *diazotization* takes place and a solution of benzene diazonium chloride is obtained:

$$C_6H_5NH_3Cl + HNO_2 = C_6H_5N_2Cl + 2H_2O$$

This important reaction is described in greater detail in the next section.

7. *Halogenation:* The amino group attached to the nucleus greatly facilitates further substitution. Thus when aniline is treated with chlorine water or bromine water at room temperature, white precipitates of trichloro- or tribromo-aniline are obtained immediately:

e.g.

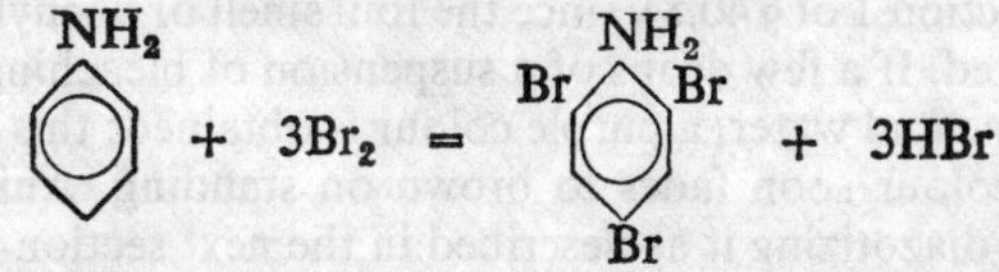

8. *Sulphonation:* When aniline is heated with concentrated sulphuric acid, sulphanilic acid is formed:

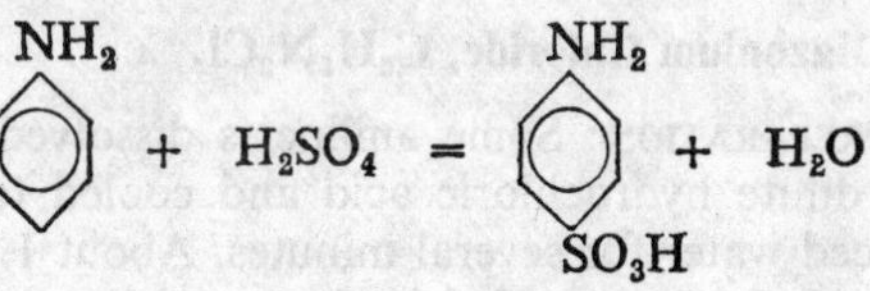

This product is important as a source of dyes and of the valuable sulphanilamide drugs.

9. *Nitration:* Aniline reacts vigorously with concentrated nitric acid, undergoing nitration and oxidation simultaneously. If the aniline is acetylated first, however, to protect the reactive amino group and then nitrated, the product gives a good yield of ortho- and para-nitroanilines when hydrolysed with hot 70% sulphuric acid, thus:

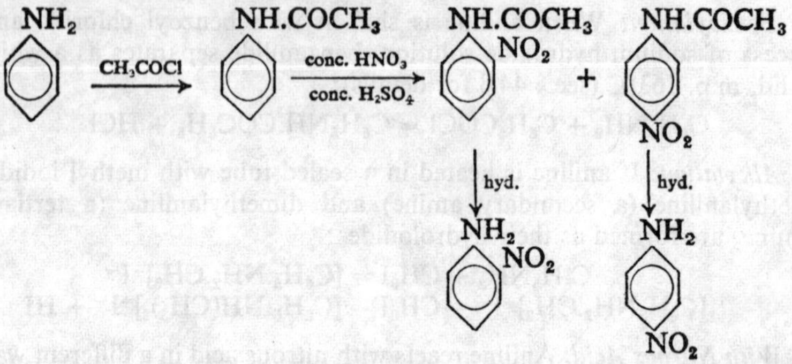

10. Oxidation: Aniline is oxidized in various ways depending upon the agent used. For example, bleaching powder gives a transient purple colour, ferric chloride solution a green colour, and chromic acid a black dye.

USES OF ANILINE: Its biggest use is in the manufacture of anti-oxidants and vulcanization accelerators for the rubber industry (§ 49.4). Large amounts are also used for making dyes, particularly azo-dyes, and for synthesizing drugs.

DETECTION OF ANILINE: It can be distinguished from secondary and tertiary amines by its response to the isocyanide or carbylamine reaction (see reaction 1 of § 40.8), since the foul smell of phenyl isocyanide is easily detected. If a few drops of a suspension of bleaching powder is added to aniline and water, a purple colour is obtained; this is a specific test, but the colour soon fades to brown on standing. Aniline is also recognized by diazotizing it as described in the next section and adding the diazonium salt to an alkaline solution of β-naphthol, when a scarlet colour is obtained.

44.6. Benzene Diazonium Chloride, $C_6H_5N_2Cl$.

LABORATORY PREPARATION: Some aniline is dissolved in about three equivalents of dilute hydrochloric acid and cooled to about 5°C by standing it in iced water for several minutes. About 1.1 equivalents of sodium nitrite solution are cooled in the same way and added a little at a time to the first solution keeping the mixture between 5°C and 10°C throughout. *Diazotization* takes place rapidly giving a colourless solution of benzene diazonium chloride, sometimes known as phenyl diazonium chloride:

$$C_6H_5NH_3Cl + HNO_2 = C_6H_5N_2Cl + 2H_2O$$

The salt is rarely prepared in the solid condition because it is explosive when dry. It has the structure $C_6H_5N_2^+Cl^-$.

REACTIONS OF BENZENE DIAZONIUM CHLORIDE SOLUTION: These are listed in Table 44.2.

TABLE 44.2. REACTIONS OF BENZENE DIAZONIUM CHLORIDE SOLUTION

BENZENE DIAZONIUM CHLORIDE SOLUTION	heated to 50°C	phenol, C_6H_5OH
	cuprous chloride in conc. hydrochloric acid	chlorobenzene, C_6H_5Cl
	cuprous bromide in conc. hydrobromic acid	bromobenzene. C_6H_5Br
	potassium cupro-cyanide soln.	benzonitrile, C_6H_5CN
	potassium iodide soln.	iodobenzene, C_6H_5I
	stannous chloride in conc. hydrochloric acid	phenylhydrazine, $C_6H_5.NHNH_2$
	phenol in caustic soda soln.	yellow dye
	β-naphthol in caustic soda soln.	scarlet dye
	ethyl alcohol	benzene, C_6H_6

1. Hydrolysis: When the solution is warmed to about 50°C, nitrogen is evolved and phenol is formed:

$$C_6H_5N_2Cl + H_2O = C_6H_5OH + N_2\uparrow + HCl$$

Side reactions also occur and if phenol is prepared in this way it is preferable to use benzene diazonium sulphate solution instead (see § 45.2).

2. With Cuprous Chloride: If the diazonium salt solution is added to a warm solution of cuprous chloride in concentrated hydrochloric acid, chlorobenzene is formed:

$$C_6H_5N_2Cl = C_6H_5Cl + N_2\uparrow$$

3. With Cuprous Bromide: The diazonium salt solution yields bromobenzene when treated with a warm solution of cuprous bromide in concentrated hydrobromic acid:

$$C_6H_5N_2Cl + CuBr = C_6H_5Br + CuCl + N_2\uparrow$$

4. With Cuprous Cyanide: Like the last two, this is another example of the *Sandmeyer reaction.* If the diazonium salt solution is treated with a warm solution of cuprous cyanide in potassium cyanide solution, benzonitrile (phenyl cyanide) is obtained:

$$C_6H_5N_2Cl + CuCN = C_6H_5CN + CuCl + N_2\uparrow$$

Benzonitrile is a colourless liquid, b.p. 191°C, with an almond smell. It is hydrolysed by boiling acids to benzoic acid, $C_6H_5.COOH$, and reduced by sodium and alcohol to benzylamine, $C_6H_5.CH_2NH_2$.

5. With Potassium Iodide: Benzene diazonium chloride solution reacts with warm potassium iodide solution to give iodobenzene:

$$C_6H_5N_2Cl + KI = C_6H_5I + KCl + N_2\uparrow$$

This is the best way of preparing iodobenzene, since it is difficult to iodinate benzene directly. After recovery by steam distillation the product is separated from water, dried with anhydrous calcium chloride, and redistilled.

6. With Stannous Chloride: A solution of stannous chloride in concentrated hydrochloric acid reduces the diazonium salt to phenylhydrazine, which is obtained as its hydrochloride:

$$C_6H_5N_2Cl + 4\bar{H} = C_6H_5NH.NH_3Cl$$

Phenylhydrazine itself is a pale yellow, very poisonous, crystalline solid, m.p. 24°C, used for identifying aldehydes, ketones, and sugars.

7. Coupling Reactions: If benzene diazonium chloride solution is added to alkaline solutions of phenols, coupling takes place and various *azo-dyes* are produced. For example, a solution of *β*-naphthol in caustic soda gives a scarlet precipitate of benzene-azo-*β*-naphthol:

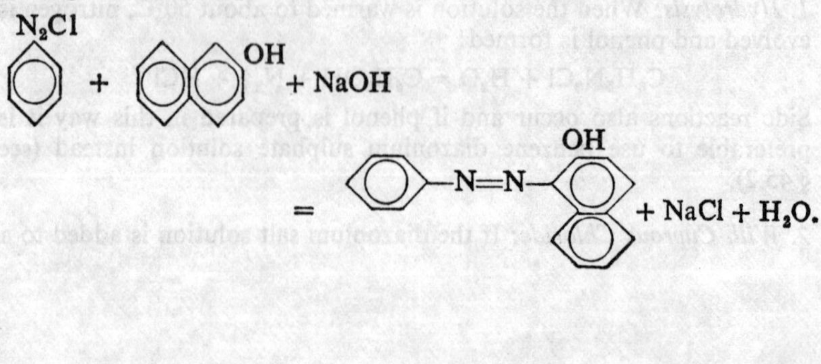

Similarly, coupling with phenol gives a yellow dye which has an orange colour in concentrated solution; this is sometimes used as a test for phenol. The diazonium salt also couples with amines in acid solution, which is why aniline must not be in excess during diazotization. Any compound with an amino group linked directly to the benzene nucleus can be diazotized and coupled with aryl amines or phenols, so a wide range of azo-dyes can be made.

8. With Alcohol: Benzene diazonium chloride is reduced to benzene when warmed with ethyl alcohol:

$$C_6H_5N_2Cl + C_2H_5OH = C_6H_6 + CH_3.CHO + N_2\uparrow + HCl$$

44.7. Acetanilide, $C_6H_5.NH.COCH_3$.

Laboratory Preparation: By heating aniline under reflux for about half an hour with a mixture of glacial acetic acid and acetic anhydride:

$$C_6H_5NH_2 + (CH_3CO)_2O = C_6H_5.NH.COCH_3 + CH_3.COOH$$

The acetanilide is precipitated when the product is poured into a large excess of cold water. It is filtered off, washed, and recrystallized from a mixture of equal volumes of glacial acetic acid and water.

Properties and Reactions of Acetanilide: It is a white crystalline solid, m.p. 114°C practically insoluble in cold water. It is hydrolysed when boiled under reflux with 70% sulphuric acid, giving aniline sulphate:

$$C_6H_5.NH.COCH_3 + H_2O + H_2SO_4 = C_6H_5NH_3HSO_4 + CH_3.COOH$$

Although unsaturated, acetanilide is ortho- and para-directing when it undergoes nuclear substitution, being one of the few common exceptions to Vörlander's rule (§ 43.5).

44.8. Benzanilide, $C_6H_5NH.COC_6H_5$.

Laboratory Preparation: By the benzoylation (§ 47.5) of aniline. Benzoyl chloride is shaken vigorously with aniline and an excess of 10% sodium hydroxide solution in a corked bottle for about twenty minutes:

$$C_6H_5NH_2 + C_6H_5COCl = C_6H_5NH.COC_6H_5 + HCl$$

The solid product is filtered off, washed thoroughly with cold water, and recrystallized from methylated spirit.

Properties and Reactions of Benzanilide: It is a white crystalline solid, m.p. 163°C, practically insoluble in cold water. Like acetanilide it is hydrolysed when boiled under reflux with 70% sulphuric acid:

$$C_6H_5NH.COC_6H_5 + H_2O + H_2SO_4 = C_6H_5NH_3HSO_4 + C_6H_5.COOH$$

44.9. Benzamide, $C_6H_5CONH_2$.

LABORATORY PREPARATION: Benzoyl chloride is shaken vigorously for about fifteen minutes with an excess of concentrated (0.88) ammonia solution in a conical flask with a well-fitting bung. When benzoylation (see § 47.5) is complete the solid benzamide is filtered off, washed repeatedly with cold water, and recrystallized from hot water.

$$C_6H_5COCl + NH_3 = C_6H_5CONH_2 + HCl$$

PROPERTIES AND REACTIONS OF BENZAMIDE: It is a colourless solid, m.p. 130°C, which dissolves readily in hot water. In its reactions it closely resembles aliphatic amides, evolving ammonia when boiled with alkalis, giving a cyanide when distilled with phosphorus pentoxide, and forming a white precipitate of benzoic acid when treated with nitrous acid.

45

Phenols

45.1. Introduction. The phenols are aromatic compounds containing one or more hydroxyl groups linked directly to the benzene ring. The simplest member is phenol itself, C_6H_5OH, and since it is typical of the others and by far the most important, it alone will be considered here.

45.2. Phenol, Hydroxybenzene, C_6H_5OH.

LABORATORY PREPARATION: 1. Aniline is diazotized by dissolving it in excess of dilute sulphuric acid and treating it with a slight excess of sodium nitrite solution at 5–10°C. On warming the benzene diazonium sulphate solution to 50°C, phenol is formed:

$$C_6H_5N_2HSO_4 + H_2O = C_6H_5OH + N_2\uparrow + H_2SO_4$$

The phenol is recovered by steam distillation and extracted with ether. The ethereal solution is dried with anhydrous potassium carbonate and distilled using an air condenser.

2. Sodium benzene sulphonate is fused with sodium hydroxide at 250°C in a nickel crucible:

$$C_6H_5.SO_3Na + 2NaOH = C_6H_5ONa + Na_2SO_3 + H_2O$$

The residue is dissolved in water and acidified, when the phenol separates as an oily layer:

$$C_6H_5ONa + HCl = C_6H_5OH + NaCl$$

It is purified as in the first method.

MANUFACTURE: Originally all phenol was obtained from coal tar, but demand has now far outstripped the supply from this source, and four other methods are also in regular use:

1. From Coal Tar: After naphthalene has been crystallized from it, the middle oil fraction (§ 43.2) is shaken with dilute sodium hydroxide solution, which dissolves the phenol as sodium phenate:

$$C_6H_5OH + NaOH = C_6H_5ONa + H_2O$$

The aqueous layer is separated and saturated with carbon dioxide:

$$2C_6H_5ONa + CO_2 + H_2O = 2C_6H_5OH\downarrow + Na_2CO_3$$

The precipitated phenol is filtered off and fractionally distilled; crystals of pure phenol separate from the distillate on cooling.

2. From Chlorobenzene: Sodium hydroxide solution is heated with chlorobenzene to 300°C under about 1.5×10^7 N m^{-2} (150 atm) pressure:

$$C_6H_5Cl + 2NaOH = C_6H_5ONa + NaCl + H_2O$$

Phenol is obtained by acidifying the product and distilling.

3. From Benzene: A mixture of benzene vapour, hydrogen chloride, and air is passed over a copper chloride catalyst at 230°C:

$$2C_6H_6 + 2HCl + O_2 = 2C_6H_5Cl + 2H_2O$$

The chlorobenzene is then hydrolysed by mixing it with steam and passing it over a silicon catalyst at 425°C:

$$C_6H_5Cl + H_2O = C_6H_5OH + HCl$$

The hydrogen chloride is recovered and used again.

4. From Sodium Benzene Sulphonate: In principle this method is similar to the laboratory preparation described above, but a temperature of about 300°C is used for the fusion.

5. From Propylene and Benzene: These two hydrocarbons, which are derived cheaply from petroleum by cracking and catalytic reforming processes, are compressed to 3×10^6 N m^{-2} (30 atm) pressure and passed over a catalyst of phosphoric acid at 250°C, when isopropylbenzene, known as *cumene*, is formed:

$$C_6H_6 + CH_3.CH : CH_2 = C_6H_5.CH.(CH_3)_2$$

The cumene is oxidized by air to its hydroperoxide, which is decomposed into phenol and acetone by treatment with sulphuric acid:

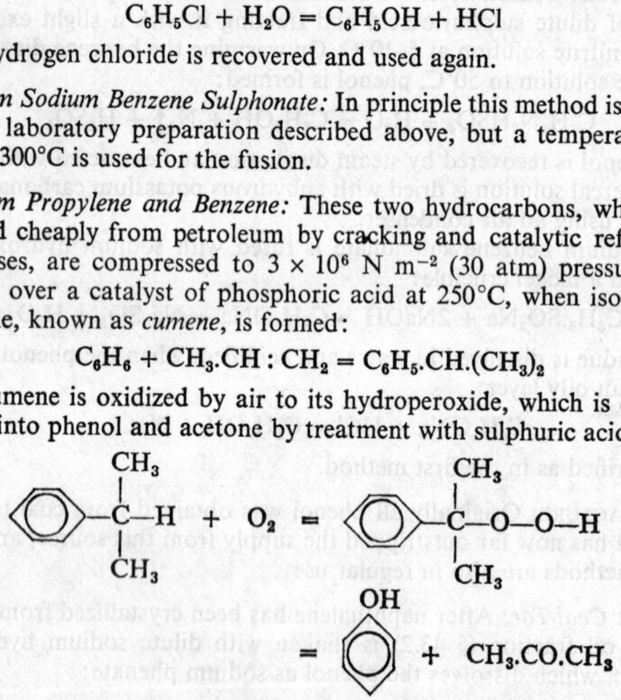

Although only in use since 1954, this process is already an important source of phenol and acetone.

Properties of Phenol: It is a colourless, hygroscopic, crystalline solid, m.p. 42°C, b.p. 181°C, with a characteristic 'carbolic' smell (it is often given the common name carbolic acid). It is very poisonous and blisters the skin on contact. Above 65.8°C phenol is miscible with water in all proportions, but below this temperature it is only partially miscible (§ 6.12). It dissolves readily in alcohol, ether, and benzene.

Reactions of Phenol: These are summarized in Table 45.1.

TABLE 45.1. REACTIONS OF PHENOL

PHENOL	sodium hydroxide soln. →	sodium phenate, C_6H_5ONa
	sodium (in ether) →	sodium phenate, C_6H_5ONa
	phosphorus pentachloride →	chlorobenzene, C_6H_5Cl
	acetyl chloride →	phenyl acetate, $CH_3.COOC_6H_5$
	benzoyl chloride →	phenyl benzoate, $C_6H_5.COOC_6H_5$
	hot zinc dust →	benzene, C_6H_6
	hydrogen (Ni catalyst) at 150°C →	cyclohexanol, $C_6H_{11}OH$
	chlorine water →	2,4,6-trichlorophenol, $HO.C_6H_2Cl_3$
	bromine water →	2,4,6-tribromophenol, $HO.C_6H_2Br_3$
	nitric acid →	nitrophenols, e.g. $HO.C_6H_4.NO_2$
	conc. sulphuric acid →	phenolsulphonic acids, e.g. $HO.C_6H_4.SO_2OH$

1. As an Acid: It is very weakly acidic, dissociating in aqueous solution thus:

$$C_6H_5OH \rightleftharpoons C_6H_5O^- + H^+$$

It dissolves readily in alkalis forming phenates, but it is too weak ($K = 1.3 \times 10^{-10}$ mol dm^{-3}) to affect litmus or to liberate carbon dioxide from carbonates (see § 50.8).

2. With Sodium: In dry ethereal solution phenol reacts with sodium forming sodium phenate:

$$2C_6H_5OH + 2Na = 2C_6H_5ONa + H_2\uparrow$$

3. With Phosphorus Pentachloride: Phenol reacts with phosphorus pentachloride at room temperature, liberating hydrogen chloride:

$$C_6H_5OH + PCl_5 = C_6H_5Cl + POCl_3 + HCl\uparrow$$

The yield of chlorobenzene is poor.

4. Acetylation: Phenol is readily acetylated by acetyl chloride or acetic anhydride, giving phenyl acetate:

$$C_6H_5OH + CH_3COCl = CH_3.COOC_6H_5 + HCl\uparrow$$

5. Benzoylation: If phenol is shaken with benzoyl chloride and excess of sodium hydroxide solution, it is benzoylated (§ 47.5) to phenyl benzoate:

$$C_6H_5OH + C_6H_5COCl = C_6H_5.COOC_6H_5 + HCl$$

6. Reduction: Phenol is reduced to benzene when its vapour is passed over red-hot zinc dust:

$$C_6H_5OH + Zn = C_6H_6 + ZnO$$

7. Hydrogenation: If hydrogen is bubbled through molten phenol at 150°C in the presence of a finely-divided nickel catalyst, cyclohexanol is produced:

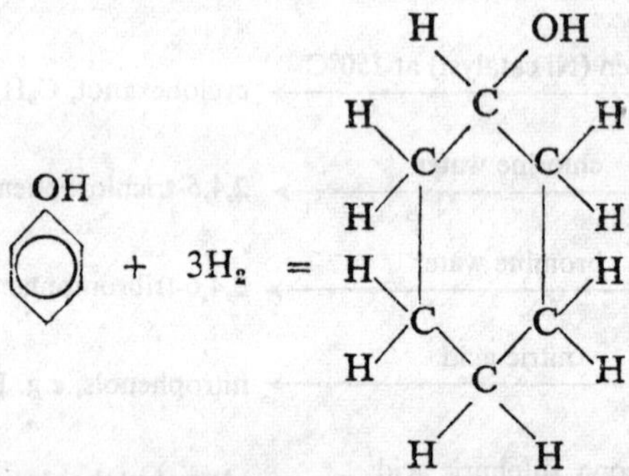

Cyclohexanol is a colourless liquid, b.p. 161°C, which smells like camphor. It is a saturated cyclic compound with none of the special properties of the benzene resonance structure. It is used to make nylon as described in § 49.3.

8. Halogenation: When an aqueous solution of phenol is treated with chlorine water or bromine water at room temperature, an immediate precipitate of trichloro- or tribromophenol is obtained:

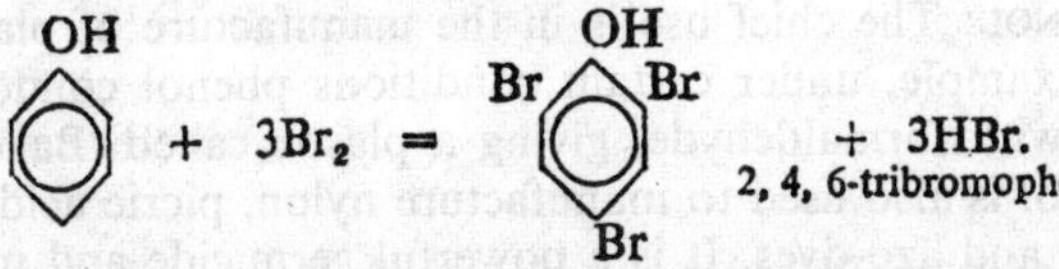

2, 4, 6-tribromophenol

Thus the hydroxyl group has the effect of facilitating substitution in the benzene nucleus.

9. Nitration: A mixture of ortho- and para-nitrophenols is obtained when phenol is treated with moderately concentrated nitric acid at room temperature. Concentrated nitric acid gives 2,4-dinitrophenol, and concentrated nitric acid in the presence of concentrated sulphuric acid forms 2, 4, 6-trinitrophenol, better known as picric acid:

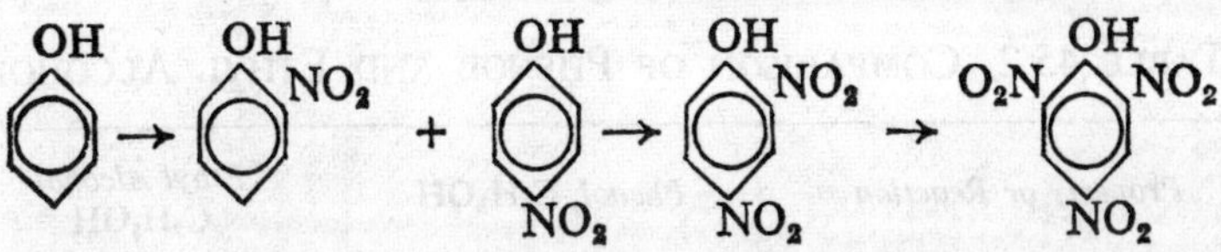

The nitro groups increase the dissociation of the OH group* and nitrophenols are more strongly acidic than phenol, e.g. they liberate carbon dioxide from sodium carbonate solution. Picric acid, m.p. 122°C, is a fairly strong acid; it is used as a yellow dye, as a high explosive, and as an antiseptic.

10. Sulphonation: Phenol gives a mixture of ortho- and para- phenolsulphonic acids when treated with concentrated sulphuric acid at room temperature.

DETECTION OF PHENOL: An aqueous solution of phenol gives a purple colour with ferric chloride solution and an immediate white precipitate with bromine water. It also couples with a solution of benzene diazonium chloride in the presence of excess of sodium hydroxide solution, giving a yellow dye, but none of these tests are specific. When phenol is heated with an equal weight of phthalic anhydride and a few drops of concentrated sulphuric acid, phenolphthalein is produced and can be recognized by its characteristic pink colour on adding an excess of alkali. Phenol can also be detected by Liebermann's nitroso reaction —a few crystals of phenol are heated gently with a little sodium nitrite and a few drops of concentrated sulphuric acid are added. A deep blue-green colour appears; on pouring the product into water the solution goes red, turning blue again when excess of alkali is added. Phenol can be distinguished from carboxylic acids by its inability to liberate carbon dioxide from sodium carbonate solution.

* An explanation of this effect is given in § 50.8.

USES OF PHENOL: The chief use is in the manufacture of plastics and resins. For example, under certain conditions phenol condenses and polymerizes with formaldehyde, giving a plastic called 'Bakelite' (see § 49.3). Phenol is also used to manufacture nylon, picric acid, salicylic acid, aspirin, and azo-dyes. It is a powerful germicide and many proprietary antiseptics contain phenol derivatives. Compounds made from phenol are used as photographic developers, detergents, selective weedkillers, and wood preservatives.

45.3. Comparison of Phenol with Ethyl Alcohol. In Table 45.2 the properties and reactions of phenol are compared with those of a typical aliphatic alcohol such as ethyl alcohol. The main differences are of three kinds. Some (e.g. nos. 1, 2, 4, 7, 8, and 9) can be ascribed to the

TABLE 45.2. COMPARISON OF PHENOL AND ETHYL ALCOHOL

Property or Reaction	*Phenol*, C_6H_5OH	*Ethyl Alcohol* C_2H_5OH
1. Solubility in water at 20°C	Only partially miscible	Miscible in all proportions
2. Acidity of solution	Weakly acidic forming salts	Neutral
3. Action of sodium	Forms sodium deriv. $+H_2\uparrow$	Forms sodium deriv. $+H_2\uparrow$
4. Action of PCl_5	Slow reaction giving poor yield of $C_6H_5Cl + HCl\uparrow$	Reacts vigorously giving $C_2H_5Cl + HCl\uparrow$
5. Action of CH_3COCl	Acetylates readily	Acetylates readily
6. Action of $C_6H_5COCl + NaOH$	Benzoylates readily	Benzoylates readily
7. Action of organic acids	None	Forms ethyl esters
8. Action of HCl, HBr, HI	None	Forms alkyl halides
9. Action of $FeCl_3$ soln.	Purple colour	No purple colour
10. Action of H_2 + Ni catalyst	Adds giving $C_6H_{11}OH$	No addition occurs
11. Action of Br_2 water	Brominates nucleus giving white ppt. immediately	None
12. Action of dil HNO_3	Nitrates nucleus	None
13. Action of conc. H_2SO_4	Sulphonates nucleus	Dehydrates to ethylene or ether
14. Action of $KMnO_4$ soln.	Gives various products	Gives aldehyde and acid
15. Coupling with $C_6H_5N_2Cl$	Yellow dye	No colour
16. Iodoform reaction	No ppt.	Gives yellow ppt.
17. Reduction by hot Zn	Gives benzene	No reduction
18. Phenolphthalein reaction	Purple colour	No colour
19. Volatility	Solid at room temperature	Liquid at room temperature

influence of the electron-attracting benzene nucleus upon the adjacent hydroxyl group. Other differences (e.g. nos. 10, 11, 12, and 13) arise from the reactions of the benzene ring itself and its ability to undergo substitution and addition reactions not shown by an alkyl radical. A third group consists of condensation or coupling reactions of which aliphatic hydroxy compounds are not capable (e.g. nos. 15 and 18) or particular reactions of ethyl alcohol not common to aliphatic alcohols in general (e.g. no. 16).

It is interesting to note that benzyl alcohol, $C_6H_5.CH_2OH$, in which the hydroxyl group is in the side chain and is not linked directly to a carbon atom of the benzene ring, closely resembles aliphatic alcohols in its reactions, as described in § 47.2. It shows no acidic properties, nor does it give a purple colour with ferric chloride solution. It can be converted in good yield to benzyl chloride by treatment with phosphorus pentachloride or hydrogen chloride, it readily forms esters with organic acids, and when oxidized it gives an aldehyde and a carboxylic acid.

46

Toluene and its Derivatives

46.1. Introduction. Toluene resembles its homologue benzene in many respects. However, its methyl group makes substitution in the benzene nucleus easier and directs it to the ortho- and para-positions. Moreover, the methyl group itself can be halogenated and oxidized, giving toluene a range of reactions not shown by benzene.

46.2. Toluene, Methylbenzene, $C_6H_5.CH_3$.

LABORATORY PREPARATION: Although rarely prepared in the laboratory, toluene can be made from benzene in two ways which are mainly of interest as general methods of preparing higher homologues:

1. By the Friedel–Craft Reaction: Methyl chloride is bubbled into a mixture of dry benzene and anhydrous aluminium chloride:

$$C_6H_6 + CH_3Cl = C_6H_5.CH_3 + HCl$$

The aluminium chloride acts as a catalyst by forming an intermediate compound which subsequently decomposes. To obtain a good yield the mass of catalyst used should be about a third that of benzene. The toluene is isolated by pouring the product into cold water, separating the upper layer, drying it with anhydrous calcium chloride, and fractionally distilling it.

2. By the Fittig Reaction: This is analogous to the Wurtz synthesis of paraffins (§ 33.3). Bromobenzene and methyl iodide in dry ethereal solution are treated with metallic sodium, when the following reaction occurs spontaneously:

$$C_6H_5Br + CH_3I + 2Na = C_6H_5.CH_3 + NaBr + NaI$$

The toluene is separated by fractional distillation.

MANUFACTURE: Toluene is obtained from two main sources. It is recovered from coal tar by fractionally distilling benzole (see Table 43.1), and it is made from petroleum, particularly in the U.S.A., by means of the catalytic reforming processes described in § 43.3.

e.g. $$\underset{\text{n-heptane}}{C_7H_{16}} = C_6H_5.CH_3 + 4H_2$$

Properties of Toluene: It is a colourless liquid, m.p. −95°C, b.p. 111°C, ρ 0.87 g cm^{-3}, with a smell like that of benzene. It is immiscible with water but it dissolves freely in organic solvents. It burns with a luminous and smoky flame.

Reactions of Toluene: These are summarized in Table 46.1.

1. Nuclear Substitution by Halogens: When toluene is treated with chlorine or bromine at room temperature in the absence of sunlight and in the presence of a halogen carrier (e.g. iodine or iron filings), substitution occurs in the nucleus giving a mixture of ortho- and para-derivatives:

e.g.

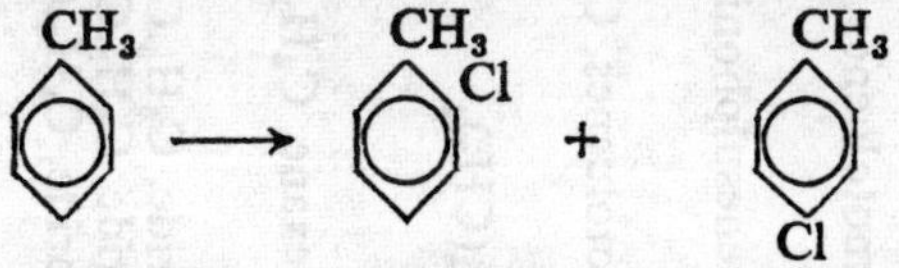

o-chlorotoluene, b.p. 159°C

p-chlorotoluene, b.p. 162°C

The chlorotoluenes resemble chlorobenzene in that the chlorine atom is firmly bound to the nucleus and cannot be replaced under laboratory conditions by —OH, —NH_2, and —CN groups.

2. Sulphonation: A mixture of ortho- and para-toluene sulphonic acids is obtained when toluene is heated with concentrated sulphuric acid:

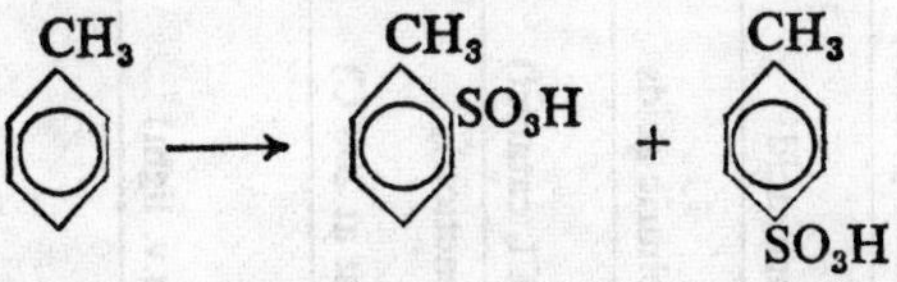

3. Nitration: When toluene is treated with equal parts of concentrated nitric acid and concentrated sulphuric acid at room temperature, a mixture of ortho- and para- nitrotoluenes is obtained:

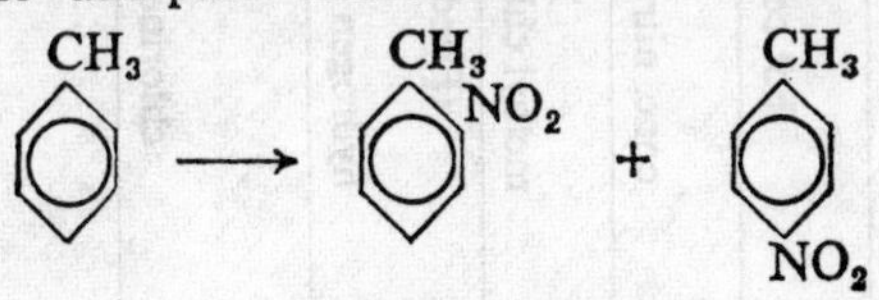

If fuming nitric acid and concentrated sulphuric acid are used, di- and tri-nitrotoluenes are obtained, as described in § 44.4.

4. The Friedel–Craft Reaction: Toluene undergoes this reaction giving alkyl derivatives, e.g. xylenes.

5. Hydrogenation: When a mixture of toluene vapour and hydrogen is

Table 46.1. Reactions of Toluene

Reagent (conditions)	Product
chlorine (halogen carrier present)	*o*- and *p*-chlorotoluenes, $CH_3.C_6H_4Cl$
bromine (halogen carrier present)	*o*- and *p*-bromotoluenes, $CH_3.C_6H_4Br$
hot conc. sulphuric acid	*o*- and *p*-toluenesulphonic acids, $CH_3.C_6H_4.SO_3H$
conc. nitric and sulphuric acids	*o*- and *p*-nitrotoluenes, $CH_3C_6H_4NO_2$
methyl chloride (Al_2Cl_6 catalyst) (Friedel-Craft reaction)	xylenes, $C_6H_4(CH_3)_2$
hydrogen (Ni catalyst at 200°C)	methylcyclohexane, $C_6H_{11}.CH_3$
chlorine (intense u.v. light)	benzyl chloride, $C_6H_5.CH_2Cl$; benzal chloride, $C_6H_5.CHCl_2$; benzotrichloride, $C_6H_5.CCl_3$
bromine (intense u.v. light)	side-chain bromine derivatives, e.g. $C_6H_5.CH_2Br$
hot potassium permanganate soln. or hot chromic acid	benzoic acid, $C_6H_5.COOH$
manganese dioxide or chromyl chloride	benzaldehyde, $C_6H_5.CHO$

All reactions start from TOLUENE.

passed over a catalyst of finely-divided nickel at 200°C, addition occurs giving methylcyclohexane:

$$C_6H_5.CH_3 + 3H_2 = C_6H_{11}.CH_3$$

6. Side-chain substitution by Halogens: If chlorine is passed into boiling toluene in strong sunlight or intense ultra-violet light in the absence of any halogen carrier, substitution occurs in the side-chain thus:

$$C_6H_5.CH_3 + Cl_2 = HCl + C_6H_5.CH_2Cl \text{ (benzyl chloride)}$$
$$C_6H_5.CH_2Cl + Cl_2 = HCl + C_6H_5.CHCl_2 \text{ (benzal chloride)}$$
$$C_6H_5.CHCl_2 + Cl_2 = HCl + C_6H_5.CCl_3 \text{ (benzotrichloride)}$$

Bromine forms the corresponding bromine derivatives under the same conditions. Reactions 1 and 6 provide yet another good example of how the products of organic reactions depend upon the conditions applied and how important it is to choose and state those conditions correctly.

7. Oxidation: If toluene is boiled under reflux for several hours with potassium permanganate solution or with chromic acid, the methyl radical is oxidized to a carboxyl group:

$$C_6H_5.CH_3 + 3\bar{O} = \underset{\text{benzoic acid}}{C_6H_5.COOH} + H_2O$$

If a weaker oxidizing agent is used (e.g. manganese dioxide or chromyl chloride), the product is benzaldehyde.

USES OF TOLUENE: Its main use is as a solvent for gums, resins, fats, oils, rubber, etc. It is preferred to benzene for this purpose because it is less volatile and not so poisonous. It is also used for manufacturing trinitrotoluene (T.N.T.), benzyl alcohol, benzaldehyde, saccharine, and a wide range of dyes, anaesthetics, and drugs.

46.3. Side-chain Derivatives. As described in reaction 6 of the preceding section, three different compounds are formed by chlorinating boiling toluene in sunlight or intense ultra-violet light in the absence of a halogen carrier. All three are colourless, pungent-smelling liquids which are immiscible with water. Their vapours cause weeping when inhaled.

In *benzyl chloride*, b.p. 179°C, the chlorine atom is even more reactive than in the alkyl halides. For example, it is readily replaced by a —OH, —NH_2, or —CN group when benzyl chloride is boiled with alkalis, alcoholic ammonia, or alcoholic potassium cyanide solution respectively. This contrasts with the inertness of the aryl halides such as chlorobenzene and the chlorotoluenes, in which the chlorine atom is linked directly to the nucleus.

Benzal chloride, b.p. 207°C, is hydrolysed to benzaldehyde when boiled with alkalis:

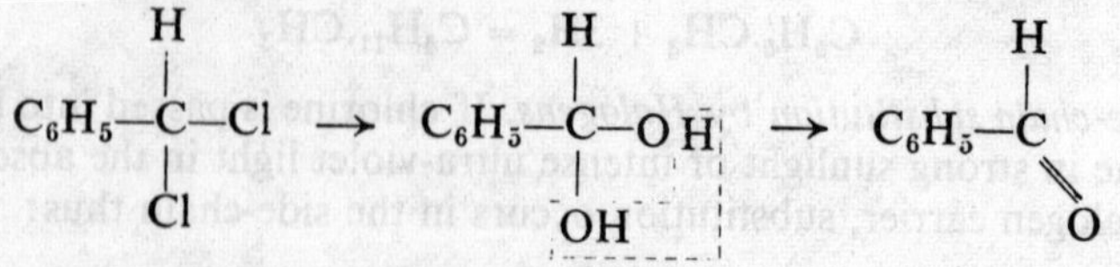

Benzotrichloride, b.p. 220°C, is hydrolysed to sodium benzoate when boiled with sodium hydroxide solution:

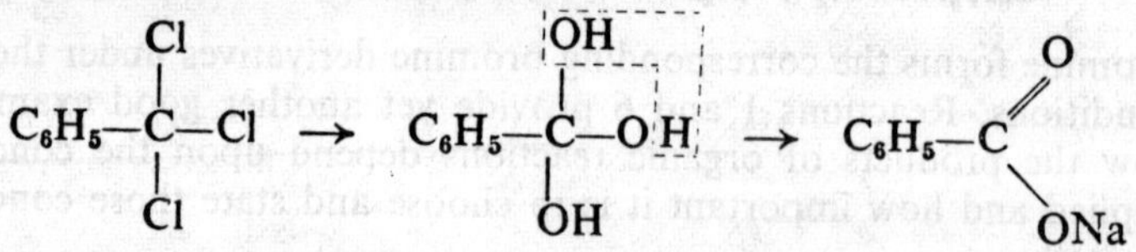

47

Benzyl Alcohol, Benzaldehyde, Benzoic Acid

47.1. Introduction. These compounds are formed by hydrolysing the side-chain chlorine derivatives of toluene. Our main interest lies in comparing them with the corresponding aliphatic compounds, and this aspect will be emphasized, therefore, throughout the chapter.

47.2. Benzyl Alcohol, $C_6H_5CH_2OH$.

LABORATORY PREPARATION: It can be made by hydrolysing benzyl chloride or benzyl esters with boiling alkali, or by reducing benzaldehyde with sodium amalgam and water, or by treating benzylamine with nitrous acid. The best method, however, is to shake benzaldehyde with concentrated potassium hydroxide solution, when Cannizzaro's reaction takes place:

$$2C_6H_5.CHO + KOH = C_6H_5.CH_2OH + C_6H_5.COOK$$

After dissolving the potassium benzoate in water, the benzyl alcohol can be extracted with ether, dried with anhydrous potassium carbonate, and distilled.

PROPERTIES OF BENZYL ALCOHOL: It is a colourless liquid, b.p. 205°C, ρ 1.05 g cm^{-3}, with a faint smell. It is almost immiscible with water, but it dissolves freely in organic solvents.

REACTIONS OF BENZYL ALCOHOL: It closely resembles aliphatic alcohols in its reactions. Thus it forms sodium benzylate and hydrogen with metallic sodium, it gives fumes of hydrogen chloride in the cold with phosphorus pentachloride, it is readily acetylated and benzoylated, and it reacts reversibly with acids to form benzyl esters. When warmed with dilute nitric acid it is oxidized to benzaldehyde, but stronger oxidizing agents (e.g. acidified potassium permanganate solution) convert it into benzoic acid.

It differs from aliphatic alcohols in undergoing nuclear substitution, forming ortho- and para-derivatives, and in reacting more slowly with aqueous reagents owing to its low solubility in water.

47.3. Benzaldehyde, $C_6H_5.CHO$.

LABORATORY PREPARATION: Benzyl chloride is boiled under reflux for several hours with a solution of cupric nitrate or lead nitrate, when hydrolysis and oxidation occur thus:

$$C_6H_5.CH_2Cl + H_2O = C_6H_5.CH_2OH + HCl$$
$$C_6H_5.CH_2OH + \bar{O} = C_6H_5.CHO + H_2O$$

A stream of carbon dioxide is usually passed through the mixture to remove oxides of nitrogen. The benzaldehyde is extracted with ether, dried with anhydrous calcium chloride, and distilled. If necessary it can be purified further by forming a crystalline addition compound with sodium bisulphite and distilling this with dilute sulphuric acid.

Benzaldehyde can also be obtained by hydrolysing benzal chloride or by oxidizing toluene with chromyl chloride (§ 25.7):

$$C_6H_5CHCl_2 + H_2O = C_6H_5.CHO + 2HCl$$

MANUFACTURE: Toluene is oxidized directly by passing its vapour, mixed with air and excess of nitrogen, over a catalyst of vanadium pentoxide or manganese dioxide at 500°C:

$$C_6H_5.CH_3 + O_2 = C_6H_5.CHO + H_2O$$

In another process toluene is oxidized by warming it with manganese dioxide and sulphuric acid. Some benzaldehyde is made by hydrolysing benzal chloride with steam or slaked lime.

PROPERTIES OF BENZALDEHYDE: It is a colourless liquid, b.p. 179°C, ρ 1.05 g cm^{-3}, with a pleasant almond smell. It dissolves readily in organic solvents, but it is only sparingly soluble in water.

REACTIONS OF BENZALDEHYDE: These are summarized in Table 47.1.

1. Oxidation: It is readily oxidized to benzoic acid:

$$C_6H_5.CHO + \bar{O} = C_6H_5.COOH$$

This reaction is brought about slowly by air (which explains the white deposit often found on the neck of a bottle of benzaldehyde), and rapidly when benzaldehyde is warmed with acidified potassium permanganate solution.

2. *Reduction:* It is reduced to benzyl alcohol by sodium amalgam and water or by sodium borohydride (§ 19.7):

$$C_6H_5.CHO + 2\bar{H} = C_6H_5CH_2OH$$

A stronger reducing agent such as Clemmensen's (amalgamated zinc and concentrated hydrochloric acid) reduces benzaldehyde to toluene:

$$C_6H_5.CHO + 4\bar{H} = C_6H_5.CH_3 + H_2O$$

TABLE 47.1. REACTIONS OF BENZALDEHYDE

Reactant	Reagent (→)	Product	
BENZALDEHYDE	acidified potassium permanganate soln.	benzoic acid, $C_6H_5.COOH$	LIKE ACETALDEHYDE
	sodium amalgam and water	benzyl alcohol, $C_6H_5.CH_2OH$	
	amalgamated zinc and conc. hydrochloric acid	toluene, $C_6H_5.CH_3$	
	hydrogen cyanide	benzaldehyde cyanohydrin $C_6H_5.CH(OH)CN$	
	conc. sodium bisulphite soln.	benzaldehyde bisulphite $C_6H_5.CH(OH).SO_3Na$	
	phenylhydrazine	benzaldehyde phenylhydrazone $C_6H_5.CH{:}NNH.C_6H_5$	
	hydroxylamine	benzaldoxime, $C_6H_5.CH{:}NOH$	
	phosphorus pentachloride	benzal chloride, $C_6H_5CHCl_2$	
	ammonia	hydrobenzamide $(C_6H_5CH)_3N_2$	UNLIKE ACETALDEHYDE
	conc. sodium hydroxide soln. (Cannizzaro's reaction)	benzyl alcohol, $C_6H_5.CH_2OH$; sodium benzoate, C_6H_5COONa	
	chlorine (no carrier present)	benzoyl chloride, $C_6H_5.COCl$	
	chlorine (halogen carrier present)	*m*-chlorobenzaldehyde, $Cl.C_6H_4.CHO$	
	conc. nitric and sulphuric acids	*m*-nitrobenzaldehyde $O_2N.C_6H_4.CHO$	
	hot conc. sulphuric acid	*m*-benzaldehyde sulphonic acid $SO_3H.C_6H_4.CHO$	

3. Addition Reactions: It forms addition compounds with hydrogen cyanide and sodium bisulphite, but not with ammonia:

e.g.

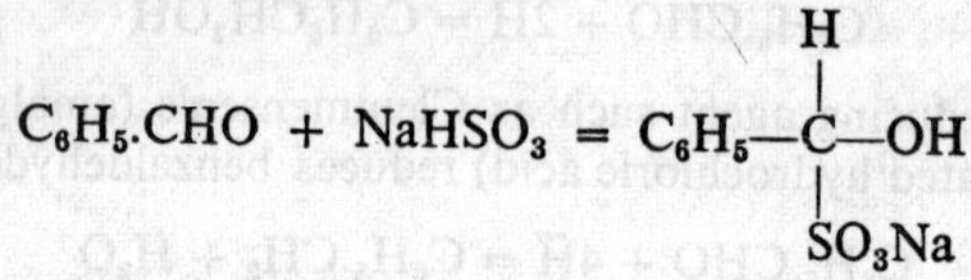

4. Condensation Reactions: Benzaldehyde condenses readily with many compounds, particularly aldehydes, ketones, and amines. With phenylhydrazine it gives yellow crystals of benzaldehyde phenylhydrazone, and with hydroxylamine it forms colourless crystals of benzaldoxime:

$$C_6H_5.CHO + C_6H_5.NH.NH_2 = C_6H_5.CH{:}N.NH.C_6H_5 + H_2O$$
$$C_6H_5.CHO + H_2NOH = C_6H_5.CH{:}NOH + H_2O$$

When benzaldehyde is shaken with concentrated ammonia solution it condenses to a solid colourless mass of hydrobenzamide, m.p. 101 °C:

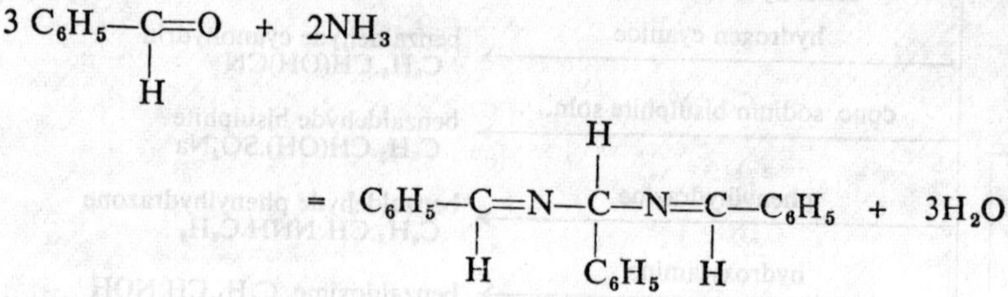

In the presence of dilute sodium hydroxide solution benzaldehyde in alcoholic solution unites with acetaldehyde to give an 'aldol' compound containing both an aldehyde and a secondary alcohol group:

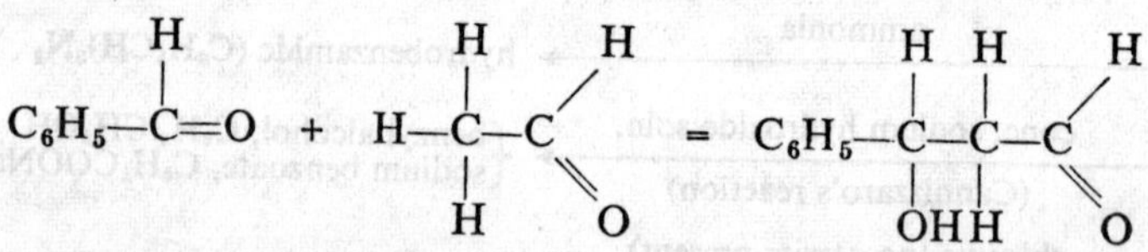

This product is unstable and changes into cinnamic aldehyde by the elimination of a molecule of water:

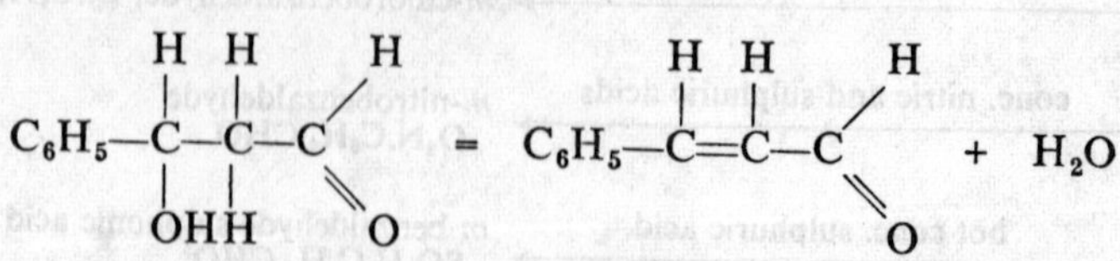

The overall condensation is an example of the Claisen reaction.

5. *Side-chain Chlorination:* Phosphorus pentachloride converts benzaldehyde into benzal chloride:

$$C_6H_5.CHO + PCl_5 = C_6H_5.CHCl_2 + POCl_3$$

If chlorine is bubbled into benzaldehyde in the absence of a halogen carrier, benzoyl chloride is obtained:

$$C_6H_5.CHO + Cl_2 = C_6H_5.COCl + HCl$$

6. *Cannizzaro's Reaction:* When shaken with a concentrated solution of potassium hydroxide, benzaldehyde is converted into benzyl alcohol and potassium benzoate:

$$2C_6H_5.CHO + KOH = C_6H_5.CH_2OH + C_6H_5.COOK$$

7. *Nuclear Substitution:* Chlorination in the presence of iron filings or iodine gives meta-chlorobenzaldehyde. Similarly, benzaldehyde gives meta-derivatives when nitrated and sulphonated.

8. *Benzoin Reaction:* When benzaldehyde is heated under reflux with a solution of potassium cyanide in aqueous alcohol, two molecules of the aldehyde unite to form benzoin:

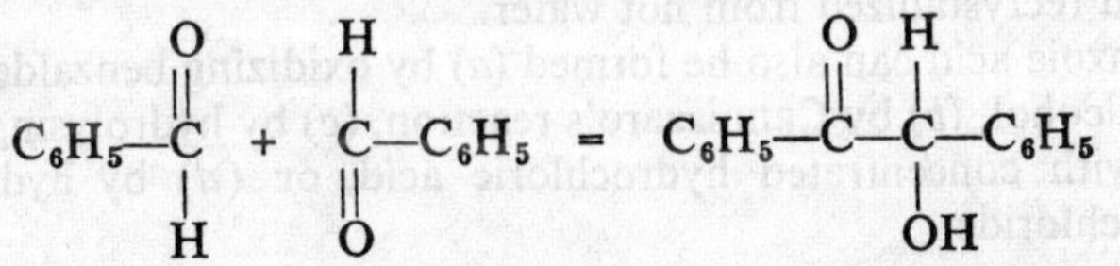

The product, which behaves as both a ketone and a secondary alcohol, forms colourless crystals melting at 137°C. Since it contains an asymmetric carbon atom, it can exist as two optically active isomers (§ 48.2).

Benzaldehyde resembles acetaldehyde in its oxidation and reduction, in its reaction with phosphorus pentachloride, and in many of its addition and condensation reactions. It differs in its reactions with ammonia, sodium hydroxide solution, and chlorine, and in undergoing nuclear substitution, and in the variety of its condensations. Benzaldehyde does not give the iodoform reaction, of course, and it is less ready than acetaldehyde to polymerize. It shows certain similarities to formaldehyde, particularly in its reactions with alkalis (Cannizzaro's reaction) and with ammonia. The reducing actions of benzaldehyde are less rapid than those of aliphatic aldehydes, presumably because of its low solubility in water.

USES OF BENZALDEHYDE: Owing to its almond-like smell it is used for scenting soap and flavouring food. It is also used for making dyes and for synthesizing the antibiotic chloromycetin.

DETECTION OF BENZALDEHYDE: It gives a pale yellow precipitate when treated with phenylhydrazine dissolved in acetic acid, and a white precipitate when shaken with a concentrated solution of sodium bisulphite, (acetaldehyde also reacts with these reagents but the products are too soluble to be precipitated). Benzaldehyde slowly restores the purple colour to Schiff's reagent, and it gives a silver mirror when heated for some time with ammoniacal silver oxide, but it has little action on Fehling's solution. When benzaldehyde is warmed with dilute nitric acid or acidified potassium permanganate solution, a white precipitate of benzoic acid is obtained.

47.4. Benzoic Acid, $C_6H_5.COOH$.

LABORATORY PREPARATION: Several methods are available:

1. Toluene can be oxidized by heating it under reflux with potassium permanganate solution or chromic acid, but the reaction is rather slow:

$$C_6H_5.CH_3 + 3\bar{O} = C_6H_5.COOH + H_2O$$

2. It is better to chlorinate toluene to benzyl chloride first and then heat this with an alkaline solution of potassium permanganate. On acidification benzoic acid is precipitated; it can be filtered off, washed well, and recrystallized from hot water.

3. Benzoic acid can also be formed (*a*) by oxidizing benzaldehyde or benzyl alcohol, (*b*) by Cannizzaro's reaction, (*c*) by hydrolysing benzonitrile with concentrated hydrochloric acid, or (*d*) by hydrolysing benzotrichloride.

PROPERTIES OF BENZOIC ACID: It forms white, plate-like crystals which melt at 122°C but which sublime when heated rapidly. It is only sparingly soluble in cold water, but it dissolves readily in hot water and in organic solvents and it is volatile in steam. Like acetic acid, it is associated into double molecules, $(C_6H_5.COOH)_2$, when dissolved in benzene (see § 6.14).

REACTIONS OF BENZOIC ACID: These closely resemble the reactions of aliphatic carboxylic acids. For example, benzoic acid is a weak, monobasic acid forming salts (benzoates) with alkalis and carbonates. It reacts reversibly with alcohols forming benzoate esters. With phosphorus pentachloride it gives an acid chloride (benzoyl chloride) and hydrogen chloride. It is decarboxylated to benzene when heated strongly with soda-lime, and it gives a good yield of benzyl alcohol when reduced with lithium aluminium hydride.

It differs from acetic acid not only in its physical properties (the benzene nucleus reduces volatility and solubility in water), but also in being a slightly stronger acid than acetic because of the influence of the electron-attracting nucleus, and in undergoing nuclear substitution. For example, it can be halogenated, nitrated, and sulphonated, giving meta-derivatives in each case.

USES OF BENZOIC ACID: It has powerful germicidal properties, and the acid and its salts are used as preservatives of processed foods and soft drinks. It is also used for making benzoyl chloride, dyes, perfumes, and rust inhibitors.

DETECTION OF BENZOIC ACID: It is recognized by its acidic properties (it dissolves readily in alkalis and liberates carbon dioxide from solutions of alkali carbonates) and by its buff precipitate with ferric chloride solution. When heated with ethyl alcohol and a little concentrated sulphuric acid it is converted into ethyl benzoate, which has a characteristic peppermint smell.

47.5. Benzoyl Chloride, C_6H_5COCl.

LABORATORY PREPARATION: By warming benzoic acid with phosphorus pentachloride or thionyl chloride:

$$C_6H_5.COOH + PCl_5 = C_6H_5.COCl + POCl_3 + HCl\uparrow$$
$$C_6H_5.COOH + SOCl_2 = C_6H_5.COCl + SO_2\uparrow + HCl\uparrow$$

The benzoyl chloride is separated by fractional distillation.

PROPERTIES AND REACTIONS OF BENZOYL CHLORIDE: It is a colourless liquid, b.p. 197°C, ρ 1.21 g cm^{-3}. Its vapour has an irritating effect upon the eyes. Unlike acetyl chloride, benzoyl chloride is only slowly hydrolysed by cold water or alkalis. Its main reaction, *benzoylation*, is similar in principle to acetylation and consists of introducing the benzoyl group C_6H_5CO— in place of a hydrogen atom in alcohols, phenols, and amines:

$$C_2H_5OH + C_6H_5.COCl = \underset{\text{ethyl benzoate}}{C_6H_5.COOC_2H_5} + HCl$$

$$C_6H_5OH + C_6H_5.COCl = \underset{\text{phenyl benzoate}}{C_6H_5.COOC_6H_5} + HCl$$

$$C_6H_5NH_2 + C_6H_5.COCl = \underset{\text{benzanilide}}{C_6H_5.NH.COC_6H_5} + HCl$$

Benzoylation is used to convert volatile compounds into crystalline derivatives which can be purified by recrystallization and identified by their melting points. It is usually carried out in the presence of excess of sodium hydroxide solution to facilitate the reaction and absorb the hydrogen chloride. It is then known as the *Schotten-Baumann* reaction.

Benzoyl chloride resembles acetyl chloride in two other reactions. When heated with sodium benzoate it is converted into benzoic anhydride $(C_6H_5CO)_2O$, and when treated with a concentrated solution of ammonia it gives benzamide, $C_6H_5.CONH_2$ (§ 44.9). When benzoyl chloride is shaken with sodium peroxide solution it gives benzoyl peroxide $(C_6H_5CO)_2O_2$. This is an important substance used for bleaching fats and flour and for catalysing many polymerization reactions.

47.6. Ethyl Benzoate, $C_6H_5.COOC_2H_5$.

LABORATORY PREPARATION: The Fischer–Speier method of esterification is used (§ 38.1). A mixture of ethyl alcohol and benzoic acid is boiled under reflux whilst a stream of dry hydrogen chloride gas is

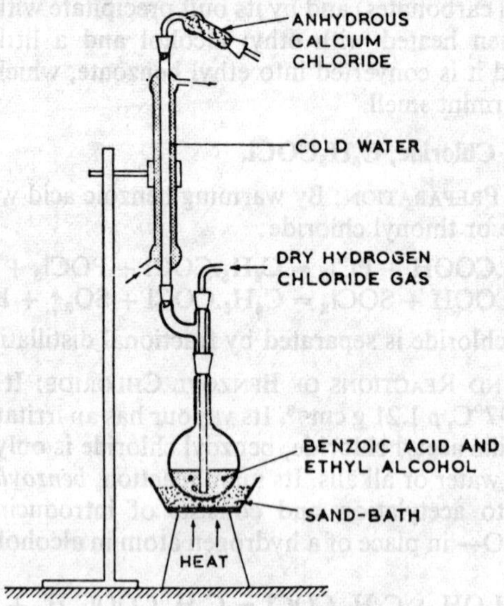

FIG. 47.1 Preparation of Ethyl Benzoate

passed through the mixture. The hydrogen chloride catalyses the esterification and helps to prevent the back reaction; it is used in preference to concentrated sulphuric acid to avoid sulphonation of the benzene ring:

$$C_6H_5.COOH + C_2H_5OH \rightleftharpoons C_6H_5.COOC_2H_5 + H_2O$$

The apparatus used is shown in Fig. 47.1. The top of the reflux condenser is fitted with a calcium chloride tube to prevent ingress of moisture. After boiling for about an hour, the contents of the flask are cooled, added to an excess of water, separated, and washed with sodium carbonate solution to remove the acid present. The ethyl benzoate is then dried with anhydrous calcium chloride and distilled.

PROPERTIES AND REACTIONS OF ETHYL BENZOATE: It is a colourless liquid, b.p. 213°C, ρ 1.05 g cm^{-3}, with a smell of peppermint. It is practically insoluble in water but it dissolves readily in organic solvents.

Ethyl benzoate is hydrolysed when boiled under reflux with sodium hydroxide solution giving sodium benzoate solution and ethyl alcohol:

$$C_6H_5.COOC_2H_5 + NaOH = C_6H_5.COONa + C_2H_5OH$$

It reacts with concentrated ammonia solution to give benzamide (§ 44.9):

$$C_6H_5.COOC_2H_5 + NH_3 = C_6H_5.CONH_2 + C_2H_5OH$$

It is reduced by lithium aluminium hydride (§ 19.14) to benzyl and ethyl alcohols:

$$C_6H_5.COOC_2H_5 + 4\bar{H} = C_6H_5CH_2OH + C_2H_5OH$$

Substitution in the nucleus can be effected by the usual reagents.

47.7. Phenyl Benzoate, $C_6H_5.COOC_6H_5$.

LABORATORY PREPARATION: By the benzoylation (§ 47.5) of phenol. The phenol is dissolved in an excess of 10% sodium hydroxide solution and shaken vigorously in a corked bottle with benzoyl chloride for about a quarter of an hour:

$$C_6H_5OH + C_6H_5.COCl = C_6H_5.COOC_6H_5 + HCl$$

The solid product is filtered off, washed thoroughly with cold water, and recrystallized from methylated spirit.

PROPERTIES AND REACTIONS OF PHENYL BENZOATE: It is a white crystalline solid, m.p. 69°C, practically insoluble in water.

Like all esters, phenyl benzoate can be hydrolysed by boiling under reflux with sodium hydroxide solution although the reaction is slower than for aliphatic esters; the products remaining in solution are sodium phenate and sodium benzoate:

$$C_6H_5.COOC_6H_5 + 2NaOH = C_6H_5.COONa + C_6H_5ONa + H_2O$$

Phenyl benzoate is reduced by lithium aluminium hydride (§ 19.14) to benzyl alcohol and phenol:

$$C_6H_5.COOC_6H_5 + 4\bar{H} = C_6H_5CH_2OH + C_6H_5OH$$

It undergoes substitution in the benzene nucleus.

48

Stereoisomerism

48.1. Introduction. As explained in chapter 32, the name isomerism is given to the phenomenon whereby *two or more different compounds can exist having the same molecular formula but different arrangements of their atoms in the molecule.* Isomerism is of two main kinds, *structural isomerism,* in which the same atoms are linked together in different ways so that the isomers have different structural formulae, and *stereoisomerism,* in which the same atoms are linked together in the same ways but are arranged differently in space. It follows that stereoisomers have identical structural formulae, since the different spatial configurations of the atoms cannot be shown in such formulae. Table 48.1 shows the various types of isomerism and refers to examples of each type. Since many instances of structural isomerism have already been encountered, this kind should now be thoroughly understood. Consequently this chapter will be devoted to the various types of stereoisomerism.

48.2. Optical Isomerism. Light may be regarded as a transverse wave motion. When it passes through certain solids it becomes plane-polarized, i.e. it vibrates at right angles to the direction of propagation in only one plane instead of in an infinite number of such planes. Fig. 48.1, which represents a cross-sectional diagram of an oncoming ray of light, should make this clear; ordinary light is shown as vibrating in all

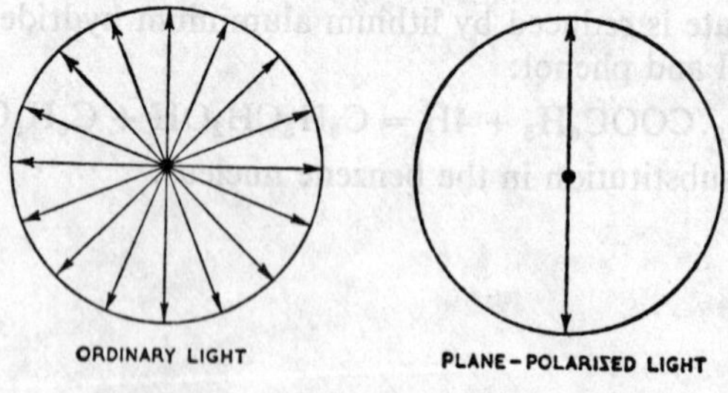

FIG. 48.1 Polarization of Light

TABLE 48.1. THE TYPES OF ISOMERISM

- ISOMERISM
  - STRUCTURAL ISOMERISM (atoms linked in different ways)
    - CHAIN ISOMERISM—different carbon chains or skeletons, e.g. the paraffins (§ 33.5)
    - METAMERISM—different radicals linked to the same functional group, e.g. the ethers (§ 39.3)
    - POSITION ISOMERISM—substituent groups in different positions on the same carbon skeleton, e.g. the propyl alcohols (§ 35.4) and o, m, + p. aromatic nuclear derivatives (§ 43.4)
    - SERIES ISOMERISM—involves compounds in different homologous series, e.g. { ethyl alcohol and dimethyl ether; acetic acid and methyl formate; acetone and propionaldehyde }
    - TAUTOMERISM—a dynamic equilibrium exists between the isomers, e.g. { glycine (§ 41.12); urea (§ 41.9) }
  - STEREOISOMERISM (atoms arranged differently in space)
    - OPTICAL ISOMERISM—occurs in compounds with asymmetric carbon atoms, e.g. { lactic acid (§ 48.3); tartaric acid (§ 48.4) }
    - GEOMETRICAL ISOMERISM—occurs in certain compounds containing double bonds, e.g. maleic and fumaric acids (§ 48.5)
    - CYCLIC ISOMERISM—occurs in certain ring molecules, e.g. { glucose (§ 42.2); benzene hexachloride (§ 43.3) }

directions, whereas light that has passed through a Nicol prism made from pieces of calcite or through a sheet of polaroid material vibrates only in one direction, say vertically up and down. If this polarized light is passed through certain substances it is found that the plane of polarization is rotated. If the rotation is in a clockwise direction as observed in the eyepiece the substance is said to be *dextro-rotatory*, and if in an anti-clockwise direction, *laevo-rotatory*. Substances which cause rotation of the plane of polarized light in this way are said to be *optically active*. They are of two kinds, those which are optically active only in the solid state, such as quartz, and those compounds which are optically active when molten or in solution as well as when crystalline. In this chapter we are concerned only with the latter kind.

The apparatus used for studying optical activity is called a polarimeter. It consists of a glass tank placed between two Nicol prisms as shown diagrammatically in Fig. 48.2. To measure the optical activity of a given substance the prism B is first adjusted to a position where minimum light is observed in the eyepiece and its setting is noted. In this position the axes of the two prisms are exactly at right angles so that none of the polarized light produced by the polarizer A is transmitted through the analyser B. The tank is then filled with a solution of

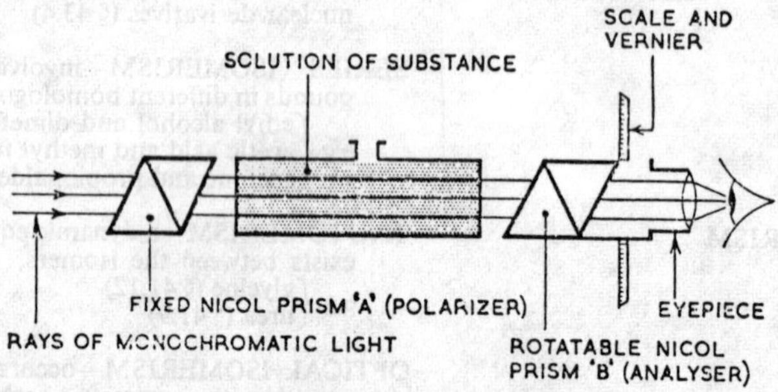

FIG. 48.2 The Polarimeter

the substance of known concentration and the prism B is rotated to the new position of maximum darkness. The change in the setting of the analyser is the rotation caused by the solution. For any given solute the rotation depends upon the length and concentration of solution, the temperature, and the wavelength of the light. Since all these variables can be measured or specified, the rotation per decimetre of a solution containing 1 g of solute per cm^3 of solution at 25°C for sodium light of D wavelength, known as the *specific rotation* [α], can be calculated for any substance. It is conventional to regard [α] as positive for dextrorotatory substances and negative for ones which are laevo-rotatory.

Only a small proportion of organic substances are optically active. When these substances are thoroughly investigated each is found to exist in two or more forms which differ in the direction in which they rotate polarized light. These forms, called *optical isomers*, are very similar in their physical and chemical properties, the main difference between them being their opposite effect upon polarized light. The two isomers are distinguished by prefixing their names by d- and l-. These prefixes stand for dextro and laevo respectively, and indicate the direction in which the plane of polarized light is rotated by a solution of the substance. They are not intended to denote the absolute configurations of the atoms in the molecule, which should be indicated by the prefixes D- and L-, and which are not considered in this book.

When the structures of these optically active substances are determined it is found that they all have one feature in common—the presence of one or more *asymmetric carbon atoms*, i.e. carbon atoms linked to four different atoms or groups. Fig. 48.3 shows a theoretical example, Cabcd, projected on to a plane, where a, b, c, and d are all different. The carbon atom itself is not asymmetrical, but the arrangement of atoms around it is, giving an asymmetry to the molecule as a whole. All molecules containing asymmetric carbon atoms give rise to

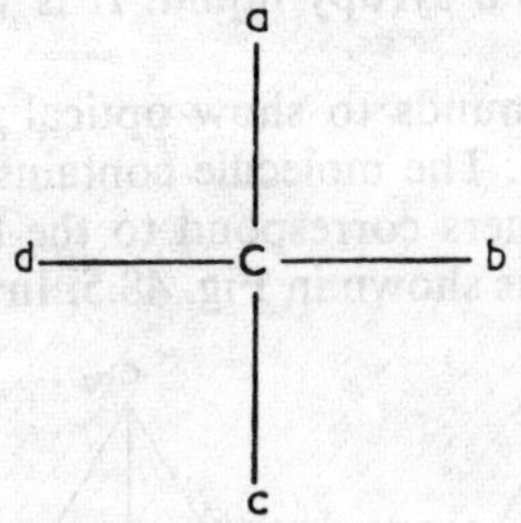

FIG. 48.3 The Asymmetric Carbon Atom

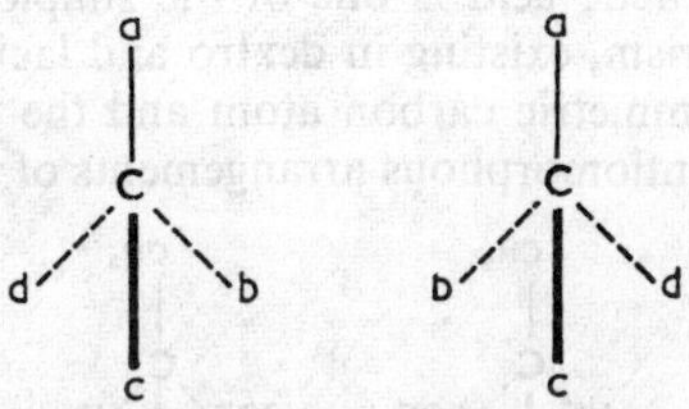

FIG. 48.4 Enantiomorphs

optical isomers, the maximum number being 2^n, where n is the number of asymmetric carbon atoms in the molecule.

The explanation of the existence of optical isomers lies in the tetrahedrally-directed valencies of the carbon atom, as suggested by van't Hoff and Le Bel in 1874. The molecule Cabcd can be regarded as a tetrahedron with the groups a, b, c, and d at the corners and the carbon atom at the centre, so that the angle between each of the bonds is 109° 28′. This view is confirmed by X-ray analysis, electron diffraction, and other physical evidence. On consideration it will be seen that two forms of such a molecule are possible, as shown in Fig. 48.4. These two forms are mirror images of each other and are known as *enantiomorphs*. They cannot be superimposed upon each other, their relationship being similar to that of a right-hand glove and a left-hand glove. These enantiomorphs are identical in their physical properties except in their effect upon polarized light, which they rotate by equal amounts in opposite directions. Thus optical isomerism is accounted for by the possible alternative spatial configurations of the atoms when four different atoms or groups are arranged tetrahedrally around a carbon atom.

Molecules of the types Ca_4 (e.g. CH_4), Ca_3b (e.g. CH_3Cl), Ca_2b_2 (e.g.

CH_2Cl_2), and Ca_2bc (e.g. $CH_2Cl.COOH$), show no optical isomerism and only one form of each is known. This is strong evidence for the tetrahedral direction of the carbon valencies, since any other spatial arrangement would inevitably give rise to more than one form of one of these molecules and some stereoisomerism would be expected.

We are now in a position to consider some important examples of optical isomerism. The first we encountered were butan-2-ol (§ 35.5) and alanine (§ 41.15).

48.3. Lactic Acid, $CH_3.CH(OH).COOH$. This compound is a colourless hygroscopic solid with a sour taste and smell, but it is so difficult to crystallize that it is usually encountered as a syrupy liquid. It is very soluble in water.

Lactic acid is one of the simplest compounds to show optical isomerism, existing in dextro and laevo forms. The molecule contains an asymmetric carbon atom and the two isomers correspond to the two enantiomorphous arrangements of the atoms shown in Fig. 48.5. In the

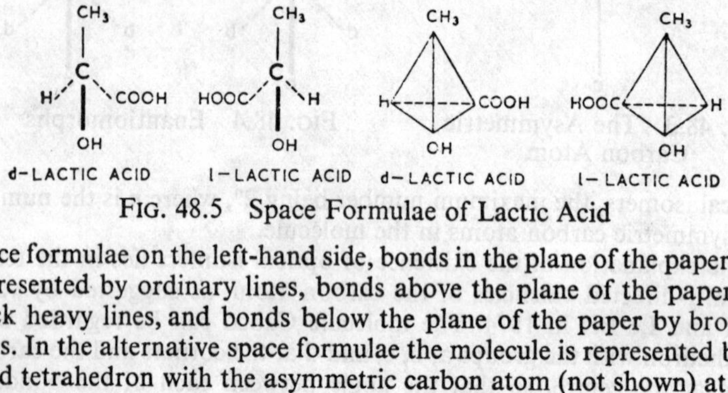

FIG. 48.5 Space Formulae of Lactic Acid

space formulae on the left-hand side, bonds in the plane of the paper are represented by ordinary lines, bonds above the plane of the paper by thick heavy lines, and bonds below the plane of the paper by broken lines. In the alternative space formulae the molecule is represented by a solid tetrahedron with the asymmetric carbon atom (not shown) at the centre and the various atoms or groups at the corners. Both types of formula show the mirror image relationship between the two isomers which accounts for their opposite effects upon polarized light, but the student should construct some three-dimensional models to satisfy himself that the difference is clearly understood.

A third form of lactic acid exists which does not rotate the plane of polarized light. This optically inactive form consists of equal amounts of the dextro and laevo forms combined together so that their effects upon polarized light cancel out. For this reason it is said to be *externally compensated* and is called a *dl compound.* Alternatively it is known as a *racemic compound* or *racemate.* When a compound capable of optical isomerism is synthesized from optically inactive substances, it is the racemic compound which is formed. For example, dl-lactic acid is obtained when lactic acid is prepared in either of the following ways:

1. 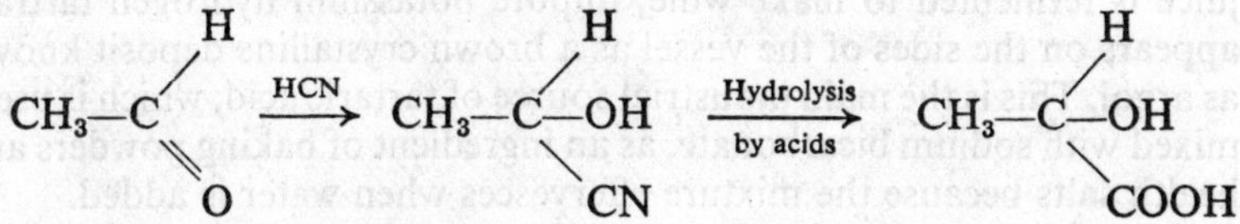

2. $$CH_3.CH_2.COOH \xrightarrow{Br_2} CH_3.CHBr.COOH \xrightarrow{hydrolysis} CH_3.CH(OH).COOH.$$

The racemic acid is also the usual product when lactic acid is obtained from sour milk.

When solutions containing equal amounts of the d- and l- forms of lactic acid are mixed and allowed to crystallize, crystals of the dl acid are obtained. The molecular weight of these crystals is twice that of the individual d- and l- forms, suggesting that the racemic compound resembles a double salt. It melts at 18°C and dissolves in water giving a solution containing equal amounts of each optically active isomer:

$$\text{dl-lactic acid} \rightleftharpoons \text{d-lactic acid} + \text{l-lactic acid}.$$

For a few substances, but not for lactic acid, the position of this equilibrium is so far over to the right-hand side that separate crystals of the two optically active isomers are deposited when the solution is concentrated, giving a *racemic mixture*. Such a mixture is optically inactive when dissolved in water because it contains equal amounts of each isomer. In rare cases the crystals in a racemic mixture can be distinguished from each other by their enantiomorphous shapes and can be separated by hand picking, as Pasteur demonstrated in his historic experiment with sodium ammonium tartrate in 1848.

d-Lactic acid, m.p. 26°C, occurs naturally in muscle and is usually prepared from meat extract. The laevo form, m.p. 26°C, does not occur naturally, but it can be prepared from the racemic compound by adding an optically active base such as d-cinchonine (which occurs naturally in Peruvian bark), or l-brucine, or l-strychnine. If d-cinchonine is used, the two salts d-lactic acid d-cinchonine and l-lactic acid d-cinchonine are formed. These salts differ in their solubility and can therefore be separated from each other by fractional crystallization. They are then reconverted into the separate d and l forms of lactic acid by acidification. This is the best general method of separating or *resolving* racemic forms into their optically active constituents. Other methods, such as the use of selective fermentation of one form by moulds, or the physical separation of enantiomorphous crystals in a racemic mixture by hand-picking, are of only very limited application and consequently of little practical importance.

48.4. Tartaric Acid, COOH.CH(OH).CH(OH).COOH. When grape-juice is fermented to make wine, impure potassium hydrogen tartrate appears on the sides of the vessel as a brown crystalline deposit known as argol. This is the main industrial source of tartaric acid, which is used, mixed with sodium bicarbonate, as an ingredient of baking powders and health salts because the mixture effervesces when water is added.

Tartaric acid is a colourless crystalline solid, very soluble in water. It decomposes when heated, leaving a black residue of carbon. Being a dibasic acid, it gives rise to two series of salts, e.g. normal potassium tartrate, $\begin{array}{l}\mathrm{CH(OH).COOK}\\ \vert\\ \mathrm{CH(OH).COOK}\end{array}$, and the acid salt potassium hydrogen tartrate $\begin{array}{l}\mathrm{CH(OH).COOK}\\ \vert\\ \mathrm{CH(OH).COOH}\end{array}$, which is commonly known as cream of tartar. This acid salt is important because it is only slightly soluble in water and can therefore be used for detecting potassium in analysis (see § 17.26). When its solution is neutralized with sodium carbonate and evaporated, crystals of sodium potassium tartrate, $\begin{array}{l}\mathrm{CH(OH).COONa}\\ \vert\\ \mathrm{CH(OH).COOK}\end{array}.4H_2O$ are obtained, known as Rochelle salt. The latter is used in the laboratory for making *Fehling's solution* by adding its alkaline solution to a solution of copper sulphate (§ 28.4).

Tartaric acid and its salts are strong reducing agents. They decolorize potassium permanganate solution and give a silver mirror with an ammoniacal solution of silver oxide. Tartrates can be distinguished from oxalates by adding calcium chloride solution because the white precipitate of calcium tartrate, unlike calcium oxalate, dissolves in acetic acid.

Two optically active isomers of tartaric acid are known which are identical in all their properties except in their effect upon the plane of polarized light, which they rotate in opposite directions. They are referred to as the d- and l- forms accordingly. In addition, two optically inactive isomers exist, the racemic compound or dl acid, and mesotartaric acid. These two forms differ from each other and from the d- and l- forms in their physical properties and in their crystalline shape. Isomers of this kind which are not enantiomorphous and differ physically are called *diastereoisomers.*

When tartaric acid is prepared from argol in the usual way, the d-form is obtained. If this acid is heated with water under pressure to about 165°C it is converted into a mixture of racemic and mesotartaric acids which can be separated by fractional crystallization. This process whereby optically active isomers change into the racemic form when heated is called *racemization*. The laevo acid is prepared from the racemic acid by resolving it with an optically active base, as for lactic acid.

The stereoisomerism of tartaric acid can be readily accounted for in terms of the three possible configurations of the atoms in a molecule which contains two asymmetric carbon atoms. These configurations are shown by means of space formulae and tetrahedral formulae in Fig. 48.6. In explaining the difference between these isomers it is helpful

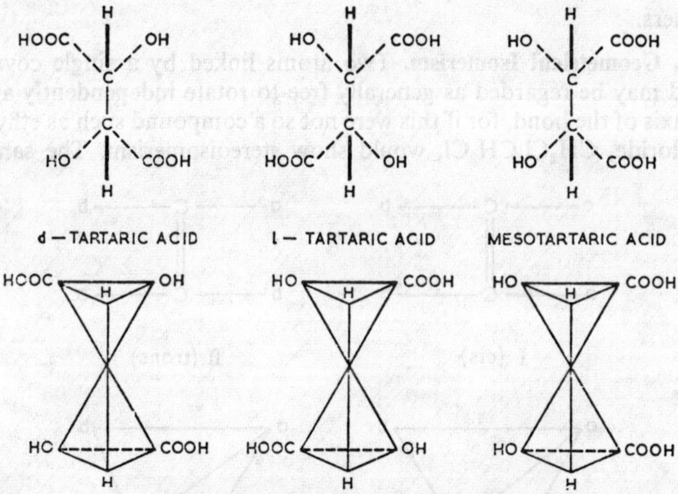

48.6 Space Formulae of Tartaric Acid

to regard the molecule of tartaric acid as composed of two similar groups linked together, each containing one asymmetric carbon atom. If the configurations of these two groups are such that they are both dextro-rotatory, for example, then the two halves of the molecule reinforce one another in their effect and the molecule as a whole will be dextro-rotatory too, and we shall have the d-form of the acid. Conversely, if the effect of both groups is to rotate the plane of polarized light in an anti-clockwise direction, then the molecule will be that of l-tartaric acid. As Fig. 48.6 makes clear, these two forms will be enantiomorphous with each other.

A third type of molecule will exist in which a group with dextro-rotatory configuration will be linked to one with laevo-rotatory properties. In this case the two parts of the molecule will have equal but opposite effects upon polarized light and the molecule as a whole will be optically inactive. Mesotartaric acid is produced in this way and is said to be *internally compensated.* It cannot be separated into the d- and l- forms because it does not contain either of them; its molecular weight is the same as that of the optically active forms.

The fourth isomer of tartaric acid, the racemic acid, is composed of equal amounts of the d- and l- forms. It is optically inactive by external compensation, therefore, and like the racemic form of lactic acid it can be separated into the d- and l- forms by the usual methods. Its melting point (206°C) is higher than that of the d- and l- forms (170°C) showing that it is a distinct compound and not merely a mixture of the two isomers.

48.5. Geometrical Isomerism. Two atoms linked by a single covalent bond may be regarded as generally free to rotate independently about the axis of the bond, for if this were not so a compound such as ethylene dichloride, $CH_2Cl.CH_2Cl$, would show stereoisomerism. The same is

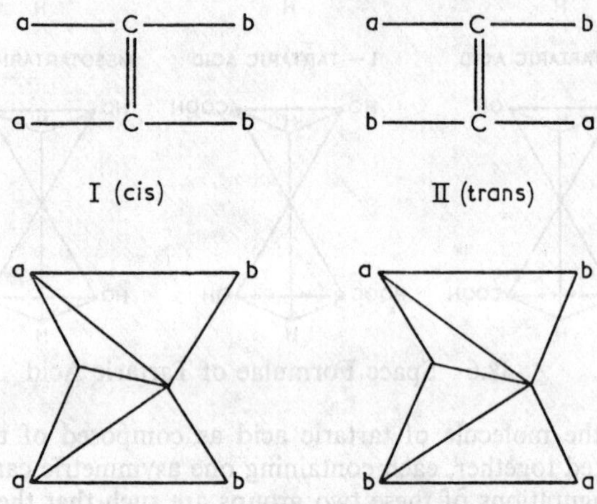

FIG. 48.7 Geometrical Isomers

not true of two atoms joined by a double bond, for this confers upon a molecule a certain rigidity giving rise to a form of stereoisomerism known as geometrical isomerism. The point is best illustrated by the hypothetical compound Cab : Cab, where a and b are different atoms or groups of atoms joined to the two carbon atoms. The configurations of the two isomers are shown in Fig. 48.7.

The compound with like groups on the same side of the double bond is known as the *cis* form, and the other as the *trans* form. Such compounds are not optical isomers because both molecules contain planes of symmetry in the plane of the paper. In addition, I has a plane of symmetry at right angles to the C=C bond and II a centre of symmetry.

One of the best-known examples of this type of isomerism is provided by maleic and fumaric acids, which may be represented by the formulae:

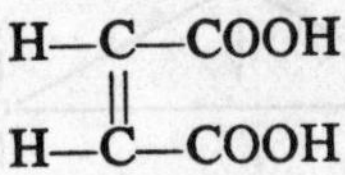

maleic acid
m.p. 130°C ρ 1.59 g cm^{-3}
solubility: 79 g in 100 g water at 25°C

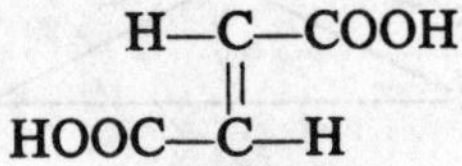

fumaric acid
m.p. 287°C ρ 1.64 g cm^{-3}
solubility: 0·7 g in 100 g water at 25°C

These two forms also differ chemically, maleic acid readily forming an anhydride when heated to 160°C owing to the proximity of the two carboxyl groups, whereas fumaric acid gives the anhydride only above 230°C:

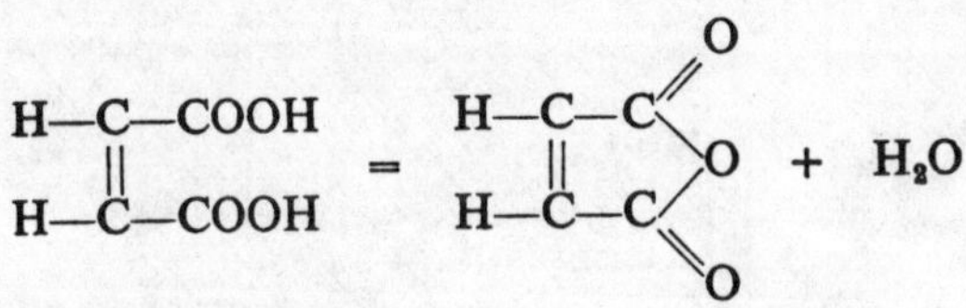

Another example is afforded by 1,2-dichloroethylene:

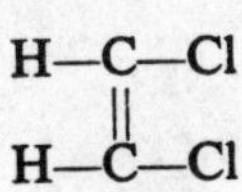

cis form
b.p. 60°C

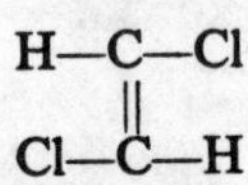

trans form
b.p. 48°C

These two isomers are readily distinguished by measuring the dipole moment (§ 4.9) which is zero for the trans form but appreciable for the cis form in which the two electronegative chlorine atoms are on the same side of the molecule.

Geometrical isomerism is not limited to compounds containing an ethylenic, $>C=C<$, bond and is also shown by benzaldoxime thus:

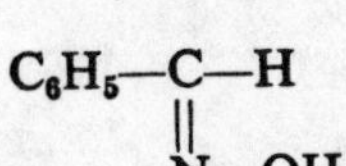

cis or syn form, m.p. 34°C

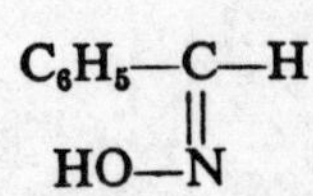

trans or anti form, m.p. 129°C

48.6. Cyclic Isomerism. This is really a special case of geometrical isomerism in which two carbon atoms are not free to rotate because they form part of a cyclic or ring compound. The simplest example is shown in Fig. 48.8, where one form is shown with all three X groups on

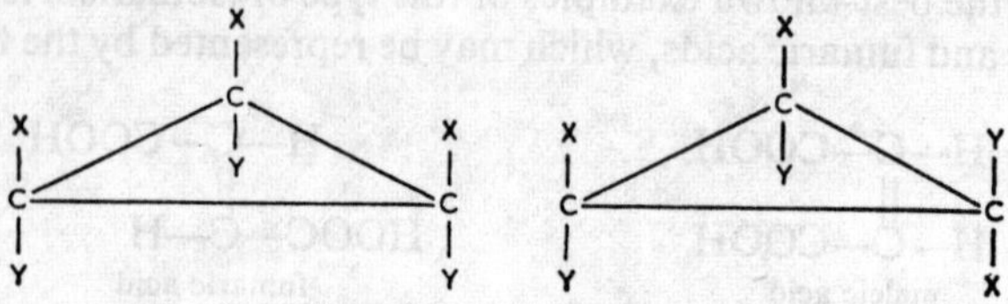

FIG. 48.8 Cyclic Isomers

one side of the ring and the other form with two X groups on one side and one on the other side. Examples of this type of isomerism include benzene hexachloride (§ 43.3, reaction 6), glucose (see Fig. 42.1) and fructose.

49

Polymers

49.1. Introduction. In the preceding chapters numerous examples have been given of *polymerization*, the process whereby relatively small and simple molecules link together to form much larger units known as *polymers*. In view of the rapidly growing chemical and industrial importance of polymers and of their already widespread use as plastics,* resins, synthetic rubbers, and fibres, it is desirable to devote a separate chapter to them and to the processes employed for making them. Some idea of the scale of production of these polymers can be gained from the fact that the world output of them has doubled in the last seven years and now exceeds the total tonnage of all the non-ferrous metals put together.

49.2. Addition Polymerization. As the name implies, polymerization of this kind proceeds by the direct addition of one molecule to another to give a new substance without forming any other product. It follows that in this type of polymerization the molecular weight of the polymer is an integral multiple of that of the *monomer*, as the original unpolymerized molecules are called.

i.e. $$n\,\mathrm{A} \rightarrow (\mathrm{A})_n$$

Addition polymerization is only possible when the monomer is unsaturated and most examples involve organic molecules with $\rangle C{=}C\langle$ double bonds (see Table 49.1).

For example, ethylene (§ 34.2) polymerizes to the plastic *polyethylene*, popularly known as *polythene*, thus:

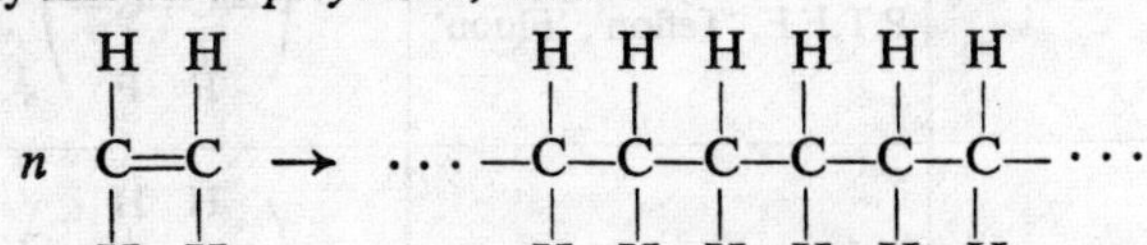

i.e. $$n\,\mathrm{CH_2{:}CH_2} \rightarrow (\mathrm{CH_2.CH_2})_n$$

* A plastic is an organic substance, usually polymetric in structure, which can be moulded into desired shapes.

Table 49.1. Some Important Addition Polymers

Monomer	*Polymer*	
Name	*Chemical name and common or trade name*	*Formula*
Ethylene (§ 34.2)	Polyethylene Polythene	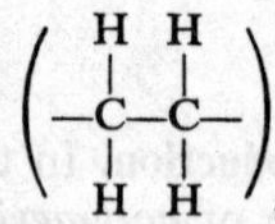
Propylene (§ 34.3)	Polypropylene 'Propathene'	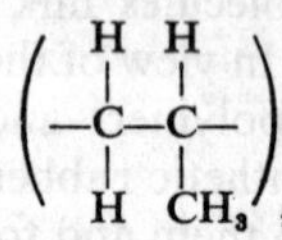
Vinyl Chloride (§ 34.4)	Polyvinyl Chloride PVC	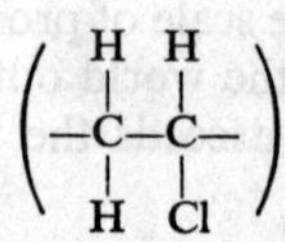
Acryonitrile (§ 34.4)	Polyacrylonitrile 'Courtelle', 'Orlon', 'Acrilan'	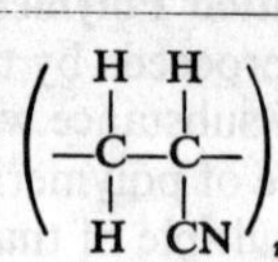
Methyl Methacrylate (§ 36.6)	Polymethylmethacrylate 'Perspex'	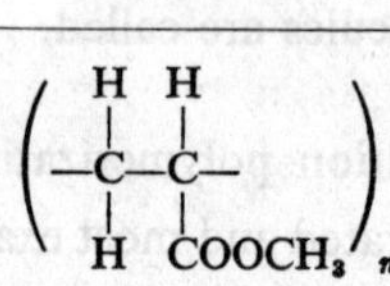
Tetrafluoroethylene (§ 40.11)	Polytetrafluoroethylene P.T.F.E. 'Teflon', 'Fluon'	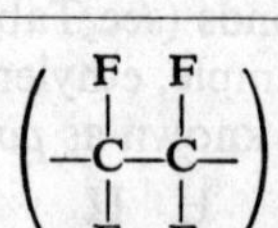
Styrene (§ 43.3)	Polystyrene —	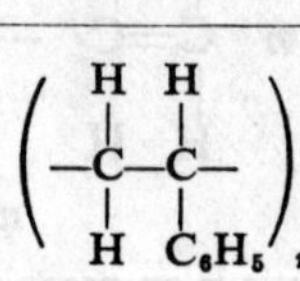

Although only six carbon atoms are shown in the simple representation, in practice the chains are often of great length consisting of many hundreds of ethylene molecules joined together. This gives the polymer a very high molecular weight and confers upon it certain characteristic physical properties. Instead of possessing a sharp and distinct melting point it tends to soften gradually over a range of temperature and to change on heating into a liquid which is viscous because the long chain molecules are still tangled together and loosely linked to each other by van der Waals forces.

Most addition polymerizations are chain reactions (§ 7.15). Once initiated they proceed very rapidly indeed (huge molecules containing thousands of atoms are formed in a fraction of a second), and as is usually the case in chain reactions they are very sensitive to catalysts capable of producing free radicals or ions and also to photochemical activation (§ 7.17). The physical properties of the polymer are considerably influenced by the catalyst or initiator used to bring about the polymerization; these substances can affect the chain length and the degree of branching and cross-linking which occurs during the formation of the polymer and also the orientation of the individual units in the chain molecule. For example, when the Ziegler catalysts (aluminium triethyl and titanium tetrachloride) discovered in 1953 are used the units of monomer tend to add to the polymer chain in a regular way with a specific orientation. This leads to linear molecules of polymer, not so contorted and folded as those produced by random polymerization, with the results that on solidifying the final product has a more orderly and partially crystalline structure with improved physical properties. Cross-linking between adjacent chains can be induced by irradiation of the polyethylene; this has the effect of giving greater rigidity and toughness and a higher softening temperature in much the same way as does vulcanization of natural rubber (§ 49.4). Another characteristic feature of a cross-linked structure is insolubility in organic solvents. Unlike linear polymers which usually dissolve readily, most cross-linked polymers are insoluble and when treated with organic liquids merely undergo slight swelling.

Similarly, the polymerization of propylene (§ 34.3) to *polypropylene* may be represented in the following way:

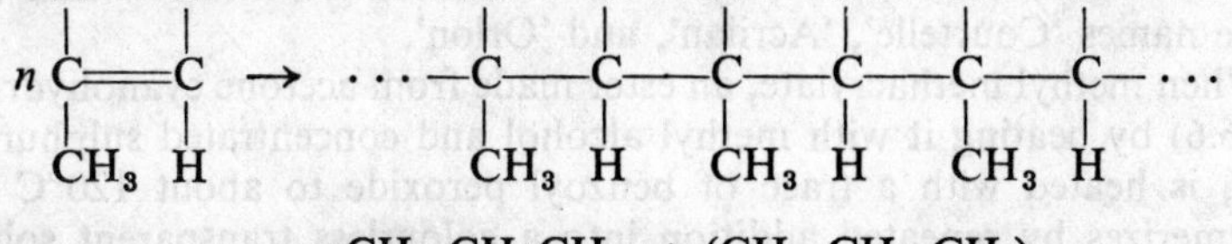

i.e. $$n\,CH_3.CH{:}CH_2 \rightarrow (CH_3.CH_2.CH_2)_n$$

Both polyethylene and polypropylene are very useful plastics now manufactured in large quantities. They are tough, flexible, and of low density. Being fully saturated they are unreactive and are not attacked by acids or alkalis, nor do they corrode or rot on prolonged exposure to the atmosphere or soil. They are excellent electrical insulators with good dielectric properties which make them very suitable for use in high frequency cables. Above all, the readiness with which they can be melted and then extruded or moulded whilst hot into sheets, pipes, or containers of any desired shape leads to their widespread use in everyday life.

Another example of addition polymerization is the formation of the important plastic *polyvinyl chloride*, commonly known by its initial letters 'PVC', from vinyl chloride (§ 34.4 and Table 34.1) thus:

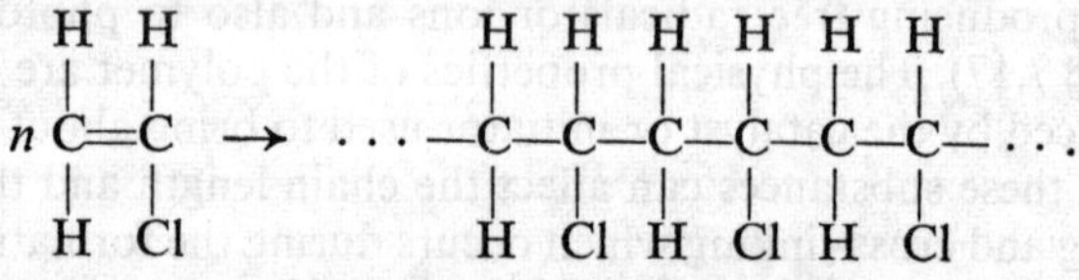

i.e. $n\,CH_2{:}CHCl \rightarrow (CH_2.CHCl)_n$

When mixed with a suitable *plasticizer* (i.e. a substance added to a plastic during manufacture to lower its softening temperature and reduce its brittleness) PVC is a tough flexible material which is used, amongst other things, for making protective clothing, floor tiles, leather-cloth, hoses and piping of all kinds, and for covering cables and electrical wiring.

Acrylonitrile or vinyl cyanide (§ 34.4) is another unsaturated molecule which is polymerized on a large scale with the aid of a peroxide catalyst:

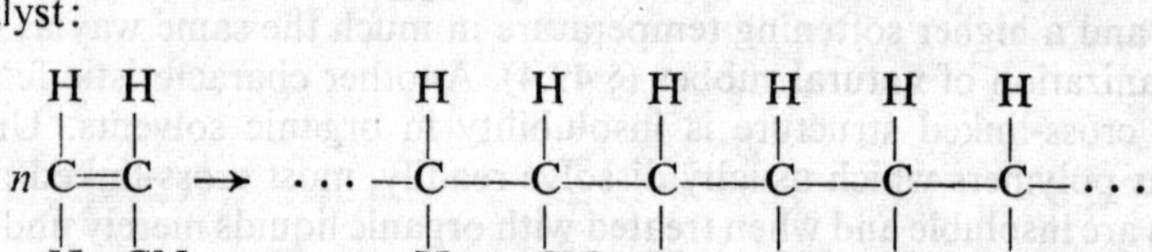

i.e. $n\,CH_2{:}CHCN \rightarrow (CH_2.CHCN)_n$

The polymer, *polyacrylonitrile*, is a valuable synthetic fibre; yarns spun from it closely resemble wool and are sold under the well known trade names 'Courtelle', 'Acrilan', and 'Orlon'.

When methyl methacrylate, an ester made from acetone cyanohydrin (§ 36.6) by heating it with methyl alcohol and concentrated sulphuric acid, is heated with a trace of benzoyl peroxide to about 120°C it polymerizes by repeated addition into a colourless transparent solid *polymethylmethacrylate*, better known under the trade name 'Perspex':

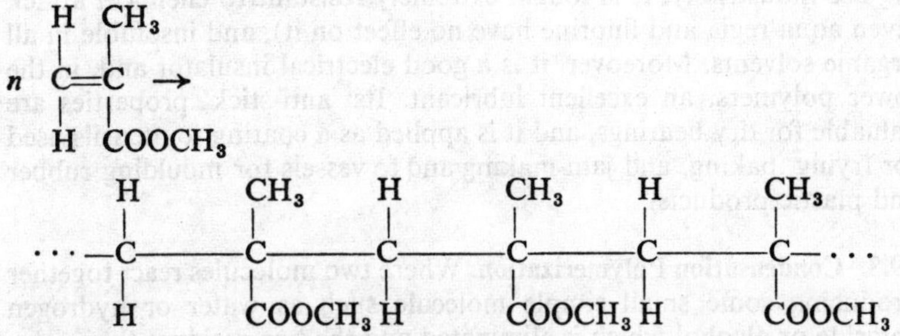

i.e. $n\ CH_2{:}CCH_3.COOCH_3 \rightarrow (CH_2.CCH_3.COOCH_3)_n$

This plastic can be moulded or machined and finds use as a substitute for glass and as a decorative material.

Another important example of addition polymerization is the production of the plastic *polystyrene* which is formed when styrene made by the catalytic dehydrogenation of ethylbenzene (§ 43.3) is heated in the presence of a peroxide catalyst:

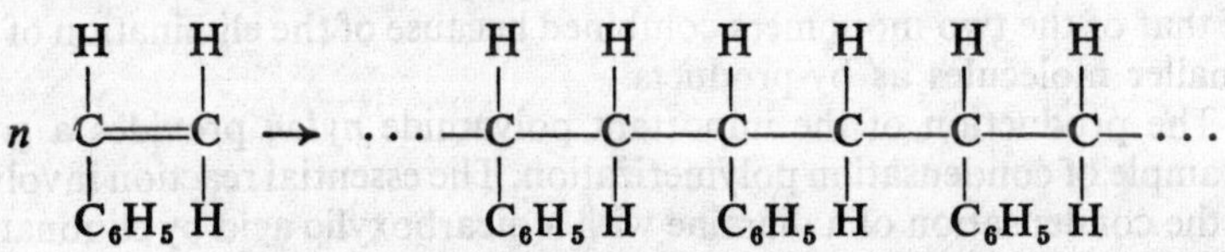

i.e. $n\ C_6H_5.CH{:}CH_2 \rightarrow (C_6H_5.CH.CH_2)_n$

Although rather brittle, polystyrene has excellent electrical insulating and dielectric properties and is highly decorative in its transparent form. It is extensively used for making parts of refrigerators, radios, toys, and household goods.

Tetrafluoroethylene (§ 40.11) provides another example of addition polymerization when it polymerizes under pressure in the presence of benzoyl peroxide to the plastic *polytetrafluoroethylene*, better known by the initial letters P.T.F.E. and sold under the trade names 'Fluon' and 'Teflon':

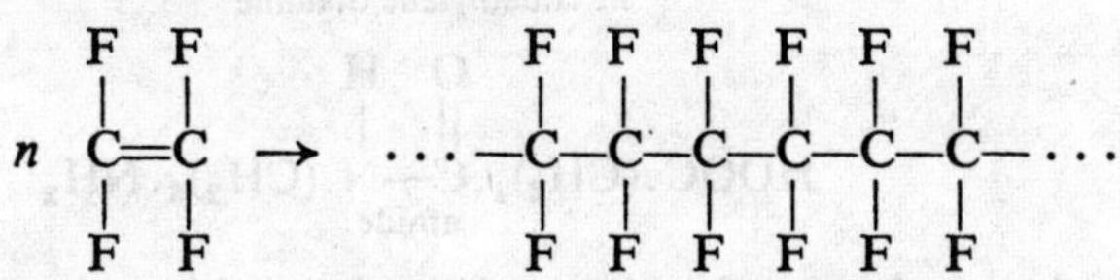

i.e. $n\ CF_2{:}CF_2 \rightarrow (CF_2.CF_2)_n$

This polymer has remarkable properties, which are leading to its increasing use industrially. It is tough, extremely resistant to chemical attack (even aqua regia and fluorine have no effect on it), and insoluble in all organic solvents. Moreover, it is a good electrical insulator and, in the lower polymers, an excellent lubricant. Its 'anti-stick' properties are valuable for dry bearings, and it is applied as a coating to utensils used for frying, baking, and jam-making and to vessels for moulding rubber and plastic products.

49.3. Condensation Polymerization. Where two molecules react together producing some small simple molecule such as water or hydrogen chloride or alcohol which is eliminated and the two residues then combine together to form a new compound of higher molecular weight than either component, the reaction is referred to as a *condensation.* If both the molecules which are condensing together in this way contain two or more suitable groups in each molecule, then the condensation process can be repeated indefinitely producing a very long chain molecule of high molecular weight. Such a process is known as *condensation polymerization.* It should be noted that in contrast to addition polymerization, the molecular weight of the polymer is not an exact multiple of that of the two monomers combined because of the elimination of the smaller molecules as by-products.

The production of the important polyamide *nylon* provides a good example of condensation polymerization. The essential reaction involved is the condensation of a diamine with a dicarboxylic acid by eliminating a molecule of water to form an amide which then condenses repeatedly with further molecules of diamine and acid to form a long chain polymer. The commonest nylon, known as nylon 66, is made from two compounds each containing six carbon atoms, hexamethylene diamine and adipic acid, which condense in the following way when heated together:

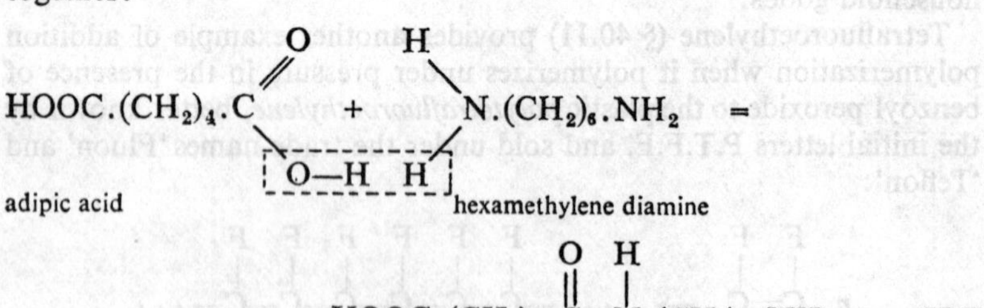

The amide then undergoes further condensations with molecules of the acid and diamine to give a polymer with the formula

$$(\text{—OC.(CH}_2)_4\text{.CONH.(CH}_2)_6\text{.NH—})_n$$

where n may be fifty or more, giving the nylon a molecular weight of about 12 000.

The adipic acid is made by oxidation of cyclohexanol (see reaction 7 of § 45.2) with nitric acid. Hexamethylene diamine is made by catalytic dehydration of adipic acid to adiponitrile $(CH_2)_4.(CN)_2$ followed by hydrogenation over a nickel catalyst.

The nylon polymer is a tough, colourless solid melting at about 250°C. It can be moulded or machined and is used for making bearing bushes and similar objects, but by far its greatest use is as a synthetic fibre in stockings, shirts, lingerie, sheets, parachutes, ropes, carpets, and brushes, where its hard wearing qualities make it economical.

Another example of condensation polymerization is the formation of the polyester *'Terylene'* from dimethyl terephthalate (obtained from *p*-xylene in coal tar, § 43.2) and ethylene gycol (§ 35.8). These two compounds condense together when heated with the elimination of a molecule of methyl alcohol in the following way:

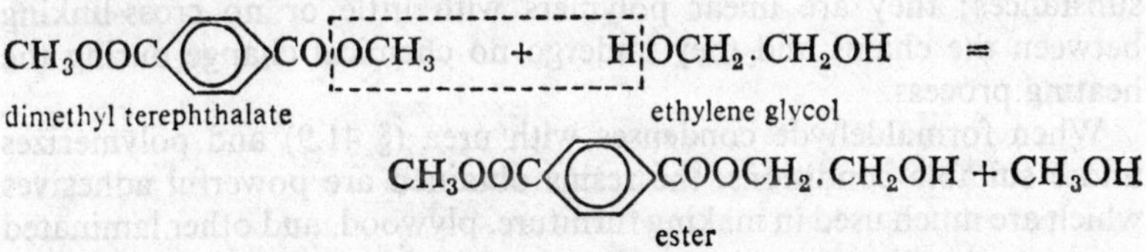

This ester, because it still contains active groups at each end of its molecule, is capable of further condensations forming a long chain polymer with the formula

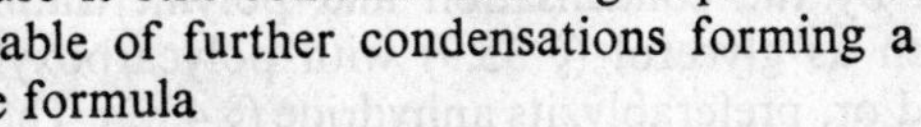

'Terylene'

and a molecular weight of about 15 000.

Terylene is much used as a synthetic fibre. It is hardwearing, resistant to creasing, moth-proof and rot-proof, and is being increasingly used for making not only clothes but also ropes, tyre-cord, safety-belts, tents, and sails. A similar polymer is made in the U.S.A. under the name of 'Dacron'.

Ethylene glycol (§ 35.8) forms a whole series of *polyester resins* by condensation polymerization. When reinforced with glass fibre these give rise to strong, light, corrosion-proof materials very suitable for making parts of vehicles and aircraft, crash helmets, and the hulls of small boats. By allowing polyester resins to react with isocyanates certain sponge-like materials known as *polyurethane foams* are obtained which are finding ready use as heat insulation in refrigerators and buildings and for packaging and upholstery, where their very low density and resistance to rot and insects are important advantages.

One of the earliest condensation polymerizations to give a commercially useful product was the formation of a plastic known as 'Bakelite' from formaldehyde (§ 36.3) and phenol (§ 45.2). This polymer has been used for making telephone receivers, electrical fittings, pens, buttons, and similar objects; it is readily moulded whilst hot but when it is cooled it sets into a hard brittle material of the required shape. Bakelite is a *thermo-setting* plastic, i.e. one which can be readily moulded or cast when heated but in the process undergoes irreversible chemical changes resulting in extensive cross-linking. It cannot be melted again by reheating and once set it forms a rigid infusible product. This should be contrasted with the majority of plastics which resemble polyethylene, polyvinyl chloride, and polystyrene in that the processes of softening on heating and hardening on cooling are entirely reversible and can be repeated indefinitely. Plastics which behave in this way and can be alternately softened and hardened at will are known as *thermoplastic* substances; they are linear polymers with little or no cross-linking between the chains and they undergo no chemical change during the heating process.

When formaldehyde condenses with urea (§ 41.9) and polymerizes under suitable conditions, the resins obtained are powerful adhesives which are much used in making furniture, plywood, and other laminated products. Another important series of resins, the *alkyd resins* or *glyptals*, is produced by the condensation and polymerization of polyhydric alcohols such as glycerol (§ 35.9) with polycarboxylic acids such as phthalic acid or, preferably, its anhydride (§ 43.2). These resins contain multiple functional groups and when baked they form extensive cross-links, changing into hard thermo-setting polymers. The alkyd resins are constituents of many modern varnishes and paints because they give a tough, high gloss protective finish.

49.4. Natural Rubber. Raw natural rubber consists of polyisoprene $(C_5H_8)_n$ and has the structure

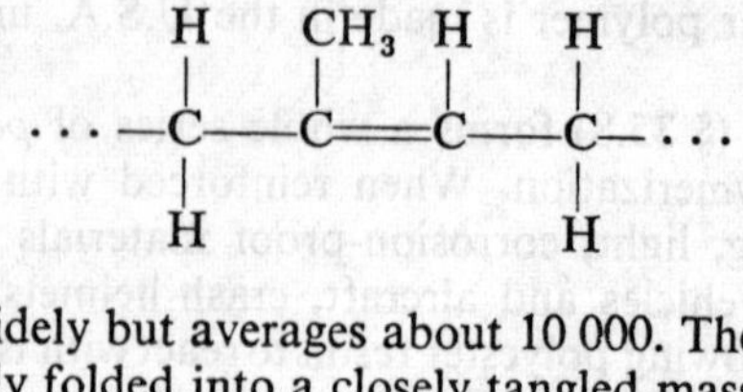

where *n* varies widely but averages about 10 000. The long chain molecules are normally folded into a closely tangled mass which is too soft and sticky to be of much practical use, particularly when warmed. It is obtained in the first instance as a milky fluid known as *latex* which exudes from the tapped rubber tree and consists of an aqueous suspension of colloidal rubber particles. This is either concentrated by centrifuging, preserved by adding ammonia, and transported as creamed latex, or, more commonly, coagulated into sheets by the addition of acetic acid and then dried and preserved by smoking. The raw rubber is converted into a tough useful product by various manufacturing processes. The most important of these is *vulcanization*, in which about 1% to 3% by weight of sulphur is added and the mixture carefully heated. As a result bridges of sulphur atoms are formed between adjacent chain molecules of polyisoprene at the double bond positions:

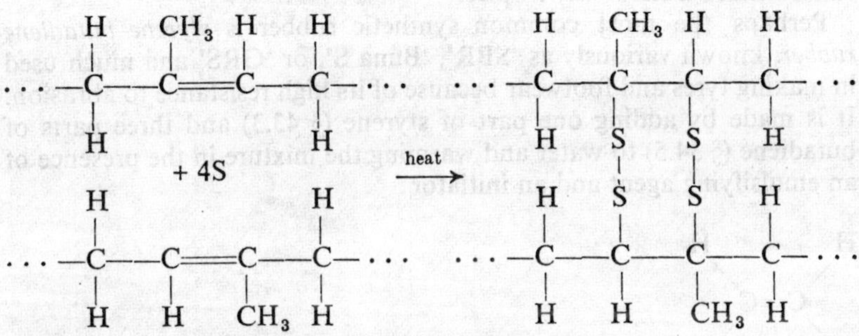

This cross-linking has a marked effect upon the properties of the rubber giving it a characteristic elasticity i.e. a tendency to regain its original shape after being stretched or compressed. This property arises from the ability of the cross-linked chain molecules which are normally extensively folded and kinked to straighten out temporarily by rotation about the single C—C bonds and to revert again to their original folded orientation when the stress is removed. Vulcanization also has the effect of making rubber less soft and sticky on warming and less soluble in organic solvents.

The physical properties of rubber can be still further changed by adding certain substances known as *fillers*, which convert it into a strong material capable of withstanding hard wear. The most widely used fillers are carbon black (§ 20.3) which can account for up to 30% of the weight of a rubber tyre, and zinc oxide (§ 29.3). Other additives include certain organic bases which have a dual role, firstly as accelerators which catalyze the chemical reactions occurring during the vulcanization process and shorten the heating period required, and secondly as antioxidants to prevent deterioration of the rubber as a result of atmospheric oxidation. When rubber is warmed it softens and is readily moulded or extruded. Nowadays this shaping process is usually combined with vulcanization in a single operation by heating the rubber to bring about vulcanization whilst it is still in the mould, so that the article retains the desired shape.

49.5. Synthetic Rubbers. Many attempts have been made to produce a polymer which would act as a substitute for natural rubber. As a result of these efforts a number of polymeric substances known as *synthetic rubbers* have been manufactured, none of them exactly like natural rubber in properties although some have proved to be very useful products despite this. Their main advantages over the natural material are economic and strategic in that they are more stable in price and they can be made outside the tropics.

Perhaps the most common synthetic rubber is *styrene–butadiene rubber*, known variously as 'SBR', 'Buna S', or 'GRS' and much used in making tyres and footwear because of its high resistance to abrasion. It is made by adding one part of styrene (§ 43.3) and three parts of butadiene (§ 34.5) to water and warming the mixture in the presence of an emulsifying agent and an initiator:

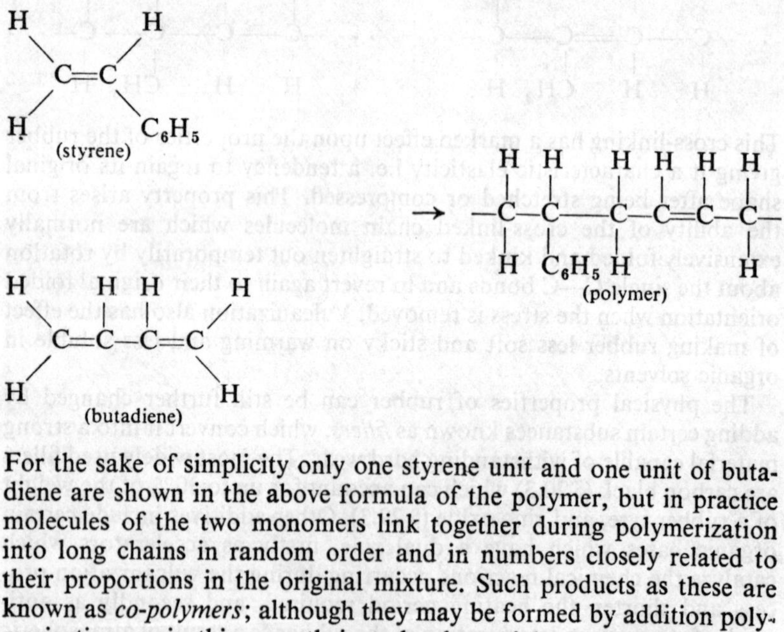

For the sake of simplicity only one styrene unit and one unit of butadiene are shown in the above formula of the polymer, but in practice molecules of the two monomers link together during polymerization into long chains in random order and in numbers closely related to their proportions in the original mixture. Such products as these are known as *co-polymers*; although they may be formed by addition polymerization, as in this case, their molecular weights are not exact multiples of those of either monomer.

Many other synthetic rubbers have been made. One of the earliest, still found useful because of its non-inflammability and resistance to oils, is *neoprene*, which is made by the polymerization of chloroprene, a compound obtained from acetylene:

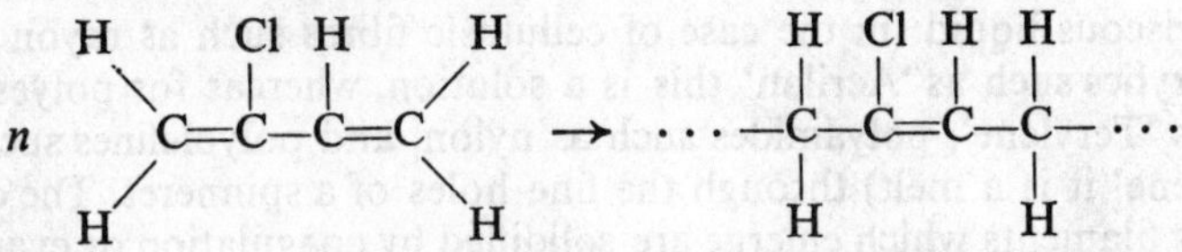

Butyl rubber, which is mainly used for making the inner tubes of tyres because of its impermeability to gases, is a polymer of isobutylene (a product of the petroleum industry) with a small proportion of isoprene. This polymerization is usually carried out in liquid methyl chloride at −95°C:

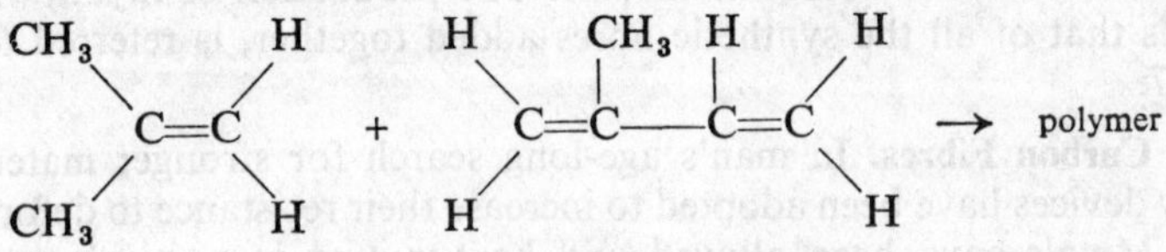

Other synthetic rubbers of value have been made by polymerizing butadiene alone and also isoprene alone. Their properties depend upon whether the polymers consist of stereoregular *cis* or *trans* forms (see § 48.5) or mixtures of these two isomers, which in turn depends upon the catalyst used and the conditions under which the polymerization is effected. Like natural rubber, synthetic rubbers are processed and vulcanized before use to increase their elasticity and toughness.

49.6. Other Polymers. As described in §§ 41.16 and 41.17, *proteins* and *polypeptides* are composed of long chains of amino-acid units linked together by the peptide linkage. In one sense they may be regarded, therefore, as condensation co-polymers of these amino-acids, although in practice it is impossible to make proteins by direct polymerization because of the difficulty of controlling the sequence in which the various units are arranged in the chain molecules of the protein. Similarly it is possible to look upon *polysaccharides* as polymers of glucose of very high molecular weight.

Polymerization is not limited to organic compounds and examples also occur in inorganic chemistry. One of the simplest is that of *sulphur*, where under appropriate conditions the individual atoms are capable of forming very long chain molecules with elastic properties as described in the section on allotropy in § 24.5. Another example is the formation of *polyphosphates* of the type $(NaPO_3)_n$ from sodium metaphosphate (§ 23.10). The important *silicone rubbers* are also polymeric in nature and are formed by repeated condensations as described in § 20.25.

49.7. Man-Made Fibres. All man-made fibres are produced by extruding a viscous liquid (in the case of cellulosic fibres such as rayon and polyacrylics such as 'Acrilan' this is a solution, whereas for polyesters such as 'Terylene', polyamides such as nylon, and polyolefines such as 'Courlene' it is a melt) through the fine holes of a spinneret. The continuous filaments which emerge are solidified by coagulation or evaporation if from a solution and by cooling if made from a melt. After being stretched these filaments are either twisted together to form a continuous filament yarn or are cut into short lengths and spun into staple yarn, often blended with other man-made or natural fibres. The methods of making many of the most important synthetic fibres have been described earlier in this chapter. The production of rayon, which equals that of all the synthetic fibres added together, is referred to in § 42.7.

49.8. Carbon Fibres. In man's age-long search for stronger materials many devices have been adopted to increase their resistance to deformation. Metals have been alloyed and heat-treated in particular ways, for example, to change the grain size of the natural crystals and so inhibit the spread of dislocations. One of the most successful methods of strengthening a given material has been to embed in it fibres or rods; the principle will be familiar from the ancient use of straw in brick-making, the steel reinforcement of concrete, and glass fibre reinforced plastics.

A recent development of great potential importance is the use of *carbon fibres*. These are made by the pyrolysis of polyacrylonitrile (e.g. Courtelle) fibres at 3000°C, when very fine threads of pure polycrystalline carbon are obtained. When these are embedded in a suitable matrix such as an epoxy resin they form a material with immense strength which retains its stiffness at high temperatures. Thanks to their low density these carbon fibres are about ten times as strong as steel, mass for mass. This high strength/mass ratio and the fact that unlike glass fibres they do not soften at high temperatures makes them extremely valuable for making components of aero engines, which at present is their principal use. Carbon fibres have the further advantage of versatility in that they are compatible with a wide range of resins and the proportion of carbon fibre can be as high as 50% in some composites. In 1969 the decision was taken to produce carbon fibres on a large scale in both Britain and America; as production increases so the price can be expected to fall and the range of uses be greatly extended.

50

Organic Reaction Mechanisms

50.1. Introduction. In the preceding chapters we have recorded many of the essential facts of organic chemistry as discovered by experiment, considering for each compound in turn its preparation, properties, and reactions. From this data we have been able to deduce the structure of each compound and ascribe to it a formula which represents this structure in a simple and conventional way. We have also been able, as our experience of organic compounds has grown, to discern the existence of certain functional groups with characteristic properties which they tend to impart to any molecule in which they occur. This has led to an appreciation of the various classes of organic compounds such as alcohols, aldehydes, acids, cyanides and amines, each of which contains one or more of these functional groups in its molecule. It has also enabled us to classify organic reactions into broad categories or types such as substitution, addition, or condensation, and to realize that in many cases these reactions are capable of proceeding simultaneously in competition with each other, so that which one prevails depends upon the conditions and environment.

This chapter is devoted to a consideration of the *mechanisms* of these organic reactions, i.e. the actual courses we believe are followed by the reactants in combining together and the various stages involved in reaching the final products. From this study we hope to arrive at an understanding of why particular compounds and functional groups react in the way they do and obtain explanations of their reactivity and their acidic or basic character. Before delving into organic mechanisms, however, we must first consider the covalent bond in greater detail, because organic reactions primarily consist of breaking and making covalent bonds and the nature of that bond has a profound and far-reaching influence upon the course of these reactions.

50.2. Polarization Effects in the Covalent Bond. There are a number of ways of describing the covalent bond; it can be portrayed as in chapter 14 as a pair of electrons positioned between two atoms, or as an overlap of their atomic orbitals, or as a region of electronic charge under the joint control and influence of two adjacent nuclei. The essential feature of all these representations is that the two electrons involved in the bond are shared between the atoms in some way to give a rigid and directional linkage. With two identical atoms the covalent bond is regarded as completely symmetrical with the electrons shared exactly equally between them. When the two atoms joined by the covalent bond are different, however, this no longer holds true and the electrons are then shared unequally between them, with the more electronegative atom having the greater affinity for electrons possessing the greater share. This effect was discussed in § 14.7, where it was explained that such an electron shift can cause a pronounced polarity in the bond and, when the molecular shape is appropriate, a high dipole moment in the molecule as a whole.

This asymmetrical distribution of electrons in bonds between atoms of different electronegativity, which is known as the *inductive effect*, can be depicted in several ways. For example, a covalent bond between carbon and chlorine can be shown with an arrow on it thus $>\mathrm{C}\!\rightarrow\!\!-\mathrm{Cl}$ to convey that the electron pair is attracted towards the electronegative chlorine atom and displaced towards it. Alternatively the notation $>\overset{\delta+}{\mathrm{C}}-\overset{\delta-}{\mathrm{Cl}}$ may be used to indicate the polarity which arises, δ^+ and δ^- standing for small ionic charges on the carbon and chlorine atoms respectively. Such conventional representations aim to make clear that as a result of the inductive effect the covalent bond in question has a partially ionic character and that there is a significant deficiency of electrons at the carbon atom and an excess of negative charge at the chlorine atom.

Similar displacement effects occur in unsaturated covalent bonds between atoms of widely different electronegativity. For example in the carbonyl group, which is conventionally portrayed as $>\mathrm{C}{=}\mathrm{O}$, there is a marked polarity because of the electron-attracting influence of the strongly electronegative oxygen atom. Such a bond is sometimes written $>\mathrm{C}\overset{\curvearrowright}{=}\mathrm{O}$ with a curved arrow to indicate electron shift, or as $>\mathrm{C}^+{-}\mathrm{O}^-$ to emphasize the very uneven electron density in the bond.

It should be noted that when a carbon atom is linked to a more electropositive atom or group, the resulting electron shift is then towards the carbon atom. For example, alkyl groups are weakly electron-donating and cause a slight displacement towards the carbon atom thus

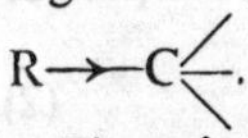

These inductive effects can be passed on to some extent to adjacent bonds in the molecule, although in saturated compounds the repercussions of a particular electron shift do not extend very far. We shall see in § 50.8 that this relay effect is of considerable importance in influencing the extent to which organic compounds containing an —OH group undergo dissociation and display acidic character.

Another notable feature of covalent bonds is their *polarizability* i.e. their readiness to adopt a temporary polarity when it is induced in them by the close proximity of a charged or highly polarized group. For example, a bromine molecule which consists of two atoms linked covalently together by a symmetrical sharing of an electron pair will develop a temporary dipole moment thus $Br^{\delta+}$—$Br^{\delta-}$ when one end of it approaches closely to a centre of high electron density such as a double or triple bond. As we shall see later, this induced polarization effect has an important bearing on the mechanism of certain organic reactions.

Yet another major factor influencing organic reactivity is the ability of certain substances, molecules or ions, to adopt a resonance structure (§ 14.11) in which any displaced electrons are *delocalized* i.e. spread over the whole structure and not concentrated entirely on one particular carbon atom. Delocalization of this kind, known as the *mesomeric effect*, leads to the formation of ions such as the carboxylate ion $R.COO^-$ and the phenoxide ion $C_6H_5O^-$ which are more stable than the alternative structural forms.

50.3. Types of Organic Reaction. When a single covalent bond between two atoms breaks it can do so in three different ways depending upon the nature of the particular atoms and the prevailing conditions. For example, the compound AB can split up in the following ways:

$$A{:}B \longrightarrow A\cdot + B\cdot \qquad (1)$$

$$A{:}B \longrightarrow A{:} + B \qquad (2)$$

$$A{:}B \longrightarrow A + B{:} \qquad (3)$$

In the first case, which is known as *homolysis* or *homolytic fission*, the bond divides symmetrically so that each atom retains one of the electrons which constituted the original bond. The products A· and B· are *free radicals*; they possess an unpaired electron and are very reactive.

In cases (2) and (3), which are known as *heterolysis* or *heterolytic fission*, the bond divides unequally so that either A or B captures both electrons and becomes a negatively charged ion whilst the other atom acquires a positive charge owing to the loss of its share in the pair of electrons:

$$A{:}B \rightarrow A^- + B^+ \qquad (2)$$

or

$$A{:}B \rightarrow A^+ + B^- \qquad (3)$$

Thus in homolytic fission the products are neutral particles, whereas in heterolysis charged ions are formed as intermediates.

50.4. Homolytic Reactions. As explained in the previous section, in homolytic reactions the covalent bond is broken symmetrically yielding two radicals each having an unpaired electron which is depicted by a single dot thus:

$$A{:}B \rightarrow A\cdot + B\cdot$$

These free radicals are extremely reactive; their energy of activation (§ 7.16) is so small, often only a few kilojoule, that they tend to combine readily with the first molecules they encounter and so have a very brief existence.

Homolytic reactions usually occur in the gaseous phase or in non-polar solvents and give rise to very rapid *chain reactions* of the type described in § 7.15. They are often initiated photochemically by ultra-violet light (§ 7.17) or by means of high temperatures or the use of a catalyst which encourages the formation of free radicals.

The halogenation of alkanes (paraffins) provides numerous examples of homolytic organic reactions. Above 250°C or in diffused daylight chlorine reacts with methane, for example, giving various substitution products as described in § 33.2. The first step involves the fission of a chlorine molecule into two chlorine radicals thus

$$Cl_2 \rightarrow Cl\cdot + Cl\cdot$$

These single chlorine atoms then rapidly combine with methane molecules as described in § 7.15 setting up a chain reaction in the mixture.

Many addition polymerizations are homolytic reactions. For example, the polymerization of vinyl chloride to polyvinyl chloride is initiated by traces of some substance such as a peroxide which decomposes thermally producing free radicals, X·. These then react with molecules of the vinyl chloride monomer to produce a succession of new radicals in the following way:

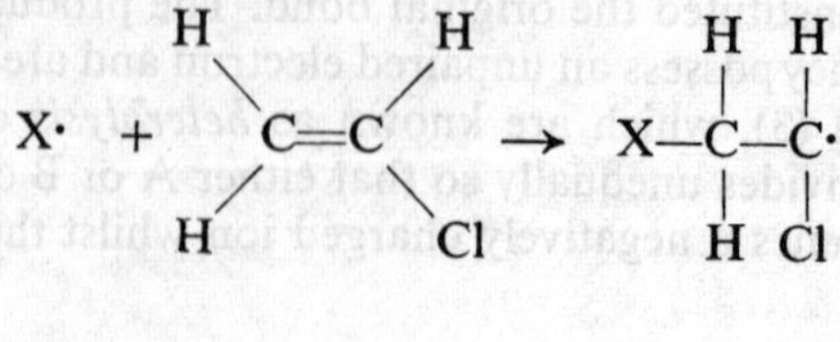

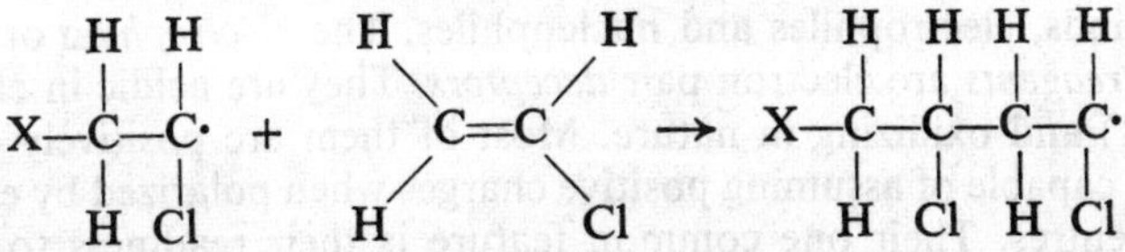

As each stage of the chain reaction is completed the size of the polymer radical grows until eventually the process is terminated by the combination of two of the free radicals to form a complete molecule. A similar mechanism applies to the polymerization of vinyl cyanide (acrylonitrile) to polyacrylonitrile and ethylene to polyethylene by the high pressure method.

The pyrolysis (i.e. thermal decomposition) of hydrocarbons is normally a homolytic reaction. For example, in the cracking of petroleum (§ 33.11) the larger molecules undergo homolytic fragmentation producing a mixture of free radicals which then combine with each other to give molecules of simpler hydrocarbons. Similarly when natural gas is raised to a high temperature it decomposes homolytically, as described in § 33.9.

These examples illustrate three common features of homolytic organic reactions, (1) their great speed, (2) their sensitivity to ultra-violet light, high temperatures, catalysts, and inhibitors (substances which react with free radicals and remove them from the reaction mixture), and (3) their indiscriminate nature, since a mixture of different products is usually obtained.

50.5. Heterolytic Reactions. This kind of reaction in which the covalent bond divides asymmetrically with the creation of two ions, can be written

$$A{:}B \rightleftharpoons A{:} + B$$

or

$$A{:}B \rightleftharpoons A^- + B^+$$

These ionic intermediates are often resonance structures which have an enhanced stability because of the delocalization of their ionic charge; benzene and its derivatives are good examples of this. A carbon atom or radical bearing a positive charge is called a *carbonium ion,* whilst one with a negative charge is known as a *carbanion.*

The tendency to heterolysis is greatest in a covalent bond between two atoms of widely different electronegativity, because in such a bond the electron pair is already displaced towards the more electronegative atom as explained in § 50.2. Most heterolytic reactions take place in solution in polar solvents because these encourage the separation of charges and the formation of ions.

It is found that reagents participating in organic reactions are of two kinds, electrophiles and nucleophiles. The *electrophiles* or *electrophilic reagents* are electron pair *acceptors*. They are acidic in character (§ 8.11) and oxidizing in nature. Most of them are positively charged or are capable of assuming positive charges when polarized by electron-rich centres. Their one common feature is their readiness to form a covalent bond by accepting a pair of electrons from another atom or ion. Examples of electrophiles are the hydrogen ion H^+, nitric acid HNO_3, sulphuric acid H_2SO_4, nitrous acid HNO_2, chlorine Cl_2, and bromine Br_2.

Nucleophiles or *nucleophilic reagents* are electron pair *donors*. They are basic in character and reducing in nature. They are usually negatively charged. Their essential characteristic is the possession of a pair of unshared electrons which they can use to form a covalent bond with another atom. Examples are the hydroxyl ion OH^-, the halide ions Cl^-, Br^-, and I^-, the bisulphite ion HSO_3^-, the cyanide ion CN^-, and ammonia NH_3.

Just as reducing agents react with oxidizing agents in redox reactions and acids with bases in neutralizations, so electrophiles and nucleophiles combine with each other in heterolytic reactions, for one compound is providing the electron pair which is to constitute the new covalent bond and the other is accepting it. To this extent it can be misleading to refer to any particular reaction as specifically electrophilic or nucleophilic, but it is conventional in discussing organic mechanisms to regard one reactant as the *reagent* and the other as the *substrate* or 'molecule under attack' and to classify a reaction in terms of the electrophilic or nucleophilic nature of the reagent, and this is the practice we shall follow in succeeding sections.

50.6. Electrophilic Reactions. As will be clear from the previous section, these are reactions in which the reagent is an electrophile i.e. it is capable of forming a new covalent bond with another atom or ion by accepting a pair of electrons donated by it. Amongst the examples we have encountered in this book are the reactions involving substitution in the benzene ring such as nitration, halogenation, sulphonation, and alkylation (Friedel–Craft reaction) and also the important processes of acetylation and benzoylation. Other examples include the numerous addition reactions across double or triple bonds such as occur in alkenes (olefines) and alkynes (acetylenes). For the sake of brevity we shall consider only four of these in detail, chosen because they are so important and typical:

(A) NITRATION OF BENZENE: There is overwhelming evidence available from cryoscopic and spectroscopic measurements that the nitrating agent is the *nitronium ion*, NO_2^+, which is produced in high concentration in a mixture of concentrated nitric and sulphuric acids:

$$2H_2SO_4 + HNO_3 \rightleftharpoons 2HSO_4^- + H_3O^+ + NO_2^+$$

A study of the kinetics of the nitration reaction shows it to be a two-stage process. In the first step the electrophilic nitronium ion is attracted to the electron-rich benzene molecule and attaches itself covalently to one of the carbon atoms by accepting an electron pair. This confers upon the intermediate product a positive charge which is spread by delocalization over the resonance structure of the benzene ring to give a stable carbonium ion:

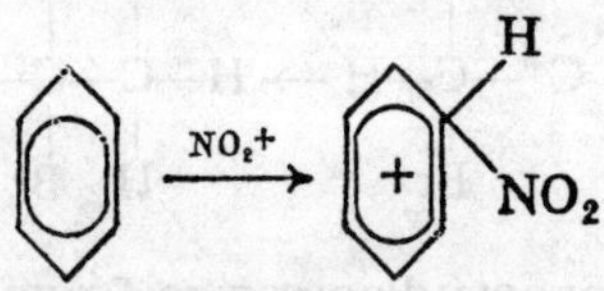

The second stage involves the loss of a proton H^+ to the base HSO_4^-, leaving a neutral molecule of nitrobenzene:

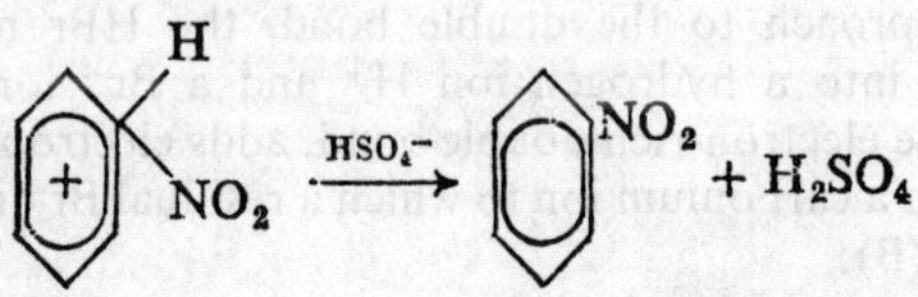

(B) Addition of Bromine to Ethylene: This reaction is believed to occur in three stages. In the first a bromine molecule approaching the ethylene becomes strongly polarized under the influence of the high electron density of the double bond as described in § 50.2 and splits heterolytically into two ions Br^+ and Br^-.

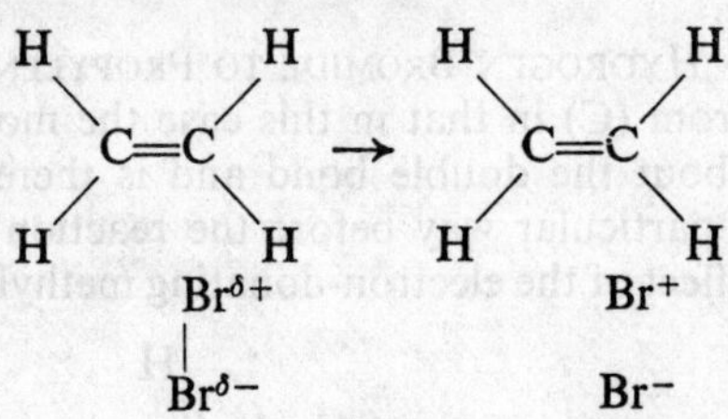

The positively-charged bromine ion is then attracted to one of the carbon atoms and combines electrophilically forming a carbonium ion as follows:

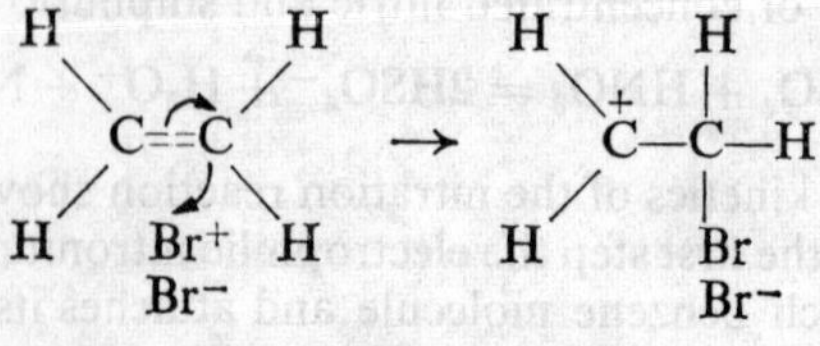

In the final stage a residual Br^- ion is attracted to the positively-charged carbon atom of the carbonium ion and combines with it to give the dibromide of ethylene, 1,2-dibromoethane:

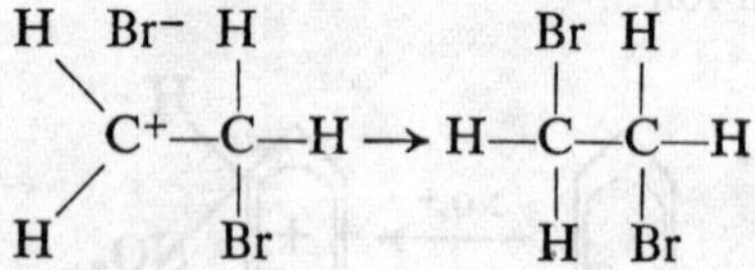

(C) Addition of Hydrogen Bromide to Ethylene: The mechanism resembles reaction (B) above, but in this case the reagent hydrogen bromide is already strongly polarized thus $H^{\delta+}$—$Br^{\delta-}$ before the reaction begins owing to the difference in electronegativity of its constituent atoms. On approach to the double bond, the HBr molecule splits heterolytically into a hydrogen ion H^+ and a Br^- ion; the former, attracted by the electron-rich double bond, adds electrophilically to the ethylene to give a carbonium ion to which a residual Br^- ion is attracted as in reaction (B):

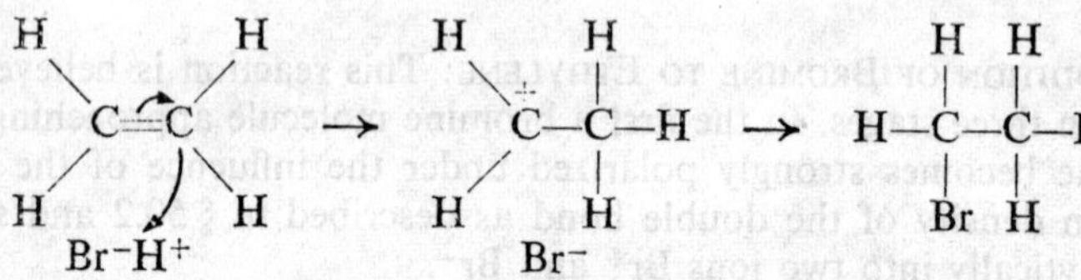

(D) Addition of Hydrogen Bromide to Propylene (Propene): This reaction differs from (C) in that in this case the molecule of alkene is unsymmetrical about the double bond and is therefore preferentially polarized in one particular way before the reaction begins as a result of the inductive effect of the electron-donating methyl group (see § 50.2):

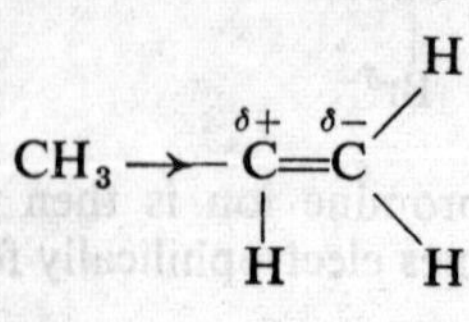

Thus when the polar hydrogen bromide molecule approaches the propylene and splits heterolytically into H^+ and Br^- ions, the positively-charged hydrogen ion is naturally attracted to the carbon atom bearing the fractional negative charge:

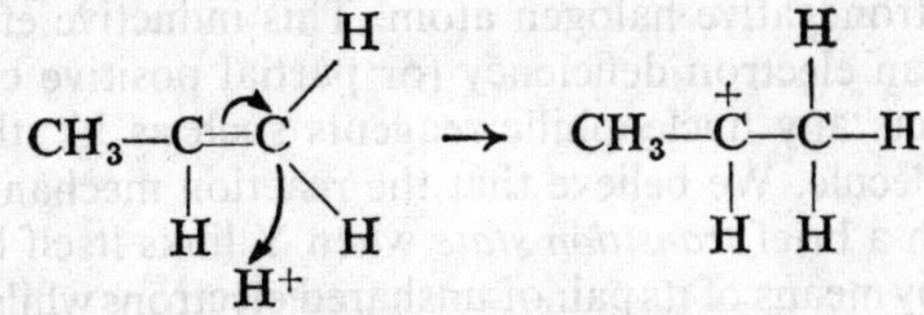

This results in a *secondary* carbonium ion (i.e. a carbonium ion in which the carbon bearing the positive charge has two alkyl groups attached to it). Not only is the formation of a secondary carbonium ion favoured by the inductive effect but it is also preferred energetically in that it is more stable than a primary carbonium ion because the two electron-donating alkyl groups partially compensate for the electron deficiency of the positively charged carbon atom. The final stage of the addition consists of the approach of a Br^- ion to the carbonium ion and its attachment to the carbon atom bearing the positive charge:

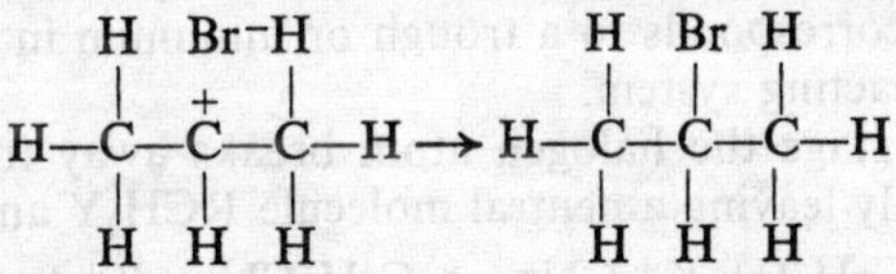

Thus the eventual product of the addition is isopropyl bromide (2-bromopropane) and not n-propyl bromide in accordance with Markownikoff's Rule (§ 34.3).

A similar mechanism explains the addition of hydrogen halides to isobutylene, $(CH_3)_2.C{=}CH_2$, the homologue of propylene. In this case there are two methyl groups exerting an inductive effect and the tertiary carbonium ion $(CH_3)_3C^+$ is formed in preference to the primary one because of its greater stability, giving tertiary butyl bromide $(CH_3)_3.CBr$ in accordance with Markownikoff's Rule.

50.7. Nucleophilic Reactions. These are reactions in which the main reagent is a nucleophile i.e. a substance possessing a pair of unshared electrons available for bond formation. The most important examples are the reactions of alkyl halides which lead to the replacement of the halogen atom by various functional groups (see § 40.4), the characteristic addition reactions of the carbonyl group (§§ 36.2 and 36.6), and the reactions of alcohols (§ 35.2), and it is these three types which will be considered in detail.

(A) Replacement Reactions of Halides: Let us consider a primary alkyl halide RCH_2X where X is an atom of chlorine, bromine, or iodine. As explained in § 50.2, the bond between the carbon atom and X will be polarized because of the pronounced electron shift towards the more electronegative halogen atom. This inductive effect gives the carbon atom an electron deficiency (or partial positive charge) which strongly attracts any nucleophilic reagents such as Y^- that approach the halide molecule. We believe that the reaction mechanism probably passes through a brief *transition state* when Y links itself loosely to the carbon atom by means of its pair of unshared electrons whilst the halogen atom still remains joined to the carbon by a loose and highly polar bond:

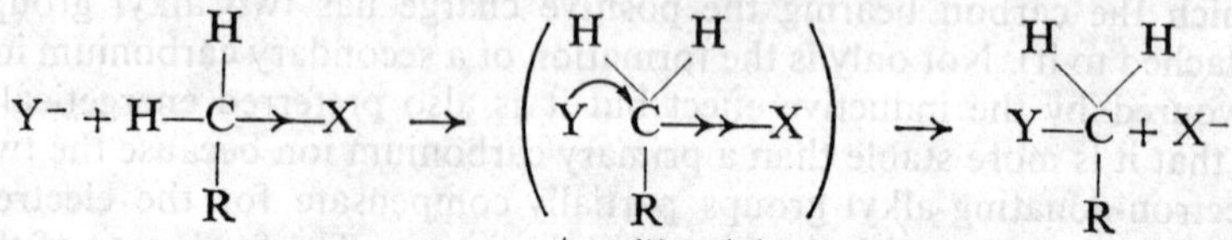

transition state

This transition state is an unstable and short-lived one and corresponds to the energy peak B in Fig. 11.5; it is not an intermediate compound, which always corresponds to a trough or minimum in the energy diagram of the reacting system.

In the final stage the halogen atom breaks away from the carbon atom completely leaving a neutral molecule RCH_2Y and an X^- ion.

e.g.
$$C_2H_5I + K^+ CN^- \rightarrow C_2H_5CN + K^+ I^-$$
$$C_2H_5Br + K^+ OH^- \rightarrow C_2H_5OH + K^+ Br^-$$

Whilst the nucleophilic reagent *Y* is usually a negatively-charged ion such as OH^- or CN^-, it may also be an uncharged molecule such as ammonia which contains a lone pair of electrons with which to attach itself to the carbon atom:

$$C_2H_5Br + NH_3 \rightarrow C_2H_5NH_3^+ + Br^-$$

A study of kinetics makes it clear that replacement reactions of tertiary alkyl halides are unimolecular (§ 7.14) and have a different mechanism involving the formation of a carbonium ion.

(B) Addition Reactions of the Carbonyl Group: The strong polarization of the carbonyl group $>C{=}O$ has already been explained in § 50.2, where the conventional notations of $>C{=}O$ or $>C^{\delta+}{=}O^{\delta-}$ were introduced. As a result of this effect a nucleophilic reagent is readily attracted to the carbon atom bearing the positive charge and attaches itself by means of its unshared electron pair:

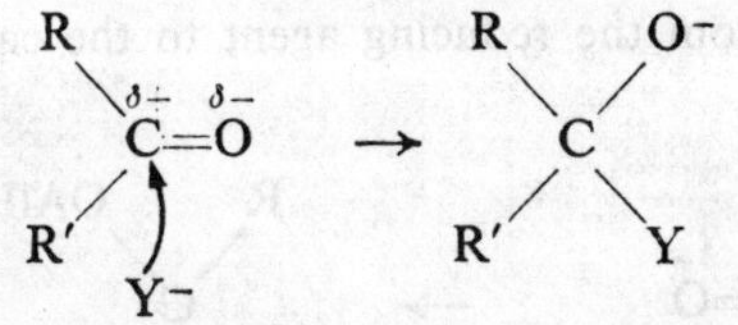

In the second stage the negative charge on the oxygen atom attracts a proton to form a hydroxyl group and so complete the course of the addition:

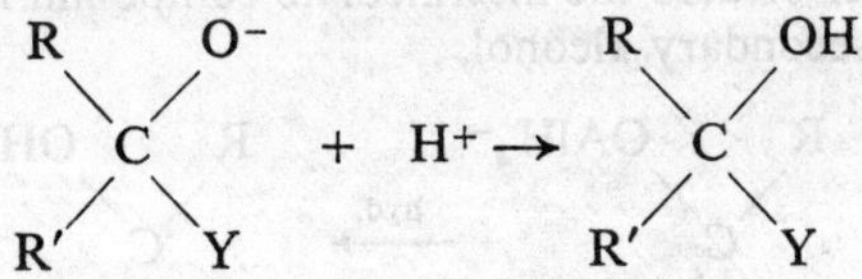

Hydrogen cyanide gives a cyanohydrin, for example:

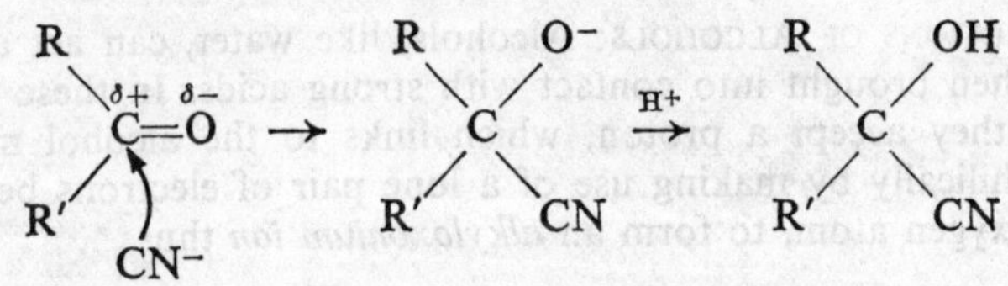

Similarly, sodium hydrogen sulphite (sodium bisulphite) and ammonia add nucleophilically as follows:

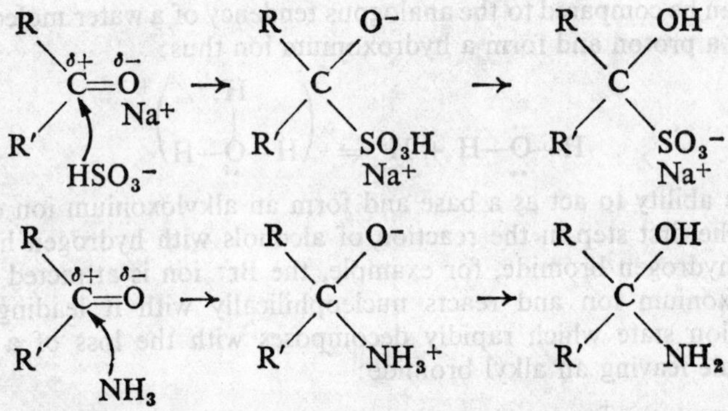

A study of the kinetics of these reactions reveals that the first step is the slow rate-determining stage, the acquisition of the proton being relatively fast.

With lithium aluminium hydride (§ 19.14) and sodium borohydride (§ 19.7), the first step in the reaction is the transfer of a nucleophilic hydride ion H^- from the reducing agent to the carbon atom of the carbonyl group:

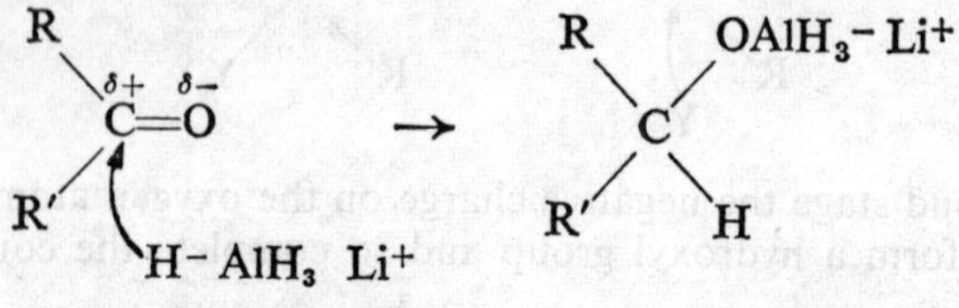

On adding water or acid the intermediate compound is rapidly hydrolysed giving a secondary alcohol.

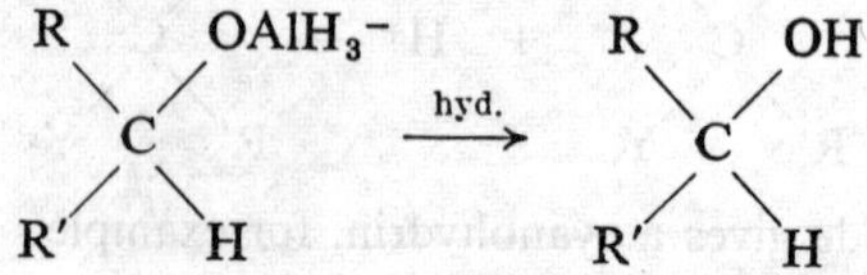

(C) Reactions of Alcohols: Alcohols, like water, can act as weak bases when brought into contact with strong acids. In these circumstances they accept a proton, which links to the alcohol molecule electrophilically by making use of a lone pair of electrons belonging to the oxygen atom, to form an *alkyloxonium ion* thus:

$$R—\ddot{\underset{\cdot\cdot}{O}}—H + H^+ \rightleftharpoons \left(R—\underset{\cdot\cdot}{\overset{\overset{H}{|}}{O}}—H\right)^+$$

This can be compared to the analogous tendency of a water molecule to accept a proton and form a hydroxonium ion thus:

$$H—\ddot{\underset{\cdot\cdot}{O}}—H + H^+ \rightleftharpoons \left(H—\underset{\cdot\cdot}{\overset{\overset{H}{|}}{O}}—H\right)^+$$

This ability to act as a base and form an alkyloxonium ion constitutes the first step in the reaction of alcohols with hydrogen halides. With hydrogen bromide, for example, the Br^- ion is attracted to the alkyloxonium ion and reacts nucleophilically with it leading to a transition state which rapidly decomposes with the loss of a water molecule leaving an alkyl bromide:

$$Br + R.\overset{+}{O}H_2 \rightarrow \underset{\text{transition state}}{(Br \frown R \rightarrow\!\!— OH_2)} \rightarrow Br—R + H_2O$$

In some secondary and most tertiary alcohols the water molecule tends to be discarded by the alkyloxonium ion first and the Br^- ion then combines nucleophilically thus:

$$R.\overset{+}{O}H_2 \longrightarrow R^+ + H_2O$$
$$R^+ + Br^- \longrightarrow R{-}Br$$

This tendency of alcohols to behave as weak bases and form alkyl-oxonium ions also plays an important part in the preparation of alkenes. For example, propan-2-ol or isopropyl alcohol reacts thus:

$$CH_3.\underset{}{\overset{\overset{\displaystyle OH}{|}}{C}}H.CH_3 + H^+ \rightleftharpoons CH_3.\overset{\overset{\displaystyle \overset{+}{O}H_2}{|}}{C}H.CH_3$$

The alkyloxonium ion then dissociates readily into a carbonium ion and water:

$$CH_3.\overset{\overset{\displaystyle \overset{+}{O}H_2}{|}}{C}H.CH_3 \rightleftharpoons CH_3.\overset{+}{C}H.CH_3 + H_2O$$

Now under certain circumstances (e.g. high temperature or the presence of a base) this carbonium ion tends to lose a proton by discarding a hydrogen ion from one of the carbon atoms adjacent to the one bearing the positive charge:

$$CH_3.\overset{+}{C}H.CH_3 \rightleftharpoons \underset{\text{(propylene)}}{CH_3.CH{=}CH_2} + H^+$$

The net result of the sequence of changes is to eliminate a molecule of water from the alcohol and create a double bond, so the reaction is essentially a dehydration and a means of preparing alkenes.

50.8. Acidity in Organic Compounds. An acid, as explained in § 8.11, is best defined in Brönsted–Lowry terms as a proton donor and a base as a proton acceptor. The acidity of carboxylic acids arises from their ability to dissociate in aqueous solution yielding hydroxonium ions thus:

$$R.COOH + H_2O \rightleftharpoons R.COO^- + H_3O^+$$

In this dissociation the hydroxyl group undergoes heterolytic fission into a proton and a negatively charged ion. The readiness with which this dissociation occurs depends upon the degree of polarity in the O—H bond. This polarity is much greater in carboxylic acids than in alcohols because of the influence of the adjacent carbonyl group, itself highly polarized by the inductive effect as described in § 50.2, which leaves the carbon atom with a small positive charge. This attracts electronic charge from the oxygen of the hydroxyl group and encourages dissociation:

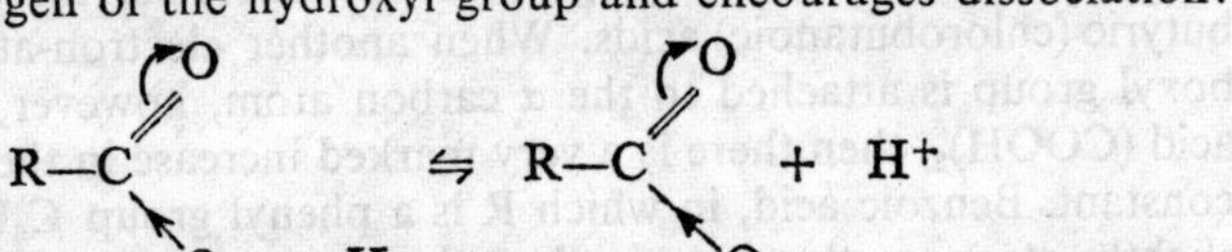

Another important factor here is the ability of the carboxylate ion R.COO⁻ to adopt a resonance structure which is a hybrid of the two canonical forms 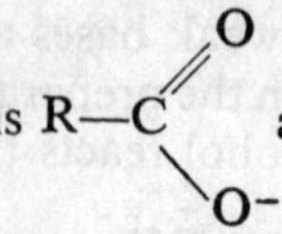 and 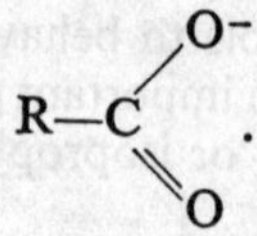. By facilitating delocalization of the negative charge this gives the resulting ion lower energy and an increased stability and hence promotes the dissociaton.

As Table 50.1 demonstrates, the strength of a carboxylic acid depends upon the nature of the radical R. There is not much difference between acetic acid and its higher homologues, but formic acid is significantly stronger because it is free of the weak electron-donating inductive effect of alkyl groups. When we consider the three chloro-substituted acetic acids the marked influence of substituent groups becomes very clear. Each substituted chlorine atom causes an electron shift away from the carbon atom to which it is attached, thereby decreasing the electron density and enhancing the dissociation of the carboxyl group. As a consequence the monochloro acid is about 80 times stronger than acetic acid and the di- and trichloro acids about 2700 and 11 000 times as strong respectively.

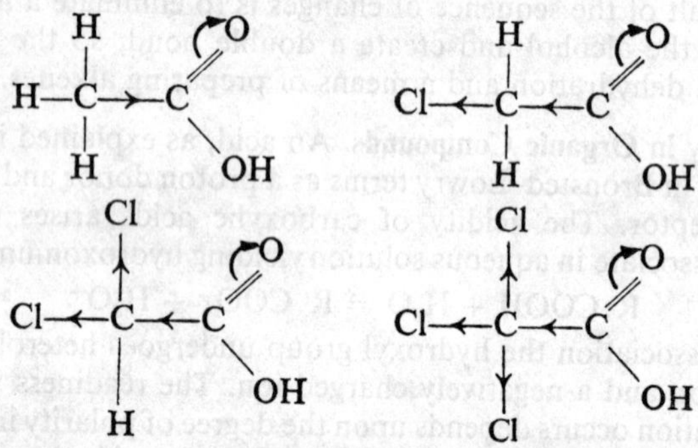

These effects are much less pronounced if the substituent groups are attached to the β and γ carbon atoms, since they are then too remote to have much inductive influence upon the dissociation of the carboxyl group, as is clear from the figures in the Table for the three chlorobutyric (chlorobutanoic) acids. When another electron-attracting carboxyl group is attached to the α carbon atom, however, as in oxalic acid $(COOH)_2$, then there is a very marked increase in the dissociation constant. Benzoic acid, in which R is a phenyl group C_6H_5—, is only slightly stronger than acetic, but the presence of substituted nitro

TABLE 50.1. THE STRENGTHS OF ORGANIC ACIDS

Acid Name	Acid Formula	Dissociation Constant K × 10^{-5} mol dm^{-3}	pK_a
Formic	H.COOH	18	3.75
Acetic	$CH_3.COOH$	1.8	4.75
Propionic	$C_2H_5.COOH$	1.3	4.89
n-Butyric	$CH_3.CH_2.CH_2.COOH$	1.5	4.82
Monochloroacetic	$CH_2Cl.COOH$	150	2.82
Dichloroacetic	$CHCl_2.COOH$	5000	1.31
Trichloroacetic	$CCl_3.COOH$	20000	0.70
α-chlorobutyric	$C_2H_5.CHCl.COOH$	140	2.85
β-chlorobutyric	$CH_3.CHCl.CH_2.COOH$	10	4.00
γ-chlorobutyric	$CH_2Cl.(CH_2)_2.COOH$	3	4.52
Oxalic	HOOC.COOH	6000	1.22
Benzoic	$C_6H_5.COOH$	6.4	4.19
o-Nitrobenzoic	*o*-$O_2N.C_6H_4.COOH$	680	2.17
Phenol	C_6H_5OH	0.00001	10.0
o-Nitrophenol	*o*-$O_2N.C_6H_4.OH$	0.006	7.22
Trinitrophenol	*2, 4, 6*-$(NO_2)_3.C_6H_2.OH$	9500	1.02

groups in the benzene ring markedly increases the degree of dissociation.

Phenol undergoes slight dissociation of its hydroxyl group thus

$$C_6H_5OH + H_2O \rightleftharpoons C_6H_5O^- + H_3O^+$$

The mild electron-attracting influence of the phenyl group facilitates this dissociation but the main reason why phenol is acidic and alcohols are not lies in the stabilizing effect on the phenoxide ion, $C_6H_5O^-$, of its resonance structure in which the negative charge is delocalized over the benzene ring. Phenol itself is a very weak acid ($K = 1.3 \times 10^{-10}$ mol dm^{-3}) but the presence of substituted groups, particularly electrophilic nitro groups, has a notable effect upon its dissociation, as the Table shows. Thus trinitrophenol, sometimes known as picric acid, is a strong acid comparable in strength to trichloroacetic and the mineral acids. The effect of substitutents in phenol is not only to increase greatly its strength as an acid but also to make the benzene ring more reactive, especially in its response to further substitution by electrophilic groups. Thus once a nitro group or bromine atom has been introduced its effect on the distribution of electronic charge is relayed around the benzene ring by the mesomeric effect and further substitution is facilitated.

The inductive effect can also be used to explain why secondary amines are stronger bases than primary amines or ammonia, since the alkyl groups are more electron-repelling than hydrogen and encourage the nitrogen atom to take up a proton. When an electron-attracting carbonyl group is introduced, however, its inductive effect acts strongly in the opposite direction and so amides are only very weakly basic substances.

50.9. Conclusion. Although substitution reactions in alkanes and arenes have a superficial resemblance, we have seen in this chapter that the mechanisms of these two reactions are entirely different. Alkane substitution consists of homolytic free-radical chain reactions, whereas aromatic substitution entails electrophilic attack by positively charged ions and highly polarized groups. Similarly, addition reactions can proceed by completely contrasting mechanisms e.g. additions to alkenes and alkynes across the double or triple bond are initiated by the approach of an electron-seeking electrophilic atom or group whereas additions across the carbonyl bond are restricted to nucleophilic reagents, most of them negatively charged which themselves provide the electron pair required to form the bond. A study of mechanism gives an appreciation of these differences and a deeper understanding of the various types of organic reaction.

Nobel Prizewinners in Chemistry

I am grateful to Abelard-Schuman, publishers of 'Nobel Prizewinners in Chemistry' by Eduard Faber, for permission to include this list. An asterisk indicates that in that year the prize was shared.

1901 Jacobus Henricus van't Hoff.
For the discovery of laws of chemical dynamics and of osmotic pressure.

1902 Emil Fischer.
For his syntheses in the groups of sugars and purines.

1903 Svante August Arrhenius.
For his theory of electrolytic dissociation.

1904 William Ramsay.
For the discovery of gaseous, indifferent elements in the air and the determination of their place in the periodic system.

1905 Adolf von Baeyer.
For his researches on organic dyestuffs and hydro-aromatic compounds.

1906 Henri Moissan.
For his research on the isolation of the element fluorine and for placing at the service of science the electric furnace which bears his name.

1907 Eduard Buchner.
For his biochemical researches and his discovery of cell-less fermentation.

1908 Ernest Rutherford.
For his investigation into the disintegration of the elements and the chemistry of radioactive substances.

1909 Wilhelm Ostwald.
For his work on catalysis and on the conditions of chemical equilibrium and velocities of chemical reactions.

1910 Otto Wallach.
For his initiative work in the field of alicyclic substances.

1911 Marie S. Curie.
For her services to the advancement of chemistry by the discovery of the elements radium and polonium, by the isolation of radium and the study of the nature and compounds of this remarkable element.

1912* Victor Grignard.
For the discovery of the so-called Grignard reagent.

1912* Paul Sabatier.
For his method of hydrogenating organic compounds in the presence of finely divided metals.

1913 Alfred Werner.
In recognition of his work on the linkage of atoms in molecules.

1914 Theodore W. Richards.
For his exact determinations of the atomic weights of a great number of chemical elements.

1915 Richard Willstätter.
For research on colouring matter in the vegetable kingdom, principally on chlorophyll.

1918 Fritz Haber.
For the synthesis of ammonia from its elements, nitrogen, and hydrogen.

1920 Walther Nernst.
For his thermochemical work.

1921 Frederick Soddy.
For his contributions to the chemistry of radioactive substances and his investigations into the origin and nature of isotopes.

1922 Francis W. Aston.
For his discovery, by means of his mass spectrograph, of the isotopes of a large number of non-radioactive elements, as well as for his discovery of the whole-number rule.

1923 Fritz Pregl.
For his invention of the method of microanalysis of organic substances.

1925 Richard Zsigmondy.
For his elucidation of the heterogeneous nature of colloid solutions and for the methods he has devised in this connection.

1926 Theodor Svedberg.
For his work on disperse systems.

1927 Heinrich Wieland.
For his research on bile acids and analogous substances.

1928 Adolf Windaus.
For his studies on the constitution of the sterols and their connection with the vitamins.

1929* Arthur Hardern and Hans von Euler-Chelpin.
For their investigations on the fermentation of sugar and of fermentative enzymes.

1930 Hans Fischer.
For his researches into the constitution of haemin and chlorophyll, especially for his synthesis of haemin.

1931* Carl Bosch and Friedrich Bergius.
In recognition of their contributions to the invention and development of chemical high-pressure methods.

1932 Irving Langmuir.
For his discoveries and investigations in surface chemistry.

1934 Harold C. Urey.
For his discovery of heavy hydrogen.

1935* Irene Joliot-Curie and Frederick Joliot.
For their synthesis of new radioactive elements.

1936 Peter J. W. Debye.
For his contributions to the study of molecular structure through his investigations on dipole moments and on the diffraction of X-rays and electrons in gases.

1937* Walter N. Haworth.
For his researches into the constitution of carbohydrates and vitamin C.

1937* Paul Karrer.
For his researches into the constitution of carotenoids, flavins, and vitamins A and B.

1938 Richard Kuhn.
For his work on carotenoids and vitamins.

1939* Adolf Butenandt.
For his work on sex hormones.

1939* Leopold Ruzicka.
For his work on polymethylenes and higher terpenes.

1943 George de Hevesy.
For his work on the use of isotopes as tracer elements in researches on chemical processes.

1944 Otto Hahn.
For his discovery of the fission of heavy nuclei.

1945 Artturi I. Virtanen.
For his researches and inventions in agricultural and nutritive chemistry, especially for his method of fodder preservation.

1946* James B. Sumner.
For his discovery that enzymes can be crystallized.

1946* John H. Northrop and Wendell M. Stanley.
For their preparation of enzymes and virus proteins in a pure form.

1947 Robert Robinson.
For research on certain vegetable products of great biological importance, particularly alkaloids.

1948 Arne Tiselius.
For his researches on electrophoresis and adsorption analysis, especially for his discoveries concerning the complex nature of the serum proteins.

1949 William F. Giauque.
For his work in the field of chemical thermodynamics, particularly concerning the behaviour of substances at extremely low temperatures.

1950* Otto Diels and Kurt Alder.
For the development of the diene synthesis.

1951* Glenn T. Seaborg and Edwin M. McMillan.
For their discoveries in the chemistry of the trans-uranic elements.

1952* Archer J. P. Martin and Richard L. M. Synge.
For their discovery of partition chromatography.

1953 Hermann Staudinger.
For his discoveries in the chemistry of macromolecular substances.

1954 Linus Pauling.
For his research into the nature of the chemical bond and its application to the elucidation of the structure of complex substances.

1955 Vincent du Vigneaud.
For his work on biologically important sulphur compounds and particularly for the first synthesis of a polypeptidic hormone.

1956* Cyril N. Hinshelwood and Nikolaj N. Semenov.
For their research on the mechanism of chemical reactions.

1957 Alexander R. Todd.
For his work on nucleotides and nucleotidic coenzymes.

1958 Frederick Sanger.
For his work on the structure of proteins, especially insulin.

1959 Jaroslav Heyrovsky.
For inventing and developing the polarographic method of analysis.

1960 Willard F. Libby.
For developing radio carbon dating techniques.

1961 Melvin Calvin.
For his work on photosynthesis.

1962* Max F. Perutz and John C. Kendrew.
For elucidation of the structures of myoglobin and haemoglobin.

1963* Karl Ziegler and Giulio Natta.
For their discoveries of chemical processes that led to new ways of making polymers and to new kinds of polymers.

1964 Dorothy C. Hodgkin.
For developing X-ray diffraction analysis and her elucidation of the structure of vitamin B_{12}.

1965 Robert C. Woodward.
For his synthesis of sterols, chlorophyll, and other naturally occurring substances.

1966 Robert S. Mulliken.
For his work on chemical bonds and the electronic structure of molecules by the molecular orbital method.

1967* Ronald G. Norrish, G. Porter, and M. Eigen.
For their work on chemical kinetics.

1968 Lars Onsager.
For his work in electrochemistry.

1969* Derek H. R. Barton and Odd Hassel.
For their work on the shapes and conformation of organic molecules.

1970 Louis F. Leloir.
For his discovery of sugar nucleotides.

1971 Gerhard Herzberg.
For his work on molecular structure.

1972* Christian B Anfinsen, Stanford Moore, and William H. Stein.
For their work on the chemistry of enzymes.

1973* Ernest O. Fischer and Geoffrey Wilkinson.
For their work on organometallic sandwich compounds.

1974 Paul J. Flory.
For his work on polymers and macromolecules.

APPENDIX I

Chemical Nomenclature

Various attempts have been made in recent years to standardize chemical nomenclature. The most important of these was the set of rules drawn up by the International Union of Pure and Applied Chemistry in 1957 which suggested systematic names for inorganic and organic chemicals.

In general these recommendations have been followed in the text of this book, although those parts of the older system of nomenclature which are still in widespread use have also been retained. Thus in many instances both the new systematic name and the older familiar one are given alongside each other, and the student is able to choose which alternative he prefers. This policy of using both names is likely to be practised in examination papers for many years to come until everyone has become thoroughly familiar with the systematic nomenclature.

In inorganic chemistry the new system involves the use of the Stock notation. In this the compounds of metals which display variable valency are distinguished from each other by including the oxidation number of the metal in the name, using Roman numerals in brackets. Thus the two chlorides of iron, previously known as ferrous and ferric, are referred to as iron(II) chloride and iron(III) chloride respectively. Again, cuprous oxide is written as copper(I) oxide, and cupric oxide (in which the copper is divalent) as copper(II) oxide. Similarly, stannous chloride and stannic chloride, in which tin is divalent and tetravalent respectively, are called tin(II) chloride and tin(IV) chloride in the Stock notation.

In organic chemistry trivial names such as ethylene, acetylene, benzene, toluene, and phenol are still approved, but most organic compounds are named systematically, as described in the text. Such names embody a syllable indicating the number of carbon atoms in the longest unbranched chain in the molecule (e.g. *eth-* for two, *prop-* for three, and *but-* for four). A second syllable indicates the linkages present (e.g. *-an* for single, *-en-* for double, and *-yne* for triple bonds). The presence of functional groups is indicated by characteristic syllables (e.g. *-ol* for the —OH group). Thus, for example, the systematic name for ethyl alcohol is *ethanol* because its molecule has two carbon atoms joined by a single bond and linked to a hydroxyl group.

In an organic compound with a branched chain, the name given is derived from that of the longest chain of carbon atoms in the molecule; the positions of the alkyl or functional groups present are conveyed by prefixing them with numbers which indicate to which carbon atom in this chain they are attached. For example, the alcohol with molecular formula C_4H_9OH and structural formula $(CH_3)_2.CH.CH_2OH$ is known as *2-methylpropan-1-ol* because its longest chain contains three carbon atoms (*propan-*) and the remaining methyl and hydroxyl groups are linked to the second and first carbon atoms respectively. The same system of numbering is employed to indicate the position of double bonds.

Complex ions are named systematically by specifying first the number and then the nature of the ligands or co-ordinated groups, followed by the name of the central metal ion, which is given the ending *-ate* in anions. Where the metal is capable of variable valency, then its valency state in the complex ion is indicated by a Roman numeral at the end. For example, the complex ion $[Fe(CN)_6]^{4-}$, previously known as ferrocyanide, is called hexacyanoferrate(II) under the new system because it contains six cyano groups linked to a central iron ion in the iron(II) state. Similarly $[Fe(CN)_6]^{3-}$, commonly referred to as the ferricyanide ion, is called hexacyanoferrate(III) because in this complex ion the iron is in the trivalent or iron(III) state. Other examples occurring in this book are $[Cu(CN)_4]^{3-}$ formerly known as cuprocyanide and now as tetracyanocuprate(I), and the complex cations $[Cu(NH_3)_4]^{2+}$ and $[Cu(H_2O)_4]^{2+}$ known as tetramminocopper(II) and tetraquocopper(II) respectively. Under the new system the ending *-ide* will be restricted to compounds involving only two constituent ions or radicals.

Some alternative names for other complex ions are given below:

Common Name	*Formula*	*Systematic Name*
argentocyanide	$[Ag(CN)_2]^-$	dicyanoargentate(I)
cuprichloride	$[Cu\,Cl_4]^{2-}$	tetrachlorocuprate(II)
cuprochloride	$[Cu\,Cl_4]^{3-}$	tetrachlorocuprate(I)
cobaltammine	$[Co(NH_3)_6]^{3+}$	hexamminocobalt(III)
cobaltinitrite	$[Co(NO_2)_6]^{3-}$	hexanitrocobaltate(III)
aluminohydride	$[Al\,H_4]^-$	tetrahydridoaluminate(III)
borohydride	$[BH_4]^-$	tetrahydridoborate(III)

APPENDIX II

Units

In recent years the system of fundamental units known as the Système International (SI) has been devised to replace the previously accepted system based on the centimetre, the gram, and the second. In this new system, which is rapidly replacing the old c.g.s. system in scientific literature and examinations and which is used exclusively throughout this book, the fundamental units of length mass and time are the metre, the kilogramme, and the second respectively.

The basic SI units are defined in the following ways:

The *metre* is defined in terms of the wavelength of a particular line in the spectrum of the krypton-86 isotope.

The *kilogramme* is defined by the mass of a piece of metal recognized as the international standard.

The *second* is defined in terms of the period of certain radiation from the caesium-133 isotope.

The *ampere*, the unit of current, is defined in terms of the force acting between two parallel conductors under specified conditions.

The *kelvin*, the unit of temperature is defined as the fraction 1/273.16 of the thermodynamic temperature of the triple point of water.

The *candela*, the unit of luminous intensity, is defined as that given from a black body under specified conditions.

The *mole*, the unit of amount of substance, is defined as that which contains the same number of atoms (or ions, molecules, or radicals, as the case may be) as there are in exactly 0.012 kilogramme of the carbon-12 isotope. This unit replaces the gramme-molecule, the gramme-ion, the gramme-atom, and the gramme-formula weight.

Some of the derived SI units of greatest importance to chemists are listed below:

Physical Quantity	*SI Unit*	*Approved Symbol*
energy	joule	J
force	newton	N
power	watt	W
electric charge	coulomb	C
electric potential difference	volt	V
electric resistance	ohm	Ω
frequency	hertz	Hz
common temperature	degree Celsius	°C

The following prefixes *inter alia* may be used to indicate decimal multiples of the basic or derived SI units:

Multiple	*Prefix*	*Symbol*	*Multiple*	*Prefix*	*Symbol*
10^{-1}	deci	d	10^{-6}	micro	μ
10^{-2}	centi	c	10^{-9}	nano	n
10^{-3}	milli	m	10^{3}	kilo	k

The following are some of the SI units most frequently used in this book:

Physical Quantity	*SI Unit*	*Symbol*
length	metre	m
length	centimetre	cm
length	nanometre	nm
volume	cubic metre	m^3
volume	cubic decimetre	dm^3
volume	cubic centimetre	cm^3
mass	kilogramme	kg
mass	gramme	g
density	kilogramme per cubic metre	kg m^{-3}
density	gramme per cubic centimetre	g cm^{-3}
pressure	newton per square metre	N m^{-2}
pressure	pascal	Pa
amount of substance	mole	mol
dissociation constant	mole per cubic decimetre	mol dm^{-3}
electrolytic conductivity	reciprocal ohm per centimetre	Ω^{-1} cm^{-1}
heat of reaction	kilojoule per mole	kJ mol^{-1}
ionization energy	kilojoule per mole	kJ mol^{-1}
bond dissociation energy	kilojoule per mole	kJ mol^{-1}
entropy	joule per kelvin	J K^{-1}

The SI equivalents of some familiar non-SI units are as follows:

Physical Quantity	*Unit*	*Equivalent*
length	ångström	10^{-10} m
force	dyne	10^{-5} N
pressure	atmosphere	101 325 N m^{-2}
energy	erg	10^{-7} J
energy	calorie	4·184 J
energy	B.t.u.	1055·06 J
volume	litre	10^{-3} m^3 (or 1 dm^3) approx.

Examination Questions

These are reprinted by permission of the Examining Bodies concerned, the source of each question being indicated in the following way:

O & C	The Oxford and Cambridge Schools Examination Board.
O	The Delegates, Oxford Local Examinations.
C	The University of Cambridge Local Examinations Syndicate.
L	The Senate of the University of London.
NJB	The Northern Universities Joint Matriculation Board.
SJB	The Southern Universities Joint Board.
O SCHOL	The Secretary, Natural Science Scholarships, Oxford.

An asterisk indicates that the question was set in an S-level paper.
I am grateful to the above authorities for permission to alter their questions so that they are expressed in SI units and in modern signs, symbols, and abbreviations. Answers to problems and numerical questions are given on page 900.

The Atomic and Molecular Theories

1. State and illustrate the Laws of Chemical Combination. What part did these laws play in the development of atomic and molecular theory? (O & C)

2. State Avogadro's hypothesis. Under what conditions would you expect it to be exactly true? Outline the experimental and theoretical considerations which led to the conclusion that the hydrogen molecule contains two, and only two, atoms. What is the significance of this in connection with the determination of (*a*) molecular weights, (*b*) atomic weights? (L)

3. State (*a*) Gay Lussac's law of gaseous volumes and (*b*) Avogadro's hypothesis. Illustrate by an example the importance of Avogadro's hypothesis in the determination of the formulae of gaseous compounds. 20 cm^3 of a mixture of ammonia and nitrogen were sparked in a eudiometer until the volume (measured at the original temperature) had increased to 28 cm^3. It was then brought into contact with sulphuric acid, when the volume decreased to 25 cm^3. A volume of 30 cm^3 of oxygen was added and the mixture sparked. After cooling, the volume of the residual gas was 37 cm^3. Explain the changes observed, and calculate the volumes of ammonia and nitrogen in the original mixture. (NJB)

4. What postulates regarding the nature of atoms did Dalton make in his Atomic Theory? What facts derived from our present knowledge of atomic structure show these postulates to have been erroneous? Why does the atomic theory still provide a sound working basis for chemistry? 1.950 g lead chloride derived from one mineral were found to contain 0.01409 g equivalents of chlorine, whereas 2.345 g lead chloride derived from another mineral were found to contain 0.01682 g-equivalents of chlorine. Calculate the atomic weight of the lead in each sample and comment very briefly on the difference in the results (Cl = 35.46). (NJB)

5.* Just over a hundred years ago the atomic weight of carbon was thought to be 6 and that of oxygen 8, and the formula for water was written HO. What arguments would you use if you were asked (*a*) why the atomic weight of carbon is now considered to be 12 and not 6, and (*b*) why the formula for water is written H_2O? What significance is there in the fact that the atomic weight of carbon is nearer 12.1 than 12? What evidence is there that the *molecular* formula of water is not H_2O? (O & C)

6. Define 'Atomic Weight' and 'Molecular Weight'. Quote *two* pieces of evidence that the atomicity of the hydrogen molecule is two and give one important application of this fact. When 0.1963 g. of an organic acid chloride is added to water the products require for complete neutralization 50.0 cm^3. 0.1M sodium hydroxide. The same weight of the compound is found to occupy approximately 74 cm^3 when in the vapour state at 100°C and 9.99×10^4 N m^{-2} pressure. Calculate the exact molecular weight of the compound and suggest its structural formula. (NJB)

7. Explain clearly why you believe in the existence of atoms. (O SCHOL)

Equivalent Weights and Atomic Weights

8. A metal *M* forms four oxides containing 23.9, 32.0, 38.6 and 44.0% of oxygen respectively. Calculate the equivalent weight of the metal in each of these oxides, and show that the figures are in accordance with a fundamental law of chemistry. If the last oxide reacts with bases to form salts isomorphous with orthophosphates, calculate the atomic weight of the metal, and deduce the formulae of the other oxides. (L)

9. Enunciate the three 'Laws of Chemical Combination'. Indicate *three* general methods other than electrolysis by which the equivalent weights of metals may be determined and describe with the chief experimental details how one of the three would be carried out in the case of copper. The passage of a constant current of 0.125 A for 20 minutes through a solution of a salt of a metal *M* caused the deposition of 0.1677 g *M*. The heat capacity of *M* is 0.232 J g^{-1} K^{-1}. What is its atomic weight? (1 Faraday = 96 500 C mol^{-1}) (NJB)

10. Explain the distinction between equivalent weight and atomic weight Summarize, giving *one* example in each case, the methods available for the determination of equivalent weights. 1.50 gramme of the chloride of a metal gave, on suitable treatment, 1.89 gramme of the sulphate. Calculate the equivalent weight of the metal. If the heat capacity of the metal is 1.029 J g^{-1} K^{-1} suggest a value for its atomic weight.

[O = 16.0; S = 32.0; Cl = 35.5.] (L)

11. What do you understand by the equivalent of (*a*) an element, (*b*) an alkali, (*c*) an oxidizing agent? Calculate the equivalents of (i) potassium permanganate ($KMnO_4$) when used as an oxidizing agent in acid solution, (ii) oxalic acid ($H_2C_2O_4$, $2H_2O$) when used as a reducing agent. Describe how you would prepare an approximately decinormal solution of potassium permanganate, and then find its exact concentration.

[H = 1; C = 12; O = 16; K = 39; Mn = 55.] (C)

12. What do you understand by the equivalent weight of a substance? Calculate the equivalent weights of the following substances and explain your calculations:

(i) oxalic acid (*a*) as an acid, (*b*) as a reducing agent;
(ii) ferrous sulphate (*a*) as a reducing agent, (*b*) in the reaction

$$CuSO_4 + Fe = Cu + FeSO_4;$$

(iii) sodium thiosulphate in the reactions
(*a*) $2Na_2S_2O_3 + I_2 = 2NaI + Na_2S_4O_6$
(*b*) $Na_2S_2O_3 + 4Cl_2 + 5H_2O = Na_2SO_4 + H_2SO_4 + 8HCl$

[H = 1; C = 12; O = 16; Na = 23; S = 32; Cl = 35.5; Fe = 56; Cu = 63.6; I = 127.] (O & C)

13.* How would you make up and standardize a decinormal solution of potassium permanganate? In estimating the concentration of a solution of sodium oxalate by means of potassium permanganate why is it necessary (*a*) to add dilute sulphuric acid, and (*b*) to heat the solution at the beginning of the titration? Why does the reaction then proceed without further heating, even if the temperature falls considerably? 25 cm^3 of a solution containing 6.1 g dm^{-3} of an oxalate of formula

$$K_xH_y(C_2O_4)_z nH_2O$$

required for titration 18 cm^3 of decinormal sodium hydroxide solution and 24 cm^3 of decinormal potassium permanganate. Calculate x, y, z, and n.

(O & C)

14. If you were supplied with clean magnesium ribbon and solutions of M hydrochloric acid and 0.1 M sodium hydroxide, describe how you would determine the equivalent weight of magnesium. Indicate how the result is calculated. State Dulong and Petit's rule. The heat capacity of magnesium is 1.046 J g^{-1} K^{-1}; how would you find the valency and the accurate atomic weight of magnesium? The sulphate of a metal *M* forms an alum with aluminium sulphate. This alum contains 23.4 per cent. of M; calculate the atomic weight of M.

[Al = 27; S = 32; H = 1; O = 16.] (O)

15.* Describe in essential detail *two* distinctly different methods by which atomic weights may be determined, and assess the accuracy and applicability of each method. An element has a metallic appearance and displays properties typical of a metal. It forms (*a*) an oxide containing 23.12 per cent. of oxygen, (*b*) a chloride containing 43.25 per cent. of chlorine, and (*c*) two sulphides containing respectively 37.55 and 25.58 per cent. of sulphur. The molar and gramme heat capacities of the element are respectively 26.8 J mol^{-1} K^{-1} and 0.145 J g^{-1} K^{-1}. What inferences can be drawn from these data?

[O = 16, S = 32, Cl = 35.5.] (NJB)

16. Give *two* methods by which the accurate atomic weight of an element may be determined. 5.21 g of the chloride of a divalent metal were dissolved in distilled water and to the solution an excess of a solution of silver nitrate was added. The mass of silver chloride obtained was 7.17 g. Calculate the atomic weight of the metal. (O & C)

17. Give an account of *three* important methods which have been used to determine the atomic weights of the elements. 0.1 gramme of a metal M of heat capacity 0.879 J g^{-1} K^{-1} dissolved in acid to give 173 cm^3 of hydrogen measured at 27°C and 8×10^4 N m^{-2} pressure. 0.1 g of its anhydrous chloride volatilized completely to 12.05 cm^3 of vapour measured under the same conditions. Calculate the atomic weight of the metal M and the formula of the chloride. (O&C)

18. State Dulong and Petit's law. Discuss its use in the determination of *exact* atomic weights, pointing out what other information is required. A metal M, whose heat capacity is 0.232 J g^{-1} K^{-1}, forms two chlorides containing 37.40 and 54.43 per cent. of chlorine respectively. The former dissolves in water but deposits a white precipitate on standing; the latter is a colourless, fuming liquid. Suggest formulae for the two chlorides, calculate the atomic weight of M and suggest, with reasons, a suitable position for it in the periodic table.

[Cl = 35.46.] (C)

19. In what ways have: (*a*) heat capacities, (*b*) vapour density, (*c*) isomorphism, been used in the determination of atomic weights? A metal (X) forms a chloride which has a vapour density of about 130 (referred to hydrogen = 1·0), and contains 45.5 per cent of the metal. Suggest a possible value for the heat capacity of the metal, and write formulae for its oxide and chloride. (L)

20. State two laws which deal with the proportions by weight in which elements may combine together to form compounds. A metal M forms three chlorides containing respectively 23.59%, 38.17% and 48.08% chlorine. The heat capacity of the metal is 0.232 J g^{-1} K^{-1}. Calculate the exact atomic wt. of the metal and give the formulae of the three chlorides (Cl = 35.5). If the heat capacity of the metal were not known, what other evidence would you seek to establish the exact atomic weight? (NJB)

21. State and explain the principles which have been used to determine the atomic weight of (*a*) a noble gas, (*b*) a solid non-metallic element, e.g. carbon or sulphur, and (*c*) a heavy metal. (O & C)

22.* Define atomic weight and describe briefly one physical and one chemical method for determining this quantity. What connection has the atomic weight of an element with the masses of the atoms of that element? Why do the isotopes of an element have the same chemical properties? (O & C)

23.* The chloride of a metal *M* is completely hydrolysed by water to give hydrochloric acid and an insoluble oxide. When 0.313 g of the chloride was hydrolysed in this way, 48.0 cm^3 of M/10 NaOH were required to neutralize the hydrochloric acid formed. When 0.230 g of the chloride was vaporized in a Victor Meyer vapour density apparatus, the volume of air expelled was 21.8 cm^3 measured at 15°C and 9.88×10^4 N m^{-2} pressure. Calculate (*a*) the equivalent weight of the metal, (*b*) the molecular weight of the chloride.

What conclusions can you draw about (i) the atomic weight of the metal, (ii) the formula of the chloride, and (iii) the heat capacity of the metal? [H = 1, O = 16, Cl = 35.5; the molar volume of hydrogen is 22.4 dm^3 mol^{-1} at s.t.p.] (NJB)

The Kinetic Theory

24.* What are the postulates of the kinetic theory of a perfect gas? From the kinetic equation, $pv = \frac{1}{3}nmc^2$, and any other necessary assumptions, deduce Boyle's Law, Charles's Law, Graham's Law, and Avogadro's hypothesis. How may the observed deviations from Boyle's Law in the behaviour of actual gases be explained? (O)

25. On what assumptions is the Kinetic Theory of Gases based? Show how an expression which embodies Avogadro's Hypothesis, Charles's Law, and Boyle's Law can be obtained from the Theory. Under what circumstances does this expression become unsatisfactory for gases such as carbon dioxide? Why is this and how has the expression been modified to render it more satisfactory? (SJB)

26. Describe the differences between solids, liquids and gases with respect to the kinetic behaviour of the molecules. What do you understand by the statement that there is a 'dynamic equilibrium' between a liquid and its vapour? Explain (*a*) the increase in the vapour pressure of a liquid as the temperature rises, (*b*) the difference between a saturated and unsaturated vapour. (NJB)

27. Explain, on the basis of the atomic and kinetic theory, (*a*) how a gas exerts pressure; (*b*) why a liquid has a definite vapour pressure at a given temperature; (*c*) the significance of heat capacities. (O SCHOL)

Molecular Weights

28.* Give a general account of the methods available for determining molecular weights, pointing out the limitations of the methods you give. Indicate how the molecular weights of acetic acid, oxalic acid, and sulphuric acid have been established. (O)

29. Outline practical methods for the determination of molecular weights which are based on (*a*) Graham's Law, (*b*) Raoult's Law, (*c*) Avogadro's hypothesis and the gas laws. Discuss briefly the importance of vapour density measurements in determining atomic weights. (L)

30. Draw a diagram of the apparatus required to find the vapour density of a liquid by V. Meyer's method. If the liquid under investigation boils at 60°C, state exactly what you would do in this determination and indicate how the result is calculated. If the vapour density of nitrogen dioxide is 30.2 at 60.2°C, what is the degree of dissociation?

[N = 14; O = 16.] (O)

31. Describe, with full experimental details, how you would determine the molecular weight of a volatile liquid. The apparent molecular weight of iodine at 550°C was found to be 165; calculate the degree of dissociation of iodine into atoms at this temperature.

[I = 127.] (C)

32. Describe how you would determine the molecular weight of ether by Victor Meyer's method. 1.0 g of the volatile chloride of a metal *M* displaced 74.2 cm^3 of air, measured at 17°C and a pressure of 101 325 $N\ m^{-2}$ (corrected for the vapour pressure of water). The chloride contained 34.46 per cent. of *M* and the heat capacity of *M* was 0.481 $J\ g^{-1}\ K^{-1}$. Find the exact atomic weight of *M* and the formula of its chloride.

[Cl = 35.46; 1 mole of a gas occupies 22.4 dm^3 at s.t.p.] (C)

33. What is the importance of vapour density measurements in the determination of molecular weights? Describe Victor Meyer's method for the determination of the vapour density of a volatile compound. 2.768 g of phosphorus pentachloride when completely vaporized at 250°C and 101 325 $N\ m^{-2}$ pressure, occupied 764.6 cm^3. What deduction can be made from this observation?

[H = 1.00; P = 31.0; Cl = 35.5; 1.0 dm^3 of hydrogen at s.t.p. weighs 0.089 g.] (L)

34. Describe an experiment to find the vapour density of benzene. Although ammonium chloride ought to have a molecular weight of 53.5, determinations from both the vapour density and the freezing point of an aqueous solution, give values of about 27. Give a detailed explanation of these facts. (SJB)

35. Give *two* applications of a knowledge of the molecular weights of substances. Describe how you would measure experimentally the molecular weight of a volatile liquid by Victor Meyer's method. When 0.65 g of aluminium chloride was heated in a Victor Meyer's apparatus at 327°C, 57.8 cm^3 of air were collected over water at 14°C and an atmospheric pressure of 102 000 N m^{-2}. When 0.25 g of aluminium chloride was heated in the same apparatus at 877°C, the volume of air, measured under the same conditions as before, was 44.5 cm^3. From these data calculate the molecular weight at each temperature, and comment on the values obtained.

[Al = 27.0. Cl = 35.5. Molar volume at s.t.p. = 22 400 cm^3 mol^{-1} vap. pressure of water at 14°C. = 1600 N m^{-2}.] (NJB)

36. Describe the experimental method of Victor Meyer for the determination of molecular weights, and make clear the theoretical principles on which it is based. When 0.10 g of formic acid is volatilized in a Victor Meyer apparatus (with a heating jacket at 120°C) 36.5 cm^3 of moist air is collected at 17°C, and 100 580 N m^{-2} pressure. Calculate the apparent molecular weight of formic acid. How would you interpret this result?

[C = 12, H = 1, O = 16, vapour pressure of water at 17°C = 1920 N m^{-2}, 1 mole of a gas occupies 22.4 dm^3 at s.t.p.] (NJB)

37. State Graham's law of diffusion and describe an experiment to show that hydrogen diffuses more rapidly than air. A given volume of a colourless, pungent gas, *A*, diffused through a porous plug in 14.1 s, while the same volume of oxygen required only 10 s. An aqueous solution of *A* decolourizes iodine and permanganate solutions. Suggest a formula for *A* and give equations for the above reactions. (C)

38. State Graham's Law of diffusion of gases, and give a qualitative explanation of it in terms of the kinetic theory. Describe an experiment to demonstrate gaseous diffusion. A certain volume of acetic acid vapour was found to diffuse in 9.68 minute, whilst the same volume of oxygen, with the same experimental conditions, took 5 minutes to diffuse. What deductions can be made as a result of this experiment. (NJB)

39. If you were provided with samples of two gases, what experiments would you carry out to compare their rates of diffusion? A volume of oxygen diffused through a porous plug in 10 second while the same volume of another

gas diffused (under the same conditions) through the plug in 9·35 second. Suggest possible formulae for the gas. Discuss some of the uses to which gaseous diffusion has been put.
(C)

40. Explain with the aid of a diagram how the addition of a non-volatile solute affects the vapour pressure, and hence the boiling-point and freezing-point of a solvent. What conclusions could be drawn from the following observations: (*a*) 0.608 g of a substance, of empirical formula weight 152, dissolved in 40 g of benzene, gave a depression of the freezing-point of 0.5 K; (*b*) 0.70 g of a substance of empirical formula weight 58.5, dissolved in 100 g of water, gave a depression of 0.4 K?

[Depression constants for 100 g of (*a*) benzene, 49.95 K mol^{-1}, (*b*) water, 18.6 K mol^{-1}.]
(O & C)

41. 1.25 g of urea dissolved in 17 cm^3 of water gave an elevation in the boiling-point of 0.65 K. The boiling-point elevation constant for 100 cm^3 of water is 5.4 K mol^{-1}. From first principles, calculate the molecular weight of urea. Draw a diagram of the apparatus you would use to carry out a determination of the above type and describe exactly what you would do. Briefly discuss the accuracy and limitations of this method of determining molecular weights.
(O)

42. How would you determine the molecular weight of a non-volatile substance by freezing-point measurements? Indicate briefly the chief sources of error in the method described. A solution containing 2.0 g of urea (mol. wt. 60) in 40 g of water freezes at − 1.55°C. Calculate the freezing-point depression constant for water (expressed as K per mole per 1000 gramme). If the density of the solution is 1.02 g cm^{-3}, what will be its osmotic pressure at 25°C?
(L)

43. Describe concisely *two* methods for determining the molecular weights of substances in solution. A solution containing 11 gramme of barium nitrate in 100 gramme of water boils at 100.46°C. Calculate (*a*) the apparent molecular weight, and (*b*) the apparent degree of dissociation of the salt, in this solution.

[The molecular elevation constant for water is 0.52 K mol^{-1} for 1 000 g of solvent. N = 14; O = 16; Ba = 137.4.]
(O & C)

44. Deduce an expression whereby the molecular weight of a solid may be calculated from a measurement of the freezing-point of its solution. Describe, with full experimental details, how you would carry out such a determination in the laboratory. What are the limitations of this method? What deductions can be drawn from the fact that a *M*/10 solution of cane sugar ($C_{12}H_{22}O_{11}$) has the same freezing-point as a solution of calcium chloride ($CaCl_2$) containing 4.44 g of solute per dm^3 of solution?

[H = 1.00; C = 12.0; O = 16.0; Cl = 35.5; Ca = 40.0; cryoscopic constant = 18.6 K mol^{-1} per 100 g.]
(L)

45. What is meant by the osmotic pressure of a solution? Describe *one* method for the direct measurement of the osmotic pressure of a solution, indicating how, from this measurement, the molecular weight of the solute can be deduced. A solution of 4.50 g of a sugar in 1·0 dm³ of water was found to have an osmotic pressure of 6320 N m^{-2} at 20°C. Calculate the molecular weight of the sugar. (L)

46. Explain what is meant by the osmotic pressure of a solution, and state the laws which govern the osmotic pressure of dilute solutions. Describe concisely how the osmotic pressure of a dilute aqueous solution of cane sugar could be measured. At what temperature would the osmotic pressure of a solution containing 18 g of glucose ($C_6H_{12}O_6$) per dm³ be 2.63×10^5 N m^{-2}? (NJB)

47. Explain clearly what you understand by the terms *osmosis*, *semi-permeable membrane*, and *osmotic pressure*. Describe *one* method of measuring osmotic pressure. To prevent the passage of water into a solution of urea through a suitable membrane, a pressure of 101 325 N m^{-2} had to be applied. The temperature of the system was 15°C, and the molecular weight of urea is 60. Calculate the weight of urea in one dm³ of the solution. (O)

48. What is meant by the statement that the osmotic pressure of a certain solution at a given temperature is 3×10^5 N m^{-2}? What are the laws which govern the osmotic pressure of dilute solutions? How do they show a similarity between the behaviour of a solution and that of a perfect gas? An aqueous solution containing 14.9 g of potassium bromide per dm³ has an osmotic pressure of 5.63×10^5 N m^{-2} at 25°C. Calculate the molecular weight of the salt in solution. Comment on the value obtained.
[K = 39.1, Br = 79.9. Molar volume at s.t.p. = 22.4 dm³ mol^{-1}.] (NJB)

49. Explain the terms (i) osmosis, (ii) dialysis, (iii) isotonic solutions. The osmotic pressure of a decimolar solution of silver nitrate is 3.80×10^5 N m^{-2} at 18°C. Calculate the apparent degree of dissociation of silver nitrate in this solution. (L)

50. 'The molecular weights of urea and acetone are well established but the molecular weight of sodium chloride cannot be found.' Outline *one* method of finding each of the molecular weights of urea and acetone and mention one alternative method for urea. Explain concisely why none of the methods you have discussed can be applied to sodium chloride. (SJB)

51. Outline, for each of four of the following, the best method for determining its molecular weight: (*a*) mercury; (*b*) water; (*c*) benzoic acid, (*d*) sulphur; (*e*) aluminium chloride. Mention briefly any interesting points that might emerge from the molecular weight determinations. (O SCHOL)

52.* How can it be shown that (*a*) iodine molecules dissociate into atoms in iodine vapour at high temperatures; (*b*) acetic acid exists as a 'dimer' (twofold polymer) in benzene solution? On what grounds would you justify the statement that water does not behave as a normal liquid? (O & C)

53. (*a*) The vapour density of nitrogen tetroxide, N_2O_4, at 26.7°C is 38.3. (*b*) The osmotic pressure of a solution of 0.0585 g sodium chloride in 22.4 cm^3 of water at 15°C is 202 650 N m^{-2}.

Discuss these data.

[N = 14, O = 16, Na = 23, Cl = 35.5.] (O & C)

54. What conclusions can be drawn from the following experimental data?

(*a*) 10 cm^3 of a gaseous hydrocarbon require 50 cm^3 of oxygen for complete combustion and 30 cm^3 of carbon dioxide are produced. [All volumes are measured at the same temperature and pressure.]

(*b*) A solution containing 54 g of glucose ($C_6H_{12}O_6$) per dm^3 has the same osmotic pressure as a solution containing 24 g of magnesium sulphate ($MgSO_4$) per dm^3.

(*c*) A solution of 60 g of acetic acid in 1 kg of benzene boils at 81.45°C.

[H = 1; C = 12; O = 16; Mg = 24; S = 32; b.p. of benzene = 80.20°C;. *K* for benzene = 25 K mol^{-1} per 100 g.] (C)

Solubility

55. Define, with reference to aqueous solutions of a solid, the terms *saturated solution*, *supersaturated solution*, and *solubility*. Describe how to determine the solubility of hydrated ferrous sulphate ($FeSO_4.7H_2O$) in water at 25°C, giving details of the analysis of the solution. Draw a rough diagram of the solubility curve of sodium sulphate and account for its abnormal shape. (O)

56. Explain fully how you would determine experimentally the solubility of a salt in water at a number of different temperatures. Illustrate your answer by reference to three particular salts and show by diagrams the different types of curves you might expect to obtain. (NJB)

57. The following figures give the solubility of a salt in water/(g per 100 g of water):

Temperature:	0°	10°	20°	30°	40°	50°	60°
Solubility	13	20	32	45	63	85	110

Plot a solubility curve from the given data and with its aid explain what happens when (*a*) a solution containing 40 g of the salt per 100 g is slowly evaporated at 30°, and (*b*) a solution containing 100 g per 100 g at 60° is cooled to 50°. (O & C)

58. Describe in detail the usual laboratory procedure for (*a*) the purification of an organic solid by recrystallization from alcohol, (*b*) the determination of its melting-point. Indicate the apparatus required at each stage and the precautions attending its use. (NJB)

59. Discuss briefly the methods available for determining the solubility of (*a*) gases in liquids, (*b*) solids in liquids. A saturated solution of a non-electrolyte, of molecular weight 180, exerted an osmotic pressure of 5.23×10^4 N m^{-2} at 22°C. What is the solubility of the substance in gramme per dm^3 of water at this temperature?

[Molar volume of a gas is 22.4 dm^3 mol^{-1} at s.t.p.] (C)

60. Explain, giving practical details and diagrams, the methods by which you would find the solubility in water of the following at the temperature and pressure of the laboratory: (*a*) nitrogen; (*b*) hydrogen chloride. Outline, without practical details, how you would find the solubility of ammonium sulphate in water at room temperature. (SJB)

61.* Describe, with experimental details, how you would determine (*a*) the solubility of oxygen in water; (*b*) the solubility of ammonia in water; (*c*) the distribution (partition coefficient) of iodine between water and chloroform. (O & C)

62. State Henry's Law and Dalton's Law, relating to the solubility of gases in liquids. In what cases are these laws not obeyed? Give reasons. What is meant by 'absorption coefficient'? If the absorption coefficients of 2 inert gases X and Y are 0.028 and 0.014 at 40°C, calculate the amounts of X and Y, expressed in cm^3 at s.t.p., dissolved in a dm^3 of water at 40°C from the following gas mixture: X at 20 250 N m^{-2}, Y at 40 500 N m^{-2}, O_2 at 19 850 N m^{-2}, and water vapour at 7 300 N m^{-2}. (NJB)

63. State briefly how Le Chatelier's principle applies to the variation with temperature of the solubility of solids in liquids and to the variation with pressure of the solubility of gases in liquids. Describe how you would measure experimentally the solubility of carbon dioxide in water at atmospheric pressure. If the absorption coefficients of oxygen and of nitrogen in water at 0°C are 0.048 and 0.024 respectively, calculate the composition by volume of the mixture of gases boiled out from water which has been saturated at 0°C with air (assumed to be $\frac{1}{5}$ of oxygen and $\frac{4}{5}$ of nitrogen by volume). (NJB)

64. A mixture containing four volumes of nitrogen and one of oxygen at a *total* pressure of 101 325 N m^{-2} was shaken with 2 dm^3 of air-free distilled water at 20°C until no more gas dissolved. After boiling the water, the gas evolved was collected and found to consist of 12.4 cm^3 of oxygen and 26.4 cm^3 of nitrogen, measured at the same temperature and barometric pressure as the original mixture. Calculate the solubilities of nitrogen and oxygen in water at 20°C and atmospheric pressure. (Part question only.) (L)

Phase Equilibria

65.* What do you understand by the term 'allotropy'? Give short accounts of the allotropy of (*a*) tin, (*b*) phosphorus, (*c*) sulphur. In what respect does the allotropy of tin and sulphur differ from that of phosphorus? How can you account for the changes observed when liquid sulphur is heated? Describe how you would prepare from crude sulphur specimens of three modifications of the element. (NJB)

66.* What is meant by (*a*) monotropy, (*b*) enantiotropy? Illustrate your answer by reference to the allotropy of phosphorus and tin. How does the allotropy of oxygen differ from that of phosphorus and tin? (O & C)

67. Give an account of the allotropy of *one* element showing monotropy and of *one* showing enantiotropy. Is it possible to apply either of these terms to the interconversion of oxygen and ozone? (O & C)

68. Define melting point and transition point as equilibrium phenomena. Describe what happens when the following cool slowly at atmospheric pressure through the temperature ranges indicated: (i) Sulphur from 105° to 85°C. (ii) Water from 10° to – 30°C. (iii) A dilute solution of sodium chloride from 10° to – 30°C. What difference would result from the application of, say, 5×10^6 N m^{-2} pressure in the second case?

[Neglect supercooling throughout this question.] (NJB)

69.* Draw temperature–concentration diagrams to illustrate the various types of vapour pressure curves relating to systems of two miscible volatile liquids. In each case discuss what will happen when a typical mixture of such liquids is fractionally distilled. (L)

70. (*a*) Describe concisely the experimental operation of distillation in steam. (*b*) Explain the theory underlying this process and show how the composition of the distillate may be calculated. Cite two substances to which the method is applicable. (*c*) When an organic compound was distilled in steam at normal atmospheric pressure, the temperature of distillation was 99°C. The vapour pressure of water at 99°C = 97 700 N m^{-2}. 80% by mass of the distillate was found to be water. Calculate the molecular weight of the organic compound.

[H = 1, O = 16.] (NJB)

71. (*a*) Describe and explain the process of distillation in steam, making clear which factors affect the composition, by mass, of the distillate. (*b*) For what kinds of substances is steam distillation especially valuable? (*c*) When bromobenzene is steam distilled at normal atmos. pressure, the temp. of distillation is 95.2°C. At 95.2°C the v.p. of water is 85 300 N m^{-2}. Calculate % by mass of bromobenzene in the distillate. (NJB)

72. A mixture of two immiscible liquids, water and nitrobenzene ($C_6H_5NO_2$), boils at 99° C, at which temperature the vapour pressures of the pure liquids are 97 725 N m^{-2} and 3600 N m^{-2} respectively. Calculate the mass ratio of nitrobenzene to water in the distillate. (Part question only). (L)

73.* Define *partition coefficient*. When an organic substance is extracted from aqueous solution with ether, successive small portions of ether are usually employed. Explain why this procedure is preferable to the use of the same total quantity of ether in one extraction. The partition coefficient of iodine between carbon disulphide and water at a given temperature is 410. 100 cm³ of a solution of iodine in water containing 0.09 g of iodine were shaken with 10 cm³ of carbon disulphide. What weight of iodine will be found in the carbon disulphide? How would you determine the weight of the iodine in the carbon disulphide? (O & C)

74. Explain what is meant by 'partition coefficient'. Phenol dissolves with its normal M.W. in both water and amyl alcohol. An aqueous solution containing 0.22 g phenol per dm³ is in equilibrium with an amyl alcohol solution containing 3.52 g phenol per dm³. What mass of phenol is extracted from 500 cm³ of an aqueous solution containing 20 g phenol per dm³ by shaking it with two successive quantities of 100 cm³ of amyl alcohol?

Give one illustration of a practical use of partition in the laboratory. (NJB)

75.* Define the term *partition coefficient*. How would you determine the partition coefficient of iodine between ether and water? Experiments on the distribution of an organic compound *X* between water and chloroform gave the following results:

Concentration in water	0.052	0.11	0.25	0.46
Concentration in chloroform	0.104	0.48	2.5	8.5

What is the molecular state of *X* in chloroform solution? (C)

76.* State the distribution law. Comment on the following results obtained in a distribution experiment:

Mass of benzoic acid in 10 cm³ water/g	0.0150	0.0195	0.0289
Mass of benzoic acid in 10 cm³ benzene/g	0.242	0.412	0.970

Given that, of the components of the equilibrium $KI + I_2 \rightleftharpoons KI_3$, only iodine is soluble in benzene, describe how, by a distribution experiment, you would determine at 25° the equilibrium constant of the reaction $KI + I_2 \rightleftharpoons KI_3$ in aqueous solution. Show how you would calculate the equilibrium constant from your results. (NJB)

77.* What is meant by the *eutectic point*? Draw a fully labelled equilibrium diagram for a system of two components showing a eutectic point, the two components being completely miscible in the liquid phase. Draw the cooling curves you would get if you cooled to solidification completely liquid systems of composition (*a*) one pure compound, (*b*) the eutectic composition, (*c*) a composition intermediate between (*a*) and (*b*). Explain the relationship between each of your cooling curves and the equilibrium diagram. (SJB)

78. Give a concise account of the methods available for the purification of chemical substances and explain the principles on which they depend. (C SCHOL)

Kinetics and Chemical Equilibrium

79.* Give an account of the factors which determine the velocity of simple chemical reactions. If the reaction is reversible, what effect do these factors have on the position of equilibrium? If dilute solutions of ethyl acetate and sodium hydroxide are mixed, the ester is slowly hydrolysed. How would you follow the rate of this reaction experimentally? (O & C)

80. What factors can affect the rate at which substances react? Illustrate your answer by examples. What explanation, if any, of the observed effects can you give in terms of the molecular kinetic theory? (O & C)

81. Describe and illustrate, with examples, the essential features of a catalyst. (O & C)

82. What are the essential characteristics of a catalyst? What general explanations can you give of catalytic action? How would you find out whether copper oxide catalysed the decomposition of potassium chlorate? (O & C)

83. What effect can a catalyst have on (*a*) the velocity of a reaction, (*b*) the equilibrium position in a reversible reaction? What types of catalyst are recognized and what theories have been advanced to explain their action? Name *three* industrial processes in which catalysts are used, stating the catalyst in each case. (O & C)

84. State *three* characteristics of a catalyst. Outline different manufacturing processes, *one* in each case, in which the following metals are used as catalysts: (*a*) iron, (*b*) platinum, (*c*) nickel. (C)

85. State the characteristic features of catalysis. Define the following terms, quoting *one* example of each: (*a*) homogeneous catalysis, (*b*) heterogeneous catalysis, (*c*) catalyst poison, (*d*) autocatalysis. What influence has a catalyst on the composition of the final equilibrium mixture of a reaction? Mentioning essential experimental details, describe concisely how you would find out whether a given mineral acid acted as a catalyst for the hydrolysis of ethyl acetate by water. (NJB)

86. State Le Chatelier's principle. State the conditions which favour the forward reactions in the following equilibria and explain why:

$$2NO_2 \rightleftharpoons 2NO + O_2;\ \Delta H = +x \text{ J mol}^{-1}$$
$$H_2 + I_2 \text{ (gaseous)} \rightleftharpoons 2HI;\ \Delta H = -y \text{ J mol}^{-1}$$
$$2SO_2 + O_2 \rightleftharpoons 2SO_3;\ \Delta H = -z \text{ J mol}^{-1}$$

The vapour density of phosphorus pentachloride at 200°C is 70; calculate the degree of dissociation. State and explain the effect of vaporising phosphorus pentachloride in an atmosphere of chlorine. (O)

87. What is the Principle of Le Chatelier? Apply this principle to deduce the effect of: (i) an increase in pressure on the thermal dissociation of ammonium chloride: (ii) an increase in temperature on the equilibrium $N_2 + 3H_2 \rightleftharpoons 2NH_3$; $\Delta H = -100$ kJ mol^{-1}, (iii) the addition of a dehydrating agent in the esterification of acetic acid by ethyl alcohol. (L)

88. Give an account of the influence of temperature, pressure and the presence of foreign bodies on (i) the rate of a chemical reaction, (ii) the position of equilibrium in a reversible reaction, (iii) the rate of attainment of equilibrium in a reversible reaction. Illustrate your answer by reference to the systems

$$CO_2 + H_2 \rightleftharpoons CO + H_2O;\ \Delta H = +50 \text{ kJ mol}^{-1}$$
$$2SO_2 + O_2 \rightleftharpoons 2SO_3;\ \Delta H = -188 \text{ kJ mol}^{-1}.$$ (L)

89. Experimental conditions for the synthesis of ammonia can be deduced from the equation:

$$N_2 + 3H_2 \rightleftharpoons 2NH_3;\ \Delta H = -100 \text{ kJ mol}^{-1}$$

Show how these deductions are applied to this reaction on the industrial scale. What other conditions are important in practice?
How is ammonia oxidized to nitric acid on the large scale? (L)

90.* Describe how the equilibrium between hydrogen, iodine, and hydrogen iodide could be measured at, say, 300°C. How would you expect this equilibrium to vary with (*a*) increasing temperature, (*b*) increasing pressure? Give your reasons. The reaction at 300°C is slightly exothermic. (O & C)

91.* State and explain Le Chatelier's principle and discuss its application to the dissociation of (*a*) nickel carbonyl, (*b*) barium peroxide, (*c*) nitric oxide. Comment on and explain the following apparent anomalies: (i) ozone is an endothermic substance but decomposes on heating to 200° C; (ii) sodium hydroxide dissolves in a moderate amount of water with considerable evolution of heat yet its solubility increases with temperature. (L)

92.* State the law of mass action, and show how it may be used to derive an expression for the equilibrium constant of the reaction

$$CH_3.COOH + C_2H_5.OH = CH_3.COOC_2H_5 + H_2O$$

Suggest one method by which (*a*) the equilibrium constant, (*b*) the heat, of this reaction may be found. (C)

93. State the Law of Mass Action and show how you would use it to derive the equilibrium constant of a reaction

$$2A + B = C + 2D;\ \Delta H = Q \text{ kJ mol}^{-1}$$

in which the reactants and resultants are all gases.

What steps could be taken (*a*) to obtain as large a yield as possible of C from a given amount of B, and (*b*) to obtain the yield as quickly as possible? (O & C)

94.* How will (*a*) variation of temperature, (*b*) variation of pressure and (*c*) addition of an inert gas affect the amounts of products formed at equilibrium in the following reactions in which all the substances involved are gases under the conditions of experiment?

(i) $A_2 + B_2 = 2AB;\ \Delta H = +x \text{ J mol}^{-1}$

(ii) $A_2 + 3B_2 = 2AB_3;\ \Delta H = -y \text{ J mol}^{-1}$

(iii) $A + B = C + D;\ \Delta H = +z \text{ J mol}^{-1}$

At 1,500°C. and atmospheric pressure carbon dioxide is 40 per cent. dissociated into carbon monoxide and oxygen. Calculate the equilibrium constant for this dissociation in terms of partial pressure. (NJB)

95. What do you understand by the term 'reversible reaction'? How would you show that the term applies to the reaction between ethyl alcohol and acetic acid? If the equilibrium constant for this reaction is 4, calculate the amount of ester present at equilibrium if one mole of acid reacts with (*a*) one mole of alcohol, (*b*) two mole of alcohol. By what means should it be possible to convert the whole of the alcohol to ester? (O & C)

96. When 13.8 gramme of ethyl alcohol and 24.0 gramme of acetic acid were allowed to react, the equilibrium mixture was found to contain 20.0 gramme of ethyl acetate. Calculate the equilibrium constant of the reaction

$$C_2H_5OH + CH_3.COOH \rightleftharpoons CH_3.COOC_2H_5 + H_2O$$

(Part question only) (L)

97. State the law of mass action for substances reacting in aqueous solution. If 1 mole of acetic acid and 1 mole of ethyl alcohol are mixed and the reaction proceeds to equilibrium, the concentrations of acid and water are found to be respectively $\frac{1}{3}$ and $\frac{2}{3}$ mole. If 1 mole of ester and 3 mole of water are mixed, how much ester is present when equilibrium is reached? ($\sqrt{52} = 7.21$.) (O & C)

98. When 60 g of acetic acid reacts with 46 g of ethyl alcohol until no further change occurs, 58.7 g of ethyl acetate is formed. When the same amount of acetic acid reacts with 23 g of alcohol, 37 g of ester is formed. Comment on, and explain, these observations.

[H = 1, C = 12, O = 16.]

(Part question only) (L)

99. State the Law of Mass Action and show how it may be applied to a reversible reaction to give an expression for the equilibrium constant. The equilibrium constant for the reaction

$$H_2 + I_2 \rightleftharpoons 2\,HI$$

is 50.0 at 444°C. If 8.0 mole of hydrogen are heated in a closed vessel at 444°C with 5.0 mole of iodine until equilibrium is attained, how many mole of hydrogen iodide will be formed? Explain what effect, if any, pressure would have on this reaction. (O & C)

100. What is meant by saying that the reaction between hydrogen and iodine is reversible? At 360°C there are present in equilibrium 1·445, 0·935, and 9.55 mole per dm^3 of hydrogen, iodine, and hydrogen iodide respectively. Calculate the equilibrium-constant for the reaction. Use this constant to find the composition of the equilibrium mixture at 360° that would be obtained from initial concentrations of 1 mole per dm^3 of hydrogen and iodine. (O & C)

101. At 200°C and 101 325 N m^{-2} pressure 208.5 g phosphorus pentachloride occupy 57.44 dm^{-3}. Calculate the equilibrium constant for the dissociation of this compound under these conditions, by way of (*a*) the vapour density, (*b*) the degree of dissociation (*c*) the conc^n in mol dm^{-3} of each species in the equilibrium mixture.

[Molar vol. at s.t.p. = 22.4 dm^3 mol^{-1}, P = 31.0 Cl = 35.5] (NJB)

102. What is meant by the term 'equilibrium' as applied to a chemical reaction? Give a definition of the equilibrium constant. One dm^3 of hydrogen iodide was kept at 300°C at atmospheric pressure until equilibrium was reached and then suddenly cooled to room temperature. The contents of the vessel were found to oxidize 38.9 c.c. $\frac{M}{10}$ sodium thiosulphate. Calculate the equilibrium constant for the decomposition of hydrogen iodide at 300°C, expressing concentrations in mol dm^{-3}. How would you expect the equilibrium constant for this reaction to depend on the pressure at a constant temperature? Give reasons.

[Molar volume of a gas at s.t.p. = 22.4 dm^3 mol^{-1}.] (NJB)

103. Distinguish between decomposition and dissociation. Give, with equations, one example of each. Explain the behaviour of calcium carbonate at high temperatures as an example of chemical equilibrium. What deductions can be drawn from the application of the law of mass action to this equilibrium?

At a given temperature, what would be the effects on this equilibrium of increasing (*a*) the mass of calcium carbonate, (*b*) the mass of calcium oxide, (*c*) the pressure of carbon dioxide? How would an increase in temperature affect the system? Explain the consequences of blowing a stream of air over the heated calcium carbonate. (NJB)

Electrochemistry

104. How does the conduction of electricity through an aqueous solution of a salt differ from its conduction through a metallic wire? State Faraday's laws of electrolysis and indicate briefly how they could be verified experimentally. An electric current is passed between Pt electrodes through the following aqueous solutions in cells arranged in series: (*a*) 0.1 M $AgNO_3$, (*b*) 0.01 M $AgNO_3$, (*c*) 0.1 M $CuSO_4$, (*d*) 0.1M$\frac{1}{2}$ H_2SO_4. If 0.36 g silver is deposited in cell (*a*), deduce the masses of silver and copper in (*b*) and (*c*) and the volume of hydrogen, measured at 17°C and 1.00×10^5 N m^{-2} pr. evolved in (*d*).

[Ag 108, Cu 63.6, 1 mole of a gas occupies 22.4 dm^3 mol^{-1} at s.t.p.] (NJB)

105. State Faraday's Laws of Electrolysis. Describe how you would determine the chemical equivalent of silver by an electrolytic method. Calculate the value of the faraday if 0.3705 g of copper were deposited from a solution of copper sulphate when a current of 0.75 A was passed for 25 minute.

[Cu = 63.6.] (L)

106. What do you understand by the term *electrochemical equivalent of an element*? How is the *electrochemical equivalent* related to the *chemical equivalent* of an element? Describe, with essential experimental details, how you would determine the electrochemical equivalent of an element such as silver. 0.406 g of a metal *X* was deposited from a solution by a current of one ampere flowing for 965 second. The metal formed a volatile chloride of vapour density 114. Calculate the atomic weight of the metal.

[1 Faraday = 96 500 coulomb; Cl = 35.5.] (O & C)

107.* Describe and explain how the products obtained by electrolysis of (*a*) sodium chloride, (*b*) sulphuric acid vary with the experimental conditions. Give details of technically important processes involving electrolysis of these substances. (L)

108.* Describe and explain what takes place when an electric current passes between platinum electrodes immersed in: (*a*) an aqueous solution of copper sulphate; (*b*) a hot aqueous solution of potassium chloride; (*c*) fused sodium acetate; (*d*) a concentrated aqueous solution of potassium bisulphate at 0°C; (*e*) a solution of silver cyanide in potassium cyanide solution. (NJB)

109. Define *conductivity* and *molar conductance* of a solution.
Describe how *you* would determine in the laboratory the *conductivity* of a solution of a strong electrolyte. The resistance of a conductivity cell containing

a 0.2 M solution of zinc sulphate at 25°C was found to be 72.2 ohm and the resistance of the same cell containing 0.02 M potassium chloride solution at 25°C was 550 ohm. If the conductivity of 0.02 M potassium chloride solution at 25°C is 0.00277 ohm, calculate: (*a*) the conductivity of the zinc sulphate solution; (*b*) the molar conductance of the zinc sulphate solution. (O & C)

110.* Explain what is meant by the molar conductance of a solution, and describe briefly how this quantity varies with concentration for different types of electrolyte. How may measurements of conductivities be used to determine the solubility of barium sulphate? (C)

111. (*a*) Define the *conductivity* and *molar conductance* of a solution of an electrolyte, and state how they are related. How is the apparent degree of ionization of an electrolyte at a particular dilution calculated from conductivity measurements? (*b*) What is meant by the pH value of a solution? What is the pH value of (i) 0.01 M hydrochloric acid, (ii) 0.01 M sodium hydroxide, (iii) the solution obtained by mixing equal volumes of (i) and (ii), (iv) the solution obtained by diluting 5 cm^3 of M hydrochloric acid with conductivity water to 1 dm^3?

$[K_W = 10^{-14} mol^2\ dm^{-6}]$ (O)

112. Define and explain the terms conductivity' and 'molar conductance.' Draw rough sketches to show how these two quantities vary during the progressive dilution of solutions of (*a*) sodium chloride, (*b*) acetic acid.

The molar conductance of a decimolar solution of acetic acid is 4.55, compared with an estimated value of 350 at infinite dilution. Calculate the pH of the decimolar solution. (L)

113.* Define the *conductivity* of a solution. How do the *molar conductances* of aqueous solutions of (*a*) sodium chloride, (*b*) acetic acid vary with the concentration? Explain the graphs obtained in these cases by plotting conductivity against dilution.

At 25°C the conductivity of a saturated solution of silver choride is $1.33 \times 10^{-6}\ \Omega^{-1}\ cm^{-1}$ and the mobilities of the silver and chloride ions are 58.1 and 68.0 respectively. Calculate the solubility of silver chloride in gramme per dm^3 at 25°C. (L)

114. Explain the terms 'molar conductance' and 'ionization constant of an acid'. The molar conductance of a 0.001 M solution of acetic acid at 25°C is 48.1 $\Omega^{-1}\ cm^2\ mol^{-1}$, compared with the calculated value of 391 at infinite dilution. Calculate the dissociation constant of the acid and the pH of the solution. (L)

115. The conductivity of M/20 acetic acid at 18°C is $4.4 \times 10^{-4}\ \Omega^{-1}\ cm^{-1}$. The mobilities of the hydrogen and the acetate ions are 310 and 77 respec-

tively at the same temperature. Find the pH of the solution and the dissociation constant of acetic acid. Mention other methods of finding pH values.
(O SCHOL)

116.* State the law of mass action, and describe how it can be verified experimentally for any ONE reaction. Deduce Ostwald's dilution law. If the dissociation constant for acetic acid is 1.8×10^{-5}, calculate the pH value of (*a*) M acetic acid, (*b*) acetic acid containing 1 mole of sodium acetate per dm^3.
(O)

117. Give *three* distinct reasons why you think that sodium chloride exists in solution in the form of ions. Strictly speaking, why is it incorrect to refer to a *molecule* of sodium chloride? If the dissociation constant of acetic acid at 18°C is 1.8×10^{-5}, what percentage of the acid is in the form of ions in a decimolar solution of the acid? Will there be a higher or lower percentage of ions present in a molar solution, and why?
(O & C)

118.* 'Urea, $CO(NH_2)_2$, is a non-electrolyte; propionic acid, $C_2H_5CO_2H$, is a weak electrolyte; sodium chloride is a strong electrolyte.' Discuss this statement with particular reference to the behaviour of aqueous solutions of the three compounds in regard to osmotic pressure, depression of the freezing point, and electrolytic conductance.
(NJB)

119. Explain what is meant by the following statements: (*a*) the pH of 0.1M acetic acid is 2.88, (*b*) the *ionic product* of water at 25°C is 1.0×10^{-14}, (*c*) the *dissociation constant* of ammonium hydroxide at 25°C is 1.8×10^{-5}.
Calculate the pH of the following solutions: (*d*) 2.0 M hydrochloric acid, (*e*) 0.1 M sodium hydroxide.
(L)

120. Define pH.
Why is an aqueous solution of sodium acetate alkaline? Explain what happens when (*a*) dilute hydrochloric acid, and (*b*) sodium hydroxide solution, are gradually added to a solution of sodium acetate in acetic acid. Why is it not possible to titrate ammonium hydroxide with acetic acid using phenolphthalein or methyl orange as indicator?
(O & C)

121. What is meant by the pH of a solution and why is it used? Explain the nature of a buffer solution.
Calculate the pH of the following solutions: (*a*) 0.01 M hydrochloric acid; (*b*) 0.01 M sodium hydroxide; (*c*) a solution where the hydrogen ion concentration is 2×10^{-6} mol dm^{-3}; (*d*) a solution where the hydroxyl ion concentration is 2×10^{-6} mol dm^{-3}.

$$([H^+][OH^-] = 1 \times 10^{-14})$$

(O & C)

122. What is a buffer solution? How does it function? What is the practical importance of such solutions? Calculate the pH of (i) a decimolar solution of acetic acid, given that the dissociation constant of the acid is 1.83×10^{-5};

(ii) a solution made by mixing equal volumes of decimolar solutions of acetic acid and sodium acetate.

[H = 1.00; C = 12.0; O = 16.0; Na = 23.0.] (L)

123. Explain the terms (i) solubility product, (ii) strength of an acid, (iii) the ionic product of water. Use the concept of solubility product to explain two procedures you have used in qualitative analysis. Briefly describe *one* experiment by which the relative strengths of two acids may be compared. (L)

124.* Discuss the principles underlying the use of hydrogen sulphide as a reagent in qualitative analysis. The solubility of silver sulphide is 1.24×10^{-14} g dm^{-3} at 25°. Calculate its solubility product at this temperature.

[Ag = 108, S = 32.] (C)

125.* Define solubility product. The solubility of silver chloride in water is 0.0015 g dm^{-3}. Calculate its solubility product. What is the maximum concentration of silver ion that should be able to exist in a decimolar solution of hydrochloric acid? Why are 'insoluble chlorides' often soluble in concentrated hydrochloric acid? How would you attempt to show the truth of your explanation?

[Cl = 35.5. Ag = 107.9.] (O & C)

126.* What is meant by the term *solubility product*? Give *two* examples of the application of this concept to qualitative analysis. The solubility of lead sulphate in water at 17°C is 0.035 g dm^{-3}. Calculate (*a*) the solubility product of lead sulphate and (*b*) the solubility of lead sulphate (in g dm^{-3}) in a 0.01 M solution of sodium sulphate at the same temperature. (Assume complete dissociation of both solutes).

[O = 16; S = 32; Pb = 207.2.] (O & C)

127.* Define *solubility product* and explain in what circumstances the term can be applied. Explain why copper sulphide is not precipitated by hydrogen sulphide from solutions of copper salts in the presence of potassium cyanide. The solubility of lead iodide (PbI_2) in water at 20°C is 0.76 g dm^{-3}. Calculate (*a*) the solubility product of lead iodide; (*b*) the solubility (in g dm^{-3}) of lead iodide in a 0.1 M solution of sodium iodide also at 20°C. (Assume complete dissociation of both solutes.)

[Na = 23.0, I = 126.9, Pb = 207.2.] (O & C)

128. Define and explain the term 'solubility product'. What is meant by the statement that the solubility product of silver chloride is 1.2×10^{-10} and that of lead iodide is 1.4×10^{-8}?

Make clear the differences between 'solubility product' and 'dissociation constant'.

Explain why hydrogen sulphide precipitates only copper sulphide from a slightly acidified solution containing copper and zinc ions; whereas if the filtrate is neutralized and made slightly alkaline with ammonia, hydrogen sulphide will now precipitate zinc sulphide.

The solubility of calcium sulphate in water at 18°C is 0.015 mol dm^{-3}. What is the value of its solubility product? Calculate how much calcium sulphate (in mole) would be precipitated from a dm^3 of its saturated solution if sodium sulphate were added to bring the total SO_4^{2-} concentration to 1.000 mol dm^{-3}. (NJB)

129. Explain the term *solubility product*. What is the *solubility* of silver chloride in g dm^{-3} of (*a*) water, (*b*) 0.01 M silver nitrate solution?

Why does the accuracy of the titration of a chloride in neutral solution with silver nitrate, using potassium chromate as indicator, depend on the concentration of potassium chromate in the solution undergoing titration?

(Solubility products of: $AgCl = 10^{-10}$, $Ag_2CrO_4 = 10^{-12}$,

Ag = 107.9, Cl = 35.5.) (SJB)

130. What do you understand by the term 'solubility product'? To what extent do you consider the principle of solubility product explains the following: (*a*) hydrogen sulphide does not precipitate zinc sulphide from a strongly acid solution of zinc sulphate; (*b*) calcium phosphate, though insoluble in water, readily dissolves in dilute hydrochloric acid; (*c*) hydrogen chloride passed into a strong solution of sodium chloride precipitates some of the sodium chloride. (C)

130.A* Explain what is meant by the statement that $\frac{M}{10}$ acetic acid has a pH value of 2·9'. Define the term 'indicator' as used in reference to acid–alkali titrations. Show by reference to pH changes during titration illustrated by suitable graphs, how the choice of indicators in acid–alkali titrations is governed. Suggest, giving reasons for your choice, suitable indicators for the titration of NaOH by (*a*) HCl, (*b*) acetic acid. (NJB)

131.* Write an essay on the principles governing the choice of indicators used in acid–alkali titrations. (O)

132. Write an account of the changes in pH which occur during the neutralization of (*a*) a strong acid and (*b*) a weak acid by sodium hydroxide, and show how these considerations affect the choice of indicators or acid–base titrations. Explain as far as you can why the range of an indicator is about two pH units. (SJB)

133.* What is a complex ion, and what do you know of methods which have been used to indicate the presence of complex ions in solution? Explain how the formation of complex ions is used in the detection of certain metals in qualitative analysis. (O)

134.* Discuss, with examples, the difference between double salts and complex salts, and describe the preparation of one typical member of each class. The freezing point of a molar solution of potassium iodide is – 3.2°C. When mercuric iodide (which is insoluble in pure water) is added in excess, the freezing point of the solution is raised to – 2.4°C. Comment on this result. (C)

135. Explain the following statements:

(*a*) It is incorrect to refer to concentrated sulphuric acid as 'strong sulphuric acid'.

(*b*) On dissolving in water hydrogen chloride ionizes and dissociates but potassium chloride only dissociates.

(*c*) Water is acid to phenolphthalein but alkaline to methyl orange.

(*d*) It is inadmissible to speak of the solubility product of acetic acid. (NJB)

136.* Explain the following:

(*a*) Addition of potassium cyanide solution in excess to a solution of cupric sulphate yields a colourless solution with which hydrogen sulphide gives no precipitate, whereas a solution of cadmium sulphate treated similarly gives a yellow precipitate.

(*b*) Electrolysis of sodium chloride solution using a steel cathode gives sodium hydroxide and hydrogen, but with a mercury cathode the product is sodium amalgam.

(*c*) Potassium chromate may be used as an indicator in titrating a solution of a chloride with silver nitrate solution, although the solubility products of silver chromate and silver chloride are respectively 2.4×10^{-12} and 1.6×10^{-10}.

(*d*) Phenolphthalein is a suitable indicator for the titration of an organic acid with a strong alkali. (O)

137. Explain what you understand by (*a*) the concentration, (*b*) the strength, (*c*) the basicity and (*d*) the dissociation constant of an acid. Describe in detail one experiment to demonstrate that sulphuric acid is dibasic and another experiment to show that hydrochloric acid is a stronger acid than acetic acid. (L)

138. What do you understand by the electrochemical series? Explain, with illustrations, how the chemical properties of an element are related to its position in this series. (L)

139. Explain what is meant by the electrochemical series of the elements. How is it arrived at, and what is its significance and uses? (O SCHOL)

140. What general methods are available for the extraction of a metal from one of its naturally occurring compounds? Illustrate by examples how the choice of method is governed by the position of the metal in the electrochemical series. (L)

Thermochemistry

Students may find it helpful to be reminded of the meaning of the sign convention used in these questions. Where heat is evolved ΔH is always shown as negative, and where heat is absorbed it is shown as positive.

e.g. $A + B = AB; \Delta H = -x$ kJ mol^{-1}

means that x kilojoule of heat is evolved when one mole of AB is formed from A and B.

141. The heat of formation of carbon dioxide is given by

$$C + O_2 = CO_2; \Delta H = -393 \text{ kJ mol}^{-1}$$

Explain fully what this statement means.

The heats of formation of carbon dioxide, water and benzene are − 393, − 286, and + 48.1 kJ mol^{-1} respectively. Calculate the heat of combustion of benzene. Outline briefly an experiment to verify your result. (SJB)

142. Define heat of formation, exothermic comp[d]., endothermic comp[d]. State Hess's Law and show, with one example, how it may be used to obtain a heat of reaction which cannot be measured directly. Given the following values of heats of combustion, calculate the heat of formation of ethylene: ethylene 1432 kJ mol^{-1}, hydrogen 272 kJ mol^{-1}, carbon 406 kJ mol^{-1}. (NJB)

143.* State the 'Law of constant heat summation' (Hess). Explain what is meant by (*a*) heat of formation, (*b*) heat of combustion. Why, and by what value, may these two quantities differ in magnitude according to whether they are measured at constant volume or at constant pressure? Distinguish between (*c*) heat of solution and (*d*) heat of dilution.

From the folowing data calculate the heat of formation of carbon bisulphide and deduce whether or not the formation of this substance from its elements is favoured by the use of a high temperature:

$$C + O_2 = CO_2; \Delta H = -405\,700 \text{ J mol}^{-1}$$
$$S + O_2 = SO_2; \Delta H = -297\,200 \text{ J mol}^{-1}$$
$$CS_2 + 3O_2 = CO_2 + 2SO_2; \Delta H = -1109\,000 \text{ J mol}^{-1}$$

(NJB)

144. Define the terms *heat of combustion, heat of formation*, and *endothermic substance*. State Hess's law and calculate the heat of formation of propane from the following data:

$$C + O_2 = CO_2; \Delta H = -406 \text{ kJ mol}^{-1}$$
$$2H_2 = O_2 = 2H_2O; \Delta H = -572 \text{ kJ mol}^{-1}$$
$$C_3H_8 + 5O_2 = 3CO_2 + 4H_2O; \Delta H = -2209 \text{ kJ mol}^{-1}$$

What is the connection between the heat of formation of a substance and its stability? (O)

145. Define (*a*) *heat of formation,* (*b*) *heat of reaction,* and (*c*) *heat of combustion.* Describe *briefly* how you would measure the heat of neutralization of molar hydrochloric acid with molar sodium hydroxide.. If the heats of combustion of carbon in the form of graphite, hydrogen, and formic acid are 405.7 kJ mol^{-1}, 286 kJ mol^{-1}, and 276.3 kJ mol^{-1} respectively, what is the heat of formation of formic acid? (C)

146.* Explain the terms: *heat of neutralization*; *heat of formation*; *heat of reaction* and *heat of combustion.* Why is the heat of neutralization of a strong acid and a strong base constant? Given that the heat of combustion of benzene is 3278 kJ mol^{-1} (evolved) and that the heats of formation of carbon dioxide and water are 393.4 and 285.8 kJ mol^{-1} (evolved) respectively, calculate the heat of formation of benzene, stating whether the heat is absorbed or evolved. (O & C)

147. Define (*a*) heat of formation, (*b*) heat of combustion. State Hess's Law of constant heat summation. If the heats of combustion of hydrogen, ethylene and ethane, are 286 kJ mol^{-1}, 1393 kJ mol^{-1}, and 1561 kJ mol^{-1}, respectively, calculate the heat of hydrogenation of ethylene to ethane. (L)

148. (*a*) State Hess's Law, and define the terms *heat of formation* and *heat of combustion.* (*b*) Calculate the heat of formation of methane from the following data: (i) heat of formation of H_2O (gas) is 286 kJ mol^{-1}, (ii) heat of formation of CO_2 is 405.7 kJ mol^{-1}, (iii) heat of combustion of methane is 886.7 kJ mol^{-1}. (*c*) Explain concisely how changes in conditions may affect (i) the thermal dissociation of calcium carbonate, and (ii) the reaction between steam and iron. (SJB)

149. Define 'heat of solution', 'heat of combustion' and 'heat of neutralization'. Describe in outline how any *one* of these quantities may be determined in the laboratory. Explain why (*a*) the heat of neutralization of any strong acid by any strong base is the same, (*b*) the heat of combustion of compounds is particularly important in organic chemistry. (L)

150. Define the terms (*a*) exothermic reaction, (*b*) heat of solution, (*c*) heat of formation. State Hess's Law. Calculate the heats of formation of ethane and acetylene from the following heats of combustion: carbon, 393.3 kJ mol^{-1}; hydrogen, 284.5 kJ mol^{-1}; acetylene, 1297 kJ mol^{-1}; ethane, 1548 kJ mol^{-1}. What information could be deduced about the relative thermal stabilities of ethane and acetylene? What heat change would be involved if 1 mole of acetylene were completely converted to ethane by being made to combine with hydrogen?

(NOTE.—For all heat changes which you calculate in this question, state precisely whether the heat is liberated or absorbed). (NJB)

151. State Hess's Law. Define the terms 'heat of formation', 'heat of neutralization' and 'heat of solution'. The heat evolved when one g of solid iodine reacts with excess of zinc dust suspended in water is 985.3 J. When one g of solid anhydrous zinc iodide dissolves in the same amount of water 160 joule are evolved. Derive the heat of formation of solid zinc iodide, showing clearly how Hess's Law is involved in your calculation.

(Zn = 65.4, I = 126.9) (L)

152. Explain and illustrate the terms (*a*) exothermic reaction, (*b*) endothermic compound. The heat of combustion of ammonia (to water and nitrogen) is 381 kJ mol^{-1} and the heat of formation of water is 214 kJ mol^{-1}. Calculate the heat of formation of ammonia and use your result to predict the effect of changes of temperature on the equilibrium between nitrogen, hydrogen, and ammonia. (O & C)

The Colloidal State

153. Describe THREE different methods by which you would prepare colloidal solutions, illustrating your answer by reference to ONE particular example for each method. What are the properties of colloidal solutions which distinguish them from true solution? Give an account of the precipitation of colloidal particles by electrolytes. (O)

154. Explain with examples how you would demonstrate three important properties of colloidal solutions. State whether the properties to which you refer are characteristic of lyophobic colloids, lyophilic colloids or both. State briefly how you would prepare in a colloidal state *one* metal and *one* non-metal or compound. (O & C)

155.* Describe in detail how you would prepare (*a*) one lyophilic and (*b*) one lyophobic colloidal solution. What experiments would you use to show three essential differences in character between the two colloidal solutions whose preparation you describe? (O & C)

156.* Describe the essential features by which you would recognize a solution as colloidal. (O & C)

157. What do the following terms signify: coagulation peptization, endosmosis, Tyndall cone? How would you make a colloidal solution of ferric hydroxide and how would you determine whether the charge on the colloidal particles was positive or negative? (O & C)

158. With regard to colloids, what is meant by: (i) peptization, (ii) Tyndall effect, (iii) adsorption, (iv) Brownian movement, (v) gold number? Give examples of some practical applications of adsorption in (vi) industry, (vii) the laboratory. (L)

The Periodic Classification

159.* Describe the principal features of the periodic classification of the elements. Give a brief explanation of the classification in terms of modern ideas of atomic structure. (O & C)

160. In the Periodic Classification, elements are arranged in (*a*) *groups* and (*b*) *periods*. Explain briefly how these arrangements can be related to the structure of the atoms concerned. What general differences are there between the reactions of the alkali metals (Group I A) and their compounds, and those of the alkaline earth metals (Group II A) and their compounds? (O & C)

161.* Explain what you understand by the terms (*a*) atom, (*b*) atomic number, (*c*) transition element. Describe briefly, giving examples, the variations of chemical and physical properties to be found in (i) a Group, (ii) a short period, of the Periodic Classification. (L)

162. Give a general account of the classification of the elements according to Mendeléeff's Periodic Law, drawing attention to any anomalies which exist. By comparison of their general characteristics, justify the inclusion in the same Group of the elements carbon and tin. (L)

163. 'Atomic number forms a more satisfactory basis for the classification of the elements than atomic weight.' Discuss this statement. (O SCHOL)

164.* Compare and contrast the electronic structures and properties of potassium and sodium, and of their important compounds, to show how their inclusion in the same group of the Periodic Table is justified. Why are the salts of potassium and sodium used to a greater extent than those of other metals, both in the laboratory and on the industrial scale? (SJB)

165. Calcium, strontium, and barium are placed in the same group of the periodic classification. Show how the properties of the hydroxides, sulphates, and other compounds justify this. How are these properties made use of in qualitative analysis? (O & C)

166. Discuss how far the classification of carbon, silicon, tin and lead in one Group of the periodic system is justified by their chemical properties. (L)

167.* What similarities are there in the chemistry of the Group IV elements, carbon, silicon, tin, and lead? Mendeléeff's prediction that an element 'eka-silicon' would be discovered to fill the place in this group between silicon and tin was fulfilled by the discovery in 1876 of germanium. What properties would you expect this element to have? (O & C)

168. Discuss the positions of (*a*) lead, (*b*) nitrogen in the Periodic Table.

To what extent do the chemical properties of these elements justify their respective positions in the Table? (L)

169.* Give a comparative account of the chemistry of nitrogen and phosphorus, with special reference to the properties of (*a*) the elements, (*b*) their hydrides, (*c*) their chlorides, (*d*) their oxides and oxyacids. (C)

170.* Phosphorus occurs in the vertical group N, P, As, Sb of the Periodic Table and in the horizontal series Al, Si, P, S, Cl. By reference to simple compounds of these elements show how P appears to be correctly placed in relation to the gradation in properties expected in the group and series respectively of the Periodic Table. Can you suggest any explanations of the gradations in properties exhibited? (NJB)

171.* Illustrate from the chemistry of the element nitrogen, its oxides and chlorides, the anomalous character of the first element in a Periodic Group. (O & C)

172.* The properties of the first member of a Periodic Group are often anomalous. Illustrate this by examples from the chemistry of fluorine and nitrogen. (O & C)

173.* Discuss (*a*) the *similarities* between oxygen and sulphur which justify classifying them as elements in the same group of the Periodic Table, (*b*) the *differences* which exist between the chlorides and hydrides of nitrogen and phosphorus in spite of their being in the same group. (SJB)

174. Illustrate from the chemistry of the halogens the nature of a Group in the Periodic Table. (O & C)

175. From a consideration of their properties, and those of their compounds, justify the positions of the following elements in the Periodic Table: (*a*) barium, (*b*) phosphorus, (*c*) iodine. (C)

176.* What general methods are used for preparing the chlorides of the non-metallic elements? Describe the chief characteristics of this class of compound. Comment on the similarities and differences between the chlorides of the elements in each of the following pairs: (*a*) B and Al, (*b*) C and Si, (*c*) N and P, (*d*) O and S. (L)

177. What are the general features of the chemistry of the transition metals? Relate these to the chemistry of any one transition metal. (O SCHOL)

178. State what you understand by the term 'transition metal'. Give *two* examples from the chemistry of each of the metals manganese, iron and copper to illustrate their transitional character. (O & C)

179.* Give examples of the use of (*a*) complex formation, (*b*) variation in valency, in the identification of any *three* transition metals. (O & C)

180. Discuss the so-called diagonal relationships in the Periodic Table, with particular reference to the elements lithium and magnesium, boron and silicon, oxygen and chlorine. (O SCHOL)

Atomic Structure and the Electronic Theory of Valency

181. Describe the modern theory of the structure of atoms, defining any particles you mention. Explain how this theory accounts for (*a*) the periodicity of elements, (*b*) the variations in valency within a short period, and (*c*) a characteristic valency in a group of the Periodic Table. Chlorine has atomic number 17. Sketch the structures of its isotopes of atomic weight 35 and 37. (SJB)

182.* Give a brief account of the present theory of atomic structure and show how it accounts for (*a*) the periodic system of the elements, (*b*) the existence of isotopes, (*c*) the principal type of valence bonds occurring in chemical compounds. (NJB)

183. What do you understand by the terms *electron*, *proton*, *neutron*, *atomic number*, *isotope*? What is the atomic structure of the chlorine isotopes? Explain why determinations of the atomic weight of chlorine give a value of 35.5. (O)

184. The element calcium has an *equivalent* of 20.04, its *atomic weight* is 40.08, its *atomic number* is 20, and there are several calcium *isotopes*. What do you understand by the terms printed in italics? Point out any ways in which these terms are connected. Give the electronic structures of calcium oxide and calcium chloride. (O)

185.* From the point of view of atomic structure, discuss the different ways in which atoms are linked in the formation of compounds. Illustrate your answer by some well-chosen examples. (O)

186. What is meant by (*a*) the electrovalent, (*b*) the covalent bond? Illustrate your answer by suitable examples. What properties do you associate with each of these bonds? (O & C)

187. Distinguish carefully between electrovalency and covalency, illustrating your answer by reference to hydrogen chloride, sodium chloride, methane, and carbon tetrachloride. (C)

188.* The principal types of bonds occurring in compounds are, according to the electronic theory of valence, described as electrovalence, covalence, and co-ordinate (dative) valences. Explain (*a*) what is meant by these terms.

(*b*) the properties conferred by such linkages on the compounds in which they occur. Give the electronic formulae of two compounds to illustrate each type of bond. (NJB)

189. Give your reasons for believing that the bonds (*a*) in silica are covalent, (*b*) in carbon tetrachloride are covalent, (*c*) in sodium chloride are ionic. What kinds of bonds do you think exist in ferrous ammonium sulphate? (O & C)

190.* Give a brief account of the bonding in the following: (*a*) sodium chloride, (*b*) carbon tetrachloride, (*c*) nitrogen, (*d*) hydrogen chloride gas, (*c*) diamond. How does the nature of the bonding affect the properties of these substances? (C)

191. Write a short account of the electron theory of valency, and discuss the types of bond found in sodium chloride, ethane, magnesium sulphate, and potassium ferrocyanide. (SJB)

192. Explain clearly the terms *electrovalency* and *covalency* giving two examples in each case. Account for the formation of the tetra-ammonia cupric ion and the compound $NH_3.BF_3$. State the more important physical properties associated with electrovalency and covalency as exemplified in the hydrides, oxides and chlorides of calcium and carbon respectively. (L)

Oxidation and Reduction

193. What do you understand by 'oxidation' and 'reduction'? Show that these two processes must always occur together and discuss the following reactions as examples of oxidation—reduction:

(*a*) $Zn + 2HCl = ZnCl_2 + H_2$

(*b*) $Cl_2 + SnCl_2 = SnCl_4$

(*c*) $2Na_2S_2O_3 + I_2 = Na_2S_4O_6 + 2NaI$

(NJB)

194. Define oxidation and reduction in terms of electron transfer and illustrate your answer by reference to the reactions which take place between the following pairs of substances in aqueous solution: (*a*) metallic zinc and copper sulphate; (*b*) bromine and ferrous sulphate; (*c*) iodine and hydrogen sulphide; (*d*) copper sulphate and potassium iodide. (L)

195. Give *four* processes to which you could apply the term *reduction*, with a reaction to illustrate each. Explain what connection there is between the processes you mention which justifies the application of the term *reduction* to them. (O & C)

196. Explain fully the meaning of the term *reducing agent*. Give *one* example of the reducing action of each of the following with a different reagent in each case: (*a*) ozone; (*b*) hydrogen sulphide; (*c*) carbon monoxide; (*d*) nitrous acid; (*e*) stannous chloride; (*f*) potassium ferrocyanide. Write equations for the reactions. (O)

Physical Terms and Laws

197. Distinguish between the following terms, either giving examples to illustrate them or stating the units in which the quantities concerned would be measured: (*a*) degree of dissociation and dissociation constant; (*b*) solubility and solubility product; (*c*) osmosis and dialysis; (*d*) conductivity and molar conductance; (*e*) drying and dehydration; (*f*) isomer and isotope. (O & C)

198. Distinguish between (*a*) velocity constant and equilibrium constant, (*b*) electrovalency and co-valency, (*c*) enantiotropy and monotropy, (*d*) diffusion and effusion. (O)

199. Explain and illustrate the essential difference between the terms in each of *four* of the following pairs: (*a*) hygroscopic and deliquescent, (*b*) 'strong' electrolyte and 'weak' electrolyte, (*c*) covalent link and co-ordinate link, (*d*) thermal dissociation and thermal decomposition, (*e*) lyophobic and lyophilic colloids. (L)

200. Describe one experiment in each case to illustrate (*a*) gaseous diffusion, (*b*) osmosis, (*c*) dialysis. What conclusions can be drawn from each experiment? Discuss the ways in which these three processes are (i) alike, (ii) different. (C)

201.* What do you understand by the terms; precipitate, semipermeable membrane, constant boiling mixture, eutectic, partition coefficient? Give examples. (O & C)

202. Explain what is meant by *four* of the following: Ostwald Dilution Law, buffer solution, isotope, triple point, complex ion. (O & C)

203. Write short notes on (*a*) deliquescence and efflorescence, (*b*) van't Hoff's factor *i*, (*c*) diffusion of gases, (*d*) Le Chatelier's principle. (C)

204. (*a*) Define precisely, with examples where possible, *four* of the following terms: empirical formula; gramme-molecular volume; molar solution; heat of combustion; freezing point; molar conductivity. (*b*) State the following laws: (i) Hess's Law. (ii) Dalton's Law of Partial Pressures. (iii) Faraday's Laws of Electrolysis. (O & C)

205. State any *five* of the following and explain their use in chemistry: Dulong and Petit's law; Mitscherlich's principle of isomorphism; Raoult's law; Henry's law; Ostwald's dilution law; Graham's law of diffusion; Gay-Lussac's law. (SJB)

206. Explain, with examples, any *three* of the following terms: common ion effect, electrode potential, eutectic point, mixed crystals, thermal dissociation. (L)

207. Explain carefully *five* of the following terms, illustrating your answer, in each case, by reference to an experiment you have seen or performed: (i) catalysis, (ii) protective colloid, (iii) allotrope, (iv) mixed crystals, (v) cataphoresis, (vi) common ion effect, (vii) transition point. (L)

208. Explain, giving examples, the meaning of *five* of the following terms: (*a*) thermal dissociation, (*b*) pH, (*c*) salt hydrolysis, (*d*) exothermic compound, (*e*) polymerization, (*f*) deliquescence. (C)

209.* Explain what is meant by the following terms: transition temperature; solubility product: critical temperature: molar conductance. In each case describe briefly an experiment by means of which the value of the property concerned could be determined. (NJB)

210. Explain what is meant by 4 of the following terms, giving one example of each: transition temperature, autocatalysis, constant boiling mixture, eutectic mixture, buffer solution. (NJB)

Miscellaneous Questions on Physical and Inorganic Chemistry

211. Describe and explain simple laboratory experiments to illustrate any *four* of the following phenomena: (*a*) the migration of ions, (*b*) the Brownian movement, (*c*) the passivity of iron, (*d*) catalysis, (*e*) adsorption, (*f*) diffusion of gases. (L)

212.* Discuss briefly the principles underlying *three* of the following: (*a*) the determination of the equivalent weight of silver; (*b*) the detection of nitrogen in organic compounds; (*c*) the purification of aniline by steam distillation; (*d*) Cannizzaro's method for the determination of atomic weights. (C)

213.* Discuss and explain the following statements: (*a*) the dehydration of copper sulphate crystals turns them white; (*b*) ethylene is prepared by the dehydration of alcohol; (*c*) aluminium sulphate solution is acid by hydrolysis; (*d*) glycerol is produced by hydrolysis of a fat; (*e*) glycine can act as an acid as well as a base; (*f*) aluminium hydroxide can act as an acid or as a base. (O & C)

214.* Comment on the following statements: (*a*) The heat capacity of carbon is 0.71 J g^{-1} K^{-1} at 20°C, but is 1.88 J g^{-1} K^{-1} at 900°C. (*b*) Quartz and sodium chlorate can have crystals which rotate the plane of polarized light, but this property disappears when the quartz is melted or the sodium chlorate is dissolved. The optical activity of organic compounds persists on melting or in solution. (*c*) Hydrogen iodide is an endothermic compound at ordinary temperatures, but has become exothermic at 400°C. (*d*) The boiling-point of methane is below that of the corresponding hydride of silicon, but the boiling points of water and ammonia are above those of hydrogen sulphide and phosphine respectively. (O & C)

215. Explain *four* of the following phenomena: (*a*) ammonium acetate is neutral in solution; (*b*) when much water is added to bismuth trichloride a turbidity is produced; (*c*) if heated very slowly, sulphur melts at 119.2°C., but if heated quickly melts at 112.8°C.; (*d*) at 400°C. the vapour density of ammonium chloride is 14.6 relative to hydrogen = 1; (*e*) the heat of neutralization of any strong acid by any strong base is 57.3 kilojoule. (L)

216.* Comment on the following topics: (*a*) The assignment of an atomic number to an element. (*b*) Isotopes and their masses. (*c*) The determination of the atomic weight of argon. (*d*) The pH range of an acid–base indicator. (SJB)

217.* Explain—(*a*) how α-particles have been used to show that atoms contain a positively-charged nucleus; (*b*) why the molar conductance of a strong electrolyte varies with dilution; (*c*) why a solution of copper sulphate to which excess of potassium cyanide solution has been added gives no precipitate with hydrogen sulphide; (*d*) the principle of the process of steam distillation used in the purification of aniline. (O)

218.* Write notes on the following: (*a*) In the production of very pure hydrogen a solution of barium hydroxide is electrolysed in preference to sodium hydroxide. Why? The hydrogen formed is passed over hot platinum and then through solid potassium hydroxide and phosphorus pentoxide. Explain the reactions and significance of this sequence. (*b*) Assuming that the formation of ozone from oxygen is reversible and endothermic, explain: (i) why ozone is destroyed when passed through a tube at 400°, and (ii) why ozone cannot satisfactorily be prepared in quantity by heating oxygen at 3 000°C. (*c*) Lead dioxide may be classified as a dioxide or an amphoteric oxide but not as a peroxide. Justify this statement by consideration of its reactions. What type of bond would you assign to the oxide? (SJB)

219. Give explanations of the following observations: (*a*) That sodium chloride is less soluble in concentrated hydrochloric acid than in water, whilst lead chloride is more soluble in conc. hydrochloric acid than in water. (*b*) That a coating of zinc on iron will continue to protect the iron from atmospheric corrosion even when the coating is partially worn away, but that this is not the case with a coating of tin on iron. (*c*) That hydrogen sulphide will precipitate arsenious sulphide from an acid soln. of arsenious oxide but not from an alkaline one, whilst it will precipitate zinc sulphide from an alkaline solution containing zinc but not from an acid soln. (*d*) That iodine, though sparingly soluble in water, is freely soluble in an aqueous solution of potassium iodide. (NJB)

220. Explain the following: (*a*) The pH of 0.1M hydrochloric acid is 1 but that of 0.1M acetic acid is 2.75. (*b*) The cations discharged when (separate) aqueous solutions of copper sulphate and sodium sulphate are electrolysed between platinum electrodes are copper and hydrogen respectively. (*c*) Pure water is acid to phenolphthalein but alkaline to methyl orange. (*d*) Nitrogen dioxide is black at 140°C.; this colour changes on cooling to red-brown at 25°C. and pale yellow at 5°C. but is discharged on heating to 600°C. (NJB)

221. Describe some methods for determining the end-points of titrations, explaining the principles on which they are based, and mention factors which limit their accuracy in practice. (O SCHOL)

222. Give explanations of three of the following: (*a*) the freezing point of a solution of potassium iodide in water is almost unaffected when some iodine is added to it; (*b*) diamond is hard, but graphite is soft and slippery; (*c*) sodium chloride is a high melting solid, but carbon tetrachloride is a volatile liquid; (*d*) silicon tetrachloride is a very reactive liquid, but carbon tetrachloride is very unreactive. (O SCHOL)

223.* Give a short description of the following processes, adding comments on the physico-chemical principles involved in each: (*a*) one method for the extraction of silver from argentiferous lead, (*b*) the production of liquid air, (*c*) the silver-plating of copper, (*d*) the preparation of a specimen of pure sodium chloride from crude salt, (*e*) the preparation of a specimen of pure ethyl alcohol from its weak aqueous solution. (NJB)

224. How and under what conditions do the following pairs of substances react together? (*a*) sodium thiosulphate and chlorine; (*b*) potassium permanganate and oxalic acid; (*c*) carbon monoxide and water; (*d*) sulphur and carbon; (*e*) calcium bicarbonate and ammonia. State briefly what use is made of these reactions. (SJB)

225. How, and under what conditions, do the following pairs of substances react with each other: (*a*) potassium iodide and potassium iodate, (*b*) ferric oxide and aluminium, (*c*) ferric chloride and potassium ferrocyanide, (*d*) ferrous sulphate and potassium permanganate? Mention *one* practical use that is made of *each* reaction. (C)

226. Describe what may be seen, and state the chemical reactions which take place, when an aqueous solution of potassium iodide is added to aqueous solutions of each of the following: (*a*) lead acetate, (*b*) mercuric chloride, (*c*) copper sulphate, (*d*) potassium dichromate acidified with sulphuric acid, (*e*) acetone and sodium hypochlorite. (C)

227. Describe and explain what happens when solutions of (i) ammonia, (ii) potassium iodide, (iii) potassium cyanide, (iv) sodium carbonate, are added to (*a*) copper sulphate solution, (*b*) silver nitrate solution. (O SCHOL)

228. How would you attempt to identify the components of *four* of the following mixtures: (*a*) sodium chloride and sodium bromide; (*b*) sodium arsenate and sodium phosphate; (*c*) zinc sulphate and magnesium sulphate; (*d*) sodium bromide and sodium nitrate; (*e*) sulphur dioxide and carbon dioxide? (L)

229. For each of the ions mentioned below describe *one* positive test, other than a flame test, which would enable you to distinguish it from the metal paired with it. State for each test the reagent(s) used, the conditions, and the results observed: (*a*) calcium; barium, (*b*) sodium; potassium, (*c*) zinc; aluminium, (*d*) silver; lead. (O & C)

230. Explain the experimental tests by which you would be able to distinguish between four of the following pairs of substances: (i) nitrogen and argon; (ii) sodium dihydrogen phosphate and disodium hydrogen phosphate; (iii) carbon monoxide and methane; (iv) chlorsulphonic acid and sulphuryl chloride; (v) acetic acid and formic acid. (O SCHOL)

231. Compare the physical and chemical properties of three of the following pairs: (i) carbon monoxide and carbon dioxide, (ii) oxygen and ozone; (iii) phosphorus trichloride and phosphorus pentachloride, (iv) acetylene and ethylene. (O SCHOL)

232. Compare and contrast the properties of three of the following pairs: (*a*) carbon dioxide and silica; (*b*) ammonia and phosphine; (*c*) boron trichloride and aluminium trichloride; (*d*) hydrogen fluoride and hydrogen chloride. (O SCHOL)

233. Discuss the influence of heat on four of the following: a mixture of nitrogen and oxygen; sulphur; ammonium cyanate; barium peroxide; phosphorus pentachloride; ammonium dichromate. (O SCHOL)

234. How would you obtain a pure sample of the first named constituent from each of the following mixtures: (*a*) ammonia and oxygen; (*b*) sodium carbonate and sodium bicarbonate; (*c*) lead dioxide and litharge; (*d*) ferrous sulphate and copper sulphate; (*e*) aluminium and zinc? (C)

235.* Give a comparative account of the chemistry of the common hydrides of nitrogen, oxygen, fluorine, phosphorus, sulphur, and chlorine. (O)

236.* Outline the production of some compounds of commercial importance using only common salt, limestone, coke, air, and water as raw materials.

It may be assumed that necessary catalysts are available but not electrical power. (O)

237.* Give an account of the use of electrical power in chemical industry. (O)

238. Classify oxides according to their chemical behaviour and constitution. Give *two* examples of each class with illustrative reactions and equations. (L)

239. Compare and contrast the properties of compounds having the formula XO_2, in which X is a metal or non-metal. (O SCHOL)

Organic Terms

240. Explain the meaning of *five* of the following terms as used in organic chemistry, giving illustrative examples and equations where possible: (*a*) polymerization, (*b*) hydrogenation, (*c*) esterification, (*d*) condensation, (*e*) isomerism, (*f*) fermentation. (L)

241. What do you understand by the terms *aromatic compound, homologous series, substitution reaction, polymerization, acetylation, diazotization*? Illustrate your answer with ONE example in each case. (O)

242. Explain what is meant by (*a*) addition, (*b*) condensation, (*c*) polymerization, (*d*) substitution, and give *one* example of each from the reactions of acetaldehyde. Describe *two* chemical tests that would enable you to distinguish between aqueous solutions of acetaldehyde and acetone. (C)

243. Write explanatory notes on *four* of the following terms: structural formula; unsaturated compound; asymmetric carbon atom; saponification; sulphonation. (O & C)

244.* Write notes on *four* of the following: (*a*) nitration; (*b*) sulphonation; (*c*) the reactions of a solution of phenyl diazonium chloride; (*d*) the biuret reaction; (*e*) the detection of a halogen in a nitrogen-containing organic compound. (O & C)

245. Explain, with *one* example in each case, the meaning of *five* of the following terms: (*a*) diazotization, (*b*) sulphonation, (*c*) metamerism, (*d*) decarboxylation, (*e*) saponification, (*f*) etherification, (*g*) optical isomerism. (L)

246. Explain clearly and concisely the meaning of the following terms in organic chemistry, and briefly indicate, by one example in each case, how the processes are carried out: (*a*) halogenation, (*b*) ether extraction, (*c*) esterification, (*d*) catalytic hydrogenation, (*e*) condensation reaction. (NJB)

247. What is meant by the term unsaturation? Give two tests which can be used to determine whether an organic substance is unsaturated. State, giving your reasons, whether the following substances may be regarded as saturated or unsaturated: ethylene, ethylene dibromide, ethyl alcohol, ethyl cyanide, acetaldehyde. (O & C)

248.* What is meant by the terms: hydrogenation; acetylation; benzoylation? State briefly how these processes may be carried out. Write down the formulae of the compounds, if any, which you would expect to obtain by the acetylation of the following compounds:

$$HO.CH_2.CO.OH; \quad HO.CH_2CH_2NH_2;$$
$$CH_3CO.NH.CH_2.CO.OH; \quad HO.OC.CH_2.CO.OH.$$

(O & C)

249. What reactions in *aromatic* chemistry do you associate with the names: Friedel-Crafts, Sandmeyer, Fittig, Schotten-Baumann, Cannizzaro? Give *one* example of each reaction, stating the essential conditions and naming the organic products formed. (SJB)

Functional Groups

250. For each of the following typical groups (i) name the class of compound of which the group is characteristic, and (ii) give *two* reactions which are characteristic of the group, stating clearly the experimental conditions under which each reaction occurs: (*a*) $>C{=}O$, (*b*) $-NH_2$, (*c*) $-CN$, (*d*) $-C{=}C-$. (NJB)

251. What main classes of compounds do you associate with the following groupings: $-CH_2OH$; $-CONH_2$; $-CN$; $>C{=}C<$? Give *two* distinctive reactions of each grouping. (O)

252. Compare and contrast the reactions of the —OH group in ethyl alcohol, acetic acid and phenol. (SJB)

253.* Compare and contrast the reactions of (*a*) the hydroxyl group in ethyl alcohol and phenol, (*b*) the bromine radical in ethyl bromide and bromobenzene. Give with some practical detail, TWO methods in each case of introducing into the benzene ring (i) —OH, (ii) —Br. (O)

254. In each of the following cases give two reactions which are characteristic of the particular type of compound mentioned, stating clearly the exptl. conditions under which each reaction occurs (*a*) an acid chloride, (*b*) an olefine, (*c*) a primary aromatic amine, (*d*) a nitrile, (*e*) a primary alcohol. (NJB)

255. With reference to methylamine, glycine (aminoacetic acid), and acetamide show what *characteristic* properties are caused by the presence of the —NH_2 group in the molecule. Give the equations and state the reagents and necessary conditions for converting acetamide into (*a*) methylamine, and (*b*) glycine. (SJB)

256.* In each pair of compounds given below, the functional group indicated may behave towards reagents in a similar manner in the two compounds or differently. Illustrate these similarities and differences by reference to the effect of *three* reagents on each of the groups concerned: (*a*) —NH_2 in ethylamine and acetamide; (*b*) —Cl in ethyl chloride and chlorobenzene; (*c*) —CHO in acetaldehyde and benzaldehyde. State briefly the conditions necessary for the reactions which you describe. (O & C)

257.* It has been stated that the reactions of organic compounds are those of certain characteristic groups. Discuss the validity of this statement on the basis of a comparison of the properties and reactions of (*a*) Acetone, acetaldehyde, and acetyl chloride, which contain the group $> C = O$; (*b*) Ethyl alcohol, phenol, and acetic acid, which contain the group —OH; (*c*) Ethylamine, aniline and acetamide, which contain the group —NH_2. (NJB)

258. A compound is believed to have the structure

$$CH_3.CH = CH.CO.OC_2H_5$$

What reactions would you expect it to show, and why? (O SCHOL)

259. What predictions can you make about the properties of the compound with the following formula:

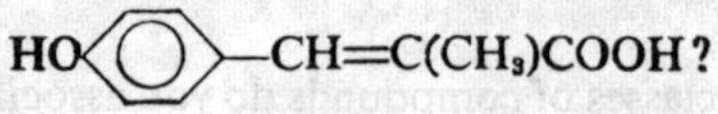

(O SCHOL)

260.* A compound has the structural formula

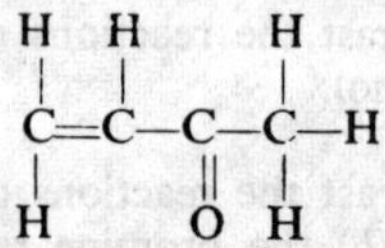

From your knowledge of the chemistry of the typical groups which it contains, give an account of the principal chemical properties which you would expect it to possess.

Which of the reactions you mention could be used as tests for the presence of the groups concerned? State exactly how each test would be carried out and what you would expect to observe. (NJB)

261.* If you were given a specimen of an unknown solid which was, in fact, *p*-nitrotoluene , explain clearly, in the correct logical order but without expt[1] details, what expts. you would make to identify the solid and to establish its structural formula. State clearly what would happen in each expt. and the conclusions you would draw from it. (NJB)

262.* The reactivity of chlorine in an organic compound *X*Cl is greatly influenced by the nature of *X*. Illustrate this statement by a comparison of the properties of the chlorine atoms in ethyl chloride, chloroacetic acid, acetyl chloride, chlorobenzene, ethylamine hydrochloride. (C)

263. What do you understand by (*a*) a saturated compound, (*b*) an unsaturated compound? In what respects does benzene differ from the usual types of both? Illustrate the effect of the aromatic nucleus on the properties of typical groups by a brief comparison of the chemical properties of (i) bromobenzene and benzyl bromide, (ii) phenol and benzyl alcohol, (iii) aniline and ethylamine. (NJB)

Distinctions between Organic Compounds

264. By what chemical tests could you distinguish between: (*a*) ethylene and acetylene; (*b*) ethyl alcohol and secondary propyl alcohol; (*c*) ethylamine and acetamide; (*d*) formic acid and acetic acid; (*e*) methyl alcohol and ethyl alcohol? (O & C)

265. Describe and explain chemical tests, *one* in each case, by which you could distinguish between: (*a*) ethylamine and aniline, (*b*) ethyl alcohol and phenol, (*c*) chlorobenzene and chloroform, (*d*) methyl alcohol and ethyl alcohol, (*e*) ethane and ethylene. (C)

266.* What are the chemical differences between the two substances in each of the following pairs of compounds: (*a*) methyl alcohol and ethyl alcohol; (*b*) formic acid and acetic acid; (*c*) acetaldehyde and benzaldehyde; (*d*) ethylamine and aniline; (*e*) chloroform and carbon tetrachloride? (O)

267.* Compare (*a*) the properties of ethyl bromide with those of monobromobenzene, (*b*) the properties of ethyl alcohol with those of phenol, (*c*) the properties of methylamine with those of aniline indicating in each case the modifying influence of the alkyl or aryl radical. (L)

268. What chemical tests would you use to distinguish between: (*a*) methyl alcohol and ethyl alcohol; (*b*) acetone and acetaldehyde; (*c*) acetamide and ammonium acetate; (*d*) formaldehyde and acetaldehyde (each in aqueous solution); (*e*) ethyl bromide and chloroform? (O & C)

269. For acetaldehyde and benzaldehyde, describe *four* reactions which are distinctive of aldehydes and apply to each of these aldehydes, and *four* reactions in which they behave differently. (SJB)

270. Outline *three* general methods by which one atom of bromine may be introduced into the molecule of an organic compound, and indicate (giving the essential reaction conditions) how far these methods are applicable to the preparation of (*a*) ethyl bromide, (*b*) bromobenzene, (*c*) benzyl bromide. Compare the behaviour of these three compounds towards aqueous potassium hydroxide, and outline a procedure by which this reagent might be used to distinguish between them. (NJB)

271.* Draw up a systematic scheme of tests which could be applied to distinguish between the five hydrocarbons ethane, ethylene, acetylene, *n*-hexane (which is a liquid paraffin) and toluene, indicating clearly the experimental observations required. How can acetylene be converted to acetaldehyde, and toluene to benzaldehyde? Compare the reactions of these two aldehydes with (*a*) ammonia and (*b*) aqueous sodium hydroxide. (NJB)

272.* On the basis of your knowledge of the reactions of (i) ethyl alcohol, benzyl alcohol and phenol, (ii) ethyl bromide, benzyl bromide and bromobenzene, and (iii) ethylene, benzene and toluene, discuss the essential differences between: (*a*) aromatic and aliphatic compounds, (*b*) nuclear and side-chain substituted derivatives of an aromatic hydrocarbon. (NJB)

Miscellaneous Questions on Organic Chemistry

273. Explain clearly and concisely what is meant by the following statements. Give examples. (*a*) Organic chemistry is the chemistry of various characteristic groups. (*b*) Esterification is a reversible reaction. (*c*) Carbonyl compounds form characteristic condensation products. (*d*) Notwithstanding its apparent high degree of unsaturation, the characteristic reactions of benzene are substitution reactions. (NJB)

274. By means of selected typical reactions briefly compare the characteristic chemical properties of ethane, ethylene and acetylene. When a given volume of a mixture of ethane and ethylene is completely oxidized, 0.440 g of carbon dioxide and 0.216 g of water are obtained. Calculate (*a*) the total volume of the original mixture at s.t.p. (*b*) its composition by volume. [H = 1, C = 12, O = 16. Molar volume = 22 400 cm^3 at s.t.p.] (NJB)

275. In each of the following cases (i) write the equations for *all* the reactions which may take place between the substances mentioned, and (ii) outline the experimental conditions under which the reactions take place (details of apparatus are not required). (*a*) Ethyl alcohol and concentrated sulphuric acid; (*b*) Benzene and a mixture of nitric and sulphuric acids; (*c*) Toluene and chlorine. (NJB)

276. Write equations, specify the necessary conditions, and name the products formed for the reactions which take place between sulphuric acid and (*a*) benzene, (*b*) ethyl alcohol, (*c*) ethylene, (*d*) methyl cyanide, (*e*) formic acid. (C)

277. What reactions can take place between potassium hydroxide and (*a*) ethyl bromide, (*b*) acetaldehyde, (*c*) benzaldehyde, (*d*) phenyl benzoate, (*e*) benzenesulphonic acid? In each case indicate how the reaction may be carried out in the laboratory and what, if anything, you would expect to *observe*. (NJB)

278. How, and under what conditions, does sodium hydroxide react with each of the following: carbon monoxide; chloroform; acetamide; ethylene dibromide; formaldehyde? (O & C)

279. Outline three reactions in which benzene behaves as if it were a saturated hydrocarbon and three in which it behaves as if it were unsaturated. In each case state clearly why you regard the reactions chosen to be suitable examples. Outline the reaction sequences (one in each case) by means of which it is possible, in the laboratory, to introduce into benzene (*a*) an amino group, (*b*) a hydroxyl group. (NJB)

280. Outline the general series of reactions (one in each case) by means of which you could (*a*) introduce a hydroxyl group into a paraffin, (*b*) introduce a hydroxyl group into the nucleus of an aromatic hydrocarbon, (*c*) introduce a carboxyl group into a paraffin, (*d*) convert a fatty acid R.COOH into the alcohol R.OH. Give, in the form of equations, one example of each type of reaction. (NJB)

281. State what reactions occur and what products are formed when (*a*) ammonium acetate is distilled and the product is treated with bromine and conc. KOH soln., (*b*) acetyl chloride is heated with an excess of anhydrous sodium acetate and the volatile product is treated with aniline; (*c*) ethylene dibromide is treated with a conc. alcoholic soln. of KOH and the resulting gas is passed into an ammoniacal solution of cuprous chloride; (*d*) methyl chloride is passed into a heated mixture of benzene and anhydrous aluminium chloride. (NJB)

282. By what series of reactions could you prepare the following? (*a*) Propylamine from ethyl iodide. (*b*) Acetone from isopropyl iodide. (*c*) Acetonitrile from acetic acid. (*d*) Benzoic acid from bromobenzene. Give the essential reaction conditions for each step. (NJB)

283. State the reagents and conditions required to perform the following syntheses, using the letters over the arrows to show the reactions to which you are referring. (Further notes are not required.)

(i) $CaC_2 \xrightarrow{a} C_2H_2 \xrightarrow{b} CH_3CHO \xrightarrow{c} CHI_3 \xrightarrow{d} H.COONa$

(ii) $CH_3COOH \xrightarrow{e} CH_2ClCOOH \xrightarrow{f} CH_2NH_2COONH_4 \xrightarrow{g}$
$CH_3CO.NH.CH_2COONH_4$

(iii) $C_6H_6 \xrightarrow{h} C_6H_5.SO_3H \longrightarrow C_6H_5ONa \longrightarrow C_6H_2(NO_2)_3OH$

Write the full equations for the reactions (*c*) and (*d*). Reaction (*c*) is used as a test. Name *two* other substances which respond to the same test. (SJB)

284. Outline how the following changes may be affected. Give equations for every reaction discussed (*a*) acetic acid ⟶ methane; (*b*) acetylene ⟶ acetic acid; (*c*) acetic acid ⟶ methyl cyanide; (*d*) ethylamine ⟶ ethyl bromide; (*e*) toluene ⟶ sodium benzoate. (C)

285. Describe the uses in organic chemistry of *five* of the following: bleaching powder; nitrous acid; phosphorus pentachloride; nitric acid; sulphuric acid; potassium cyanide. (O & C)

286. Describe the uses of *four* of the following reagents in organic chemistry: sodium; sodium hypobromite; soda-lime; ferric chloride; bromine. (O & C)

287. Give an account of the uses of the following reagents in organic chemistry: sodium; nitric acid; ammonia; bromine; potassium cyanide. (O & C)

288. Give an example of the use of each of the following reagents in organic chemistry: (*a*) iodine and caustic soda, (*b*) conc H_2SO_4, (*c*) mercury–aluminium couple, (*d*) phenylhydrazine, (*e*) metallic sodium. State briefly how each reagent is used and for what purpose. (NJB)

289.* For each of *four* of the reagents named below, describe in concise detail one important use in organic chemistry, and if there is an alternative reagent for the same operation explain any special advantages obtained by using the compound mentioned in the question (*a*) thionyl chloride, (*b*) benzoyl chloride, (*c*) dry hydrogen chloride, (*d*) moist silver oxide, (*e*) hydroxylamine hydrochloride, (*f*) chloroplatinic acid, H_2PtCl_6. (NJB)

Organic Problems

290. Give two methods for preparing ethane in the laboratory.

A mixture of 10 cm³ of a gaseous hydrocarbon and 100 cm³ of oxygen (excess) was exploded. The volume after explosion was 75 cm³, and this was reduced to 35 cm³ on treatment with potassium hydroxide solution. Deduce the molecular formula of the hydrocarbon and give its possible structural formulae. (All measurements were made at the same temperature and atmospheric pressure.) (O & C)

291. 75 cm³ of oxygen were mixed with 12.5 cm³ of a gaseous hydrocarbon *X*. After exploding and cooling to room temperature, 50 cm³ of gas were left and when this was shaken with potassium hydroxide solution, 12.5 cm³ of oxygen remained. Calculate the formula of *X*. (All volumes were measured at the same temperature and pressure.)

Explain the evidence on which the structural formula of ethylene is based. Briefly describe *one* preparation of ethylene and explain how it could be converted into any *two* other hydrocarbons. (SJB)

292.* Indicate briefly how samples of ethane, ethylene and acetylene could be obtained from ethyl alcohol. 8 cm^3 of a gaseous hydrocarbon were exploded with 70 cm^3 of oxygen; the resulting gas-volume, measured under the same conditions of temperature and pressure, was 58 cm^3 and this, on shaking with a concentrated solution of potassium hydroxide, decreased further to 26 cm^3. The hydrocarbon yielded a red precipitate when passed into an ammoniacal solution of cuprous chloride. Write its structural formula. (NJB)

293. Name *three* organic compounds, other than hydrocarbons, which are gases at room temperature (18°C).

Outline *three* methods for the preparation of ethane, giving the necessary reagents and temperature conditions.

16 cm^3 of a mixture of methane and ethane were exploded in a eudiometer with 50 cm^3 of oxygen. The volume of the residual gas was 32 cm^3, which was reduced on shaking with potassium hydroxide solution to 12 cm^3. (All volumes were measured at the same room temperature and pressure.) Calculate the composition by volume of the hydrocarbon mixture. (NJB)

294. Excess of oxygen was added to 100 cm^3 of a mixture of methane and acetylene. After exploding and cooling to room temperature, the residual volume was read. On adsorption with potassium hydroxide solution, the volume decreased by 170 cm^3 and the remaining gas was shown to be oxygen. (All gases were measured at the same temperature and pressure.) Calculate the original composition of the mixture. Describe briefly, with essential practical details, how you would make acetylene from ethylene and ethylene from acetylene. What happens to acetylene (*a*) when heated to 400°, (*b*) when placed with hydrogen chloride? (SJB)

295.* A compound *A*, of low boiling point, contained carbon, hydrogen and oxygen. After 10 cm^3 of the vapour *A* were exploded with 50 cm^3 of oxygen (excess), the volume was 45 cm^3 and this was reduced to 25 cm^3 by treatment with potash, all volumes being measured at 25°C and 10^5 N m^{-2}. When 0.88 g of *A* was dissolved in water, oxidized and distilled, the distillate required 20 cm^3 M sodium hydroxide for neutralization. What is the structural formula of *A* and what happens when it is oxidized? [H = 1, C = 12, O = 16.] (C)

296. Three substances, *A*, *B* and *C*, were found to have the following percentage composition:

A. C = 58.54; H = 7.32; N = 34.14
B. C = 40.68; H = 8.48; O = 27.12; N = 23.73
C. C = 31.18; H = 9.09; O = 41.57; N = 18.18

The molecular formula in each case was the same as the empirical formula. Give a brief description of each of the three compounds, which were easily convertible into one another, and state how this interconversion could be effected. (o & c)

297. The percentage composition of an aliphatic compound was found to be C = 20.0, H = 6.7, O = 26.7, N = 46.6. A solution of 0.25 gramme of the compound in 20 gramme of water froze at – 0.39°C. What was the compound and what would be the action of heat on it?

(Molecular depression constant for water is 18.6 K mol^{-1} per 100 gramme.)

[H = 1; C = 12; N = 14; O = 16; Na = 23; Cl = 35.45.] (o & c)

298. An organic compound *A*, containing 85.1 per cent. of bromine) reacted with moist silver oxide to yield *B*, $C_2H_6O_2$. *B* was oxidized to an acid *C*, which evolved carbon monoxide and dioxide on warming with concentrated sulphuric acid. Substance *A*, on boiling with alcoholic potash, yielded a gas *D* which, on being passed through ammoniacal cuprous chloride solution, deposited a red solid containing 84.1 per cent. of copper.

Deduce structures for *A*, *B*, *C* and *D*, and represent the above information in the form of a reaction scheme.

[H = 1, C = 12, O = 16, Cu = 63.5, Br = 80.] (c)

299. An organic compound *A* of empirical formula C_2H_2N was reduced to a compound *B* of empirical formula C_2H_6N. On boiling *A* with dilute sulphuric acid an acid *C* was obtained. *C* was esterified with ethyl alcohol and gave an ester *D* which had a vapour density of 87 (H = 1). Deduce the formulae for *A*, *B*, *C*, and *D*. Account for the above reactions and suggest how *A* could be prepared.

[H = 1, C = 12, N = 14.] (o & c)

300.* An aliphatic compound *A* contained C = 21.2 per cent., H = 1.8 per cent., Cl = 62.8 per cent. On being treated with water in the cold it gave a compound *B* containing C = 25.4 per cent., H = 3.15 per cent., Cl = 37.6 per cent. When *B* was heated with sodium carbonate solution and subsequently acidified, an acid *C* of molecular formula $C_2H_4O_3$ was isolated. What were *A*, *B*, and *C*? Account for the above reactions.

[H = 1, C = 12, O = 16, Cl = 35.5.] (o & c)

301.* A colourless, solid, aromatic compound *X*, which was soluble in hot water but sparingly soluble in cold, was found to contain C, 71.2 per cent.; H, 6.67 per cent.; N, 10.37 per cent. After boiling with dilute sulphuric acid for some time and then distilling, an acid was obtained which gave no reaction with potassium permanganate solution and which, after neutralization, gave a red colouration with neutral ferric chloride solution. On adding an excess of a concentrated solution of sodium hydroxide to the residue in the distilling flask, an oil separated which readily redissolved in an acid and which gave a purple coloration with a solution of bleaching powder. Deduce the nature of *X*, explain the reactions described, and say how *X* could be prepared.

[H = 1; C = 12; N = 14; O = 16.] (O & C)

302. A primary monoacid organic base has the composition C = 77.42%, H = 7.52%, N = 15.05%. On ignition 1.4924 g of its platinichloride left a residue of 0.4880 g of platinum. Calculate the molecular formula of the base and state how it would react with (*a*) bromine, (*b*) nitrous acid.

[H = 1, C = 12, N = 14, Cl = 35.5, Pt = 195.] (L)

303.* Assuming that you know the formula of monochloroacetic acid to be $Cl.CH_2.CO_2H$, what principal chemical characteristics would you expect it to show?

An organic acid crystallized in a hydrated form, which on analysis gave C = 34.29%, H = 4.76%, O = 60.95%. It was found that 0.70 g of the hydrated acid, dissolved in water, neutralized 20 cm^3 of 0.5 M alkali. The anhydrous silver salt and the ethyl ester were prepared; 0.57 g of the salt on ignition left a residue of 0.36 g of silver, and 12 g of the ester, dissolved in 50 g of ethyl alcohol, raised the boiling point of the solvent by about 1 °C. Deduce (*a*) the basicity of the acid, (*b*) the molecular formula of the *anhydrous* acid.

[H = 1, C = 12, O = 16, Ag = 108. The molecular elevation constant for alcohol, for 1 mole of solute in 1 000 g of solvent, is 1.15 K mol^{-1}.] (NJB)

304.* A neutral aliphatic liquid *A* was found to contain C = 49.3; H = 6.85 per cent. On shaking *A* with aqueous ammonia a compound *B* was precipitated which contained 31.82 per cent of nitrogen. On heating *B* with a solution of sodium hydroxide, ammonia was evolved and the residual solution after neutralization gave a white precipitate with a solution of calcium chloride which was insoluble in acetic acid. Deduce the structural formulae of *A* and *B* and explain the above reactions. Describe how *A* could be synthesized from acetylene.

[H = 1, C = 12, N = 14, O = 16.] (O & C)

305.* An organic compound *A* of molecular formula C_8H_8O gave the iodoform reaction and on oxidation yielded a compound *B* of molecular formula $C_7H_6O_2$, which was sparingly soluble in cold water and which, after neutralization, gave a buff precipitate with ferric chloride. Treatment of *B* with phosphorus pentachloride gave a compound *C* and on treatment of this with aqueous ammonia a compound *D* of molecular formula C_7H_7ON was precipitated. Treatment of *D* with phosphorus pentoxide gave a compound *E*, which on reduction gave a compound *F* of molecular formula C_7H_9N. On reduction with sodium amalgam *A* gave a compound *R* of molecular formula $C_8H_{10}O$, which also gave the iodoform reaction but did not give a coloration with ferric chloride. Deduce the structure of *A* and give equations to explain the above reactions. (O & C)

Isomerism

306. What is meant by (*a*) empirical formula, (*b*) molecular formula, (*c*) structural isomerism? Write down the structural formula of all the isomers of $C_4H_8O_2$ which are either acids or esters. Describe briefly how you would obtain an aqueous solution of ethyl alcohol from one of the esters. (C)

307. What is meant by isomerism? Describe *four* different ways in which isomerism can arise in organic compounds. Where possible, illustrate your answer by examples of each type. (L)

308.* Discuss, with examples, the various types of isomerism which you have encountered in organic compounds. Which of these types do you think might occur in inorganic compounds and why? (O & C)

309. Two substances *A* and *B* both possess the same molecular formula $C_2H_4O_2$, when treated with a cold solution of sodium hydrogen carbonate *A* remains unchanged, but *B* dissolves with liberation of carbon dioxide. When *A* is boiled under reflux with NaOH soln. it yields an alcohol, which does not give the iodoform reaction, and the sodium salt of an acid which decolourizes a warm solution of $KMnO_4$ acidified with dil. H_2SO_4. *Deduce* and write down the names and full structural formulae of *A* and *B* and state clearly what reactions occur in all the above expts. What would be the action (if any) of warm conc. H_2SO_4 on *A* and on *B*? (NJB)

310.* An aqueous solution of an organic compound, *X*, molecular formula $C_3H_{10}NCl$, gave an instantaneous precipitate with aqueous silver nitrate. Write the possible structures for *X*, and describe how the various possibilities could be distinguished by their chemical reactions. (C)

311. A neutral liquid only partially miscible with water, gave the following composition: 48.65 per cent C, 8.11 per cent H, 43.24 per cent O. Its vapour density was 37. Draw structural formulae for the compounds which would satisfy these facts, and give practical details of tests to distinguish the isomers. State briefly how *one* of them may be prepared. (SJB)

312. An organic compound was found on analysis to contain 31.86 per cent C, 5.31 per cent H, and 62.83 per cent Cl. In a Victor Meyer determination, 0.113 g of this compound displaced 23.75 cm^3 of air measured over water, at 15°C and 102 500 N m^{-2}. Calculate the molecular formula of the compound and suggest possible constitutional formulae. How would these compounds behave on hydrolysis? (L)

313.* An acid *X* containing 40.90 per cent C and 4.55 per cent H and of molecular weight about 90, on reduction gave an acid *Y* containing 40.00 per cent C and 6.67 per cent H. Write structural formulae for *X* and *Y*, and an equation for the reaction involved. Write an account of the isomerism exhibited in *Y*, pointing out where the physical and chemical properties of the isomers are similar and where they are different. (SJB)

314. *X* is an ester of a monocarboxylic acid and a monohydric alcohol. When one gramme of *X* is boiled with 25 cm^3 M potassium hydroxide solution until saponification is complete, it is found that 13.65 cm^3 M hydrochloric acid are required to neutralize the excess alkali. Calculate the molecular weight of the ester. Name and give the structural formulae of all such esters which have this molecular weight. Describe the preparation of *one* of them in the laboratory. [H = 1, C = 12, O = 16.] (C)

315.* Name the isomers which have the formula C_7H_7Cl and write their structural formulae. One of the isomers, *A*, when refluxed with potassium carbonate solution forms *B*, formula C_7H_8O. Identify *A* and describe how it can be converted into *pure C*, formula C_7H_6O.

How does *C* react with (*a*) ammonia, (*b*) potassium hydroxide? How can *C* be converted into *D*, formula $C_{13}H_{12}N_2$. (O)

316.* Explain the following terms: polarized light; optically active compound; asymmetric carbon atom; stereoisomerism.

Give an example of an optically active alcohol and indicate how EITHER this alcohol OR lactic acid can be synthesized.

Describe ONE method of separating the isomers of an optically active acid. (O)

317.* Describe precisely how each of the following structural formulae can represent two compounds,

(*a*) CH_3.CH(OH).COOH,

(*b*) HOOC.CH:CH.COOH.

Write down possible compounds having the empirical formulae C_3H_8O. Comment on the number of possible compounds of empirical formula CH_2. (C)

318.* Distinguish between optical and geometrical isomerism. In what ways does the optical isomerism of tartaric acid differ from that of lactic acid? (O & C)

319. Write an essay on one of the following subjects: Isotopes, The Noble Gases, Stereoisomerism, Fuels, Carbohydrates, Petroleum, Coal, Plastics, Fertilizers, Hydrolysis, Catalysis, Allotropy, Colloids, Osmosis, Valency, Isomerism, Corrosion, Electrolysis, The Atmosphere, Chromatography, Order of Reaction, Polymers, Natural Gas.

Answers to Examination Questions

3. 11 cm^3 of ammonia; 9 cm^3 of nitrogen
4. 205.9; 206.4
6. 78.52
8. Equiv. wts. are 25.48, 17.0, 12.73, 10.18; At. wt. is 51; Formulae are MO, M_2O_3, MO_2, M_2O_5
9. 107.9
10. Equiv. wt. is 12.56; at. wt. is probably 25.12
11. (i) 31.6; (ii) 63
12. (i) (*a*) 45, (*b*) 45; (ii) (*a*) 152, (*b*) 76; (iii) (*a*) 158, (*b*) 19.75
13. $x = 1$, $y = 3$, $z = 2$, $n = 2$
14. 132.9
15. At. wt. is 186.2; the element exerts valencies of 4 and 7
16. 137.4
17. 27.1
18. MCl_2, MCl_4; 118.7
19. 0.226 J g^{-1} K^{-1}; XO_2, XCl^4
20. 115; MCl, MCl_2, MCl_3
23. (*a*) 29.7; (*b*) 256; (i) 118.8; (ii) MCl_4; (iii) 0.222 J g^{-1} K^{-1}
30. 52.3%
31. 54%
32. 55.95; M_2Cl_6
33. $\alpha = 34\%$
35. 265.2; 132.5
36. 67
37. *A* is SO_2
38. Acetic acid exists as dimers (mol. wt. 120) in the vapour
39. N_2, CO, C_2H_4
40. (*a*) Mol. wt. is 152; (*b*) $\alpha \simeq 80\%$
41. 61
42. 1.86 K per mole per 1000 gramme; 2×10^6 N m^{-2}
43. (*a*) 124.4; (*b*) 55%
44. The calcium chloride appears to be 75% dissociated
45. 173.4
46. 44°C
47. 2.54 g
48. 65.5
49. 57%
53. (*a*) The apparent degree of dissociation of the sodium chloride is 89.6%; (*b*) The tetroxide is 20.1% dissociated at that temperature
54. (*a*) C_3H_8; (*b*) $\alpha = 50\%$ for the magnesium sulphate; (*c*) Acetic acid is 100% associated in benzene
59. 3.84 g dm^{-3}
62. 6.44 cm^3 of each
63. $\frac{1}{3}$ oxygen, $\frac{2}{3}$ nitrogen
64. 0.0165 cm^3 nitrogen and 0.0310 cm^3 oxygen in 1 cm^3 of water
70. 122
71. 62%
72. 1:3.97
73. 0.088 g
74. 9.43 g
75. *X* is dimerised (M_2) in chloroform
76. The acid is associated into dimers in benzene
94. $K_d = 2/27$
95. (*a*) $\frac{2}{3}$ mole; (*b*) 0·845 mole

96. 4.12
97. 0.465 mole
98. $K = 4$
99. 9.06 mole
100. $K = 67.2$; 0.195 mole H_2 and I_2, 1.61 mole HI.
101. $K = 0.0082$
102. $K = 0.0124$
104. 0.36 g, 0.106 g, 40.2 cm^3
105. 96 550 coulomb
106. 121.8
109. (*a*) 0.0211 $\Omega^{-1}\ cm^{-1}$; (*b*) 105.5 $\Omega^{-1}\ cm^{-2}\ mol^{-1}$
111. (*b*) (i) 2, (ii) 12, (iii) 7, (iv) 2.3
112. 2.89
113. 1.60×10^{-3} g dm^{-3}
114. 1.73×10^{-5} mol dm^{-3}; 3.91
115. 2.9; 2.65×10^{-5} mol dm^{-3}
116. (*a*) 2.37; (*b*) 4.74
117. 1.34%
119. (*d*) −0.3; (*e*) 13.0
121. (*a*) 2; (*b*) 12; (*c*) 5.7; (*d*) 8.3
122. (i) 2.87; (ii) 4.74
124. 5×10^{-49} $mol^3\ dm^{-9}$
125. 1.09×10^{-10} $mol^2\ dm^{-6}$; 1.09×10^{-9} mol dm^{-3}
126. (*a*) 1.33×10^{-8} g dm^{-3}; (*b*) 4.04×10^{-4} g dm^{-3}
127. (*a*) 1.79×10^{-8} $mol^3\ dm^{-9}$; (*b*) 8.26×10^{-4} g dm^{-3}
128. 2.25×10^{-4} $mol^2\ dm^{-6}$; 1.48×10^{-2} mole
129. (*a*) 1.43×10^{-3} g dm^{-3}; (b) 1.43×10^{-6} g dm^{-3}
141. −3265 kJ mol^{-1} (evolved)
142. +67 kJ mol^{-1} (absorbed)
143. +108.8 kJ mol^{-1} (absorbed)
144. −149 kJ mol^{-1} (evolved)
145. −416 kJ mol^{-1} (evolved)
146. +59.91 kJ mol^{-1} (absorbed)
147. −118.8 kJ mol^{-1} (evolved)
148. −91.00 kJ mol^{-1} (evolved)
150. −79.5 kJ mol^{-1} (evolved); +226 kJ mol^{-1} (absorbed); −305 kJ mol^{-1} (evolved)
151. −199 kJ mol^{-1} (evolved)
152. −52.3 kJ mol^{-1} (evolved)
274. (*a*) 112 cm^3; (*b*) 60% ethylene, 40% ethane
290. C_4H_{10}
291. C_3H_8
292. $HC{\vdots}C.C_2H_5$
293. 12 cm^3 methane, 4 cm^3 ethane
294. 30 cm^3 methane, 70 cm^3 acetylene
295. $CH_3.CHO$
296. *A* is CH_3CN; *B* is $CH_3.CONH_2$; *C* is $CH_3.COONH_4$
297. Urea, $NH_2.CO.NH_2$
298. *A* is $\begin{array}{c}CH_2Br\\|\\CH_2Br\end{array}$; *B* is $\begin{array}{c}CH_2OH\\|\\CH_2OH\end{array}$; *C* is $\begin{array}{c}COOH\\|\\COOH\end{array}$; *D* is $HC{\equiv}CH$
299. *A* is $\begin{array}{c}CH_2CN\\|\\CH_2CN\end{array}$; *B* is $\begin{array}{c}CH_2CH_2CN\\|\\CH_2CH_2CN\end{array}$; *C* is $\begin{array}{c}CH_2.COOH\\|\\CH_2.COOH\end{array}$; *D* is $\begin{array}{c}CH_2COOC_2H_5\\|\\CH_2COOC_2H_5\end{array}$
300. *A* is $ClCH_2.COCl$; *B* is $ClCH_2.COOH$; *C* is $HOCH_2.COOH$
301. *X* is $C_6H_5NH.COCH_3$
302. C_6H_7N
303. (*a*) 3; (*b*) $C_6H_8O_7$
304. *A* is $\begin{array}{c}COOC_2H_5\\|\\COOC_2H_5\end{array}$; *B* is $\begin{array}{c}CONH_2\\ \\CONH_2\end{array}$
305. *A* is $C_6H_5.CO.CH_3$
309. *A* is $H.COOCH_3$; *B* is $CH_3.COOH$
311. Molecular formula is $C_3H_6O_2$
312. Molecular formula is $C_3H_6Cl_2$
313. *X* is $CH_2{:}C(OH)COOH$; Y is $CH_3.CH(OH).COOH$ or $CH_2OH.CH_2.COOH$
314. 88
315. *A* is benzyl chloride, $C_6H_5CH_2Cl$

Index

Approximate Atomic Weights

Aluminium	27	Manganese	55
Barium	137	Mercury	200
Boron	11	Nitrogen	14
Bromine	80	Oxygen	16
Calcium	40	Phosphorus	31
Carbon	12	Platinum	195
Chlorine	35.5	Potassium	39
Chromium	52	Silicon	28
Copper	63.5	Silver	108
Hydrogen	1	Sodium	23
Iodine	127	Strontium	88
Iron	56	Sulphur	32
Lead	207	Tin	119
Magnesium	24	Zinc	65

The Naturally Occurring Isotopes of Some Common Elements

(mass numbers arranged in order of abundance)

Aluminium	27	Mercury	202, 200, 199, 201, 198, 204, 196
Barium	138, 137, 136, 135, 134 130, 132	Nitrogen	14, 15
Boron	11, 10	Oxygen	16, 18, 17
Bromine	79, 81	Phosphorus	31
Calcium	40, 44, 42, 48, 43, 46	Platinum	195, 194, 196, 198, 192, 190
Carbon	12, 13		
Chlorine	35, 37	Potassium	39, 41, 40
Chromium	52, 53, 50, 54	Silicon	28, 29, 30
Copper	63, 65	Silver	107, 109
Hydrogen	1, 2	Sodium	23
Iodine	127	Strontium	88, 86, 87, 84
Iron	56, 54, 57, 58	Sulphur	32, 34, 33, 36
Lead	208, 206, 207, 204	Tin	120, 118, 116, 119, 117 124, 122, 112, 114, 115
Magnesium	24, 25, 26		
Manganese	55	Zinc	64, 66, 68, 67, 70

Precise Atomic Weights

The following atomic weights are based on the exact number 12 for the carbon isotope 12, as agreed between the International Unions of Pure and Applied Physics and of Pure and Applied Chemistry.

Name	Symbol	At. No.	At. wt.
Actinium	Ac	89	—
Aluminium	Al	13	26.9815
Americium	Am	95	—
Antimony	Sb	51	121.75
Argon	Ar	18	39.948
Arsenic	As	33	74.9216
Astatine	At	85	—
Barium	Ba	56	137.34
Berkelium	Bk	97	—
Beryllium	Be	4	9.0122
Bismuth	Bi	83	208.980
Boron	B	5	10.811*
Bromine	Br	35	79.909
Cadmium	Cd	48	112.40
Cæsium	Cs	55	132.905
Calcium	Ca	20	40.08
Californium	Cf	98	—
Carbon	C	6	12.01115*
Cerium	Ce	58	140.12
Chlorine	Cl	17	35.453
Chromium	Cr	24	51.996
Cobalt	Co	27	58.9332
Copper	Cu	29	63.54
Curium	Cm	96	—
Dysprosium	Dy	66	162.50
Einsteinium	Es	99	—
Erbium	Er	68	167.26
Europium	Eu	63	151.96
Fermium	Fm	100	—
Fluorine	F	9	18.9984
Francium	Fr	87	—
Gadolinium	Gd	64	157.25
Gallium	Ga	31	69.72
Germanium	Ge	32	72.59
Gold	Au	79	196.967
Hafnium	Hf	72	178.49
Helium	He	2	4.0026
Holmium	Ho	67	164.930
Hydrogen	H	1	1.00797*
Indium	In	49	114.82
Iodine	I	53	126.9044
Iridium	Ir	77	192.2
Iron	Fe	26	55.847
Krypton	Kr	36	83.80
Lanthanum	La	57	138.91
Lawrencium	Lw	103	—
Lead	Pb	82	207.19
Lithium	Li	3	6.939
Lutetium	Lu	71	174.97
Magnesium	Mg	12	24.312
Manganese	Mn	25	54.9380
Mendelevium	Md	101	—
Mercury	Hg	80	200.59
Molybdenum	Mo	42	95.94
Neodymium	Nd	60	144.24
Neon	Ne	10	20.183
Neptunium	Np	93	—
Nickel	Ni	28	58.71
Niobium	Nb	41	92.906
Nitrogen	N	7	14.0067
Nobelium	No	102	—
Osmium	Os	76	190.2
Oxygen	O	8	15.9994*
Palladium	Pd	46	106.4
Phosphorus	P	15	30.9738
Platinum	Pt	78	195.09
Plutonium	Pu	94	—
Polonium	Po	84	—
Potassium	K	19	39.102
Praseodymium	Pr	59	140.907
Promethium	Pm	61	—
Protoactinium	Pa	91	—
Radium	Ra	88	—
Radon	Rn	86	—
Rhenium	Re	75	186.2
Rhodium	Rh	45	102.905
Rubidium	Rb	37	85.47
Ruthenium	Ru	44	101.07
Samarium	Sm	62	150.35
Scandium	Sc	21	44.956
Selenium	Se	34	78.96
Silicon	Si	14	28.086*
Silver	Ag	47	107.870
Sodium	Na	11	22.9898
Strontium	Sr	38	87.62
Sulphur	S	16	32.064*
Tantalum	Ta	73	180.948
Technetium	Tc	43	—
Tellurium	Te	52	127.60
Terbium	Tb	65	158.924
Thallium	Tl	81	204.37
Thorium	Th	90	232.038
Thulium	Tm	69	168.934
Tin	Sn	50	118.69
Titanium	Ti	22	47.90
Tungsten	W	74	183.85
Uranium	U	92	238.03
Vanadium	V	23	50.942
Xenon	Xe	54	131.30
Ytterbium	Yb	70	173.04
Yttrium	Y	39	88.905
Zinc	Zn	30	65.37
Zirconium	Zr	40	91.22

* These atomic weights are known to be variable because of natural variations in isotopic composition.